Electromagnetic Interactions in Nuclear and Hadron Physics

Proceedings of the International Symposium

Electromagnetic Interactions in Nuclear and Hadron Physics

Osaka, Japan   4 – 7 December 2001

editors

Mamoru Fujiwara
Tatsushi Shima

Research Center for Nuclear Physics, Osaka University, Japan

World Scientific
New Jersey • London • Singapore • Hong Kong

Published by

World Scientific Publishing Co. Pte. Ltd.

P O Box 128, Farrer Road, Singapore 912805

USA office: Suite 1B, 1060 Main Street, River Edge, NJ 07661

UK office: 57 Shelton Street, Covent Garden, London WC2H 9HE

British Library Cataloguing-in-Publication Data
A catalogue record for this book is available from the British Library.

ELECTROMAGNETIC INTERACTIONS IN NUCLEAR AND HADRON PHYSICS
Proceedings of the International Symposium

ISBN 981-238-044-2

International Symposium on

Electromagnetic Interactions in Nuclear and Hadron Physics (EMI2001)

RCNP, Osaka, Japan December 4-7, 2001

International advisory committee

Sam M. Austin (NSCL), I.T. Cheon (Yonsei), K. de Jager (JLab),
S. Gales (Orsay), M.N. Harakeh (KVI), K. Hicks (Ohio),
H. Horiuchi (Kyoto), H. Kamitsubo (SPring-8),
T.-S.H. Lee (Argonne), B. Mecking (JLab), Y. Nagai (RCNP),
S. Nagamiya (KEK), A. Sandorfi (BNL), C. Schaerf (Rome),
H. Stroeher (Jeulich), Z. Suikowski (Swierk), I. Tanihata (RIKEN),
H. Toki (RCNP), M. Urin (Moscow)

Organizing committee

M. Fujiwara (RCNP/JAERI ; Chairperson), T. Nakano (RCNP),
H. Shimizu (Yamagata), H. Utsunomiya (Konan),
T. Kishimoto (Osaka), T. Sato (Osaka), T. Morii (Kobe),
J. Kasagi (Tohoku), Y. Ohashi (SRing-8), S. Date (Spring-8),
H. Sagawa (Aizu), T. Shima (RCNP ; Scientific secretary),
K. Takahisa (RCNP)

Host Institute
Research Center for Nuclear Physics (RCNP)
Osaka University

PREFACE

The International Symposium on Electromagnetic Interactions in Nuclear and Hadron Physics (EMI2001) was held at Osaka University, Japan from December 4th to 7th, 2001. The symposium is a topical one to discuss the fundamental problems in nuclear and hadron physics developed with "electromagnetic probes". When a decision was made nearly two years ago to organize this symposium, several movements towards new developments of physics with electromagnetic probes were expected to be common in the world. Actually, interesting and hot results from many laboratories in the world have been presented and discussed in the symposium.

This symposium is supported by Ministry of Education, Culture, Sport, Science and Technology (Monbu-Kagaku-shou) under COE (Center of Excellence) Program, and is hosted by the Research Center for Nuclear Physics (RCNP). The symposium belongs to a series of international meetings at RCNP. This time, we would like to organize the symposium to discuss nuclear and hadron physics with emphasis on real and virtual photon interactions. It is timely to discuss the subjects mentioned above since we expect to have some preliminary results from the LEPS (Laser-Electron-Photon Spectrometer) facility at SPring-8 using the photon beam with an energy of 1.5–2.4 GeV, and since other new technical developments will be delivered for future experiments.

Subjects discussed in the symposium are

1. Meson and hadron productions by real and virtual photon interaction with nucleons and nuclei
2. Astrophysics studies via photoreactions and hadron reactions
3. New technologies for the electromagnetic (E.M.) probes and the detector development
4. Nuclear structure studied with E.M. probes
5. Fundamental symmetries with E.M. probes and related problems

The symposium is intended also for celebrating the 30 years anniversary of RCNP, Osaka University. The ceremony and reception of the 30 years RCNP anniversary is held on December 3rd, 2001, before the symposium.

In addition to invited talks, we arranged a poster session and a special session where the students working at RCNP presented their latest results for the attendant experienced scientists. The number of the participants was about 130 including 80 scientists from outside Japan. We could have lively discussions with ample time.

On behalf of the organizing committee, we thank many students and young researchers of RCNP for their great help during the symposium. We appreciate the RCNP official staff for their considerable help to make the present symposium successful. They include A. Nagasawa, S. Yoshida, Y. Inui, M. Wada, S. Kinjo, T. Murakami, A. Nakazawa, Y. Ogawa, S. Hyodo, Y. Ito, C. Sano, M. Nishimoto and M. Matsuura.

We would like to acknowledge the generous offer of the Welding Research Institute of Osaka University for allowing the use of the Arata Memorial Hall as the main meeting hall of the EMI2001 symposium.

Finally, we thank the members of the International Advisory Committee and the Organizing Committee for their sound advice which helped us to prepare the symposium. One of our scientific goals was to publish this EMI2001 Proceedings in a relatively short time after the symposium that is important to encourage the subsequent further scientific work. We wish to thank all the speakers for submitting their contribution papers.

Editors: *Mamoru Fujiwara*
Tatsushi Shima

CONTENTS

Proceedings of the International Symposium

Electromagnetic Interactions in Nuclear and Hadron Physics

NUCLEON ELECTROMAGNETIC FORM FACTORS

KEES DE JAGER

Jefferson Lab, 12000 Jefferson Avenue, Newport News, VA 23606, USA
E-mail: kees@jlab.org

A review of data on the nucleon electromagnetic form factors in the space-like region is presented. Recent results from experiments using polarized beams and either polarized targets or nucleon recoil polarimeters have yielded a significant improvement on the precision of the data obtained with the traditional Rosenbluth separation. Future plans for extended measurements are outlined.

1 Introduction

The nucleon electromagnetic form factors (EMFF) are of fundamental importance for the understanding of the nucleon's internal structure. These EMFF are generally measured through elastic electron-nucleon scattering. In Plane Wave Born Approximation (PWBA) the cross section can be expressed in terms of the Dirac and Pauli form factors F_1 and F_2, respectively,

$$\frac{d\sigma}{d\Omega} = \sigma_M [(F_1^2 + \kappa^2 \tau F_2^2) \cos^2(\frac{\theta_e}{2}) + 2\tau(F_1^2 + \kappa F_2^2)^2 \sin^2(\frac{\theta_e}{2})] \qquad (1)$$

where $\tau = Q^2/4(m_N^2)$, Q is the four-momentum transfer, m_N the mass of the nucleon, σ_M the Mott cross section for scattering off a point-like particle, κ the nucleon anomalous magnetic moment, θ_e the electron scattering angle and E_e the electron energy. F_1 parametrizes the distribution of the nucleon charge and of the normal part of the nucleon magnetic moment, F_2 that of the anomalous part of the magnetic moment. These two form factors can be expressed in the electric and magnetic Sachs form factors G_E and G_M, respectively,

$$G_E = F_1 - \tau\kappa F_2$$
$$G_M = F_1 + \kappa F_2 \qquad (2)$$

leading to the so-called Rosenbluth[1] formula

$$\frac{d\sigma}{d\Omega} = \sigma_M [\frac{G_E^2 + \tau G_M^2}{1 + \tau} + 2\tau G_M^2 \tan^2(\frac{\theta_e}{2})] \qquad (3)$$

This equation shows that G_E and G_M can be determined separately by performing cross-section measurements at fixed Q^2 over a range of (θ_e, E_e) combinations (Rosenbluth separation). In the non-relativistic Breit frame the Sachs form factors can be identified with the Fourier transform of the nucleon charge and magnetization density distributions, such that their derivative at $Q^2 \to 0$ is related to the charge and magnetization radius, respectively.

Through the middle of the previous decade practically all available proton EMFF data had been collected using the Rosenbluth separation technique. This experimental procedure requires an accurate knowledge of the electron energy and the total luminosity. In addition, since the contribution to the elastic cross section from the magnetic form factor is weighted with Q^2, data on G_E^p suffer from increasing systematic uncertainties at higher Q^2-values. Data for the neutron resulted mainly from quasi-elastic scattering off the deuteron, since a free neutron target is not available in nature. This additional constraint caused large uncertainties, especially on the data for G_E^n.

These restrictions are clearly presented in the review paper by Bosted et al.[2] The then available world data set was compared to the so-called dipole parametrization G_D, which corresponds to exponentially decreasing radial charge and magnetization densities:

$$G_D = (\frac{\Lambda^2}{\Lambda^2 + Q^2}) \text{ with } \Lambda = 0.84 \ GeV/c \text{ and } Q \text{ in } GeV/c$$

$$G_E^p = G_M^p \approx \mu_p G_D$$

$$G_M^n \approx \mu_n G_D \qquad\qquad G_E^n \approx 0 \qquad\qquad (4)$$

Accurate data were available for G_M^p up to Q^2-values of over 20 $(GeV/c)^2$, whereas for G_E^n no significant deviation from zero was measured[44]. For all four EMFF the available data agreed with the dipole parametrization to within 20 %. Both the G_E^p and the G_M^p data could be fitted adequately with an identical parametrization. However, the limitation of the Rosenbluth separation is evident from fig. 1, which shows all available data on G_E^p. Different data sets deviate from each other by up to 50 % at higher Q^2-values, way beyond the already sizeable estimate of the experimental uncertainty.

2 Theory

A frequently used framework to describe the EMFF is that of Vector Meson Dominance (VMD)[6], in which one assumes that the virtual photon - after having become a quark-antiquark pair - couples to the nucleon as a vector meson. The EMFF can then be expressed in terms of coupling strengths

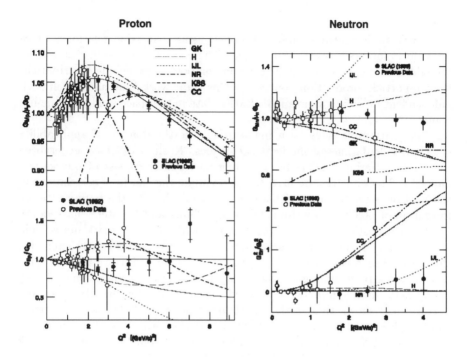

Figure 1. World data on nucleon EMFF available in 1995. The full circles are from ref. [3].
The theory curves are (from the top): GK[5], H[6], IJL[7], NR[8], KSS[9] and CC[10].

between the virtual photon and the vector meson and between the vector
meson and the nucleon, summing over all possible vector mesons. In some
cases additional terms are included to account for the effect of unknown or
lesser known mesons. A common restriction of VMD models is that they
do not predict a correct behaviour of the EMFF at high Q^2-values. The
quark-dimensional scaling framework[11] predicts that only valence quark states
contribute at sufficiently high Q^2-values. Under these conditions the EMFF
Q^2-dependence is determined simply by the number of gluon propagators,
causing the Dirac and Pauli form factors to be proportional to Q^{-4} and Q^{-6},
respectively, whereas any VMD-model will predict a Q^{-2} behaviour at large
Q^2-values. Gari and Krümpelmann[4,5] have constructed a hybrid (EVMD)
model which combines the low Q^2-behaviour of the VMD model with the
asymptotic behaviour predicted by pQCD. In their first paper they considered
only coupling to the ρ and ω mesons, whereas later the ϕ meson was also

included.

VMD models form a subset of models using dispersion relations, which relate form factors to spectral functions. These spectral functions can also be thought of as a superposition of vector meson poles, but include contributions from n-particle production continua. This framework thus allows a model-independent fit[12,13] to all available EMFF data in the space- and the time-like region.

Many attempts have been made to enlarge the domain of applicability of pQCD calculations to moderate Q^2-values. Kroll et al.[9] have generalized the hard-scattering scheme by assuming nucleons to consist of quarks and diquarks. The diquarks are used to approximate the effects of correlations in the nucleon wave function. This model is equivalent to the hard-scattering formalism of pQCD in the limit $Q^2 \to \infty$. Chung and Coester[10] have developed a relativistic constituent quark model with effective quark masses and a confinement scale as free parameters. Lu et al.[15] have recently expanded the cloudy bag model, whereby the nucleon is described as a bag containing three quarks, by including an elementary pion field coupled to them, in such a way that chiral symmetry is restored. As mentioned earlier, Brodsky and Farrar[11] had shown that in the pQCD limit the ratio F_2/F_1 is proportional to Q^{-2}. Recently, Ralston et al.[16] predicted this ratio to scale with Q^{-1} if they assumed the contributions to the proton wavefunction from the quark angular momentum $L_q = 0$ or 1 to be equal.

Recent developments[17] within the Generalized Parton Distribution formalism indicate a relation between the EMFF behaviour at larger Q^2-values and the nucleon spin. One should keep in mind, however, that all presently available theories are at least to some extent effective (or parametrizations). Only lattice QCD theory could provide a truely ab initio calculation, but reliable results for the EMFF are still several years away.

3 Polarization Instrumentation

Over 20 years ago Akhiezer and Rekalo[25] and Arnold et al.[26] showed that the accuracy of EMFF measurements could be increased significantly by scattering polarized electrons off a polarized target (or by equivalently measuring the polarization of the recoiling nucleon). In the early nineties a series of measurements[45,46,47,48,35,36] at the MIT-Bates facility demonstrated the feasibility of this new measurement principle, as shown in figs. 2 and 3.

Since then, significant technological advances have resulted in a large number of new data with a significantly improved accuracy. Polarized electron beams[27,28] are now reliably available with a polarization close to 80 %

at currents of up to 100 μA. The beam polarization is measured with either Møller[29] or Compton[30] polarimeters with an accuracy approaching 1 %. The dynamical polarization technique[31] provides polarized hydrogen or deuterium targets with an average polarization of 80 or 20 %, respectively, while polarized helium targets are available with a polarization close to 50 % , either through spin[32] or metastability[33] exchange at a density of 10 atm. Finally, the polarization of recoiling or knocked-out reaction products can be measured with focal-plane[34] or neutron[54] polarimeters.

4 Neutron Magnetic Form Factor

Significant progress has been made in measurements of G_M^n at low Q^2-values by measuring the ratio of quasi-elastic neutron and proton knock-out from a deuterium target. This method is insensitive to nuclear binding effects and to fluctuations in the luminosity and detector acceptance. The basic set-up used in all such measurements is very similar: the electron is detected in a magnetic spectrometer with coincident neutron/proton detection in a large scintillator array. The main technical difficulty in such a ratio measurement is the absolute determination of the neutron detection efficiency. For the measurements at Bates[36] and ELSA[37] the efficiency was measured in situ using the $D(\gamma, p)n$ or the $p(\gamma, \pi^+)$ reaction with a bremsstrahlung radiator up stream of the experimental target. The hadron detectors used in the experiments at NIKHEF[39] and Mainz[40,41] were calibrated at the PSI neutron beam using the kinematically complete $p(n, p)n$ reaction. Figure 2 shows the results of those four experiments. The Mainz G_M^n data are 8-10 % lower than those from ELSA, at variance with the quoted uncertainty of appr. 2 %. This discrepancy would require a 16-20 % error in the detector efficiency. The contribution from electroproduction in the ELSA set-up, caused by the electron contamination in the bremsstrahlung beam, which could result in a loss of events due to the three-body kinematics in electroproduction, has been extensively investigated[38]. Thus far, the detection inefficiency due to electroproduction has been established at less than 5 %, clearly much smaller than required to explain the discrepancy in the data.

The high accuracy of the Mainz data set by itself has been used to extract a precise value of the magnetic radius of the neutron $< r_M^2 >_n^{1/2} = 0.89 \pm 0.07 \, fm$, in excellent agreement with the proton charge and magnetic radius values 0.86 ± 0.01 and $0.86 \pm 0.06 \, fm$, respectively.

None of the available theoretical predictions[12,14,15,22] shown in fig. 2 provides an accurate description of the data. The dispersion theory of Mergell et al.[12] and the old VMD prediction of Höhler et al.[6] are in reasonable agree-

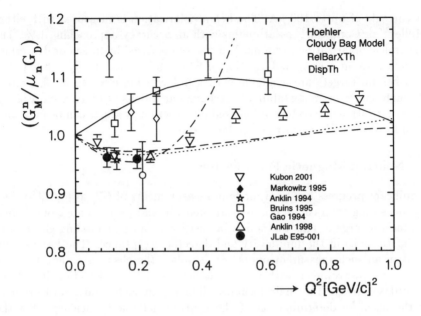

Figure 2. The neutron magnetic form factor G_M^n, in units of the dipole parametrization μG_D, as a function of Q^2. The data are from refs.[41,36,39,37,35,40,42]. The theory curves are dotted[6], dashed[12], dot-dashed[14] and solid[15].

ment with the general trend of the data from Mainz.

Recently, inclusive quasi-elastic scattering of polarized electrons off a polarized ^3He target was measured[42] in Hall A at JLab in a Q^2-range from 0.1 to 0.6 $(GeV/c)^2$. This experiment has provided an independent accurate measurement of G_M^n at Q^2-values of 0.1 and 0.2 $(GeV/c)^2$, in excellent agreement with the Mainz data. At the higher Q^2-values the nuclear medium corrections[23,24] have not yet been expanded with relativistic effects. A study of G_M^n at Q^2-values up to 5 $(GeV/c)^2$ has recently been completed at JLab by measuring the neutron/proton quasi-elastic cross-section ratio using the CLAS detector[43].

5 Neutron Electric Form Factor

Since a free neutron target is not available, one has to use neutrons bound in nuclei to study the neutron EMFF. The most precise data on G_E^n prior to any spin-dependent experiment were obtained from the elastic electron-deuteron scattering experiment by Platchkov *et al.*[49] The deuteron elastic form factor $A(Q^2)$ contains a term of the form $G_E^n G_E^p$. However, in order to extract G_E^n from the data, one has to calculate the deuteron wave function, which requires a choice of the nucleon-nucleon potential. Figure 3 shows the values extracted from the Platchkov data with the Paris potential, while the grey band indicates the range of G_E^n-values extracted with the Nijmegen, AV14 and RSC potentials. Evidently, the choice of NN-potentials results in a systematic uncertainty of appr. 50 % in G_E^n. Recently, Schiavilla and Sick[50] have extracted G_E^n from the data available on the quadrupole form factor G_Q of the deuteron (by combining data on A, B and T_{20}). The results shown in fig. 3 indicate that the model dependence in G_E^n extracted from G_Q is significantly smaller than from A.

In the last decade a series of spin-dependent measurements have provided new accurate data on G_E^n by utilizing the fact that the ratio of the beam-target asymmetry with the target polarization perpendicular and parallel to the momentum transfer is directly proportional to the ratio of the electric and magnetic form factors:

$$\frac{G_E^n}{G_M^n} = \frac{A_\perp}{A_{//}} \sqrt{\tau + \tau (1 + \tau) \tan^2 (\theta_e/2)} \tag{5}$$

A similar relation can be derived for the reaction $^2H(\vec{e}, e'\vec{n})$ when one measures the polarization of the recoiling neutron directly and after having precessed the neutron spin over 90° with a dipole magnet.

Figure 3 shows results obtained through the reaction channels $^2\vec{H}(\vec{e}, e'n)$[52], $^2H(\vec{e}, e'\vec{n})$[53,54] and $^3\vec{He}(\vec{e}, e'n)$[55,56]. At low Q^2-values corrections for nuclear medium and rescattering effects can be sizeable: 65 % for ^2H at 0.15 $(GeV/c)^2$ and 50 % for $^3\vec{He}$[23] at 0.35 $(GeV/c)^2$. These corrections are expected to decrease significantly with increasing Q, although no reliable results are presently available for $^3\vec{He}$ above 0.5 $(GeV/c)^2$. Thus, there are now data from a variety of reaction channels available in a Q^2-range up to 0.6 $(GeV/c)^2$ with an overall accuracy of appr. 20 %, which are in mutual agreement. However, neither the VMD[4] nor the dispersion relation[14] calculation agrees with the data. Only the Galster parametrization[57], basically a modified version of the dipole form factor, is able to describe the data adequately.

Shown on the horizontal axis in fig. 3 are results expected in the near future, from the $^2\vec{H}(\vec{e}, e'n)$[62] and $^2H(\vec{e}, e'\vec{n})$[60] channels at JLab and from the $^2H(\vec{e}, e'\vec{n})$[58] channel at Mainz. Recently an experiment[63] has been approved at JLab to measure G_E^n at Q^2-values of 2.4 and 3.4 $(GeV/c)^2$ using the $^3\vec{He}(\vec{e}, e'n)$ reaction. In addition, the BLAST facility[61] at MIT is expected to

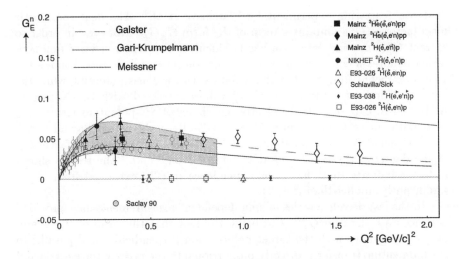

Figure 3. The neutron electric form factor G_E^n as a function of Q^2. The data are indicated by the following symbols: solid - squares[55,56], diamonds[51], triangles[53,54], and circle[52]-, open - circles[49], triangle[59], diamonds[50] - and small diamonds[60]. Theory curves are dotted[14] and solid[5], while the Galster parametrization[57] is presented by the dashed curve. Symbols on the zero axis indicate the Q^2-value and the size of the error bars of ongoing experiments[60,58].

provide highly accurate data on G_E^n in a Q^2-range from 0.1 to 0.8 $(GeV/c)^2$. Thus, within a couple of years G_E^n data with an accuracy of 10 % or better will be available up to a Q^2-value of 3.4 $(GeV/c)^2$.

6 Proton Electric Form Factor

Arnold *et al.*[26] have shown that the systematic error in a measurement of G_E^p, inherent to the Rosenbluth separation, can be significantly reduced by scattering longitudinally polarized electrons off a hydrogen target and measuring the ratio of the transverse to longitudinal polarization of the recoiling proton.

$$\frac{G_E^p}{G_M^p} = -\frac{P_t}{P_l}\frac{E_e + E_e}{2m_N}\tan\left(\frac{\theta_e}{2}\right) \tag{6}$$

This ratio of the two polarization components P_t and P_l can be measured in a focal plane polarimeter, while neither the beam polarization nor the polarimeter analyzing power need be known. This method was first used by

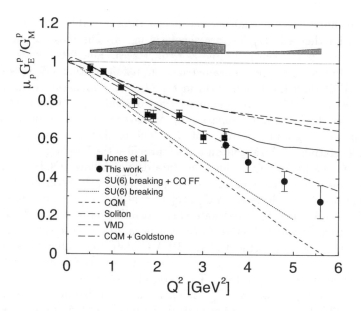

Figure 4. The ratio $\mu G_E^p/G_M^p$ from ref.[66](open squares) and ref.[67](solid circles), compared to various theoretical calculations: dash-dotted[13], thin dashed[19], solid[18], dashed[20], dotted[18] and short dashed[21]. The systematic errors for both experiments are shown as a band at the top of the figure.

Milbrath $et\ al.$[64] at MIT-Bates to measure the ratio G_E^p/G_M^p at low Q^2.

This technique has been used in two experiments[66,67], performed in Hall A at JLab, to measure the ratio G_E^p/G_M^p in a Q^2-range from 0.5 to 5.6 $(GeV/c)^2$. Longitudinally polarized electrons, with a polarization of up to 70 % and energies between 0.9 and 4.6 GeV were scattered in a 15 cm long liquid hydrogen target. Elastic ep events were selected by detecting electrons and protons in coincidence in the two identical HRS spectrometers. At the four highest Q^2-values a lead-glass calorimeter was used to detect the scattered electrons. The polarization of the recoiling proton was determined with a Focal Plane Polarimeter (FPP) in the hadron HRS, consisting of two pairs of straw chambers with a carbon or polyethylene analyzer in between. Instrumental asymmetries are cancelled by taking the difference of the azimuthal distributions of the protons scattered in the analyzer for positive and negative beam helicity. A Fourier analysis of this difference then yields the transverse and normal polarization components at the FPP. The data were analyzed in bins of each of the target coordinates. No dependence on any of these variables was observed. The results for the ratio G_E^p/G_M^p are shown in fig. 4. The most striking feature of the data is the sharp decline as Q^2 increases. Since it is known that G_M^p closely follows the dipole parametrization, it follows that G_E^p falls more rapidly with Q^2 than the dipole form factor G_D. A comparison with fig. 1 confirms the expected improvement in accuracy of such a spin-dependent measurement over the Rosenbluth separation. All theoretical calculations shown in fig. 4 predict a gradual decrease of G_E^p. The prediction by Holzwarth[20], who used a relativistic chiral soliton model, provides the best agreement with the data. At the highest Q^2-values the data do not follow pQCD scaling[11], but rather the $1/Q$ behaviour in F_2/F_1, predicted by Ralston et $al.$[16]. If one assumes the linear decrease in G_E^p/G_M^p to continue, G_E^p would cross zero at $Q^2 \approx 7.7$ $(GeV/c)^2$. To investigate this possibility an extension[68] of this experiment to a Q^2-value of 9.6 $(GeV/c)^2$ has been approved to run in Hall C.

The symmetric arrangement of the BLAST detector[61] allows a simultaneous measurement of two cross-section asymmetries at the same Q^2-value in the left and right detector sectors. The ratio of these two asymmetries then provides a measurement of G_E^p/G_M^p, independent of the beam and target polarizations. This technique[69] will provide data on G_E^p with an unprecedented accuracy of ≈ 0.3 % in a Q^2-range from 0.07 to 0.9 $(GeV/c)^2$.

7   Proton Magnetic Form Factor

No new data have become available on the magnetic form factor of the proton G_M^p since ref.[3]. Brash et $al.$[70] have reanalyzed the world data set on G_M^p imposing the form factor ratio G_E^p/G_M^p as measured in refs.[66,67] as a constraint. As a result the data lie appr. 2 % higher than the original parametrization of Bosted et $al.$[2]

8   Summary

Recent advances in polarized electron sources, polarized nucleon targets and nucleon recoil polarimeters have made it possible to accurately measure the spin-dependent elastic electron-nucleon cross section. New data on nucleon electro-magnetic form factors with an unprecedented precision have (and will continue to) become available in an ever increasing Q^2-domain. These data will form tight constraints on models of nucleon structure and will hopefully incite new theoretical efforts. In addition they will significantly improve the accuracy of the extraction of strange form factors from parity-violating experiments.

Acknowledgments

The author expresses his gratitude to Haiyan Gao, Franz Gross, Mark Jones and Richard Madey for fruitful discussions and for receiving their results prior to publication. The Southeastern Universities Research Association (SURA) operates the Thomas Jefferson National Accelerator Facility for the United States Department of Energy under contract DE-AC05-84ER40150.

References

1. M.N. Rosenbluth, *Phys. Rev.* **79**, 615 (1950).
2. P. Bosted *et al.*, *Phys. Rev.* C **51**, 409 (1995).
3. L. Andivahis, *Phys. Rev.* D **50**, 5491 (1994).
4. M.F. Gari and W. Krümpelmann, *Z. Phys.* A **322**, 689 (1985).
5. M.F. Gari and W. Krümpelmann, *Phys. Lett.* B **274**, 159 (1992).
6. G. Höhler *et al.*, *Nucl. Phys.* B **114**, 505 (1976).
7. F. Iachello, A. Jackson and A. Lande, *Phys. Lett.* B **43**, 191 (1973).
8. V.A. Nesterenko and A.V. Radyushkin, *Phys. Lett.* B **128**, 439 (1983).
9. P. Kroll, M. Schürmann, and W. Schweiger, *Z. Phys.* A **338**, 339 (1991); private communication (1998).
10. P.L. Chung and F. Coester, *Phys. Rev.* D **44**, 229 (1991); private communication (1999).
11. S.J. Brodsky and G. Farrar, *Phys. Rev.* D **11**, 1309 (1975).
12. P. Mergell *et al.*, *Nucl. Phys.* A **596**, 367 (1996).
13. E.L. Lomon, *Phys. Rev.* C **64**, 035204 (2001).
14. U.-G. Meissner, *Nucl. Phys.* A **666**, 51 (2000).
15. D.H. Lu, A.W. Thomas and A.G. Williams, *Phys. Rev.* C **57**, 2628 (1998).
16. J. Ralston *et al.*, Proc. 7th Int. Conf. on Intersection of Particle and Nuclear Physics, Quebec City (2000), p. 302.
17. X. Ji, *Phys. Rev.* D **55**, 7114 (1997); *Phys. Rev. Lett.* **78**, 610 (1997).
18. F. Cardarelli and S. Simula, *Phys. Rev.* C **62**, 065201 (2000).
19. R.F. Wagenbrunn *et al.*, *Phys. Lett.* B **511**, 33 (2001).
20. G. Holzwarth, *Z. Phys.* A **356**, 339 (1996).
21. M.R. Franck, B.K. Jennings and G.A. Miller, *Phys. Rev.* C **54**, 920 (1995).
22. F. Schlumpf, *J. Phys.* G **20**, 237 (1994).
23. J. Golak *et al.*, *Phys. Rev.* C **63**, 034006 (2001).
24. W. Glöckle, private communication (1999).
25. A.I. Akhiezer and M.P. Rekalo, *Sov. J. Part. Nucl.* **3**, 277 (1974); based

on J.H. Scofield, *Phys. Rev.* **141**, 1352 (1966).

26. R. Arnold, C. Carlson and F. Gross, *Phys. Rev.* C **23**, 363 (1981).
27. K. Aulenbacher *et al.*, *Nucl. Instrum. Methods* A **391**, 498 (1997).
28. M. Poelker *et al.*, Proc. of the 14th Int. Symp. on High-Energy Spin Physics, Osaka (2000), p. 943.
29. M. Hauger *et al.*, *Nucl. Instrum. Methods* A **462**, 382 (2001).
30. N. Falletto *et al.*, *Nucl. Instrum. Methods* A **459**, 412 (2001).
31. D. Crabb and D. Day *et al.*, *Nucl. Instrum. Methods* A **356**, 9 (1995); T.D. Averett *et al.*, *Nucl. Instrum. Methods* A **427**, 440 (1999).
32. P.L. Anthony *et al.*, *Phys. Rev. Lett.* **71**, 959 (1993); *Phys. Rev.* D **54**, 6620 (1996); J.S. Jensen, Ph.D. Thesis, CalTech, 2000 (unpublished).
33. R. Surkau *et al.*, *Nucl. Instrum. Methods* A **384**, 444 (1997).
34. M.K. Jones *et al.*, AIP Conf. Proc. 412, ed. T.W. Donnelly, p. 342 (1997); L. Bimbot *et al.*, to be submitted to *Nucl. Instr. Meth.*
35. H. Gao *et al.*, *Phys. Rev.* C **50**, R546 (1994).
36. P. Markowitz *et al.*, *Phys. Rev.* C **48**, R5 (1993).
37. E.E.W. Bruins *et al.*, *Phys. Rev. Lett.* **75**, 21 (1995).
38. B. Schoch, private communication (1999).
39. H. Anklin *et al.*, *Phys. Lett.* B **336**, 313 (1994).
40. H. Anklin *et al.*, *Phys. Lett.* B **428**, 248 (1998).
41. G. Kubon *et al.*, *Phys. Lett.* B **524**, 26 (2002).
42. W. Xu *et al.*, *Phys. Rev. Lett.* **85**, 2900 (2000).
43. W. Brooks and M.F. Vineyard, JLab experiment E94-017.
44. A.F. Lung *et al.*, *Phys. Rev. Lett.* **70**, 718 (1993).
45. T. Eden *et al.*, *Phys. Rev.* C **50**, R1749 (1994).
46. C.E. Woodward *et al.*, *Phys. Rev. Lett.* **65**, 698 (1990).
47. C.E. Jones-Woodward *et al.*, *Phys. Rev.* C **44**, R571 (1991).
48. A.K. Thompson *et al.*, *Phys. Rev. Lett.* **68**, 2901 (1992).
49. S. Platchkov *et al.*, *Nucl. Phys.* A **510**, 740 (1990).
50. R. Schiavilla and I. Sick, *Phys. Rev.* C **64**, 041002 (2001).
51. M. Meyerhoff *et al.*, *Phys. Lett.* B **327**, 201 (1994).
52. I. Passchier *et al.*, *Phys. Rev. Lett.* **82**, 4988 (1999).
53. C. Herberg *et al.*, *Eur. Phys. Jour.* A **5**, 131 (1999).
54. M. Ostrick *et al.*, *Phys. Rev. Lett.* **83**, 276 (1999).
55. J. Becker *et al.*, *Eur. Phys. Jour.* A **6**, 329 (1999).
56. D. Rohe *et al.*, *Phys. Rev. Lett.* **83**, 4257 (1999).
57. S. Galster *et al.*, *Nucl. Phys.* B **32**, 221 (1971).
58. H. Schmieden *et al.*, MAMI proposal A1/2-99.
59. H. Zhu *et al.*, *Phys. Rev. Lett.* **87**, 081801 (2001).
60. B.D. Anderson, S. Kowalski and R. Madey, JLab experiment E93-038.

61. Bates Large Acceptance Spectrometer Toroid, http://mitbates.mit.edu /blast.
62. D. Day, JLab proposal E93-026.
63. B. Wojtsekhowski *et al.*, JLab proposal E02-013.
64. B. Milbrath *et al.*, *Phys. Rev. Lett.* **80**, 452 (1998); erratum, *Phys. Rev. Lett.* **82**, 2221 (1999).
65. T. Pospischil *et al.*, *Eur. Phys. Jour.* A **12**, 125 (2001).
66. M.K. Jones *et al.*, *Phys. Rev. Lett.* **84**, 1398 (2000).
67. O. Gayou *et al.*, accepted for *Phys. Rev. Lett.*
68. C.F. Perdrisat *et al.*, JLab experiment E01-109.
69. H. Gao, J.R. Calarco and H. Kolster, MIT-Bates proposal 01-01.
70. E. Brash *et al.*, hep-ex/0111038.

SEARCH FOR NEW BARYON RESONANCES

Bijan Saghai

Service de Physique Nucléaire, DAPNIA - CEA/Saclay,
F-91191 Gif-sur-Yvette Cedex, France
E-mail: bsaghai@cea.fr

Zhenping Li

Physics Department, Peking University, Beijing 100871, P.R. China

Within a chiral constituent quark formalism, allowing the inclusion of all known resonances, a comprehensive study of the recent η photoproduction data on the proton up to $E_\gamma^{lab} \approx 2$ GeV is performed. This study shows evidence for a new S_{11} resonances and indicates the presence of an additional missing P_{13} resonance.

1 Introduction

For several decades, the baryon resonances have been investigated [1] mainly through partial wave analysis of the "pionic" processes $\pi N \to \pi N$, ηN, $\gamma N \to \pi N$, and to less extent, from two pion final states.

Recent advent of high quality electromagnetic beams and sophisticated detectors, has boosted intensive experimental and theoretical study of mesons photo- and electro-production. One of the exciting topics is the search for new baryon resonances which do not couple or couple too weakly to the πN channel. Several such resonances have been predicted [2,3,4] by different QCD-inspired approaches, offering strong test of the underlying concepts.

Investigation of η-meson production *via* electromagnetic probes offers access to fundamental information in hadrons spectroscopy. The properties of the decay of baryon resonances into γN and/or $N^* \to \eta N$ are intimately related to their internal structure [3,5,6]. Extensive recent experimental efforts on η photoproduction [7,8,9,10,11] are opening a new era in this topic.

The focus in this manuscript is to study all the recent $\gamma p \to \eta p$ data for $E_\gamma^{lab} < 2$ GeV ($W \equiv E_{total}^{cm} < 2.2$ GeV) within a chiral constituent quark formalism based on the $SU(6) \otimes O(3)$ symmetry. The advantage of the quark model for meson photoproduction is the ability to relate the photoproduction data directly to the internal structure of the baryon resonances. Moreover, this approach allows the inclusion of all of the known baryon resonances. To go beyond the exact $SU(6) \otimes O(3)$ symmetry, we introduce [12,13] symmetry breaking factors and relate them to the configuration mixing angles generated by the gluon exchange interactions in the quark model [5].

2 Theoretical Frame

The chiral constituent quark approach for meson photoproduction is based on the low energy QCD Lagrangian [14]

$$\mathcal{L} = \bar{\psi} \left[\gamma_\mu (i\partial^\mu + V^\mu + \gamma_5 A^\mu) - m \right] \psi + \ldots \tag{1}$$

where ψ is the quark field in the $SU(3)$ symmetry, $V^\mu = (\xi^\dagger \partial_\mu \xi + \xi \partial_\mu \xi^\dagger)/2$ and $A^\mu = i(\xi^\dagger \partial_\mu \xi - \xi \partial_\mu \xi^\dagger)/2$ are the vector and axial currents, respectively, with $\xi = e^{i\Pi f}$; f is a decay constant and Π the Goldstone boson field.

The four components for the photoproduction of pseudoscalar mesons based on the QCD Lagrangian are:

$$\mathcal{M}_{fi} = \mathcal{M}_{seagull} + \mathcal{M}_s + \mathcal{M}_u + \mathcal{M}_t \tag{2}$$

The first term in Eq. (2) is a seagull term. It is generated by the gauge transformation of the axial vector A_μ in the QCD Lagrangian. This term, being proportional to the electric charge of the outgoing mesons, does not contribute to the production of the η-meson. The second and the third terms correspond to the s- and u-channels, respectively. The last term is the t-channel contribution and contains two parts: $i)$ charged meson exchanges which are proportional to the charge of outgoing mesons and thus do not contribute to the process $\gamma N \to \eta N$; $ii)$ ρ and ω exchange in the η production which are excluded here due to the duality hypothesis [13].

The u-channel contributions are divided into the nucleon Born term and the contributions from the excited resonances. The matrix elements for the nucleon Born term are given explicitly, while the contributions from the excited resonances above 2 GeV for a given parity are assumed to be degenerate so that their contributions could be written in a compact form [15].

The contributions from the s-channel resonances can be written as

$$\mathcal{M}_{N^*} = \frac{2M_{N^*}}{s - M_{N^*}\left[M_{N^*} - i\Gamma(q)\right]} e^{-\frac{k^2 + q^2}{6\alpha_{ho}^2}} \mathcal{A}_{N^*}, \tag{3}$$

where $k = |\mathbf{k}|$ and $q = |\mathbf{q}|$ represent the momenta of the incoming photon and the outgoing meson respectively, $\sqrt{s}$ is the total energy of the system, $e^{-(k^2+q^2)/6\alpha_{ho}^2}$ is a form factor in the harmonic oscillator basis with the parameter α_{ho}^2 related to the harmonic oscillator strength in the wave-function, and M_{N^*} and $\Gamma(q)$ are the mass and the total width of the resonance, respectively. The amplitudes $\mathcal{A}_{N^*}$ are divided into two parts [15,16]: the contribution from each resonance below 2 GeV, the transition amplitudes of which have been translated into the standard CGLN amplitudes in the harmonic oscillator basis,

Table 1: Resonances included in our study with their $SU(6)\otimes O(3)$ configuration assignments.

Resonance	$SU(6) \otimes O(3)$ State	C_{N^*}	Resonance	$SU(6) \otimes O(3)$ State	C_{N^*}
$S_{11}(1535)$	$N(^2P_M)_{\frac{1}{2}-}$	1	$S_{11}(1650)$	$N(^4P_M)_{\frac{1}{2}-}$	0
$P_{11}(1440)$	$N(^2S'_S)_{\frac{1}{2}+}$	1	$P_{11}(1710)$	$N(^2S_M)_{\frac{1}{2}+}$	1
$P_{13}(1720)$	$N(^2D_S)_{\frac{3}{2}+}$	1	$P_{13}(1900)$	$N(^2D_M)_{\frac{3}{2}+}$	1
$D_{13}(1520)$	$N(^2P_M)_{\frac{3}{2}-}$	1	$D_{13}(1700)$	$N(^4P_M)_{\frac{3}{2}-}$	0
$F_{15}(1680)$	$N(^2D_S)_{\frac{5}{2}+}$	1	$F_{15}(2000)$	$N(^2D_M)_{\frac{5}{2}+}$	1
			$D_{15}(1675)$	$N(^4P_M)_{\frac{5}{2}-}$	0

and the contributions from the resonances above 2 GeV treated as degenerate, since little experimental information is available on those resonances.

The contributions from each resonance to η photoproduction is determined by introducing [12] a new set of parameters C_{N^*} and the following substitution rule for the amplitudes $\mathcal{A}_{N^*}$,

$$\mathcal{A}_{N^*} \to C_{N^*}\mathcal{A}_{N^*}, \tag{4}$$

so that

$$\mathcal{M}_{N^*}^{exp} = C_{N^*}^2 \cdot \mathcal{M}_{N^*}^{qm}, \tag{5}$$

where $\mathcal{M}_{N^*}^{exp}$ is the experimental value of the observable, and $\mathcal{M}_{N^*}^{qm}$ is calculated in the quark model [16]. The $SU(6) \otimes O(3)$ symmetry predicts $C_{N^*} = 0$ for $S_{11}(1650)$, $D_{13}(1700)$, and $D_{15}(1675)$ resonances, and $C_{N^*} = 1$ for other resonances in Table 1. Thus, the coefficients C_{N^*} give a measure of the discrepancies between the theoretical results and the experimental data and show the extent to which the $SU(6) \otimes O(3)$ symmetry is broken in the process investigated here.

One of the main reasons that the $SU(6) \otimes O(3)$ symmetry is broken is due to the configuration mixings caused by the one gluon exchange [5]. Here, the most relevant configuration mixings are those of the two S_{11} and the two D_{13} states around 1.5 to 1.7 GeV. The configuration mixings can be expressed in terms of the mixing angle between the two $SU(6) \otimes O(3)$ states $|N(^2P_M) >$ and $|N(^4P_M) >$, with the total quark spin 1/2 and 3/2,

$$\begin{pmatrix} |S_{11}(1535) > \\ |S_{11}(1650) > \end{pmatrix} = \begin{pmatrix} \cos\theta_S & -\sin\theta_S \\ \sin\theta_S & \cos\theta_S \end{pmatrix} \begin{pmatrix} |N(^2P_M)_{\frac{1}{2}-} > \\ |N(^4P_M)_{\frac{1}{2}-} > \end{pmatrix}, \tag{6}$$

and

$$\begin{pmatrix} |D_{13}(1520) > \\ |D_{13}(1700) > \end{pmatrix} = \begin{pmatrix} \cos\theta_D & -\sin\theta_D \\ \sin\theta_D & \cos\theta_D \end{pmatrix} \begin{pmatrix} |N(^2P_M)_{\frac{3}{2}-} > \\ |N(^4P_M)_{\frac{3}{2}-} > \end{pmatrix}. \tag{7}$$

To show how the coefficients C_{N^*} are related to the mixing angles, we express the amplitudes $\mathcal{A}_{N^*}$ in terms of the product of the meson and photon transition amplitudes:

$$\mathcal{A}_{N^*} \propto < N|H_m|N^* >< N^*|H_e|N >, \tag{8}$$

where H_m and H_e are the meson and photon transition operators, respectively. Using Eqs. (6) to (8), for the resonance $S_{11}(1535)$ one finds

$$\mathcal{A}_{S_{11}(1535)} \propto \left[< N|H_m|N(^2P_M)_{\frac{1}{2}-} > \cos\theta_S - < N|H_m|N(^4P_M)_{\frac{1}{2}-} > \sin\theta_S \right]$$
$$\left[< N(^2P_M)_{\frac{1}{2}-}|H_e|N > \cos\theta_S - < N(^4P_M)_{\frac{1}{2}-}|H_e|N > \sin\theta_S \right] (9)$$

Due to the Moorhouse selection rule, the amplitude $< N(^4P_M)_{\frac{1}{2}-}|H_e|N >$ vanishes [16] in our model. So, Eq. (9) becomes

$$\mathcal{A}_{S_{11}(1535)} \propto \left[< N|H_m|N(^2P_M)_{\frac{1}{2}-} >< N(^2P_M)_{\frac{1}{2}-}|H_e|N > \right]$$
$$\left[\cos^2\theta_S - \sin\theta_S \cos\theta_S \frac{< N|H_m|N(^4P_M)_{\frac{1}{2}-} >}{< N|H_m||N(^2P_M)_{\frac{1}{2}-} >} \right]. \tag{10}$$

where $< N|H_m|N(^2P_M)_{\frac{1}{2}-} >< N(^2P_M)_{\frac{1}{2}-}|H_e|N >$ determines [16] the CGLN amplitude for the $|N(^2P_M)_{\frac{1}{2}-} >$ state, and the ratio

$$\mathcal{R} = \frac{< N|H_m|N(^4P_M)_{\frac{1}{2}-} >}{< N|H_m|N(^2P_M)_{\frac{1}{2}-} >}, \tag{11}$$

is a constant determined by the $SU(6) \otimes O(3)$ symmetry. Using the same meson transition operator H_m from the Lagrangian as in deriving the CGLN amplitudes in the quark model, we find $\mathcal{R} = -1$ for the S_{11} resonances and $\sqrt{1/10}$ for the D_{13} resonances. Then, the configuration mixing coefficients can be related to the configuration mixing angles

$$C_{S_{11}(1535)} = \cos\theta_S(\cos\theta_S - \sin\theta_S), \tag{12}$$

$$C_{S_{11}(1650)} = -\sin\theta_S(\cos\theta_S + \sin\theta_S), \tag{13}$$

$$C_{D_{13}(1520)} = \cos\theta_D(\cos\theta_D - \sqrt{1/10}\sin\theta_D), \tag{14}$$

$$C_{D_{13}(1700)} = \sin\theta_D(\sqrt{1/10}\cos\theta_D + \sin\theta_D). \tag{15}$$

3 Results and Discussion

Our effort to investigate the $\gamma p \to \eta p$ process has gone through three stages.

In our early work[12], we took advantage of the differential cross section data from MAMI[7] (100 data points for $E_\gamma^{lab} = 0.716$ to 0.790 GeV) and polarization asymmetries measured with polarized target at ELSA[8] (50 data points for $E_\gamma^{lab} = 0.746$ to 1.1 GeV) and polarized beam at GRAAL[9] (56 data points for $E_\gamma^{lab} = 0.717$ to 1.1 GeV). Those data allowed us to study the reaction mechanism in the first resonance region and led to the conclusion[12] that the $S_{11}(1535)$ plays by far the major role in this energy range.

Later, differential cross section data were released by the GRAAL collaboration[10] (244 data points for $E_\gamma^{lab} = 0.732$ to 1.1 GeV). Using the four data sets we extended our investigations to the second resonance region and performed a careful treatment of the configuration mixing effects. This work[13] led us to the conclusion that the inclusion of a new S_{11} resonance was needed to interpret those data.

Finally, the third resonance region has just been covered by the CLAS g1a cross section measurements[11] (192 data points for $E_\gamma^{lab} = 0.775$ to 1.925 GeV).

Within our approach, we are in the process of interpreting all available experimental results and report here our preliminary findings.

Below, we summarize the main ingradients of the starting point and the used procedure leadind to the models $\mathcal{M}_1$, $\mathcal{M}_2$, and $\mathcal{M}_3$ (see Table 2 and Figs. 1 to 3):

- **Mixing angles:** Our earlier works[12,13] have shown the need to go beyond the exact $SU(6) \otimes O(3)$ symmetry. To do so, we used the relations (12) to (15) for the S_{11} and D_{13} resonances and left the mixing angles θ_S and θ_D as free parameters.

- **Model $\mathcal{M}_1$:** This model includes explictly all the eleven known relevant resonances (Table 1) with mass below 2 GeV, while the contributions from the known excited resonances above 2 GeV for a given parity are assumed to be degenerate and hence written in a compact form[15].

- **Model $\mathcal{M}_2$:** Because of the poor agreement between the model $\mathcal{M}_1$ and the data above $E_\gamma^{lab} \approx 1$ GeV, as explained below, and given our previous

Figure 1: Differential cross section for the process $\gamma p \to \eta p$: angular distribution for E_γ^{lab} = 0.775 GeV to 1.725 GeV. The curves come from the models $\mathcal{M}_1$ (dotted), $\mathcal{M}_2$ (dashed), and $\mathcal{M}_3$ (full). Data are from Refs. [7] and [10].

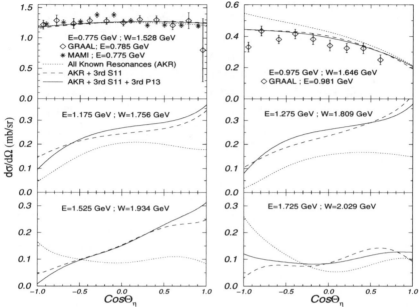

findings[13], we introduce a third S_{11} resonance with three free parameters (namely the resonance mass, width, and strength).

- **Model $\mathcal{M}_3$:** To improve further the agreement between our results and the data, we introduce a third P_{13} missing resonance with three additional adjustable parameters.

- **Fitting procedure:** The free parameters of all the above three models have been extracted using the MINUIT minimization code from the CERN Library. The fitted data base contains roughly 650 values: differential cross-sections from MAMI[7], GRAAL[10], and JLab[11], and the polarization asymmetry data from ELSA[8] and GRAAL[9].

In the following, we compare the results of our models with different measured observables[a].

---

[a]The differential cross sections from JLab[11] were kindly provided to us by B. Ritchie and M. Dugger and were included in our fitted data-base. However, given that these data have not yet been published by the CLAS Collaboration, we do not reproduce them here.

Table 2: Results of minimizations for the models as explained in the text.

parameter		$\mathcal{M}_1$	$\mathcal{M}_2$	$\mathcal{M}_3$
Mixing angles:				
	Θ_S	$-37°$	$-34°$	$-34°$
	Θ_D	$8°$	$11°$	$11°$
Third S_{11}	Mass (GeV)		1.795	1.776
	Width (MeV)		350	268
Third P_{13}	Mass (GeV)			1.887
	Width (MeV)			225
	$\chi^2_{d.o.f}$	6.5	3.1	2.7

Figure 1 shows our results at six of the CLAS data energies. At the lowest energies we compare our results with data from GRAAL and MAMI. As already mentioned, at the lowest energy the reaction mechanism is dominated by the first S_{11} resonance and the data are equally well reproduced by the three models. At the next shown energy, $E_\gamma^{lab}=0.975$ GeV, we are already in the second resonance region and the model $\mathcal{M}_1$ overestimates the data, while the models $\mathcal{M}_2$ and $\mathcal{M}_3$ improve equally the agreement with the data.

At $E_\gamma^{lab}=1.175$ and 1.275 GeV, the model $\mathcal{M}_1$ underestimates the unshown JLab data (see footnote a). This is also the case at the two depicted highest energies, except at backward angles, where again the model $\mathcal{M}_1$ overestimates the JLab data. The reduced χ^2 for this latter model is 6.5, see Table 1.

The most dramatic improvement is obtained by introducing a new S_{11} resonance (Fig. 1, model $\mathcal{M}_2$), which brings down the reduced χ^2 by more than a factor of 2.

Finally the introduction of a new P_{13} resonance (Fig. 1, model $\mathcal{M}_3$) gives the best agreement with the data, though it does not play as crucial a role as the third S_{11} resonance.

Predictions of those models for the total cross section, as well as results for the fitted polarizations observables are depicted in Figures 2 and 3, respectively. These theory/data comparisons bolster our conclusions about the new resonances, without providing further selectivity between models $\mathcal{M}_2$ and $\mathcal{M}_3$.

Here, we would like to comment on the values reported in Table 1.

Mixing angles : The extracted values for mixing angles θ_S and θ_D are identical for models $\mathcal{M}_2$ and $\mathcal{M}_3$ and differ by $3°$ from those of the model $\mathcal{M}_1$. These values are in agreement with angles determined by Isgur-Karl model [5]

Figure 2: Total cross section as a function of total center-of-mass energy for the process $\gamma p \rightarrow \eta p$; curves and data as in Fig. 1.

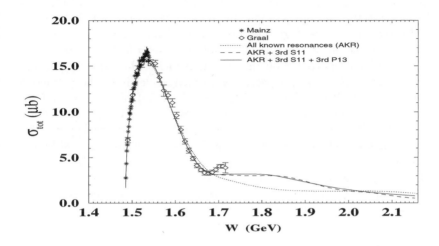

Figure 3: Single polarization asymmetries angular distribution for the process $\gamma p \rightarrow \eta p$; curves as in Fig. 1. Data are from Refs. [8] and [9].

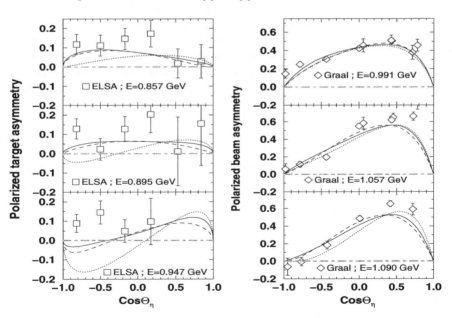

and by large-N_c approaches [20].

Third S_{11} resonance: The extracted values for the mass and width of a new S_{11} are close to those predicted by the authors of Ref. [17] (M=1.712 GeV and Γ_T=184 MeV), and our previous findings [13]. Moreover, for the one star $S_{11}(2090)$ resonance [1], the Zagreb group coupled channel analysis [18] produces the following values M = 1.792 ± 0.023 GeV and $\Gamma_T = 360 \pm 49$ MeV. The BES Collaboration reported [19] on the measurements of the $J/\psi \to p\bar{p}\eta$ decay channel. In the latter work, a partial wave analysis leads to the extraction of the mass and width of the $S_{11}(1535)$ and $S_{11}(1650)$ resonances, and the authors find indications for an extra resonance with $M = 1.800 \pm 0.040$ GeV, and $\Gamma_T = 165^{+165}_{-85}$ MeV. Finally, a recent work [4] based on the hypercentral constituent quark model predicts a missing S_{11} resonance with M=1.861 GeV.

Third P_{13} resonance: The above mentioned hypercentral CQM predicts also three P_{13} resonances with M=1.816, 1.894, and 1.939 GeV. Finally a relativized pair-creation $^3P_0$ approach [3] predicts four missing P_{13} resonances in the relevant energy region with masses betwenn 1.870 and 2.030 GeV.

4 Concluding remarks

We reported here on a study of the process $\gamma p \to \eta p$ for E_γ^{lab} between threshold and ≈ 2 GeV, using a chiral constituent quark approach.

We show how the symmetry breaking coefficients C_{N^*} are expressed in terms of the configuration mixings in the quark model, thus establish a direct connection between the photoproduction data and the internal quark gluon structure of baryon resonances. The extracted configuration mixing angles for the S and D wave resonances in the second resonance region using a more complete data-base are in good agreement with the Isgur-Karl model [5] predictions for the configuration mixing angles based on the one gluon exchange, as well as with results coming from the large-N_c effective field theory based approaches [20].

However, one of the common features in our investigation of η photoproduction at higher energies is that the existing S-wave resonances can not accommodate the large S-wave component above $E_\gamma^{lab} \approx 1.0$ GeV region. Thus, we introduce a third S-wave resonance in the second resonance region suggested in the literature [17]. The introduction of this new resonance, improves significantly the quality of our fit and reproduces very well the cross-section increase in the second resonance region. The quality of our semi-prediction for the total cross-section and our results for the polarized target and beam asymmetries, when compared to the data, gives confidence to the presence of a third S_{11} resonance, for which we extract some static and dynamical properties: $M \approx$

1.776 GeV, $\Gamma_T \approx 268$ MeV. These results are in very good agreement with those in Refs. [4,17,19], and compatible with ones in Ref. [18]. Finally, we find indications for a missing P_{13} resonace[3,4] with $M \approx 1.887$ GeV, $\Gamma_T \approx 225$ MeV.

Acknowledgments

One of us (BS) wishes to thank the organizers for their kind invitation to this very stimulating symposium. We are intebted to Barry Ritchie and Michael Dugger for having provided us with the CLAS g1a data prior to publication.

References

1. Particle Data Group, *Eur. Phys. J.* A **15**, 1 (2000).
2. S. Capstick, W. Roberts, *Prog. Part. Nucl. Phys.* **45**, 5241 (2000); R. Bijker, F. Iachello, A. Leviatan, *Ann. Phys.* **236**, 69 (1994); *Ann. Phys.* **284**, 89 (2000); L.Ya. Glozman, D.O. Riska, *Phys. Rep.* **268**, 263 (1996); M. Anselmino *et al.*, *Rev. Mod. Phys.* **65**, 1199 (1993).
3. S. Capstick, *Phys. Rev.* D **46**, 2864 (1992); S. Capstick, W. Roberts, *Phys. Rev.* D **49**, 4570 (1994).
4. M.M. Giannini, E. Santopinto, A. Vassalo, nucl-th/0111073.
5. N. Isgur, G. Karl, *Phys. Lett.* B **72**, 109 (1977); N. Isgur, G. Karl, R. Koniuk, *Phys. Rev. Lett.* **41**, 1269 (1978).
6. R. Bijker, F. Iachello, A. Leviatan, *Phys. Rev.* C **54**, 1935 (1996).
7. B. Krusche *et al.*, *Phys. Rev. Lett.* **74**, 3736 (1995).
8. A. Bock *et al.*, *Phys. Rev. Lett.* **81**, 534 (1998).
9. J. Ajaka *et al.*, *Phys. Rev. Lett.* **81**, 1797 (1998).
10. F. Renard *et al.*, submitted to *Phys. Lett.* B, hep-ex/0011098.
11. M. Dugger, B. Ritchie *et al.* (CLAS Collaboration), to be submitted to *Phys. Rev. Lett.*; B. Ritchie, private communication (Oct. 01).
12. Z. Li, B. Saghai, *Nucl. Phys.* A **644**, 345 (1998).
13. B. Saghai, Z. Li, *Eur. Phys. J.* A **11**, 217 (2001).
14. A. Manohar, H. Georgi, *Nucl. Phys.* B **234**, 189 (1984).
15. Z. Li, *Phys. Rev.* D **48**, 3070 (1993); *Phys. Rev.* D **50**, 5639 (1994); *Phys. Rev.* C **52**, 1648 (1995).
16. Z. Li, H. Ye, M. Lu, *Phys. Rev.* C **56**, 1099 (1997).
17. Z. Li, R. Workman, *Phys. Rev.* C **53**, R549 (1996).
18. M. Batinić *et al.*, *Phys. Scripta* **58**, 15 (1998); A. Švarc, S. Ceci, nucl-th/0009024.
19. J.Z. Bai *et al.* (BES Collaboration), *Phys. Lett.* B **510**, 75 (2001).
20. D. Pirjol, T.-M. Yan, *Phys. Rev.* D **57**, 5434 (1998); C.E. Carlson *et al.*, *Phys. Rev.* D **59**, 114008 (1999).

OVERVIEW OF LASER-ELECTRON PHOTON FACILITY AT SPring-8

T. NAKANO

FOR THE LEPS COLLABORATION

RCNP, Osaka University,
10-1 Mihogaoka, Ibaraki, Osaka 567-0047, JAPAN
E-mail: nakano@rcnp.osaka-u.ac.jp

The GeV photon beam at SPring-8 is produced by backward-Compton scattering of laser photons from 8 GeV electrons. Polarization of the photon beam will be ∼100 % at the maximum energy with fully polarized laser photons. We report the status of the new facility and the prospect of hadron physics study with this high quality beam. Preliminary results from the first physics run are presented.

1 LEPS FACILITY

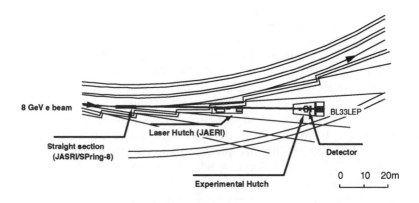

Figure 1. Plan view of the Laser-Electron Photon facility at SPring–8 (LEPS).

The Spring-8 facility is the most powerful third-generation synchrotron radiation facility with 62 beamlines. We use a beamline, BL33LEP (Fig. 1), for the quark nuclear physics studies. The beamline has a 7.8-m long straight section between two bending magnets. Polarized laser photons are injected from a laser hutch toward the straight section where Backward-Compton scattering (BCS) [1] of the laser photons from the 8 GeV electron beam takes place. The BCS photon beam is transferred to the experimental hutch, 70 m downstream of the straight section. The maximum energy of the BCS photon is expressed by

$$k_2 = \frac{4k_1 E_e^2}{m_e^2 + 4k_1 E_e},$$
(1)

where k_1 is the energy of the laser photon, E_e is the energy of the electron, and m_e is the electron mass. For a 351-nm (3.5 eV) Ar laser and a 8-GeV electron beam, the maximum energy is 2.4 GeV well above the threshold of ϕ-phtoproduction from a nucleon (1.57 GeV).

If laser lights are 100 % polarized, a backward-Compton-scattered photon is highly polarized at the maximum energy. The polarization drops as the photon energy decreases. However, an energy of laser photons is easily changed so that the polarization remains reasonably high in the energy region of interest.

The incident photon energy is determined by measuring the energy of a recoil electron with a tagging counter. The tagging counter locates at the exit of the bending magnet after the straight section. It consists of multi-layers of a 0.1 mm pitch silicon strip detector (SSD) and plastic scintillator hodoscopes. Electrons in the energy region of 4.5 – 6.5 GeV are detected by the counter. The corresponding photon energy is 1.5 – 3.5 GeV. The position resolution of the system is much better than a required resolution. The energy resolution (RMS) of 15 MeV for the photon beam is limited by the energy spread of the electron beam and an uncertainty of a photon-electron interaction point.

The operation of the BCS beam at SPring–8 started in July, 1999. Ar laser at 351-nm wave length is used, and the intensity of the beam is about 2.5×10^6 photons/sec for a 5 W laser-output.

2 PHYSICS MOTIVATION

2.1 ϕ photoproduction near threshold

Since a ϕ meson is almost pure $s\bar{s}$ state, diffractive photo-production of a ϕ meson off a proton in a wide energy range is well described as a pomeron-exchange (multi gluon-exchange) process in the framework of the Regge theory and of the Vector Dominance Model (VDM) [2]; a high energy photon converts into a ϕ meson and then it is scattered from a proton by an exchange of the pomeron [3,4,5] while the meson-exchange is suppressed by the OZI rule. However, at low energies other contributions arising from meson (π, η)-exchange [5], a scaler (0^{++} glueball)-exchange [6], and $s\bar{s}$ knock-out [7] are possible. These contributions fall off rapidly as the incident γ-ray energy increases, and can be studied only in the low energy region near the production threshold. Linearly polarized photons are an ideal probe to decompose these contributions. For natural-parity exchange such as pomeron and 0^{++} glueball exchanges, the decay plane of $K^{+}K^{-}$ is concentrated in the direction of the photon polarization vector. For unnatural-parity exchange processes like π and η exchange processes, it is perpendicular to the polarization vector. Jlab [8] and LEPS will greatly contribute to this field.

2.2 K^{+} photoproduction

Recent measurements for $K^{+}\Lambda$ photoproduction at SAPHIR indicated a structure around $W = 1.9$ GeV in the total cross-section [9]. It attracted theorist's interest to study missing nucleon resonances in this process. Mart and Bennhold [10] showed that the SAPHIR data can be reproduced by inclusion of a new D_{13} resonance which have large couplings both to the photo and the $K\Lambda$ channels according to the quark model calculation [11]. Although it is difficult to draw a strong conclusion on the existence of D_{13} resonance from the cross-section measurements, the photon polarization asymmetry is very sensitive to the missing nucleon resonance. The LEPS collaboration will measure the asymmetry in the energy region of 1.5–2.4 GeV [12].

2.3 ω photoproduction

Missing nucleon resonances could couple to the π-nucleon channel weakly but couple to the ω-nucleon channel strongly [13]. The LEPS can measure a forward proton from backward ω photoproduction. In this kinematic region u-channel and s-channel contributions dominate and the effects of missing resonances would be large. The old differential cross section data for backward ω photoproduction [14] shows a structure around $u = -0.15$ GeV, which is very hard to be reproduced by model calculations [15]. New precise measurements are awaited to confirm the structure.

3 Detector

The LEPS detector (Fig. 2) consists of a plastic scintillator to detect charged products after a target, an aerogel Cerenkov counter with a refractive index of 1.03, charged-particle tracking counters, a dipole magnet, and a time-of-flight TOF wall. The design of the detector is optimized for a ϕ photo-production

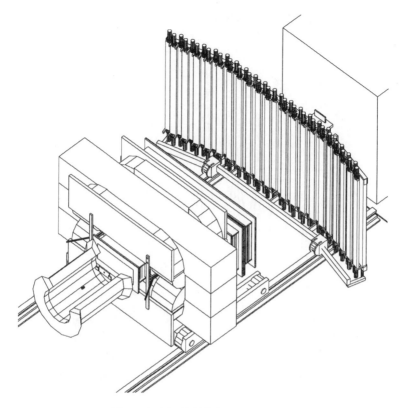

Figure 2. The LEPS detector setup.

at forward angles.

The opening of the dipole magnet is 135-cm wide and 55-cm height. The length of the pole is 60 cm, and the field strength at the center is 1 T. The vertex detector consists of 2 planes of single-sided SSDs and 5 planes multi-wire drift chamber, which are located upstream of the magnet. Two sets of MWDCs are located downstream of the magnet.

The identification of momentum analyzed particles is performed by measuring a time of flight from the target to the TOF wall. The start signal for the TOF measurement is provided by a RF signal from the 8-GeV ring, where electrons are bunched at every 2 nsec with a width (σ) of 16 psec. A stop signal is provided by the TOF wall consisting of 40 2m-long plastic scintillation bar with a time cross resolution of 150 psec.

4 PRELIMINARY RESULTS

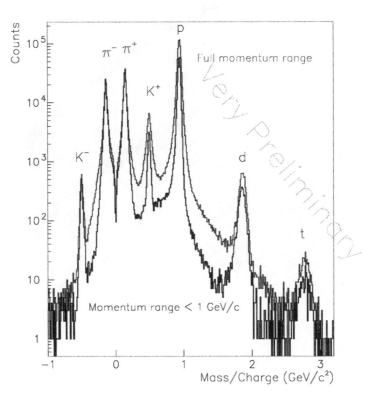

Figure 3. A mass distribution of charged particles reconstructed from momentum and TOF information.

The physics run with a 5-cm long liquid H_2 target wad carried out during December, 2000 to June, 2001. The trigger required a tagging counter hit, no charged particle before the target, charged particles after the target, no signal in the aerogel cerenkov counter, at least one hit on the TOF wall. A typical trigger rate was about 20 counts per second.

Figure 3 shows a preliminary mass distribution of charged particles reconstructed from momentum and TOF information. The events of ϕ mesons are successfully identified through the reconstruction of the K^+K^- invariant mass as shown in Figure 4.

Figure 5 shows a missing mass distribution of the (γ, K^+) reactions. The Λ and Σ peaks are clearly identified, but the $\Lambda(1405)$ is not separated from

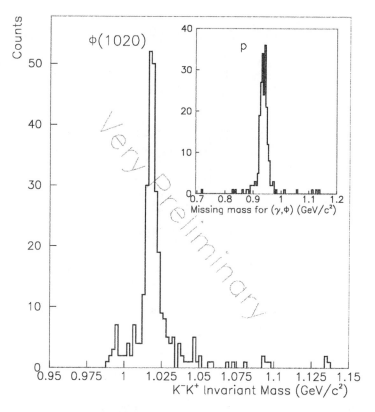

Figure 4. A two-kaon invariant mass distribution. The ϕ peak is clearly identified.

the $\Sigma(1385)$. A time-projection chamber to decompose the two resonances by detecting their decay products will be constructed in 2002.

Figure 6 shows a missing mass distribution of the (γ, p) reactions. The ω, η, and η' peaks are observed.

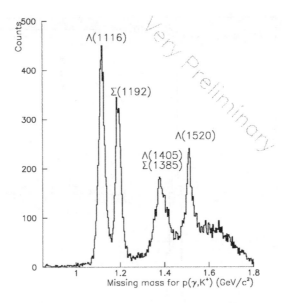

Figure 5. A missing mass distribution of (γ, K^+) reactions.

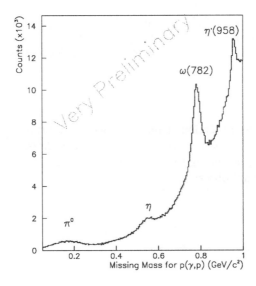

Figure 6. A missing mass distribution of (γ, p) reactions.

References

1. R.H. Milburn, Phys. Rev. Lett. 10, 75 (1963).
2. J.J. Sakurai, Ann. Phys. 11, 1 (1960); J.J. Sakurai, Phys. Rev. Lett. 22, 981 (1969).
3. T.H. Bauer et al., Rev. Mod. Phys. 50, 261 (1978).
4. A. Donnachie and P.V. Landshoff, Nucl. Phys. B267, 690 (1986).
5. M.A. Pichowsky and T.-S. H. Lee, Phys. Rev. D56, 1644 (1997).
6. T. Nakano and H. Toki, in Proc. of Intern. Workshop on Exciting Physics with New Accelerator Facilities, SPring-8, Hyogo, 1997, World Scientific Publishing Co. Pte. Ltd., 1998, p.48.
7. A.I. Titov, Y. Oh, and S.N. Yang, Phys. Rev. Lett. 79, 1634 (1997); A.I. Titov, Y. Oh, and S.N. Yang, Phys. Rev. C58, 2429 (1998).
8. D.J. Tedeschi, in these proceedings; P.L. Cole, in these proceedings.
9. M.Q. Tran et al., Phys. Lett. B 445, 20 (1998).
10. T. Mart and C. Bennhold, Phys. Rev. C61, (R)012201 (2000).
11. S. Capstick and W. Roberts, Phys. Rev. D58, 074011 (1998).
12. R.G.T. Zegers, in these proceedings.
13. Q. Zhao, in these proceedings; A.I. Titov, in these proceedings.
14. R.W. Clifft et al., Phys. Lett. B 72 (1977) 144.
15. T.-S. H. Lee, private communication.

WEAK NUCLEON FORM FACTORS

Paul A. Souder[a]

Syracuse University, Syracuse, NY 13244

E-mail: souder@physics.syr.edu

Experiments at JLab, Mainz, and MIT-Bates are probing weak nucleon factors by measuring the parity-violating asymmetry in the scattering of polarized electrons from nucleons. The goal of this work is the extraction of the contribution of strange quarks to nucleon form factors.

1 Introduction

The role that antiquarks in general and strange quarks in particular play in the structure of the nucleon is a key issue. Strange quarks contribute approximately 3% of the momentum of a nucleon, [1] and analyses of data on spin structure functions [2,3,4,5] suggest that strange quarks also carry some of the spin. Kaplan and Manohar [6] suggested that strange quarks might also contribute to elastic nucleon form factors and that this contribution can be isolated by measuring weak form factors.

The main technique for observing the weak form factors is measuring the parity violating asymmetry in the scattering of polarized electrons from an unpolarized target: [7,8,9]

$$A^{PV} = \frac{\sigma_R - \sigma_L}{\sigma_R + \sigma_L}, \qquad (1)$$

where $\sigma_{L(R)}$ is the cross section for the scattering of left(right) handed electrons. There has been much progress in measuring weak form factors. [10,11] Completed [12,13,14,15] or approved [16,17,18,19,20,21] experiments in the field are listed in Table 1.

2 Theory Elastic Form Factors and Strange Quarks

The cross section for unpolarized targets is the sum of three amplitudes, the electromagnetic amplitude f^γ, and the weak amplitudes f^Z_V, and f^Z_A shown in Fig. 1. The parity violating asymmetry isolates the weak amplitudes. In the Standard Model, the axial coupling of the electron to the Z is large, so f^Z_V is also large and provides a useful tool for measuring the weak vector form factors of the target. On the other hand, g^V for the electron and thus f^Z_A is small.

[a]Work supported by the DOE under contract number DE-FG02-84ER40146

Table 1: Survey of parity experiments measuring weak nucleon form factors

Experiment	Reaction	Physics Goals	A^{PV}
Completed Experiments			
SAMPLE-P [12]	$\vec{e}P$ Elastic	$G_M^s(0) = \mu_s$	10^{-5}
SAMPLE-D [13]	$\vec{e}D$ Elastic	G_A^e	10^{-5}
HAPPEX(JLab) [14,15]	$\vec{e}P$ Elastic	$G_M^s + 0.39 G_E^s$	10^{-5}
Approved Experiments			
Mainz [16]	$\vec{e}P$ Elastic	G_M^s, G_E^s	10^{-5}
G^0(JLab) [17]	$\vec{e}P$ Elastic	G_M^s, G_E^s	10^{-5}
^{4}He(JLab) [18]	$\vec{e}\ ^4$He Elastic	G_E^s	10^{-5}
HAPPEX II(JLab) [19]	$\vec{e}P$ Elastic	$G_M^s + 0.39 G_E^s$	10^{-6}
SAMPLE-D [20]	$\vec{e}D$ Elastic	G_A^e	10^{-5}
HAPPEX ^{4}He [21]	$\vec{e}\ ^4$He Elastic	G_E^s	10^{-5}

The *electromagnetic* cross section for elastic scattering from the nucleon with four-momentum transfer Q^2 is

$$\frac{d\sigma}{d\Omega} = \left(\frac{d\sigma}{d\Omega}\right)_{Mott} \left[(F_1^{\gamma N})^2 + \tau(F_2^{\gamma N})^2 + 2\tau(F_1^{\gamma N} + F_2^{\gamma N})^2 \tan^2(\theta/2) \right], \quad (2)$$

where θ is the scattering angle, $\tau = Q^2/4M_N^2$, and M_N is the mass of the nucleon. The quantities $F_{1,2}^{\gamma N}$ are the Dirac elastic electromagnetic form factors, which are functions of Q^2. The form factors may be written in terms of contributions from individual quarks as follows:

$$F_{1,2}^{\gamma p} = \frac{2}{3}F_{1,2}^u - \frac{1}{3}F_{1,2}^d - \frac{1}{3}F_{1,2}^s; \quad F_{1,2}^{\gamma n} = \frac{2}{3}F_{1,2}^d - \frac{1}{3}F_{1,2}^u - \frac{1}{3}F_{1,2}^s \quad (3)$$

where the second relation follows from charge symmetry. Weak scattering from the proton, involving the Z-boson, is governed by the weak form factor

$$F_{1,2}^{Zp} = F_{1,2}^u - F_{1,2}^d - F_{1,2}^s - 4\sin^2\theta_W F_{1,2}^{\gamma p}. \quad (4)$$

The strange form factors $F_{1,2}^s$ may be determined from the above equations after the electromagnetic form factors for both the proton and neutron as well as the weak form factors F_i^{Zp} are measured.

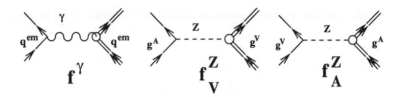

Figure 1: Feynman diagrams for electroweak scattering.

The parity-violating asymmetry for the proton is given in terms of the Sachs form factors: $G_E^i = F_1^i - \tau F_2^i$, and $G_M^i = F_1^i + F_2^i$. by:[22]

$$A^{PV} = \left[\frac{-G_F M_p^2 \tau}{\pi\alpha\sqrt{2}}\right]\left\{(1 - 4\sin^2\theta_W) - \right. \tag{5}$$

$$\frac{[\varepsilon G_E^{p\gamma}(G_E^{n\gamma} + G_E^s) + \tau G_M^{p\gamma}(G_M^{n\gamma} + G_M^s)]}{\varepsilon(G_E^{p\gamma})^2 + \tau(G_M^{p\gamma})^2} -$$

$$\left.\frac{(1 - 4\sin^2\theta_W)\sqrt{\tau(1+\tau)}\sqrt{1 - \varepsilon^2}G_M^{p\gamma}(-G_A^{(1)} + \frac{1}{2}F_A^s)}{\varepsilon(G_E^{p\gamma})^2 + \tau(G_M^{p\gamma})^2}\right\}.$$

Here $\varepsilon = [1 + 2(1 + \tau)\tan^2(\theta/2)]^{-1}$ and F_A^s is the strange quark contribution to the axial form factor.

The kinematics of the scattering determines the sensitivity of the asymmetry to various form factors. For small values of both τ and θ, the asymmetry is sensitive only to the fundamental weak interaction. A^{PV} is primarily sensitive to G_E^s and G_M^s for small θ and large τ, and primarily to G_M^s and G_A^e for large values of θ. This fact motivates the wide range of kinematics evident in Table 1.

The radiative correction for the axial term has substantial uncertainties.[24] The SAMPLE collaboration defines a quantity $G_A^e(T = 0)$, which includes all radiative corrections and replaces $G_A^{(1)}$ in the above equation. On the other hand, the radiative corrections for the vector form factors terms are well known.[23]

A number of papers have made predictions for the sizes of the form factors. [25,26,27,28,29,30,31,32,33,34,35,36,37,38] The predictions may be expressed in terms of the parameters ρ_s and μ_s, which are the low Q^2 limits

$$G_E^s \to \tau\rho_s : \quad G_M^s \to \mu_s. \tag{6}$$

Table 2: Some published theoretical estimates for ρ_s and μ_s.

Method	ρ_s	μ_s	Author
Pole fits	-2.1 ± 1.0	-0.31 ± 0.09	Jaffe[25]
	-2.9 ± 0.5	-0.24 ± 0.03	Hammer[26]
Kaon Loops	0.2	-0.03	Koepf[27]
	0.5 ± 0.1	-0.35 ± 0.05	Ramsey-Mulolf[28]
	0.3	-0.12	Ito[29]
Unquenched Quarks	0.6	0.04	Geiger[30]
Meson Exchange	0.03	0.002	Meissner[31]
Meson Cloud	–	$0.-0.066$	Ma[32]
NJL	3.0 ± 0.08	-0.15 ± 0.10	Weigel[33]
Skyrme	1.6	-0.13	Park[34]
	-0.7	-0.05	Park[35]
Chiral Bag Model	–	0.37	Hong[36]
Lattice QCD	1.7 ± 0.7	-0.36 ± 0.20	Dong[37]
Dispersion Relations	0.99	-0.42	Hammer[38]

The predictions given in Table 2 also give an estimate of reasonable sixes for the strange form factors. More reliable calculations are the goal of ongoing work with lattice calculations providing an especially promising approach.

3 Experimental Techniques

Many experimental features are common to all polarized electron parity experiments. A typical apparatus is shown in Fig. 2. Photoemission of circularly polarized laser light from a GaAs cathode produces the polarized electrons. The helicity of the electron beam is determined by the helicity of the light, which is in turn determined by the voltage on a Pockels cell. The electrons are accelerated and then pass through a beam line which is highly instrumented to measure any small correlations between beam parameters and helicity. The beam then strikes a high-power target. The scattered electrons are detected by a spectrometer. A diversity of spectrometers may be used to detect the scattered electrons for particular experiments. A computer, which monitors all the signals and records the data, may also control the Pockels cell to null any intensity asymmetry. Coils may be used to calibrate the sensitivity of both the monitors and the spectrometers to beam parameters.

Since the measured asymmetries are small, it is essential to keep the effects

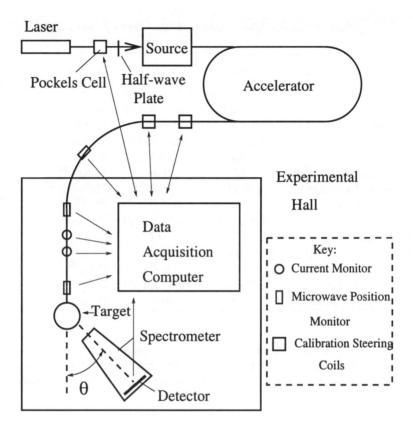

Figure 2: Generic Polarized Electron Parity Experiment

of helicity correlations on the cross section due to any other beam parameter small. Innovative techniques have been developed [39] to maintain the helicity correlations at acceptable levels. The effect of the beam parameters on the number of detected events may be calibrated by dithering steering coils in the beam line.

Recently, impressive progress has been made in the technology required for these challenging experiments. Polarized electron sources with intensities of 100μA and polarizations of $> 70\%$ have been achieved.

The helicity of the beam is reversed rapidly, typically between 15 and 300 times a second. For pulsed accelerators, including MIT-Bates and SLAC, the signals are integrated to accommodate the high instantaneous rates. For continuous wave accelerators including JLab and Mainz, either integrating or counting methods may be chosen.

4 Published Results

The HAPPEX collaboration at JLab used the high-resolution spectrometers in Hall A. Electrons of energy 3.5 GeV scattered elastically at 12.5° were detected with negligible background. The average value of Q^2 was 0.477 $(GeV/c)^2$. The experiment took place in two runs. The first was in 1998,[14] and the second in 1999.[15] The 1998 run used a bulk GaAs crystal which provided a polarization of ∼40% and a current on target of ∼ 100μA. For the 1999 data, a strained GaAs crystal was used which produced a beam with ∼70% polarization and a current of typically 40μA.

The experimental asymmetry for the combined 1998 and 1999 was $A^{PV} = -15.1 \pm 1.0(\text{stat}) \pm 0.6(\text{sys})$ ppm.[15] Combining the asymmetry data with published data on electromagnetic form factors [40,41,42,43,44,45,46,47,48] gives the result

$$(G_E^s + 0.39 G_M^s) = 0.025 \pm 0.020 \pm 0.014 \tag{7}$$

where the first error is the experimental uncertainties added in quadrature, and the second error arises from uncertainties in the electromagnetic form factor data. This number is consistent with negligible strange form factors. The allowed band, together with theoretical predictions for $Q^2 \sim 0.5$, are shown in Figure 3.

The SAMPLE collaboration at the MIT-Bates Laboratory is in the later stages of a program of experiments measuring the asymmetry for electrons scattered from both hydrogen and deuterium by an angle of ∼ 130° and $Q^2 \sim 0.1$ $(GeV/c)^2$. An array of ellipsoidal mirrors that focus the Čerenkov light produced by the scattered electrons in the air onto phototubes provides a detector with a large solid angle of 1.5 sr. Since the beam energy is only 200 MeV, inelastic events are below threshold.

The result from the SAMPLE group is $A^{PV} = -6.79 \pm 0.64 \pm 0.55$ ppm for the proton[12] and $A^{PV} = -4.92 \pm 0.61 \pm 0.73$ ppm for the deuteron.[13] Here the first errors are statistical, and the second are systematic. By using Eqn. 5, the data can be interpreted in terms of G_M^s and the radiatively corrected axial form factor G_A^e as shown in Figure 4. The region where the bands intersect gives $\mu_s = 0.01 \pm 0.29(\text{stat}) \pm 0.31(\text{sys}) \pm 0.07(\text{theor.})$ and $G_A^e(T = 1) = 0.22 \pm 0.45(\text{stat}) \pm 0.39(\text{sys})$. The latter value is a bit off from the theoretical value $G_A^e = -0.83 \pm 0.26$.[24] A new SAMPLE run will provide deuterium data with a lower energy beam in the year 2002.

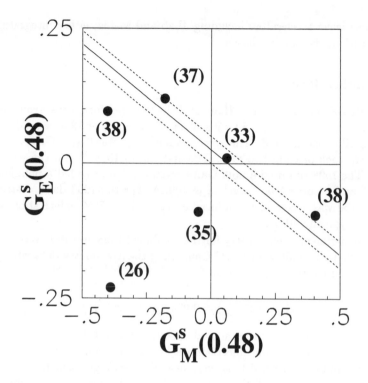

Figure 3: Band shows allowed region in the $G_E^s - G_M^s$ parameter space based on the HAPPEX data. Points with references are theoretical predictions for $Q^2 = 0.48$ (GeV/c)^2.

5 Upcoming Experiments

A number of upcoming experiments will provide more data on strange form factors. An experiment at Mainz by the A4 collaboration [16] will measure elastic scattering from hydrogen at $\theta = 35°$ and $Q^2 = 0.23$ (GeV/c)^2. A unique feature of the experiment is the detector. It will count individual events with an array of 1022 tapered PbF$_2$ crystals. Identification of elastic events is achieved by the excellent resolution of the calorimeter. Initial data-taking is completed and results are expected soon.

The G_0 experiment [17] is a major program at JLab to study strange form factors over a wide kinematic range. The unique feature is a large supercon-

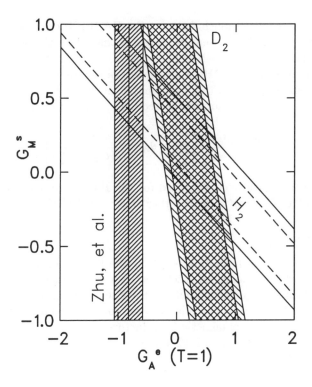

Figure 4: Results from the SAMPLE collaboration in terms of G_M^s and the radiatively corrected axial form factor $G_A^e(T = 1)$. The bands labeled H$_2$ and D$_2$ are extracted from the data. The vertical band is a theoretical estimate for $G_A^e(T = 1)$.

ducting toroidal magnet that will serve as the spectrometer. In one configuration, the recoil protons from elastic scattering with $62° < \theta < 78°$ will be detected with plastic scintillators. This kinematics, with a beam energy of 3 GeV, corresponds to electrons with $15° > \theta > 5°$ and $0.16 < Q^2 < 0.95$ $(\text{GeV/c})^2$. In another configuration, electrons scattered with large angles will be detected. Runs with various beam energies between 0.3 and 0.9 GeV will cover the same Q^2 range as the forward data. The goal is to determine G_E^s and G_M^s separately over a large range of Q^2. Commissioning is scheduled for the year 2002.

An extension of the HAPPEX experiment,[19] will measure elastic scattering from a hydrogen target at $Q^2 \sim 0.1(\text{GeV/c})^2$. The new run is motivated by the possibility that the strange form factors are proportional to Q^2 only at low values of Q^2, but fall off much faster at the HAPPEX kinematics. The small angle will be attained with septum magnets that allow the spectrometers in Hall A to reach angles as small as 6°.

Elastic scattering from ^{4}He is also sensitive to strange form factors. However, with a spinless target, there are neither magnetic form factors nor axial hadronic contributions to radiative corrections. The asymmetry is given by[9,49]

$$A^{PV} = \frac{G_F Q^2}{\pi \alpha \sqrt{2}} \left[\sin^2 \theta_W + \frac{G_E^s}{2(G_E^p + G_E^n)} \right].$$

(8)

Two experiments have been approved for the Hall A spectrometers at JLab, one at $Q^2 \sim 0.6$ (GeV/c)^2 and the other at $Q^2 \sim 0.1$ (GeV/c)^2.

6 Conclusions

Great progress has been made recently in the field of measuring weak form factors by scattering polarized electrons. Several precise results have been published which set significant limits on the size of strange elastic form factors of the nucleon. More precise data is expected soon. I should mention here that improved data on electromagnetic form factors will also be required. Several such experiments at JLab and MIT-Bates are in progress.

References

1. A. O. Bazarko *et al.*, *Z. Phys.* C **65**, 189 (1995).
2. K. Abe *et al.*, *Phys. Lett.* B **405**, 180 (1997).
3. B. Adeva *et al.*, *Phys. Rev.* D **58**, 112002 (1998).
4. G. Alterelli *et al.*, Acta Phys. Polon. B **29**, 1145 (1998).
5. E. Leader *et al.*, *Phys. Lett.* B **462**, 189 (1999).
6. D. B. Kaplan and A. Manohar, *Nucl. Phys.* B **310**, 527 (1988).
7. R. D. McKeown, *Phys. Lett.* B **219**, 140 (1989).
8. E. J. Beise and R. D. McKeown, Comments Nucl. Part. Phys. **20**, 105 (1991).
9. D. H. Beck, *Phys. Rev.* D **39**, 3248 (1989).
10. K. S. Kumar and P. A. Souder, Prog. Part. Nucl. Phys. **45** S333 (2000).
11. D. H. Beck and B. R. Holstein, Int. J. Mod. Phys. E**10**, 1 (2001).
12. D. T. Spayde *et al.*, *Phys. Rev. Lett.* **84**, 1106 (2000).
13. R. Hasty, *et al.*, *Science* **290**, 2117 (2000).
14. K. Aniol *et al.*, *Phys. Rev. Lett.* **82**, 1096 (1999).
15. K. A. Aniol *et al.*, *Phys. Lett.* B **509**, 211 (()2001).
16. Mainz proposal A4/1-93 (D. von Harrach, spokesperson).
17. JLab experiment 91-017 (D. Beck, spokesperson).
18. JLab experiment 91-004 (E. J. Beise, spokesperson).
19. JLab experiment 99-115 (K. S. Kumar and D. Lhuillier, spokespersons).

20. MIT-Bates experiment 00-04, T. M. Ito, spokesperson.
21. JLab experiment 00-114 (D. S. Armstrong and R. Michaels, spokespersons).
22. M. J. Musolf *et al.*, *Phys. Rep.* **239**, 1 (1994), and references therein.
23. Particle Data Group, C. Caso *et al.*, *Eur. Phys. J.* C **3**, 1 (1998).
24. S. -L. Zhu *et al.*, *Phys. Rev.* D **62**, 033008 (2000).
25. R. L. Jaffe, *Phys. Lett.* B **229**, 275 (1989).
26. H. -W. Hammer, Ulf-G. Meissner, and D. Drechsel, *Phys. Lett.* B **367**, 323 (1996).
27. W. Koepf, E. M. Henley, and J. S. Pollock, *Phys. Lett.* B **288**, 11 (1992).
28. M. J. Musolf and M. Burkhardt, *Z. Phys.* C **61**, 433 (1994).
29. H. Ito, *Phys. Rev.* C **52**, R1750 (1995).
30. P. Geiger and N. Isgur, *Phys. Rev.* D **55**, 299 (1997).
31. Ulf-G. Meissner *et al.*, *Phys. Lett.* B **408**, 381 (1997).
32. B.-Q Ma, *Phys. Lett.* B **408**, 387 (1997).
33. H. Weigel *et al.*, *Phys. Lett.* B **353**, 20 (1995).
34. N. W. Park, J. Schecter, and H. Weigel, *Phys. Rev.* D **43**, 869 (1991).
35. N. W. Park and H. Weigel, *Nucl. Phys.* A **541**, 453 (1992).
36. S-T. Hong, B-Y. Park, and D-P. Min, *Phys. Lett.* B **414**, 229 (1997).
37. S. J. Dong, K. F. Liu, and A. G. Williams, *Phys. Rev.* D **58**, 074504 (1998).
38. H.-W. Hammer and M. J. Ramsey-Musolf, *Phys. Rev.* C **60**, 045205 (1999).
39. T. Averett *et al.*, *Nucl. Instrum. Methods* **438**, 246 (1999)
40. M. K. Jones *et al.*, *Phys. Rev. Lett.* **84**, 1398 (2000).
41. R. C. Walker *et al.*, *Phys. Rev.* D **49**, 5671 (1994).
42. H. Anklin *et al.*, *Phys. Lett.* B **428**, 248 (1998).
43. E. E. W. Bruins *et al.*, *Phys. Rev. Lett.* **75**, 21 (1995).
44. C. Herberg *et al.*, *Eur. Phys. Jour.* A **5**, 131 (1999).
45. M. Ostrick *et al.*, *Phys. Rev. Lett.* **83**, 276 (1999).
46. I. Passchier *et al.*, *Phys. Rev. Lett.* **82**, 4988 (1999).
47. D. Rohe *et al.*, *Phys. Rev. Lett.* **83**, 4257 (1999).
48. H. Zhu *et al.*, *Phys. Rev. Lett.* **87**, 081801 (2001).
49. M. J. Musolf, R. Schiavilla, and T. W. Donnelley *Phys. Rev.* C **50**, 2173 (1994).

PARITY VIOLATING ELECTRON-PROTON SCATTERING EXPERIMENTS*

WILLEM T.H. VAN OERS

(FOR THE G0 COLLABORATION)

Department of Physics and Astronomy, University of Manitoba
Winnipeg, MB, R3T 2N2, Canada
and
TRIUMF, 4004 Wesbrook Mall, Vancouver, BC, V6T 2A3, Canada
Email: vanoers@triumf.ca

High energy electron beams can act as sensitive probes of the structure of nuclear matter at very small scales. Electron scattering experiments have given information on the distribution of electric charge and magnetization inside the proton and neutron, represented by the electric and magnetic form factors, the so called Sachs form factors. These are functions of the four-momentum transferred by the virtual photon, which mediates the electromagnetic scattering process. But the weak interaction also contributes to electron scattering through the exchange of the Z^0 boson, and one can probe entirely different aspects of the proton and neutron structure. With the Z^0 contribution so small one has to make use of the fact that the weak interaction exhibits parity violation. At the present relatively little is known about the analogous weak form factors of the proton and neutron. Of particular interest is the possibility of disentangling the strange quark contributions to these form factors, which can be accomplished by combining measurements of the parity violating asymmetry (longitudinal analyzing power) with measurements of the electric and magnetic form factors. This presents the objective of the G0 experiment at Jefferson Laboratory. A status report is given in light of recent results of the kinematically more restrictive SAMPLE (MIT-Bates), HAPPEX (Jefferson Laboratory), and PV-A4 (Mainz) experiments. The first one of these showed the importance of the nucleon anapole moment contribution to the axial form factor. The anapole moment is discussed with reference to the weak-meson nucleon couplings determined in hadronic parity violation experiments.

1 Introduction

The first glimpse of the internal structure of the nucleon appeared as the anomalous magnetic moment of the proton as measured in 1933 by Otto Stern and coworkers. They measured the magnetic moment of the proton to be 2.5 times (with an error of 10%) the magnetic moment of a Dirac proton[1]. This was followed by a series of measurements at Stanford of the Sachs electric and magnetic form factors of the proton by Hofstadter and coworkers in the 1950's, showing the nucleons to be of finite size[2]. The first direct evidence for point-like constituents in the nucleons came from the observation of scaling phenomena in deep-inelastic scattering experiments at SLAC[3]. The point-like charged constituents, called partons, were found to be consistent with spin-1/2 fermions. However, valence quarks alone do not suffice, with the evidence that quark-antiquark pairs belonging to the sea form part of the nucleon structure, as corroborated by the charge distribution of the neutron, which has a positive core and a negative shell.

The Quantum Chromo Dynamics (QCD) picture of the proton has it consisting of three valence quarks (uud), gluons mediating the strong force, and quark- antiquark pairs belonging to the sea. One expects the lightest of the second generation of quarks, the strange quark, to be present as strange-antistrange pairs exclusively belonging to the sea. There have been various indications of the presence of strange quarks in the nucleon. Neutrino deep-inelastic scattering (measuring the axial current) has shown the longitudinal momentum of the $\bar{s}$ quarks to be given by $\kappa = 2\bar{s}/(\bar{u}+\bar{d}) = 0.42\pm0.07 \pm 0.06$, with no significant difference between the s(x) and $\bar{s}$(x) distributions and x the fraction of longitudinal momentum carried by the quark[4]. Electron and muon spin-dependent deep-inelastic scattering measurements (sensitive to the vector-axial current) have shown a significant contribution of the strange quarks to the spin of the nucleon[5]. But it is to be noted that this deduction is model dependent and assumes SU(3) flavor symmetry. A third indication may stem from the πN Σ term as deduced from phase shift analyses of π-N scattering and the pion decay constant. It gives the mass of the nucleon carried by the strange quarks; theory gives a contribution of 51 $\pm$ 7 MeV, whereas experiment gives values ranging between 70 and 90 MeV[6]. Parity violating electron-nucleon scattering gives direct information on the strange vector current.

2 Electron-Nucleon Scattering

Assuming the nucleons to be composed of the three lightest quarks (u, d, and s quarks), and assuming isospin symmetry (equating the u quark distribution of the proton with the d quark distribution of the neutron, and vice versa), the nucleon electromagnetic form factors can be expressed in terms of the quark form factors:

$$G_{E,M}^{\gamma,p} = (2/3)G_{E,M}^{u} - (1/3)G_{E,M}^{d} - (1/3)G_{E,M}^{s}$$

$$G_{E,M}^{\gamma,n} = (2/3)G_{E,M}^{d} - (1/3)G_{E,M}^{u} - (1/3)G_{E,M}^{s}$$

These equations give four experimental observables expressed in six unknown quark form factors. By measuring the parity violating longitudinal analyzing power in electron-proton scattering one can determine the weak proton form factors to provide two additional observables. In completely analogous fashion one can express the proton weak form factors in terms of the quark form factors:

$$G_{E,M}^{Z,p} = (1/4 - (2/3)\sin^2\theta_W)G_{E,M}^u$$
$$- (1/4 - (1/3)\sin^2\theta_W)G_{E,M}^d - (1/4 - (1/3)\sin^2\theta_W)G_{E,M}^s$$

The strange quark form factors can then be deduced from the above set of six equations:

$$G_{E,M}^s = (1 - 4\sin^2\theta_W)G_{E,M}^{\gamma,p} - G_{E,M}^{\gamma,n} - 4G_{E,M}^{Z,p}$$

The parity violating analyzing power in electron-proton scattering is related to the electromagnetic and weak form factors as follows:

$$A_z(Q^2) = \frac{-(1/P_z)G_F Q^2}{\pi\alpha\sqrt{2}} \frac{(\epsilon G_E^{\gamma,p}G_E^{Z,p} + \tau G_M^{\gamma,p}G_M^{Z,p} + \eta G_M^{\gamma,p}G_A^e)}{(\epsilon(G_E^{\gamma,p})^2 + \tau(G_M^{\gamma,p})^2)},$$

where ϵ, τ, and η are kinematical parameters:

$$\epsilon = [1 + 2(1+\tau)\tan^2\theta/2]^{-1}$$
$$\tau = Q^2/(4M^2)$$
$$\eta = -(1 - 4\sin^2\theta_W)[\tau(1+\tau)(1-\epsilon^2)^{1/2} ,$$

and with P_z the longitudinal polarization of the electron beam, G_F the Fermi coupling constant, α the fine structure constant, and G_A^e the axial form factor. Note that the term containing the latter is multiplied with $(1 - 4\sin^2\theta_W) = 0.075$. However, there are uncertainties in calculating G_A^e and consequently it needs to be determined experimentally. A separation of the electric and magnetic form factors requires both forward angle and backward angle measurements at the same Q^2 (Rosenbluth like separation). A determination of the axial form factor requires a further measurement from the $T = 0$ target deuterium at the same Q^2 backward angles. Consequently the G0 experiment at Jefferson Laboratory will consist of three phases: a forward angle mode, and a backward angle mode with both hydrogen and deuterium targets.

3   Previous Experimental Results

Early electron-nucleus parity violating experiments had as their objective testing the Standard Model and determining $\sin^2\theta_W$. The pioneering experiment was a measurement of A_z in electron-deuteron scattering at SLAC using a longitudinally polarized electron beam from a bulk Ga-As electron source[7]. The value for the Weinberg angle obtained is $\sin^2 \theta_W = 0.224 \pm 0.020$, the most precise value at that time. This first experiment was followed by two lower energy experiments: a measurement of A_z in quasi-elastic scattering of electrons from ^{9}Be at Mainz giving a result of $\sin^2 \theta_W = 0.221 \pm 0.014 \pm 0.004$ (with the first error giving the statistical uncertainty an the second error the systematic uncertainty)[8], and a measurement of A_z in elastic scattering of electrons from the $T = 0$ target ^{12}C at MIT-Bates giving a result $\gamma = 0.136 \pm 0.032 \pm 0.009$ (in the Standard Model $\gamma = (2/3) \sin^2 \theta_W$)[9].

With the effective Weinberg angle now known to three significant figures ($\sin^2 \theta_W = 0.23147 \pm 0.00016$), the objectives of the more recent experiments have shifted from testing the Standard Model to assuming its correctness and determining the nucleon quark form factors, and in particular the strange quark form factors. Three experiments have reported results: SAMPLE at MIT-Bates, HAPPEX at Jefferson Laboratory, and PV-A4 at MAMI-Mainz.

The SAMPLE experiment has used a 200 MeV longitudinally polarized electron beam with an intensity of 40 μA and a polarization of about 0.37 (from a bulk Ga-As polarized electron source) incident on 0.40 m long LH$_2$ and LD$_2$ targets. Beam pulses have a width of 25 μs with a repetition rate of 600 Hz. The scattered electrons at backward angles (from $130°$ to $170°$) with an average Q^2 value of 0.1 $(\mathrm{GeV}/c)^2$ were detected by the Čerenkov light produced in air as observed by ten 0.20 m diameter photomultipliers tubes via ten mirrors positioned around the beam axis and covering a solid angle of 1.4 sr. By measuring A_z for both elastic scattering from hydrogen and quasi-elastic scattering from deuterium, both G_M^s and G_A^e can be deduced. The result obtained[10] is $G_M^s(Q^2 = 0.1(\mathrm{GeV}/c)^2) = 0.14 \pm 0.29 \pm 0.31$ which, using chiral perturbation theory, gives a contribution to the magnetic moment due to strange quarks $\mu_s = 0.01 \pm 0.29 \pm 0.31 \pm 0.07$ nm, a result not inconsistent with zero. Many theoretical predictions for μ_s have been made with predominantly negative values. The value for the e-N isovector part of the axial vector form vector is $G_A^e(T = 1) = +0.22 \pm 0.45 \pm 0.39$ while theoretical predictions give values of -0.71 ± 0.20 (see Ref. 11) and -0.83 ± 0.26 (see Ref. 12), a difference of about 1.5 σ with experiment (see Fig. 1 for G_M^s versus $G_A^e(T = 1)$). As shown in Fig. 2 there are three important contributions to G_A^e: G_A^Z the neutral weak axial form factor determined from neutron β decay and neutrino scattering and multiplied by $(1 - 4\sin^2 \theta_W)$, R^e which represents the electroweak radiative corrections to e-N scattering, and ηF^A with F^A the nucleon anapole moment, the parity violating electromagnetic moment of the nucleon[13]. It is the latter which is

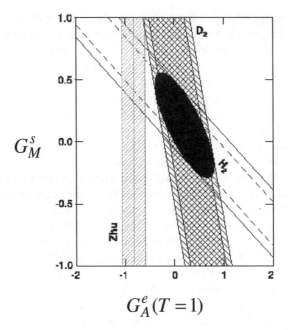

Figure 1.   Results from the SAMPLE experiments on hydrogen and deuterium plotted as G_M^s versus $G_A^e(T = 1)$(see Ref. 10). Inner error bands are for statistical errors only; outer error bands have the systematic errors added in quadrature. The vertical band is a theoretical prediction for $G_A^e(T = 1)$(see Ref. 12).

most uncertain since it depends on the weak meson-nucleon coupling constants like h_ρ, which can only be determined in precise proton-proton parity violation experiments[14]. Consequently, it is essential to determine both G_M^s and G_A^e experimentally; a comparison with theoretical predictions may shed light on the nucleon anapole moment. A further SAMPLE experiment is taking data at an energy of 125 MeV at the same backward angles corresponding to $Q^2 = 0.04$ $(\text{GeV/c})^2$. The experiment aims for an uncertainty given by $\delta G_A^e(T = 1) = \pm 0.61(\text{stat}) \pm 0.25(\text{sys})$.

The HAPPEX experiment has used 3.35 GeV longitudinally polarized electron beams incident on a 0.15 m long LH_2 target. Scattered electrons were observed in two magnetic spectrometers placed at 12.3° left and right with respect to the incident electron beam with Lead/Lucite sandwich Čerenkov detectors in the focal planes. In the first experiment the electron beam had an intensity of 100 μA and an average polarization of 0.388 ± 0.027 (from a bulk Ga-As polarized electron source), while in the second experiment the beam intensity was 35 μA with a polarization of about 0.70 (from a "strained" Ga-As polarized electron source) and a Compton polarimeter was used to continuously monitor the electron beam polarization[15]. The combined result from the two experiments is $A_z = (-14.60 \pm 0.94 \pm 0.54) \times 10^{-6}$. From this, one deduces a combination of G_E^s and G_M^s at $Q^2 = 0.477$ $(\mathrm{GeV/c})^2$: $(G_E^s + 0.392 G_M^s)/(G_M^{\gamma,p}/\mu_p) = 0.091 \pm 0.054 \pm 0.039$, where the first error is experimental and the second one arises from the uncertainties in the electromagnetic form factors. Again the result deduced is not inconsistent with zero. A further HAPPEX experiment has been scheduled for data taking at the more forward angles of 6° corresponding to $Q^2 = 0.1 (\mathrm{GeV/c})^2$ to complement the SAMPLE backward angle measurement.

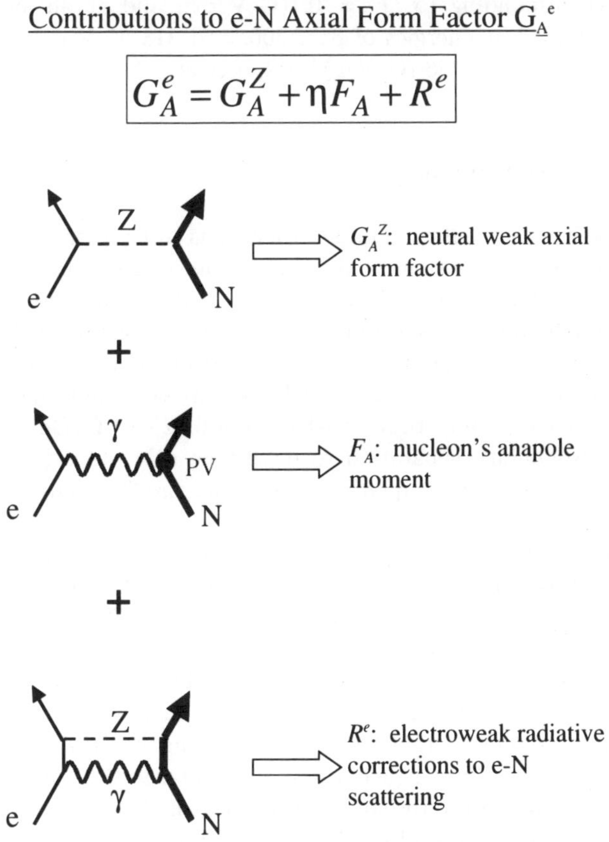

Contributions to e-N Axial Form Factor G_A^e

$$G_A^e = G_A^Z + \eta F_A + R^e$$

G_A^Z: neutral weak axial form factor

+

F_A: nucleon's anapole moment

+

R^e: electroweak radiative corrections to e-N scattering

Figure 2. Contributions to the e-N axial form factor G_A^e: G_A^Z the neutral weak axial form factor, F_A the nucleon anapole moment, and R^e electroweak radiative corrections.

The PV-A4 experiment at MAMI-Mainz uses a 855 MeV longitudinally polarized electron beam incident on a 0.10 long LH_2 target. Scattered electrons in the angular range $30° - 40°$ corresponding to $Q^2 = 0.213$ $(GeV/c)^2$ are observed in a calorimeter consisting of 512 PbF_2 crystals (one half of the final number of crystals). The electron beam intensity is 20 μA with a polarization of about 0.80 (from a "strained" Ga-As polarized electron source). The polarization is measured with a combination of a Moeller polarimeter, Compton laser back scattering polarimeter and transmission Compton polarimeter. The experiment determines the combination $G_E^s + 0.218 G_M^s$. The results of this first experiment can be found elsewhere in these proceedings[16]. Further measurements are planned at 570 MeV also at an average angle of $35°$ corresponding to $Q^2 = 0.1(GeV/c)^2$, and at an average backward angle of $145°$ at energies of 817, 506, and 318 MeV corresponding to $Q^2 = 0.94$, 0.47, and 0.23 $(GeV/c)^2$, respectively.

4   The G0 Experiment

The G0 experiment at Jefferson Laboratory[17] has three research objectives: i) to determine the contributions of the three lightest quarks to the charge and magnetization distributions of the nucleons; ii) to determine the axial coupling of the photon to the nucleon and the contribution of the nucleon anapole moment; and iii) to measure the weak neutral transition current in the region of the Δ-resonance. The experiment will measure the longitudinal analyzing power at a range of momentum transfers from 0.12 to 1.0 $(GeV/c)^2$ and will be using a Rosenbluth separation to determine G_E^s and G_M^s. Theoretical calculations without strange quark contributions predict analyzing powers in the range of -3 to -35 $\times 10^{-6}$; it is planned to measure these with statistical uncertainties of $\Delta A_z/A_z = 0.05$ and systematic uncertainties related to helicity correlated effects of $\Delta A \leq 2.5 \times 10^{-7}$. The electron beam with an intensity of 40 μA and a polarization of about 0.70 (from a "strained" Ga-As electron source) will be scattered from a 0.20 m long LH_2 target. The detection of scattered particles is based on a 1.6 Tm superconducting toroidal magnet with eightfold symmetry for momentum analysis and uses sets of 16 pairs of scintillators per octant which are placed in the focal plane at various Q^2 contours. The bending angle in the spectrometer is $35°$; collimators placed within the enclosure of the toroidal magnet in each of the octants prevent the scintillators the direct view of the LH_2 target. Fig. 3 presents a cross section of the target, toroidal spectrometer with collimators, and detectors in one of the eight symmetry planes.

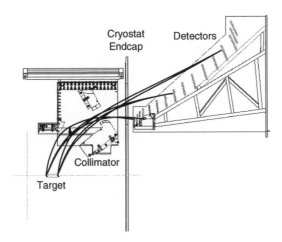

Figure 3. Cross section of the G0 target, toroidal magnetic spectrometer with collimators, and detectors in one of the eight symmetry planes between a pair of "coil pancakes (forward angle measurements)".

In the first phase of the experiment recoil protons from elastic e-p scattering at an incident energy of 3 GeV will be detected in the range of $70° \pm 10°$ (corresponding to forward scattered electrons at about 7°). Since time-of-flight needs to be incorporated over a 32 ns flight path to select the elastic events, a pulsed beam at 31.25 MHz will be used. In the second phase of the experiment the toroidal spectrometer will be turned around in order to detect scattered electrons in the range of $110° \pm 10°$. In this mode each incident electron energy corresponds to one given Q^2 value. Since both the strange quark magnetization distribution and the isovector axial form factor need to be determined, measurements will be made with hydrogen and deuterium targets at the three Q^2 values (0.3, 0.5, and 0.8 $(\text{GeV/c})^2$). The second phase experiment will reduce the focal plane detector array to a single layer of 16 scintillators and add arrays of 9 "cryostat exit detector" scintillators (CEDs) in coincidence mode. To reject the π^- background (from the entrance and exit foils of the liquid target and from deuterium), there will be a Čerenkov detector for each octant placed between the CEDs and the focal plane detectors. The expected uncertainties for the strange quark charge and magnetization distributions and the axial vector form factor at three values of Q^2, together with selected theoretical predictions, are shown in Fig. 4.

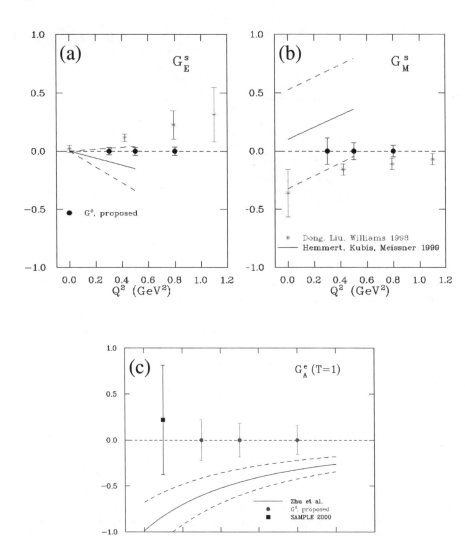

Figure 4. Expected uncertainties for the strange quark charge and magnetization distributions and axial form factor at three values of Q^2 0.3. 0.5, and 0.8 (GeV/c)^2 with theoretical predictions of Dong, Liu, and Williams [19], Hemmert, Kubis, and U.-G.Meiszner[20], and Zhu *et al.* [12]

The detector packages for each octant have been completed and are installed in the eight-fold symmetric Ferris wheel, placed in the experimental hall (Hall-C) for testing with cosmic rays and parasitic beam. The electronics has been installed in the counting room and on the beam floor. The superconducting toroidal magnet and the target are undergoing acceptance tests. The CEDs and Čerenkov detectors for the second phase experiment are being fabricated. The current schedule asks for installation of the magnet and target on the beam line starting February 2002, with the first engineering and commissioning run in the summer of 2002 and the second commissioning run and first data taking run in 2003. The turning of the toroidal spectrometer (so that the entrance and exit are interchanged) and backward angle data taking for the first Q^2 point (at $0.8(\text{GeV/c})^2$) with both LH_2 and LD_2 targets is tentatively scheduled for 2004 and data taking for the remaining two Q^2 points for 2005.

5 Future Experimentation

Very-high precision measurements of parity violating electron scattering from the proton at very low Q^2 and at forward angles allow in turn to challenge the predictions of the Standard Model and search for new physics. A unique experiment to carry out the first measurement of the weak charge of the proton $Q_W^p = 1 - 4\sin^2\theta_W$ has been proposed at JLab[18] It will measure the parity violating longitudinal analyzing power in elastic electron proton scattering at $Q^2 = 0.03(\text{GeV/c})^2$ employing a 1.165 GeV 180 μA electron beam with a polarization of 0.80 incident on a 0.35 m long LH_2 target. The experiment intends to measure the weak charge of the proton with better than 4% combined statistical and systematic errors. The Standard Model makes a firm prediction of the weak charge of the proton based on the running of the Weinberg angle $\sin^2\theta_W$ from the Z-pole to lower energies, corresponding to a 10 σ effect at the kinematic conditions of the experiment. In the absence of physics beyond the Standard Model, the experiment will provide an approximately 0.3% measurement of $\sin^2\theta_W$, a very competitive measurement indeed.

6 Conclusions

Precision parity violating electron-proton scattering experiments, like G0, allow for the measurent of a new fundamental property of the proton: $G_{E,M}^{p,Z}$, the weak current distribution of the proton. This enables a decomposition of the ground state matrix elements in the quark charge and magnetization distributions; in particular those of the strange quark will provide a direct measurement of the quark sea. Experiments performed to date (SAMPLE,

HAPPEX, and PV-A4) have given results for the strange quark distributions not inconsistent with zero. The additional measurements with a deuterium target will give information on the nucleon anapole moment, of great current interest. A next generation of very high precision measurements of $G_E^{p,Z}$ opens the door to determine the weak charge of the proton and $\sin^2(\theta_W)$ as a search for new physics, beyond the Standard Model, at the TeV scale.

References

[*] Work supported in part by the Natural Sciences and Engineering Research Council of Canada.
 1. R. Frisch and O. Stern, Z. Phys. **85**, 4 (1933); I. Estermann and O. Stern, *ibid. 17.*
 2. R.E. Hofstadter, Ann. Rev. Nucl. Sci. **7**, 231 (1957).
 3. E.D. Bloom *et al.*, Phys. Rev. Lett. **23**, 930 (1969); M. Breidenbach *et al., ibid.* 935.
 4. T. Adams *et al.*, (NuTeV collaboration), hep-ex/9906037.
 5. D. Adams *et al.*, Phys. Rev. **D56**, 5330 (1997).
 6. U.-G. Meiszner and G. Smith, Report FZJ-IKP-TH-2000-27.
 7. C.Y. Prescott *et al.*, Phys. Lett. **B77**, 347 (1978); Phys. Lett. **B84**, 524 (1979).
 8. W. Heil *et al.*, Nucl. Phys. **B327**, 1 (1979).
 9. P.A. Souder *et al.*, Phys. Rev. Lett. **65**, 694 (1990).
10. R. Hasty *et al.*, Science **290**, 2117 (2000).
11. M.J. Ramsey-Musolf and B.R. Holstein, Phys. Lett. **B242**, 461 (1990).
12. S.-L. Zhu *et al.*, Phys. Rev. **D62**, 033008 (2000).
13. Ya.B. Zel'dovich, JETP **9**, 682 (1959).
14. A.R. Berdoz *et al.*, Phys. Rev. Lett. **87**, 272301 (2001).
15. K.A. Aniol *et al.*, Phys. Rev. Lett. **82**, 1096 (1999); K.A. Aniol *et al.*, Phys. Lett. **B509**, 211 (2001).
16. F. Maas, these Proceedings.
17. JLab Proposals E-99-016, E-00-006, E-01-116, The G0 Experiment, spokesperson D.H.Beck; JLab Proposal E-01-115, Measurement of the Parity Violating Asymmetry in the N-Δ Transition, spokespersons S.P. Wells and N.Simicevic.
18. JLab Proposal E-02-020, The Qweak Experiment, spokesperson R. Carlini.
19. S.J. Dong, K.F. Liu, and A.G. Williams, Phys. Rev. **D58**, 074504 (1998).
20. T.R. Hemmert, B. Kubis, and U.-G. Meiszner, Phys. Rev. **C60**, 045501 (1999).

PARITY VIOLATION EXPERIMENT AT MAINZ

J. VAN DE WIELE AND M. MORLET

Institut de Physique Nucléaire/IN2P3, F-91406 Orsay Cedex, France
E-mail: vandewi@ipno.in2p3.fr, morlet@ipno.in2p3.fr

AND THE A4 COLLABORATION[(A,B,C,D)]*

Measurement of asymmetry in polarized electron-nucleon elastic scattering provides information on the weak charge and current distributions in the nucleon through the interference between photon and Z^0 exchange graphs. This asymmetry is due to the parity violation in the weak interaction. The expansion of weak form factors in term of flavor (u,d,s) form factors leads to information on quark structure of the nucleon, especially on the strangeness contribution to the vector form factors. The measurement of the parity violating asymmetry in elastic scattering of polarized electrons on protons at MAMI facility in Mainz is presented. At the MAMI current maximum energy of 0.855 GeV, the electrons scattered out a 10 cm liquid hydrogen target are detected between 30° and 40°. Under these kinematic conditions, the value of the transfer momentum Q^2, weighted by the elastic cross section in this angular range, is $Q^2 = 0.233$ (GeV/c)^2. This experiment measures the $G_E^s + 0.218 G_M^s$ combination of the electric and magnetic strange form factors. The detection is done using a calorimeter built of 511 Čerenkov crystals in the phase I of the experiment. The experimental set-up with its main characteristics are shown. The origin and magnitude of possible systematic and helicity correlated false asymmetries are discussed.

1 Introduction

As already mentioned in this symposium, parity violation in the scattering of polarized electrons provides an observable which is sensitive to the vector matrix element $< p|\bar{\psi}_s \gamma_\mu \psi_s|p >$. As the weak interaction violates the discrete symmetry of parity, the scattering of polarized electrons on unpolarized protons produces an asymmetry in the cross section. In this contribution,

[(A)] P. ACHENBACH, K. AULENBACHER, S. BAUNACK, B. FIEDLER, K. GRIMM, T. HAMMEL, D. VON. HARRACH, E. KABUß, R. KOTHE, K. W. KRYGIER, A. LOEPES GINJA, F. E. MAAS, E. SCHILLING, G. STEPHAN, C. WEINRICH, H. SCHMIEDEN, P. BARSTCH, M. SEIDL. INSTITUT FÜR KERNPHYSIK, UNIVERSITÄT MAINZ, BECHERWEG 45, D-55099 MAINZ [(B)] J. ARVIEUX, B. COLLIN, S. ESSABAA, R. FRASCARIA, M. GUIDAL, H. GULER, R. KUNNE, D. MARCHAND, M. MORLET, S. ONG, L. ROSIER, J. VAN DE WIELE. INSTITUT DE PHYSIQUE NUCLÉAIRE/IN2P3, F-91406 ORSAY CEDEX, FRANCE [(C)] I. ALTAREV. ST. PETERSBURG, INSTITUT FOR NUCLEAR RESEARCH, RAS, ST PETERSBURG 183550, RUSSIA [(D)] S. KOWALSKI. M.I.T. LABORATORY FOR NUCLEAR SCIENCE, CAMBRIDGE

we are interested by the elastic scattering of polarized electrons off unpolarized protons in the kinematics governed by the MAMI facility in Mainz. The transition amplitude $\mathcal{T}(p'e', ep)$ of ep elastic electroweak scattering is the sum of the electromagnetic amplitude $\mathcal{T}^{EM}(p'e', ep)$ due to the exchange of a virtual photon and of the neutral current amplitude $\mathcal{T}^{NC}(p'e', ep)$ due to the exchange of the Z^0. As the neutral current amplitude is small compared to the electromagnetic amplitude ($\approx 10^{-5}$), the measured asymmetry defined as $A = \frac{\sigma_+ - \sigma_-}{\sigma_+ + \sigma_-}$ is practically due to the interference of the electromagnetic and neutral currents. In this expression σ_+ is the cross section associated with incident electron with an helicity equal to $1/2$ and σ_- is the cross section associated with incident electron with an helicity equal to $-1/2$.

2   The PVA4 Experiment at MAMI

2.1   The experimental conditions

The experiment is performed at the MAMI facility at IKP Mainz University using the polarized electron beam at the present maximum energy of 0.855 GeV. The square of the relative statistical uncertainty is proportional to the inverse of the Factor of Merit (FOM) defined as the product of the square of the asymmetry A by the differential cross section $d\sigma/d\Omega$. The elastic differential cross section is calculated using the electromagnetic form factors parameterized as usual [2]. The asymmetry, used in the FOM, is calculated

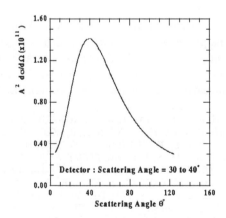

Figure 1. Figure Of Merit (FOM) for the A4 experiment at 0.855 GeV

without any strange form factors in the proton and without the electro weak radiative corrections, with the recent value $\sin^2\theta_W = 0.2311$ [3]. The values of FOM versus the scattering angle is displayed in Fig.1. The maximum is located around $35° \pm 5°$. The value of the transfer momentum Q^2 at $\theta = 35°$ is equal to 0.227 $(\text{GeV}/c)^2$, while the value, weighted by the elastic cross section between 30° and 40°, is $<Q^2> = 0.233$ $(\text{GeV}/c)^2$. At 35°, the value of A_0, the asymmetry without explicit strangeness, is expected to be equal to -6.39 ppm under the preceding assumptions. The value of this asymmetry, weighted by the differential cross section over the angular range, is $< A_0 > = -6.78$ ppm. Using the parameterization of Hammer [4], the predicted values of A_s (including strange form factors but no radiative electroweak corrections) are -9.44 ppm and $< A_s > = -9.99$ ppm respectively. For the central angle of 35°, this experiment measures a combination of electric and magnetic strange form factors given by: $G_E^s + 0.218G_M^s$.

2.2 Experimental Setup

The experimental set up is shown in Fig. 2. Polarized electrons with a

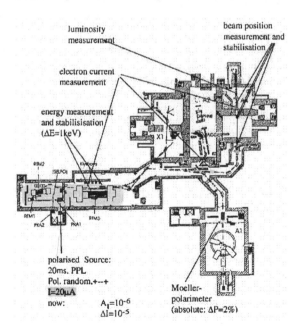

Figure 2. Overview of the experimental setup

polarization between 75% and 80% are extracted from the strained AsGa source and accelerated in the MAinz MIcrotron in three stages to an energy of 0.855 GeV. The 20 μA extracted electron beam is then focused on a liquid hydrogen target in the A4 hall. The necessary luminosity $\simeq 5.10^{37}\ cm^{-2}s^{-1}$ is achieved with a high cooling power, high flow liquid hydrogen target [5] of 10 cm length. At this intensity, the electron beam deposits 80 W of heating power in the target. The polarization of the beam is regularly measured with a Möller polarimeter by switching the beam in the hall A1. The electron current, energy, beam position and angle are continuously measured on beam line using a combination of resonant cavities. The energy of the particles scattered out of the target is absorbed in a PbF_2 Čerenkov shower calorimeter. The luminosity is measured at small angle after the target. The beam is then caught in a water cooled beam dump.

2.3 Polarized Electron Source and associated uncertainties

A laser is used to provide a linearly polarized light which is converted into a circular polarization using a $\lambda/2$ plate and a Pockels cell. Polarized electrons are produced by illuminating a strained GaAs photocathode with this laser light, the polarization of the light being transmitted to the electrons [6]. The helicity of the electrons is pseudo randomly reversed in 100 μs by switching the laser light polarization from right to left (or reverse). This is done by changing the sign of the applied high voltage to the Pockels cell. Each polarization state is maintained during a measurement gate time of nominally 20 ms in order to integrate out any fluctuation due to the European 50 Hz frequency. The time needed to reverse the polarization introduces a necessary random phase shift compared to the nominal frequency. The measurement gate lengths are continuously monitored for the two polarization states, and the false asymmetry introduced by this gate is measured, found lower than 10^{-9} and is therefore negligible. An intensity correlation at the electron source can induce an energy correlation. This energy correlation can propagate into a position and angle correlation of the electron beam on the target. It is therefore important to reduce the intensity correlation at the polarized source over the duration of the experiment. Therefore the angle of the $\lambda/2$ plate is adjusted during the measurements to obtain the same luminosity for each helicity state. The luminosity for each helicity state is measured continuously during the data acquisition. The absolute value of the polarization may be different in each helicity state. A second $\lambda/2$ plate may be introduced in the laser light before the Pockels cell to reverse the global sign of beam polarization and to allow the minimization of this difference. Half of measurements are done with this plate in, and half with the plate out. The introduction or removal of this second plate is performed regularly.

2.4 Calorimeter and electronics

The angular range covered by this experiment goes from 30° to 40° where the FOM is maximum for the chosen energy (0.855 GeV). The counting rate,

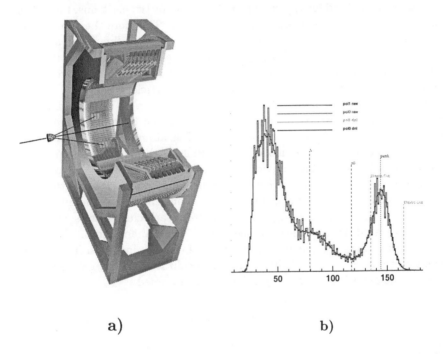

a)   b)

Figure 3. a):Schematic of the Čerenkov calorimeter b):Typical spectrum showing the elastic peak and the inelastic events. The shoulder is due to the Δ excitation.

including elastic scattered electrons and background events, is very high ($\approx 10^8$/s). Therefore the calorimeter must be segmented and the material must be very carefully selected. The electronics has been especially developed for this experiment [5]. Systematic changes in the measured asymmetry due to pile-up as well as to dilution of elastic scattering events with inelastic processes nearest in energy have to be avoided. Therefore a very compact material with a fast response time (< 20 ns) is needed which in addition can stand the high dose of ionizing radiation collected during the whole measurement time. Associated with the good energy resolution needed for the separation of elastic events, we found that Leadfluoride (PbF_2) single Čerenkov crystals is the best material and has been chosen [7]. The calorimeter is built of 1022 Čenrenkov crystals distributed in 7 rings of $\Delta\theta=1.43°$. Each crystal corresponds to $\Delta\phi=2.47°$. For the present experiment two sectors corresponding only to one half of the calorimeter has been built. A schematic view of the detector is shown in fig. 3 illustrating the principle of the detection. In order to obtain the needed energy resolution for the detected electrons, the collection of all the electromagnetic shower is needed. Therefore the sum of nine measured signals in a matrix of 3x3 crystals is done on line. The achieved energy resolution is about $4\%/\sqrt{E}$. A typical spectrum is displayed in fig. 3. The signal of the Čerenkov calorimeter is substantially different from high energy electrons or photons and heavier charged particles where the Čerenkov photon yield is suppressed. The radiation damage can be healed completely by shining visible light on the material. Taking into account of the very high counting rate ($\simeq$100 kHz per crystal), a parallelization in the electronics and data acquisition system is needed to eliminate each kind of pile-up. Each cluster of 3x3 modules has individual, self triggering electronics. To avoid the spatial pile-up, we inhibit a second hit in neighborhood of the crystal hit in first. To eliminate the temporal pile-up, a second hit within the integration time is also inhibited. The signals coming from each cluster are summed up then integrated using analogical electronics. A total acquisition dead time of 20ns, including the pile-up rejection, has been obtained. To improve the energy resolution. After digitization, the localization in the calorimeter as well as the deposit energy in crystal are histogrammed.

3 Correction of the helicity correlated false asymmetry

3.1 Possible systematic uncertainties and their correction

The experimental asymmetry is defined by $A_{exp} = (N_e^+ - N_e^-)/(N_e^+ + N_e^-)$, $N_e^\pm$ being the detector counting rates in the two helicity states. It is given at the first order by :

$$A_{exp} = |P_e| \, A_{PV} + A_{I_e} + A_\Sigma \qquad (1)$$

where A_{PV} is the searched physical asymmetry. Its value, weighted over the angular range by the differential cross section, is expected equal to $-6.78 \ 10^{-6}$ if we neglect the strangeness contribution and the electroweak radiative corrections. Using the Musolf's estimations [1], the electroweak corrections in our kinematic conditions are small and decrease the absolute value of A_{PV} of 7% $\pm$ 3%. The term A_{I_e} is the helicity correlated asymmetry in the incident intensity defined by $A_{I_e} = (I_e^+ - I_e^-)/(I_e^+ + I_e^-)$ where $I_e^\pm$ are the intensities of the electron beam in the two helicity states respectively. The term A_Σ is the false asymmetry due to correlated fluctuations in energy, positions and angles of electron beam. In order to reduce the helicity correlated intensity fluctuations, the beam intensity is stabilized and the use of the second $\lambda/2$ plate allows, as previously indicated, to minimize the value of A_{I_e}. When the first $\lambda/2$ plate is introduced, the sign of the beam polarization $|P_e|$ as well as the helicity correlated false asymmetries are reversed but not the sign of the non helicity correlated false asymmetries. Therefore the difference between the values obtained for A_{exp} with and without this plate does no longer contain the non helicity correlated asymmetries. As an example, the target density fluctuations are not correlated to helicity and will disappear with this procedure. The luminosity measurement is continuously done using height integrating water Čerenkov counters regularly laid out on a cone and viewing the target at small scattering angle ($4°$ to $10°$) where the PV asymmetry is expected to be very weak ($2.5 \ 10^{-8}$). The target density fluctuations (boiling, temperature effects,...) appear as a broadening in the distribution of the counting rates yielded in these luminosity monitors. The electron current is continuously measured at three positions in the beam line (see fig.2) and permits the calculation of A_{I_e} for each run ($\simeq 5$ mn) as well as the pertaining corrections. Stabilization of the current is done using a feed back loop on the intensity monitors. The false asymmetry A_Σ is defined by :

$$A_\Sigma = \frac{\Sigma^+ - \Sigma^-}{\Sigma^+ + \Sigma^-} \quad \text{with} \quad \Sigma^\pm = \int_{\Omega^\pm} \rho^\pm \ell^\pm \frac{d\sigma(E^\pm, \omega)}{d\omega} d\omega \tag{2}$$

$\Omega^\pm$, which are the solid angles of the detector seen from the target for the two helicity states, depend on the beam position (x, y) and angle $(\dot{x}, \dot{y})$ which may be helicity correlated. The target thickness in the two helicity states $\ell^\pm$ are depending also on (x, y) and $(\dot{x}, \dot{y})$. The target density $\rho^\pm$, being non helicity correlated, can be replaced by its mean value ρ and disappears in equation 2. The distribution of the counting rate yielded by the luminosity monitors during a run gives some information on the ρ distribution. An increase of the width of such a distribution is an indication of a possible boiling of the target. The elastic cross section may be correlated to helicity via the beam energy $E^\pm$. Energy, beam position and angle, are continuously measured using a combination of resonant cavities on the beam line (See Fig. 2). A feed back loop permits to stabilize these quantities using steering magnets on the beam line for the beam positions and angles. The width of the beam energy distribution is typically of 2 keV with the energy stabilization, and the positions are stabilized within about 50 μm. During the experiment the helicity correlated position difference is typically 20 nm, and the helicity correlated energy difference is of about 13.6 eV. As these differences are small, it may be possible to evaluate A_Σ neglecting the helicity correlations in Ω and ℓ . Using this approximation, the asymmetry then called A_ϵ depends only on the elastic cross section and his value is a lower limit of A_Σ. Using the typical experimental ΔE value of 13.6 eV, we find $|A_\epsilon| \simeq 2.67\ 10^{-8}$.

The small values of the positions and angle fluctuations may be taken into account using a linear χ^2 procedure.

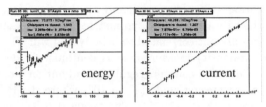

Figure 4. Example of helicity correlated parameters

The linear dependence of the luminosity fluctuations on the beam parameters clearly appears in fig. 4. Numerical determination, using a general matrix method for solving the multiple regression problem, is presently in progress. We give some very preliminary results in the section 4. We notice that the false asymmetry term $A_{P_e} = (|P_e^+| - |P_e^-|)/(|P_e^+| + |P_e^-|)$ corresponding to the helicity correlated fluctuation in the value of the beam polarization does not contribute in the first order to the experimental asymmetry.

3.2   Background effect and extracted preliminary results

When a background is present under the elastic peak, the preceding formulas have to be modified. Under the assumptions of *(i)* a small background contribution to the elastic cross section and of *(ii)* an asymmetry associated to this background of the same order or weaker than the PV asymmetry, the equation 1 becomes at the first order:

$$A_{exp} = \frac{|P_e|\ A_{PV} + A_{I_e}}{1 + \frac{\Sigma^b}{\Sigma}} + A_{\Sigma+\Sigma^b} \tag{3}$$

where $\Sigma = (\Sigma^+ + \Sigma^-)/2$ is the mean value of $\Sigma^\pm$ and Σ^b is the same quantity associated to the background. The asymmetry $A_{\Sigma+\Sigma^b}$ has the same definition as A_Σ by replacing $\Sigma^\pm$ by $\Sigma^\pm + \Sigma^{b\ \pm}$. In figure 3b) the typical spectrum shows a large elastic peak with some additional events above the pion threshold.

Besides the contribution of the elastic radiative tail, photons coming from high energy π^0 decay, contribute to the small background under the elastic peak. A simulation was done to generate such a background and the associated asymmetry [8]. This asymmetry is only due to the events where a photon and the scattered electron hit the same crystal. It has been found to have a negligible contribution to the PV asymmetry in the elastic peak. The dilution factor coming from the background is also weak: the ratio of the number of photons with an energy greater than 600 MeV to the elastic electrons is about 0.4%.

4   Preliminary results

A first analysis of the data obtained in 600 h of beam at 0.855 GeV, leads to the following preliminary value for the experimental PV asymmetry:

$$A_{PV} = -8.5 \pm 0.9 \pm \text{(not yet) ppm}$$

This experimental value has to be compared with the theoretical prediction for $< A_s >$. In this analysis the corrections to obtain the physical asymmetry have been partially done and the calculation of the systematic uncertainties is still in progress.

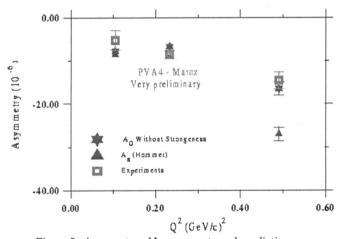

Figure 5. Asymmetry: Measurements and predictions.

The systematic uncertainty is expected to be smaller than the statistical one ($\simeq$ 5% to be compared to the 10% of the statistical ones). A more judicious choice of the limits taken for the elastic peak as well as the use of the

Table 1. Effect of the strangeness in PV experiments

Experiment	Q^2	δ_s^{exp}	δ_s^{th}
SAMPLE	0.105	-0.324	0.105
PVA4	0.233	0.278	0.346
HAPPEX	0.491	-0.114	0.641

data coming from the external rings of the calorimeter (which have been not presently included), will increase the statistical accuracy. The comparison with the results of the SAMPLE and HAPPEX experiments is done in fig.5. The theoretical predictions for $< A_0 >$ and $< A_s >$ are displayed for the three experiments. In these predictions the electroweak radiative corrections are included. The asymmetry $< A_0 >$ has been calculated in the frame of the standard model assuming no strangeness in the weak form factors. The asymmetry $< A_s >$ includes the strange electric and magnetic form factors with the parameterization given by Hammer [4]. To evaluate the effect of these form factors, we can calculate their relative contribution defined as:

$$\delta_s^{th} = \frac{< A_s > - < A_0 >}{< A_0 >} \qquad \delta_s^{exp} = \frac{A_{PV} - < A_0 >}{< A_0 >}$$

This quantity is given for the three experiments and compared to the theoretical predictions of Hammer in the table 1.

References

1. M.J. Musolf et al., *Phys. Rep.* **239**, 1 (1994).
2. P.E. Bosted, *Phys. Rev.* C **51**, 409 (1995), and references therein.
   S. Galster et al., *Nucl. Phys.* B **32**, 221 (1971).
3. D.E. Groom et al., The European Pysical Journal **C15**,1 (2000).
4. H.W. Hammer et al., Phys. Lett. **B367**, 323 (1996).
5. F.E.Maas and the A4 coll., Parity Violation in Atoms and Polarized Electron Scattering. Word Scientific, 491 (1999) Ed. by B. Frois and M.A. Bouchiat.
6. K. Aulenbacher, C. Nachtigall et al., NIM **A391**, 498 (1997).
7. P. Achenbach et al., NIM **A416**, 357 (1998).
8. S. Ong, M.P. Rekalo and J. Van de Wiele, Eur. Phys. J. **A6**, 215 (1999).

Parity Violating Measurements of Neutron Densities: Implications for Neutron Stars

C. J. Horowitz

Dept. of Physics and Nuclear Theory Center, Indiana University, Bloomington, IN 47405 USA

E-mail: horowitz@iucf.indiana.edu

J. Piekarewicz

Department of Physics, Florida State University, Tallahassee, FL 32306

E-mail: jorgep@scri.fsu.edu

Parity violating electron scattering can measure the neutron density of a heavy nucleus accurately and model independently. This is because the weak charge of the neutron is much larger then that of the proton. The Parity Radius Experiment (PREX) at Jefferson Laboratory aims to measure the root mean square neutron radius of ^{208}Pb with an absolute accuracy of 1% (± 0.05 Fm). This is more accurate then past measurements with hadronic probes, which all suffer from controversial strong interaction uncertainties. PREX should clearly resolve the neutron-rich skin. Furthermore, this benchmark value for ^{208}Pb will provide a calibration for hadronic probes, such as proton scattering, which can then be used to measure neutron densities of many exotic nuclei. The PREX result will also have many implications for neutron stars. The neutron radius of Pb depends on the pressure of neutron-rich matter: the greater the pressure, the larger the radius as neutrons are pushed out against surface tension. The same pressure supports a neutron star against gravity. The Pb radius is sensitive to the equation of state at normal densities while the radius of a 1.4 solar mass neutron star also depends on the equation of state at higher densities. Measurements of the radii of a number of isolated neutron stars such as Geminga and RX J185635-3754 should soon improve significantly. By comparing the equation of state information from the radii of both Pb and neutron stars one can search for a softening of the high density equation of state from a phase transition to an exotic state. Possibilities include kaon condensates, strange quark matter or color superconductors.

1 Introduction

The size of a heavy nucleus is one of its most basic properties. However, because of a neutron skin of uncertain thickness, the size does not follow from measured charge radii and is relatively poorly known. For example, the root mean square neutron radius in ^{208}Pb, R_n is thought to be about 0.2 Fm larger then the proton radius $R_p \approx 5.45$ Fm. An accurate measurement of R_n would provide the first clean observation of the neutron skin in a stable heavy nucleus. This is thought to be an important feature of all heavy nuclei.

Ground state charge densities have been determined from elastic electron scattering, see for example ref.[1]. Because the densities are both accurate and model independent they have had a great and lasting impact on nuclear physics. They are, quite literally, our modern picture of the nucleus.

In this paper we discuss future parity violating measurements of neutron densities. These purely electro-weak experiments follow in the same tradition and can be both *accurate* and *model independent*. Neutron density measurements have implications for nuclear structure, atomic parity nonconservation (PNC) experiments, isovector interactions, the structure of neutron rich radioactive beams, and neutron rich matter in astrophysics. It is remarkable that a single measurement has so many applications in atomic, nuclear and astrophysics.

Donnelly, Dubach and Sick[2] suggested that parity violating electron scattering can measure neutron densities. This is because the $Z-$boson couples primarily to the neutron at low Q^2. Therefore one can deduce the weak-charge density and the closely related neutron density from measurements of the parity-violating asymmetry in polarized elastic scattering.

Of course the parity violating asymmetry is very small, of order a part per million. Therefore measurements were very difficult. However, a great deal of experimental progress has been made since the Donnelly *et. al.* suggestion, and since the early SLAC experiment [3]. This includes the Bates ^{12}C experiment[4], Mainz ^{9}Be experiment [5], SAMPLE [6] and HAPPEX [7]. The relative speed of the HAPPEX result and the very good helicity correlated beam properties of CEBAF show that very accurate parity violation measurements are possible. Parity violation is now an established and powerful tool.

It is important to test the Standard Model at low energies with atomic parity nonconservation (PNC), see for example the Colorado measurement in Cs [8,9]. These experiments can be sensitive to new parity violating interactions such as additional heavy $Z-$bosons. Furthermore, by comparing atomic PNC to higher Q^2 measurements, for example at the Z pole, one can study the momentum dependence of Standard model radiative corrections. However, as the accuracy of atomic PNC experiments improves they will require increasingly precise information on neutron densities[10,11]. This is because the parity violating interaction is proportional to the overlap between electrons and neutrons. In the future the most precise low energy Standard Model test may involve the combination of an atomic PNC measurement and parity violating electron scattering to constrain the neutron density.

There have been many measurements of neutron densities with strongly interacting probes such as pion or proton elastic scattering, see for example ref. [12]. Unfortunately, all such measurements suffer from potentially serious theoretical systematic errors. As a result no hadronic measurement of neutron densities has been generally accepted by the field.

Relative measurements of isotope differences of neutron radii can be more accurate. See for example [13]. Therefore one can use a single parity violating measurement of the neutron radius of ^{208}Pb to "calibrate" hadronic probes. Then these hadronic probes can be used to measure neutron radii of many other stable and unstable nuclei. For example, (^{3}He,T) measurements of neutron radii differences for Sn isotopes were based on measuring spin dipole strength and a spin dipole sum rule along with assuming a theoretical Hartree Fock radius for ^{120}Sn [14].

Finally, there is an important complementarity between neutron radius measurements in a finite nucleus and measurements of the neutron radius of a neutron star. Both provide information on the equation of state (EOS) of dense matter. In a nucleus, R_n is sensitive to the EOS at normal nuclear densities. The neutron star radius depends on the EOS at higher densities. In the near future, we expect a number of improving radius measurements for nearby isolated neutron stars such as Geminga [15] and RX J185635-3754 [16].

We now present general considerations for neutron density measurements, discusses possible theoretical corrections, outline an approved Jefferson Laboratory experiment on ^{208}Pb and then relate this Pb measurement to ongoing measurements of neutron star radii.

2 General Considerations

In this section we illustrate how parity violating electron scattering measures the neutron density and discuss the effects of Coulomb distortions and other corrections. These corrections are either small or well known so the interpretation of a measurement is clean.

2.1 Born Approximation Asymmetry

The weak interaction can be isolated by measuring the parity-violating asymmetry in the cross section for the scattering of left (right) handed electrons. In Born approximation the parity-violating asymmetry is,

$$A_{LR} = \frac{G_F Q^2}{4\pi\alpha\sqrt{2}} \left[4\sin^2\theta_W - 1 + \frac{F_n(Q^2)}{F_p(Q^2)} \right], \tag{1}$$

with G_F the Fermi constant and θ_W the weak mixing angle. The Fourier transform of the proton distribution is $F_p(Q^2)$, while that of the neutron distribution is $F_n(Q^2)$, and Q is the momentum transfer. The asymmetry is proportional to $G_F Q^2/\alpha$ which is just the ratio of Z^0 to photon propagators.

Since $1-4\sin^2\theta_W$ is small and $F_p(Q^2)$ is known we see that A_{LR} directly measures $F_n(Q^2)$. Therefore, A_{LR} provides a practical method to cleanly measure the neutron form factor and hence R_n.

2.2 Coulomb distortions

By far the largest known correction to the asymmetry comes from coulomb distortions. By coulomb distortions we mean repeated electromagnetic interactions with the nucleus remaining in its ground state. All of the Z protons in a nucleus can contribute coherently so distortion corrections are expected to be of order $Z\alpha/\pi$. This is 20 % for ^{208}Pb.

Distortion corrections have been accurately calculated in ref. [18]. Here the Dirac equation was numerically solved for an electron moving in a coulomb and axial-vector weak potentials. From the phase shifts, all of the elastic scattering observables including the asymmetry can be calculated.

Other theoretical corrections from meson exchange currents, parity admixtures in the ground state, dispersion corrections, the neutron electric form factor, strange quarks, the dependence of the extracted radius on the surface shape, etc. are discussed in reference [17]. These are all small. Therefore the interpretation of a parity violating measurement is very clean.

3 Parity Radius experiment

The Parity Radius Experiment (P-ReX) will measure the parity violating asymmetry for elastic electron scattering from ^{208}Pb [19]. This Jefferson Laboratory Hall A experiment will use 850 MeV electrons scattered at six degrees. The planned 3% accuracy in the approximately 0.7 parts per million asymmetry will allow one to deduce the neutron root mean square radius R_n to 1% ($\approx \pm 0.05$ Fm). The neutron radius R_n is expected to be about 0 to 0.3 Fm larger then the proton radius R_p. Therefore PREX should cleanly resolve the neutron skin $R_n - R_p$.

The target will be a thick foil, enriched in ^{208}Pb, sandwiched between two thin diamond foils. The very high thermal conductivity of the diamond keeps the Pb from melting and allows a high beam current of order 100 microamps. Note, the thin diamond foils introduce only a few percent background. Furthermore, the asymmetry from ^{12}C can be calculated with high accuracy so this background is not a problem for the interpretation of the experiment.

PREX requires some improvements in the helicity correlated beam properties and an improvement in the measurement of the absolute beam polarization in Hall A. This is presently of order 3% and needs to be improved to 1-2%. It should take about 30 days of beam time to get the 3% statistics.

4 Implications of the ^{208}Pb neutron radius for neutron stars

It is an exciting time to study neutron stars. These gigantic atomic nuclei are more massive then the Sun and yet have a radius of only about 10 kilometers. New telescopes, operating at many different wave lengths, are finally turning these theoretical curiosities into detailed observable worlds. The structure of a neutron star depends only on the equation of state (EOS) of neutron rich matter together with the know equations of General Relativity. The equation of state gives the pressure as a function of (energy) density. Densities in neutron stars are comparable to, or greater, then the densities in atomic nuclei. The central density of a 1.4 solar mass neutron star is expected to be a few times greater then the saturation density of nuclear matter, $\rho_0 \approx 0.16$ nucleons per Fm^3.

Likewise the neutron radius of a conventional atomic nucleus such as ^{208}Pb also depends on the equation of state of neutron rich matter. Higher pressures lead to greater neutron radii and thicker neutron skins as neutrons are pushed out against surface tension. Indeed, Alex Brown finds a strong correlation between the pressure of neutron matter at $\rho \approx 0.1$ Fm^{-3} and the neutron radius in ^{208}Pb [20]. This correlation is valid for many different nonrelativistic and relativistic effective interactions. The density $\rho = 0.1$ Fm^{-3} is about 2/3 of ρ_0 and represents some average over the interior and surface density of the nucleus.

Therefore, the neutron radius in Pb has many implications for the structure of neutron stars and several other areas of astrophysics. The common unknown is the equation of state of neutron rich matter. Information on the EOS from a measurement of R_n for ^{208}Pb could be very important for astrophysics.

4.1 Neutron Skin versus Neutron Star Crust

The Pb radius constrains the EOS at normal densities ≈ 0.1 Fm^{-3}. Neutron stars are expected to undergo a phase transition near this density from a solid crust to a liquid interior. We have shown that the Pb radius is strongly correlated with the liquid to solid transition density [21]. A high pressure for neutron rich matter more quickly favors the uniform liquid over the nonuniform solid. Therefore, a large neutron radius in Pb implies a low transition density for the crust. Thus, a measurement of the thickness of the neutron rich skin in Pb helps determine the thickness of the solid crust of a neutron star. We note that both the skin of a heavy nucleus and the crust of a neutron star are made of neutron rich matter at similar densities. Many neutron star observables such as glitches in the rotational period, gravitational waves from quadrupole deformations, and the surface temperature depend on the thickness of the crust. Adrian Cho has written a short popular article on using this pression measurement in Pb to learn about neutron star crusts [22].

4.2 Pb radius versus Neutron Star Radius

In general the radius of a neutron star R_* is related to the neutron radius in Pb R_n. A large R_n implies a high pressure for the EOS of neutron rich matter and this same pressure supports a star against gravity. Therefore, a larger R_n might imply a larger R_*. However, the radius of a neutron star R_* depends on the EOS of neutron rich matter over a range of densities from near ρ_0 to higher densities. In contrast, R_n only depends on the EOS at ρ_0 and lower densities. Thus, R_n only constrains the low density EOS. Models with different high density behavior can have the same R_n but predict different R_*. Therefore, we find no unique relationship between R_* and R_n [23].

One way to characterize the different information on the EOS contained in R_n compared to R_* is to consider low mass neutron stars. Most, well measured, neutron stars have masses near 1.4 solar masses. These stars have central densities significantly above ρ_0. Instead, 0.5 Solar mass neutron stars are expected to have central densities only slightly greater then ρ_0. We find a sharp correlation between R_* for 0.5 solar mass neutron stars and R_n [24]. This is because, now, both R_* and R_n depend on the EOS at similar densities.

Note, such low mass neutron stars probably don't exist. This is because conventional stars with cores near 0.5 solar masses are not expected to collapse. Thus, R_n contains unique information on the low density EOS that can not be obtained by measuring neutron star radii directly. Furthermore, *measuring both R_n in Pb and R_* for a neutron star [a] provides important information on the density dependence of the EOS.*

For example, if R_n is measured to be relatively large, this implies a stiff (high pressure) EOS at normal nuclear densities. If R_* is also measured to be relatively small, say near 10 km, this implies a soft (low pressure) high density EOS. This softening of the EOS with increasing density could be strong evidence for a phase transition of neutron rich matter to some exotic phase. Note, an exotic phase that increases the pressure would not be thermodynamicly favored. There is much speculation on possible high density exotic phases for neutron rich matter. Examples include: kaon condensation, strange quark matter, or color superconductivity.

Neutron stars provide essentially the only way to study cold very dense matter. Relativistic heavy ion collisions can reach high energy densities but not at low temperatures. Thus, if an exotic phase exits at high density it is very

---

[a] That is probably near 1.4 solar masses.

hard to find experimental evidence. This comparison of the EOS information from R_n and R_* may be one of the few sharp signals. Therefore, it is important to measure *both* the neutron radius of a heavy nucleus R_n, and the radius of a neutron star R_*.

4.3 Measurements of Neutron Star Radii

There are several ongoing measurements of the radius of neutron stars. Most of these are based on measuring the stars luminosity, distance and surface temperature T. If the star were a black body, the luminosity would be σT^4 times the surface area. Thus, the surface area $4\pi R_\infty^2$ and effective radius R_∞ can be deduced. Corrections for non black body behavior can be made with model atmospheres.

Because of the curvature of space in the Star's very strong gravitational field, the effective radius R_∞ is somewhat larger then the coordinate radius R_*,

$$R_\infty = R_*/(1 - 2GM/R_*)^{1/2}, \qquad (2)$$

where M is the Star's mass and G is Newton's constant. Some of the light from the far side of the star is bent by gravity and still reaches an observer. This makes the star appear larger. Note, the light is also gravitationally red shifted so that the actual surface temperature is about 30% higher then the apparent temperature deduced from the observed spectrum.

The Stony Brook group has fit the visible and X-ray spectrum of the isolated nearby neutron star RX J185635-3754 with a model Fe atmosphere [25]. Their fit combined with a preliminary parallax distance to RX J185635 of 61 parsecs (about 180 light years) yields a very small radius of $R_* = 6$ km! This radius is smaller then that predicted by any present neutron matter EOS and seems unrealistic. However Kaplan et al. [26] have questioned the parallax distance. They reanalyze the same Hubble optical images and infer a larger distance. For their distance, $R_\infty = 15 \pm 6$ km. Note for a 1.4 solar mass star, $R_\infty = 15$ km corresponds to $R_* \approx 13$ km. This new value is fully consistent with many neutron matter EOSs. However, the error is still large.

Sanwal et al. [27] fit the X-ray spectrum of the well known Vela pulsar with a high energy power law, from the pulsar mechanism, and a thermal component from a magnetized hydrogen atmosphere. They deduce $R_\infty = 15.5 \pm 1.5$ km assuming a distance to Vela of 250 parsec. Finally Rutledge et al. analyze the X-ray spectra of a neutron star in the globular star cluster NGC 5139, for which the distance to the cluster is well known [28]. Their preliminary result is $R_\infty = 14.3 \pm 2.5$ km.

In the near future, we should have better distance measurements to RX J185635 and better X-ray spectra. Also, there will be better measurements on other nearby isolated neutron stars such as Geminga, and more measurements of neutron stars in globular clusters. This will allow checks on neutron star radii measured for stars with different surface temperatures and magnetic field strengths. A reasonable near term goal is a number of neutron star radius measurements accurate to about one km.

5 Conclusion

With the advent of high quality electron beam facilities such as CEBAF, experiments for accurately measuring the weak density in nuclei through parity violating elastic electron scattering (PVES) are feasible. From parity violating asymmetry measurements, one can extract the neutron density of a heavy nucleus accurately and model independently. This is because the weak charge of a neutron is much larger then that of a proton. Therefore, the Z^0 boson couples primarily to neutrons (at low momentum transfers).

These neutron density measurements allow a direct test of mean field theories and other models of the size and shape of nuclei. They can have a fundamental and lasting impact on nuclear physics. Furthermore, PVES measurements have important implications for atomic parity nonconservation (PNC) experiments. Atomic PNC measures the overlap of atomic electrons with neutrons. High precsion PNC experiments will need accurate neutron densities. In the future, it may be possible to combine atomic PNC experiments and PVES to provide a precise test of the Standard Model at low energies.

The Parity Radius Experiment at Jefferson Laboratory aims to measure the neutron radius in ^{208}Pb to 1% with parity violating elastic electrons scattering. This will provide unique information on the equation of state (EOS) of neutron rich matter at normal nuclear densities. The EOS describes how the pressure depends on the density. This information has many astrophysical implications.

The structure of a neutron star depends only on the EOS. There are many ongoing measurements of the radius of neutron stars. These are sensitive to the EOS at greater then nuclear densities. By comparing the EOS information from the ^{208}Pb neutron radius measurement with that from neutron star measurements one can deduce the density dependence of the EOS. This allows one to search for a softening (lower pressure) of the high density EOS from a possible phase transition to an exotic phase for neutron rich matter. Possible phases include kaon condensates, strange quark matter and color superconductors.

Acknowledgments

The work on parity violating measurements of neutron densities was done in collaboration with Robert Michaels, Steven Pollock and Paul Souder. We acknowledge financial support from DOE grants: DE-FG02-87ER40365 and DE-FG05-92ER40750.

References

1. B. Frois et. al., Phys. Rev. Lett. **38**, 152 (1977).
2. T.W. Donnelly, J. Dubach and Ingo Sick, Nuc. Phys. **A 503**, 589 (1989).
3. C. Y. Prescott *et al.*, Phys. Lett. **84B**, 524 (1979).
4. P. A. Souder *et al.*, Phys. Rev. Lett. **65**, 694 (1990).
5. W. Heil *et al.*, Nucl. Phys. **B327**, 1 (1989).
6. B. Mueller *et al.*, Phys. Rev. Lett. **78**, 3824 (1997).
7. K. A. Aniol *et al.*, Phys. Rev. Lett. **82**, 1096 (1999).
8. C. S. Wood et al, Science **275**, 1759 (1997).
9. S. C. Bennett and C. E. Wieman, Phys. Rev. Lett.**82**, 2484 (1999).
10. S. J. Pollock, E. N. Fortson, and L. Wilets, Phys. Rev. C **46**, 2587 (1992), S.J. Pollock and M.C. Welliver, Phys. Lett. **B 464**, 177 (1999).
11. P. Q. Chen and P. Vogel, Phys. Rev. **C 48**, 1392 (1993).
12. L.Ray and G.W.Hoffmann, Phys. Rev. C **31**, 538 (1985). L. Ray, Phys. Reports **212**, 223 (1992).
13. L. Ray, Phys. Rev. **C27**, 2143 (1983).
14. A. Krasznahorkay et al., Phys. Rev. Lett. **82**, 3216 (1999).
15. Patrizia A. Caraveo et. al., ApJ **461**, L91 (1996); A. Golden and A. Shearer, astro-ph/9812207.
16. F. Walter, S. Wolk and R. Neuhauser, Nature **379**, 233 (1996); Bennett Link, Richard I. Epstein and James M. Lattimer, PRL **83**, 3362 (1999).
17. C.J. Horowitz, S. Pollock, P.A. Souder and R. Michaels, Phys Rev **C63**, 025501 (2001).
18. C.J. Horowitz, Phys. Rev. **C57**, 3430 (1998).
19. Jefferson Laboratory Experiment 00-003, R. Michaels, P. A. Souder and G. Urciuoli spokespersons.
20. B. Alex Brown, Phys. Rev. Lett. **85**, 5296 (2000).
21. C. J. Horowitz and J. Piekarewicz, Phys. Rev. Lett. **86**, 5647 (2001).
22. Adrian Cho, Newscientist **2294**, 11 (2001).
23. C. J. Horowitz and J. Piekarewicz, Phys. Rev. **C64**, 062802 (2001).
24. C. J. Horowitz and J. Piekarewicz, to be published.
25. J. A. Pons et al., asto-ph/0107404.
26. D. L. Kaplan, M. H. van Kerkwijk and J. Anderson, astro-ph/0111174.

27. Divas Sanwal et al., astro-ph/0112164.
28. Robert E. Rutledge et al., astro-ph/0105405.

HIGH ENERGY APPROACHES TO LOW ENERGY PHENOMENA IN ASTROPHYSICS

SAM M. AUSTIN

NSCl and Department of Physics and Astronomy
Michigan State University, East Lansing MI 48824 USA
E-mail: Austin@nscl.msu.edu

Studies in nuclear astrophysics have long been associated with long runs at small accelerators, measuring ever-decreasing cross sections as one approached (but rarely reached) the energy of reactions in stars. But in recent years pioneering studies have shown that studies at high-energy accelerators can often yield the same information, and in some important cases, provide information not otherwise available. This is particularly so for studies of the properties and reactions of the short-lived radioactive nuclei that play a crucial role in explosive phenomena such as novae, supernovae, and neutron stars. I'll give an overview of some of the possibilities, and then concentrate on two extended examples: measurements of the rates of radioactive capture reactions using Coulomb breakup reactions, and the relationship of charge exchange cross sections and beta-decay strength for L = 1 transitions.

1 Introduction

It is a common perception that experimental nuclear astrophysics involves long measurements of small cross sections at lower and lower energies, so as to permit a reliable extrapolation to actual astrophysical energies. This perception is only partially correct. Recent developments, especially of radioactive beams, often permit one to obtain equivalent information with higher energy beams. The high energy experiments commonly yield higher event rates and sometimes yield information not available in the classical approach.

An important advantage is that one can confidently reduce the overall uncertainty in the determination of an astrophysical reaction rate by combining various approaches, with assurance that the systematic uncertainties are independent. For example, if a reaction rate is needed to 5%, four independent approaches at the 10% level will be sufficient, and may lead to a more reliable overall result.

1.1 Energy scales

The energy scale for nuclear reactions is given by the location of the Gamow Peak that is determined by the product of the Maxwell Boltzmann distribution of particle velocities at a given temperature and the rapidly increasing barrier penetrability leading to a nuclear reaction. For reactions in the sun, the Gamow-peak energies are in the 5-25 keV range.

1.2   Resonant and non-resonant processes.

If a resonance occurs in the Gamow Peak it typically dominates the total reaction rate. For a (p, γ) reaction the rate is $\propto [\Gamma_p\Gamma_\gamma/(\Gamma_p + \Gamma_\gamma)]\exp(-E_r/kT)$; it is only necessary to measure the relevant nuclear widths and the energy of the resonance to specify the rate. If the reaction is non-resonant it is characterized by the S factor, $S = \sigma E\exp(bE^{-1/2})$ and one needs to determine S in the region of the Gamow Peak.

High-energy reactions have commonly been applied to determination of resonant, rates and particularly for the measurement of resonance energies E_r. But it is only recently, with the development of radioactive beams, that it has been possible to obtain non-resonant rates using high-energy techniques. In this paper I'll describe a selection of high-energy approaches, some briefly with selected references to the recent literature, and two others in more detail.

2   Some approaches, briefly

2.1   Measurements of the Asymptotic Normalization Coefficient (ANC)

At low energies radiative capture reactions such as (p, γ) and (α, γ) are dominated by processes occurring far outside the nuclear radius. The cross section then depends on the square of a wave function whose radial dependence, governed by the Coulomb force, is given by the Whittaker function, and whose normalization is given by the ANC[1,2]; the value of the ANC is then sufficient to specify the value of the S factor at E = 0. The most active program is at Texas A&M University where ANCs have been measured by studying transfer reactions at low energies. S-factors have been obtained, for example, for the ^{13}C(p, γ)^{14}N, ^{16}O(p, γ) ^{17}F, and ^{7}Be(p, γ)^{8}B reactions. The ^{7}Be(p, γ)^{8}B reaction produces the high-energy neutrinos that dominate the response of solar neutrino detectors. The transfer reactions ^{10}B(^{7}Be,^{8}B)^{9}Be and ^{14}N(^{7}Be,^{8}B)^{13}C were studied [3] to determine the S factor, S_{17}, that describes the rate of this reaction. The results are given in Fig.5.

The major issue in this approach is the uncertainty in the optical model potentials (OMPs) necessary to carry out the necessary DWBA calculations. A systematic effort has been made to determine the relevant OMPs with the result that overall precisions of about 10% are possible. The ANC results for the ^{16}O(p,γ) reaction agree with the direct observation at this level.

2.2   The Trojan-Horse Method

The Trojan-Horse (THM) method [4,5] can be regarded an adaptation of the quasi-free knockout method that has been applied in the past to study the momentum distribution of nucleons in   nuclei using   (p,p') and (α,α') reactions.   The essence of the THM approach can be obtained from an examination of an example case: study of the low energy cross section of the ^{7}Li(p, α)α reaction using the ^{2}H(^{7}Li, αα)n reaction.   This is diagrammed in Fig. 1. The reaction is induced by ^{7}Li ions with an energy above the Coulomb barrier, and the kinematics of the detected α's are chosen so the neutron is essentially a spectator to the reaction between ^{7}Li and the proton that occurs at the circled vertex. The Fermi energy of the proton in ^{2}H can (partially) compensate for the energy of the ^{7}Li, so that low relative ^{7}Li-p energies can be studied.

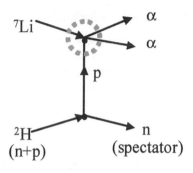

Fig. 1:   Trojan-Horse Method for the ^{7}Li (p,α)α reaction.

Results   obtained   [6]   for   the   reaction diagrammed in Fig. 1 are shown in Fig. 2. Because   of   the   relatively   simple   theory employed to connect the three-body and two-body cross sections, absolute cross sections are unreliable; it is necessary to normalize   to   the   directly measured cross sections at higher energies. At lower energies the THM cross section deviates from the   direct   measurements, presumably because of electron screening corrections: the low-E reactions take place at such large distances from the nucleus that   the   nuclear   charge   is somewhat   shielded   by   atomic electrons   and   the   Coulomb barrier   is   reduced.   These screening effects are not present in the THM cross sections. If one accepts the extrapolation of the THM   cross   section   to   low energies, then the difference can

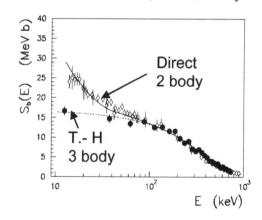

Fig. 2:   Direct and Trojan Horse Method S-factors for the ^{7}L(p,α)α reaction. From Ref. [6].

be ascribed to screening effects and compared to established theories. The surprise is that the effects are much larger than anticipated. In this and the other case studied in detail (6Li(d,αα) for ^{2}H(α, ^{6}Li), Ref. [7]) the effective screening potential is about twice that expected.

These studies are in an early state, and their accuracy is not yet clear. But the TH technique has promise for shedding light on screening effects [8]. It may also be possible to extend the technique to radiative capture reactions.

2.3   Gamow-Teller strength

It is now well established that Hadronic charge exchange (CEX) reactions such as (p,n) or (n,p) provide an accurate measure of GT strength B(GT), especially for strong transitions [9,10].   Taking advantage of this capability,   (n,p) reactions on nuclei in the iron range have been used to measure the strength of electron capture reactions expected to be important in describing supernova core collapse.   Recent shell model calculations [11] are in reasonable agreement with these results, but higher resolution data and data on odd-odd and higher mass nuclei, some radioactive, are necessary to establish a truly reliable experimental database. And calculations of supernova evolution using the new theoretical rates show that radioactive nuclei also play an important role in supernovae[12]. As a result of all these considerations, it seems that additional measurements are required, and that they will have to employ CEX reactions involving heavier projectiles.

This raises an important issue. These heavier projectiles are often strongly absorbed and strong absorption means that a limited region of space contributes to the cross section.   One would anticipate that by the uncertainty principle, a large range of momenta in the transition form factor would contribute to the CEX cross section. In contrast, B(GT) depends only on the form factor at q≈0.   In the cases that have been investigated, however, it has been found that even strongly absorbed projectiles (as in the (^{12}C,^{12}N) reaction [13])   yield accurate B(GT).   Osterfeld *et al.* [14] have shown that this surprising result follows from two facts: that the CEX reaction is only sensitive to the form factor over a limited (although fairly large) range of q, and that, (in the cases studied) the transition form factors have essentially the same q dependence in this region.

This conclusion makes it possible to use (^{3}He,t), (t, ^{3}He), and (^{7}Li,^{7}Be) reactions on stable nuclei to obtain GT$_+$ and GT$_-$ strengths with   high resolution;   to use the (^{6}Li,^{6}Li* (0^+ T=1)) reaction to measure GT$_0$ strength by inelastic scattering; and to use some of these reactions in inverse kinematics to study GT strength for radioactive nuclei using radioactive beams.   It is not yet clear what techniques will prove most useful for these inverse kinematics studies, because the outgoing light particles have such low energies. But at a minimum, low-energy-resolution data will be obtainable by catching and identifying the heavy product in high-resolution spectrometers such as the NSCL S800.

3 Coulomb dissociation-a detailed example

Only recently has Coulomb dissociation (CD) been employed to study the breakup of radioactive nuclei of interest for astrophysics and thereby to obtain information about the related radiative capture process. The principle is simple and we illustrate it by considering the CD of ^{8}B and how it can be used to determine the S-factor, S_{17}, for the ^{7}Be(p, γ)^{8}B reaction. Upon passing a high-Z nucleus a fast ^{8}B projectile absorbs a virtual photon, and breaks up into ^{7}Be and a proton. The energy of the absorbed photon is determined by measuring the momenta of the products and the cross section for the (p,γ) reaction is determined by detailed balance. The event rate is larger for CD than for (p,γ) because thick targets can be used and because the momentum ratio in detailed balance favors CD. There is however a complexity: detailed balance applies only to gamma rays of a given multipolarity, but the contributions of various multipoles differ for the two processes. For ^{8}B, L=1 photons dominate both processes, but E2 photons, entirely negligible for (p,γ), can contribute significantly to CD. This was recognized early, but still remains a source of significant controversy and some uncertainty.

The first experiments involving radioactive nuclei were those of Motobayashi, *et al.* [15] at RIKEN, on the breakup of ^{14}O and of ^{8}B[16]. More recently measurements on ^{8}B have been done at higher energies at GSI[17] and ^{8}B[18,19,20] and ^{9}Li [21] at MSU/NSCL. The results for ^{14}O have been confirmed by (p,γ) experiments and other CD measurements. The results for ^{8}B are also in general agreement, as we shall see in Fig. 5.

3.1 Inclusive Experiments

Because of the concern about E2 contributions for ^{8}B, all experiments have attempted to assess its contributions. The RIKEN and GSI experiments obtained limits on E2 values that were significantly smaller than theoretical estimates (see summary in [20]). At MSU/NSCL, an attempt was made to obtain a more accurate value by observing the interference between E1 and E2 amplitudes in the inclusive distribution of ^{7}Be longitudinal momenta; the ^{7}Be ions were observed in the high acceptance S800 spectrograph. The results, shown in Fig. 3, are strongly asymmetric; such an asymmetry can occur only if E2 amplitudes contribute. For B(E2) = 0, the distribution would be symmetric in first order perturbation theory (PT) while higher order effects would produce the opposite asymmetry.

We have analyzed these data in PT and obtain a value of E2 strength about half that predicted by theoretical estimates; if higher order effects were included in the analysis, one would expect to obtain a somewhat larger value [22]. We have checked this inclusive result by examining the angular distribution of protons in the c.m. of ^{8}B as observed in the exclusive experiment [20]. It is consistent with the

78

inclusive result, but with lower statistical precision. The reason for the disagreement with the earlier data is not understood, but may possibly result from the contributions of nuclear processes in the RIKEN and GSI experiments.

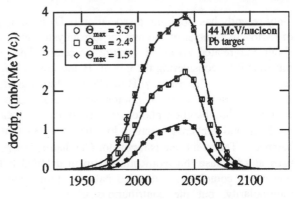

3.2 Exclusive measurement

Exclusive studies of ^{8}B breakup were carried out at 83 MeV/nucleon at MSU/NSCL. The breakup products were separated in a magnetic field and detected and identified by an array of multiwire drift chambers

Fig. 3: Measured longitudinal momentum distributions of ^{7}Be fragments from the CD of 44 MeV/nucleon ^{8}B on Pb; several maximum ^{7}Be angle cuts are shown. Also shown are first-order perturbation theory calculations convoluted with the experimental resolution (5 MeV/c). From [18, 20]. See the text for details.

and plastic scintillators [20]. The results shown in Fig. 4 are for a scattering angle cutoff of 1.77 deg, or a classical impact parameter around 30 fm. First order PT calculations show that the E2 contribution dominates at low E_{rel}, but is less that 10% above 130 keV. The Continuum Discretized Coupled Channel (CDCC) calculations show that nuclear contributions are small. After small corrections for E2 effects and the contributions of breakup leading to ^{7}Li (478 keV), we obtain S_{17} = 17.8 (+1.4/-1.2) keV b. This value is consistent with most other recent results with the single exception of the recent and as yet unpublished result of Junghans, et al. [26]. See Fig. 5. for details and a weighted mean.

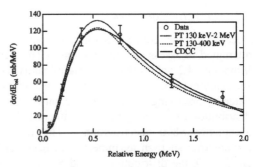

The result of Junghans, et al. [26] has not been included in the weighted average, because it is inconsistent with most recent results. Yet [26] describes a carefully done experiment that must be taken very seriously. A

Fig. 4: Measured cross sections for breakup of 83 MeV/nucleon ^{8}B ions with ^{8}B scattering angles of ≤ 1.77 deg. PT calculations normalized to the data over the regions noted, and CDCC calculations are also shown. The lowest energy point was excluded from the fit because E2 contributions are large at low relative energy. From [19,20].

major goal of future work to understand this difference. It is perhaps worth noting that both the CD and ANC methods have been checked against particular direct measurements to good accuracy. In the case of the MSU/NSCL measurements care was taken to choose the conditions of observation to minimize the confounding effects (E2, nuclear) of the CD method.

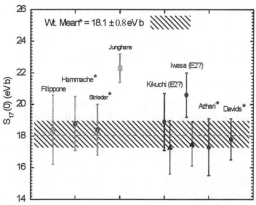

Fig. 5: Results for S_{17}. The results to the left of the horizontal break are the direct results, while those to the right have been measured by CD or from the ANC (See refs [23, 24, 25, 26, 16, 17, 3, 20] in horizontal order, left to right). The cross hatched area is the weighted mean of the results marked with *. The CD results of Kikuchi et al. and Iwasa, et al. have been excluded because of the present author's concern about E2 contributions. If they are included the weighted mean is 18.7 eV b. Earlier work is reviewed in [27].

Another issue involves possible E2 contributions in the RIKEN and GSI measurements. We believe that the MSU measurements of the E2 effects are the most sensitive, because they observe interference phenomena. For that reason, we have made a rough estimate of the effects of E2 amplitudes on the GSI/RIKEN results. The estimated changes are shown by the triangles near the bottom of the cross hatched area.

4 Can charge exchange reaction measure L=1 electroweak strength?

It appears that forbidden electroweak transitions can play an important role in astrophysics. Neutrinos can excite the spin-dipole resonance in abundant alpha-particle nuclei; part of the time the excited nucleus will be unbound and can emit nucleons, leading to the production of rare nuclides such as ^{7}Li, ^{11}B, and ^{19}F. Similar phenomena can modify the distribution of nuclides formed in the r-process. Neutrino reactions also play a role in the transport of energy by neutrinos in supernova explosions and are the basis for the observation of supernova neutrinos in terrestrial detectors. Yet direct observation of neutrino induced reactions will be experimentally feasible in only a very few cases.

In this circumstance, it will be necessary to rely on hadronic reactions to measure the required strengths, following the approach used for determination of allowed (GT) strength that was discussed in section 3.3. It is not known whether this procedure will work reliably, for the reasons discussed in section 2.3 and because the qualitative conditions are still less favorable: larger momentum transfers and contributions of tensor forces will be a confounding factor. We [28] have

80

investigated two questions. (1) Are σ (CEX) and B(L=1) proportional when both are calculated with the same wave functions. A positive answer to this questions is a necessary condition for CEX to be useful. (2) What range of momentum transfer is important in the transition form factor F(q').

We considered a simple case: $^{12}C(p, n)^{12}N$ at $E_p = 135$ MeV [29]. A simple eikonal model, taking into account the real and imaginary parts of the OMP, was used to provide insight into the reaction mechanism. This permitted the evaluation of a sensitivity function S(q, q') which characterizes the range of q' in the transition form factor which contributes at a given asymptotic momentum transfer q. Specifically, $T(q) = T_{PW}' + \int dq' S'(q,q') F(q')$ Where T_{PW} and $T(q)$ are the plane wave and total (p,n) reaction amplitudes. The approximations involved were checked against the distorted wave impulse approximation; the two predictions were in good agreement with each other and with angular distributions for J=1⁻ and 2⁻ states.

The results are shown in Figs. 6.

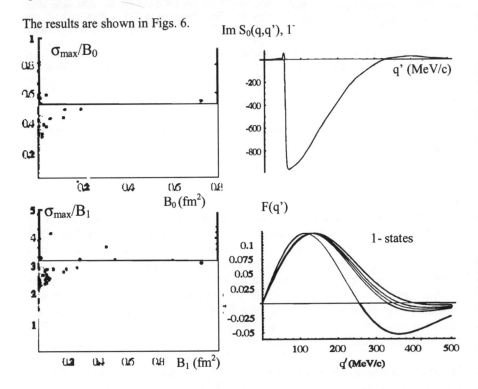

Fig. 6: The left panel shows the ratio of calculated cross sections (at maxima) and calculated B(L=1) for 0⁻ and 1⁻ states. The right panel shows the sensitivity functions (imaginary part) for a 1⁻ state, and the q' dependence of the transition form factors for various 1⁻ states.

The conclusions we draw from Fig. 6 are twofold. For $B_J > 0.1$ fm^2, $B_J \propto \sigma(CEX)$ to within 10-15%, about the same level of proportionality observed for GT transitions. And the location of the peaks in the sensitivity function $S(q,q')$ indicates that the (p,n) reaction amplitude is dominated by q' for which the transition form factors for various states are proportional. This gives us some confidence that CEX reactions can be used to measure the bulk of $L = 1$ electroweak strength. It remains to generalize to other systems: to determine whether $S(q, q')$ is localized for heavier systems and more strongly absorbed articles, and whether the $F_J(q')$ are similar for the important q'.

5 Final comments

5.1 Techniques discussed

For techniques involving high-energy particles we find:

- In inverse kinematics at high beam energies, thicker targets and large solid angle coverage yield large event rates.
- For low energies, screening effects are absent, an advantage for some reactions
- It remains to reduce the systematic uncertainties that may still be present in these new techniques. But they already approach in accuracy a typical experiment using standard techniques.
- A combination of techniques with independent systematic errors may offer the best chance of obtaining accurate results in difficult cases.

5.2 Other measurements

Use of high energy beams may give a larger reach toward the drip line even for experiments that can be done at low energy. For example:

- Identification of beam particles at high energy and then stopping them to measure β-decay half lives and to perform spectroscopy of β-delayed neutrons and gamma rays near the r-process path.
- Mass measurements with moderate (but probably sufficient) accuracy near the r-process and rp-process paths.

References

[1] A. M. Mukhamedzhanov, *et al.* Phys. Rev. C **56**, 1302 (1997).

[2] A. M. Mukhamedzhanov, C. A. Gagliardi, and R. E. Tribble, Phys Rev C **63**, 024612 (2001)

82

[3] A. Azhari *et al.*, Phys. Rev. C **63**, 055803 (2001).

[4] G. Baur, Phys. Lett. **B178**, 135 (1986).

[5] S. Typel and H. H. Wolter, Few-Body Syst. **29**, 75 (2000)

[6] M. Lattuada, *et al.*, Ap.J. **562**, 1076 (2001).

[7] C. Spitaleri *et al.*, Phys. Rev. C **63**, 055801 (2001)

[8] K. Rolfs, Nucl. Phys. News **11**, No. 3, 5 (2001)

[9] W. P. Alford and B. M. Spicer, Adv. Nucl. Phys. **24**, 1 (1998).

[10] J. Rapaport and E. Sugarbaker, Ann. Rev. Nucl. Part. Sci. **44**, 109 (1994).

[11] E. Caurier, K. Langanke, G. Martinez-Pinedo, and F. Nowacki, Nucl. Phys. **A653**, 439(1999).

[12] A. Heger, S. E. Woosley, G. Martinez-Pinedo, and K. Langanke Ap. J. **560**,307(2001).

[13] N. Anantaraman, J. S. Winfield, S. M. Austin, J. A. Carr, C. Djalali, A. Gillibert, W. Mittig, J. A. Nolen, Jr., and Z. W. Long, Phys. Rev. C **44**, 398 (1991).

[14] F. Osterfeld, N. Anantaraman, S. M. Austin, J. A. Carr, and J. S. Winfield, Phys. Rev. C **45**, 2854 (1992).

[15] T. Motobayashi *et al.*, Phys. Rev. Lett. **73**, 2680 (1994); Phys. Lett. **B64**, 259 (1991).

[16] T. Kikuchi *et al.*, Phys. Lett. **B391**, 261 (1997); T. Kikuchi et al., Eur. Phys. J. A3, 213 (1998).

[17] N. Iwasa *et al.*, Phys. Rev. Lett. **83**, 2910 (1999).

[18] B. Davids, *et al.*   Phys. Rev. Lett. **81**, 2209 (1998).

[19] B. Davids, *et al.*   Phys. Rev. Lett. **86**, 2750 (2001).

[20] B. Davids, Sam M. Austin, D. Bazin, H. Esbensen, B. M.Sherrrill, I.J.Thompson, and J. A. Thompson, Phys. Rev. C **63**, 065806 (2001).

[21] P. D. Zecker *et al.*, Phys. Rev. C **57**, 959 (1998)

[22] H. Esbensen and G. F. Bertsch, Nucl. Phys. **A600**, 37 (1996).

[23] B. W. Filippone,A. J. Elwyn, C. N. Davids, and D. D. Koetke, Phys. Rev. C **28**, 2222 (1983).

[24] F. Hammache, *et al.*, Phys. Rev. Lett. **80**, 928 (1998); **86**, 3985 (2001)

[25] F. Strieder, *et al.* Nucl. Phys. **A696**, 219 (2001).

[26] A. R. Junghans, *et al.* Nucl-ex/0111014.

[27] E. G. Adelberger *et al.*, Rev. Mod. Phys. **70**, 1265 (1998).

[28] V. F. Dmitriev, V. Zelevinsky, and Sam M. Austin, Phys. Rev. C, to be published.

[26] B. D. Anderson, *et al.* Phys. Rev. C **54**, 237 (1996).

NUCLEAR ASTROPHYSICS EXPERIMENTS WITH LASER-ELECTRON MEV PHOTONS

H. UTSUNOMIYA, S. GOKO, H. AKIMUNE, K. YAMASAKI, T.YAMAGATA,
AND M.OHTA

*Department of Physics, Konan University, Okamoto 8-9-1, Higashinada,
Kobe 658-8501, Japan, E-mail: hiro@konan-u.ac.jp*

H. OHGAKI

*Institute of Advanced Energy, Kyoto University, Gokanosho, Uji,
Kyoto 611-0011,JAPAN*

H. TOYOKAWA

*National Institute for Advanced Industrial Science and Technology (AIST),
1-1-1 Umezono, Tsukuba, Ibaraki 305-8568 Japan*

Y.-W. LUI

Cyclotron Institute, Texas A & M University, College Station, Texas 77843, USA

T. HAYAKAWA AND T. SHIZUMA

*Advanced Photon Research Center, Japan Atomic Energy Research Institute,
Tokai, Ibaraki, 319-1195, Japan*

K. SUMIYOSHI

Numazu College of Technology, Ooka 3600, Numazu, Shizuoka 410-8501, Japan

T. KAJINO

*Division of Theoretical Astrophysics, National Astronomical Observatory,
Osawa 2-21-1, Mitaka, Tokyo 181-8588, Japan*

Photonuclear reactions have direct impact on the nucleosynthesis of the p-process nuclei and play a role of probing radiative capture processes. Laser-electron MeV photon beams (LEMPBs) developed at AIST have been used for nuclear astrophysics experiments. We discuss two topics of the supernova nucleosynthesis of ^{9}Be and the threshold behavior of ^{181}Ta(γ,n) cross sections in conjunction with the origin of the p-process nuclei.

1 Introduction

Photonuclear reactions play an important role in nuclear astrophysics. They have direct impact on the nucleosynthesis of the p-process nuclei which lie on the proton-rich side of the valley of β stability with small abundance.[1] The p-process nuclei may be produced in the photodisintegration of pre-existing s-

and r-process nuclei in the O/Ne-rich layers of Type II supernovae.[2] The effective energy window for the photodisintegration is defined immediately above the neutron threshold with a typical width 1 MeV.[3] It is therefore important to elucidate the threshold behavior of the GDR cross section for nuclear astrophysics. This is also the case for destruction of the p-process nuclei. Further, it is important to study not only total but also partial cross sections in photodisintegration. For example, the nature's rarest isotope, ^{180}Ta, which has survived as a long-lived isomer ($T_{1/2} > 1.2 \times 10^{15}$ y) at the 75 keV excitation energy, may be produced by (γ, n) reaction on ^{181}Ta. On the contrary, ^{180}Ta in the ground state has become extinct with $T_{1/2} = 8.1$ h.

Photodisintegration plays a role of probing radiative-capture processes. This role is important particularly for light nuclei in which the level density is not very high. For example, (γ,n) reactions on D and ^{9}Be are inverse reactions of p(n,γ)D in the early universe[4] and of $\alpha(\alpha n, \gamma)^9$Be in the core-collapse supernovae,[5] respectively. The $\alpha(\alpha n, \gamma)^9$Be reaction bridges the mass gap at A = 8 in the supernova nucleosynthesis and is followed by (α,n) reaction to reach ^{12}C. The synthesis of ^{9}Be cannot directly be investigated because both ^{8}Be and ^{5}He that capture a neutron and an α particle, respectively, to form ^{9}Be are particle-unbound. The inverse reaction, ^{9}Be(γ,n)$\alpha\alpha$, however, can be investigated provided that a suitable real photon source is available.

For medium to heavy nuclei with a high level density, (n,γ) and (γ,n) reactions are not *equivalent*. The entrance channel (the ground state γ transition) of (γ,n) corresponds to a tiny portion of the exit channel (complicated cascade γ transitions) of (n,γ). However, the exit channel of (γ,n) provides useful information on the entrance channel of stellar (n,γ) reactions which take place on low-lying states populated in a stellar photon bath. This feature will lead to a large number of applications, including the s-process contribution, ^{186}Os(9.8 keV)(n,γ)^{187}Os, to the Re/Os cosmochronometer.[6] Thus, the breakdown of the total (γ,n) cross section into partial cross sections is important.

In this paper, we focus on photodisintegration of ^{9}Be which probes the $\alpha(\alpha$ n, $\gamma)^9$Be reaction in core-collapse supernovae and the threshold behavior of the ^{181}Ta(γ,n) cross section in conjunction with the origin of the nature's rarest isotope ^{180}Ta.

2 LEMPBs at AIST

Real photon sources that have widely been used in nuclear physics are radioactive isotopes[7,8], (n,γ) reactions,[9,10] bremsstrahlung,[11,12] and positron annihilation in flight.[13,14] Real photon beams in the MeV region have been developed at the National Institute for Advanced Industrial Science and Technology

(AIST) [16] in head-on collisions of laser photons on relativistic electrons.[17] It is straightforward to use the laser-electron MeV photon beam (LEMPB) for nuclear astrophysics experiments. The LEMPB has recently become available at HIGS facility of Duke University based on the FEL technology.[18] Further, the LEMPB is under development at SPring-8 with far-infrared laser photons at the wavelength 118 μm.[19]

In the present work, the LEMPB was produced at AIST with a Nd:YLF Q-SW laser in the first ($\lambda = 1053$ nm) and second harmonics. The energy of an electron beam in the accumulator ring TERAS was tuned in the 300 - 800 MeV region to vary the LEMPB energy. The LEMPB naturally with 100 % linear polarization can be unpolarized by using an optical element *depolarizer*.

Figure 1 shows an energy spectrum of the LEMPB measured with a high-purity Ge detector. The spectrum was simulated with the Monte Carlo code EGS4.[20] The original energy distribution of the LEMPB which best reproduced the response of the Ge detector is shown by the solid line in the figure. The LEMPB is purely quasi-monochromatic and has a characteristic low-energy tail that is determined by the beam emittance of electrons and laser photons in the interaction region and the size of a γ-ray collimator. This is a great advantage over the position annihilation in flight which has a bremsstrahlung component in addition to the annihilation component.

A BGO detector and a NaT(Tl) detector in a later stage of the measurements were used as a flux monitor of the LEMPB. The energy distributions of pile-up signals and single-photon signals from the flux monitor were used to determine the average number of γ rays per beam pulse. The average number was multiplied by the beam frequency (1 kHz) and the data acquisition time to obtain the total number of γ rays. When the LEMPB energy is close to the neutron threshold, the γ flux above the threshold was determined with help of the EGS4 code.

3   Neutron detection

A neutron detector consisting of four BF_3 counters embedded in a polyethylene moderator was used in the photoneutron cross section measurement on ^{9}Be. The neutron detection efficiency was measured with a source of ^{24}NaOH + D_2O and calibrated ^{252}Cf and Am/Be sources. The ^{24}NaOH was produced by irradiating a NaOH sample with thermal neutrons at the Research Reactor Institute of Kyoto University. The dependence of the detection efficiency on neutron energy was determined with the Monte Carlo code MCNP.[21] Some more details are seen in Ref. [22].

Another neutron detector was used for the measurement on ^{181}Ta. Eight

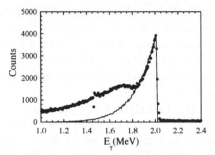

Figure 1: An energy spectrum of the LEMPB measured with a high-purity Ge detector. The energy distribution of the LEMPB which best reproduces the response of the Ge detector to the LEMPB is shown by the solid line.

BF$_3$ counters were embedded in a polyethylene cube, four each located in a concentric ring at 11 cm and 13 cm from the beam axis, respectively. In this double-ring arrangement, the detector has nearly constant detection efficiency for neutrons in the 600 - 3000 keV region. The average neutron energy was determined by the ratio of neutron events detected in the inner and the outer rings.

A 4 cm thick ^{9}Be rod was irradiated with both linearly-polarized and un-polarized LEMPBs and two 2 mm thick ^{181}Ta disks by unpolarized LEMPBs. The distribution of the neutron moderation time in the polyethylene was mea-sured with the LEMPB at the frequency 1 kHz. Background neutrons that randomly arrived at BF$_3$ counters were separated from reaction neutrons with a good signal-to-noise ratio.

4 ^{9}Be(γ,n) cross sections

Figure 2 shows photoneutron cross sections for ^{9}Be. The flux correction was made for the data at energies from the neutron threshold to 2 MeV. These data are plotted at the average energies by the solid circles in the figure. Note that the three data points at 1.69, 1.71, and 1.73 MeV which covered the peak region of the $1/2^+$ state are consistent with the data of Fujishiro et al.[23]

Including all the data with the flux correction, a new least-squares fit was performed to deduce the Breit-Wigner parameters for resonance states in ^{9}Be. The same fitting procedure as in Ref. [22] was followed. The best fit is shown by the solid lines. The best-fit parameters are listed in TABLE 1.

The present B(E1↓) for the $1/2^-$ state is lager by a factor of two than the results of electron scattering (B(E1↓) = 0.050 ± 0.020 [24] and B(E1↓) = 0.054 ± 0.004 [25]). Barker [26] previously pointed out a potential problem of

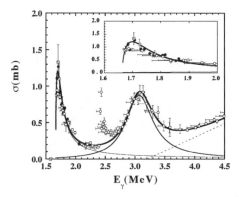

Figure 2: Photoneutron cross sections (open and solid circles) for ^{9}Be. The solid circles represent cross sections with the flux correction made below 2 MeV. The data of Ref. [23] are shown by the slashed boxes.

Table 1: Resonance parameters deduced in the present experiment.

I^π	$X\lambda$	E_R [MeV]	$B(X\lambda \downarrow)$ E1:$[e^2 fm^2]$ M1:$[(e\hbar/2Mc)^2]$	Γ_n [keV]
$1/2^+$	E1	1.735 ± 0.003	0.104 ± 0.002	225 ± 12
$5/2^-$	M1	2.43	0.295 ± 0.072	
$5/2^+$	E1	3.077 ± 0.009	0.0406 ± 0.0007	549 ± 12

background subtraction in electron scattering which weakly excited the $1/2^-$ state adjacent to the strongly excited $5/2^-$ state.

5 $\alpha(\alpha n, \gamma)^9$Be reaction rate

A new reaction rate for $\alpha(\alpha n, \gamma)^9$Be was evaluated with the present (γ,n) cross sections. We followed the same method as that derived for the triple α process by Nomoto et al. [27] The contribution of the $5/2^-$ state through the ^{8}Be + n channel was taken into account assuming a 7 % decay branch,[28,29] whereas the contribution through the ^{5}He + ^{4}He channel was neglected.

Figure 3 shows the new reaction rate $N_A^2 < \alpha\alpha n >$ in the temperature range $T_9 = 0.1$ - 10.

The CF88 rate [30] (dashed line) differs from the present rate by a factor of

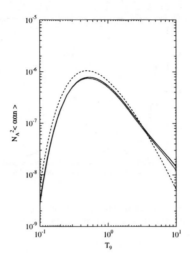

Figure 3: The reaction rate for $\alpha(\alpha\,n,\,\gamma)^9$Be (solid line) evaluated with the present data. The CF88 rate and the NACRE rate are shown by the dashed and the dotted lines, respectively.

two. In contrast, the NACRE rate [31] (dotted line) is consistent with the new rate to within 20 %. The present reaction rate will be given in a tabular form and in a fitting formula in a forthcoming paper [32] over a wide temperature range of $T_9 = 10^{-3}$ - 10^1.

6   p-processing of ^{180}Ta

The origin of the nature's rarest isotope ^{180}Ta has remained to be a mystery. ^{180}Ta is off the main path of the s-process and shielded from the r-process by a stable ^{180}Hf nucleus. In addition to the exotic s-processing of ^{180}Ta, ^{179}Hf$(e^-\,\nu)^{179}$Ta$(n,\gamma)^{180}$Ta, at high temperatures in ABG stars [6,34,35] and the ν-processing, ^{181}Ta$(\nu,n)^{180}$Ta,[36] the p-processing in the O/Ne-rich layers of Type II supernovae may make a major contribution.[2] Both the production and the destruction of ^{180}Ta should be treated dynamically during the expansion and freeze-out of supernova layers.[33] ^{180}Ta(9^-) may be produced in the photodisintegration of the pre-existing ^{181}Ta (an s- and r-process nucleus) and destructed by two mechanisms: (γ,n) and ^{180}Ta$(9^-)(\gamma,\gamma')^{180}$Ta(gs) via intermediate states above 1 MeV.[35] On the production side, it is necessary to measure the branching to the long-lived isomeric state, ^{180}Ta at 75 keV with the spin-parity 9^-.

The branching is experimentally determined by

$$\sigma(9^-) = \sigma^{total} - \sigma(gs), \tag{1}$$

where $\sigma(9^-)$ and $\sigma(gs)$ are partial (γ,n) cross sections leading to the 9^- state and the ground state, respectively, and σ^{total} is the total photoneutron cross section. The σ^{total} is determined by direct photoneutron counting, while the $\sigma(gs)$ with an activation technique[37] for the ground-state decay with a half-life of 8.1 h.

Figure 4 shows σ^{total} for ^{181}Ta near the neutron threshold. In the figure, the energy window is shown for temperatures $T_9 = 2.0$, 2.5, and 3.0 on an arbitrary scale. Clearly, cross sections near the neutron threshold are of astrophysical significance.

None of the Livermore data [13] and the Saclay data [14] are consistent with the present data. The Livermore data do not really cover the neutron threshold region. While the Saclay data agree with the present data above $E_\gamma = 9$ MeV, they exhibits non-vanishing cross sections with large uncertainties below the neutron threshold energy of Audi and Wapstra (7.58 MeV).[15] Both the Livermore and the Saclay groups used γ-ray beams resulting from the positron annihilation in flight. As a result, the contribution of the bremsstrahlung radiation had to be subtracted by using electron beams under the same experimental condition. It is most plausible that the subtraction especially near the neutron threshold resulted in subtracting a large number from a large number.

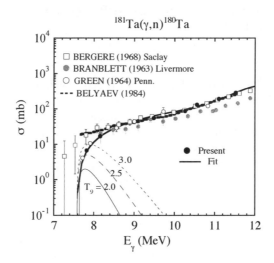

Figure 4: Photoneutron cross sections for ^{181}Ta.

The present data agree well with the result of Green and Donahue [10] who used monochromatic γ rays from thermal neutron capture reactions. In view of the fact that radioactive isotopes produced in the (n,γ) reaction usually emit a few discrete γ rays with low intensities, it is much straightforward to use the LEMPB for the photoneutron cross section measurement.

The threshold behavior of the photoneutron cross section is expressed in the form of

$$\sigma^{total} = \sum_{\ell} \sigma_{\ell} = \sum_{\ell} \sigma_{\ell}^{o} \left(\frac{E_\gamma - S_n}{S_n} \right)^{\ell + \frac{1}{2}} = \sigma^{o} \left(\frac{E_\gamma - S_n}{S_n} \right)^{p}, \qquad (2)$$

where ℓ is the neutron orbital angular momentum, σ_{ℓ} is the partial cross section for a given ℓ, and S_n is the neutron separation energy. The third part of Eq.(2) explicitly shows the ℓ dependence [39] of σ_{ℓ}. The last part is a phenomenological expression for σ^{total}. The present data for ^{181}Ta shows that $p \sim 1.1$. In contrast, the data [40] for ^{197}Au shows that $p \sim 0.5$.

The threshold behavior for ^{197}Au indicates s-wave nature of σ^{total}, whereas that for ^{181}Ta clearly indicates mixed-ℓ nature. In a naive picture of the excitation of the GDR in ^{A}X near the neutron threshold and its decay into the ground state of $^{A-1}$X + n, a dominant ℓ is expected to be 0 for ^{197}Au and 1 for ^{181}Ta. Note that $I^{\pi} = 3/2^{+}$ for ^{197}Au, 2^{-} for ^{196}Au, $7/2^{+}$ for ^{181}Ta, and 1^{+} for ^{180}Ta. The mixed-ℓ nature for ^{181}Ta can be associated with the branching to low-lying states in ^{180}Ta after the GDR excitation and/or the excitation with M1 and E2 multipoles.

7 Conclusion

Laser-electron MeV photon beams (LEMPBs) developed at the National Institute for Advanced Industrial Science and Technology (AIST) were used for nuclear astrophysics experiments. Photoneutron cross sections for ^{9}Be and ^{181}Ta are presented and discussed in the context of nuclear astrophysics. A variety of subjects in nuclear astrophysics can be studied with the LEMPB by measuring not only total but also partial photoneutron cross sections from light to heavy nuclei. It is important to create a nuclear astrophysics database of photoneutron cross sections with focus on the neutron-threshold behavior.

Acknowledgments

This work is supported by the Japan Private School Promotion Foundation, the HIRAO TARO Foundation of Konan University, and the Japan Society for the Promotion of Science under the Grant-in-Aid Program for Scientific Research (C).

References

1. D.L. Lambert, *Astron. Astrophys. Rev.* **3**, 201 (1992).
2. M. Rayet, N. Prantzos, M. Arnould, *Astron. Astrophys.* **227**, 271 (1990).
3. P. Morh *et al.*, *Phys. Lett.* B **488**, 127 (2000).
4. R.V. Wagoner, W.A. Fowler, and F. Hoyle, *Astron. Astrophys. Rev.* **3**, 201 (1992).
5. S.E. Woosley and R.D. Hoffman, *Astrophys. J.* **395**, 202 (1992).
6. K. Yokoi, K. Takahashi, and M. Arnould, *Astrophys. J.* **117**, 65 (1983).
7. B. Russel *et. al.*, *Phys. Rev.* **73**, 545 (1948).
8. B. Hamermesh and C. Kimball, *Phys. Rev.* **90**, 1063 (1953).
9. R.E. Welsh and D.J. Donahue, *Phys. Rev.* **61**, 880 (1961).
10. L. Green and D.J. Donahue, *Phys. Rev.* **64**, B701 (1964).
11. M.K. Jakobson, *Phys. Rev.* **123**, 229 (1961).
12. B.L. Berman, R.L. Van Hemert, and C.D. Bowman, *Phys. Rev.* **163**, 163 (1967).
13. R.L. Bramblett, J.T. Caldwell, G.F. Auchampaugh, and S.C. Fultz, *Phys. Rev.* **129**, 2723 (1963).
14. R. Bergère, H. Beil, and A. Veyssière, *Nucl. Phys.* **A121**, 463 (1968).
15. G. Audi and A.H. Wapstra, *Nucl. Phys.* **A565**, 1 (1993).
16. reorganized from the Electrotechnical Laboratory (ETL).
17. H. Ohgaki *et al.*, *IEEE Trans. Nucl. Sci.* **38**, 386 (1991).
18. H.R. Weller, this symposium.
19. Y. Aritomo, Y. Sato, H. Ohkuma, S. Suzuki, K. Tamura, and S. Okajima in *LEPS2000*, ed. M. Yosoi and H. Shimizu (Research Center for Nuclear Study, Osaka University, 2001)
20. W.R. Nelson, H. Hirayama and W.O. Roger, *The EGS4 Code Systems*, SLAC-Report-265 (1985).
21. J.F. Briesmeister, *MCNP*, A General Monte Carlo N-Particle Transport Code, Version 4B, Las Alamos National Laboratory (1997).
22. H. Utsunomiya, Y. Yonezawa, H. Akimune, T. Yamagata, M. Ohta, M. Fujishiro, H. Toyokawa and H. Ohgaki, *Phys. Rev.* **C63**, 018801 (2001).
23. M. Fujishiro *et. al.*, *Can. J. Phys.* **60**, 1672 (1982); ibid. **61**, 1579 (1983).
24. H.-G. Clerc *et. al.*, *Nucl. Phys.* **A120**, 441 (1968).
25. G. Kuechler *et. al.*, *Z. Phys.* **A326**, 447 (1987).
26. F.C. Barker, *Can. J. Phys.* **61**, 1371 (1983).
27. K. Nomoto, F.-K. Thielemann and S. Miyaji, *Astron. Astrophys.* **149**, 239 (1985).
28. P.R. Christensen, and C.L. Cocke, *Nucl. Phys.* **89**, 656 (1966).

29. Y.S. Chen, T.A. Tombrello, and R.W. Kavanagh, *Nucl. Phys.* **A146**, 136 (1970).
30. G. R. Caughlan and W. A. Fowler, *At. Data and Nucl. Data Tables,* **40**, 283 (1988).
31. C. Angulo et al., *Nucl. Phys.* **A656**, 3 (1999).
32. K. Sumiyoshi, H. Utsunomiya, S. Goko, and T. Kajino, in preparation.
33. M. Arnould, private communication.
34. Zs. Németh, F. Käppeler, and G. Reffo, *Astrophys. J.* **392**, 277 (1992).
35. D. Belic *et al., Phys. Rev. Lett.* **83**, 5242 (1999).
36. S.E. Woosley and W.M. Haward, *Astrophys. J. Lett.* **354**, L21 (1990).
37. A.P. Tonchev and J.F. Harmon, *Applied Radiation Isotopes* **52**, 873 (2000).
38. S.N. Belyaev *et al., Yad. Fiz.* **42**, 1050 (1985).
39. A.M. Lane and R.G. Thomas, *Rev. Mod. Phys.* **30**, 257 (1958).
40. P. Mohr, this symposium.

OPEN QUESTIONS IN STELLAR HELIUM BURNING STUDIED WITH PHOTONS *

MOSHE GAI

Laboratory for Nuclear Science, University of Connecticut,
2152 Hillside Rd., U3046, Storrs, CT 06269-3046, USA
E-mail: gai@uconn.edu, http://www.phys.uconn.edu

The outcome of helium burning is the formation of the two elements, carbon and oxygen. The ratio of carbon to oxygen at the end of helium burning is crucial for understanding the final fate of a progenitor star and the nucleosynthesis of heavy elements in Type II supernova. While an oxygen rich star is predicted to end up as a black hole, a carbon rich star leads to a neutron star. Type Ia supernovae (SNIa) are used as standard candles for measuring cosmological distances with the use of an empirical light curve-luminosity stretching factor. It is essential to understand helium burning that creates the carbon/oxygen white dwarf and thus the initial stage of SNIa. Since the triple alpha-particle capture reaction, ^{8}Be$(\alpha, \gamma)^{12}$C, the first burning stage in helium burning, is well understood, one must extract the cross section of the ^{12}C$(\alpha, \gamma)^{16}$O reaction at the Gamow peak (300 keV) with high accuracy of approximately 10% or better. This goal has not been achieved, despite repeated strong statements that appeared in the literature. Constraints from the beta-delayed alpha-particle emission of ^{16}N were shown to not sufficiently restrict the p-wave cross section factor; e.g. low values can not be rulled out. Measurements at low energies, are thus mandatory for measuring the ellusive cross section factor for the ^{12}C$(\alpha, \gamma)^{16}$O reaction. We are constructing a Time Projection Chamber (TPC) for use with high intensity photon beams extracted from the HIγS-TUNL facility at Duke University to study the ^{16}O$(\gamma, \alpha)^{12}$C reaction, and thus the direct reaction at low energies, as low as 0.7 MeV. This work is in progress.

1 Introduction: Oxygen Formation in Helium Burning and The ^{12}C$(\alpha, \gamma)^{16}$O Reaction

The outcome of helium burning is the formation of the two elements, carbon and oxygen [1,2,3]. The ratio of carbon to oxygen at the end of helium burning is crucial for understanding the fate of Type II supernovae and the nucleosynthesis of heavy elements. While an oxygen rich star is predicted to end up as a black hole, a carbon rich star leads to a neutron star [2]. At the same time helium burning is also very important for understanding Type Ia supernovae (SNIa) that are now being used as a standard candle for cosmological distances [4]. All thus far luminosity calibration curves and the stretching factor are based on empirical observations without fundamental understanding the

---

*WORK SUPPORTED BY USDOE GRANT NO. DE-FG02-94ER40870.

relation between the time characteristics of the light curve and the maximum magnitude of Type Ia supernova. Since the first burning stage in helium burning, the triple alpha-particle capture reaction ($^8\text{Be}(\alpha,\gamma)^{12}\text{C}$), is well understood [1], one must extract the p-wave [$S_{E1}(300)$] and d-wave [$S_{E2}(300)$] cross section of the $^{12}\text{C}(\alpha,\gamma)^{16}\text{O}$ reaction at the Gamow peak (300 keV) with high accuracy of approximately 10% or better to completly understand stellar helium burning, and better understand Type II and Type Ia supernova.

1.1 Beta-Delayed Alpha-Particle Emission of ^{16}N

Early hopes [5,6] for extracting the astrophysical E1 S-factor [$S_{E1}(300) = \sigma_{E1} \times E \times e^{2\pi\eta}$] of the $^{12}\text{C}(\alpha,\gamma)^{16}\text{O}$ reaction through constraints imposed by new data on the beta-delayed alpha-particle emission of ^{16}N [7,8,9,10] were examined in detail over the last few years and a few observations were made. The original Yale data [7,8] were improved [11,12] in a phase II experiment of the Yale-UConn group, and were found to be in disagreement with the TRIUMF data [10] but consistent with the unpublished data of the Seattle group [13] (see Fig. 1). In addition, an independent R-matrix analysis [14] of the world data including the beta decay of ^{16}N data was found to not rule out a small S-factor solution (10-20 keV-b). It is thus doubtful that one can extract the p-wave cross section factor with a reasonable accuracy as stated in Ref. [10]. The confusion in this field mandates a direct measurement of the cross section of the $^{12}\text{C}(\alpha,\gamma)^{16}\text{O}$ reaction at low energies. A recent measurement at lower energies [15] suggests a d-wave cross section factor that is at least twice larger than "the accepted value", and the their low energy data point(s) measured with low precision can not rule out a small p-wave cross section factor.

2 The Proposed $^{16}\text{O}(\gamma,\alpha)^{12}\text{C}$ Experiments

For determination of the cross section of the $^{12}\text{C}(\alpha,\gamma)^{16}\text{O}$ at very low energies, as low as $E_{cm} = 700$ KeV, considerably lower than measured till now [15], it is advantageous to have an experimental setup with larger (amplified) cross section, high luminosity and low background. It turns out that the use of the inverse process, the $^{16}\text{O}(\gamma,\alpha)^{12}\text{C}$ reaction may indeed satisfy all three conditions. The cross section of $^{16}\text{O}(\gamma,\alpha)^{12}\text{C}$ reaction (with polarized photons) at the kinematical region of interest (photons approx 8-8.5 MeV) is larger by a factor of approximately 100 than the cross section of the direct $^{12}\text{C}(\alpha,\gamma)^{16}\text{O}$ reaction. Note that the linear polarization of the photons yields an extra factor of two in the enhancement due to detailed balance. Thus for the lowest thus far measured data point at 0.9 MeV with the direct cross section of $\approx$60

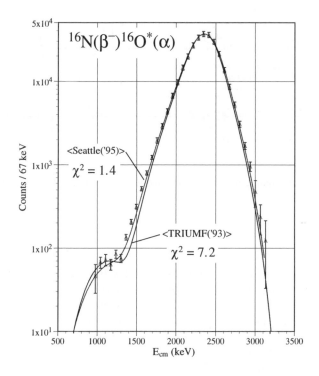

Figure 1. The Yale-UConn data on the beta-delayed alpha-particle emission of ^{16}N [11,12] compared to the TRIUMF [10] and Seattle results [13]. The TRIUMF and Seattle data are averaged over the energy resolution of the Yale-UConn experiment and are shown by continuous lines. The unpublished Seattle data are listed (by permision) in the appendix of Ref. [12].

pb, the photodissociation cross section is 6 nb. It is evident that with similar luminosities and lower background, see below, the photodissociation cross section can be measured to yet lower center of mass energies, as low as 0.7 MeV, where the direct ^{12}C$(\alpha, \gamma)^{16}$O reaction cross section is of the order of 1 pb. A very small contribution (less than 5%) from cascade gamma decay can not be measured in this method, but appears to be negligible and below the design goal accuracy of our measurement of ±10%.

The High Intensity Gamma Source (HIγS) [16], shown in Fig. 2, has already achieved many of its design milestones and is rapidly approaching its

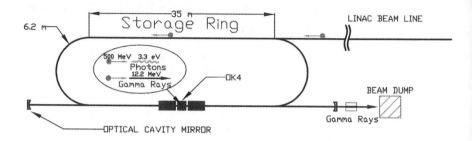

Figure 2. Schematic diagram of the HIγS facility [16] for the production of intense MeV gamma beams.

design goal for 2-200 MeV gammas. For 9 MeV gammas we expect an energy resolution of 0.1% and intensity of order 10^9 /sec. Currently achieved intensities are of the order of 10^8 /sec with energy resolution of ≈0.5%.

The backscattered photons of the HIγS facility are collimated (3 mm diameter) and enter the target/detector TPC setup as we discuss below. With a Q value of -7.162, our experiment will utilize gammas of energies approximately 8 to 10 MeV. Note that the emitted photons are linearly polarized [17] and the emitted particles are primarily in a horizontal plane (parallel to target room floor) with a $sin^2\phi$ azimuthal angular dependence [19], thus simplifying the tracking of particles in this experiment. The pulsed photon beam (0.1 ns every 180 nsec with at most 500 gammas per pulse) provides additional trigger for removing background. The image intensified CCD camera is triggered by light detected in the PMT, see below, and the time projection information from the drift chamber yields the azimuthal angle of the event of interest. The scattering angle is measured with high accuracy using the (8 cm long) alpha tracks and (2 cm long) carbon tracks. Background events from contaminants carbon, oxygen, and fluorine isotopes are discriminated using the TPC as a calorimeter with a 2% energy resolution. Time of flight techniques, and flushing of the CCD between two events will also be used. To reduce noise, the CCD will be cooled. We note that similar research program with high intensity photon beams and a TPC already exists at the RCNP at Osaka, Japan [18], proving that tracks from low energy light ions can be identified in the TPC with a managable electron background.

An $^{16}O(e, e'\alpha)^{12}C$ experiment with virtual photons proposed at the MIT-Bates accelerator [20] is useful to extract only the d-wave astrophysical cross section factor and thus it complements our experiment proposed for the HIγS-TUNL facility.

2.1 Proposed Time Projection Chamber (TPC)

We are constructing an Optical Readout Time Projection Chamber (TPC), similar to the TPC constructed in the Physikalisch Technische Bundesanstalt, (PTB) in Braunschweig, Germany and the Weizmann Institute, Rehovot, Israel [21], for the detection of alphas and carbon, the byproduct of the photodissociation of ^{16}O. Since the range of available alphas is approximately 8 cm (at 100 mbars) the TPC is 40 cm wide and up to one meter long. We first construct a 40 cm long TPC for initial use at the HIγS beam line at TUNL/Duke. The TPC is largely insensitive to single Compton electrons, and the large compton electron flux, if a problem, can be blocked using a standard beam blocker placed between the drift chamber volume and the Multi Wire Proportional Counter of the TPC. The TPC allows for tracking of both alphas and carbons emitted almost back to back from the beam position in time correlation. The very different range of alphas and carbons (approximately a factor of 4), and differences in the lateral ionization density, will aid us in particle identification. The TPC also allow us to measure angular distributions with respect to the polarization vector of the photon thus seperating the E1 and E2 components of the $^{12}C(\alpha, \gamma)^{16}O$ reaction. The excellent energy resolution of the TPC (approx. 2%) allows us to exclude events from the photodissociation of nuclei other than ^{16}O, including isotopes of carbon, oxygen and fluorine, that are present in the gas. In Fig. 3, taken from Titt et al. [21], we show a schematic diagram of the Optical Readout TPC.

The photon beam enters the TPC through an entrance window in the drift chamber part of the TPC and mainly produce background e^+e^- pairs and a smaller amount of Compton electrons, as well as the photodissociation of various nuclei present in the CO_2 + Ar gas mixture, including ^{16}O. The charged particle byproducts of the photodissociation create delta electrons that create secondary electrons that drift in the chamber electric field with a total time of the order of 1 μs per 5 cm. The time projection of the drift electrons allows us to measure the inclination angle (ϕ) of the plane of the byproducts, and the tracks themselves allow for measurement of the scattering angle (θ), both with an angular resolution better than two degrees. The electrons that reach the multi-wire chamber are multiplied (by approx. a factor of 10^5) and interact with a small (3%) admixture of triethylamine (TEA) [21] or CF_4 [22] gas to produce UV or visible photons, respectively. The light detected in the photomultiplier tube, see Fig. 3, triggers the Image

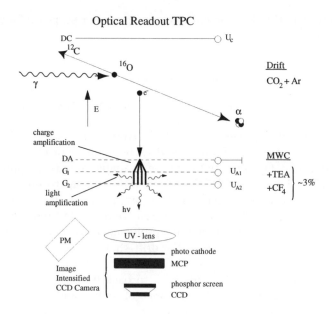

Figure 3. Schematic diagram of the Optical Readout TPC [21].

Intensifier and CCD camera which takes a picture of the visible tracks. The picture is downloaded to a PC and analyzed for recognition of the two back-to-back alpha-carbon tracks originating from the beam position. The background electrons lose approximately 0.5 keV/cm in the TPC and are removed by an appropriate threshold in the trigger Photo Multiplier Tube (PMT). Events from the photodissociation of nuclei other than ^{16}O are removed by measuring the total energy (Q-value) of the event with a resolution of 2%.

2.2 Design Goals

The luminosity of our proposed $^{16}O(\gamma, \alpha)^{12}C$ experiment can be very large. For example, with a 30 cm long fiducial length target with 30% CO_2 at a pressure of 76 torr (100 mbar) and a photon beam of 2×10^9 /sec, we obtain a luminosity of 10^{29} sec^{-1}cm^{-2}, or a day long integrated luminosity of 10 nb^{-1}. Thus a measurement of the photodissociation of ^{16}O with cross section of 0.1 nb, yields one count per day. Hence it is conceivable that a facility with such luminosity and low background together with the high efficiency TPC will allow us to measure the photodissociation cross section down to 100 pb and thus approaching 1 pb for the direct $^{12}C(\alpha, \gamma)^{16}O$ reaction, corresponding to energies as low as 700 keV. The construction of the TPC is in progress at UConn, CERN, Brookhaven, and the Weizmann Institute.

3 Acknowledgements

The author would like to acknowledge the help of Amos Breskin and Racehl Chechik (Weizmann), Uwe Titt (Mass General), Angelo Gandi (CERN), Veljko Radeka (BNL) as well as Ralph H. France III (UConn) and James E. McDonald (UHartford), in the design and construction of the TPC. This work is in collaboration with Henry Weller (HIγS-TUNL) and Tatsushi Shima (RCNP).

References

1. W.A. Fowler, Rev. Mod. Phys. **56**, 149 (1984).
2. T.A. Weaver, and E. Woosley, Physics Report, **227**, 65 (1993).
3. M. Gai, From the Sun to the Great Attractor, Lecture Notes in Physics, Springer, 2000, p. 49.
4. A.G. Riess *et al.*, Astron. J. **116** 1009 (1998).
5. B.G. Levi, Search and Discovery, Physics Today, July 1993, p.23.
6. G.F. Bertsch, Physics News, American Physical Society, 1993.
7. Zhiping Zhao, Ph.D. thesis, Yale University, 1993.
8. Z. Zhao, R.H. France III, K.S. Lai, S.L. Rugari, M. Gai, and E.L. Wilds, Phys. Rev. Lett. **70**, 2066 (1993)2066; ER **70**, 3524 (1993).
9. L. Buchmann *et al.*, Phys. Rev. Lett. **70**, 726 (1993).
10. R.E. Azuma *et al.*, Phys. Rev. **C50** 1194 (1994).
11. Ralph H. France III, Ph.D. thesis, Yale University, 1996, http://www.phys.uconn.edu/~france/thesis.html.
12. R.H. France III, E.L. Wilds, N.B. Jevtic, J.E. McDonald, and M. Gai, Nucl. Phys. **A621**, 165c (1997).
13. Z. Zhao, L. Debrackeleer, and E.G. Adelberger, 1995, *Private Communication*.
14. G.M. Hale, Nucl. Phys. bf A621, 177c (1997).
15. R. Kunz *et al.*, Phys. Rev. Lett. **86**, 3244 (2001).
16. A proposal: "A FREE-ELECTRON LASER GENERATED GAMMA-RAY BEAM FOR NUCLEAR PHYSICS", W. Tornow, R. Walter, H.R. Weller, V. Litvinenko, B. Mueller, P. Kibrough, Duke/TUNL, 1997.

17. V.N. Litvinenko *et al.*, Phys. Rev. Lett. **78** ,4569 (1997).
18. T. Shima, Y. Nagai, T. Kii, T. Baba, T. Takahashi, and H. Ohgaki, Nucl. Phys. **A629**, 475c (1998).
19. E.C. Schreiber *et al.*, Phys. Rev. **C61**, 061604(R) (2000).
20. I. Tsentalovitch *et al.*, Bates proposal, 1999.
21. U. Titt, A. Breskin, R. Chechik, V. Dangendorf, H. Schmidt-Bocking, H. Schuhmacher, Nucl. Instr. Meth. **A416**, 85 (1998); U. Titt, V. Dangendorf, H. Schuhmacher; Nucl. Phys. **B Supp. 78**, 444 (1999); U. Titt, Dissertation zur Erlangung des Doktorgrades. (Ph.D. thesis), J.W. Goethe Universitat Frankfurt, 1999.
22. A. Pausky *et al.*, Nucl. Instr. Meth. **354**, 262 (1995).

PHOTO-NUCLEAR REACTIONS IN THE BIG-BANG AND SUPERNOVAE

T. KAJINO, M. ORITO, K. OTSUKI AND M. TERASAWA

National Astronomical Observatory, Mitaka, Tokyo 181-8588, Japan

Recent observation of the power spectrum of Cosmic Microwave Background Radiation has exhibited that the flat cosmology is most likely. This suggests too large universal baryon-density parameter $\Omega_b h^2 \approx 0.022 \sim 0.030$ to accept a theoretical prediction, $\Omega_b h^2 \leq 0.017$, in the homogeneous Big-Bang model for primordial nucleosynthesis. Theoretical upper limit arises from the sever constraints on the primordial ^{7}Li abundance. We propose two cosmological models in order to resolve the descrepancy; lepton asymmetric Big-Bang nucleosynthesis model, and baryon inhomogeneous Big-Bang nucleosynthesis model. In these cosmological models the nuclear processes are similar to those of the r-process nucleosynthesis in gravitational collapse supernova explosions. Massive stars $\geq 10 M_\odot$ culminate their evolution by supernova explosions which are presumed to be the most viable candidate site for the r-process nucleosynthesis. Even in the nucleosynthesis of heavy elements, initial entropy and density at the surface of proto-neutron stars are so high that nuclear statistical equilibrium favors production of abundant light nuclei. In such explosive circumstances many neutron-rich radioactive nuclei of light-to-intermediate mass as well as heavy mass nuclei play the significant roles.

1 Big-Bang Cosmology

Recent progress in cosmological deep survey has clarified progressively the origin and distribution of matter and evolution of Galaxies in the Universe. The origin of the light elements among them has been a topic of broad interest for its significance in constraining the dark matter component in the Universe and also in seeking for the cosmological model which best fits the recent data of cosmic microwave background (CMB) fluctuations. This paper is concerned with neutrinos during Big-Bang nucleosynthesis (BBN). In particular, we consider new insights into the possible role which degenerate neutrinos may have played in the early Universe [1].

There is no observational reason to insist that the universal lepton number is zero. It is possible, for example, for the individual lepton numbers to be large compared to the baryon number of the Universe, while the net total lepton number is small $L \sim B$. It has been proposed recently [2] that models based upon the Affleck-Dine scenario of baryogenesis might generate naturally lepton number asymmetry which is seven to ten orders of magnitude larger than the baryon number asymmetry. Neutrinos with large lepton asymmetry and masses ~ 0.07 eV might even explain the existence of cosmic rays with energies in excess of the Greisen-Zatsepin-Kuzmin cutoff [3]. It is, therefore, important for both particle physics and cosmology to carefully scrutinize the limits which cosmology places on the allowed range of both the lepton and baryon asymmetries.

1.1 Lepton Asymmetric Big-Bang Model

Although lepton asymmetric BBN has been studied in many papers [4] (and references therein), there are several differences in the present work: For one , we have included finite temperature corrections to the mass of the electron and photon [5]. Another is that we have calculated the neutrino annihilation rate in the cosmic comoving frame, in which the Møller velocity instead of the relative velocity is to be used for the integration of the collision term in the Boltzmann equations [6,7].

Neutrinos and anti-neutrinos drop out of thermal equilibrium with the background thermal plasma when the weak reaction rate becomes slower than the universal expansion rate. If the neutrinos decouple early, they are not heated as the particle degrees of freedom change. Hence, the ratio of the neutrino to photon temperatures, T_ν/T_γ, is reduced. The biggest drop in temperature for all three neutrino flavors occurs for $\xi_\nu \sim 10$. This corresponds to a decoupling temperature above the cosmic QCD phase transition. ξ_ν is the neutrino degeneracy parameter defined by $\xi_\nu = \mu_\nu/T_\nu$, where μ_ν is the chemical potential and T_ν is the neutrino temperature. Finite ξ_ν leads to a lepton asymmetric ($L \neq 0$) Universe.

Non-zero lepton numbers affect nucleosynthesis in two ways. First, neutrino degeneracy increases the expansion rate. This increases the ^{4}He production. Secondly, the equilibrium n/p ratio is affected by the electron neutrino chemical potential, $n/p = \exp\{-(\Delta M/T_{n\leftrightarrow p}) - \xi_{\nu_e}\}$, where ΔM is the neutron-proton mass difference and $T_{n\leftrightarrow p}$ is the freeze-out temperature for the relevant weak reactions. This effect either increases or decreases ^{4}He production, depending upon the sign of ξ_{ν_e}.

A third effect emphasized in this paper is that T_ν/T_γ can be reduced if the neutrinos decouple early. This lower temperature reduces the energy density of neutrinos during BBN, and slows the expansion of the Universe. This decreases ^{4}He production.

Figure 1 highlights the main result of this study, where we take $\xi_{\nu_\mu} = \xi_{\nu_\tau}$. For low $\Omega_b h_{50}^2$ models, only the usual low values for ξ_{ν_e} and $\xi_{\nu_{\mu,\tau}}$ are allowed. Between $\Omega_b h_{50}^2 \approx 0.188$ and 0.3, however, more than one allowed region emerges. For $\Omega_b h_{50}^2 \gtrsim 0.4$ only the large degeneracy solution is allowed. Neutrino degeneracy can even allow baryonic densities up to $\Omega_b h_{50}^2 = 1$.

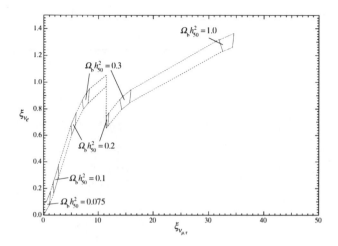

Figure 1. Allowed values of ξ_{ν_e} and $\xi_{\nu_{\mu,\tau}}$ for which the constraints from light element abundances are satisfied for values of $\Omega_b h_{50}^2 = 0.075$, 0.1, 0.2, 0.3 and 1.0 as indicated. Note that $\Omega_b h_{50}^2 = 4 \times \Omega_b h^2$, and $h = h_{100}$.

1.2 Cosmic Microwave Background

Several recent works [8,9,10] have shown that neutrino degeneracy can dramatically alter the power spectrum of the CMB. However, only small degeneracy parameters with the standard relic neutrino temperatures have been utilized. Here, we have calculated the CMB power spectrum to investigate effects of a diminished relic neutrino temperature.

The solid line on Fig. 2 shows a $\Omega_\Lambda = 0.4$ model for which $n = 0.78$, where n is the power index of primordial fluctuations. This fit is marginally consistent with the data at a level of 5.2σ. The dotted line shows the matter dominated $\Omega_\Lambda = 0$ best fit model with $n = 0.83$ which is consistent with the data at the level of 3σ. The main differences in the fits between the large degeneracy models and our adopted benchmark model are that the first peak is shifted to slightly higher l value and the second peak is suppressed. One can clearly see that the suppression of the second acoustic peak is consistent with our derived neutrino-degenerate models. In particular, the MAXIMA-1 results are in very good agreement with the predictions of our neutrino-degenerate cosmological models. It is clear that these new data sets substantially improve the goodness of fit for the neutrino-degenerate models [9]. Moreover, both data sets seem to require an increase in the baryonic contribution to the closure density as allowed in our neutrino-degenerate models [1,13].

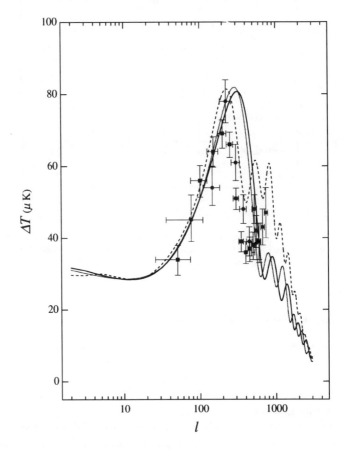

Figure 2. CMB power spectrum from BOOMERANG [11] (squares) and MAXIMA-1 [12] (circles) binned data compared with calculated $\Omega = 1$ models.

1.3   Baryon Inhomogeneous Big-Bang Model

The biggest advantage of the baryon inhomogeneous Big-Bang nucleosynthesis model [14,15,16] is to allow larger $\Omega_b h^2 \leq 0.05$, which well covers the constraint from recent CMB data $\Omega_b h^2 \approx 0.022 \sim 0.030$, still satisfying the light element abundance constraints.

Let us consider what kind of observational signature is expected in this model. Nuclear reaction flow stops at the $A = 7$ nuclear systems in the homogeneous Big-Bang nucleosynthesis model because of the instability of ^{8}Be. In the baryon inhomogeneous model, however, the nucleosynthesis occurs in an environment of proton-neutron segregated inhomogeneous distribution. Therefore, the radioactive nuclear reactions play the significant roles in the production of intermediate-to-heavy mass elements via unstable nuclei ^{8}Li(838 ms), ^{9}Li(178.3 ms), ^{7}Be(53.29 d),
^{10}Be($1.51 \times 10^6 y$), ^{8}B(770 ms), ^{12}B(20.20 ms), ^{11}C(20.385 m), ^{14}C(5730 y), ^{15}C(2.449 s), ^{13}N(9.965 m), ^{16}N(7.13 s), ^{14}O(70.606 s), etc.

$$^4\text{He}(^3\text{H}, \gamma)^7\text{Li}(n, \gamma)^8\text{Li}(\alpha, n)^{11}\text{B}(n, \gamma)^{12}\text{B}(\beta \ \nu) \tag{1}$$
$$^{12}\text{C}(n, \gamma)^{13}\text{C}(n, \gamma)^{14}\text{C}...,$$

$$^7\text{Li}(n, \gamma)^8\text{Li}(n, \gamma)^9\text{Li}(\beta \ \nu)^9\text{Be}(n, \gamma)^{10}\text{Be}(n, \gamma) \tag{2}$$
$$^{11}\text{Be}(\beta \ \nu)^{11}\text{B}...,$$

$$^7\text{Li}(^3\text{H}, n)^9\text{Be}(^3\text{H}, n)^{11}\text{B}. \tag{3}$$

The two reaction chains (1) and (2) play the key roles in the production of heavy neutron-rich isotopes [14,16] in the neutron-rich zones. Since ^{7}Li is the heaviest element to be created in the homogeneous Big-Bang nucleosynthesis model, all nuclear reactions for the production of heavier elements have ever been ignored in the previous calculations. We however found that the reaction chain (3) is extremely important for the production of ^{9}Be in the baryon inhomogeneous Big-Bang nucleosynthesis models [17,18,19]. With these reactions being included in the network of the baryon inhomogeneous Big-Bang models, the ^{9}Be abundance increases by three orders of magnitude to $N(\text{Be})/N(\text{H}) \approx 10^{-14}$ which approaches the current observational level.

2 Supernova Explosion

Stars with various masses provide a variety of production sites for intermediate-to-heavy mass elements. Very massive stars $\geq 10 M_\odot$ culminate their evolution by supernova (SN) explosions which are presumed to be most viable candidate for the still unknown astrophysical site of r-process nucleosynthesis. We discuss in this section the neutrino-driven winds from Type II SN explosion of very massive stars. Although there is still a room for the prompt explosion [20] to account for one part of the r-process nucleosynthesis, we concentrate on the gravitaional core-collpase Type II SNe here.

Even in the nucleosynthesis of heavy elements, initial entropy and density at the surface of proto-neutron stars are so high that nuclear statistical equilibrium (NSE) favors production of abundant light nuclei. In such explosive circumstances of so called hot-bubble scenario, not only heavy neutron rich nuclei but light unstable nuclei play a significant role.

The study of the origin of r-process elements is also critical in cosmology. It is a potentially serious problem that the cosmic age of the expanding Universe derived from cosmological parameters may be shorter than the age of the oldest globular clusters. Since both age estimates are subject to the uncertain cosmological distance scale, an independent method has long been needed. Thorium, which is a typical r-process element and has half-life of 14 Gyr, has recently been detected along with other elements in very metal-deficient stars. If we model the r-process nucleosynthesis in these first-generation stars, thorium can be used as a cosmochronometer completely independent of the uncertain cosmological distance scale.

2.1 Neutrino-Driven Winds in Type-II Supernovae

Recent measurements using high-dispersion spectrographs with large Telescopes or the Hubble Space Telescope have made it possible to detect minute amounts of heavy elements in faint metal-deficient ([Fe/H] $\leq$ -2) stars [21]. The discovery of r-process elements in these stars has shown that the relative abundance pattern for the mass region

$120 \leq A$ is surprisingly similar to the solar system r-process abundance independent of the metallicity of the star. Here metallicity is defined by [Fe/H] = log[N(Fe)/N(H)] - log[N(Fe)/N(H)]$_\odot$. It obeys the approximate relation $t/10^{10}$yr $\sim 10^{[Fe/H]}$. The observed similarity strongly suggests that the r-process occurs in a single environment which is independent of progenitor metallicity. Massive stars with $10M_\odot \leq M$ have a short life

$\sim 10^7$ yr and eventually end up as violent supernova explosions, ejecting material into the intersteller medium early on quickly from the history of the Galaxy. However, the iron shell in SNe is excluded from being the r-process site because of the observed metallicity independence.

Hot neutron stars just born in the gravitational core collapse SNeII release most of their energy as neutrinos during the Kelvin-Helmholtz cooling phase. An intense flux of neutrinos heat the material near the neutron star surface and drive matter outflow (neutrino-driven winds). The entropy in these winds is so high that the NSE favors a plasma which consists of mainly free nucleons and alpha particles rather than composite nuclei like iron. The equilibrium lepton fraction, Y_e, is determined by a delicate balance between $\nu_e + n \rightarrow p + e^-$

and $\bar{\nu}_e + p \rightarrow n + e^+$, which overcomes the difference of chemical potential between n and p, to reach $Y_e \sim 0.45$. R-process nucleosynthesis occurs because there are plenty of free neutrons at high temperature. This is possible only if seed elements are produced in the correct neutron-to-seed ratio before and during the r-process.

Although Woosley et al. [22] demonstrated a profound possibility that the r-process could occur in these winds, several difficulties were subsequently identified. First, independent non relativistic numerical supernova models [23] have difficulty producing the required entropy in the bubble S/k $\sim$ 400. Relativistic effects may not be enough to increase the entropy dramatically [24,25,26]. Second, even should the entropy be high enough, the effects of neutrino absorption $\nu_e + n \rightarrow p + e^-$ and $\nu_e + A(Z,N) \rightarrow A(Z+1, N-1) + e^-$ may decrease the neutron fraction during the nucleosynthesis process. As a result, a deficiency of free neutrons could prohibit the r-process [27].

In order to resolve these difficulties, we have studied [26,28] neutrino-driven winds in a Schwarzschild geometry under the reasonable assumption of spherical steady-state flow. The parameters in the wind models are the mass of neutron star, M, and the neutrino luminosity, L_ν. The entropy per baryon, S/k, in the asymptotic regime and the expansion dynamic time scale, τ_{dyn}, which is defined as the duration time of the α-process when the temprature drops from T ≈ 0.5 MeV to 0.5/e MeV, are calculated from the solution of hydrodynamic equations. Then, we carried out r-process nucleosynthesis calculations in our wind model. We found [26] that the general relativistic effects make τ_{dyn} much shorter, although the entropy increases by about 40 % from the Newtonian value of S/k $\sim$ 90. By simulating many supernova explosions, we have found some interesting conditions which lead to successful r-process nucleosynthesis, as to be discussed in the following sections.

2.2 R-process Nucleosynthesis

Previous r-process calculations [22,30] had complexity that the seed abundance distribution was first calculated by using smaller network for light-to-intermediate mass elements, and then the result was connected further to another r-process network in a different set of the computing run. For this reasaon it was less transparent to interpret the whole nucleosynthesis process. This inconvenience happened because it was numerically too heavy to run both α-process and r-process in a single network code for too huge number of reaction couplings among $\sim$ 3000 isotopes. Our nucleosynthesis calculation [26,28] is completely free from this complexity because we exploited fully

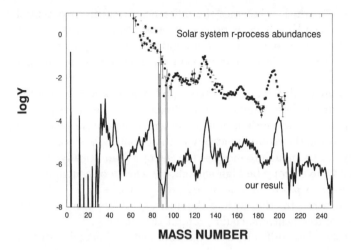

MASS NUMBER

Figure 3. R-process abundance [26] (solid line) as a function of atomic mass number A compared with the solar system r-process abundance (filled circles) from Käppeler, Beer, & Wisshak [29]. The neutrino-driven wind model used is for $L_\nu = 10^{52}$ ergs/s and $M = 2M_\odot$. The solar system r-process abundance is shown in arbitrary unit.

implicit single network code which is applied to a sequence of the whole processes of NSE - α-process - r-process.

Let us remind the readers that there were at least three difficulties in the previous theoretical studies of the r-process. The first difficulty among them is that an ideal, high entropy in the bubble S/k ~ 400 [22] is hard to be achieved in
the other simulations [23,24,25,26].

The key to resolve this difficulty is found with the short dynamic time scale $\tau_{dyn} \sim 10$ ms in our models [26,28] of the neutrino-driven winds. As the initial nuclear composition of the relativistic plasma consists of neutrons and protons, the α-burning begins when the plasma temperature cools below T ~ 0.5 MeV. The ^{4}He$(\alpha\alpha, \gamma)^{12}$C reaction is too slow at this temperature, and alternative nuclear reaction path

^{4}He$(\alpha n, \gamma)^9$Be$(\alpha, n)^{12}$C triggers explosive α-burning to produce seed elements with A ~ 100 [31]. Therefore, the time scale for nuclear reactions is regulated by the ^{4}He$(\alpha n, \gamma)^9$Be. It is given by $\tau_N \equiv \left(\rho_b^2 Y_\alpha^2 Y_n \lambda(\alpha\alpha n \rightarrow^9 Be)\right)^{-1}$. If the neutrino-driven winds fulfill the condition $\tau_{dyn} < \tau_N$, then fewer seed nuclei are produced during the α-process with plenty of free neutrons left over when the r-process begins at T $\sim$ 0.2 MeV. The high neutron-to-seed ratio, $n/s \sim 100$, leads to appreciable production of r-process elements, even for low entropy S/k $\sim$ 130, producing both the 2nd ($A \sim 130$) and 3rd ($A \sim 195$) abundance peaks and the hill of rare-earth elements ($A \sim 165$) (Figure 3).

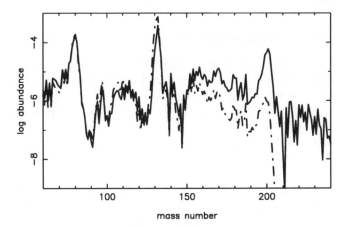

Figure 4. The same as those in Figure 3, but for the neutrino-driven wind model of $L_\nu =$ 5×10^{52} ergs/s. Solid line respresents the result by using the Woosley & Hoffman rate [31] of the ^{4}He$(\alpha n, \gamma)^9$Be reaction, and long-dashed line for the rate multiplied by factor 2, as suggested by the recent experiment of Utsunomiya et al. [32].

The three body nuclear reaction cross section for ^{4}He$(\alpha n, \gamma)^9$Be is one of the poorly determined nuclear data which may alter the r-process nucleosynthesis yields. The inverse process has recently been studied experimentally by Utsunomiya et al. [32], and photodisintegration cross section of ^{9}Be has been measured with better precision than those of the previous experiments. Applying the principle of the detailed balance to this process, one can estimate the cross section for ^{4}He$(\alpha n, \gamma)^9$Be. They found that the thermonuclear reaction rate is almost twice as big as that of Woosley and Hoffman [31] but in resonable agreement with recent compilation of Angulo et al. [33]. However, there still remain several questions on the consistency between their result and electron-scattering experiments, on the contribution from the narrow resonance $J^\pi = 5/2^-$ (2.429 MeV), etc. It is also a theoretical challenge to understand the reaction mechanism and the resonance structure because two different channels, ^{8}Be + n and ^{5}He + α, contribute to this process.

Therefore, we show two calculated results in Figure 4: The solid line displays the result obtained by using the Woosley and Hoffman cross section [31], assuming a ^{8}Be + n structure for ^{9}Be. We also calculated the r-process by multiplying this cross section by factor of 2 (long-dashed line). This makes a drastic change in the r-process yields in the 3rd ($A \sim 195$) abundance peak. More theoretical and experimental studies of the ^{4}He$(\alpha n, \gamma)^9$Be reaction are highly desired.

2.3 Neutrino-nucleus interactions

Neutrino interactions with nucleons and nuclei take the key to resolve
the second difficulty which was pointed out in sect. 1. The difficulty is that the effects of neutrino absorptions $\nu_e + n \to p + e^-$ and
$\nu_e + A(Z, N) \to A(Z + 1, N - 1) + e^-$ during the α-process may induce the deficiency of free neutrons
and break down the r-process conditions [27]. These two types of neutrino interactions control most sensitively the electron fraction and the neutron fraction, as well, in a neutron-rich environment. In order to resolve this difficulty, we have updated the electron-type neutrino capture rates for all nuclei and electron-type anti-neutrino capture rate for free proton [34,35].

The new r-process calculation proves to be almost invariant. One can understand this robustness of the succesful r-process in the following way: The specific collision time for neutrino-nucleus interactions is given by

$$\tau_\nu \approx 201 \times L_{\nu,51}^{-1} \times \left(\frac{\epsilon_\nu}{\text{MeV}}\right) \left(\frac{r}{100\text{km}}\right)^2 \left(\frac{\langle\sigma_\nu\rangle}{10^{-41}\text{cm}^2}\right)^{-1} ms, \qquad (4)$$

where $L_{i,51}$ is the individual neutrino or antineutrino luminosity in units of 10^{51} ergs/s, $\epsilon_i = < E_i^2 > / < E_i >$ in MeV ($i = \nu_e$, $\bar{\nu}_e$, etc.),
and $\langle\sigma_\nu\rangle$ is the averaged cross section over neutrino energy spectrum. At the α-burning site of r $\approx$ 100 km for $L_{\nu,51} \approx 10$, $\epsilon_{\nu_e} = 12$ MeV, and $\langle\sigma_\nu\rangle \approx 10^{-41} cm^2$, τ_{ν_e}(r=100 km) turns out to be $\approx$ 240 ms. This collision time is larger than the expansion dynamic time scale; $\tau_{dyn} \approx 10$ ms $\ll \tau_{\nu_e}$(r=100 km) $\approx$ 240 ms. Because there is not enough time for ν_e's to interact with n's in such rapidly expanding neutrino-driven wind, the neutron fraction is insensitive to the neutrino absorptions.

One might wonder if our dynamic time scale $\sim$ 10 ms is too short for the wind to be heated by neutrinos. Careful comparison between proper expansion time and specific collision time for the neutrino heating is needed in order to answer this question. Otsuki et al. [26] have found that the supernova neutrinos transfer their kinetic energy to the wind most effectively just above the neutron star surface at 10km $\leq r <$ 20km. Therefore, one should refer the duration time for the wind to reach the α-burning site, τ_{heat}, rather than

τ_{dyn}. One can estimate this heating time scale

$$\tau_{\text{heat}} = \int_{r_i}^{r_f} \frac{dr}{u},$$ (5)

where u is the fluid velocity of the wind. By setting the radius of neutron star surface $r_i = 10$ km and $r_f = 100$ km, we get $\tau_{\text{heat}} \approx 30$ ms. The collision time τ_ν is given by Eq. (4) by setting $L_{\nu,51} \approx 10$, $\epsilon_\nu = (\epsilon_{\nu_e} + \epsilon_{\bar{\nu}_e})/2 = (12 + 22)/2 = 17$ MeV, r $\approx$10 km, and $\langle \sigma_\nu \rangle \approx 10^{-41} cm^2$. Let us compare τ_{heat} and τ_ν to one another: $\tau_\nu \approx 3.4$ms $\ll \tau_{\text{heat}} \approx 30$ms. We can thus conclude that there is enough time for the expanding wind to be heated by neutrinos even with short dynamic time scale for the α-process, $\tau_{\text{dyn}} \sim 10$ ms.

2.4 Roles of Light Neutron-Rich Nuclei

The r-process is thought to proceed after the pile up of seed nuclei produced in the α-process at higher temperatures $T_9 \approx 5 \sim 2.5$. Since charged-particle reactions, which reassemble nucleons into α-particles and α-particles into heavier nuclei (i.e. α-process), are faster than the neutron-capture flow which is regulated by beta-decays, the light-mass neutron-rich nuclei were presumed to be unimportant.

However, Terasawa et al. [36] have recently found that even light neutron-rich nuclei progressively play the significant roles in the production of seed nuclei. Nuclear reaction network used in the previous studies [22,30] includes only limited number of light unstable nuclei, ^{3}H, ^{7}Be, ^{8,9}B, ^{11,14}C, ^{13}N, ^{15}O, ^{18,20}F, ^{23,24}Ne, and so on. We therefore need to extend the network code so that it covers all radioactive nuclei to the neutron-drip line. We take the rates of charged particle reactions from those used in the Big-Bang nucleosynthesis calculations [16,18,19] and the NACRE compilation [33].

Let us briefly discuss preliminary result of the r-process calculation, using the extended reaction network [36]. At early epoch of the wind expansion, t $\leq$ a few dozens ms, both temperature and density are so high that the charged particles interact with one aother to proceed nucleosynthesis around the β-stability line in the light-mass region A $\leq$ 20. There are plenty of protons and α-particles as well as neutrons at this epoch, and the main reaction flow is triggered by ^{4}He$(\alpha n, \gamma)^9$Be [22,31]:

$$^4\text{He}(\alpha n, \gamma)^9\text{Be}(\alpha, n)^{12}\text{C}(n, \gamma)^{13}\text{C}(\alpha, n)$$
$$^{16}\text{O}(n, \gamma)^{17}\text{O}(\alpha, n)^{20}\text{Ne} \quad \text{or} \quad ^{16}\text{O}(\alpha, \gamma)^{20}\text{Ne}...$$ (6)

However, at relatively later epoch even after the α-rich freeze out, a new reaction path [36]

$$^3\text{H}(\alpha,\gamma)^7\text{Li}(n,\gamma)^8\text{Li}(\alpha,n)^{11}\text{B}(n,\gamma)^{12}\text{B}(n,\gamma) \tag{7}$$
$$^{13}\text{B}(n,\gamma)^{14}\text{B}(n,\gamma)^{15}\text{B}(e^-\nu)^{15}\text{C}...$$

also takes some appreciable flux of baryon number to continuously supply the seed nuclei. The classical r-process like flow, (n,γ) followed by beta decay, has already started from light nuclei. This is a very different result from the previous picture that the r-process starts from only intermediate-mass seed nuclei $A \approx 100$.

Since we do not have much information of $(2n,\gamma)$ reactions, we did not include ^6He, ^8He, ^{11}Li, ^{14}Be, $^{17,19}\text{Be}$, ^{22}C, etc. The yields of even the most neutron-rich isotopes were found to be abundant in this calculation [36], and we plan to study the possible role of the $(2n,\gamma)$ reactions. There are several branching points between (n,γ) and (α,n) reactions. They are at ^{18}C, ^{24}O, ^{36}Mg, etc. Exerimetal studies to measure these reaction cross sections are highly desirable.

3 Quest for Nuclear Data and Astrophysics Data

Our result in Fig. 3 reproduces fairly well the observed second abundance peak (A $\approx$ 130), the hill of rare-earth elements (A $\approx$ 165), and the third peak (A $\approx$ 195). However, there are several defects, too. The first defect is a shift of the third abundance peak around A $\approx$ 195 by a couple of mass units. This is a common feature found in the previous r-process calculations [22,26,30], too. These elements are the beta-decay products of extremely neutron-rich unstable nuclei on the neutron magic N = 126. Peak position depends on the timing of freezeout of the r-process. Therefore, a particular combination of environmental evolution of neutron-number density, N_n, and temperature, T_9, as well as the expansion dynamic time scale, τ_{dyn}, might match the freeze out so that it results in the right position of abundance peak [36].

The second defect is the deficiency of abundance right above or below the peak elements, i.e. at A $\approx$ 90, 120, 150, 190, and 210. These deficiencies seem related to yet unseen effects of deformation or strucure change of unstable nuclei surrounding the neutron magic numbers N = 50, 82, and 126. Further extensive theoretical studies and observational challenge to determine the masses, lives, and beta Q-values of these nuclei are highly desired.

The third defect is the underproduction of actinoid elements, Th-U-Pu (A = 230 $\sim$ 240), by more than one order of magnitude. The observed high abundance level of these nuclei might suggest an existence of a new magic number around $N = 150 \sim 160$: Xenon $^{129}Xe_{75}$ and platinum $^{195}Pt_{117}$ are the typical

r-process elements on the second and third abundance peaks, which are decay products from extremely neutron-rich unstable nuclei with neutron magic numbers $N = 82$ and 126, respectively. ¿From these observations we estimate that the waiting point nucleus is located by shifting $\Delta N \approx 7$ or 9 units from the peak element. Applying the same shift $\Delta N \approx 7$ or 9 to $^{232}Th_{142}$ and $^{238}U_{146}$, we could assume a new magic number around $N = 150 \sim 160$ which may lead to the fourth abundance peak at $A = 230 \sim 240$. Actually, in the very light nuclear systems, a new magic number $N = 16$ was found [37] in careful experimental studies of the neutron

separation energies and interaction cross sections of extremely neutron-rich nuclei. Since these possibilities were not taken into account in the present and previous calculations, the deficiency of actinoids might be improved by modernizing nuclear mass formula including such effects. Another possibility is to make actinoid elements in neutron star mergers or the mergers of the neutron star and black hole binaries which have extremely small lepton fraction, $Y_e \leq 0.2$ [38]. However, these processes do not virtually produce any intermediate-mass nuclei including iron, which contradicts with the fact that the observed iron abundance is proportional to the r-process and actinoid elements over the entire history of Galactic evolution $\sim 10^{10}$ yr.

Since we discuss only material ejected from the proto-neutron star behind the shock, it does not make any serious problem to see the underproduction in mass region $A \leq 90$. Most of these intermediat-mass nuclei are ejected from the exploded outer shells in supernovae.

Acknowledgment

This paper has been supported in part by the Grant-in-Aid for Science Researches 10044103, 10640236, 12047233 and 13640313 of the Ministry of Education, Science, Sports, and Culture of Japan.

References

1. M. Orito, T. Kajino, G. J. Mathews, and R. N. Boyd, ApJ, submitted, (2000), astro-ph/0005446; Nucl. Phys. A., **688**, 17c, (2001)
2. A. Casas, W. Y. Cheng, & G. Gelmini, Nucl. Phys. B., **538**, 297, (1999)
3. G. Gelmini & A. Kusenko Phys. Rev. Lett., **82**, 5202, (1999)
4. H. Kang & G. Steigman Nucl. Phys. B., **372**, 494, (1992)

114

5. N. Fornengo, C. W. Kim, & J. Song, Phys. Rev. D., **56**, 5123, (1997)
6. P. Gondolo, & G. Gelmini, Nucl. Phys. B., **360**, 145, (1991)
7. K. Enqvist, K, K. Kainulainen, & V. Semikoz, Nucl. Phys. B., **374**, 392, (1992)
8. W. K. Kinney & A. Riotto, Phys. Rev. Lett., **83**, 3366, (1999)
9. J. Lesgourgues & S. Pastor, Phys. Rev. D., **60**, 103521, (1999)astro-ph/0004412
10. S. Hannestad, Phys. Rev. Lett., **85**, 4203, (2000)
11. P. Bernardls, et al. (Boomerang Collaboration) Nature., **404**, 955, (2000)
12. S. Hanany, et al. (MAXIMA-1 Collaboration), ApJ, **545**, L5, (2000)
13. M. Orito, G. J. Mathews, T. Kajino, and Y. Wang, Phys. Rev. D., submitted, (2001)
14. J. H. Applegate, C. J. Hogan and R.J. Scherrer, Phys. Rev. D., **35**, 1151, (1987)
15. C. R. Alcock, G. M. Fuller and G. J. Mathews, ApJ, **320**, 439, (1987)
16. T. Kajino, G.J. Mathews and G.M. Fuller, ApJ, **364**, 7, (1987)
17. R.N. Boyd and T. Kajino, ApJ, **336**, L55, (1989)
18. T. Kajino and R.N. Boyd, ApJ, **359**, 267, (1990)
19. Orito, M., Kajino, T. & Mathews, G. M., ApJ, **488**, 515, (1997)
20. K. Sumiyoshi, M. Terasawa, G. J. Mathews, T. Kajino, S. Yamada, and H. Suzuki, ApJ, **562**, 880, (2001)astro-ph/0106407
21. Sneden, C., McWilliam, A., Preston, G. W., Cowan, J. J., Burris, D. L., & Armosky, B. J., ApJ, **467**, 819, (1996)
22. Woosley, S. E., Wilson, J. R., Mathews, G. J., Hoffman, R. D., & Meyer, B. S., ApJ, **433**, 229, (1994)
23. Witti, J., Janka, H.-Th. & Takahashi, K., A&A, **286**, 842, (1994)
24. Qian, Y. Z. & Woosley, S. E., ApJ, **471**, 331, (1996)
25. Cardall, C. Y. & Fuller, G. M., ApJ, **486**, L111, (1997)
26. Otsuki, K., Tagoshi, H., Kajino, T. & Wanajo, S., ApJ, **533**, 424, (2000)
27. Meyer, B. S., ApJ, **449**, L55, (1995)
28. Wanajo, S., Kajino, T., Mathews, G. J., & Otsuki, K., ApJ, **554**, 578, (2001)
29. Käppeler, F., Beer, H., & Wisshak, K., Rep. Prog. Phys., **52**, 945, (1989)
30. Meyer, B. S., Mathews, G. J., Howard, W. M., Woosley, S. E., & Hoffman, R. D., ApJ, **399**, 656, (1992)
31. Woosley, S. E. & Hoffman, R. D., ApJ, **395**, 202, (1992)
32. Utsunomiya, H., Yonezawa, Y., Akimune, H., Yamagata, T., Ohta, M., Fujishiro, M., Toyokawa, H., & Ohgaki, H., Phys. Rev. C., **63**, 018801, (2001)
33. Angulo, C., et al. (NACRE collaboration), Nucl. Phys. A., **656**, 3, (1999)
34. Qian, Y.-Z., Haxton, W. C., Langanke, K., & Vogel, P., Phys. Rev. C., **55**, 1533, (1997)
35. Meyer, B. S., McLaughlin, G. C., & Fuller, G. M., Phys. Rev. C., **58**, 3696, (1998)
36. Terasawa, M., Sumiyoshi, K., Kajino, T., Mathews, G. J., & Tanihata, I., ApJ, **562**, 470, (2001)astro-ph/0107368
37. Ozawa, A., Kobayashi, T., Suzuki, T., Yoshida, K., & Tanihata, I., Phys. Rev. Lett., **84**, 5493, (2000)
38. Freiburghaus, C., Rosswog, S., & Thielemann, F.-K., ApJ, **525**, L121, (1999)

THE FUNDAMENTAL $\gamma N \to \pi N$
PROCESSES AT JLAB ENERGIES

H. GAO

Laboratory for Nuclear Science and Department of Physics,
Massachusetts Institute of Technology,
Cambridge, MA 02139, USA
E-mail: haiyan@mit.edu

The $\gamma n \to \pi^- p$ and $\gamma p \to \pi^+ n$ reactions are essential probes of the transition from meson-nucleon degrees of freedom to quark-gluon degrees of freedom in exclusive processes. The cross sections of these processes are also advantageous, for investigation of the oscillatory behavior around the quark counting prediction, since they decrease relatively slower with energy compared with other photon-induced processes. Moreover, these photoreactions in nuclei can probe the QCD nuclear filtering effects. In this talk, I discuss the preliminary results on the $\gamma n \to \pi^- p$ process at a center-of-mass angle of 90° from Jefferson Lab experiment E94-104. I also discuss a new proposal in which singles $\gamma p \to \pi^+ n$ measurement from hydrogen, and coincidence $\gamma n \to \pi^- p$ measurements at the quasifree kinematics from deuterium and ^{12}C for photon energies between 2.25 GeV to 5.8 GeV in fine steps at a center-of-mass angle of 90° are planned. The proposed measurement will allow the detailed investigation of the oscillatory scaling behavior in the photopion production differential cross-section and the study of the nuclear dependence of rather mysterious oscillations with energy that previous experiments have indicated. The various nuclear and perturbative QCD approaches, ranging from Glauber theory, to quark-counting, to Sudakov-corrected independent scattering, make dramatically different predictions for the experimental outcomes.

1 Introduction

Exclusive processes are essential to studies of transitions from the non-perturbative to the perturbative regime of QCD. The differential cross-section for many exclusive reactions [1] at high energy and large momentum transfer appear to obey the quark counting rule [2]. The quark counting rule was originally obtained based on dimensional analysis of typical renormalizable theories. The same rule was later obtained in a short-distance perturbative QCD approach by Brodsky and Lepage[3]. Despite many successes, a model-independent test of the approach, called the hadron helicity conservation rule, tends not to agree with data in the similar energy and momentum region. The presence of helicity-violating amplitudes indicates that the short-distance expansion cannot be the whole story. In addition some of the cross-section data can also be explained in terms of non-perturbative calculations [4].

In recent years, a renewed trend has been observed in deuteron photodisintegration experiments at SLAC and JLab [5,7]. Onset of the scaling behavior has been observed in deuteron photo-disintegration [7] at a surprisingly low momentum transfer of 1.0 (GeV/c)^2 to the nucleon involved. However, a recent polarization measurement on deuteron photo-disintegration [8], carried out in Hall A at JLab, shows disagreement with hadron helicity conservation in the same kinematic region where the quark counting behavior is apparently observed. These paradoxes make it essential to understand the exact mechanism governing the early onset of scaling behavior.

Moreover, it is important to look closely at claims of agreement between the differential cross section data and the quark counting prediction. Historically, the elastic proton-proton (pp) scattering at high energy and large momentum transfer has played a very important role. In fact, the re-scaled 90° center-of-mass pp elastic scattering data, $s^{10}\frac{d\sigma}{dt}$ show substantial oscillations about the power law behavior. With new high luminosity experimental facilities such as CEBAF, these oscillatory scaling behavior can be investigated with significantly improved precision. This will help identify the exact nature and the underlying mechanism responsible for the scaling behavior.

Oscillations are not restricted to the pp sector; they are also seen in πp fixed angle scattering [9]. The situation for meson photoproduction is unsettled and needs urgent investigation. Rough power-law dependence of meson photoproduction seems to agree with the constituent quark counting rule prediction [10] within experimental uncertainties, for example in the case of the $\gamma p \to \pi^+ n$ process at a center-of-mass angle of 90°. Yet it is not clear whether the $\gamma p \to \pi^0 p$ process follows the correct counting rule prediction because discrepancies exist between different measurements. For the $\gamma n \to \pi^- p$ process, no cross section data exist above a photon energy of 2.0 GeV prior to the recent Jefferson Lab E94-104 [11] experiment, in which cross section measurements of this process from a deuterium target up to a photon energy of 5.6 GeV have been carried out. Preliminary results indicate the constituent counting rule behavior in this channel at center-of-mass angle of 90°, for photon energies above $\sim$ 3 GeV. In addition to the $\frac{1}{s^7}$ scaling behavior, these preliminary results suggest an oscillatory behavior. These hints need to be investigated carefully, because the energy settings of E94-104 were designed only to investigate the global constituent quark counting rule behavior, and thus chosen to be rather coarse.

Oscillatory behavior is also suggested by the existing data on the $\gamma p \to \pi^+ n$ channel, though large uncertainties preclude any conclusive statement. Thus, it is essential to confirm and map out such oscillatory scaling behavior, which will provide insight on the origins of the scaling behavior.

The energy and nuclear dependence of such oscillatory behavior is also crucial in the search for signatures of the nuclear filtering effect. The nuclear transparency of the $\gamma\, n \to \pi^-\, p$ process can be studied by taking the ratio of the pion photoproduction yield from a nuclear target such as ^{12}C to the yield from ^{2}H. By finely mapping out the nuclear transparency over the scaling region it should be possible to test the nuclear filtering effect in a new regime.

We also note that the traditionally accepted "Glauber approximation" might be tested in the reactions under study. If the oscillations are a persistent feature of hard nucleon scattering, then established methods for obtaining the expected attenuation in nuclear targets exist. The qualitative nature of this approach is dramatically different from nuclear filtering. One way or the other, then, the experimental outcome of the photopion production from a nuclear target is expected to be of interest to a wide audience.

2 Constituent Counting Rule and Oscillations

The constituent counting rule predicts the energy dependence of the differential cross section at fixed center-of-mass angle for an exclusive two-body reaction at high energy and large momentum transfer as follows:

$$d\sigma/dt = h(\theta_{cm})/s^{n-2}, \tag{1}$$

where s and t are the Mandelstam variables, s is the square of the total energy in the center-of-mass frame and t is the momentum transfer squared in the s channel. The quantity n is the total number of elementary fields in the initial and final states, while $h(\theta_{cm})$ depends on details of the dynamics of the process. In the case of pion photoproduction from a nucleon target, the quark counting rule predicts a $\frac{1}{s^7}$ scaling behavior for $\frac{d\sigma}{dt}$ at a fixed center-of-mass angle. The quark counting rule was originally obtained based on dimensional analysis under the assumptions that the only scales in the system are momenta and that composite hadrons can be replaced by point-like constituents. Implicit in these assumptions is the approximation that the class of diagrams, which represent on-shell independent scattering of pairs of constituent quarks (Landshoff diagrams) [12], can be neglected. This counting rule was also confirmed within the framework of perturbative QCD analysis up to a logarithmic factor of α_s and are believed to be valid at high energy, in the perturbative QCD region. Such analysis relies on the factorization of the exclusive process into a hard scattering amplitude and a soft quark amplitude inside the hadron.

Although the quark counting rule agrees with data from a variety of exclusive processes, the other natural consequence of pQCD: the helicity conservation selection rule, tends not to agree with data in the experimentally tested region. Hadron helicity conservation arises from quark helicity conservation at high energies and the vector gluon-quark coupling nature of QCD, by neglecting the higher angular momentum states of quarks or gluons in hadrons. The same dimensional analysis which predicts the quark counting rule also predicts hadron helicity conservation for exclusive processes at high energy and large momentum transfers. If hadron helicity conservation holds, the induced polarization of the recoil proton in the unpolarized deuteron photo-disintegration process is expected to be zero. A polarization measurement [8] in deuteron photo-disintegration has been carried out recently by the JLab E89-019 collaboration. While the induced polarization does seem to approach zero around a photon energy of 1.0 GeV at 90° center-of-mass angle, the polarization transfer data are inconsistent with hadron helicity conservation.

The entire subject is very controversial. Isgur and Llewellyn-Smith [4] argue that if the nucleon wave-function has significant strength at low transverse quark momenta $(k_\perp)$, then the hard gluon exchange (essential to the perturbative approach) which redistributes the transfered momentum among the quarks, is no longer required. The applicability of perturbative techniques at these low momentum transfers is in serious question. There are no definitive answers to the question- *what is the energy threshold at which pQCD can be applied?* Indeed the exact mechanism governing the observed quark counting rule behavior remains a mystery. Thus, it is crucial to also look for other QCD signatures.

Apart from the early onset of scaling and the disagreement with hadron helicity conservation rule, several other striking phenomena have been observed in pp elastic scattering. One such phenomena is the oscillation of the differential cross-section about the scaling behavior predicted by the quark counting rule $(s^{-10}$ for pp scattering), first pointed out by Hendry [13] in 1973.

Secondly, the spin correlation experiment in pp scattering first carried out at Argonne by Crabb et al. [14] shows striking behavior: it is ~ 4 times more likely for protons to scatter when their spins are both parallel and normal to the scattering plane than when they are anti-parallel, at the largest momentum transfers $(p_T{}^2 = 5.09$ (GeV/c)^2, $\theta_{c.m.} = 90°$). Later spin-correlation experiments [15] confirm the early observation by Crabb et al. [14]. Theoretical interpretation for such an oscillatory behavior $(s^{10}\frac{d\sigma}{dt})$ and the striking spin-correlation in pp scattering was attempted by Brodsky, Carlson, and Lipkin [16] within the framework of quantum chromodynamic quark and gluon interactions, where interference between hard pQCD short-distance and long-distance (Landshoff) amplitudes was discussed for the first time. The Landshoff amplitude arises due to multiple independent scattering between quark pairs in different hadrons. Although each scattering process is itself a short distance process, different independent scatterings can be far apart, limited only by the hadron size. Moreover, gluonic radiative corrections give rise to a phase to this amplitude which is calculable in pQCD [17]. This effect is believed to be analogous to the coulomb-nuclear interference that is observed in low-energy charged-particle scattering. It was also shown that at medium energies this phase (and thus the oscillation) is energy dependent [18], while becoming energy independent at asymptotically high energies [18,19].

Lastly, Carroll *et al.* [20] reported the anomalous energy dependence of nuclear transparency from the quasi-elastic A(p,2p) process: the nuclear transparency first rises followed by a decrease. This intriguing result was confirmed recently at Brookhaven [21] with improved experimental technique in which the final-state was completely reconstructed. Ralston and Pire [22] explained the free pp oscillatory behavior in the scaled differential cross section and the A(p,2p) nuclear transparency results using the ideas of interference between the short-distance and long-distance amplitudes and the QCD nuclear filtering effect. Carlson, Chachkhunashvili, and Myhrer [23] have also applied such an interference concept to the pp scattering and have explained the pp polarization data.

It was previously thought that the oscillatory $s^{10}\frac{d\sigma}{dt}$ feature is unique to pp scattering or to hadron induced exclusive processes. However, it has been suggested that similar oscillations should occur in deuteron photo-disintegration [24], and photo-pion productions at large angles [25]. The QCD re-scattering calculation of the deuteron photo-disintegration process by Frankfurt, Miller, Sargsian and Strikman [24] predicts that the energy dependence of the differential cross-section, $s^{11}\frac{d\sigma}{dt}$ arises primarily from the $n-p$ scattering in the final state. If these predictions are correct, such oscillatory behavior may be a general feature of high energy exclusive photoreactions. Thus it is very important to experimentally search for these oscillations in photoreactions.

Farrar, Sterman and Zhang [26] have shown that the Landshoff contributions are suppressed at leading-order in large-angle photoproduction but they can contribute at subleading order in $\frac{1}{Q}$ as pointed out by the same authors. In principle, the fluctuation of a photon into a $q\bar{q}$ in the initial state can contribute an independent scattering amplitude at sub-leading order. However, the vector-meson dominance diffractive mechanism is already suppressed in vector meson photoproduction at large values of t [27]. On the other hand such independent scattering amplitude can contribute in the final state if more than one hadron exist in the final state, which is the case for both the deuteron photo-disintegration and nucleon photo-pion production reactions. Thus, an unambiguous observation of such an oscillatory behavior in exclusive photoreactions with hadrons in the final state at large t may provide a signature of QCD final state interaction. The most recent data on $d(\gamma, p)n$ reaction [7] show that the oscillations, if present, are very weak in this process, and the rapid drop of the cross section ($\frac{d\sigma}{dt} \propto \frac{1}{s^{11}}$) makes it impractical to investigate such oscillatory behavior.

Given that the nucleon photo-pion production has a much larger cross-section at high energies ($\frac{d\sigma}{dt} \propto \frac{1}{s^7}$), it is very desirable to use these reactions to verify the existence of such oscillations. In fact, the existing data on $\gamma p \to \pi^+ n$ suggest oscillatory behavior but the statistical uncertainties are poor at higher energies. The preliminary $\gamma n \to \pi^- p$ (E94-104) data with high statistical accuracy show hints of oscillation in the scaled differential cross-section. However, the rather coarse beam energy settings prevent a conclusive statement about the oscillatory behavior. Thus, to verify any structure in the scaled cross-section of photo-pion production processes, it is imperative that one does a fine scan of the scaling region for the $\gamma p \to \pi^+ n$ and the $\gamma n \to \pi^- p$ processes at a 90° center-of-mass angle.

2.1 Nuclear Filtering

Nuclear filtering refers to the suppression of the long distance amplitude (Landshoff amplitude) in the strongly interacting nuclear environment. Large quark separations tend not to propagate in the nuclear medium while small quark separations propagate with small attenuation. This leads to suppression of the oscillation phenomena arising from interference of the long distance amplitude with the short distance amplitude (as seen in pp scattering, mentioned earlier). Nuclear transparency measurements in A(p,2p) experiments carried out at Brookhaven [20] have shown a rise in transparency for $Q^2 \approx 3$ - 8 $(\text{GeV}/c)^2$, and a decrease in the transparency at higher momentum transfers. A more recent experiment [21], completely reconstructing the final-state of the A(p,2p) reaction, confirms the validity of the earlier Brookhaven experiment. If the oscillatory behavior of the cross-section is suppressed in nuclei one would expect to see oscillations in the transparency, which are 180^0 out of phase with the oscillations in the free pp cross-section. This is because the transparency is formed by dividing the A(p,2p) cross-section by the pp cross-section scaled by the proton number Z of the nuclear target. Brodsky and de Teramond [28] claimed that the structure seen in $s^{10}\frac{d\sigma}{dt}(pp \to pp)$, the A_{NN} spin correlation at $\sqrt{s} \sim 5$ GeV (around center-of-mass angle of 90°) [14,15], and the $A(p, 2p)$ transparency result can be attributed to $c\bar{c}uuduud$ resonant states. The opening of this channel gives rise to an amplitude with a phase shift similar to that predicted for gluonic radiative corrections.

While interpretations of the elastic $pp \to pp$ cross section, the analyzing power A_{NN} and the transparency data remain controversial, the ideas of nuclear filtering effect and the interference between the hard pQCD short-distance and the long-distance Landshoff amplitudes by Ralston and Pire [22] are able to explain both the $s^{10}\frac{d\sigma}{dt}(pp \to pp)$ oscillatory behavior and the Brookhaven A(p,2p) transparency data. Carlson, Chachkhunashvili, and Myhrer [23] have also applied such an interference concept to explain the pp polarization data.

Recently, a first complete calculation of "color transparency" and 'nuclear filtering' in perturbative QCD has been carried out for electro-production experiments [29]. These calculations show that the nuclear filtering effect is complementary to color transparency (CT) effect. Color transparency, first conjectured by Mueller and Brodsky [30] refers to the suppression of final (and initial) state interactions of hadrons with the nuclear medium in exclusive processes at high momentum transfers. The phenomenon of CT occurs when exclusive processes proceed via the selection of hadrons in the so-called point-like-configuration (PLC) states. Furthermore this small configuration should be "color screened" outside its small radius and the compact size should be maintained while it traverses the nuclear medium. While nuclear filtering uses the nuclear medium actively, in CT large momentum transfers select out the short distance amplitude which are then free to propagate through the passive nuclear medium. The expansion time relative to the time to traverse the nucleus is an essential factor for the observation of the CT effect, based on the quantum diffusion model by Farrar, Liu, Frankfurt and Strikman [31]. Thus, while one expects to observe the onset of CT effect sooner in light nuclei compared to heavier nuclei, the large A limit provides a perturbatively calculable limit for the nuclear filtering effect. This makes ^{12}C a good choice as the target for the proposed nuclear transparency measurement. The experimental verification of the nuclear filter effect would be a very interesting confirmation of this QCD based approach in the transition region. For a detailed discussion on the nuclear filtering effect and related subjects, we refer to a review article on the subject [32].

As mentioned in the introduction, oscillations are suggested by the existing cross-section data in photo-pion production reactions. Such oscillatory behavior is predicted from QCD if the independent scattering (Landshoff) regions are important. Nuclear filtering in turn should remove these regions. Thus one can investigate the nuclear filtering effect by measuring transparency in the pion photoproduction reactions as well. A calculation based on the two-component model of Jain, Kundu, and Ralston [25] was carried out for the nuclear transparency of the $\gamma p \to \pi^+ n$ reaction and predicted the following features: (i) an overall slow increase in the transparency as $\sqrt{s}$ increases; (ii) an oscillatory behavior with amplitude depending strongly on the relative phase between the effective nuclear potential of the short-distance and the long-distance amplitudes in the nuclear medium. While a relative phase angle of ZERO predicts the smallest amplitude for the oscillation, it is disfavored by the Brookhaven A(p,2p) transparency data [33].

3 JLab experiment E94-104 and future extension

Experiment E94-104 was proposed to carry out coincidence measurements of the exclusive $\gamma + n \to \pi^- + p$ process to investigate the onset of the constituent counting rule behavior in this unexplored region of $\sqrt{s} > 2.0$ GeV for this reaction. Furthermore, a coincidence measurement of the exclusive reaction $\gamma + n \to \pi^- + p$ in ^{4}He target to study the nuclear transparency of this fundamental process was also proposed for the first time.

The experiment was performed in JLab Hall A using unpolarized electrons incident upon a copper radiator, generating bremsstrahlung photons. Final state protons and pions from the photopion process of interest were detected in the two high resolution spectrometers in Hall A, with the left arm configured for optimum π^+/proton separation and the right arm for optimum π^-/electron separation. While the left arm spectrometer was used for the proton detection, the right arm was employed for the π^- detection for the coincidence measurement. The photon energy was reconstructed from the measured momenta and scattering angles of the detected particles after well defined particle identification cuts. This experiment was carried out in the spring of 2001. The preliminary results [34] on the $\gamma n \to \pi^- p$ differential cross-section at the 90° center-of-mass angle suggest a general agreement with the constituent quark counting rule prediction. The data also suggest hints of oscillatory scaling behavior as discussed previously.

Recently a new experiment [35] was proposed to carry out a measurement of the photo-pion production cross-section for the fundamental $\gamma n \to \pi^- p$ process from a ^{2}H and ^{12}C target and for the $\gamma p \to \pi^+ n$ process from a hydrogen target at a center-of-mass angle of 90°, at $\sqrt{s} \sim 2.25$ GeV to 3.41 GeV in steps of approximately 0.07 GeV. The nuclear transparency for the $\gamma n \to \pi^- p$ process will be formed by taking the ratio of the production cross-section from ^{12}C to that from ^{2}H. The calculations of Jain et al. [25] in this region predict oscillations of the order of 30% for ^{12}C. The new experiment will make individual cross-section measurements with a 2% statistical uncertainty and point-to-point systematic uncertainties of < 3%, which will allow the test of the oscillatory behavior in the scaled free cross section. The systematic uncertainties for the transparency measurement will be greatly reduced when one takes the ratio of carbon to ^{2}H. Thus, the proposed transparency measurement is expected to have combined statistical and systematic uncertainties of $< 5\%$, which should be sufficient to provide evidence for or against the nuclear filtering effect in nuclear photo-pion production processes. With combined statistical and systematic uncertainties of $<5\%$, it should be possible to confirm the Glauber predictions as well. The proposed experiment will only be possible with the unique JLab capability of high luminosity and such an experiment will be carried out in Hall A at JLab.

In summary, the preliminary E94-104 results in a rather coarse step of $\sqrt{s}$, seem to suggest oscillatory behavior in $s^7 \frac{d\sigma}{dt}$. Thus, it is essential to confirm such oscillatory behavior in finer step of $\sqrt{s}$ in the $\gamma p \to \pi^+ n$ and the $\gamma n \to \pi^- p$ processes. Furthermore, a nuclear transparency measurement of the $\gamma n \to \pi^- p$ process from a ^{12}C target will allow the investigation of the nuclear filtering effect. Such an experiment is currently planned at JLab.

Acknowledgement

I acknowledge stimulating discussions with D. Dutta, R.J. Holt, P. Jain, G.A. Miller, J.P. Ralston, M. Sargsian. This work is supported by the U.S. Department of Energy under contract number DE-FC02-94ER40818.

References

1. G. White *et al.*, Phys. Rev. **D49**, 58 (1994).
2. S.J. Brodsky and G.R. Farrar, Phys. Rev. Lett.**31**, 1153 (1973); Phys. Rev. D **11**, 1309 (1975); V. Matveev *et al.*, Nuovo Cimento Lett. **7**, 719 (1973);
3. G.P. Lepage, and S.J. Brodsky, Phys. Rev. D **22**, 2157 (1980).
4. N. Isgur and C. Llewelyn-Smith, Phys. Rev. Lett. **52**, 1080 (1984).
5. J. Napolitano *et al.*, *Phys. Rev. Lett.* **61**, 2530 (1988); S.J. Freedman *et al.*, **48**, 1864 (1993); J.E. Belz *et al.*, *Phys. Rev. Lett.* **74**, 646 (1995).
6. C. Bochna *et al.*, *Phys. Rev. Lett.* **81**, 4576 (1998).
7. E.C. Schulte, *et al.*, *Phys. Rev. Lett.* **87**, 102302 (2001);
8. K. Wijesooriya, *et al.*, Journal*Phys. Rev. Lett.*86, 2975 (2001).
9. D. P. Owen *et al.*, Phys. Rev. **181**, 1794 (1969); K. A. Jenkins *et al.*, Phys. Rev. D **21**, 2445 (1980); C. Haglin *et al.*, Nucl. Phys. B **216**, 1 (1983).
10. R.L. Anderson *et al.*, Phys. Rev. **D14**, 679 (1976).
11. Jefferson Lab Experiment E94-104, Spokespersons: H. Gao, R.J. Holt.

124

12. P. V. Landshoff, Phys. Rev. D **10**, 1024 (1974).
13. A.W. Hendry, Phys. Rev. D **10**, 2300 (1974).
14. D.G. Crabb *et al.*, Phys. Rev. Lett. **41**, 1257 (1978).
15. G.R. Court *et al.*, Phys. Rev. Lett. **57**, 507 (1986), T.S. Bhatia *et al.*, Phys. Rev. Lett. **49**, 1135 (1982), E.A. Crosbie *et al.*, Phys. Rev. D **23**, 600 (1981).
16. S.J. Brodsky, C.E. Carlson, and H. Lipkin, Phys. Rev. D **20**, 2278 (1979).
17. A. Sen, Phys. Rev. D **28**, 860 (1983).
18. J. Botts and G. Sterman, Nucl. Phys. **B325**, 62 (1989).
19. A. H. Mueller, Phys. Rep. **73**, 237 (1981).
20. A.S. Carroll *et al.*, Phys. Rev. Lett. **61**, 1698 (1988).
21. Y. Mardor *et al.*, Phys. Rev. Lett. **81**, 5085 (1998); A. Leksanov *et al.*, Phys. Rev. Lett. **87**, 212301-1 (2001).
22. J.P. Ralston and B. Pire, Phys. Rev. Lett. **61**, 1823 (1988), J.P. Ralston and B. Pire, Phys. Rev. Lett. **65**, 2343 (1990).
23. C.E. Carlson, M. Chachkhunashvili, and F. Myhrer, Phys. Rev. D **46**, 2891 (1992).
24. L.L. Frankfurt, G.A. Miller, M.M. Sargsian, and M.I. Strikman, Phys. Rev. Lett. **84**, 3045 (2000), M.M. Sargsian, private communication.
25. P. Jain, B. Kundu, and J. Ralston, hep-ph/0005126.
26. G.R. Farrar, G. Sterman, and H. Zhang, Phys. Rev. Lett. **62**, 2229 (1989).
27. E. Anciant *et al.*, Phys. Rev. Lett. **85**, 4682 (2000).
28. S. J. Brodsky, and G. F. de Teramond, Phys. Rev. Lett. **60**, 1924 (1988).
29. B. Kundu, J. Samuelsson, P. Jain and J.P. Ralston, Phys. Rev. D **62**, 113009 (2000).
30. S.J. Brodsky and A.H. Mueller, Phys. Lett. **B 206**, 685 (1988).
31. G.R. Farrar, H. Liu, L.L. Frankfurt, and M.I. Strikman, Phys. Rev. Lett. **61**, 686 (1988).
32. P. Jain, B. Pire, and J.P. Ralston, Phys. Rep. **271**, 67 (1996).
33. P. Jain, private communications.
34. L.Y. Zhu, private communications.
35. Jefferson Lab Proposal PR02-010, Spokespersons: D. Dutta, H. Gao and R.J. Holt.

THE NEW CRYSTAL BALL EXPERIMENTAL PROGRAM

W.J. BRISCOE

Department of Physics and Center for Nuclear Studies
The George Washington University, Washington, DC 20052, USA
E-mail: briscoe@gwu.edu

THE CRYSTAL BALL COLLABORATION
Abilene Christian University, Argonne National Laboratory,
Arizona State University, Brookhaven National Laboratory,
University of California at Los Angeles, University of Colorado,
George Washington University, Universität Karlsruhe,
Kent State University, University of Maryland,
Petersburg Nuclear Physics Institute, University of Regina,
Rudjer Boskovic Institute and Valparaiso University

The Crystal Ball Spectrometer is being used at Brookhaven National Laboratory in a series of experiments which study all neutral final states of $\pi^- p$ and $K^- p$ induced reactions. We report about the experimental set up and progress in obtaining new results for the radiative capture reactions $\pi^- p \to \gamma n$ and $K^- p \to \gamma \Lambda$, charge exchange $\pi^- p \to \pi^\circ n$, two π° production $\pi^- p \to \pi^\circ \pi^\circ n$, and η production $\pi^- p \to \eta$ reactions. Data have also been obtained on the decays of N^*, Δ, λ, and Σ resonances. Threshold η production has been studied in detail for both $\pi^- p$ and $K^- p$. Sequential resonance decays have been studied by studying the $2\pi^\circ$ production mechanism both in the fundamental interaction and in nuclei. In addition, we have used the ηs produced near threshold to make precision measurements searching in particular for rare and forbidden η decays.

1 Introduction

The major goal of Nuclear Physics is to understand the strong interaction. The best candidate theory is Quantum Chromodynamics, QCD, which attempts to explain the strong interaction in terms of underlying quark and gluon degrees of freedom. The study of the structure of baryons and their excitations in terms of the elementary quark and gluon constituents is thus pivotal to our understanding of nuclear matter within QCD. Within this goal our motivations for the particular reactions of interest are described briefly in the following paragraphs.

The radiative decay of a resonance provides the ideal laboratory for testing theories of the strong interaction, gives us insight into the fundamental interactions between mesons and nucleons, and allows us to probe into the structure of the nucleon itself. In particular, these data are important in the study of the radiative decay of the neutral Roper resonance. They can be combined with recent JLab Hall B data for the reactions $\gamma p \to \pi^+ n$ and $\gamma p \to \pi^\circ p$ which study the mesonic decays of the charged Roper in the incident photon energy region from 400 to 700 MeV. In addition, comparison of our data to the new JLab data taken on the inverse reaction $\gamma n \to \pi^- p$, using a deuteron target, tests extrapolation techniques for the deuteron correction and study medium effects within the deuteron.

The elusive charge exchange process has been the weakest link in partial-wave and coupled-channel studies. The accurate data that we obtained in this momentum region will help in improving the determinations of the isospin-odd s-wave scattering length, the πNN coupling constant, and the π-N σ term. In addition, better charge exchange data helps in evaluating the mass splitting of the Δ and the charge splitting of the P_{33} resonance and may result in new values for the $P_{11}(1440)$ mass and width.

Two pion production provides a means of studying sequential pion resonant decays; with neutral pions we study the $\pi\pi$ interaction in the absence of final-state Coulomb effects and owing to isospin considerations there is no contribution of ρ decay. The study of this process on the proton $(\pi^- p \to \pi^\circ \pi^\circ n)$ is the subject of a recently completed Ph.D. thesis. [1] We have also published measurements on this process in the nuclear medium. [2]

Near threshold η-production measurements provides data useful in verifying models of η-meson production and are also necessary for extraction of the η-N scattering length. Precise η production data are necessary to resolve ambiguities in the resonance properties of the $S_{11}(1535)$ and in the η photoproduction helicity amplitudes.

Using the Crystal Ball Detector, we have the ability of selecting pure isospin states. For example in the reactions $K^- p \to \eta\Lambda$ [3] and $K^- p \to \pi^\circ \Sigma^\circ$ we select a pure $I = 0$ Λ^* and in the reaction $K^- p \to \pi^\circ \Lambda$ we select a $I = 1$ Σ^*. Figure 1 shows the production cross section of the former reaction. A significant part of our program is geared toward the study of $K^- p$ reactions are described in a recent publication. [4]

The large production cross section of tagged ηs allow us to search for breaking of fundamental symmetries (e.g. C and CP invariance), and test Chiral Perturbation Theory as well as other theoretical models. We have published an article on $\eta \to 4\pi^\circ$ [5] which presents a new upper limit for this branching ratio $(B \leq 6.9 \times 10^{-7})$ of the CP forbidden decay at the 90% confidence level. This value of B puts a 2% limit on CP in quark-family-conserving interactions.

Another article has been published (since my presentation at the conference) on the rare $\eta \to 3\pi^\circ$ [6] which presents our determination of the quadratic

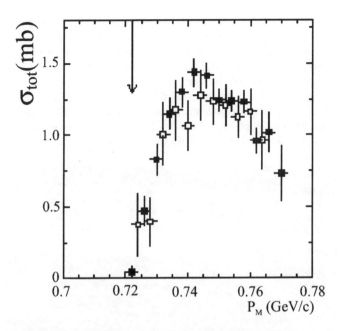

Figure 1. Total cross section for the reaction $K^-p \to \eta\Lambda$. Solid squares show cross section derived from $\eta \to \gamma\gamma$ and the open squares are derived from the $3\pi^o$ decay mode. The threshold is indicated by the arrow.

slope parameter α for that decay. The value obtained ($\alpha = -0.031 \pm 0.004$) disagrees significantly with current theory. Since this published material is now readily available, I will discuss our recent and yet unpublished work on the $\pi^o\gamma\gamma$ decay of the η.

2   Experimental Considerations

2.1   The Crystal Ball

We used the SLAC Crystal Ball to make these measurements at the C6 line at the Brookhaven National Laboratory, BNL, Alternating Gradient Synchrotron, AGS, with pion momenta from 147 MeV/c to 760 MeV/c. Data are taken simultaneously on all reactions which helps ensure that background events are accurately subtracted. Data taking using the Crystal Ball began in July 1998 and continued until late November 1998.

128

The Crystal Ball is a segmented, electromagnetic calorimetric spectrometer, covering 94% of 4π steradians. It was built at SLAC and used for meson spectroscopy measurements there for three years. It was then used at DESY for five years of experiments and put in storage at SLAC from 1987 until 1996 when it was moved to BNL by our collaboration.

The Crystal Ball is constructed of 672 hygroscopic NaI crystals, hermetically sealed inside two mechanically separate stainless steel hemispheres. The crystals are viewed by photomultipliers, PMT. There is an entrance and exit tunnel (see Fig. 2) for the beam, LH_2 target plumbing, and veto counters.

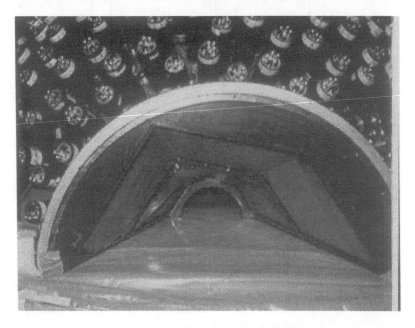

Figure 2. A view into the entry tunnel of the Crystal Ball. PMTs can be seen attached to individual crystals.

The crystal arrangement is based on the geometry of an icosahedron (20 triangular faces or "major-triangles" arranged to form a spherical shape). Each "major-triangle" is subdivided into four "minor-triangles", which in turn consist of nine individual crystals. Each crystal is shaped like a truncated triangular pyramid, points towards the interaction point, is optically isolated, and is viewed by a PMT which is separated from the crystal by a glass window.

The beam pipe is surrounded by 4 scintillators covering 98% of the target tunnel (these scintillators form the veto-barrel).

This high degree of segmentation provides excellent resolution. Electromagnetic showers in the ball are measured with an energy resolution of $\sigma/E = 2.7\%/E[GeV]^{1/4}$. Shower directions are measured with a resolution in θ of $\sigma = 2°$–$3°$ for energies in the range 50–500 MeV; the resolution in ϕ is $2°/\sin\theta$. Typically, 98% of the deposited energy of each photon is contained in a cluster of thirteen crystals (a crystal with its twelve nearest neighbors). The thickness of the NaI amounts to nearly one hadron interaction length resulting in two-thirds of the charged pions interacting in the detector. The minimum ionization energy deposited is 197 MeV; the length of the counters corresponds to the stopping range of 233 MeV for $\mu^{\pm}$, 240 MeV for $\pi^{\pm}$, 341 MeV for $K^{\pm}$, and 425 MeV for protons. The preliminary energy calibration is performed using the 0.661 MeV γ's from a ^{137}Cs source. The final energy calibration is done using three reactions: i) $\pi^- p \rightarrow \gamma n$ at rest, yielding an isotropic, monochromatic γ flux of 129.4 MeV; ii) $\pi^- p \rightarrow \pi° n$ at rest, yielding a pair of photons in the energy range 54.3—80 MeV, almost back to back; and iii) $\pi^- p \rightarrow \eta n$ at threshold, yielding two photons, about 300 MeV each, in coincidence almost back to back. The PMT analog pulses are sent to ADCs for digitization. Analog sums of the signals from each minor-triangle are available for trigger purposes.

In addition to the expected high efficiency for photons, the Crystal Ball is also fairly responsive to neutrons. We were able to measure the response of the NaI(Tl) to neutrons by using the reaction $\pi^- p \rightarrow \pi° n$ and kinematics to determine the efficiency (as high as 40%) as a function of energy, see Fig. 3.[7]

2.2 Beam Line

Figure 4 shows the experimental setup at BNL on the C6 beam line. The final stages of the C6 beamline consist of four quadrupoles and a dipole that form a beam momentum spectrometer. Wire chambers are located on both sides of the dipole to track the particles through the spectrometer. The momentum resolution is 0.3%. The scintillators located up and down stream of the dipole provide TOF information and the coincidence trigger for the beam. Scintillators surround the LH2 target to provide a charged particle veto. Two columns of scintillator neutron counters are located downstream of the Crystal Ball. A beam veto scintillator is located further downstream. A concrete shield wall located upstream of the beam stop shields the Crystal Ball from low energy photons from the stop. A Cerenkov counter is located just after this wall to monitor electron contamination in the beam.

130

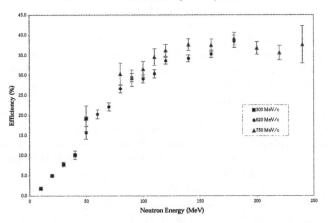

Figure 3. Efficiency of NaI(Tl) for neutron detection as a function of neutron energy.

The usual trigger consists of: a beam coincidence trigger, no downstream beam veto, and a total energy-over-threshold signal from the Crystal Ball. The Crystal Ball trigger is normally a total energy trigger. A trigger based on the distribution of the energy in different regions of the Crystal Ball was also used to provide a more restrictive trigger in certain cases.

2.3 Simulations

The primary reason so few data are available for the radiative capture reaction is the difficulty in separating its contribution from other reactions. This is mainly due to the significant background from $\pi^- p \to \pi^\circ n$ whose cross-section is about 50 times larger. The geometry of the Crystal Ball provides the capability of discriminating against multiple γ-rays that arise from the decay of π°s and ηs. However, because of the large entrance and exit tunnels there is a 20% chance that one of the two γ's from say π° decay is missed, resulting in a fake one-photon event.

The separation of signal for $\pi^- p \to \gamma n$ from 'background' was investigated with GEANT. A full-fledged Monte-Carlo simulation of the Crystal Ball including all 672 NaI crystals, the hydrogen target, its mechanical support and the down-stream neutron counters was done. Electromagnetic show-

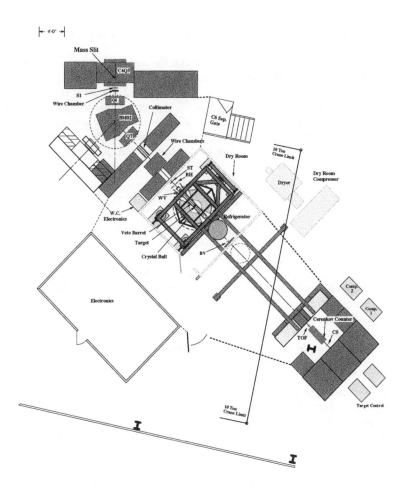

Figure 4. Experimental setup at BNL AGS C6 line.

ers are propagated by EGS within GEANT. Our Monte-Carlo includes such subtleties as secondaries from photon and pion breakup of a nucleus in the NaI, photon split-offs (a single photon cluster split into two) and backward Compton scattering. It reproduces and improves upon the photon energy and angular resolution that were measured in the course of the Crystal Ball's eight-year tenure at SLAC and DESY.

The large solid angle acceptance of the Crystal Ball and the additional use of the forward neutron detector wall lead to a rejection factor of 40-150 for the background events from $\pi^- p \to \pi^0 n$. Since the total cross-section ratio is approximately 50 at 700 MeV/c, we expect a signal to background of about 2 to 1 for the measurement of $\pi^- p \to \gamma n$ within the angular range of about 70° to about 140°. This background and thus the angular range of the radiative capture measurement can be further reduced by using endcaps to increase the solid angle.

3 Rare Decays of the η Meson

As we have alluded to above, by studying the rare and forbidden decay modes of the η meson, we are able to test the limits of such fundamental symmetries as C and CP invariance and G parity. Additionally, these also provide a laboratory in which we can test chiral perturbation theory and other recently proposed models. As mentioned above some of these measurements are already in the literature.[5] [6]

To measure the η-decay processes, we took a series of dedicated $\pi^- p$ runs at 720MeV/c which was at the maximum in eta production and yet close enough to threshold that the ηs were essential going forward in the lab. In effect we produced an η beam.

For the decay $\pi^- p \to \eta n \to \pi^0 \gamma \gamma n$ we looked at events in which 4 photons were detected. Even with this restriction, we still had backgrounds due to $3\pi^0$ decay and direct $2pi^0$ production. These backgrounds were reduced by a series of kinematic checks which not only required that the 4 photons satisfied the kinematics of the desired final state, but also eliminated any events with even a small probability (0.1%) of satisfying the kinematics of possible background processes.

In addition to the above kinematic restrictions we made various target and detector cuts that tested our abilities to Monte Carlo the acceptance and detection efficiencies. In all cases we obtained results consistent within our statistical and estimated systematic uncertainties.

Our preliminary result for the $\eta \to \pi^0 \gamma \gamma$ decay branching ratio is $3.2 \pm 0.9_{tot} \times 10^{-4}$.[8] This is less than half of the current Particle Data Group value of $7.1 \pm 1.4 \times 10^{-4}$ and disagrees by about 3-4 standard deviations. However, our experimental value does agree with the latest chiral perturbation theory calculations.

Using a similar analysis procedure, one of our colleagues has just reported an upper limit of the $\eta \to \pi^o\pi^o\gamma$ branching ratio of 5×10^{-4} at the 90% confidence level.[9]

4 Summary

The Crystal Ball Program at BNL has producing a large number of high-quality result in a very short time period after data taking. We have put severe constraints on tests of chiral perturbation theory and the limitations of fundamental symmetries. Full and short reports as well as downloads of publications and conference contributions are available to the public on our Crystal Ball Collaboration web site - URL http://bmkn8.physics.ucla.edu. While hoping to complete our planned experimental program at BNL, members of the collaboration are currently making plans for the future which include bring the Ball back to Europe - in particular to MAMI at Mainz.

5 The Collaboration

The new Crystal Ball Collaboration consists of B. Draper, S. Hayden, J. Huddleston, D. Isenhower, C. Robinson and M. Sadler, *Abilene Christian University*, C. Allgower and H. Spinka, *Argonne National Laboratory*, J. Comfort, K. Craig and A. Ramirez, *Arizona State University*, T. Kycia (deceased), *Brookhaven National Laboratory*, M. Clajus, A. Marusic, S. McDonald, B. M. K. Nefkens, N. Phaisangittisakul and W. B. Tippens, *University of California at Los Angeles*, J. Peterson, *University of Colorado*, W. Briscoe, A. Shafi and I. Strakovsky *George Washington University*, H. Staudenmaier, *Universität Karlsruhe*, D. M. Manley and J. Olmsted, *Kent State University*, D. Peaslee, *University of Maryland*, V. Abaev, V. Bekrenev, N. Kozlenko, S. Kruglov, A. Kulbardis, I. Lopatin and A. Starostin, *Petersburg Nuclear Physics Institute*, N. Knecht, G. Lolos and Z. Papandreou, *University of Regina*, I. Supek, *Rudjer Boskovic Institute* and A. Gibson, D. Grosnick, D. D. Koetke, R. Manweiler and S. Stanislaus, *Valparaiso University*.

Acknowledgements

The members of the Crystal Ball Collaboration are supported in part by the United States Department of Energy, the United States National Science Foundation, the National Sciences and Engineering Research Council of Canada, the Russian Ministry of Sciences, Volkswagen Stiftung and the George Washington University Research Enhancement Fund and Virginia Campus.

References

1. Craig, K., Ph.D. Thesis, Arizona State University (2001).
2. Starostin, A., *et al.*, Phys. Rev. Lett. **85**, 5539 (2000).
3. Starostin, A. *et al.*, Phys Rev. C **64**, 055205 (2001).
4. Manley, D.M., **in press**, Phys. Rev. Lett. (2001).
5. Prakhov, S., *et al.*,Phys. Rev. Lett. **84**, 4802 (2000).
6. Tippens, W.B. *et al.*, Phys. Rev. Lett. **87**, 192001 (2001).
7. Stanislaus, T.D.S., *et al.*, Nucl. Instrum. Methods A **462**, 463 (2001).
8. Prakhov, S., Proceedings of the III International Conference on Non-Accelerator New Physics (2001); and Crystal Ball Report CB-01-008 (2001).
9. Prakhov, S., Crystal Ball Report CB-01-009 (2001).

THE FIRST PION PHOTOPRODUCTION RESULTS FROM POLARIZED HD AT LEGS

C. STEVEN WHISNANT[1], K. ARDASHEV[2], V. BELLINI[3], M. BLECHER[4], C. CACACE[5], A.CARACAPPA[5], A. CICHOCKI[6], C. COMMEAUX[7], I. DANCHEV[8], A. D'ANGELO[3], J. P. DIDELEZ[7], R. DEININGER[2], C. GIBSON[8], K. HICKS[2], S. HOBLIT[5], A. HONIG[9], T. KAGEYA[4], M. KHANDAKER[10], O. KISTNER[5], A. KUCZEWSKI[5], F. LINCOLN[5], R. LINDGREN[6], A. LEHMANN[1], M. LOWRY[5], M. LUCAS[2], J. MAHON[2], H. MEYER[4], L. MICELI[5], D. MORICCIANNI[3], B. M. PREEDOM[8], B. NORUM[6], T. SAITOH[4], A. M. SANDORFI[5], C. SCHAERF[3], D. J. TEDESCHI[8], C. THORN[5], K. WANG[6], AND X. WEI[5].
(THE LEGS SPIN COLLABORATION)

[1] *James Madison U.*,* [2] *Ohio U.*, [3] *U. di Roma and INFN-Sezione diRoma*, [4] *Virginia Tech.*, [5] *Brookhaven National Lab.*, [6] *U. Virginia*, [7] *Orsay*, [8] *U. of South Carolina*, [9] *Syracuse U.*, [10] *Norfolk St. U*

A solid, polarized *HD* target has been developed for the measurement of double-polarization observables in the Δ resonance region. We report here the inaugural data obtained with this target. This new polarized target technology, combined with the high quality LEGS photon beam and the large acceptance spectrometer, SASY, provides a unique facility for studying the spin structure of the nucleon below 500 MeV. Pion production data collected on a longitudinally polarized target with six γ-ray polarization states provides the first simultaneous measurement of $\frac{d\sigma}{d\Omega}$, as well as the Σ, G and E asymmetries. With the future addition of magnetic analysis to SASY, a complete set of pion production observables on the proton and the deuteron (neutron) will be obtained.

1 Introduction

Considerations of forward Compton scattering leads to the Gerasimov Drell-Hearn[1] (GDH) and the forward spin polarizability[2] sum rules.

$$ GDH = \int\limits_{\omega_0}^{\infty} \frac{\sigma_{1/2} - \sigma_{3/2}}{\omega} d\omega \quad \text{and} \quad \gamma_0 = \frac{1}{4\pi^2} \int\limits_{\omega_0}^{\infty} \frac{\sigma_{1/2} - \sigma_{3/2}}{\omega^3} d\omega \quad (1) $$

The latter is rapidly convergent, while the GDH sum rule requires the additional assumption that the Compton spin-flip amplitude vanishes at infinite energy[3].

---

*e-mail: whisnacs@jmu.edu

Due to the Δ dominance in the helicity $\frac{1}{2}$ (photon and target spins parallel in the center of mass) and the helicity $\frac{3}{2}$ (spins anti-parallel) reaction cross sections, and the energy weighting of the integral, these sum rules are dominated by single pion production. Moreover, it has been shown that the key physics issues are in the proton - neuton differences[3]. The investigation of the proton and neutron under identical conditions, thereby minimizing the systematic uncertainties, is central to the accurate determination of this physics.

With the development of the Strongly Polarized Hydrogen deuteride ICE target (SPHICE), LEGS (Laser Electron Gamma Source) begins a program of double-polarization measurements. This unique target is complemented by the high quality polarized Compton backscattered photon beam at LEGS and the large acceptance Spin ASYmmetry (SASY) detector system constructed for these experiments. With the combination of SPHICE and SASY at LEGS, we are beginning the detailed study of pion photoproduction and the nucleon spin structure.

2 Beam Characteristics

The γ-ray beam at LEGS is produced by Compton backscattering of laser light from the 2.8 GeV electron beam at the National Synchrotron Light Source (NSLS) at Brookhaven National Laboratory. This electron energy, combined with a new frequency quadrupled laser and a conventional Ar-Ion laser, permits the production of tagged photons from π threshold up to 471 MeV.

The combination of the frequency quadrupled laser and the existing Ar-Ion laser produces γ distributions with Compton edges ranging from below pion threshold up to 471 MeV. The polarization of the γ-ray beam is determined by the polarization of the incoming laser beam. A polarimeter sampling the laser beam just before the laser-electron interaction region determines the polarization to $\pm 1\%$. Because the incident beam may be linearly, circularly, or elliptically polarized, the complete specification of the laser polarization is obtained by the measurement of the Stokes vector, $S_l = \{Q_l, U_l, V_l\}$. The three components of this vector are Q_l, the amplitude of $0°$ polarized beam, U_l, the amplitude of the beam polarized at $+45°$, and V_l, the amplitude of right circularly polarized beam. The γ-ray beam Stokes vector, $S_\gamma = \{Q_\gamma(E_\gamma), U_\gamma(E_\gamma), V_\gamma(E_\gamma)\}$ is obtained from S_l, the polarization transfer function, $\mathcal{P}_{linear}$ or $\mathcal{P}_{circular}$ obtained from the Klein-Nishina cross section and a Monte Carlo of the overlap and divergence of the laser and electron beams, and P_{brem}, the depolarization due to the bremsstrahlung produced from the residual gas in the storage ring. The γ polarization essentially equals that of the laser at the Compton edge and falls slowly with decreasing energy for all polarization states. The variety of laser energies available ensure that the γ polarization is $\geq 70\%$ throughout the tagging range.

With this specification of the photon polarization, the most general expression for the cross section on a longitudinally polarized target can be written as

$$\frac{d\sigma}{d\Omega}(\theta, \phi; E_\gamma) = \frac{d\sigma}{d\Omega}(\theta; E_\gamma) \cdot \tag{2}$$
$$\{ 1 + [Q_\gamma(E_\gamma) \cdot \Sigma(\theta; E_\gamma) - P_z \cdot U_\gamma(E_\gamma) \cdot G(\theta; E_\gamma)] \cos(2\phi)$$
$$+ [P_z \cdot Q_\gamma(E_\gamma) \cdot G(\theta; E_\gamma) + U_\gamma(E_\gamma) \cdot \Sigma(\theta; E_\gamma)] \sin(2\phi)$$
$$- P_z \cdot V_\gamma(E_\gamma) \cdot E(\theta; E_\gamma) \}$$

where $\Sigma(\theta; E_\gamma)$ is the $0°/90°$ beam polarization asymmetry on an unpolarized target, $G(\theta; E_\gamma)$ is the $\pm 45°$ beam polarization asymmetry with longitudinal target polarization, $E(\theta; E_\gamma)$ is the helicity cross section asymmetry, and P_z is the target polarization.

To disentangle these asymmetries from data obtained with beam polarizations less than 100% requires measurement with four linear $(0°, 90°, \pm 45°)$ as well as left and right circular polarizations. This is readily done at LEGS by randomly cycling the laser polarization through all six states. Data collected in this way with a longitudinally polarized target permits the extraction of $\frac{d\sigma}{d\Omega}$, $\Sigma(\theta; E_\gamma)$, $G(\theta; E_\gamma)$, and $E(\theta; E_\gamma)$ *simultaneously*.

3 SPHICE

SPHICE represents a new technology utilizing molecular HD in the solid state[4]. These targets are polarized at low temperature (15-20 mK) and high field (15-17 T) in a dilution refrigerator. The spin-lattice coupling which permits polarization (and depolarization) of these targets is affected by a small $(\approx 10^{-4})$ concentration of ortho-H_2 (molecular rotational angular momentum, $J = 1$). Since these molecules decay at low temperature to the magnetically inert $J = 0$, para-H_2 with a time constant of $\approx$ 6 days, the spin-lattice relaxation time, T_1, increases as a function of the time. Once polarized, the spins are "frozen in" by holding the target in the polarizing conditions. Protons are polarized by permitting the target to equilibrate at low-temperature and high-field. Because the initial T_1 is short, the proton polarization may be transferred to the deuteron using rf techniques and the proton repolarized. By this technique, a proton polarization of 80% and a deuteron polarization of about 50% are achievable.

Holding the target in the polarizing conditions for 30-50 days (depending on whether the goal is $\vec{H}D$, $H\vec{D}$ or $\vec{H}\vec{D}$) increases T_1 permitting target extraction using a specially designed transfer cryostat[5] (TC). The frozen spin targets can then be stored/transported in a storage dewar (4 K/8 T) or inserted into the in-beam cryostat (IBC) (1.25 K/0.65 T).

3.1 Target Production and Characteristics

Targets are produced in a top-loading dilution refrigerator containing a superconducting solenoid capable of producing up to three 2.5 cm diameter × 5 cm long targets simultaneously. Once the spins are frozen in, the field is reduced, the system is warmed to 2 K, and the TC is inserted. This device, containing both LN_2 and LHe jackets, has a central portion that can translate and rotate allowing it to screw into the target mount and extract the target. While in the TC, polarization is maintained with a small magnet providing a few hundred Gauss field. For this experiment, the target was inserted directly into the IBC. The IBC is a pumped ^{4}He system operating at 1.25 K. The 0.65 T polarization holding field is provided by a small superconducting solenoid inside the IBC.

The target used for the experiments we report here was polarized at $\approx$18 mK and 15 T for 40 days. The T_1 for $\vec{H}$ measured in-beam (1.25 K/0.65 T) is 13 days and for $\vec{D}$ it is 36 days. The initial polarization obtained for hydrogen was 70% in this first target. However, due to the numerous tests and manipulations done on this first target, the hydrogen polarization was 33 ± 2% when in-beam. This large drop in polarization is the result of, among other things, the equivalent of five transfers of the target between the dilution refrigerator and the IBC, and a detailed mapping of the relaxation time as a function of temperature and field.

The target is shown in the IBC schematically in Figure 1. Aside from the thin CTFE target cell, the only material in the photon beam other than HD is the high purity (99.999%) Al cooling wires. Because HD at 18 mK is a poor thermal conductor and the ortho- to para-H_2 conversion generates heat throughout the target, these wires are required to ensure a uniform low temperature while polarizing.

The 2050 50 μm wires represent 20% of the target by weight. Because these wires and the 0.6 mm thick CTFE target cell are the only sources of unpolarized background, the sprectra are quite clean. As seen in the missing energy spectrum for $\pi^{\pm}$ in Figure 2, the empty target contribution results in a 5:1 integrated full/empty ratio. The extended tails on the free proton peak are caused by the Fermi motion of the bound proton in the deuteron.

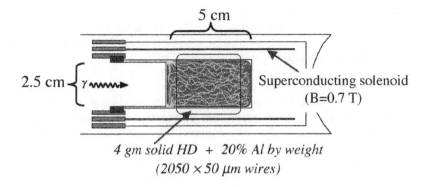

Figure 1. A cross sectional view of the SPHICE target in the IBC.

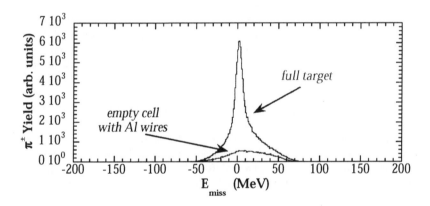

Figure 2. A full and empty target spectrum for $\pi^{\pm}$ obtained with an unpolarized HD target.

4 Experiment

SASY, the Spin-ASYmmetry detector array, determines angle, energy, and particle identity for all reactions induced by photons on hydrogen and deuterium over the entire LEGS energy range. The major calorimetry subsystems are instrumented: the crystal box (an array of 432 NaI(Tl) crystals), and a forward wall of plastic scintillator ($\approx$ 30% neutron efficiency) and Pb-Glass

Čerenkov counters. Atomic events are rejected by a gas Čerenkov at $0°$ and an Aerogel detector covering angles out to $30°$. The space between the target cryostat and the crystal box is filled by a scintillator azimuthally segmented in 32 sections, extending the neutron coverage to $90°$.

The photon beam was produced by backscattering UV Ar-Ion laser light (333-363 nm) to produce tagged photons in the energy range from 230 MeV to 380 MeV. The beam polarization was randomly cycled through the six polarization states. An average of 100 sec was spent with linear beam polarization at $0°/90°$ with respect to the reaction plane, 200 sec in the $\pm 45°$ states, 500 sec with circular polarization, and 25 sec with laser off (bremsstrahlung) per cycle. The laser was typically 98% polarized, or better. However, it is technically very difficult to reach circular polarizations larger than 98%, and a 98% circular beam has 20% linear polarization, $(0.98)^2 + (0.20)^2 = 1$. *Thus, to separate* $\Sigma(\theta; E_\gamma)$, $G(\theta; E_\gamma)$, *and* $E(\theta; E_\gamma)$ *with imperfectly polarized beams requires the simultaneous measurement of all six polarization states.* We report here the first such measurement.

Using the measured π^+ angle, the tagged photon energy, correcting for the energy lost in the cryostat, and subtracting the mass of the proton gives the missing energy spectrum. This is shown for π^+ in Figure 3 for left and right circular polarization of the beam. These accidental corrected and background subtracted spectra correspond to $E_\gamma = 300 \pm 20$ MeV and $\theta_\pi^{cm} = (80 \pm 10)°$. These sprectra represent two days of beam time.

From equation 2 we see that with an unpolarized target the $0°/90°$ beam asymmetry is expected to have a $\cos 2\phi$ and the $\pm 45°$ asymmetry will have a $\sin 2\phi$ dependence with an amplitude proportional to $\Sigma(\theta; E_\gamma)$. However, with a longitudinally polarized target a non-zero $G(\theta; E_\gamma)$ mixes these two ϕ dependences and produces a phase shift in both asymmetries. The effect of this asymmetry is seen in the top and center panels of Figure 4 as a small systematic phase shift in the data.

The circular polarization asymmetry is more complicated. Although the $E(\theta; E_\gamma)$ asymmetry is independent of ϕ, because the beam is not perfectly polarized, there will also be an arbitrary admixture of the other two polarization states. Hence, we see in the bottom panel of Figure 4 an offset from zero due to the $E(\theta; E_\gamma)$ asymmetry and a linear combination of the $\cos 2\phi$ and $\sin 2\phi$ dependence due to the linear polarizations present in the beam.

5 Conclusions

We have presented here the first simultaneous measurement of the $\Sigma(\theta; E_\gamma)$, $G(\theta; E_\gamma)$, and $E(\theta; E_\gamma)$ asymmetries using a longitudinally polarized *HD* target at LEGS. This first result obtained with a 33% polarized $\vec{H}D$ target in the beam for two days demonstrates the high quality data that can be achieved using this new technology.

A systematic program to measure the GDH and γ_0 sum rule integrals on the proton will run in the spring of 2002. Subsequently, a major upgrade to the SASY detector is planned. A large, warm-bore superconducting magnet that will fit inside the crystal box has been constructed. Its installation will provide larger holding fields for the polarized target and tracking of $\pi^{\pm}$ in the magnetic field. This will permit the removal the solenoid in the target cryostat, thereby lowering the threshold for charged particles and enable the discrimination between π^+ and π^- without requiring the detection of the nucleon in coincidence. This capability is necessary to extend the charged pion measurements into the critical region near threshold and for detailed measurements on the neutron.

Acknowledgments

This work is supported by US Dept. of Energy under contract DE-AC02-98CH10886 and by the US National Science Foundation.

References

1. S. B. Gerasimov. Sov. J. Nucl. Phys. **2**, 430 (1966), S. D. Drell and A. C. Hearn. Phys. Rev. Lett. **16**, 908 (1966).
2. V. Bernard, N. Kaiser, J. Kambor, and Ulf-G. Meissner. Nucl. Phys. **B38**, 315 (1992).
3. A. M. Sandorfi, C.S. Whisnant, and M. Khandaker. Phys. Rev. **D50**, R6681 (1994).
4. A. Honig, Q. Fan, X. Wei, A. M. Sandorfi and C. S. Whisnant. Nucl. Inst. Meth. **A356**, 39 (1995).
5. N. Alexander, J. Barden, Q. Fan, and A. Honig. Rev. Sci. Instrum. **62**, 2729 (1991).

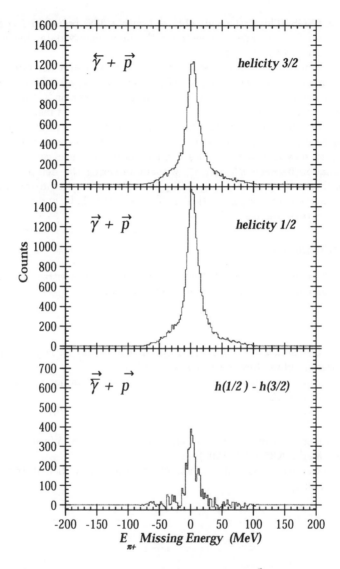

Figure 3. The missing mass spectra for π^+ obtained with an $\vec{H}D$ target and a circularly polarized beam. The top panel is total helicity $\frac{3}{2}$ (anti-parallel spins in the laboratory frame), the middle panel is helicity $\frac{1}{2}$ (parallel spins), and the bottom panel is the difference.

$$E_\gamma = 300 \text{ MeV}, \; \theta_\pi^{cm} = 80°$$

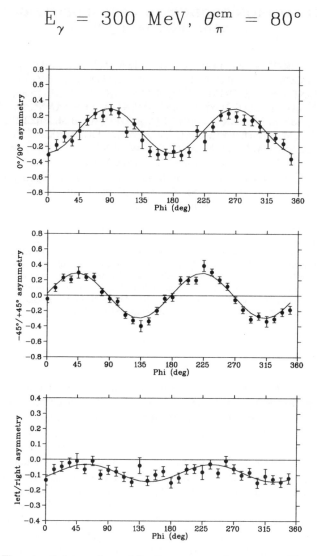

Figure 4. The azimuthal dependence of the three asymmetries constructed from the data. The top panel shows the 0°/90° beam polarization asymmetry, the center panel shows the asymmetry for ±45° linear polarization and the bottom panel contains the circular polarization asymmetry. The curves on the top and center panels are $\cos 2\phi$ and $\sin 2\phi$, respectively. In the bottom panel it is a fit to the data.

NEUTRAL PION PRODUCTION FROM DEUTERIUM AT THE LEGS FACILITY

K.H. HICKS[1*], K. ARDASHEV[1], M. BLECHER[2], A. CARACAPPA[3], A. CICHOCKI[4], C. COMMEAUX[5], A. D'ANGELO[6], J.-P. DIDILEZ[5], R. DEININGER[1], S. HOBLIT[3], M. KHANDAKER[7], O. KISTNER[3], A. KUCZEWSKI[3], F. LINCOLN[3], R. LINDGREN[4], A. LEHMANN[8], M. LOWRY[3], M. LUCAS[1], H. MEYER[2], L. MICELI[3], A. OPPER[1], B.M. PREEDOM[8], B. NORUM[4], A.M. SANDORFI[3], C. SCHAERF[6], H. STRÖHER[9], C.E. THORN[3], J. TONNISON[2], K. WANG[4], X. WEI[3], C.S. WHISNANT[10], AND D. WILLITS[1]
(THE LEGS COLLABORATION)

1) Ohio U., 2) Va Tech, 3) Brookhaven National Lab, 4) U. Virginia, 5) Orsay, 6) Roma, 7) Norfolk St., 8) U. So. Carolina, 9) Jülich, and 10) James Madison
* e-mail: hicks@ohio.edu

Neutral pion photoproduction from a liquid deuterium target was measured in the energy region near 300 MeV at the LEGS facility of Brookhaven National Laboratory. The inclusive cross sections from deuterium are in agreement with measurments from Mainz, yet the exclusive cross sections and spin asymmetries for neutral pion production in coincidence with a detected nucleon are much smaller than expected from a quasi-free approximation. This may indicate that substantial final state interactions play a significant role, which will complicate the extraction of the desired amplitudes that would be measured if a free neutron target could be used.

1 Introduction

The importance of deuterium as a source of quasi-free neutrons is often used to motivate experimental studies that aim to measure reactions that could be used to extract "free" neutron scattering observables. Because the deuteron is bound by slightly more than 2 MeV, deuterium is easily broken apart in reactions where the incident beam has energies of several hundred MeV. As a result, the struck nucleon emerges at nearly the same kinematics as if it had been free (or stationary in the lab) rather than bound (in which case it had some initial momentum with a Fermi distribution). However, for some reactions the beam may interact with the deuteron as a whole with a higher probability than one might expect. This is the case for neutral pion photoproduction, where coherent production is the dominant reaction mechanism at forward pion angles, and incoherent (quasifree) production becomes strong only at backward pion angles. The fact that π^0 photoproduction from deuterium with photon beam energies in the range of 300-400 MeV is far from the quasifree approximation has far-reaching consequences.

Pion photoproduction from hydrogen at beam energies of 300-400 MeV is dominated by the Δ resonance. Non-resonant Born terms also contribute to the reaction amplitudes, and are particularly important for observables where interference between amplitudes has a big effect. With the precise data available today[1 2] from the free proton, the reaction mechanism is fairly well understood although multipole analysis shows that more data is needed to fully determine the individual amplitudes[1]. As shown later, data from the neutron is necessary to separate amplitudes that come from the scalar and vector parts of the photon field. However, no neutron target exists and we are forced to use deuterium or some other nuclear target. Hence, final-state interactions and coherent production must be investigated in order to understand the corrections in going from bound neutron measurements to free neutron amplitudes.

Neutral pions offer a small advantage over charged pions in this regard. For π^0 production, the pole terms (t-channel) diagrams are suppressed and so the Born terms are much smaller when compared with charged pion production. Also, the π^0 decays quickly into two photons, which easily escape from the liquid hydrogen target whereas charged pions may have significant energy loss or be stopped in the target. Although detection of the $\pi 0$ is complicated by the necessity of reconstructing the $\pi 0$ from 2 gamma-rays, the large solid angle detectors present at many facilities allows measurements of this reaction even at very low π^0 kinetic energies. This, coupled with the dominance of the Δ resonance at medium beam energies and the symmetry between production from protons and neutrons, makes π^0 photoproduction a good place to investigate the role of final-state interactions present in measurements from deuterium. Again, our goal is to get to the free neutron amplitudes and knowledge of corrections induced by final-state interactions is crucial to achieve this end.

2   Experimental Details

The present experiment was carried out at LEGS (laser electron gamma source)[3] at Brookhaven National Laboratory, using the SASY (spin-asymmetry) detector. The heart of SASY is the crystal box (XTAL) made up of 432 NaI crystals, each 25 cm long with a square face 6.4 cm along the edge, in a rectangular geometry, see Fig. 1. The two γ's from π^0 decay were detected as clusters in XTAL and the invariant mass of the π^0 was calculated based on the opening angle between the gammas and the energy of the higher-energy cluster. The XTAL covers an angular region of about 45° to 135° in the lab. This enabled detection of π^0 decays with a reasonable efficiency for

Figure 1. The SASY (spin asymmetry) detector at the LEGS facility.

center-of-mass angles from 75° to 155°. In addition, nucleons were detected at forward angles by using an array of plastic scintillator bars, each with a square face 10 cm on an edge and 160 cm long. Preceding the scintillator bars was a thin wall of scintillator paddles which were used to determine whether the particle hitting the bars was charged or not. Following the bars was an array of lead glass cerenkov detectors, which were not included in the present analysis. Downstream from the target and centered along the beam was a gas cerenkov detector used to veto electrons or positrons produced from atomic events.

The photon beam was produced by Compton backscattering of laser light from electrons of energy 2.58 GeV in the NSLS (National Synchrotron Light Source) storage ring. The photon beam ranged in energy from 213 to 333 MeV, with high linear polarization[1] using a standard Argon laser.

The trigger consisted of a hit anywhere in the XTAL above a threshold of 10 MeV in coincidence with a hit in the tagger spectrometer which measured the energy of the Compton scattered electron. The trigger also required no signal in the gas cerenkov detector used to veto atomic e^+e^- pairs. The data was recorded using standard electronics giving a livetime of typically in the range 85-95

The target was a cylindrical 10.0 cm long cell of liquid D_2, 5 cm in diameter with rounded end caps. The target walls were made from 0.12 mm thick electroformed Nickel.

The reconstruction of the π^0 angle and energy is described elsewhere[4]. The efficiency for π^0 detection was modeled by Monte Carlo simulations using the GEANT package[5]. The nucleon detection efficiency was also determined from computer simulations, which was particularly important for the neutron detection efficiency. The GEANT efficiencies were also checked against the STANTON code developed extensively to match medium-energy neutron time-of-flight measurements at LAMPF and IUCF. There was good agreement between these calculated efficiencies.

3 Results and Discussion

The multipoles for π^0 photoproduction from a nucleon target can be put in terms of definite isospin,

$$A_{\gamma p \to \pi^0 p} = (A^{(0)} + \frac{1}{3}A^{(1)}) + \frac{2}{3}A^{(3/2)} \tag{1}$$

$$A_{\gamma n \to \pi^0 n} = -(A^{(0)} - \frac{1}{3}A^{(1)}) + \frac{2}{3}A^{(3/2)} \tag{2}$$

where the first bracketed term is the isospin $1/2$ multipole, broken into isoscalar (0) and isovector (1) parts of the photon field, and the second term is the isospin $3/2$ component. The isospin $3/2$ component is dominated by the Δ resonance at the present photon energies and the isospin $1/2$ term is dominated by Born diagrams which, as noted earlier, are suppressed for neutral pion channels. For π^0 photoproduction the photon coupling produces different isospin amplitudes for proton and neutron targets in the isospin $1/2$ amplitude.

In the extreme limit where the the $A^{(3/2)}$ amplitude completely dominates, the cross sections for π^0 photoproduction from a free proton is simply related to that from a free neutron by the ratio of magnetic moments[6]. However, as noted earlier, final state interactions for photoproduction from a bound neutron and a bound proton will interfere with both isospin $1/2$ and $3/2$ terms, and although the FSI diagrams may be invariant under isospin, the calculated observables will not be. To first order, one might ignore the smaller contribution from the isospin $1/2$ channel, but it is unlikely this approximation will be sufficiently accurate to describe the precise measurements available today.

An additional complication with a deuteron target, as noted earlier, is coherent photoproduction. It is not clear what reaction mechanism leads to such strong coherent production at forward angles. Also, meson exchange currents (MEC's), which are not isospin invariant, may play a large role in photoproduction from the deuteron. If these effects are significant, it may be difficult to extract the free neutron amplitudes without a better understanding of how coherent production and MEC's affect the calculated observables.

As a first indication of the strength of π^0 photoproduction from the

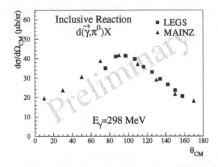

Figure 2. Preliminary inclusive pi^0 photoproduction cross sections at $E_\gamma = 298$ MeV as a function of the π^0 angle. The data from Mainz using the TAPS setup is also shown.

deuteron, the inclusive cross section is shown in Fig. 2 which includes the present data from LEGS and previously published data using the TAPS detector and the photon tagging facility at Mainz[2]. There is good agreement (5-10%) between the two data sets. We note in passing that the inclusive cross sections are within 20% of a naive model where the free proton multipoles are used to predict (via isospin symmetry) the neutron photoproduction cross section, and both proton and neutron cross sections are simply added together. The LEGS data are labeled preliminary because final corrections have not yet been applied in the data analysis. The comparison with Mainz data is nonetheless quite good.

Next we show exclusive cross sections for a π^0 detected in coincidence with a proton or a neutron in Fig. 3. The nucleon was detected in the forward-angle scintillator bars. As one can easily see, the measured exclusive cross sections are far below the naive prediction using the SAID multipoles for free nucleons. Clearly, the quasifree approximation is not good for π^0 photoproduction from the deuteron in the angular region shown, where the summed neutron plus proton cross sections are up to a factor of 3 smaller than the SAID curve. We note that the Mainz data, when separated into coherent and incoherent parts, show a similar magnitude for the incoherent component[2]. At $\theta_{CM} = 87°$, less than half of the total inclusive cross section is seen in the exclusive channel. Presumably, the difference is due to coherent photoproduction where the deuteron remains intact (in agreement with the Mainz result). At a photon energy of 298 MeV, there is still about 100 MeV

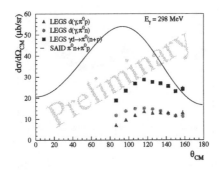

Figure 3. Preliminary exclusive pi^0 photoproduction cross sections at $E_\gamma = 298$ MeV as a function of the π^0 angle. The prediction from SAID multipoles is also shown.

left for kinetic energy after subtraction of the pion mass. Considering the momentum transfer at $\theta_{CM} \simeq 90°$, it may be surprising that over half of the cross section is due to coherent production.

In Fig. 4 are shown preliminary results for the spin asymmetries Σ for linear polarization. There are no previously published asymmetry data for the exclusive neutron channel (top plot). The photon energy in the figure (298 MeV) is just a sample of about 10 energy bins in the full data set. The exclusive proton data (middle plot) are similar to the neutron channel, and significantly different from the free proton spin asymmetry represented by the SAID curve. The spin asymmetries are often assumed to be less sensitive to FSI effects, because Σ is a ratio of cross sections. These data indicate again that the quasifree approximation is not sufficient to explain the deuteron reaction mechanism. The apparent agreement between the inclusive data (bottom plot) and the SAID multipole predictions is accidental and suggests that Σ for coherent photoproduction on deuterium is larger than that for hydrogen.

In conclusion, data has been presented for inclusive and exclusive π^0 photoproduction from a deuterium target. The inclusive data from LEGS are in reasonable agreement with similar data from Mainz. The exclusive LEGS data, when combined with the inclusive data, suggest that a significant part of the total cross section comes from coherent photoproduction. Both exclusive data and spin asymmetries indicate that significant final state interactions are present.

150

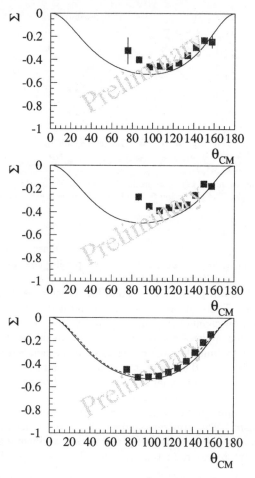

Figure 4. Preliminary spin asymmetries for pi^0 photoproduction at $E_\gamma = 298$ MeV as a function of the π^0 angle. The plot are for exclusive neutron (top), exclusive proton (middle) and inclusive (bottom) measurements. The prediction from SAID multipoles is also shown.

References

1. G. Blanpied *et al.*, Phys. Rev. C **64**, 25203 (2001).
2. U. Siodlaczek, doctoral thesis, University of Tübingen, 2000; U. Siodlaczek *et al.*, Nucl. Phys. **A663** (2000) 428c.
3. C.E. Thorn, *et al.*, Nucl. Instr. Meth. **A45**, 447 (1989).
4. H. Ströher *et al.*, Nucl. Instr. Meth. **A269** (1988) 568.
5. GEANT software package, CERN Program Library Writeup W5013.
6. X. Li, L.E. Wright and C. Bennhold, Phys. Rev. C **48**, 816 (1993).

TRANSITION PROPERTIES OF LOW-LYING RESONANCES IN A RELATIVISTIC QUARK MODEL WITH A MESON CLOUD EFFECT

Y. B. DONG[1], AMAND FAESSLER[2] AND K. SHIMIZU[3]

Institute of High Energy Physics, The Chinese Academy of Sciences,
P. O. B. 918(4-1), Beijing 100039, P. R. China[1]
Institute of Theoretical Physics, Tuebingen University, Germany[2]
Dept. of Physics, Sophia University, Tokyo, Japan[3]

Pion cloud effect on nucleon structure and on nucleon resonance transition properties is studied based on a relativistic quark model approach. We display our calculated results for the electromagnetic form factors of nucleon, for the electroproduction amplitudes of the $\Delta(1232)$ and $P_{11}(1440)$ resonances and moreover, for the strong decay width of the $\Delta(1232)$ resonance. Our results indicate the important effect of the pion cloud on those physical observables in the low energy region.

1 Introduction

For a long time constituent quark models, such as Isgur-Karl model[1] and its improved versions, have been extensively used to study nucleon resonance spectra, nucleon form factors and its resonance transition properties. Most of the constituent quark models are based on nonrelativistic framework but with relativistic corrections. For decades, the simple nonrelativistic constituent quark models are successful in explaining some simple structures of the resonances and even the nucleon-nucleon interaction. However, some detailed properties of the resonances still deserve to be investigated further, because more and more new and precise data are coming out recently. For example, the new data of the pion photoproduction from MAINZ and from LEGS, the new data of the spin structure functions from E143 of SLAC and from Jefferson Lab and the new measurement for the low-lying resonance transition properties from the Jefferson Lab. have been published. It is expected that in the near future even more precise measurement of the resonance transition properties and of the nucleon spin structure functions in the resonance region will be obtained at different photon- and electron-facilities, for instant, at Jefferson Lab., SLAC and Mainz. These experiments include the measurements of the photon- and electroproduction of the $\Delta(1232)$ resonance, the $P_{11}(1440)$ (Roper) resonance and other low-lying resonances, and of the test of Gerasimov-Drell-Hearn sum rule. Many of the existing and forthcoming data request an improved understanding for the nucleon structures and for nucleon resonance properties in this resonance region where nonperturbative QCD is dominant.

Besides the well-known Isgur-Karl model[1], the nonrelativistic chiral constituent quark model[2] is also widely employed to re-study the baryon resonance spectra and the transition properties recently. In this model, the Goldstone boson degrees of freedom are considered. Better results for the baryon spectra are obtained than the Isgur-Karl model, particularly, for the Roper resonance $P_{11}(1440)$ spectrum and for the spin-orbit splitting of the baryons.

In fact, meson cloud effect on some of the resonance properties has been investigated for a long time by the cloudy bag model[3], by the chiral quark model[4] for the nucleon spin content. Our recent calculations[5] based on a relativistic independent quark potential model approach, particularly, with explicit inclusion of the pion cloud both in the nucleon and its resonance wave functions and in the electromagnetic current, is another approach to show and to confirm the important effect of the meson cloud on the physical observables in the low energy region.

It should be mentioned that the cloudy bag model and our approach are different from the chiral constituent quark model[2] because the intermediate quark state in the latter one is in plane wave which is unlike the bound state in the former two approaches. Moreover, in our approach and the cloudy bag model, the Goldstone bosons can be emitted and absorbed by the same quark[6].

2 Relativistic potential model approach

In our relativistic independent quark potential model approach, one assumes that the single quark as a Dirac particle moving in a mean field of a confinement potential

$$V_{conf}(\vec{r}) = \frac{a_c}{2}(1 + \gamma_0)\vec{r}^{\,2} = V_s(\vec{r}) + V_v(\vec{r}), \tag{1}$$

where a_c is the strength of the confinement and $V_{s,v}$ stands for the scalar and vector confinements, respectively. The confinement does not have Klein paradox problem. There are two reasons to use the scalar-vector harmonic oscillator confinement. One is that the wave function of the Dirac equation can be solved analytically. Another is from the recent work of Page et al.[7]. In their work, they found that the observation of the spin-orbit degeneracy in heavy-light quark mesons can be explained by a relativistic symmetry of the Dirac Hamiltonian which occurs when the vector and scalar potentials exerted on the light quark by the heavy anti-quark differ approximated by a constant. They argued that this kind of confinement may occurs in QCD in the region of light quark wave function dominance. It should be pointed out that QCD yields asymptotically a linear confinement. However previous investigation in the past decades has shown that as long as the baryons have about the same radius, their properties do not sensitively depend on the radial dependence of the confinement potential, if the confinement strength is adjusted to reproduce the correct root mean charge radius.

The total Lagrangian of the system is

$$\mathcal{L}_{tot} = \mathcal{L}_q + \mathcal{L}_{GB} + \mathcal{L}_I, \tag{2}$$

where $\mathcal{L}_{GB}$ and $\mathcal{L}_I$ are the Lagrangians of the Goldstone bosons and of the interaction between Goldstone bosons and quarks, respectively. The two terms will be explicitly discussed in the next section. Here, we focus on the three-quark core part $\mathcal{L}_q$. The corresponding Lagrangian density for quarks is

$$\mathcal{L}_q = \frac{i}{2}\bar{\psi}(\vec{r})\gamma^\mu \partial_\mu \psi(\vec{r}) - \bar{\psi}(\vec{r})[V_{conf.}(\vec{r}) + m_q]\psi(\vec{r}). \tag{3}$$

Single quark wave function can be obtained by solving the following Dirac equation

$$(\vec{\alpha} \cdot \vec{p} + \gamma_0(V_{conf}(\vec{r}) + m_q) - E_q^{nl})\psi_q^{nl}(\vec{r}) = 0, \tag{4}$$

where, m_q is the quark mass, $n, l = 0, 1, 2, ...$ are radial and orbital angular momentum quantum numbers of the large component of the Dirac spinor, respectively. The relations among the energy eigenvalue E_q^{nl} of the Dirac equation, the confinement strength a_c and the harmonic oscillator constant α_q^{nl} are

$$a_c = \frac{(E_q^{nl} - m_q)^2(E_q^{nl} + m_q)}{4(2n + l + 3/2)^2}, \quad \alpha_q^{nl} = \sqrt{a_c(E_q^{nl} + m_q)} = \frac{1}{(b_q^{nl})^2}, \tag{5}$$

where b_q^{nl} is the harmonic oscillator length of quark q. The single quark ground-state $(n, l = 0)$ wave function is

$$\psi_q^{0s}(\vec{r}) = (\frac{\alpha_q^{0s}}{\pi})^{3/4}(1 + \frac{E_q^{0s} - m_q}{2(E_q^{0s} + m_q)})^{-1/2} \left(\begin{array}{c} i \\ -\frac{\alpha_q^{0s}}{E_q^{0s}+m_q}\vec{\sigma} \cdot \vec{r} \end{array} \right) e^{(-\frac{\alpha_q^{0s}\vec{r}^{\,2}}{2})}. \tag{6}$$

Moreover, the analytical expression of the wave function for the first radial excited state ($n = 1, l = 0$) is

$$\psi_q^{1s}(\vec{r}) = (\frac{\alpha_q^{1s}}{\pi})^{3/4}(1 - \frac{2m_q}{3(E_q^{1s} + m_q)})^{-1/2} \tag{7}$$

$$\times \left(\begin{array}{c} i(1 - \frac{2}{3}\alpha_q^{1s}\vec{r}\,^2) \\ -\frac{\alpha_q^{1s}}{E_q^{1s}+m_q}\vec{\sigma}\cdot\vec{r}(\frac{7}{3} - \frac{2}{3}\alpha_q^{1s}\vec{r}\,^2) \end{array} \right) e(-\frac{\alpha_q^{1s}\vec{r}\,^2}{2}),$$

It should be mentioned that the parameter α_q^{nl} depends on n and l. Thus, it is not identical for all the solutions as in the non-relativistic case. It is simple to verify the orthogonal relation between the two single quark wave functions in eqs. (6) and (7) i.e. $< 1s \mid 0s >= 0$. The general form of the eigen values of the 0s and 1s states is

$$E_q^{0s,1s} = \frac{m_q}{3} + [-\frac{x}{2} + (\frac{x^2}{4} + \frac{y^3}{27})^{1/2}]^{1/3} + [-\frac{x}{2} - (\frac{x^2}{4} + \frac{y^3}{27})^{1/2}]^{1/3}, \tag{8}$$

where

$$x = \frac{16}{27}m_q^3 - a_c \left(\begin{array}{c} 9(0s\ state) \\ 49(1s\ state) \end{array} \right); \qquad y = -\frac{4}{3}m_q^2. \tag{9}$$

We first treat the nucleon as a three-quark bound state. Then, the nucleon wave functions are in the simplest approach: the product of the three individual single quark wave functions. However, the center of mass motion should be removed. To this end, we use the Peierls-Yoccoz[8] method which was also employed by Tegen, Brockmann and Weise[9] and by many others. In this approach one projects the system on a good total momentum. There are several other approaches to consider the center of mass correction in the relativistic framework. Detailed discussion about the center of mass correction is referred to recent work by Shimizu et al.[10]. It should be mentioned that all methods can only approximately remove the center of mass motion in a relativistic many-body system.

If no any center of mass correction (NCMC) is considered, the nucleon and Δ wave functions are

$$\Psi_{N,\Delta}^{NCMC}(\vec{r}_1, \vec{r}_2, \vec{r}_3) = \psi_{u,d}(\vec{r}_1)\psi_{u,d}(\vec{r}_2)\psi_{u,d}(\vec{r}_3). \tag{10}$$

After removing the center of mass motion, we can express it as

$$\Psi_{\vec{P}}^{B}(\vec{r}_1, \vec{r}_2, \vec{r}_3) = \frac{N_B(\vec{P})}{(2\pi)^9} \int d^3p_1 d^3p_2 d^3p_3 e^{i\vec{p}_1 \cdot \vec{r}_1 + i\vec{p}_2 \cdot \vec{r}_2 + i\vec{p}_3 \cdot \vec{r}_3} \Psi_{\vec{P}}^{B}(\vec{p}_1, \vec{p}_2, \vec{p}_3) \tag{11}$$

where

$$\Psi_{B\vec{P}}^{PY}(\vec{p}_1,\vec{p}_2,\vec{p}_3) = (2\pi)^3 \delta^3(\vec{p}_1 + \vec{p}_2 + \vec{p}_3 - \vec{P})\Psi_B^{NCMC}(\vec{p}_1,\vec{p}_2,\vec{p}_3) \qquad (12)$$

based on the Peierls-Yoccoz method. In eq. (12), the δ function indicates that $\vec{P} = \vec{p}_1 + \vec{p}_2 + \vec{p}_3$, and $\vec{P}$ is the total three momentum of the three-quark core and $N_B(\vec{P})$ is a normalization constant for the baryon B. In the relativistic approach the intrinsic three-body wave function $\Psi_{B\vec{P}}^{PY}(\vec{p}_1,\vec{p}_2,\vec{p}_3)$ and the normalization constant $N_B(\vec{P})$ depend on the total momentum $\vec{P}$ and on the method to remove the center of mass motion within the relativistic quark model with a self-consistent potential. If one projects always on the total momentum zero and then boosts the wave function to the center of mass momentum $\vec{P}$ with the above method, one obtains a slightly different intrinsic wave function than when one projects immediately on the momentum $\vec{P}$. Thus one gets different values for the root mean square radius and the magnetic moment for the nucleon in the two cases. In the latter case (direct projection on $\vec{P}$) the limits of the total momentum $\vec{P}$ to zero for the form factors and other observable yield different values than, if one projects immediately to $\vec{P} = 0$. These differences reflect the fact that one is not able to remove the center of mass motion in a completely covariant way from a relativistic many body wave function involving a common potential.

3   Goldstone boson degrees of freedom

To include the Goldstone boson degrees of freedom explicitly as shown in eq. (2), we have the Lagrangian for the Goldstone bosons as

$$\mathcal{L}_{GB} = \frac{1}{2}(\partial_\mu \hat{\Phi})^2, \qquad (13)$$

and the interactive Lagrangian density between quarks and Goldstone bosons in eq. (2) as

$$\mathcal{L}_I = -ig_0 \bar{\psi}_q(\vec{r})(U_s(\vec{r}) + m_q)\gamma^5 \frac{\hat{\Phi}}{F}\psi_q(\vec{r}). \qquad (14)$$

Where $F = 88 MeV$ is the pion decay constant in the chiral limit from the chiral perturbation theory, g_0 is the strength of the interaction which is regarded as a free parameter and the pseudo-scalar Goldstone boson field $\hat{\Phi}$ is

$$\frac{\hat{\Phi}}{\sqrt{2}} = \sum_{i=1}^{8} \frac{\Phi_i \lambda_i}{\sqrt{2}} = \Phi, \qquad (15)$$

with

$$\Phi = \begin{pmatrix} \frac{\pi^0}{\sqrt{2}} + \frac{\eta}{\sqrt{6}} & \pi^+ & K^+ \\ \pi^- & -\frac{\pi^0}{\sqrt{2}} + \frac{\eta}{\sqrt{6}} & K^0 \\ K^- & \bar{K}^0 & -\frac{2\eta}{\sqrt{6}} \end{pmatrix}. \tag{16}$$

The pseudo-scalar Goldstone boson quark interaction in eq. (14) is generated from the following chiral transformation

$$\psi(\vec{r}) \to \psi(\vec{r}) - i\gamma_5 \frac{\vec{\tau} \cdot \vec{\pi}}{2f_\pi} \psi(\vec{r}). \tag{17}$$

It should be mentioned that the present pseudo-scalar interaction in eq. (14) comes from the scalar term $-\bar{\psi}(\vec{r})(V_s(\vec{r}) + m_q)\psi(\vec{r})$ in the quark Lagrangian in eq.(3)[11]. The remaining terms in the same Lagrangian are chiral invariant.

After including of the pion meson cloud, one can re-write the baryon wave function as

$$| \tilde{B} >= \sqrt{Z^B} | \Psi^B > + \sum_{B'=N,\Delta} C^B_{B'\pi} | (\Psi^{B'}\pi)_B >, \tag{18}$$

where, the second term on the right hand side of eq. (18) indicates the Goldstone boson degrees of freedom explicitly in the wave function. In the above equation Z^B is a renormalization constant for the baryon resonance B, and B' is always restricted to be only N and $\Delta(1232)$ resonance for the intermediate states like cloudy bag model. The explicit expressions for the coefficients in eq. (18) are

$$| C^B_{B'\pi} |^2 = Z^B | C'^B_{B'\pi} |^2, \tag{19}$$

$$| C'^B_{B'\pi} |^2 = \int \frac{v^{*BB'}_{0j}(\vec{q}) v^{BB'}_{0j}(\vec{q})}{(E - \omega_{\vec{q}} - M_{B'})^2} \frac{d^3q}{(2\pi)^3},$$

$$Z^B = \frac{1}{1 + \sum_{B'=N,\Delta} | C'^B_{B'\pi} |^2}, \tag{20}$$

where E is initial baryon energy, ω_q and $\vec{q}$ are the energy and the three-momentum of the Goldstone bosons, respectively. In eq. (20), $v^{BB'}_{0j}$ stands for the interactive vertex.

Then, we use above wave function in eq. (18) to calculate the transition properties of baryon resonances. It should be pointed out that since we include the pion meson cloud explicitly, we have not only the impulse photo-quark current

$$j_{q\gamma}^{\mu} = \sum_i e_i \Psi^+ \gamma_0(i) \gamma^{\mu}(i) \exp(i\vec{q} \cdot \vec{r}_i) \Psi, \tag{21}$$

but also the photo-pion current

$$j_{\gamma\pi}^{\mu} = ie[\phi_+^{\dagger}(\pi)\partial^{\mu}\phi(\pi) + -\phi_+(\pi)\partial^{\mu}\phi_+^{\dagger}(\pi)], \tag{22}$$

where $\phi_+ = \frac{1}{\sqrt{2}}[\phi_1 + i\phi_2]$. Then, the total electromagnetic current is

$$J^{\mu} = j_{q\gamma}^{\mu} + j_{\pi\gamma}^{\mu}. \tag{23}$$

In our numerical calculation, we only have three free parameters. They are g_0, m_q and E_q^{0s}. The first parameter g_0 can be determined from the empirical value of $g_{\pi NN} \sim 14.0$ with

$$\frac{g_{\pi NN}}{2M_N} F(\vec{q})\vec{\sigma}_N \cdot \vec{q}\vec{\tau}_N \leftrightarrow < \Psi_N \mid \sum_{i=1}^{3} \frac{U_s(r_i) + m_q}{f_\pi} \gamma_0(i)\gamma^5(i)\vec{r}(i)e^{i\vec{q}\cdot\vec{r}_i} \mid \Psi_N >, \tag{24}$$

where $F(\vec{q})$ is the form factor and $F(0) = 1$. Moreover, since the averaged mass of nucleon and $\Delta(1232)$ resonance is about 1080MeV, we select $E_q^{0s} = 540MeV$, then the calculated averaged nucleon and $\Delta(1232)$ mass is about 100MeV heavier than $2E_{0s} \sim 1080MeV$ after removal of the center of mass motion. It is expected that Goldstone boson exchange and one-gluon exchange potentials can suppress this value. The quark mass can vary from 0 to 300MeV or even heavier. However, our previous analyses indicate that small quark mass is favored by our model. In the following, the small quark mass $m_q \sim 0$ is selected.

4   Results and conclusions

Based on the relativistic quark potential model approach and on the consideration of explicit pion cloud, one can calculate the electromagnetic form factors of nucleon, the helicity amplitudes for the resonances, such as two low-lying resonances $\Delta(1232)$ and $P_{11}(1440)$ and the transition rate for the $\Delta(1232)$ resonance ($\Delta^{++} \to \pi^+ p$). Moreover, the spin content of the nucleon can be also estimated within our model. Our calculated results for the charge radii and magnetic moments (in unit of nucleon magneton) of the proton and neutron are

$$\sqrt{< r_{Ep}^2 >} = 0.827(fm), \quad < r_{En}^2 >= -0.0958fm^2 \tag{25}$$

$$\mu_p = 2.491, \quad \mu_n = -1.785.$$

The recent data[12] for the electromagnetic form factors of the nucleon are

$$\sqrt{< r_{Ep}^2 >} = 0.847 \pm 0.008(fm), \quad < r_{En}^2 >= -0.113 fm^2 \quad (26)$$

$$\mu_p = 2.7928, \quad \mu_n = -1.913.$$

Our above results in eq. (25) are in good agreement with the data. Moreover, we find that if the quark mass m_q increasing, the obtained results become slightly worse than the case with smaller quark mass. Therefore, the small quark mass is chosen in this work. The better results for the neutron charge mean square radii come from the pion cloud. If we consider only the three-quark core component of the nucleon, where we have only impulse quark-photon interaction and all three quarks are in S-wave, we cannot get a good description for the neutron properties, particularly, for the neutron charge distribution. After including the pion cloud explicitly, we have not only the quark-photon interaction, but also the pion-photon interaction and moreover, we have $| \pi^- p >$ and $| \pi^0 n >$ components in the neutron wave function because of the fluctuation. Consequently, the explanation for the neutron charge distribution becomes good. Our calculations confirm the important role of the pion cloud.

Our calculations need small quark mass as an input. In fact, in the nonrelativistic constituent quark model approach, the nucleon mass comes from quark mass and usually the constituent quark mass is selected to be $M_N/3 \simeq 313 MeV$. However, in our relativistic approach, the nucleon mass is contributed by the kinetic energy term $\sum_i \vec{\alpha}_i \cdot \vec{p}_i$, by the confinement term $\sum_i \frac{a_c}{2}(1 + \gamma_0(i))(\vec{r}_i - R_c)^2$ (R_c is center of mass coordinate of the system) and by the quark mass term $\sum_i \beta_i m_q$, respectively.

The helicity amplitudes of the resonances $\Delta(1232)$ and $P_{11}(1440)$ of our model are (in unit of $10^{-3} GeV^{-1/2}$)

$$Re(A_{1/2}) = -147, \quad Re(A_{3/2}) = -277; \quad for \quad \Delta(1232) \quad (27)$$

$$Re(A_{1/2}^p) = -76.6, \quad Re(A_{1/2}^n) = 34.6 \quad for \quad N_{11}(1440).$$

The corresponding experimental measurements[12] are (also in unit of $10^{-3} GeV^{-1/2}$)

$$Re(A_{1/2}) = -135 \pm 6, \quad Re(A_{3/2}) = -255 \pm 8; \quad for \quad \Delta(1232) \quad (28)$$

$$Re(A_{1/2}^p) = -75 \pm 4, \quad Re(A_{1/2}^n) = 40 \pm 10 \quad for \quad N_{11}(1440).$$

As we all know, the nonrelativistic constituent quark model approach[13] can only give about 70% for the helicity amplitudes of the experimental data of the $\Delta(1232)$ resonance and it gives a large discrepancy between the data and the theoretical prediction for the next low-lying resonance $P_{11}(1440)$ as well[14]. The Roper resonance is even believed to be a hybrid state $\mid 3qg >$[15] in order to get a good results for its helicity amplitudes. In our calculation, we find that after including the pion meson cloud explicitly, we can get good results comparing with the data because the pion cloud plays an extremely important role. Moreover, we also calculate the electroproduction amplitudes of the two resonances. We find that for the $P_{11}(1440)$ resonance, our predictions are different from the results of the constituent quark model and from the predictions of the hybrid model[15] which regards the Roper resonance as a $\mid 3qg >$. This is because the evolution of our electroproduction amplitudes of the Roper resonance with respect to the Q^2 has a crossing point. It is expected that the new data from Jefferson Lab. can tell from the different model estimates[16].

Finally, our model is also applied to calculate the decay width of the $\Delta(1232)$ resonance, such as $\Delta^{++} \rightarrow \pi^+ p$. It has been reviewed[17] that the predication of the conventional nonrelativistic chiral constituent quark model is about 30% underestimated for this observable. If we consider only the three-quark core part in the wave functions of the nucleon and $\Delta(1232)$ resonance, we got

$$g_{\Delta\pi N} = 11.36 GeV^{-1}, \quad \Gamma_{\Delta\pi N} = 62.3 MeV. \tag{29}$$

These results are underestimated comparing with the empirical value and the data

$$g_{\Delta\pi N} = (15.15 \pm 0.07) GeV^{-1}, \quad \Gamma_{\Delta\pi N} = (111.0 \pm 1.0) MeV. \tag{30}$$

Moreover, we consider the coupling of the three-quark core with pion and also the same coupling when we have pion cloud and the pion is in flight. Our result shows that the pion cloud can also improve theoretical estimates for these two observables as

$$g_{\Delta\pi N} = 15.60 GeV^{-1}, \quad \Gamma_{\Delta\pi N} = 117.4 MeV. \tag{31}$$

Thus, we reach that the three-quark core components in the nucleon and $\Delta(1232)$ resonance wave functions are too simple and not enough. The pion cloud should be included. In this way, we can get a good description both for the coupling $g_{\pi NN}$ and for the decay width of $\Delta^{++} \rightarrow \pi^+ p$, simultaneously.

To summarize our calculations, we find that the effect of the pion cloud should be considered in the low energy region. This conclusion is nature because in this region the incoming photons or leptons are easy to probe the pion cloud surrounding the bare three-quark core. The important role of the pion cloud in some physical observables, such as in the neutron charge distribution, in the helicity amplitudes of the $\Delta(1232)$ and $P_{11}(1440)$ resonances and moreover, in the decay width of $\Delta(1232)$, points out that one has to take into account the Goldstone boson degrees of freedom explicitly in the nucleon and its resonance wave functions as well as in the interactions.

References

1. N. Isgur and G. Karl, *Phys. Rev.* D **19**, 2653 (1979); *Phys. Rev.* D **19**, 1191 (1979).
2. L. Ya. Glozman and D. O. Riska, *Phys. Rept.* **268**, 251 (1999).
3. S. Theberge and A. W. Thomas, *Nucl. Phys.* A **383**, 252 (1983).
4. T. P. Cheng and L. F. Li, *Phys. Rev. Lett.* **74**, 2872 (1995).
5. Y. B. Dong, Amand Faessler and K. Shimizu, *Eur. Phys. J.* A **6**, 203 (1999); *Phys. Rev.* C **60**, 035203 (2000); *Nucl. Phys.* A **689**, 889 (2001).
6. T. W. Thomas and G. Krein, *Phys. Lett.* B**456**, 5 (1999).
7. P. R. Page, T. Goldman and J. N. Ginocchio, *Phys. Rev. Lett.* **86**, 204 (2001).
8. R. E. Peierls and J. Yoccoz, *Proc. Phys. Soc.* **30**, 381 (1957).
9. R. Tegen, R. Brockmann and W. Weise, *Z. Phys.* A **307**, 339 (1982).
10. K. Shimizu, Y. B. Dong, Amand Faessler and A. J. Buchmann, *Phys. Rev.* C **63**, 025212 (2001).
11. N. Barik and B. K Dash, *Phys. Rev.* D **33**, 1925 (1986).
12. C. Caso et al., Review of Particle Properties, *Eur. Phys. J.* C **3**, 1 (1998).
13. Z. P. Li and F. E. Close, *Phys. Rev.* D **42**, 2194 (1990).
14. S. Capstick, *Phys. Rev.* D **46**, 1965 (1992).
15. Z. P. Li, *Phys. Rev.* D **44**, 2841 (1999).
16. V. D. Burkert et al., The CLAS Collaboration, *Nucl. Phys.* A **684**, 16 (2001).
17. D. O. Riska and G. E. Brown, *Nucl. Phys.* A **679**, 577 (2001).

Studies of the Dilepton Emission from Nucleon-Nucleon Interactions [a]

J. C.S. Bacelar

Kernfysisch Versneller Instituut Zernikelaan 25, 9747AA Groningen The Netherlands

The real- and virtual-photon emission during interactions between few-nucleon systems have been investigated at KVI with a 190 MeV proton beam. Here I will concentrate the discussion on the results of the virtual-photon emission for the proton-proton system and proton-deuteron capture. Predictions of a fully-relativistic microscopic-model of the proton-proton interaction are discussed. For the proton-deuteron capture process the data is compared with predictions of a relativistic gauge-invariant impulse approximation and a Faddeev calculation. For the virtual photon processes, the nucleonic electromagnetic response functions were obtained for the first time and are compared to model predictions.

1 Introduction

The study of photon emission during nucleon-nucleon collisions provides a sensitive and unobtrusive way to study the nucleon-nucleon interaction. The photon couples directly to the electromagnetic currents associated with the dynamics of the collision. Recent accurate measurements [1,2] of this process provided exciting developments [3,4,5,6] in the theoretical attempts to describe the interacting nucleons. Most of the precise data is associated with proton-proton collisions, for obvious experimental simplicities.

At KVI we have performed [2] the most accurate measurements to date on the reaction $pp \rightarrow pp\gamma$. Here I discuss the virtual-bremsstrahlung, $pp \rightarrow pp e^+ e^-$, yields [7,8] The proton beam is polarized, with an energy of 190 MeV. A liquid H_2 target is used to reduce background. The forward angle hodoscope SALAD (Small Angle Large Acceptance Detector) was used to measure the energies and the tracks of the two protons, within scattering polar angles of 6^0 and 28^0. It consists of two wire chambers (with a total of five wire planes) and two stacked planes of scintillator detectors (see fig. 1). This detector is equiped with a specially designed trigger system which allows it to deal with the unwanted 14 MHz rate of the elastic channel at these small angles. The virtual photon was detected with TAPS, an array of 384 BaF_2 crystals which for the data discussed here were placed in the median plane around the target in blocks of 8x8 crystals. The experimental setup shown in this figure is common to all the experiments discussed in this paper.

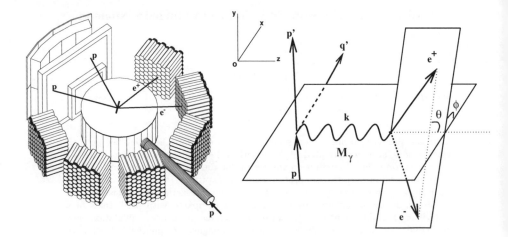

Figure 1: Left: Schematic view of the experimental setup consisting of SALAD (two wire-chambers and 50 plastic scintillators) and TAPS (384 hexagonal BaF$_2$ crystals). Not drawn are the thin plastic scintillators placed in front of each BaF$_2$ crystal. An example of a ppe^+e^- event originating from the liquid-hydrogen target is indicated. On the right panel the chosen coordinate system to describe the dynamics of these events is given.

2 Proton-proton bremsstrahlung

During this experiment we measured [7,8] for the first time the virtual-photon emission process, which in the laboratory means the measurement of the positron-electron pair. Such an event is shown in fig. 1. The added information provided by the virtual photon is the decomposition of the electromagnetic response of the nucleon in components related to the polarizations of the photon. The six nucleonic response functions are related to specific angular distributions of the two leptons. Following reference [9] the reaction cross section for the virtual bremsstrahlung process, $pp \to ppe^+e^-$, is proportional to the following amplitude:

$$|A|^2 = \frac{1}{M_\gamma^2} \left\{ W_T \left(1 - \frac{\ell^2}{2M_\gamma^2} \sin^2 \theta \right) + W_L \left(1 - \frac{\ell^2}{k_0^2} \cos^2 \theta \right) \right.$$
$$+ \frac{\ell^2 \sin^2 \theta}{2M_\gamma^2} \left(W_{TT} \cos 2\phi + W'_{TT} \sin 2\phi \right)$$
$$\left. + \frac{\ell^2 \sin 2\theta}{2k_0 M_\gamma} \left(W_{LT} \cos \phi + W'_{LT} \sin \phi \right) \right\}, \tag{1}$$

where M_γ and k_0 are the invariant mass and the energy of the virtual photon, respectively. The polar (θ) and azimuthal (ϕ) angles of the momentum-difference vector, ℓ, of the two leptons are shown in fig. 1. The cross-section for the virtual bremsstrahlung process as a function of the invariant mass of the photon, as well as its angular distribution are plotted in fig. 2.

The data are compared with predictions of the fully-relativistic microscopic-calculation[6] (solid line) as well as a low energy (LET) calculation[9,10] (dashed line). The microscopic-model, based on the Fleischer-Tjon potential[11], includes the off-shell dynamics of the interacting protons. It includes virtual Δ-isobar excitations, meson exchange currents, negative energy states and rescattering diagrams explicitly. The LET calculation uses an expansion procedure for the on-shell T-matrix in order to account for the off-shell dynamics of the interacting protons. Furthermore, it only takes into account the rescattering contributions, meson-exchange currents and the virtual Δ-isobar excitation by enforcing charge and current conservation. The microscopic calculation overpredicts the virtual photon data by approximatelly 30% over the full acceptance of the experimental setup. The LET calculation on the other hand reproduces the data rather well. We note that the cross-section depicted in fig. 2, is integrated over the leptonic angles and is therefore a measure of the transverse (W_T) and longitudinal (W_L) nucleonic response functions. The interference terms cancel out in the integration over the leptonic phase-space (see equation 1). A similar discrepancy was observed for the real photon differential cross-sections. In many regions of the phase space, the microscopic model overpredicts the data, whereas the LET calculation is in all regions of the phase space covered in the experiment, in excellent agreement with the data. We note that whereas the LET calculations use a proton-proton potential which is fitted to the worlds elastic data (phase-shifts), the microscopic model uses a Fleischer-Tjon relativistic potential which does not describe the elastic data with the same degree of precision. Work is at the moment in progress to fit this potential to the worlds proton-proton elastic data. The different amplitudes of the orthogonal cosine and sine functions of the leptonic dihedral angle ϕ (see fig. 1 and equation 1) were extracted from the data. They are directly related to each of the four interference nucleonic response functions: W_{TT}, W_{TL}, W'_{TT} and W'_{TL}. The experimentally extracted response functions are shown in fig. 2. They are integrated over the full acceptance of our experimental setup, and shown for two regions of the invariant mass of the photon. The predictions of the fully relativistic calculations are also shown as solid lines. Within the accuracy of the data the model predictions are in general good. These response functions are sensitive to the different diagrams included in the calculations, and with an accurate dataset one will be able to

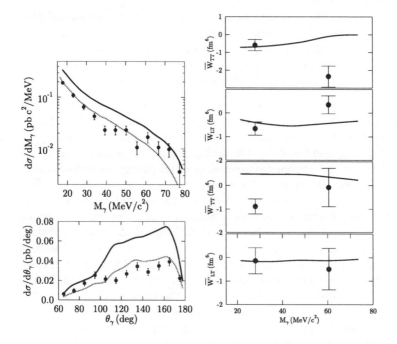

Figure 2: Top Left: differential cross section as a function of the invariant mass (M_γ) of the virtual photon integrated over the entire detector acceptance. Bottom Left: virtual-photon angular distribution in the laboratory frame for invariant masses integrated from 15 to 80 MeV/c². Right: The experimental values of the proton electromagnetic interference response functions $\overline{W}_{TT}$, $\overline{W}_{LT}$, $\overline{W'}_{TT}$ and $\overline{W'}_{LT}$. The solid lines present the results of a microscopic model and the dotted lines the results of a LET calculation.

disentangle the effects of individual diagrams.

3 Proton-Deuteron capture process

Another reaction investigated was proton-deuteron scattering. All exit channels were measured. Here I will only discuss the process $pd \rightarrow {}^3$He both for real as well as virtual photon capture [12]. The experimental setup is the same as in the proton-proton studies. The experiment was performed with an integrated luminosity of 370±40 pb^{-1}. We have identified 12000 real-photon capture events and 300 events from the $pd \rightarrow {}^3$He$+e^+e^-$ reaction. Only events with invariant mass $M_\gamma > 15$ MeV were selected [7,8].

The differential cross-sections as function of the photon angle are shown for the real and virtual photon capture in fig. 3. For the real photon capture our data are compared with previously published data from IUCF[13] measured at a proton beam energy of 150 MeV, and an experiment performed at TSL[14] at 176 MeV. Furthermore, the data are compared to calculations with a relativistic gauge-invariant model[15] (dotted curve) and a recent Faddeev calculation[16] (not shown). The latter includes initial-state interactions as well as π and ρ meson exchange currents explicitly, with the method outlined in Ref.[17]. The solid lines are discussed below.

In the impulse approximation, radiation from all external legs are included, and a contact term, constructed by applying gauge-invariance, is used to account for other contributions such as meson-exchange currents, etc.

The discrepancy noted for the real-photon capture cross section is also observed for the virtual-photon capture process in the angular range $70° < \theta_\gamma^{CM} < 140°$ (see dotted lines in the figure).

In an attempt at fitting the real-photon cross sections, Ref.[18] introduces an ad-hoc parameter (α) which enhances the magnetic contribution. Calculations of this model with $\alpha = 1.2$ are shown as solid lines in fig. 3. This value of α is chosen such that it best fits the real-photon capture cross sections at 190 MeV. The large discrepancy between theoretical predictions and experimental data for both the real- and virtual-photon capture cross sections implies enhanced transverse radiation which could be attributed to magnetic contributions. An example of such a process is the virtual excitation of the Δ. This process is not taken into account in the theoretical predictions shown as dotted lines in fig. 3, and also not in the Faddeev calculations[16].

4 Future plans at KVI

At KVI this whole program is going to enter its second phase, whereby the Plastic Ball detector[19], originally developed at GSI, is used as a photon/dilepton spectrometer. This detector has an almost complete coverage of 4π for the electromagnetic radiation, and increases the efficiency of the experiments reported here by a factor of 25. It is highly granular, with a total of 654 phoswich elements. At KVI this detector has been modified (see fig. 4) to operate with a liquid H_2 target. Furthermore, modification are underway to include an inner-shell of Cherenkov-detectors which will provide an highly selective trigger for dilepton events. The electromagnetic response functions of the proton will be studied with a precision such that the effects of different diagrams entering in the nucleon-nucleon interaction can be disentangled.

I would like to acknowledge all my KVI collaborators: M.J. van Goethem, M.N. Harakeh, M. Hoefman, H. Huisman, N. Kalantar-Nayestanaki, H. Löhner,

J.G. Messchendorp, R. Ostendorf, S. Schadmand, R. Turissi, M. Volkerts, H.W. Wilschut, A. van der Woude, and the TAPS collaboration. This work was supported in part by the "Stichting voor Fundamenteel Onderzoek der Materie" (FOM) with financial support from the "Nederlandse Organisatie voor Wetenschappelijk Onderzoek" (NWO), by GSI, the German BMBF, and by the European Union HCM network under contract HRXCT94066.

References

1. K. Michaelian et al., *Phys. Rev.* D **41**, 2689 (1990).
2. H. Huisman et al., *Phys. Rev. Lett.* **83**, 4017 (1999).
3. F. de Jong and K. Nakayama, *Phys. Lett.* B **385**, 33 (1996).
4. J.A. Eden and M.F. Gari, *Phys. Rev.* C **53**, 1102 (1996).
5. G.H. Martinus, O. Scholten and J.A. Tjon, *Phys. Rev.* C **58**, 686 (1998).
6. G.H. Martinus, O. Scholten and J.A. Tjon, *Few-Body Sys.* B **26**, 197 (1999).
7. J.G. Messchendorp et al., *Phys. Rev.* C **61**, 064007 (2000).
8. J.G. Messchendorp et al., *Phys. Rev. Lett.* **82**, 2649 (1999), and *Phys. Rev. Lett.* **83**, 2530 (1999).
9. A.Yu. Korchin and O. Scholten, *Nucl. Phys.* A **581**, 493 (1995)
10. A.Yu. Korchin, O. Scholten and D. Van Neck, NPA **602**, 423 (1996).
11. J. Fleischer and J.A. Tjon, *Nucl. Phys.* B **84**, 375 (1975); and *Phys. Rev.* D **21**, 87 (1980).
12. J.G. Messchendorp et al., *Phys. Lett.* B **481**, 171 (2000).
13. M.J. Pickar et al., *Phys. Rev.* C **35**, 37 (1987).
14. R. Johansson et al., *Nucl. Phys.* A **641**, 389 (1998).
15. A.Yu. Korchin et al., *Phys. Lett.* B **441**, 17 (1998).
16. J. Golak, private communication; and W. Glöckle et al., *Physics Reports* **274**, 107 (1996).
17. R. Schiavilla et al.,*Phys. Rev.* C **40**, 2294 (1989).
18. A.Yu. Korchin and O. Scholten, *Phys. Rev.* C **59**, 1890 (1999).
19. A. Baden et al., *Nucl. Instrum. Methods* **203**, 189 (1982).

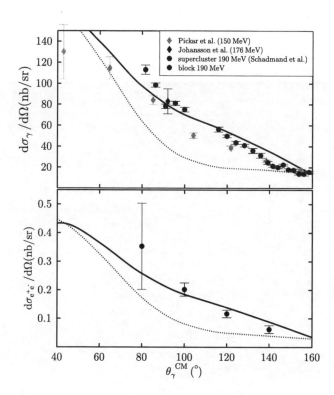

Figure 3: The real- (top) and virtual-photon (bottom) capture cross sections of the $pd \to {}^3\mathrm{He} + \gamma^{(*)}$ reaction as a function of $\theta_\gamma^{\mathrm{CM}}$. Present data are shown as full dots. Empty dots, and full and empty diamonds are data from Refs. 5, 6 and 7, respectively. The lines are the results of calculations using a relativistic gauge-invariant model (solid and dotted curves) as discussed in the text. For the virtual-photon case the data and calculations are integrated for $M_\gamma > 15$ MeV.

Figure 4: The Plastic Ball and SALAD detectors at KVI.

PION ELECTROPRODUCTION OFF ^{3}HE AND SELF ENERGIES OF THE PION AND THE Δ ISOBAR IN THE MEDIUM

A. RICHTER

Institut für Kernphysik, Technische Universität Darmstadt,
Schlossgartenstr. 9, D-64289 Darmstadt, Germany
E-mail: richter@ikp.tu-darmstadt.de

The differential coincident pion electroproduction cross section of the ^{3}He(e,e$'\pi^+$)^{3}H reaction in the excitation region of the Δ resonance has been measured with the high resolution three-spectrometer facility at the Mainz Microtron MAMI. It was the aim of the experiment to study the influence of the nuclear medium on the properties of the pion and the Δ(1232) resonance. Two experimental methods have been applied. For fixed four–momentum transfers $Q^2 = 0.045$ [0.100] (GeV/c)^2 with the pions detected in parallel kinematics, the incident energy was varied between 555 and 855 MeV in order to separate the longitudinal (L) and transverse (T) structure functions. In the second case the emitted pions with respect to the momentum transfer direction were detected over a large angular range at fixed incident energy $E_0 = 855$ MeV and the two fixed four–momentum transfers. From the angular distributions the LT interference term has been extracted. The experimental data are compared to model calculations which are based on the elementary pion production amplitude that contains besides the Born terms also the excitation of the Δ and higher resonances. Moreover, three-body Faddeev wave functions are used and the final state interaction of the outgoing pion is taken into account. The experimental cross sections are reproduced only after additional medium modifications of the pion and the Δ isobar have been considered in terms of self energies. In the framework of Chiral Perturbation Theory the pion self energy is related to a reduction of the π^+ mass of $\Delta m_{\pi^+} = \left(-1.7 \, {}^{+ \, 1.7}_{- \, 2.1}\right)$ MeV/c^2 in the neutron-rich nuclear medium at a density of $\rho = \left(0.057 \, {}^{+ \, 0.085}_{- \, 0.057}\right)$ fm^{-3}. This result is fully consistent with the one obtained within a two-loop approximation of ChPT. It is also interesting to compare the determined negative mass shift Δm_{π^+} with a positive mass shift Δm_{π^-} of 23 to 27 MeV/c^2 derived recently from deeply bound pionic states in ^{207}Pb and ^{205}Pb. Both pion mass shifts observed in complementary approaches may be understood within the Tomozawa-Weinberg scheme of isovector dominance of the πN interaction, which provides a strong guidance for the understanding of the pion modifications in the nuclear medium. The Δ self energy as extracted from previous data of π^0 photoproduction from ^{4}He is compatible with the results of the present experiment. Subsequently, the mass and the decay width of the Δ resonance in the nuclear medium are increased by $\Delta M_\Delta = (40-50)$ MeV/c^2 and $\Delta\Gamma_\Delta = (60-70)$ MeV in the considered kinematical region.

1 Introduction

A fundamental research field in mediate energy physics concerns the properties of hadrons as they are embedded in a nuclear medium. Such medium effects are commonly treated in terms of self energies from which effective masses and decay widths are deduced. Electroproduction of charged pions from ^{3}He represents an excellent testing ground to study the influence of the nuclear medium on the production and propagation of mesons and nucleon resonances such as the pion and the Δ resonance. While measurements with real photons probe only the transverse structure of the hadronic system, the electroproduction process is also sensitive to longitudinal components. Separate access to these components is achieved by an appropriate choice of kinematical settings which allows a detailed study of of the dynamics.

As part of the scientific program of the A1 collaboration at MAMI, studies of the ^{3}He(e,e′$\pi^\pm$) reactions were initiated. The motivation for this work has been many-fold. An important impetus was to learn about possible Δ isobar components in the ^{3}He ground state from the comparison of the π^+- to the π^--production in the break-up channels.[1] Such meson exchange currents are closely related to the question of a pion excess in nuclei which has been examined very recently with pion electroproduction at Jefferson Lab.[2] With the excellent missing mass resolution at the MAMI three-spectrometer set-up,

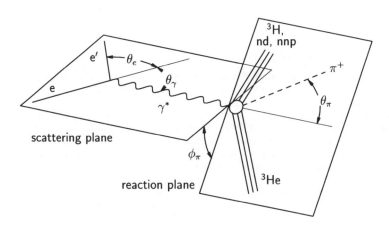

Figure 1. Kinematics of the ^{3}He(e,e′π^+) reaction. The variables in the scattering and reaction planes are indicated.

the two-body final state (^{3}Hπ^+) could be clearly separated from the three-
and four-body (dnπ^+, pnnπ^+) final states. This allowed to examine in a
kinematically complete measurement the ^{3}Hπ^+-channel, which is of particular
theoretical interest. As a simple composite nucleus, ^{3}He is amenable to precise
microscopic calculations of the ground state properties. In comparison with
heavier nuclei, the effects of final state interaction are expected to be much
smaller. Moreover, the mass-three nucleus may already be considered as a
medium. It is the aim of the present paper to report on the results of the
^{3}He(e,e$'\pi^+$)^{3}H reaction. This presentation is based on and draws much from
our previous works.[3,4,5,6] The experimental results are presented including a
discussion of the self energies of the pion and the Δ isobar from the analysis
of the longitudinal and transverse cross section components, respectively.

2 Experiment

2.1 Differential Cross Section

Pion electroproduction off the ^{3}He nucleus can be visualized as in Fig. 1.
The direction of the scattered electron and of the virtual photon define the
scattering plane; their angles with respect to the incoming electron are denoted
by θ_e and θ_γ, respectively. The pion direction is specified by its angle with
respect to the virtual photon, θ_π, and by the out-of-plane angle ϕ_π. The
three-fold differential cross section can be written as[7]

$$\frac{d^3\sigma}{d\Omega_{e'}\,dE_{e'}\,d\Omega_\pi} = \Gamma \frac{d\sigma_V}{d\Omega_\pi}(W, Q^2, \theta_\pi; \epsilon, \phi_\pi) \tag{1}$$

with

$$\frac{d\sigma_V}{d\Omega_\pi} = \frac{d\sigma_T}{d\Omega_\pi} + \epsilon \frac{d\sigma_L}{d\Omega_\pi} + \sqrt{2\epsilon(1+\epsilon)}\,\cos\phi_\pi \frac{d\sigma_{LT}}{d\Omega_\pi} + \epsilon\cos 2\phi_\pi \frac{d\sigma_{TT}}{d\Omega_\pi}. \tag{2}$$

Here the quantities ϵ and Γ denote the polarization and flux of the virtual
photon. The first two terms in Eq. (2) are the transverse (T) and longitudinal (L) cross sections. The third and fourth term, respectively, describe
the longitudinal-transverse (LT) and transverse-transverse (TT) interference
structure (or response) functions. They vanish in parallel kinematics ($\theta_\pi = 0°$)
since they are proportional to $\sin\theta_\pi$ and $\sin^2\theta_\pi$, respectively.

The four structure functions in Eq. (2) depend on the total energy W of
the photon–^{3}He system, on the four-momentum transfer Q^2 and on the polar
angle θ_π of the emitted pion. All four structure functions are independent of
the azimuthal angle ϕ_π which solely occurs explicitly in the coefficients of the
interference terms.

From Eq. (2) it is clear that a separation of the tranverse and longitudinal response can be done in parallel kinematics ($\theta_\pi = 0°$) for fixed W and Q^2 by varying the virtual photon polarization ϵ, whereas an extraction of the LT and TT interference terms is possible by their respective azimuthal dependencies. Especially the LT interference is accessible by an in-plane measurement of the pion, i.e. $\phi_\pi = 0°$ or $180°$, respectively. For the extraction of the TT interference an out-of-plane experiment is necessary.

2.2 Measurements

The measurements were carried out at the high-resolution three-spectrometer facility of the A1 collaboration at the Mainz Microtron MAMI[8] at two four-momentum transfers $Q^2 = 0.045$ and 0.100 $(\mathrm{GeV/c})^2$, referred to as kinematics 1 and 2, respectively. The energy transfer in the laboratory frame has been chosen at $\omega = 400$ and 394 MeV, respectively, i.e. in the Δ resonance region. At each Q^2, three measurements in parallel kinematics with various values of ϵ were made to determine the L and T cross sections (Rosenbluth separation). In addition, the in-plane pion angular distribution (i.e. $\phi_\pi = 0°$ or $180°$, respectively) for the second kinematics at $\epsilon = 0.74$ has been measured to determine the LT interference term. The experiment was performed with a cryogenic ^{3}He gas target which was cooled with liquid hydrogen at 20 K. Due to a novel target cell made from a solid block of aluminum with a wall

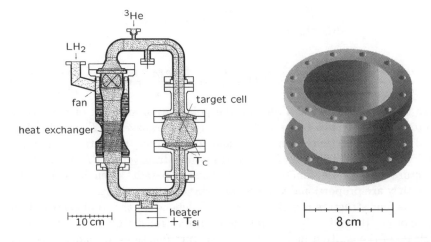

Figure 2. L.h.s.: Schematic view of the target-gas loop. R.h.s.: The new target cell made from aluminum.

thickness of 250 μm, it was possible to operate the target at a pressure of 20 bar, resulting in a thickness of $\approx$ 250 mg/cm^2.

In Fig. 3 the double differential cross section for the ^{3}He(e,e'π^+) reaction is shown as a function of the missing mass (with the ^{3}H mass of the ground state subtracted). The narrow peak at origin refers to the ^{3}Hπ^+ coherent channel. It is clearly separated from the continuum of the three- and four-body break-up channels, indicating the excellent missing mass resolution of $\approx$ 700 keV (FWHM) of the set-up.

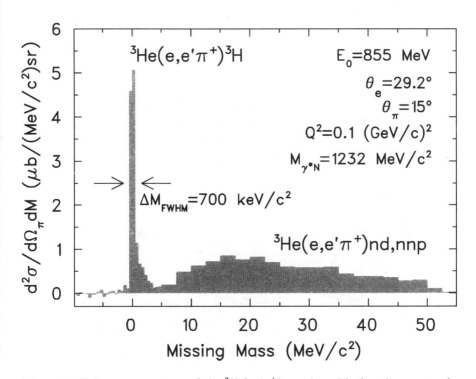

Figure 3. Missing mass spectrum of the ^{3}He(e,e'π^+) reaction with the triton mass subtracted from the abscissa. The narrow peak at $M_{\text{miss}} = 0$ refers to the two-body (^{3}Hπ^+) final state and is well separated from the three- and four-body break-up continuum states.

3 Results

3.1 Rosenbluth Separation

In the present contribution we discuss the coherent pion production to the two-hadron final state, viz. $^3H\pi^+$. Figure 4 shows the measured cross sec-

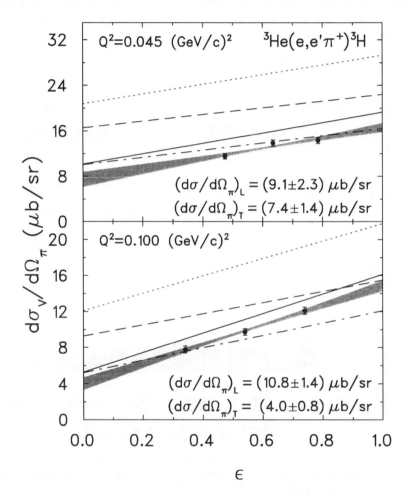

Figure 4. Rosenbluth separations of the differential cross sections. The data are given by the solid dots. The shaded area is identified with a one standard deviation error band of a straight line fit to the data. The extracted values of the longitudinal and transverse components with their statistical errors are shown in the plot. The PWIA and DWIA calculations are shown dotted and dashed, respectively. The dash-dotted line results after accounting for the Δ self energy. The solid line represents the full calculation including the pion and the Δ self energies in the medium (from [6]).

tions plotted as a function of ϵ. The transverse cross section is given by the value on the ordinate at $\epsilon = 0$ and the slope of the straight line is identified with the longitudinal cross section. Also shown are the fit results for the L and T components with statistical errors. The systematic errors (mainly due to normalization) amount to 10 % (8 %) for kinematics 1 (2), respectively. The experimental data are compared to results of theoretical calculations which are based on the most recent elementary pion production amplitude in the framework of the so-called Unitary Isobar Model.[7,9] The amplitude includes the Born terms as well as Δ- and higher resonance terms. In Plane-Wave Impulse Approximation (PWIA), three-body Faddeev wave functions are employed for the mass-three nuclei. The final state interaction due to pion rescattering is included in the Distorted-Wave (DWIA) calculations.[10] As is seen in Fig. 4, the DWIA calculations underestimate the longitudinal component and overestimate the transverse component, each by about a factor of two. Since the longitudinal component is dominated by the pion-pole term and a large part of the transverse part arises from the Δ resonance excitation, a modification of both the pion and the Δ propagators is suggested by the data. In parallel kinematics, the pion-pole and the Δ contribution essentially decouple in the longitudinal and transverse channel and can therefore be studied separately.

3.2 Modification of the Pion in the Medium

The DWIA underestimate of the longitudinal response (cf. Fig. 4) is remedied by replacing the free pion propagator in the t-channel pion-pole term of the elementary amplitude, $[\omega_\pi^2 - \vec{q}_\pi^2 - m_\pi^2]^{-1}$, by a modified one, $[\omega_\pi^2 - \vec{q}_\pi^2 - m_\pi^2 - \Sigma_\pi(\omega_\pi, \vec{q}_\pi)]^{-1}$, where $\Sigma_\pi(\omega_\pi, \vec{q}_\pi)$ denotes the pion self energy in the nuclear medium.[11] For the two values of Q^2, the energy ω_π and the momentum $\vec{q}_\pi$ of the virtual pion are fixed as $\omega_\pi = 1.7$ (4.1) MeV and $|\vec{q}_\pi| = 80.9$ (141.2) MeV/c, such that two experimental numbers for Σ_π can be determined from a fit to the respective longitudinal cross sections. The best-fit values result in $\Sigma_\pi = -(0.22 \pm 0.11)\, m_\pi^2$ for kinematics 1 and $\Sigma_\pi = -(0.44 \pm 0.10)\, m_\pi^2$ for kinematics 2 and are shown in Fig. 5. For $\omega_\pi \approx 0$ the pion self energy can be written as

$$\Sigma_\pi(0, \vec{q}_\pi) = -\frac{\sigma_N}{f_\pi^2}(\rho_p + \rho_n) - \vec{q}_\pi^2 \, \chi(0, \vec{q}_\pi), \qquad (3)$$

where ρ_p and ρ_n denote the proton and neutron densities, $\sigma_N = 45$ MeV the πN sigma term,[12] $f_\pi = 92.4$ MeV the pion decay constant, and $\chi(0, \vec{q}_\pi)$ the p-wave pionic susceptibility. Utilizing local density approximation, the susceptibility can be shown to be constant below the Fermi surface. With the two values for Σ_π given above, we immediately obtain $\chi = 0.31 \pm 0.22$. How-

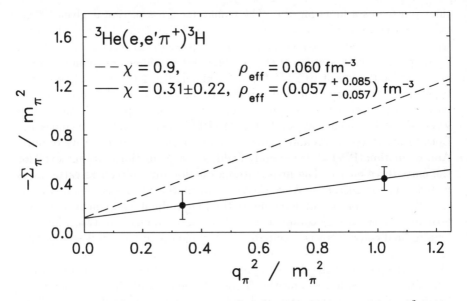

Figure 5. The pion self energy as a function of the virtual pion momentum $\vec{q}_\pi^2$ in the vicinity $\omega_\pi \approx 0$. The two data points with errors are resulting from fits to the longitudinal cross sections in the two kinematics. The dashed line corresponds to $\chi = 0.9$ from the Lindhard function with $\rho = 0.06$ fm^{-3}, while the solid line represents the best fit of χ and ρ_{eff} to the data according Eq. (3) with the values given in the plot (from [6]).

ever, a standard nuclear matter calculation of the Lindhard function results already at the small density $\rho = 0.06$ fm^{-3} in a much higher value for the susceptibility of $\chi = 0.9$, which is understood to reflect the conceptual limitation of the local density approximation. For further steps, the above value from experiment is used, which allows an extrapolation of the self energy to $\vec{q}_\pi = 0$ and to determine the mean density experienced by the virtual pion, with the result $\rho = \rho_p + \rho_n = \left(0.057 \, {}^{+\,0.085}_{-\,0.057}\right)$ fm$^{-3} \approx \frac{1}{3}\rho_0$ ($\rho_0 = 0.17$ fm^{-3} being the saturation density), albeit with a large error. The self energy corresponding to the best fit is displayed in Fig. 5.

The effective mass of the π^+ can be obtained from an extrapolation of the pion self energy to the mass shell. In the framework of Chiral Perturbation Theory up to second order in ω_π and m_π, the self energy of a charged pion in homogeneous, spin-saturated, but isospin-asymmetric nuclear matter in the vicinity of $\omega_\pi \approx m_\pi$ and for $\vec{q}_\pi = 0$ is given by the expansion

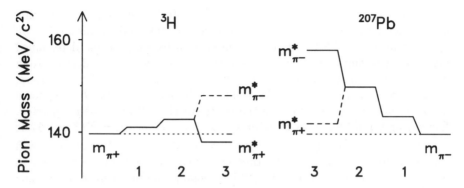

Figure 6. L.h.s.: The charged-pion mass shifts in the triton at the effective density of the present experiment. Starting from the bare mass m_π the first and the second isoscalar term in Eq. (4) are repulsive while the isovector third term is repulsive for π^- and attractive for π^+ and causes the mass splitting. R.h.s.: Applying Eq. (4) to the case of a deeply bound π^- in ^{207}Pb, all three terms are repulsive with a resulting repulsion $U_{\pi^-}(0) = \Sigma_\pi^{(-)}(m_\pi, 0)/(2m_\pi) \approx 18$ MeV.

$$\Sigma_\pi^{(\pm)}(\omega_\pi, 0) = \left(-\frac{2\,(c_2 + c_3)\omega_\pi^2}{f_\pi^2} - \frac{\sigma_N}{f_\pi^2} \right) \rho \tag{4}$$

$$+ \frac{3}{4\pi^2} \left(\frac{3\pi^2}{2} \right)^{1/3} \frac{\omega_\pi^2}{4f_\pi^4} \rho^{4/3} \pm \frac{\omega_\pi}{2f_\pi^2} (\rho_p - \rho_n) + \dots,$$

where the $+/-$ signs refer to the respective charge state of the pion. The low-energy constants (LEC's) c_2 and c_3 are fixed as $(c_2 + c_3) \times m_\pi^2 = -26$ MeV.[12] The pion self energy in Eq. (4) consists of two isoscalar parts proportional to ρ and $\rho^{4/3}$, respectively, and an isovector part proportional to $(\rho_p - \rho_n)$. The latter is known as the "Tomozawa-Weinberg term".[13] Based on PCAC arguments, it reflects the isovector dominance of the πN interaction at $\omega_\pi = m_\pi$, where the isoscalar πN scattering length as given by the first coefficient in Eq. (4) vanishes at leading order. The second isoscalar term proportional to $\rho^{4/3}$ is caused by s-wave pion scattering from correlated nucleon pairs.[14] The sign of the Tomozawa-Weinberg term depends on the isospin asymmetry of the nuclear medium. In the present case of a virtual π^+ propagating in a triton-like medium with $\rho_p - \rho_n = -\frac{1}{3}\rho$, the isovector term becomes attractive. The effective π^+ mass $m_{\pi^+}^*$ is deduced from the pole of the pion propagator at $\vec{q}_\pi = 0$ which is determined by the solution of $\omega_\pi^2 - m_\pi^2 - \Sigma_\pi(\omega_\pi, 0) = 0$ with the self energy as given by Eq. (4). Using $\rho = \left(0.057 \, ^{+\,0.085}_{-\,0.057}\right)$ fm^{-3}, one obtains a mass shift $\Delta m_{\pi^+} = m^*_{\pi^+} - m_\pi = \left(-1.7 ^{+\,1.7}_{-\,2.1}\right)$ MeV/c^2 for a π^+ propagating in ^{3}H. This result is fully consistent with the one obtained within a two-loop approximation of ChPT.[15] It is also interesting to compare the determined negative mass shift Δm_{π^+} with a positive mass shift Δm_{π^-} derived from deeply bound pionic states [16,17] in ^{207}Pb and ^{205}Pb with $N/Z \simeq 1.5$. Here, a strong repulsion of 23 to 27 MeV due to the local potential $U_{\pi^-}(r)$ for a deeply bound π^- in the center of the neutron-rich ^{207}Pb nucleus is reported. Evaluating Eq. (4) for this case with $\rho_p + \rho_n = \rho_0$ and $\rho_n/\rho_p = N/Z \simeq 1.5$ one calculates $U_{\pi^-}(0) = \Sigma_\pi^{(-)}(m_\pi, 0)/(2m_\pi) \approx 18$ MeV in good agreement with the two-loop calculation,[15] indicating a significant increase of the π^- mass in neutron-rich nuclear matter. Yet, there remains the problem of a "missing repulsion" in the interpretation of the pionic atom data.

Figure 6 shows the contributions to the pion mass shift in ^{3}H and ^{207}Pb: The two isoscalar contributions to Σ_π are both repulsive and increase the pion mass. For a neutron-rich nucleus the isovector πN interactions are attractive (repulsive) for $\pi^+ (\pi^-)$ giving rise to a splitting of the mass shifts (contribution 3 in Fig. 6). In ^{3}H, the isoscalar and isovector terms are compensating each other to a large extent, resulting in the very small decrease of the π^+ mass. In the case of ^{207}Pb, all three terms in Eq. (4) are repulsive and responsible for the strong increase of the π^- mass.

3.3 Modification of the Δ in the Medium

The DWIA overestimate in the transverse channel (cf. Fig. 4) suggests a medium modification of the Δ isobar. The in-medium Δ propagator is written[7] as $[\sqrt{s} - M_\Delta + i\Gamma_\Delta/2 - \Sigma_\Delta]^{-1}$, where one introduces a complex self-energy term Σ_Δ in the free Δ propagator. The Δ self energy is parametrized as $\Sigma_\Delta = V_1(E_\gamma)F(q_A^2)$. Here the quantities E_γ and q_A denote the photon energy and the nuclear recoil momentum, respectively. For the momentum dependence of Σ_Δ, an s-shell harmonic oscillator form factor $F(q_A^2)$ is applied. The potential parameter $V_1(E_\gamma)$ has been deduced from an energy-dependent fit to a large set of π^0 photoproduction data[18] from ^{4}He. The original fitting procedure[7] has been redone with the unitary phase excluded from the propagator in accordance with prescriptions often used in the Δ-hole model.[19] The resulting Δ self energy exhibits an energy dependence which is shown in Fig. 7. Evaluated for the kinematics 1 and 2, the real and imaginary parts are $Re\,\Sigma_\Delta \approx 50$ and 39 MeV and $Im\,\Sigma_\Delta \approx -36$ and -29 MeV. As a result, the agreement with the transverse cross section is significantly improved, although the experimental values are still overestimated by about 30%. The remaining discrepancy may be due to additional theoretical uncertainties. For example,

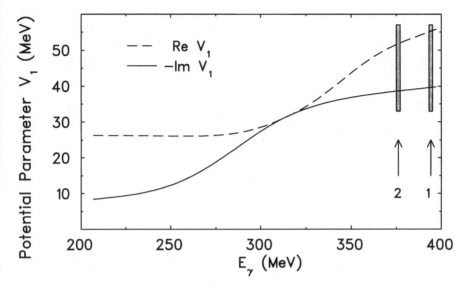

Figure 7. The potential parameter V_1 as a function of the photon energy, resulting from a fit to previously measured π^0 photoproduction cross sections.[18] The photon equivalent energies of the two kinematics of the present experiment are marked.

the Fermi motion of the nucleons is effectively accounted for by a factorization ansatz.[7] An exact treatment might reduce the prediction of the transverse cross section by about 10%. A second uncertainty of the order of 10% concerns the knowledge of the elementary π^+ production amplitude at $\theta_\pi = 0°$. This kinematical region is not probed in photoproduction but may be accessible in the future with appropriate electroproduction data from the proton. Attributing the entire Δ self energy to a mass shift ΔM_Δ and a width change $\Delta\Gamma_\Delta$, we deduce an increase by 40 to 50 MeV and 60 to 70 MeV, respectively. These values seemingly differ from our earlier results,[3] where we have employed a parameterization[20] which did not include the Δ-hole interaction, giving $Re\,\Sigma_\Delta \approx -14$ MeV for a mean ^{3}He density of $\rho = 0.09$ fm^{-3}. On the other hand, the self-energy term of the present work is an effective parameter which incorporates the influence of the Δ-spreading potential, Pauli- and binding effects as well as the Δ-hole interaction including the Lorentz-Lorenz correction. This finally leads to the positive sign.

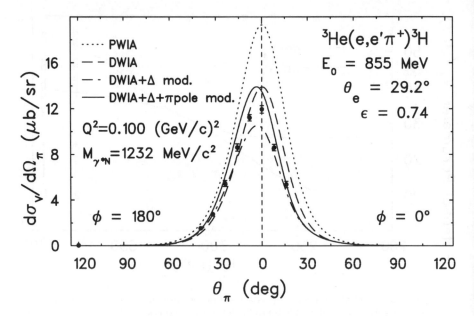

Figure 8. Pion angular distribution in the photon-^{3}He cm system for kinematics 2. The solid dots are the measured data. Additionally the results of the model calculations are shown. The labeling of the curves is the same as in Fig. 4 (from [6]).

3.4 Pion Angular Distribution and LT Interference Term

The pion angular distribution measured in kinematics 2 is illustrated in Fig. 8 along with the model calculations. The differential cross section is shown as a function of the polar angle θ_π, with the azimuthal angles centered around $\phi_\pi = 0°$ and $180°$, respectively. Most of the observed strength is restricted to a forward angle region up to $60°$ where also the medium effects on the angular distribution are concentrated. As is seen in Fig. 8, the shape of the data agrees well with the calculations involving medium effects, although the absolute height is not reproduced (at $\theta_\pi = 0°$ this discrepancy is already seen in Fig. 4 at the highest value of ϵ). More information is provided by the azimuthal dependence ϕ_π of the cross section. In Fig. 9 the differential cross section is shown as a function of the azimuthal angle ϕ_π for three bins of the polar angle θ_π.

The coefficient B of the fit function in Fig. 9 is proportional to the LT interference term according to Eq. (2). Also the TT interference term may be extracted if the out-of-plane acceptance is large enough. This is the case for

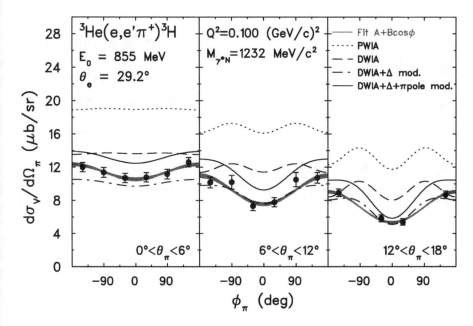

Figure 9. The differential cross section as a function of the azimuthal angle ϕ_π for three bins of the polar angle θ_π. The shaded area corresponds to a one sigma error band of a fit of the function $A + B\cos\phi_\pi$ to the data. Additionally the results of the model calculations are shown. The labeling of the curves is the same as in Fig. 4.

the two most forward bins, however no significance was found. The extracted LT interference term is finally plotted in Fig. 10 as a function of the pion emission angle θ_π. Comparing it with the model calculations, it is obvious again that only the full calculation, incorporating the medium modifications in the pion and Δ propagators, is able to reproduce the data.

4 Summary

The present experimental investigation of charged pion production in the $^3\mathrm{He}(e,e'\pi^+)^3\mathrm{H}$ reaction for invariant photon-nucleon masses slightly above and at the Δ resonance has for the first time unequivocally demonstrated that the longitudinal, transverse and longitudinal-transverse structure functions can only be described by model calculations in which medium effects due to the finite density of $^3\mathrm{He}$ are included. The measurements show evidence for self-energy corrections in both the pion and Δ-isobar propagators

182

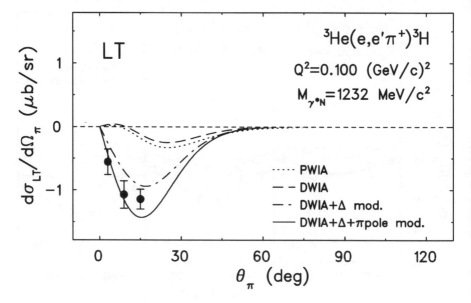

Figure 10. Extracted *LT* interference term for kinematics 2. Additionally the results of the model calculations are shown. The labeling of the curves is the same as in Fig. 4 (from [6]).

and complement the large body of previous results from pion-nucleus data. The pion modification in the nuclear medium is primarily understood within the Tomozawa-Weinberg scheme of isovector dominance of the πN interaction. Despite uncertainties in the transverse channel, the medium modification of the Δ isobar is evident.

Acknowledgement

This work has been performed within the A1 collaboration at MAMI. It is a great pleasure to thank in particular M. Kohl, C. Rangacharyulu, G. Schrieder, M. Urban, J. Wambach and A. Wirzba for their numerous contributions to the scientific material presented in this text and for enlightening discussions. I am particularly grateful to M. Kohl for his tremendous effort in helping me to produce this manuscript. Within his thesis he has also carried the biggest load of the experiment. Finally, I thank M. Fujiwara for having organized a most interesting symposium EMI2001.

This work has been supported by the Deutsche Forschungsgemeinschaft under contract No. Ri 242/15-2 and through a travel grant 446JAP113/267/0-1.

References

1. K.I. Blomqvist *et al.*, *Phys. Rev. Lett.* **77**, 2396 (1996).
2. D. Gaskell *et al.*, *Phys. Rev. Lett.* **87**, 202301 (2001).
3. K.I. Blomqvist *et al.*, *Nucl. Phys.* **A626**, 871 (1997).
4. A. Richter in *Proc. Intern. Workshop XXVIII on Gross Properties of Nuclei and Nuclear Excitations*, M. Buballa *et al.*, eds., (GSI Darmstadt, ISSN 0720−8715), p. 40.
5. M. Kohl in *Proc. Intern. Conference XVI*th *IUPAP on Few-Body Problems in Physics*, *Nucl. Phys.* **A684**, 454c (2001).
6. M. Kohl *et al.*, nucl-ex/0104004, submitted to Phys. Lett. B.
7. D. Drechsel *et al.*, *Nucl. Phys.* **A660**, 423 (1999).
8. R. Neuhausen, *Nucl. Phys. (Proc. Suppl.)* **B44**, 695 (1995); K. I. Blomqvist *et al.*, *Nucl. Instr. and Meth.* **A403**, 263 (1998).
9. D. Drechsel *et al.*, *Nucl. Phys.* **A645**, 145 (1999).
10. S. Kamalov, L. Tiator, and C. Bennhold, *Phys. Rev.* **C47**, 941 (1993).
11. T. Ericson, W. Weise, *Pions and Nuclei* (Clarendon Press, Oxford, 1988).
12. V. Thorsson, A. Wirzba, *Nucl. Phys.* **A589**, 633 (1995); M. Kirchbach, A. Wirzba, *Nucl. Phys.* **A604**, 695 (1996).
13. Y. Tomozawa, *Nuovo Cim.* **46A**, 707 (1966); S. Weinberg, *Phys. Rev. Lett.* **17**, 616 (1966).
14. M. Ericson, T. Ericson, *Ann. Phys. (NY)* **36**, 323 (1966).
15. W. Weise in *Proc. Intern. Workshop on Nuclei and Nucleons*, *Nucl. Phys.* **A690**, 98 (2001); N. Kaiser and W. Weise, *Phys. Lett.* **B512**, 283 (2001).
16. T. Yamazaki *et al.*, *Z. Phys.* **A355**, 219 (1996); *Phys. Lett.* **B418**, 246 (1998).
17. K. Itahashi *et al.*, *Phys. Rev.* **C62**, 25202 (2000).
18. F. Rambo *et al.*, *Nucl. Phys.* **A660**, 69 (1999).
19. J.H. Koch *et al.*, *Ann. Phys. (NY)* **154**, 99 (1984).
20. R.C. Carrasco, E. Oset, *Nucl. Phys.* **A536**, 445 (1992); C. García-Recio *et al.*, *Nucl. Phys.* **A526**, 685 (1991).

THE DEUTERON ELECTROMAGNETIC FORM FACTORS

GERASSIMOS G. PETRATOS

Department of Physics, Kent State University, Kent, OH 44242, USA
E-mail: gpetrato@kent.edu

A review of the deuteron electromagnetic form factors, as measured in unpolarized and polarized elastic electron-deuteron scattering experiments is presented along with a brief description of theoretical models. Particular emphasis is given to measurements at large momentum transfers. The experimental data are compared to theoretical calculations based on Propagator and Hamiltonian Dynamics and predictions of perturbative Quantum Chromodynamics.

Electron scattering from the deuteron has long been a crucial tool in understanding the internal structure and dynamics of the nuclear two-body system[1]. In particular, the deuteron electromagnetic form factors, measured in elastic scattering, offer unique opportunities to test models of the short-range nucleon-nucleon interaction and meson-exchange currents[2] as well as the possible influence of explicit quark degrees of freedom[3].

The cross section for elastic electron-deuteron scattering is described by the Rosenbluth formula:

$$\frac{d\sigma}{d\Omega} = \sigma_M \left[A(Q^2) + B(Q^2) \tan^2\left(\frac{\theta}{2}\right) \right], \qquad (1)$$

where $\sigma_M = \alpha^2 E' \cos^2(\theta/2)/[4E^3 \sin^4(\theta/2)]$ is the Mott cross section. Here E and E' are the incident and scattered electron energies, θ is the electron scattering angle, $Q^2 = 4EE' \sin^2(\theta/2)$ is the four-momentum transfer squared and α is the fine-structure constant. The elastic electric and magnetic structure functions $A(Q^2)$ and $B(Q^2)$ are given in terms of the charge, quadrupole and magnetic form factors of the deuteron $F_C(Q^2)$, $F_Q(Q^2)$ and $F_M(Q^2)$:

$$A(Q^2) = F_C^2(Q^2) + \frac{8}{9}\tau^2 F_Q^2(Q^2) + \frac{2}{3}\tau F_M^2(Q^2), \qquad (2)$$

$$B(Q^2) = \frac{4}{3}\tau(1+\tau) F_M^2(Q^2), \qquad (3)$$

where $\tau = Q^2/4M_d^2$, with M_d being the deuteron mass.

Separation of the two elastic structure functions is accomplished by cross section measurements at different scattering angles. Forward angle scattering[4] yields $A(Q^2)$, while backward angle scattering[5] allows for the determination of $B(Q^2)$. Separation of all three elastic form factors is achieved by measuring a polarization observable in a single- or double-scattering experiment. Single-scattering experiments have used unpolarized electron beams and tensor-polarized deuteron targets[6]. Double-scattering experiments have used unpolarized electron beams and recoil deuteron tensor polarimeters[7]. The polarized deuteron scattering cross section is given, in general, by[8]:

$$\sigma(\theta, \phi) = \sigma_0(\theta)\left(1 + t_{20}T_{20} + 2t_{21}T_{21}\cos\phi + 2t_{22}T_{22}\cos 2\phi\right), \qquad (4)$$

where $\sigma(\theta)$ is the unpolarized cross section. In a single (double) scattering experiment, t_{kl} are the known deuteron polarization parameters (measured deuteron tensor moments), T_{kl} are the measured analyzing powers of the reaction (known analyzing powers of the polarimeter) and ϕ is the angle between the normal to the reaction plane and the magnetic field of reference (angle between the two scattering planes). Due to time reversal symmetry, $t_{kl} = (-1)^{k+l}T_{kl}$. Asymmetry measurements with electron beams and a tensor-polarized target or a deuteron tensor polarimeter result in the extraction of the deuteron tensor polarization observables t_{20}, t_{21}, t_{22}, which are given in terms of the deuteron form factors as:

$$t_{20} = -\frac{1}{\sqrt{2}S}\left[\frac{8}{3}\tau F_C F_Q + \frac{8}{9}\tau^2 F_Q^2 + \frac{1}{3}\tau f(\theta)F_M^2\right], \qquad (5)$$

$$t_{21} = \frac{2}{\sqrt{6}S}\tau\sqrt{\tau(1 + f(\theta))}F_M F_Q \sec\frac{\theta}{2}, \qquad (6)$$

$$t_{22} = -\frac{1}{2\sqrt{3}S}\tau F_M^2, \qquad (7)$$

where $S = A + B\tan^2(\theta/2)$ and $f(\theta) = 1 + 2(1+\tau)\tan^2(\theta/2)$. Since t_{22} provides the same information as an unpolarized backward angle measurement and because of the small magnitude of F_M in the expression of t_{21}, the observable of choice to extract F_C and F_Q is the t_{20} tensor polarization. It is customary to neglect the small contribution of F_M in t_{20} and use the alternate quantity $\tilde{t}_{20}$, defined as:

$$\tilde{t}_{20} = \sqrt{2}\frac{y(2+y)}{1+2y^2}, \qquad y = \frac{2\tau F_Q}{3F_C}. \qquad (8)$$

The advantage of this quantity is that it is independent of the nucleon electromagnetic form factors, as they cancel in the ratio F_Q/F_C (see below).

In the non-relativistic impulse approximation (IA), the electron interacts with one of the two moving nucleons in the deuteron and the two-body bound state is solved using the Schrödinger equation with a realistic nucleon-nucleon potential. The deuteron form factors are then described in terms of the deuteron wave function and the electromagnetic form factors of the nucleons[2]:

$$F_C = (G_E^p + G_E^n)C_E, \tag{9}$$

$$F_Q = (G_E^p + G_E^n)C_Q, \tag{10}$$

$$F_M = \frac{M_d}{M}((G_M^p + G_M^n)C_S + \frac{1}{2}(G_E^p + G_E^n)C_L), \tag{11}$$

where $G_E^{p(n)}$ and $G_M^{p(n)}$ are the electric and magnetic form factors of the proton (neutron) and M is the nucleon mass. The factors C_E, C_Q, C_S and C_L give the distribution of the proton and neutron point currents inside the deuteron as determined by the deuteron wave function. They are integrals of quadratic combinations of the S- and D-state wave functions $u(r)$ and $w(r)$ of the deuteron, with r being the internucleon separation, expressed as:

$$C_E = \int_0^\infty \left[u^2(r) + w^2(r)\right] j_0(k)dr, \tag{12}$$

$$C_Q = \frac{3}{\sqrt{2}\tau} \int_0^\infty w(r) \left[u(r) - \frac{w(r)}{2\sqrt{2}}\right] j_2(k)dr, \tag{13}$$

$$C_S = \int_0^\infty \left[u^2(r) - \frac{w^2(r)}{2}\right] j_0(k) + \frac{w(r)}{2} \left[\sqrt{2}u(r) + w(r)\right] j_2(k)dr, \tag{14}$$

$$C_L = \frac{3}{2} \int_0^\infty w^2(r)[j_0(k) + j_2(k)]dr, \tag{15}$$

where $j_0(k)$ and $j_2(k)$ are spherical Bessel functions, with $k = Qr/2$.

Theoretical calculations based on the IA approach using various nucleon-nucleon potentials and parametrizations of the nucleon form factors underestimate the $A(Q^2)$ data and fail to reproduce the positions of the first diffraction minimum and the height of the secondary maximum of the charge and magnetic form factor data[1]. It has long been known that the form factors of the deuteron are very sensitive to the presence of meson-exchange currents[2] (MEC) in the deuteron. The inclusion of MEC to the impulse approximation brings the theory into better agreement with the data but still fails to describe at the same time all available form factor data for moderate and large momentum transfer measurements[1]. It should be noted that some calculations of the deuteron form factors show also sensitivity to the possible presence of six-quark[9] and isobar configurations in the deuteron[10], but the magnitude of these configurations is essentially unknown.

The failure of the non-relativistic calculations with increasing momentum transfers dictates the need for relativistic calculations. There are two relativistic approaches: Hamiltonian Dynamics[11] and Propagator Dynamics[12]. In the first approach, the basic dynamics contains a finite number of particles and has a corresponding Hilbert space when quantized. The quantization is performed along constant time surfaces (instant form dynamics), along spacelike surfaces with constant interval (point form dynamics) or along the light cone (light front dynamics). The advantage of the Hamiltonian Dynamics framework is that it can lead to equations of motion of the same form as the two-body Schrödinger equation, where it is possible to use non-relativistic nucleon-nucleon potentials without modification. The disadvantage of this formalism is, in general, the loss of locality and manifest covariance. Also no consensus has been reached concerning consistent techniques for the construction of electromagnetic currents in this framework[1].

The second approach is based on a field theory description of two interacting nucleons using three-dimensional reductions of the Bethe-Salpeter (BS) equation[13], which is a four-dimensional integral equation with a complicated analytical structure. The three-dimensional reduction to so called quasipotential equations is accomplished by replacing the free propagator in the BS equation with a new one chosen to include a constraint in the form of a delta function involving either the relative energy[14,15] or time[16] of the interacting nucleons. The advantage of the propagator dynamics is the retainment of locality and manifest covariance. The disadvantage of this approach is the inclusion of negative energy states in the particle propagators, which tends to make the calculations technically more difficult and their physical interpretation harder.

At sufficiently large momentum transfers the deuteron form factors are expected to be calculable in terms of only quarks and gluons within the framework of Quantum Chromodynamics (QCD). The first attempt at a quark-gluon description of the deuteron form factors was based on quark dimensional scaling (QDS)[17]: the underlying dynamical mechanism during electron-deuteron scattering is the rescattering of the constituent quarks via the exchange of hard gluons, which implies that $\sqrt{A(Q^2)} \sim (Q^2)^{-5}$. This prediction was later substantiated in the framework of perturbative QCD (pQCD), where it was shown[18] that to leading-order:

$$\sqrt{A(Q^2)} = [\frac{\alpha_s(Q^2)}{Q^2}]^5 \sum_{m,n} d_{mn}[\ln(\frac{Q^2}{\Lambda^2})]^{-\gamma_n-\gamma_m}, \tag{16}$$

where $\alpha_s(Q^2)$ and Λ are the QCD strong coupling constant and scale parameter, and $\gamma_{m,n}$ and d_{mn} are QCD anomalous dimensions and constants.

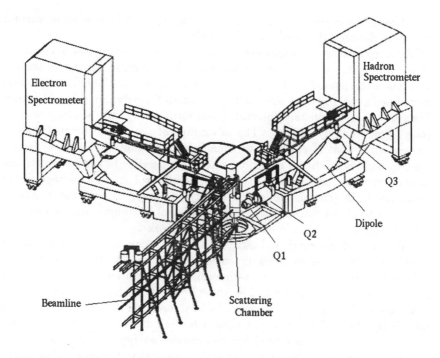

Figure 1. Plan view of the JLab Hall A Facility as used for the recent measurement of the deuteron $A(Q^2)$ at large momentum transfers[4]. Shown are the target scattering chamber, and the two identical high resolution spectrometers, each consisting of three quadrupole Q_1, Q_2, Q_3 and a dipole magnet (see text).

The unique features of the Continuous Electron Beam Accelerator and Hall A Facilities of the Jefferson Laboratory (JLab) offered recently the opportunity to extend the kinematical range of $A(Q^2)$ and to resolve inconsistencies in previous data sets from different laboratories by measuring[4] forward scattering angle elastic electron-deuteron scattering for $0.7 \leq Q^2 \leq 6.0$ (GeV/c)^2. Backward angle measurements in conjunction with the forward measurements enabled also determination of $B(Q^2)$ in the kinematical range $0.7 \leq Q^2 \leq 1.4$ (GeV/c)^2. Electron beams were scattered off a liquid deuterium target. Scattered electrons were detected in the electron High Resolution Spectrometer. To suppress backgrounds and separate elastic from inelastic processes, recoil deuterons were detected in coincidence with the scattered electrons in the hadron High Resolution Spectrometer. A schematic of the Hall A Facility as used in this experiment is shown in Figure 1.

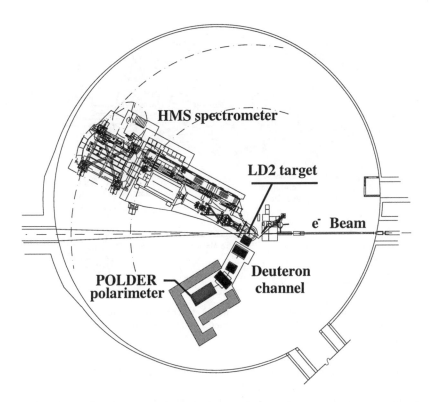

Figure 2. Plan view of the JLab Hall C Facility as used for the recent measurement[7] of the deuteron $t_{20}(Q^2)$. Shown are the electron High Momentum Spectrometer, the deuteron cryotarget, and the recoil deuteron spectrometer with the POLDER polarimeter (see text).

Recent advances in deuteron polarimetry and the high luminosity of the JLab Hall C Facility enabled also large Q^2 measurements[7] of $t_{20}(Q^2)$ for $0.7 \leq Q^2 \leq 1.7$ (GeV/c)^2. The experiment used the Hall C High Momentum Spectrometer to detect scattered electrons and a special deuteron channel to detect and spin-analyze recoil deuterons in coincidence with the electrons (see Figure 2). The deuteron channel consisted of a large solid angle magnetic spectrometer comprised of three quadrupole and one dipole magnets, and the POLDER deuteron tensor polarimeter, calibrated previously at the Saturn Laboratory. The polarimeter was based on the charge exchange reaction $H(\vec{d}, pp)n$ induced by the recoil deuterons in a liquid hydrogen cell. Deuterons incident on the polarimeter were identified with two scintillator trigger planes and two sets of wire chambers. The pairs of protons from

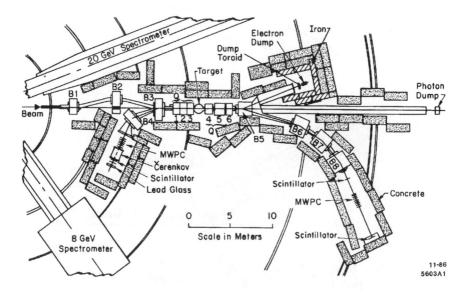

Figure 3. Plan view of the 180° electron and 0° deuteron magnetic spectrometer systems used for the SLAC End Station A measurement[5] of the deuteron $B(Q^2)$ (see text).

the charge exchange reactions were identified with two pairs of scintillator hodoscopes and a wire chamber set.

The highest Q^2 measurements of $B(Q^2)$ were performed[5] in the mid-1980's with a special double-arm spectrometer system[19] in the SLAC End Station A, as shown in Figure 3. Electron beams from the Nuclear Physics Injector were directed on a deuteron target through a chicane of dipole magnets (B1, B2 and B3). Electrons scattered at 180° were transported through a large solid angle spectrometer system consisting of three quadrupole (Q1, Q2, and Q3) and two dipole (B3 and B4) magnets to a set of detectors. Deuterons recoiling at 0° were transported through a spectrometer system consisting of three quadrupole (Q4, Q5, and Q6) and four dipole (B5, B6, B7 and B8) magnets to another set of detectors. The primary (unscattered) beams were directed onto a beam dump located inside the End Station. Electron-deuteron elastic coincidence events were identified by the time-of-flight method.

The experimental data for $A(Q^2)$, $B(Q^2)$ and $\tilde{t}_{20}(Q^2)$ are shown in Figures 4, 5 and 6, together with theoretical calculations (see below). A complete compilation of all available data and an extended overview of the theory are given in Ref. 1. The data for $A(Q^2)$ indicate a smooth fall-off with no apparent diffractive structure. The data for $B(Q^2)$ indicate the presence of a diffraction

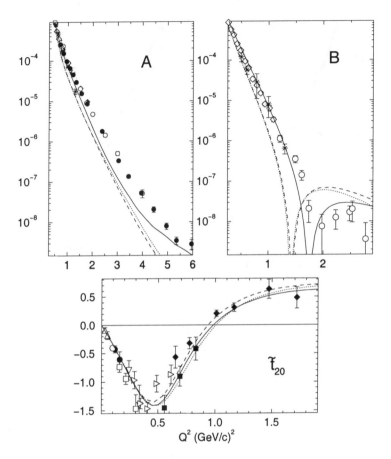

Figure 4. Experimental data on the deuteron $A(Q^2)$, $B(Q^2)$ and $\tilde{t}_{20}$ compared with theoretical calculations based on quasipotential equations (see text).

minimum in the vicinity of $Q^2 = 1.8$ $(\text{GeV/c})^2$. The $t_{20}(Q^2)$ data together with the $A(Q^2)$ and $B(Q^2)$ data show that the charge form factor exhibits a diffractive structure with a minimum at $Q^2 = 0.7$ $(\text{GeV/c})^2$ and that the quadrupole form factor falls off exponentially in the measured Q^2 region[7].

Shown also in Fig. 4 are theoretical calculations of $A(Q^2)$, $B(Q^2)$ and $\tilde{t}_{20}(Q^2)$ based on the Propagator Dynamics approach [sometimes referred to as the relativistic impulse approximation (RIA)] by three different groups. The solid, dotted and dashed curves represent the RIA calculations of Van Orden, Devine and Gross (VDG)[14], Hummel and Tjon (HT)[15], and Phillips,

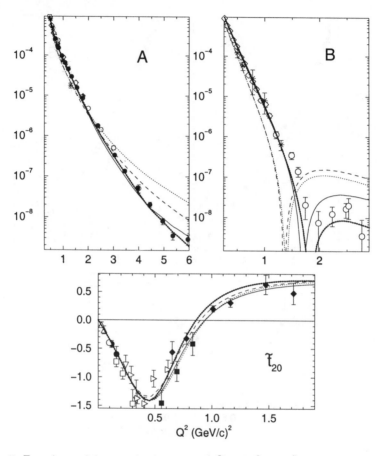

Figure 5. Experimental data on the deuteron $A(Q^2)$, $B(Q^2)$ and $\tilde{t}_{20}$ compared with theoretical calculations based on quasipotential equations, including the $\rho\pi\gamma$ contribution (see text).

Wallace and Devine (PWD)[16], respectively. The VDG curve is based on the Gross quasipotential equation[20] with a one-boson-exchange interaction and assumes that the electron interacts with an off-mass-shell nucleon or a nucleon that is on-mass-shell right before or after the interaction. The HT curve is based on a one-boson-exchange quasipotential approximation of the Bethe-Salpeter equation[21] where the two nucleons are treated symmetrically by putting them equally off their mass-shell with zero relative energy. The PWD curve is based on an also one-boson-exchange interaction but with a single-time equation that constraints the relative time to be zero.

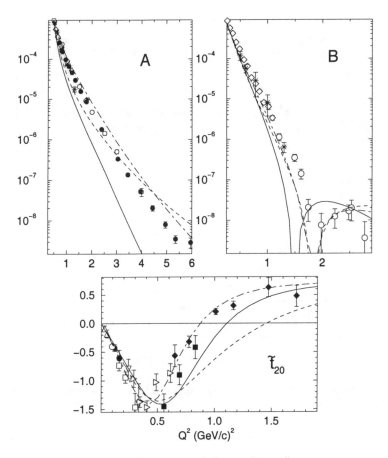

Figure 6. Experimental data on the deuteron $A(Q^2)$, $B(Q^2)$ and $\tilde{t}_{20}$ compared with selected theoretical calculations based on Hamiltonian Dynamics (see text).

In all three cases the RIA appears to fail to describe the data. All three theory groups have augmented their models by including the $\rho\pi\gamma$ MEC contribution as shown in Figure 5. The magnitude of this contribution depends on the $\rho\pi\gamma$ coupling constant and vertex form factor choices. The VDG model (solid curve) uses a soft $\rho\pi\gamma$ form factor. The HT model (dotted curve) uses a Vector Dominance Model hard form factor. The PWD model (dashed curve) uses an intermediate form factor. The inclusion of the $\rho\pi\gamma$ MEC contribution has a small effect on $B(Q^2)$ and $\tilde{t}_{20}(Q^2)$ but increases dramatically $A(Q^2)$. The $A(Q^2)$ data favour use of the softest possible form factor. The thick dotted curve is an alternate calculation by the VDG group with a different

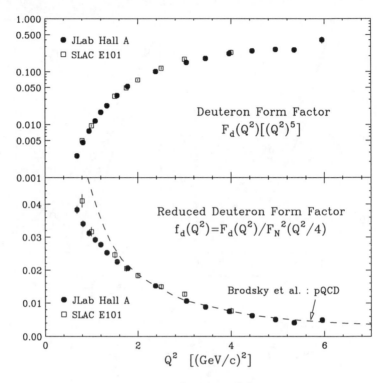

Figure 7. The deuteron form factor $F_d(Q^2)$ times $(Q^2)^5$ (top) and the reduced deuteron form factor $f_d(Q^2)$ (bottom) from SLAC and JLab. The curve is the asymptotic pQCD prediction for $\Lambda_{QCD} = 100$ MeV, arbitrarily normalized to the data at $Q^2 = 4$ (GeV/c)^2.

nucleon form factor choice in the off-shell modification of the single nucleon current required to insure current conservation. Although the difference in the models is indicative of the size of theoretical uncertainties and ambiguities, it appears that the relativistic impulse approximation can, as in the case of the alternate VDG model, reproduce the deuteron form factor data fairly well.

Figure 6 shows a comparison of the experimental data with selected recent Hamiltonian Dynamics calculations. The solid curve is a point form calculation by Allen, Klink and Polyzou[22] using the Argonne v_{18} potential. The dashed curve is a front form calculation by Lev, Pace and Salmè[23] using the Nijmegen potential. The dot-dashed curve is an instant form calculation by Forest and Schiavilla[24] using the Argonne v_{18} potential. These three calculations employ different approaches for the construction of the electromagnetic current. It is evident that none of these admittedly promising approaches is able, at this stage, to describe simultaneously all three $A(Q^2)$, $B(Q^2)$ and $\tilde{t}_{20}$ deuteron observables.

Figure 7 (top) shows values for the "deuteron form factor" $F_d(Q^2) \equiv \sqrt{A(Q^2)}$ multiplied by $(Q^2)^5$. It is evident that the data exhibit a behavior consistent with the power law of QDS and pQCD. Figure 7 (bottom) shows values for the "reduced" deuteron form factor[25] $f_d(Q^2) \equiv F_d(Q^2)/F_N^2(Q^2/4)$ where the two powers of the nucleon form factor $F_N(Q^2) = (1 + Q^2/0.71)^{-2}$ remove in a minimal and approximate way the effects of nucleon compositeness[25]. The $f_d(Q^2)$ data appear to follow, for $Q^2 > 2$ (GeV/c)^2, the asymptotic Q^2 prediction of pQCD[18]: $f_d(Q^2) \sim [\alpha_s(Q^2)/Q^2][\ln(Q^2/\Lambda^2)]^{-\Gamma}$. Here $\Gamma = -(2C_F/5\beta)$, where $C_F = (n_c^2 - 1)/2n_c$, $\beta = 11 - (2/3)n_f$, with $n_c = 3$ and $n_f = 2$ being the numbers of QCD colors and effective flavors. Although several authors have questioned the validity of QDS and pQCD at the momentum transfers of the JLab experiment[26], similar scaling behavior has been reported in deuteron photodisintegration at moderate photon energies[27].

In summary, there is a wealth of data on the deuteron form factors from 50 years of experimental work at numerous laboratories. Recent relativistic impulse approximation calculations with the inclusion of $\rho\pi\gamma$ MEC contributions are able to describe the data fairly well. Future experiments at JLab[28] are expected to extend the deuteron elastic structure functions measurements to even larger momentum transfers and improve the precision of the $B(Q^2)$ data around the region of its first diffraction minimum. These measurements will be critical in testing the validity of the apparent scaling behavior of $A(Q^2)$ and are expected to provide severe constraints to the ingredients of the recent theoretical relativistic calculations.

Acknowledgements

The author would like to acknowledge the support of the U.S. National Science Foundation in this work and to thank the EMI2001 Symposium organizers of Osaka University for their kind invitation and hospitality.

References

1. R. Gilman and F. Gross, arXiv:nucl-th/0111015 (2001); M. Garcon and J.W. Van Orden, *Advances in Nucl. Phys.* **26**, 293 (2001).
2. J. Carlson and R. Schiavilla, *Rev. Mod. Phys.* **70**, 743 (1998).
3. C.E. Carlson, J.R. Hiller and R.J. Holt, *Annu. Rev. Nucl. Part. Sci.* **47**, 395 (1997).

196

4. L.C. Alexa *et al.*, *Phys. Rev. Lett.* **82**, 1374 (1999); and references therein.
5. P.E. Bosted *et al.*, *Phys. Rev.* C **42**, 38 (1990); and references therein.
6. M. Bouwhuis *et al.*, *Phys. Rev. Lett.* **82**, 3755 (1999); and references therein.
7. D. Abbott *et al.*, *Phys. Rev. Lett.* **84**, 5053 (2000); and references therein.
8. J. Arvieux and J.M. Cameron, *Advances in Nucl. Phys.* **18**, 107 (1987).
9. T-S. Cheng and L.S. Kisslinger, *Phys. Rev.* C **35**, 1432 (1987); H. Dijk and B.L.G. Bakker, *Nucl. Phys.* A **494**, 438 (1989).
10. R. Dymarz and F.C. Khanna, *Nucl. Phys.* A **516**, 549 (1990); A. Amghar, N. Aissat and B. Desplanques, *Eur. Phys. J.* A **1**, 85 (1998).
11. B.D. Keister and W.N. Polyzou, *Advances in Nucl. Phys.* **20**, 225 (1991); J. Carbonell *et al.*, *Phys. Rep.* **300**, 215 (1998).
12. F. Gross, in *Modern Topics in Electron Scattering*, ed. B. Frois and I. Sick (World Scientific, Singapore, 1991).
13. E. Salpeter and H. Bethe, *Phys. Rev.* **84**, 1232 (1951).
14. J.W. Van Orden, N. Devine and F. Gross, *Phys. Rev. Lett.* **75**, 4369 (1995).
15. E. Hummel and J.A. Tjon, *Phys. Rev.* C **42**, 423 (1990).
16. D.R. Phillips, S.J. Wallace and N.K. Devine, *Phys. Rev.* C **58**, 2261 (1998).
17. S.J. Brodsky and G.R. Farrar, *Phys. Rev. Lett.* **31**, 1153 (1973); V.A. Matveev, R.M. Muradyan and A.N. Tavkhelidze, *Lett. Nuovo Cimento* **7**, 719 (1973).
18. S.J. Brodsky, C-R. Ji and G.P. Lepage, *Phys. Rev. Lett.* **51**, 83 (1983).
19. A.T. Katramatou *et al.*, *Nucl. Instrum. Meth.* A **267**, 448 (1988);
20. F. Gross, *Phys. Rev.* C **26**, 2203 (1982); and references therein.
21. R. Blankenbecler and R. Sugar, *Phys. Rev.* **142**, 1051 (1966); A.A. Logunov and A.N. Tavkhelidze, *Nuovo Cimento* **29**, 380 (1963).
22. T.W. Allen, W.H. Klink and W.N. Polyzou, *Phys. Rev.* C **62**, 054002 (2000).
23. F.M. Lev, E. Pace and G. Salmè, *Phys. Rev.* C **62**, 064004 (2000).
24. J.L. Forest and R. Schiavilla, to be published.
25. S.J. Brodsky and B.T. Chertok, *Phys. Rev.* D **14**, 3003 (1976).
26. N. Isgur and C.H. Llewellyn Smith, *Phys. Lett.* B **217**, 535 (1989); G.R. Farrar, K. Huleihel and H. Zhang, *Phys. Rev. Lett.* **74**, 650 (1995).
27. E.C. Schulte *et al.*, *Phys. Rev. Lett.* **87**, 102302 (2001).
28. G.G. Petratos, in *Proceedings of the Workshop on JLab Physics and Instrumention with 6-12 GeV Beams*, Newport News, Virginia (1998).

SEARCHING FOR QUARKS IN THE DEUTERON

R. GILMAN

Rutgers University, 126 Frelinghuysen Rd, Piscataway, NJ 08855 USA
and
Jefferson Lab, 12000 Jefferson Ave, Newport News, VA 23606 USA
E-mail: gilman@jlab.org

The deuteron has been studied in numerous elastic scattering and photodisintegration experiments over a wide range of momentum transfers. A primary motivation is investigating the validity of theories based on hadronic vs. quark degrees of freedom. It is often thought that hadronic approaches will break down at large momentum transfers, and it will be necessary to use quark degrees of freedom. I will very briefly review the status of our understanding of the underlying theoretical issues, the world data sets for elastic scattering and photodisintegration, and the ability of hadronic and quark based theories to explain the data. Hadronic theories appear to provide good explanations for elastic scattering, but photodisintegration is very difficult to understand.

1 Introduction

The structure of the deuteron is the preeminent problem of few body nuclear physics. It provides the most stringent tests of our understanding of nuclear structure with microscopic models. The deuteron lacks some of the complications of heavier nuclei, and has been investigated experimentally over a large kinematic range.

A prime interest in the recent past, motivating experimental studies in the GeV range, is whether one should use quark or hadronic degrees of freedom to describe the deuteron. It is possible to formulate the hadronic theory for the reaction in many different frameworks, some of which will be discussed below. Various approaches should in principle give equivalent correct answers, if the theory is formulated completely and consistently and the calculation converges. Approximations and truncations lead to variations between the theories, and between theories and (presumably correct) data. Using quark degrees of freedom in this view is merely an alternate formulation, that will provide better convergence and agreement with data in some kinematic regimes. Thus, a better question might be: in what kinematic regimes does each theory provide a concise and accurate description of the data?

In this proceedings, we will briefly cover how well various approaches agree with elastic scattering and photodisintegration data. This presentation is based on a significantly longer work [1], from which several of the figures

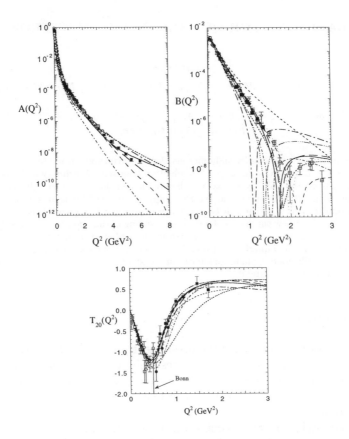

Figure 1. Experimental data for A, B, and t_{20} compared to eight calculations, described in the text. The calculations, in order of the Q^2 of their mimima in B, are: CK (long dot-dashed line), PWDM (dashed double-dotted line), AKP (short dot-dashed line), VDG full calculation (solid line), VDG in RIA (long dashed line), LPS (dotted line), DB (widely spaced dotted line), FSR (medium dashed line), and ARW (short dashed line).

have been taken.

2   Elastic scattering

Figure 1 shows several theoretical calculations, to be described below, which provide generally good descriptions of t_{20}, A, and B. It is impressive that theories can reasonably describe the deuteron as the structure functions decrease over about 8 orders of magnitude. The large variations in the predictions for B near the apparent minimum make it clear that these data need to be improved, to firmly determine that the minimum exists, as well as to better determine its location. The disagreement between the higher Q $(= \sqrt{Q^2})$ A data and the theories is not as significant. At higher Q, the calculations are sensitive to poorly defined aspects of the theory, such as the $\rho\pi\gamma$ exchange current and the form factor associated with the additional term in the nucleon current, required by current conservation for the bound nucleon.

Figure 1 is not useful to see precisely how well we do understand the deuteron. Is theory good to 2% or 20%? The 10% differences between the Jefferson Lab Hall A and Hall C data are not visible in this plot, and the slight systematic differences between some of the t_{20} data sets also do not seem to be very important. (There is insufficient space in these proceedings to reference the complete data set; please see instead recent fits [2] and reviews by Garçon and Van Orden [3], by Sick [4], and by Gilman and Gross [1].)

We now turn to a more precise examination of the A data at low Q. Here, pQCD and quark models are of little concern, but there are effective field theory (EFT) and chiral perturbation theory (χPT) calculations, as well as conventional theories. Recent χPT calculations [5] provide very good descriptions of B up to Q about 0.3 Gev, and of A and t_{20} up to Q about 0.5 GeV. This may indicate a relatively larger short range contribution to B. However, as we shall demonstrate below, there are possible questions about the exact size of A at low Q. Due to limited space and because they predict the structure functions over a larger range reasonably well, the focus will be on conventional calculations of the low Q A data.

Figure 2 shows the data for A at low Q. The data have been divided by a "fit" function, which depends on Q and some constants, to remove much of the Q dependence and to enable differences to be seen. The data sets generally overlap well, except for the 10% disagreement between the Mainz [6] and Saclay [7] data sets at $Q = 0.2 - 0.4\,\text{GeV}$. This is difficult to understand from comparison of experiments, since both experiments overlap well with each other and with Monterrey [8] at lower Q, both are cross checked with the very precise Mainz ep measurements [9] at these Qs, and the Saclay measurements agree well with the hgher Q, although less precise, experiments. The underlying origin of the difference is cross sections at $50°$, $60°$, $80°$, and $90°$ (lab) taken at 298.9 MeV

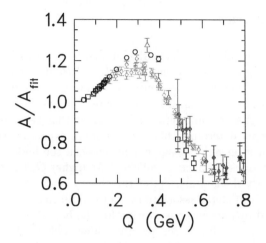

Figure 2. The data for A at low and moderate Q, divided by a fit function to take out much of the Q dependence of A, so that data sets may be compared.

(300 MeV) at Mainz (Saclay). For these data $\sigma/\sigma|_{NS} = A + B\tan^2(\theta/2)$ differs by almost 10%. The correction for the slight difference in Q is only about 1%. It is difficult to understand this large a disagreement for these two high precision experiments.[a]

This difference may appear small and unimportant, but, to put it in perspective, a 10% uncertainty in A in a region in which its magnitude is 10^{-1} – 10^{-2} is much greater than a 100% uncertainty in B in a region in which its magnitude is 10^{-7}. Furthermore, ambiguities in the theory are much reduced at lower Q, and the difficulty of getting cancellations exactly right, for precise reproduction of the position of minima, is well known. Thus, we conclude that the potential problems with A at low Q are at least as significant in assessing how well we understand the deuteron as the question of the minimum in B.

A set of conventional nonrelativistic calculations is shown in Fig. 3. The calculations (in order of decreasing magnitude at $Q = 0.1$ GeV) use W16 (long dot-dashed), CD Bonn (short dashed), AV18 (solid), IIB (short dot-dashed), and Paris (long dashed) wave functions. The W16 and IIB models use the S and D wave functions of a relativistic model, neglecting the P-state components. The models vary by about $\pm 2\%$, and are generally larger (smaller)

---

[a]Note that the tabulated Mainz cross sections in [6] have some typographical errors. For three of the data points, the tabulated cross section is a factor of 10 different from the tabulated $\sigma/\sigma|_{NS}$, which appears to be correct. One assumes a typographical error. For a fourth point, the difference is a factor of 8.1, and $\sigma/\sigma|_{NS}$ again appears to be correct.

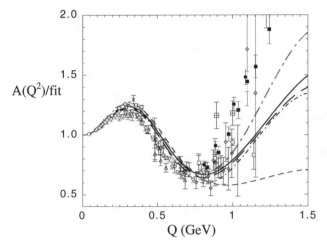

Figure 3. The data for A at low and moderate Q, divided by the fit function, compared to five nonrelativistic calculations described in the text.

than the Saclay (Mainz) data. Along with the assumption that relativistic corrections are small and under control at low Q, this variation suggests that a $\approx 2\%$ theoretical description of the form factors should be possible. Note that there is a long standing discrepancy of this size in conventional calculations with the quadrupole moment. In the pionless EFT approach, there is a short distance four nucleon one photon coupling, which is unknown and can be adjusted to give the quadrupole moment.

It is important to assess how accurately the nucleon form factors are known, in considering how precise the calculations might be. In a pure wave function model, the structure function results from a sum of deuteron body form factors multiplied by isoscalar nucleon form factors. The first problem is that the ep data base is largely made up of 4 – 5% cross sections, although these are some more precise experiments. The second point is that the Platchkov data were used, with a Paris potential calculation plus calculated corrections, and the knowledge of G_{ep}, to extract G_{en}. These G_{en} values were the major input to the "MMD" fits [10] to the nucleon form factors. The MMD form factor parameterization was then an input to these calculations. Thus, the logic is somewhat circular, as the Platchkov A data affect the input to the theoretical A calculations. However, the MMD paramaterization

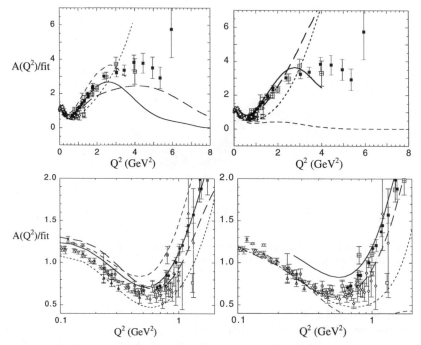

Figure 4. The data for $A(Q)$, compared to eight relativistic calculations. Left panels show the propagator and instant-form results: FSR (solid line), VDG in RIA approximation (long dashed line), ARW (medium dashed line), and PWDM (short dashed line). Right panels show the front-form CK (long dashed line) and LPS (short dashed line), the point-form AKP (medium dashed line) and the quark model calculation DB (solid line).

appears to get the isoscalar electric form factor about correct, slightly underestimating G_E^n and overestimating G_E^p. A refit to the proton and neutron data base, omitting the Platchkov G_{En} values, would be desirable.

Figure 4 compares seven relativistic calculations and one quark model calculation to the $A(Q)$ data. All of these calculations were done without the $\rho\pi\gamma$ meson-exchange current, which is not under good control, and but appears in modern calculations to have a negligible effect at low Q. The calculations include both propagator formulations and Hamiltonian dynamics (instant, point, and front form calculations), as well as a nonrelativistic quark model calculation. While there have been many investigations concerning the implications of pQCD and helicity conservation on the deuteron properties,

Table 1. The eight theories shown in Figures 1 and 4.

model	J?	A	B	t_{20}
Carbonell and Karmanov (CK) [11] front-form, dynamical light front	no	Saclay	no	OK
Phillips, Wallace, Devine, and Mandelzweig (PWDM) [12] propagator "equal-time"	no	low	no	OK
Allen, Klink, and Polyzou (AKP) [13] point form	no	Saclay	no	no
Van Orden, Devine, and Gross (VDG) [14] propagator, Gross equation	yes	Mainz	OK	OK
Lev, Pace, and Salmé (LPS) [15] front-form, fixed light front	no	Saclay	OK	no
Dijk and Bakker (DB) [16] nonrelativistic quark compound bag	yes	high	OK	OK
Forest, Schiavilla and Riska (FSR) [17] instant form, no v/c expansion	yes	Mainz	OK	OK
Arenhövel, Ritz and Wilbois (ARW) [18] instant form, v/c expansion	yes	Mainz	no	OK

we consider these limits to not be of concern here.

Table 2 summarizes the calculations, ordered by the Q^2 of the minimum for B. The column labelled J assesses which calculations have more complete currents [1], while the columns labelled A, B, and t_{20} assess whether the agreement with these observables is good, though only up to about 1 GeV^2 for A. The more technically complete calculations tend to agree better with B and t_{20} than do the others, and they also favor the Mainz data rather than the Saclay data. Note that ARW calculation involves a v/c expansion which is suggested by the authors to not be valid much beyond 1 GeV.

Comparing the W16 and IIB models in Figure 3 with the VDG model in Figure 4, from which they come, indicates that the relativistic corrections are likely small. However, if we compare the nonrelativistic calculations to the data, it is not clear what to conclude about the size and sign of the relativistic corrections. If the Mainz data are correct, both the VDG and FSR calculations provide a good account of the full data set, and the ARW calculation is good over its claimed range. However, if the Saclay data are correct, it appears that *no* conventional calculation is entirely satisfactory.

We conclude that there are several hadronic theories that provide good qualitative agreement with the deuteron electromagnetic structure. It is possible that the agreement is actually as good as ≈2%, but problems in the ep and ed databases prevent this precise an assessment. There is no apparent need to use anything but nucleon and meson degrees of freedom to understand the elastic scattering, and there is no satisfactory quark based model. Understanding A at low Q^2 is an often unnoticed but important issue.

3 Photodisintegration

The kinematics for photodisintegration and elastic scattering are actually quite different, despite similar beam energies; quark models may indeed be quite appropriate for photodisintegration. In elastic scattering, the energy and momentum transfer are matched so that in the deuteron rest frame $W = m_d$, independent of the beam energy or Q^2. The two nucleon system is never far off shell, and there is little need to consider $\Delta\Delta$, quark, or other exotic components to the wave function.

In photodisintegration, $W >> m_d$, and the two nucleon system is far off shell. As $E_\gamma \to 4$ GeV, all known baryon resonance can be excited on shell, and 286 distinct combinations of 2 on-shell known (four star) baryons occur. With the uncertainties in the baryon's masses, widths, decay channels, and interactions, and with the numerical problems inherent in such a coupled channels calculation, one cannot be optimistic about the prospects for a microscopic hadronic theory. Quark degrees of freedom may provide an appropriate averaging over all the combinations of resonances.

The agreement of the best microscopic theory available with data starts to break down by $E_\gamma \approx 400$ MeV, just above the Δ resonance. Observables coming from imaginary interferences, such at p_y and T, appear more difficult to describe than real interferences, such as Σ and C_x, or sums of amplitudes squared, such as C_z and $d\sigma/d\Omega$. Figure 5 compares calculations to the Σ asymmetry and p_y, both of which have extensive data sets. Conventional hadronic calculations by Schwamb and Arenhövel (SA) [19] and Kang et al. [20] reproduce Σ at least qualitatively, but neither reproduces p_y well at all. The SA calculation is similarly detailed to the recent elastic scattering work; the Kang calculation is much simpler, except for including 17 well established resonances, and it reasonably reproduces the cross section up to 1.6 GeV.

Two quark model calculations, the QCD rescattering model [21] and the quark gluon string model [22] show reasonable agreement with the cross sections above about 1 GeV. The QCD rescattering model relates the photodisintegration to nucleon nucleon scattering. It relies on the dominance of quark exchange diagrams at higher energies, and assumes the photon is absorbed on the exchanged quarks. The prediction that p_y is small at high energies appears correct. Estimates in this model of the polarization transfers are also qualitatively correct. The quark gluon string model [22] uses a Reggie general-

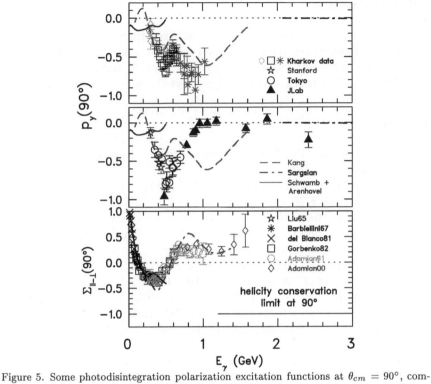

Figure 5. Some photodisintegration polarization excitation functions at $\theta_{cm} = 90°$, compared to theoretical calculations. Top panel: Kharkov data for the induced polarization p_y. There were experimental difficulties for the higher energy data sets. Middle panel: Other world data for p_y. Lower panel: Several data sets for the linearly polarized photon asymmetry Σ; several other data sets with only a few points each were omitted.

ization of the nucleon exchange Born term. It is based on the dominance of planar graphs in QCD and has been applied successfully to other high energy reactions. Results from polarization calculations are expected soon.

We conclude that construction of a conventional microscopic hadronic calculation appears to be not possible in the near future. Promising quark models exist, with reasonable agreement with the high energy data – though certainly not the several percent agreement of elastic scattering. However, the models are not yet sufficiently tested to conclude that our understanding of high energy deuteron photodisintegration is satisfactory.

Acknowledgments

Some of the figures in this work have been taken from [1]. I am particularly indebted to Franz Gross for many useful dscussions on the structure of the deuteron. I also thank H. Gao, M. Garçon, R. Holt, X. Jiang, K. McCormick, G.G. Petratos, S. Strauch, and K. Wijesooriya for many helpful discussions.

References

1. R. Gilman and F. Gross, submitted to *J. Phys. G*; preprint nucl-th/0111015.
2. D. Abbott *et al.*, *Eur. Phys. J* A **7**, 421 (2000).
3. M. Garçon and J.W. Van Orden, *Advances in Nucl. Phys.* **26**, 293 (2001).
4. I. Sick, submitted to *Progress in Theoretical Physics*.
5. D.R. Phillips, preprint nucl-th/0108070, to be published in *Proceedings of Conference on Mesons and Light Nuclei*.
6. G.G. Simon *et al.*, *Nucl. Phys.* A **364**, 285 (1981).
7. S. Platchkov *et al.*, *Nucl. Phys.* A **510**, 740 (1990).
8. R.W. Berard *et al.*, *Phys. Lett.* B **47**, 355 (1973).
9. G.G. Simon *et al.*, *Nucl. Phys.* A **333**, 381 (1980).
10. P. Mergell, U.G. Meissner and D. Drechsel, *Nucl. Phys.* A **596**, 367 (1996).
11. J. Carbonell and V.A. Karmanov, *Eur. Phys. J* A **6**, 9 (1999).
12. D.R. Phillips, S.J. Wallace, and N. Devine, *Phys. Rev.* C **58**, 2261 (1998).
13. T.W. Allen, W.H. Klink and W.N. Polyzou, *Phys. Rev.* C **63**, 034002 (2001).
14. J.W. Van Orden, N. Devine and F. Gross, *Phys. Rev. Lett.* **75**, 4369 (1995).
15. F.M. Lev, E. Pace and G. Salmé, *Phys. Rev.* C **62**, 064004 (2000).
16. H. Dijk and B.L.G. Bakker, *Nucl. Phys.* A **494**, 438 (1988).
17. J. Forest and R. Schiavilla 2001 (to be published).
18. H. Arenhövel, F. Ritz and T. Wilbois, *Phys. Rev.* C **61**, 034002 (2000).
19. M. Schwamb and H. Arenhövel, preprint nucl-th/0105033; *Nucl. Phys.* A **690**, 682 (2001); *Nucl. Phys.* A **690**, 647 (2001).
20. Y. Kang, P. Erbs, W. Pfeil, and H. Rollnik, *Abstracts of the Particle and Nuclear Intersections Conference*, (MIT, Cambridge, MA, 1990); Y. Kang, Ph.D. thesis, Bonn (1993).
21. L.L. Frankfurt, G.A. Miller, M.M. Sargsian and M.I. Strikman, *Phys. Rev. Lett.* **84**, 3045 (2000); M Sargsian, private communication.
22. V.Yu. Grishina *et al.*, *Eur. Phys. J* A **10**, 355 (2001).

PHOTOREACTIONS IN NUCLEAR ASTROPHYSICS

P. MOHR, M. BABILON, W. BAYER, D. GALAVIZ, T. HARTMANN,
C. HUTTER, K. SONNABEND, K. VOGT, S. VOLZ, AND A. ZILGES

Institut für Kernphysik, Technische Universität Darmstadt,
Schlossgartenstraße 9, D-64289 Darmstadt, Germany
E-mail: mohr@ikp.tu-darmstadt.de

Photon-induced reactions are the key reactions for the nucleosynthesis of neutron-deficient p-nuclei. We present a new technique which simulates a thermal photon bath at typical temperatures of several billion Kelvin during a supernova explosion. We are able to determine astrophysical reaction rates for the relevant photon-induced reactions. Additionally, we give some examples for the relevance of photon-induced reactions in the astrophysical s-process.

1 Introduction

The nucleosynthesis of light nuclei ($A \leq 60$) proceeds mainly by fusion reactions of charged particles along the $N \approx Z$ line. At the nuclei around $A \approx 60$ which have the highest binding energy per nucleon fusion reactions stop, and the nucleosynthesis of heavier nuclei requires a different mechanism. It is well-known that the bulk of heavy nuclei with $A \geq 60$ have been synthesized by neutron capture reactions and subsequent β decays, and already in the famous B[2]FH paper in 1957 the two dominating neutron capture processes have been characterized [1] which are the slow neutron capture process (s-process) and the rapid neutron capture process (r-process).

At s-process conditions moderate neutron densities of the order of 10^8 neutrons per cm^3 and moderate temperatures of about $kT \approx 8 - 30\,\text{keV}$ lead to a nucleosynthesis path along the valley of stability because the time between two neutron captures is significantly larger than the β^- decay half-life. The astrophysical sites for the s-process are helium shell burning in low mass AGB stars for its main component and core helium burning in massive stars for its weak component [2].

In the r-process much higher neutron densities of more than 10^{20} neutrons per cm^3 lead to a series of neutron captures before β^- decay can occur. The r-process runs around $15 - 20$ mass units away from the the valley of stability on the neutron-rich side of the chart of nuclides. Here, typical neutron separation energies are of the order of $2\,\text{MeV}$ [3].

Both processes together are able to produce about $99\,\%$ of the heavy nuclei, and both processes contribute almost equally in strength to the solar abundance distribution. However, there are a some 35 neutron-deficient nuclei which cannot be produced by neutron capture reactions. These so-called p-nuclei have been synthesized in supernova explosions mainly by photon-induced reactions, i.e., (γ,n), (γ,p), and (γ,α) reactions [4,5,6,7,8,9,10,11,12,13]. This nucleosynthesis process requires that heavy neutron-rich nuclei which have been synthesized earlier in the s-process and the r-process are available as seed nuclei. A common property of all 35 p-nuclei is their low relative isotopic abundance of the order of 1 % or less compared to the isotopes made in the s-process and the r-process.

The astrophysical process for the nucleosynthesis of the p-nuclei has been called (i) γ-process [7] or (ii) p-process [8]. Note that the "p" in "p-process" should be interpreted as "photodisintegration" [14]. To clarify this confusing nomenclature, we use the terminus "p-process" for the full nucleosynthesis network for the production of the 35 p-nuclei which includes more than 1000 nuclei and 10000 reaction rates [12]. The γ-process consists mainly of (γ,n), (γ,p), and (γ,α) reactions; it is a good approximation for the full p-process network for masses above $A \approx 140$.

Possible astrophysical sites for this process are still under discussion. Typical conditions of $T_9 \approx 2 - 3$ (T_9 is the temperature in billion degrees, corresponding $kT \approx 172 - 259\,\text{keV}$), densities of $\rho \approx 10^6\,\text{g/cm}^3$, and timescales $\tau \approx 1\,\text{s}$ occur in the oxygen- and neon-rich layers of type II supernovae [7,8,9,12,13]. Alternatively, type Ia supernovae have also been proposed [10] where similar thermodynamical conditions may occur.

All calculations of the p-process nucleosynthesis show a common trend: the general agreement between the calculated overproduction factors and the solar abundances of the p-nuclei is good. However, there is a systematic overproduction of heavy p-nuclei and a systematic underproduction of light p-nuclei [7,8,9,10,12]. Recently, Costa et al. [13] have suggested that an enhanced $^{22}\text{Ne}(\alpha,\text{n})^{25}\text{Mg}$ reaction rate leads to an enhanced weak component of the s-process which occurs in the progenitor stars of type II supernovae. The enhancement of these s-process seed nuclei leads to a stronger production of light p-nuclei. However, the enhanced $^{22}\text{Ne}(\alpha,\text{n})^{25}\text{Mg}$ reaction rate of Ref. [13] was ruled out by a recent experiment [15].

Almost no experimental data are available at astrophysically relevant energies for the nuclear reactions in the p-process. All nucleosynthesis calculations have to rely on theoretical calculations within the framework of the statistical model [16,17,18]. Therefore, we started a program to measure (γ,n) cross sections and reaction rates using a quasi-thermal photon bath at typical temperatures of $T_9 = 2 - 3$. This quasi-thermal photon bath which is produced by the superposition of bremsstrahlung spectra with different endpoint energies will be discussed in detail.

This paper is organized as follows. In Sect. 2 the reaction rate of a photon-induced reaction in a photon bath at a given temperature T is defined. In Sect. 3 our experimental set-up is presented, and in Sect. 4 we give our results for the nucleus ^{197}Au. In addition, we present some examples where photon-induced reactions are relevant for the nucleosynthesis in the s-process (Sect. 5). Finally, some conclusions are drawn in Sect. 6.

2 Reaction rates in a thermal photon bath

The reaction rate λ of a photodisintegration reaction $B(\gamma,x)A$ is given by

$$\lambda(T) = \int_0^\infty c\, n_\gamma(E,T)\, \sigma_{(\gamma,x)}(E)\, dE \qquad (1)$$

with the speed of light c and the cross section of the γ-induced reaction $\sigma_{(\gamma,x)}(E)$. Obviously, λ is also the production rate of the residual nucleus A. The photon density $n_\gamma(E,T)$, the number of γ-rays at energy E per volume and per energy interval, is given by the Planck distribution:

$$n_\gamma(E,T) = \left(\frac{1}{\pi}\right)^2 \left(\frac{1}{\hbar c}\right)^3 \frac{E^2}{\exp\left(E/kT\right)-1} \qquad (2)$$

The integrand of Eq. (1) is given by the product of the photon density n_γ which decreases with the energy E and the photodisintegration cross section σ which increases with the energy E. The integrand has a sharp maximum at energies of about $kT/2$ above the neutron threshold with a typical width of about 1 MeV. A measurement of the cross section in this narrow window is already sufficient to derive the astrophysical reaction rate. The properties of this Gamow-like window for (γ,n) reactions have been presented in Ref. [19], and they have been analyzed in detail in Ref. [20].

3 Experimental set-up of a quasi-thermal photon bath

3.1 Experimental procedure

Our experiments are performed at the injector of the superconducting linear electron accelerator S-DALINAC installed at Technische Universität Darmstadt [21,22]. The electron beam with typical currents up to about $50\,\mu$A and energies up to 10 MeV is completely stopped in a massive copper radiator. A continuous so-called white bremsstrahlung spectrum is produced with an endpoint energy E_0 equal to the kinetic energy of the incoming electrons. The bremsstrahlung photons are mainly emitted in forward direction. The photons are collimated using a massive copper collimator with a length of about 1 m. Behind the collimator the photons hit the target with typical masses less than 1 g of natural isotopic composition corresponding to several milligrams of the relevant p-nuclei. The target is usually sandwiched between two thin layers of boron for normalization of the incoming photons. A detailed description of the set-up can be found in Ref. [23].

During the irradiation of the target the incoming photon flux is determined by photon scattering using the ^{11}B(γ,γ') reaction. Two large volume HPGe detectors are placed at distances of 26 cm and at angles of 90° and 130° relative to the incoming photons [24]. A good description of the photon flux is obtained from slightly modified GEANT simulation calculations [25].

After irradiation the (γ,n) cross section is determined by the photoactivation method. The activity of the target is measured using a third HPGe detector with an energy resolution of about 2 keV. A typical spectrum from the ^{197}Au$(\gamma,n)^{196}$Au experiment is shown in Fig. 1. The excellent sensitivity of the photoactivation method allows one to measure (γ,n) cross sections with small amounts of target material of natural isotopic composition. In a previous experiment the half-life of ^{196}Au has been measured: $T_{1/2} = 6.1669 \pm 0.0006$ d [26], and the uncertainty of this value has been reduced significantly.

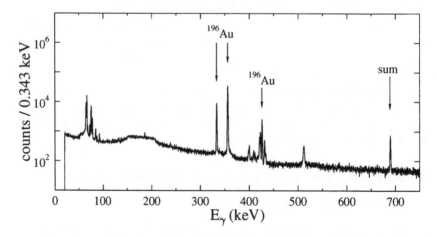

Figure 1. Photon spectrum of the gold target after irradiation with bremsstrahlung photons. The decay lines at $E_\gamma = 333.0$ keV, 355.7 keV, and 426.1 keV from the decay of ^{196}Au are clearly visible. Additional lines stem from X-rays after the electron capture ^{196}Au $\rightarrow$ ^{196}Hg, and summing effects in the detector.

3.2 Simulation of a thermal photon bath

Several irradiations were performed with photon endpoint energies of 8325, 8550, 8775, 9000, 9450, and 9900 keV. The intensity of the bremsstrahlung spectra decreases steeply close to the endpoint energy. This behavior is similar to the thermal photon distribution in Eq. (2). Therefore, a quasi-thermal photon spectrum Φ^{qt}_{brems} can be obtained in a given energy range by a careful superposition of bremsstrahlung spectra $\Phi_{brems}(E_0)$ with different endpoint energies $E_{0,i}$:

$$c n_\gamma(E, T) \approx \Phi^{qt}_{brems}(T) = \sum_i a_i(T) \times \Phi_{brems}(E_{0,i}) \qquad (3)$$

where the $a_i(T)$ are strength coefficients which can be adjusted for any temperature T relevant for the γ process. The superposition is shown in Fig. 2 for a temperature of $T_9 = 2.5$, i.e. for $T = 2.5$ billion degrees.

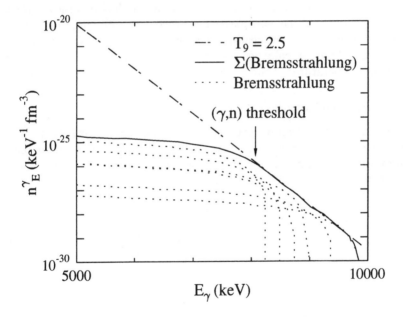

Figure 2. The superposition of several bremsstrahlung spectra $\Phi^{qt}_{brems}(T)$ (full line) with different endpoint energies E_0 is compared to the thermal Planck spectrum $n_\gamma(E, T)$ (dashed line) at a temperature of $T_9 = 2.5$. A good agreement is found in the relevant energy range between the neutron separation energy of ^{197}Au ($S_n = 8071$ keV, marked by an arrow) up to about 10 MeV. The six contributing bremsstrahlung spectra $\Phi_{brems}(E_{0,i})$ are shown as dotted lines.

The experimental yield Y_i per target nucleus is given by

$$Y_i = \int \Phi_{\text{brems}}(E_{0,i}) \, \sigma_{(\gamma,x)}(E) \, dE \qquad . \tag{4}$$

A comparison of Eq. (4) with Eqs. (1) and (3) relates the astrophysical decay rate λ directly to the experimental yields Y_i by

$$\lambda(T) = \sum_i a_i(T) \times Y_i \qquad . \tag{5}$$

The average deviation between the thermal Planck distribution [Eq. (3), *l.h.s.*] and the bremsstrahlung-approximated quasi-thermal distribution [Eq. (3), *r.h.s.*] is less than 10% in the relevant energy region around $E_{\text{eff}} \approx S_{\text{n}} + \frac{1}{2}kT$ (see also Fig. 2).

The astrophysical reaction rate $\lambda(T)$ can be derived directly from the experimental yield of the irradiations with different endpoint energies. The temperature T of the quasi-thermal spectrum can be chosen in the full relevant temperature range of the astrophysical γ-process by adjusting the coefficients $a_i(T)$. However, the experimentally measured laboratory reaction rates have to be corrected theoretically because of thermal excitations of the target nuclei at typical temperatures of the γ-process.

3.3 Conventional analysis

The direct determination of reaction rates in the previous section avoids the problems of the determination of the cross section. However, for comparison with existing (γ,n) data for ^{197}Au it is necessary to provide the cross section $\sigma(E)$ instead of the reaction rate $\lambda(T)$. From Eq. (4) it is obvious that one needs an assumption on the energy dependence of the cross section. Close above the threshold the cross section may be parametrized by

$$\sigma_{\text{conv}}(E) = \sigma_0 \times \sqrt{\frac{E - S_{\text{n}}}{S_{\text{n}}}} \tag{6}$$

4 Results for ^{197}Au$(\gamma,n)^{196}$Au

In Fig. 3 the result of the conventional analysis of our ^{197}Au$(\gamma,n)^{196}$Au experiment is compared to several data available in literature [27,28,29,30,31]. We find excellent agreement with the most recent data point of Ref. [31] and about 5 % lower values than Veyssiere et al. [28]. Within the error bars, this is in agreement with the recommendations of Berman et al. [30] to reduce the Veyssiere data by 8 %. The Fultz data [27] seem to be obsolete [30].

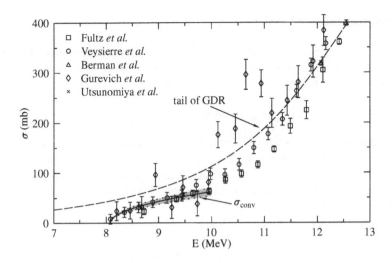

Figure 3. Cross section of the reaction ^{197}Au(γ,n)^{196}Au close above the threshold (S_n = 8.071 MeV) [27,28,30,29,31]. Berman et al. [30], Veyssiere et al. [28], and Fultz et al. [27] used quasi-monochromatic photons from positron annihilation in their experiments; Gurevich et al. [29] used electron bremsstrahlung. Recently, Utsunomiya et al. [31] used Laser-Compton backscattered photons with $E_\gamma = 9.47$ MeV. The Lorentzian parametrization of the GDR from [30] is shown as dashed line. The Lorentzian parametrization overestimates the cross section close to the reaction threshold. Additionally, the result of the conventional analysis of the present experiment is shown (full line from threshold to 10 MeV, uncertainties gray shaded). The numerical data were taken from the compilation [32].

Our recent results of the ^{197}Au(γ,n)^{196}Au experiment are summarized in Tab. 1. The good agreement between λ_{conv} from the conventional analysis and λ_{qt} from the quasi-thermal superposition shows that the assumption in Eq. (6) is realistic for ^{197}Au. This fact is confirmed by the observation that all σ_0 values which have been determined at all measured endpoint energies agree within their uncertainties. The quoted value of $\sigma_0 = 127 \pm 16$ mb is the weighted average of all values.

Table 1. Summary of the results of the gold activation experiment: the constant σ_0 from Eq. (6) and the decay rates λ from the conventional analysis, from the quasi-thermal analysis, and from a statistical model calculation [33] are given. All reaction rates λ are given for $T_9 = 2.5$.

nucleus	S_n (keV)	σ_0 (mb)	λ_{conv} (s^{-1})	λ_{qt} (s^{-1})	λ_{theo}
^{197}Au	8071	127±16	5.90	6.17	4.81

5 Photon-induced reactions in the s-process

Of course, (n,γ) reaction rates are the most important ingredient for nucleosynthesis calculations in the s-process. But here we present some examples for the relevance of photon-induced reactions.

5.1 The s-process branching at ^{185}W

If the half-life of an unstable nucleus in the s-process path is sufficiently long, this nucleus may capture a neutron before the β^- decay occurs. From the analysis of these so-called s-process branchings one can determine the neutron density during the s-process. For the branching at ^{185}W a neutron density of about $4 \times 10^8/\mathrm{cm}^3$ was derived [34].

A direct measurement of the (n,γ) cross section of unstable ^{185}W is difficult. We have measured the inverse reaction ^{186}W(γ,n)^{185}W at energies close above the threshold. A theoretical calculation agrees with our preliminary data at low energies and with (γ,n) data from literature at higher energies [35]. Therefore, this calculation of the ^{185}W(n,γ)^{186}W cross section is also reliable, and the result confirms the adopted value of about 700 mb [36].

5.2 The nucleosynthesis of ^{180}Ta

The nucleosynthesis of nature's rarest isotope ^{180}Ta is still an unresolved puzzle. Several astrophysical sites have been proposed for its production including the s-process. However, because the $J^\pi = 1^+$ ground state of ^{180}Ta is unstable and all ^{180}Ta is in the $J^\pi = 9^-$ isomeric state, the synthesized ^{180}Ta may be destroyed by photon-induced transitions between the isomeric state, an intermediate state, and the ground state [37].

Recently, ^{180}Ta has been irradiated at the photoactivation set-up at Universität Stuttgart, and the destruction of $J^\pi = 9^-$ isomer could be observed down to energies of about 1 MeV. The effective half-life of ^{180}Ta under s-process conditions is reduced by many orders of magnitude. Realistic s-process models [38] still allow ^{180}Ta to survive because of convection which transports freshly synthesized ^{180}Ta into cooler regions of the low mass AGB star [37].

6 Conclusions and outlook

Because of its high sensitivity photoactivation is an excellent tool to study photon-induced reactions under astrophysical conditions. The simulation of a quasi-thermal photon bath at typical temperatures during a supernova explosion enables us to measure astrophysical reaction rates directly. These experimental results should be used to test theoretical calculations which are necessary because of the huge number of reaction rates in the γ-process [8]. Further exciting experimental possibilities arise from the development of high-intensity monochromatic γ-ray beams from Laser-Compton backscattering [31].

Acknowledgments

We want to thank the S–DALINAC group around H.-D. Gräf for the reliable beam during the photoactivation and U. Kneissl, A. Mengoni, T. Rauscher, and A. Richter and for valuable discussions. This work was supported by the Deutsche Forschungsgemeinschaft (contracts Zi 510/2-1, FOR 272/2-1).

References

1. E. M. Burbidge, G. R. Burbidge, W. A. Fowler, F. Hoyle, Rev. Mod. Phys. **29**, 547 (1957).
2. F. Käppeler, Prog. Part. Nucl. Phys. **43**, 419 (1999).
3. G. Wallerstein et al., Rev. Mod. Phys. **69**, 995 (1997).
4. D. L. Lambert, Astron. Astrophys. Rev. **3** (1992) 201.
5. K. Ito, Prog. Theor. Phys. **26**, 990 (1961).
6. R. L. Macklin, Astroph. J. **162**, 353 (1970).
7. S. E. Woosley and W. M. Howard, Astrophys. J. Suppl. **36**, 285 (1978).
8. M. Rayet, N. Prantzos, and M. Arnould, Astron. Astroph. **227**, 271 (1990).
9. N. Prantzos, M. Hashimoto, M. Rayet, and M. Arnould, Astron. Astroph. **238**, 455 (1990).
10. W. M. Howard, B. S. Meyer, and S. E. Woosley, Astroph. J. **272**, L5 (1991).
11. B. S. Meyer, Ann. Rev. Astron. Astrophys. **32**, 153 (1994).
12. M. Rayet, M. Arnould, M. Hashimoto, N. Prantzos, and K. Nomoto, Astron. Astroph. **298**, 517 (1995).
13. V. Costa, M. Rayet, R. A. Zappalà, M. Arnould, Astron. Astroph. **358**, L67 (2000).
14. M. Arnould, private communication.
15. M. Jaeger, R. Kunz, J. W. Hammer, G. Staudt, K.-L. Kratz, and B. Pfeiffer, Phys. Rev. Lett. **87**, 202501 (2001).
16. T. Rauscher and F.-K. Thielemann, At. Data Nucl. Data Tables **75**, 1 (2000).
17. T. Rauscher and F.-K. Thielemann, At. Data Nucl. Data Tables **79**, 47 (2001).

18. S. Goriely, Proc. 10th Int. Symp. Capture Gamma-Ray Spectroscopy, ed. S. Wender, AIP Conference Proceedings **529**, 287 (2000).

19. P. Mohr, K. Vogt, M. Babilon, J. Enders, T. Hartmann, C. Hutter, T. Rauscher, S. Volz, and A. Zilges, Phys. Lett. B **488**, 127 (2000).

20. P. Mohr, M. Babilon, J. Enders, T. Hartmann, C. Hutter, K. Vogt, S. Volz, and A. Zilges, Nucl. Phys. **A688**, 82c (2001).

21. A. Richter, Proc. 5th European Particle Accelerator Conference, Barcelona 1996, ed. S. Myers et al., IOP Publishing, Bristol, 1996, p. 110.

22. A. Richter, Prog. Part. Nucl. Phys. **44**, 3 (2000).

23. P. Mohr, J. Enders, T. Hartmann, H. Kaiser, D. Schiesser, S. Schmitt, S. Volz, F. Wissel, and A. Zilges, Nucl. Instr. Meth. Phys. Res. A **423**, 480 (1999).

24. T. Hartmann, J. Enders, P. Mohr, K. Vogt, S. Volz, and A. Zilges, Phys. Rev. Lett. **85**, 274 (2000).

25. K. Vogt, P. Mohr, M. Babilon, J. Enders, T. Hartmann, C. Hutter, T. Rauscher, S. Volz, A. Zilges, Phys. Rev. C **63**, 055802 (2001).

26. K. Lindenberg, F. Neumann, D. Galaviz, T. Hartmann, P. Mohr, K. Vogt, S. Volz, and A. Zilges, Phys. Rev. C **63**, 047307 (2001).

27. S. C. Fultz, R. L. Bramblett, T. J. Caldwell, and N. A. Kerr, Phys. Rev. **127**, 1273 (1962).

28. A. Veyssiere, H. Beil, R. Bergere, P. Carlos, and A. Lepretre, Nucl. Phys. **A159**, 561 (1970).

29. G. M. Gurevich, L. E. Lazareva, V. M. Mazur, S. Yu. Merkulov, G. V. Solodukhov, V. A. Tyutin Nucl. Phys. **A351**, 257 (1981).

30. B. L. Berman, R. E. Pywell, S. S. Dietrich, M. N. Thompson, K. G. McNeill, and J. W. Jury, Phys. Rev. C **36**, 1286 (1987).

31. H. Utsunomiya et al., private communication and to be published.

32. I. N. Boboshin, A. V. Varlamov, V. V. Varlamov, D. S. Rudenko, and M. E. Stepanov, The Centre for Photonuclear Experiments Data (CDFE) nuclear data bases, http://depni.npi.msu.su/cdfe, INP Preprint 99-26/584, Moscow, 1999.

33. T. Rauscher, private communication.

34. F. Käppeler, S. Jaag, Z. Y. Bao, and G. Reffo, Astroph. J. **366**, 605 (1991).

35. A. Mengoni, private communication.

36. Z. Y. Bao, H. Beer, F. Käppeler, F. Voss, and K. Wisshak, At. Data Nucl. Data Tables **76**, 70 (2000).

37. D. Belic et al., Phys. Rev. Lett. **83**, 5242 (1999).

38. R. Gallino et al., Astroph. J. **497**, 388 (1998).

WEAK INTERACTION, GIANT RESONANCES AND NUCLEAR ASTROPHYSICS

K. LANGANKE

Institute for Physics and Astronomy, University of Aarhus, Aarhus, Denmark
E-mail: langanke@ifa.au.dk

G. MARTÍNEZ-PINEDO

Departement für Physik und Astronomie der Universität Basel, Basel, Schweiz
E-mail: martinez@quasar.physik.unibas.ch

The manuscript reviews some astrophysically important weak-interaction processes. These include electron captures, beta-decays and neutrino-induced reactions in a core-collapse supernova, and β-decays and neutrino-nucleus processes in r-process nucleosynthesis. Giant resonance response plays an essential role in all these applications.

1 Introduction

Reactions mediated by the weak interaction play decisive roles in many astrophysical problems. Often the relevant reaction rates under astrophysical conditions are dominated by giant resonance contributions. Prominent examples are the electron captures and beta-decays which occur during the presupernova phase in the core collapse of a massive star. Both processes are mainly given by Gamow-Teller (and Fermi) transitions [1] and thus, their reliable determination requires the accurate description of the Gamow-Teller strength distribution in nuclei around the iron mass range and in heavier nuclei. Another important example are neutrino-nucleus reactions which occur at several stages during a type II supernova. Related to differences in the neutrino spectra, the neutrino-nucleus reactions induced by ν_e and $\bar{\nu}_e$ neutrinos are dominated by allowed transitions, while the neutral-current reactions induced by ν_μ and ν_τ neutrinos and their antiparticles are mainly given by first-forbidden transitions.

There has been quite impressive experimental progress in studying Gamow-Teller distributions in pf-shell nuclei, e.g. [2], and the nuclear shell model appears to be able to describe the data well. Thus one now has a theoretical tool in hand which allows to calculate stellar weak interaction rates. This goal has been achieved recently and we will discuss their consequences for the core collapse in section 2. For many nuclei, the allowed contribution to the neutrino-nucleus reaction cross section can be derived on the basis of the shell model too. However, the forbidden transitions have still to be evaluated within more simplified theoretical models like the random phase approximation. Section 3 discusses the calculation of neutrino-induced reactions for core-collapse supernovae, while section 4 focusses on neutrino-induced reactions on very neutronrich nuclei relevant for the r-process. This section also discusses the status of β halflife calculations for r-process waiting-point nuclei.

2 Electron captures and beta-decays in a core-collapse supernova

A detailed description of the current core-collapse supernova picture is given in [3]. During hydrostatic oxygen and silicon burning the density and electron chemical potential in the stellar core have grown sufficiently to make electron captures energetically favorable. The onset of electron capture accelerates the collapse as it reduces the amount of electron degeneracy pressure available to counteract gravity and leads to energy losses by the emission of neutrinos. Furthermore, the matter composition, which approximately reaches nuclear statistical equilibrium conditions during silicon burning, is driven more neutronrich. The relevant captures occur on nuclei in the iron mass range which, during silicon burning, become abundant in the core. The respective stellar weak-interaction rates have been recently evaluated [4,5] on the basis of large-scale shell model calculations [6], supplemented by experimental data whereever available. The shell model electron capture rates turned out to be systematically and significantly smaller than the previous estimates based on the independent particle model and the data then available [7]. This difference leads to noticeable changes in the presupernova evolution of massive stars (until the central density reaches about 10^{10} g/cm^3), which are examplified in Fig. 1 in terms of the three decisive quantities for the collapse: i) the central electron-to-baryon ratio Y_e increases with the new rates. This suggests a larger homologous core size after neutrino trapping; ii) the iron core masses are reduced. Together with i) this implies that the shock wave has less material to traverse reducing its energy losses; iii) for stars with $M \leq 20 M_\odot$ the entropy is smaller. As a consequence, the abundance of free protons is smaller. This reduces the electron capture rate in the successive collapse stage, which is dominated by capture on free protons. Details of the presupernova studies with the shell model weak-interaction rates can be found in [8,9]. We note that these studies also show that beta-decays, an additional source for stellar cooling, become competative during a short period in silicon burning.

Although the shell model calculations agree nicely with the available data for Gamow-Teller strength distributions and halflives for nuclei in the iron mass range [6], further experimental information is asked for. The shell model calculations [4] predict that, due to pairing effects, the centroid of the GT$_+$ distribution resides at about 2 MeV (4 MeV) higher energies in the daughter

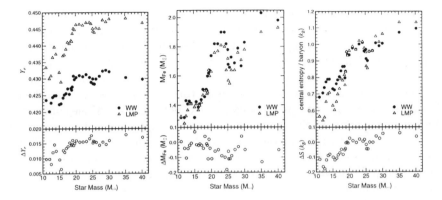

Figure 1: Comparison of the center values of Y_e (left), the iron core sizes (middle) and the central entropy (right) for $11 - 40 M_\odot$ stars for presupernova models based on the weak-interaction rates derived from the independent particle model (WW) and from the shell model (LMP)[8].

nucleus than for odd-A (even-even) nuclei. This should be experimentally verified. Furthermore these calculations also indicate a dependence of the GT_+ centroid energy on the neutron excess[4]. The presupernova models with the new shell model rates evolve along stellar trajectories where nuclei like ^{55,57}Fe and ^{54,56,58}Mn contribute significantly to the electron capture and β-decay rates. The corresponding GT distributions can be measured at future radioactive ion-beam facilities.

As electron captures drive the matter more neutronrich, the matter composition in the collapse phase, post the presupernova models, will be dominated by nuclei with neutron numbers $N > 40$ and proton numbers $Z < 40$. In the simple independent particle model, which is employed to estimate the electron capture rates in the collapse phase, GT transitions are then completely Pauli-blocked. As a consequence electron capture on nuclei ceases out in these simulations[10,11]. However, the Pauli blocking of the GT transitions will be overcome by thermal excitations[12] and correlation effects[13]. To verify the unblocking by correlations and to guide theoretical models it would be very helpful, if the GT_+ strength were determined for nuclei like ^{72,74,76}Ge.

3 Neutrino-induced processes during a supernova collapse

While the neutrinos can leave the star unhindered during the presupernova evolution, neutrino-induced reactions become more and more important during the subsequent collapse stage due to the increasing matter density and neutrino energies; the latter are of order a few MeV in the presupernova models, but increase roughly approximately to the electron chemical potential[10,14]. Elastic neutrino scattering off nuclei and inelastic scattering on electrons are the two most important neutrino-induced reactions during the collapse. The first reaction randomizes the neutrino paths out of the core and, at densities of a few 10^{11} g/cm^3, the neutrino diffusion time-scale gets larger than the collapse time; the neutrinos are trapped in the core for the rest of the contraction. Inelastic scattering off electrons thermalizes the trapped neutrinos then rather fastly with the matter and the core collapses as a homologous unit until it reaches densities slightly in excess of nuclear matter, generating a bounce and launching a shock wave which traverses through the infalling material on top of the homolgous core. In the currently favored explosion model, the shock wave is not energetic enough to explode the star, it gets stalled before reaching the outer edge of the iron core, but is then eventually revived due to energy transfer by neutrinos from the cooling remnant in the center to the matter behind the stalled shock.

Neutrino-induced reactions on nuclei, other than elastic scattering, can also play a role during the collapse and explosion phase[15]. We note that during the collapse only ν_e neutrinos are present. Thus, charged-current reactions $A(\nu_e, e^-)A'$ are strongly blocked by the large electron chemical potential[11,14]. Inelastic neutrino scattering on nuclei can compete with (ν_e, e^-) scattering at higher neutrino energies $E_\nu \geq 20$ MeV[11]. Here the cross sections are mainly dominated by first-forbidden transitions. Finite-temperature effects play an important role for inelastic $\nu + A$ scattering below $E_\nu \leq 10$ MeV. This comes about as nuclear states get thermally excited which are connected to the ground state and low-lying excited states by modestly strong GT transitions and increased phase space. As a consequence the cross sections are significantly increased for low neutrino energies at finite temperature and might be comparable to inelastic $\nu_e + e^-$ scattering[17]. Examples are shown in Fig. 2. A reliable estimate for the inelastic neutrino-nucleus cross sections requires the knowledge of the GT$_0$ strength. Shell model predictions (see Fig. 3) imply that the GT$_0$ centroid resides at excitation energies around 10 MeV and is independent of the pairing structure of the ground state[18,17]. This should be experimentally verified. Finite temperature effects become unimportant for stellar inelastic neutrino-nucleus cross sections once the neutrino energy is large enough to reach the GT$_0$ centroid, i.e. for $E_\nu \geq 10$ MeV.

The trapped ν_e neutrinos will be released from the core in a brief burst shortly after bounce. These neutrinos can interact with the infalling matter just before arrival of the shock and eventually preheat the matter requiring less energy from the shock for dissociation[15]. The relevant preheating processes

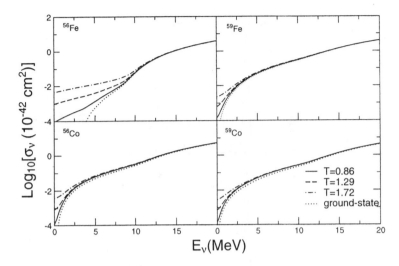

Figure 2: Cross sections for inelastic neutrino scattering on nuclei at finite temperature. The temperatures are given in MeV [17]

are charged- and neutral-current reactions on nuclei in the iron and also silicon mass range. So far, no detailed collapse simulation including preheating has been performed. The relevant cross sections can be calculated on the basis of shell model calculations for the allowed transitions and RPA studies for the forbidden transitions [18]. The main energy transfer to the matter behind the shock, however, is due to neutrino absorption on free nucleons. The efficiency of this transport depends strongly on the neutrino opacities in hot and very dense neutronrich matter [19]. It is likely also supported by convective motion, requiring multidimensional simulations [20]. Finally, elastic neutrino scattering off nucleons, mainly neutrons, also influences the efficiency of the energy transfer. A non-vanishing strange axialvector formfactor of the nucleon [21] will likely reduce the elastic neutrino-neutron cross section.

4 Weak-interaction processes in r-process nucleosynthesis

It is known that approximately half of the heavy elements with mass number $A > 70$ and all of the transuranics are formed by the process of rapid neutron capture, the r-process. The astrophysical site where the required conditions occur - neutron number densities in excess of $\sim 10^{20}$ cm^{-3} and temperatures of $\sim 10^9$ K lasting for on the order of 1 s [22] - has been a matter of speculation for almost four decades. R-process simulations indicate that the process

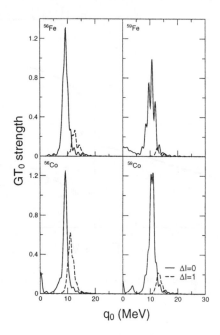

Figure 3: Distribution of the GT$_0$ strength built on the ground-state for four selected nuclei as a function of excitation energy q_0. The full (dashed) lines show the $\Delta I = 0$ ($\Delta I = 1$) contributions to the strength.

occurs approximately in $(n, \gamma) \leftrightarrow (\gamma, n)$ equilibrium implying, that for fixed neutron density and temperature, the r-process path runs along a line of about constant neutron separation energies, $S_N \sim 2 - 3$ MeV. The currently favored r-process site is the neutronized atmosphere just above the supernova core, i.e. the neutrino-driven wind model [23,24]. In this model neutron density and temperature are dependent on time. Thus also the r-process path changes with time. In particular the matter cools during the r-process as it moves further outwards in the star. This shifts the r-process path to smaller neutron separation energies and hence more neutronrich nuclei. As the peaks of the elemental r-process abundances are related to the magic neutron numbers ($N = 50, 82, 126$), more neutronrich paths will shift these peaks to smaller mass numbers [25]. Thus, in the neutrino-driven wind model the abundance peak positions reflect the competition of the speed with which the matter moves to cooler regions and the halflives of the nuclei along the r-process path, which determine how long it takes to reach the magic neutron numbers. R-process halflives are difficult to calculate as they usually depend on the rather weak tail of the GT_ distribution. Most important are the nuclei with magic neutron numbers as their halflives are long compared with non-magic nuclei and hence determine the overall time-scale of the r-process matter-flow. Fig. 4 compares halflives calculated in various nuclear models for such waiting point nuclei with magic neutron numbers $N = 82$ and 126, showing a rather unsatisfying spread between the various models. Importantly the halflives for three waiting point nuclei with $N = 82$ have been determined experimentally[26] indicating that the halflives in most models, except for the shell model calculations, are too long. For waiting point nuclei with $N = 126$ no data exist so far leaving the models untested. The situation will change dramatically once the future radioactive ion-beam facilities are operational as they will allow the determination of the halflives for basically all $N = 82$ and a few key $N = 126$ waiting point nuclei. Although the shell model calculations appear as the most reliable tool to calculate halflives, the model spaces needed for studies of non-semimagic nuclei become prohibitively large and make such calculations impossible. Thus, other models have to be improved to calculate the required halflives along the r-process path. We further note that the unknown Q_β values are another important source of uncertainty in the calculations. These Q_β values are often large enough that β-delayed neutron emission occurs. To estimate this one needs precise values for the neutron separation energies and reliable low-energy GT_ distributions. Other open questions are the importance of forbidden transitions and of isomeric states for the r-process halflives.

Due to the immense neutrino fluxes, (ν_e, e^-) reactions can compete with β-decays at rather modest distances above the neutron star (typical (ν_e, e^-) halflives are shown in Fig. 5) and speed-up the r-process matter-flow to heavier nuclei. We note, however, that the average energy of supernova ν_e neutrinos is about 12 MeV. Thus, in contrast to β-halflives, neutrino absorption on neutronrich nuclei can be rather accurately calculated as the capture is dominated by allowed transitions which are governed by sumrules (the Ikeda sumrule fixes the GT_ strength in very neutronrich nuclei as the GT_+ strength basically vanishes) and the neutrino energies are large enough to reach the IAS state and the GT_ centroids. Capture to the low-energy tail of the GT_ distribution has been estimated to introduce an uncertainty of about a factor 2 into the cross sections[29]. Current calculations adjust the energy scales by invoking the standard parametrization of the IAS state[30] which has been derived for nuclei near stability. We stress that absorption cross sections for supernova neutrinos do not pronounce magic neutron numbers[31]. Thus, the occurence of the peaks in the r-process elemental abundance distribution signals that at freeze-out β-decays have to be faster than neutrino absorption[32]. This puts constraints on the neutrino flux at the time and position of the freeze-out. These constraints

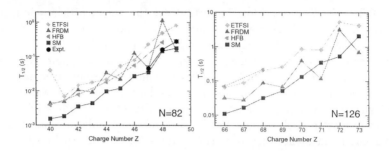

Figure 4: Comparison of halflives for r-process $N = 82$ and $N = 126$ waiting point nuclei calculated in various models (see [27,28]). The experimental data for the $N = 82$ nuclei are from Kratz *et al.* [26].

are more stringent if $\nu_e \leftrightarrow \nu_{\mu,\tau}$ oscillations occur between supernova neutrinos as the latter have larger average energies ($E_\nu \sim 25$ MeV) reducing the absorption halflives (see Fig. 5). For these higher energy neutrinos forbidden transitions contribute significantly to the absorption cross section.

In both ν_e-induced charged-current and $\nu_{\mu,\tau}$-induced neutral-current reactions the final nucleus will be in an excited state and most likely decay by the emission of one or several neutrons. If these processes occur after freeze-out the neutrons will not be recaptured and the r-process abundance is changed due to this neutrino postprocessing[33]. Neutrino postprocessing can be relevant for the peak distributions where it removes some abundance from the top of the peaks and shifts it to the wings at smaller mass numbers. Considering that the postprocessing contribution cannot be more than the observed abundance for the wing nuclides allows one to put constraints on the neutrino fluence in the neutrino-driven wind scenario[34].

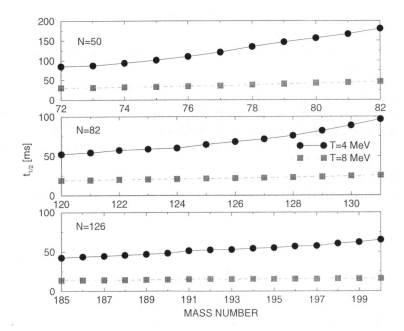

Figure 5: Halflives for (ν_e, e^-) absorption on $N = 50, 82$ and 126 waiting point nuclei. The calculations have been performed assuming a Fermi-Dirac spectrum for the neutrinos with zero chemical potential and $T = 4$ MeV (describing supernova ν_e neutrinos) and $T = 8$ MeV (assuming complete $\nu_e \leftrightarrow \nu_{\mu,\tau}$ oscillations). The luminosity was set to 10^{51} erg/s at a radius of 100 km [33].

Acknowledgement

The work has been in part supported by the Danish Research Council.

1. H.A. Bethe, G.E. Brown, J. Applegate and J.M. Lattimer, Nucl. Phys. A324 487 (1979)
2. Y. Fujita *et al.*, Phys. Lett. B365 29 (1996)
3. H.A. Bethe, Rev. Mod. Phys. 62 801 (1990)
4. K. Langanke and G. Martinez-Pinedo, Nucl. Phys. A673 481 (2000)
5. K. Langanke and G. Martinez-Pinedo, At. Data Nucl. Data Tables 79 1 (2001)
6. E. Caurier, K. Langanke, G. Martinez-Pinedo and F. Nowacki, Nucl. Phys. A653 439 (1999)
7. G.M. Fuller, W.A. Fowler and M.J. Newman, ApJS 42 447 (1980) ; 48 279 (1982); ApJ 252 715 (1982); 293 1 (1985)

8. A. Heger, K. Langanke, G. Martinez-Pinedo and S.E. Woosley, Phys. Rev. Lett. 86 1678 (2001)

9. A. Heger, S.E. Woosley, G. Martinez-Pinedo and K. Langanke, Ap.J. 560 307 (2001)

10. S.W. Bruenn, Ap.JS 58 771 (1985)

11. S.W. Bruenn and W.C. Haxton, ApJ. 376 678 (1991)

12. J. Cooperstein and J. Wambach, Nucl. Phys. A420 591 (1984)

13. K. Langanke, E. Kolbe and D.J. Dean. Phys. Rev. C63 032801 (2001)

14. K. Langanke, G. Martinez-Pinedo and J.M. Sampaio, Phys. Rev. C64 055801 (2001)

15. W.C. Haxton, Phys. Rev. Lett. 60 (1988) 1999

16. J.M. Sampaio, K. Langanke and G. Martinez-Pinedo, Phys. Lett. B511 11 (2001)

17. J.M. Sampaio, K. Langanke, G. Martinez-Pinedo and D.J. Dean, submitted to Phys. Lett.

18. J. Toivanen et al., Nucl. Phys. A694 395 (2001)

19. S. Reddy, M. Prakash, J.M. Lattimer and S.A. Pons, Phys. Rev C59 2888 (1999)

20. A. Mezzacappa, Nucl. Phys. A688 158c (2001) and references therein

21. M.J. Musolf et al., Phys. Rep. 239 1 (1994); J. Ashman et al., Nucl. Phys. B328 1 (1989)

22. G.J. Mathews and J.J. Cowan, Nature 345 491 (1990) and references therein

23. K. Takahashi, J. Witti and H.-T. Janka, Astron. Astrophys. 286 857 (1994)

24. S.E. Woosley et al., Ap.J. 433 229 (1994)

25. I.N. Borzov and S. Goriely, Phys. Rev. C62 035501 (2000)

26. K.-L.Kratz, Nucl. Phys. A688 308c (2001) and references therein

27. G. Martinez-Pinedo and K. Langanke, Phys. Rev. Lett. 83 4502 (1999)

28. G. Martinez-Pinedo, Nucl. Phys. A688 357c (2001)

29. R. Surman and J. Engel, Phys. Rev. C58 2526 (1998)

30. K. Langanke and E. Kolbe, At. Data Nucl. Data Tables, in print

31. A. Hektor, E. Kolbe, K. Langanke and J. Toivanen, Phys. Rev. C61 055803 (2000)

32. B.S. Meyer, G. McLaughlin and G.M. Fuller, Phys. Rev. C58 3696 (1998)

33. Y.-Z. Qian, W.C. Haxton, K. Langanke and P. Vogel, Phys. Rev. C55 1532 (1997)

34. W.C. Haxton, K. Langanke, Y.-Z. Qian and P. Vogel, Phys. Rev. Lett. 78 2694 (1997)

THE HIγS FACILITY – A FREE-ELECTRON LASER GENERATED GAMMA-RAY BEAM FOR NUCLEAR PHYSICS

H.R. WELLER

Triangle Universities Nuclear Laboratory, Duke University
Durham, NC 27708-0308, USA

The High Intensity Gamma Ray Source (HIγS), a collaborative project between TUNL and the Duke Free Electron Laser Laboratory at Duke University, is described. The results of some initial experiments and plans for the future research program are discussed.

1 The High Intensity γ-Ray Source (HIγS)

There are at present a number of facilities which produce polarized γ rays for nuclear physics studies. All of those facilities which employ Compton backscattering techniques operate by scattering conventional laser light from electrons circulating in a storage ring. In our scheme, however, intracavity scattering of the UV-FEL light will produce a γ-flux enhancement of approximately 10^3 over the existing sources. The Duke storage ring was designed to operate at energies from 250 MeV to about 1 GeV. The range of operating energies has been extended to between 200 MeV and 1.1 GeV as demonstrated in 1995. Minor upgrades will extend this range up to 1.2 GeV. The Duke OK-4 storage ring XUV FEL can presently produce FEL photons up to 6.4 eV, with an upgrade underway which will increase this to 12.5 eV. This will allow for the production of γ rays up to an energy of about 225 MeV having an average flux in excess of 10^7 /sec/MeV. This can be obtained with a modest beam an average-stored current of only about 100-150 mA. If necessary, the flux of γ rays can be increased by increasing the current in the target bunch and/or by operating eight electron bunches. The present OK-4 FEL will soon be upgraded to a helical undulator system (OK-5). This new system has many advantages over the present one including making switchable linear and circularly polarized beams available, an increase in power and a decrease in mirror-damaging radiation.

The Duke University Free-Electron Laser Laboratory (DFELL) is a 5000 m^2 facility which presently houses a 1.1 GeV electron storage ring and related support facilities. The major components of the HIγS facility are the OK-4 XUV FEL and the Duke 1.1 GeV electron storage ring with its 280 MeV linac injector. A diagram showing the γ-ray production scheme is presented in Fig. 1. In the case shown here, the two electron bunches contained in the ring can be thought of as a lasing bunch and a scattering bunch. The lased photons from one electron bunch are reflected from the downstream mirror, then collide head on with the second electron bunch, producing gamma rays. The OK-4 XUV FEL and its projected performance are described in other publications.[1,2]

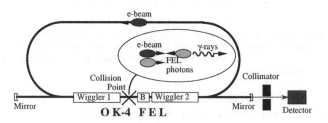

Figure 1. Schematic diagram of the HIγS – the DFELL-TUNL γ-ray facility.

There are a number of distinct advantages of this system as compared to presently available ones. The first is the high flux which is already allowing us to measure nuclear processes with low cross sections with good precision in realistic times. The second advantage is the fact that tagging is not needed. The high-quality of the electron beam permits energy definition by the use of collimation alone. This means that, unlike many tagged sources, there will be NO untagged high-energy γ rays. Another way of saying this is that since all of our γ rays are untagged, the tagging efficiency is 100%. In addition, the energy of these beams can be tuned from about 2 to 225 MeV. The energy resolution of these beams will be exceptional. For example, a 1 mm radius collimator located 30 meters from the collision point should allow us to produce 100 MeV γ rays having a FWHM energy spread of less than 400 keV. These γ-ray beams will be essentially 100% linearly polarized, and the beam environment will be exceptionally clean; backgrounds will be negligibly small.

2 The Duke 1.1 GeV Electron Storage Ring and the Booster Injector

The present system can support the HIγS operation in the "no loss" mode up to 20 MeV. Total beam intensities on the order of 10^8 γ/s are available now between 2 and 20 MeV. For γ-ray energies above 20 MeV, the electrons are knocked out of the ring. However, beams with intensities of 10^6 γ/s are presently available up to 48 MeV. Higher energy γ-ray production requires that the electrons be replaced at full-energy.

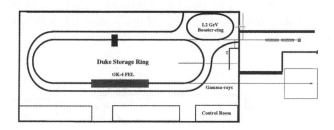

Figure 2. Proposed layout of the HIγS facility with full-energy booster injector.

The current electron source for the ring is a 280 MeV S-band radio-frequency linear accelerator. The present system consists of eleven accelerator sections and a RF gun, all driven by three klystrons. This system is adequate to support the initial stage of operation during the construction of a 1.2 GeV booster injector. The construction of the booster will not interfere with normal operation of the HIγS facility. The booster injector utilizes standard technology and will provide for efficient injection at any chosen operating energy of the storage ring from 200 MeV to 1.2 GeV. Parts of the existing linac injector will be used for injection into the booster. The transition period from the existing linac injector to the full-energy booster injector will take about two months of time once the booster is completed. The proposed layout of the HIγS facility with the full-energy booster injector is shown in Fig. 2.

The booster injector has been designed to have a maximum energy of 1.2 GeV, adequate to provide injected beams when the maximum storage ring energy is upgraded from the present 1.1 to 1.2 GeV. Operating with 1–3 Hz repetition rate, this injector will be capable of supporting any of the projected operating modes of the HIγS facility.

3   Experimental Area for Nuclear Physics

A target room having dimensions of 32 ft by 32 ft and 17 ft high, as shown in Fig. 3, has been constructed for the HIγS research program. This room was provided with all services necessary for the HIγS program, including proper radiation shielding. In designing this room, particular attention was paid to requirements associated with the installation of the LAMPF neutral meson spectrometer (NMS). The final design is compatible with the space and other physical requirements of the NMS. The new building was completed in May, 1999. The first nuclear physics experiment (photodisintegration of the deuteron near threshold) was performed in the new HIγS target room in the spring of 2000.

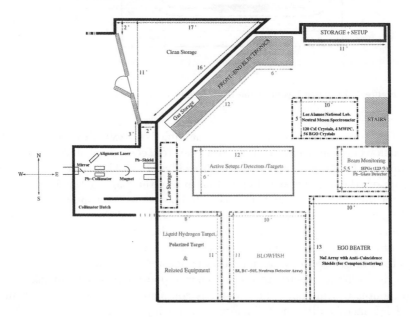

Figure 3. The HIγS target room. Present layout of various experimental setups are indicated.

A beam dump approximately 5 feet deep was constructed at the rear of the HIγS target room. The square opening of the dump is $20'' \times 20''$, and it has been filled with shielding material to contain the radiation created in the dump. A secondary flux monitor detector (a lead-glass detector) has been installed at the rear of the beam dump.

4   The Physics Program at HIγS

4.1   Studies of the nucleon

The availability of high-intensity beams of polarized γ rays below 225 MeV will allow for a series of measurements which will test some of the most basic theories of nuclear and particle physics. At present, Quantum Chromodynamics (QCD) is the most fundamental theory of the strong force.[3] However, a treatment which allows for a direct solution of the QCD Lagrangian for the nucleus has not been discovered. Recently, however, a great deal of progress has been made in understanding low-energy interactions of pions, nucleons, and photons using Chiral Perturbation Theory (CPT). Chiral Perturbation Theory is an effective field theory which exploits the chiral symmetry of QCD in order to make rigorous contact with low-energy nuclear and particle-physics phenomenology. Calculations done within this framework allow one to test whether low-energy descriptions of strong-interaction phenomena are consistent with CPT, or more generally, with the standard model. Working within this framework assures us that the results will connect to the direct solution of the QCD Lagrangian, if and when it is discovered.

In the limit of vanishing u, d and s quark masses, the QCD Lagrangian admits a global chiral symmetry: $SU(3)_L \times SU(3)_R$. This symmetry is broken spontaneously, which implies the existence of eight pseudoscalar massless Goldstone bosons. Furthermore, since the quark masses are finite (but small), these Goldstone bosons acquire a small mass and are identified with the pions, the kaons and the eta. It is the fact that the interaction of these Goldstone bosons with themselves or matter fields, e.g., the nucleons, is weak, which allows for a systematic low-energy expansion in terms of small momenta and quark masses: Chiral Perturbation Theory. Chiral Perturbation Theory is a low-energy procedure, expected to be able to deal with reactions below about 500 MeV.[4] Consequently, CPT allows us to test our understanding of the spontaneous and explicit chiral-symmetry breaking and isospin-symmetry breaking contained in QCD.

The quark masses are important input parameters in the Standard Model. The determination of the light quark masses has, however, proven to be extremely difficult. CPT offers the framework to precisely determine the quark mass ratios.[3] The currently accepted values (at a renormalization scale of 1 GeV) are:

$$m_u \approx 5 \text{ MeV and } m_d \approx 9 \text{ MeV}$$

with $m_d/m_u \approx 2$. This large ratio might lead one to expect large isospin-violating effects. These effects are, however, efficiently masked since (m_d −

$m_u)/\Lambda$ is, with $\Lambda = 1$ GeV, less than or equal to about 0.01. Note that Λ is the typical scale of the hadronic (chiral) interactions. For reactions involving only pions, effects related to the quark-mass difference $m_u - m_d$ cannot appear in leading order (G-parity). This is different in the three flavor case (e.g., in $\eta \to 3\pi$) or in the presence of nucleons. In this latter case, the HIγS facility could provide an important contribution. To be specific, pion-photo-production offers two ways of observing effects due to isospin violation. The first one consists of a precise measurement of all four photo-pion production s-wave amplitudes $\gamma p \to \pi^+ n$, $\gamma p \to \pi^0 p$, $\gamma n \to \pi^- p$, and $\gamma n \to \pi^0 n$, where the latter two cases would be observed via coherent production from the deuteron. A recent TRIUMF experiment (E643) attempted to extract the s-wave amplitude for the process $\gamma n \to \pi^- p$ from the measurement of the total and differential cross sections of the inverse reaction $\pi^- p \to \gamma n$. To be quantitative, the s-wave amplitude for the charged-particle channels should be determined to within an accuracy of 1%, and to within 5% for the neutral channels. Accurate predictions for all four channels exist [5] making use of the conventional isospin symmetric basis of three independent amplitudes. The theoretical framework of consistently including operators related to the quark-mass difference and to virtual photons is currently being developed at Bonn.[6] A determination of the (absolute) total cross section for the charged channels with a 2% accuracy appears to be sufficient to give a consistency check on the quark-mass ratio m_d/m_u extracted from mesonic processes. A more quantitative assessment of the effect is still lacking, but further study is underway.[6]

Recently, there have been two independent claims that isospin has been violated at the $\approx$7% level in medium energy πN scattering.[7] As pointed out by Bernstein,[8] a very precise determination of the phase of $\gamma p \to \pi^0 p$ below the secondary $\pi^+ n$ threshold would allow for a determination of the s-wave πN scattering length $a_{\pi N}(\pi^0 p)$ via a generalized three-channel Fermi-Watson analysis. As shown by Weinberg,[9] the difference $a_{\pi N}(\pi^0 p)$ - $a_{\pi N}(\pi^0 n)$ is very sensitive to the quark-mass difference $m_u - m_d$. A measurement of Im (E_{0+}) is equivalent to a measurement of the corresponding πN phase shift. It is important to map out this quantity in the region of the so-called unitary cusp (150–170 MeV)[8]: the discontinuity in E_{0+} which results from the fact that the $\pi^0 p$ and $\pi^+ n$ thresholds are different (a result of isospin breaking). It is clear that a measurement of Im (E_{0+}), performed using a polarized target, can be obtained using the intensity and energy resolution of the HIγS facility with an accuracy which will easily display isospin violation if it is present at the level claimed above. These important experiments can only be performed at the level of accuracy needed using a facility such as HIγS.

These experiments involve the use of polarized-proton targets, a technology at which TUNL excels.

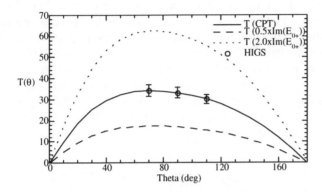

Figure 4. CPT prediction for $T(\theta)$ for a polarized proton target at $E_\gamma = 158$ MeV for various values of $\text{Im}(E_{0+})$. The data points indicate the uncertainity which is expected from the HIγS beam in 90 hours of running time.

The analyzing power for a spin-1/2 polarized target will be called $T(\theta)$. The prediction from CPT for $T(\theta)$ at $E_\gamma = 158$ MeV is shown as the solid curve in Fig. 4.[5,6] The other two curves in Fig. 4 are the predicted values for $T(\theta)$ when the $(\text{Im } E_{0+})$ amplitude is halved or doubled, respectively.

We have made a count-rate estimate for this study based on the expected γ-ray flux from the proposed facility and the modified TUNL dynamically polarized proton target. The three data points shown in Fig. 4 represent the accuracy which we would obtain if all three angles were measured simultaneously in just 90 hours of beam time. A 4π spectrometer would, of course, allow us to obtain a full angular distribution in the same amount of time.

Chiral Perturbation Theory (CPT) predicts a number of observables which the HIγS facility will be ideally suited to testing.[10] These include the polarizabilities of the nucleons.[10] The electric and the magnetic polarizabilities are defined as the first non-trivial moments in the energy expansion of the Compton scattering amplitude:

$$T(\gamma N \to \gamma N) = \frac{e^2 Z^2}{4\pi_m} + \bar{\alpha}\omega'\omega\vec{\epsilon}' \cdot \vec{\epsilon} + \bar{\beta}(\vec{\epsilon}' \times \vec{\kappa}) \cdot (\vec{\epsilon} \times \vec{\kappa}) + O(\omega^4), \quad (1)$$

where the first term is the well-known Thomson limit which is only sensitive to global parameters like the charge Z and the mass m of the particle from which the photon scatters. The non-trivial structure information is encoded in the polarizabilities $\bar{\alpha}$ and $\bar{\beta}$. In a non-relativistic picture, these quantities measure the response of a system to external electric and magnetic fields: the ability to induce an electric and a magnetic dipole moment, respectively. They have been determined to date by Compton scattering from the free proton and the quasi-free neutron (bound in deuterium) as well as by scattering slow neutrons in the field of a heavy nucleus.[11,12,13] The results have large uncertainties and depend upon the use of a sum rule for the value of $\bar{\alpha} + \bar{\beta}$ which is derived from the optical theorem. The proposed intense polarized beam will allow for a new precise determination of $\bar{\alpha}$ and $\bar{\beta}$ which is independent of the dispersion sum rule for $\bar{\alpha} + \bar{\beta}$. The equations below illustrate how Compton scattering data with polarized γ rays will provide a direct determination of $\bar{\alpha}$ and $\bar{\beta}$.

$$\left[\frac{d\sigma_\perp}{d\Omega} - \frac{d\sigma_\perp^{pt}}{d\Omega}\right]^{\frac{1}{2}} - \cos\theta \left[\frac{d\sigma_\parallel}{d\Omega} - \frac{d\sigma_\parallel^{pt}}{d\Omega}\right]^{\frac{1}{2}} = +\bar{\alpha}\sin^2\theta\left(\frac{E_\gamma}{hc}\right)^2 \qquad (2)$$

$$\cos\theta\left[\frac{d\sigma_\perp}{d\Omega} - \frac{d\sigma_\perp^{pt}}{d\Omega}\right]^{\frac{1}{2}} - \left[\frac{d\sigma_\parallel}{d\Omega} - \frac{d\sigma_\parallel^{pt}}{d\Omega}\right]^{\frac{1}{2}} = -\bar{\beta}\sin^2\theta\left(\frac{E_\gamma}{hc}\right)^2, \qquad (3)$$

where $\perp$ and $\parallel$ refer to having the photon polarization vector perpendicular or parallel to the reaction plane, respectively. σ^{pt} is the exact Born cross section for a nucleon with an anomalous magnetic moment, but no other structure.

The intensity and quality of the proposed polarized beam will allow for an order of magnitude improvement in our knowledge of these quantities. The measurement for the proton using the proposed γ-ray source was simulated using the following parameters:

- $E_\gamma = 120$ MeV (and others)

- target thickness $= 80$ mg/cm^2

- Flux $= 10^7$ γ/sec

- Running time $= 280$ hours

It was assumed that the detector had a 2π coverage in ϕ, the azimuthal angle, and covered from $20°$ to $160°$ in θ, the polar angle. The pseudo-data were divided into 5 degree bins in both θ and ϕ, thus making for a total of 2016 bins. The cross sections and asymmetries are both smoothly varying, so no smearing due to instrumental resolution was deemed necessary. It was assumed that the detection efficiency was 100% and that no background subtraction was required. The 120 CsI detectors (each of which is 30 cm long and 10 cm × 10 cm at the back face) of the NMS spectrometer, which we expect to be able to use for this experiment, can be configured to approximate these operating conditions, if not at the 100% level, then certainly to within a factor of two with respect to the product of efficiency and solid angle assumed in our simulation.

For each bin, the number of events to be "expected" was computed from L'vov's formalism[14] which contained the Born term, the pion-exchange term, and the polarizability terms. The number of events assigned to that bin was then determined using a Poisson random number generator with the expected value as the mean. The total number of events in each simulation was approximately 10^6.

The data were then fit using the L'vov calculation as the fitting function. The parameters allowed to vary independently were (1) the overall cross section, (2) $\bar{\alpha}$, and (3) $\bar{\beta}$. No sum-rule constraint was used. The fit was done using a routine that maximized the probability using Poisson statistics. Accordingly, the simulations reflect only the statistical contribution to the possible errors in the experiment. Note that $\bar{\alpha}$ and $\bar{\beta}$ are presented in units of 10^{-4} fm^3.

The plot presented in Fig. 5 shows the results for the 120 MeV simulation. The error envelope for the HIGS measurement is the 1 σ limit. It is important to note once again that this result is independent of the sum rule!

4.2 The Gerasimov-Drell-Hearn Sum Rule for the Deuteron (and ^{3}He)

Measurements of the Gerasimov-Drell-Hearn integral on the proton and neutron to test the Gerasimov-Drell-Hearn (GDH) Sum Rule have been and are being performed at Mainz, LEGS and later at GRAAL and TJNAF.[15,16] For the LEGS measurements the SPHICE target composed of molecular HD will be used, with the deuteron providing the "neutron target." The GDH sum rule for a nucleon is

$$\int_{k_\pi}^{\infty} (\sigma_p(k) - \sigma_a(k)) \frac{dk}{k} = 2\pi^2 \alpha \left(\frac{\kappa_N hc}{2\pi M_N c^2} \right)^2 , \qquad (4)$$

where k_π corresponds to the threshold energy for pion production from the nucleon, $\sigma_P (\sigma_A)$ is the total inelastic photon cross section when the nucleon and the circularly polarized photon spins are parallel (anti-parallel), and κ_N is the anomalous magnetic moment of the nucleon. Significantly, this sum rule is based upon very general principals: causality, unitarity, gauge and Lorentz invariance. In addition, it involves the reasonable assumptions that an

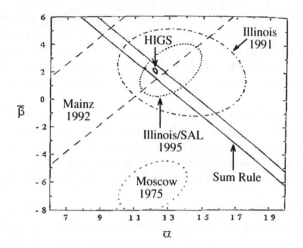

Figure 5. Error contours in the $\bar{\alpha}\bar{\beta}$ plane as determined by previous experiments, as well as a possible envelope for a similar experiment run at HIGS. The HIGS result is independent of the sum rule.

unsubtracted dispersion relation can be used in calculating the forward-angle contribution from Compton scattering and that the LETs give the correct low-energy limits. It was subsequently pointed out by Hosada and Yamamoto[17,18] and Gerasimov[16] that these arguments could be applied equally well to the deuteron. That is, the deuteron could be treated as the object of the sum rule rather than simply as a source of neutrons. The resultant "GDH" sum rule is given by

$$\int_{k_2}^{\infty} (\sigma_p(k) - \sigma_a(k)) \frac{dk}{k} = 4\pi^2 \alpha \left(\frac{\kappa_d h c}{2\pi M_d c^2} \right)^2, \tag{5}$$

where k_2 corresponds to the threshold not for pion production ($\approx$ 145 MeV) but for photodisintegration ($\approx$ 2.2 MeV), and m_d (κ_d) is the mass (anomalous magnetic moment) of the deuteron. The sum rule values for the proton, neutron, and deuteron are given in the table below:

The GDH integral for the deuteron can be separated into two pieces

$$\int_{k_2}^{\infty} GDH_d = \int_{k_2}^{k_\pi} GDH_d + \int_{k_\pi}^{\infty} GDH_d = 4\pi^2 \alpha \left(\frac{\kappa_d h c}{2\pi M_d c^2} \right)^2 = 0.6\mu\text{b}. \tag{6}$$

Target	κ	$\int GDH$
p	1.79	204.0 μb
n	-1.91	232.0 μb
d	-0.14	0.6 μb

The second term on the right hand side will be measured at LEGS and GRAAL. Its value can be estimated from the GDH integrals for the nucleons under the assumption that the impulse approximation is valid (this assumption is useful to obtain this estimate but is not significant to the basic argument since what will actually be measured at Mainz, LEGS, GRAAL, and TJNAF is the GDH integral for the deuteron). Thus,

$$\int_{k_\pi}^{\infty} GDH_d = (204\mu b) + (232\mu b) = 436\mu b. \tag{7}$$

Therefore, the previous equation predicts that

$$\int_{k_2}^{k_\pi} GDH_d = -436\mu b. \tag{8}$$

That is, if the assumptions underlying the GDH sum rules are generally valid, then asymmetries in the total cross sections dominated by high energy pion production and resonance excitation processes give a firm prediction for asymmetries in a very low-energy process, namely photodisintegration below pion threshold. This is a connection that merits testing.

The connection between the GDH sum rule for the nucleons and that for the deuteron is potentially more interesting. There are suggestions that the GDH sum rule may be violated for nucleons. If this turns out to be true, then the immediate question will be, "What is the origin of the violation?" Possible explanations include a failure of the GDH integral to converge, the need to use a subtracted dispersion relation in calculating the forward-angle Compton amplitudes, or a breakdown of the LETs. Measurements on the deuteron would help to differentiate among these possibilities. For example, if the discrepancy in the GDH sum rule for the nucleons arises from a failure of the integral to converge as E_γ approaches infinity, then one would conclude that some very short range phenomena have not been accounted for accurately, a potentially very exciting result. On the basis of the nucleon data alone one could not determine whether this was indeed the cause. However, noting that at very high energies the deuteron can be treated as a free proton plus a free neutron (with very small corrections) one can use the nucleon data to determine a "correction" to the GDH sum rule for the deuteron. If one were to find that this "corrected" sum rule was satisfied for the deuteron when integrated from photodisintegration threshold up to the same cutoff as the nucleon data, then one could reasonably infer that the source of the discrepancy for the nucleons is in the high-E_γ behavior of the integrands. Note that, since the experiments at Mainz, LEGS, GRAAL and TJNAF will use a deuterium target to measure the sum rule for the neutron, deuteron data above pion threshold will be available. These data, when combined with the lower energy data of the present proposal, will provide data over the complete energy range needed. In summary, the measurement of the GDH integral for the deuteron will be a valuable complement to measurements on the nucleons which are currently in preparation.

The cross section, total and differential at a few angles, for low-energy photodisintegration of the deuteron has been studied since the 1950's (see Arenhövel and references contained therein).[19] Very few measurements involving polarization observables have been made. Moreover, only the quantities P_y (nucleon polarization), Σ^l (photon linear polarization), and A_y (analyzing power) have been extracted. No measurement of the asymmetries contained in the GDH integrals has ever been attempted.

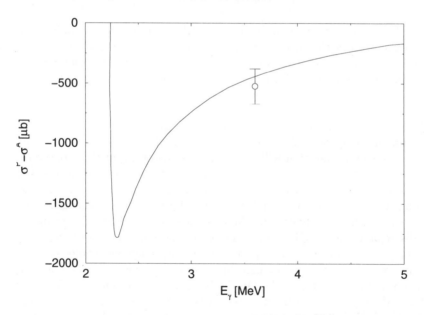

Figure 6. The theoretical prediction for $\sigma_P - \sigma_A$ [20] as a function of E_γ for the deuteron along with the preliminary result deduced from recent HIGS data.

Figure 6 shows a recent calculation by Arenhövel et al. of the integrand of the GDH integral for the deuteron.[20] The dominant contribution to the integral comes from the near threshold $^1S_0$ resonance. A measurement of the integral between threshold and $E_\gamma = 10$ MeV will be the crucial component of the total measurement. The dotted curve includes only the contribution of the nucleon-nucleon potential [N], the dash-dot curve adds meson-exchange currents to this [N+MEC], the dashed curve adds isobar currents [N+MEC+IC], and the solid curve includes relativistic contributions as well [N+MEC+IC+RC]. Little sensitivity is seen to the inclusion or omission of these individual contributions. Figure 7 shows the difference between the total cross section when the helicity of the incident photon and the spin of the target are parallel ($\bullet \sigma_P$) and that when they are antiparallel ($\bullet \sigma_A$) divided by the unpolarized total cross section. This asymmetry shows a surprisingly large sensitivity to relativistic effects, a result meriting serious investigation.

The HIGS facility will be ideally suited to the task of obtaining these measurements. In fact, since the major part of the GDH integral as well as the predicted onset of signicant relativistic effects both occur at energies below 60 MeV, these will constitute the initial program of experiments at HIGS. Intense beams having energies below 40 MeV should be available prior to the injector upgrade. The production of circularly polarized FEL photons, and hence circularly polarized γ rays, will be accomplished using the helical undulator (OK-5), which has been previously discussed.

The polarized target will be obtained by modifying an existing TUNL polarized target. Initial measurements of the asymmetries predicted in Fig. 6 will entail detecting neutrons with kinetic energies of 1 to 10 MeV or more. For these measurements we will use detectors currently available at TUNL. Measuring the GDH integral near threshold will involve the detection of neutrons with kinetic energies as low as 50 to 100 keV. Preliminary calculations indicate that a statistical accuracy of better than 5% in the GDH integral up to an energy of 50 MeV can be obtained in less than 200 hours of running.

The HIGS facility has been used to perform measurements of the photodisintegration of the deuteron in the threshold region using 100% linearly polarized γ rays on an unpolarized target. Our initial measurements were performed using four neutron detectors.[21] The analyzing power was measured at $E_\gamma = 3.58$ MeV, and used to determine the percentage M1 strength at this energy. Our results indicated that the M1 contribution to the total cross section was $9.2 \pm 1.8\%$, which is in good agreement with the theoretical prediction of Arenhövel.[20] More recently, we have made a high precision measurement of these analyzing powers for E_γ between 2.6 and 10 MeV using the 88-neutron detector array BLOWFISH. This detector system is shown in Fig. 7. The data from this experiment are currently being analyzed.

Deuteron Photodisintegration

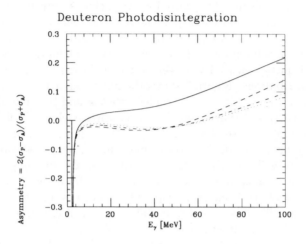

Figure 7. The asymmetry $2(\sigma_P - \sigma_A)/(\sigma_P + \sigma_A)$ in the total cross section for deuteron photodisintegration where $\sigma_P(\sigma_A)$ is the total cross section when the photon helicity and target spin are parallel (antiparallel). Dotted curve includes N-N potential only, dot-dash adds MEC, dashed adds isobar currents, and solid adds relativistic effects as well.

The term in the integrand of the GDH sum rule, $(\sigma_P - \sigma_A)$, can be written in terms of the contributing transition matrix elements. In the threshold region, there are only 5 matrix elements which are expected to contribute. Three of these are the E1 matrix elements. If we assume, as expected, that the reduced E1 matrix elements are not j-dependent, then they do not contribute to the value of $(\sigma_P - \sigma_A)$. There are also two M1 matrix elements. One of these has the same quantum numbers as the ground state, and is therefore expected to not contribute to the M1 strength due to orthogonality. This result permits us to relate the M1 strength measured in Schreiber et al.[21] to the quantity $(\sigma_P - \sigma_A)$. The resulting value of $(\sigma_P - \sigma_A)$ obtained from our experiment is in excellent agreement with the calculated value at this energy (3.58 MeV) - (see Fig. 6). It is clearly very important to extend these measurements to lower energies.

Figure 8. The BLOWFISH neutron detector array at HIGS. This array was built and commissioned for the UVA-USASK HIGS-GDH collaboration.

4.3 Nuclear-Structure Studies

The proposed high-intensity gamma-ray source at the Duke Free-Electron Laser Laboratory will present unique opportunities for the study of nuclear structure. As will be shown below, the high intensity beams of essentially 100% linearly polarized γ rays in the energy range $2 \le E_\gamma \le 10$ MeV are ideal for the development of a program to study low-spin, collective excitations of nuclei.

The study of excited nuclear states by NRF has for many years made important contributions to the understanding of nuclear structure. Recently, a great deal of excitement has followed the discovery of the so-called magnetic dipole "scissors" mode (see the review article by Kneissl[22] and references therein). Up to the present time, the majority of these studies have been performed using electron bremsstrahlung beams. While bremsstrahlung photon sources have many useful characteristics, they suffer from the serious drawback that the photon energy spectrum is continuous and has an exponentially increasing background with decreasing photon energy. It is clear that there are a number of significant advantages to NRF studies offered by the HIγS facility. In particular, the nearly monochromatic photon energies will allow the selective population of resonant states of interest with much less uncertainty due to feeding from higher-lying states. The background from nonresonant scattering will be substantially lower in the energy region where branches to lower-lying excited states would occur. Furthermore, due to the essentially 100% linear polarization of the photon beam, there will be much greater sensitivity for parity measurements.

Among the open questions that we plan to investigate using the NRF technique at the HIγS facility are:

- Magnetic dipole strength and fragmentation in odd-A nuclei.

- Deformation dependence of M1 strength in transitional even-even nuclei.

- Two-phonon dipole excitations and decay to one-phonon states, coupling with particle excitations.

- Electric dipole transitions in deformed nuclei, octupole degree of freedom.

Much of this program can be accomplished using the available TUNL detector systems. In particular, we currently have two large volume HPGe detectors (128% and 140%) along with four smaller (60%) HPGe detectors. These detectors will allow for the measurement of the resonantly scattered photons. Detector arrangements similar to those used at the Stuttgart and Darmstadt NRF facilities[22] will be used to measure the angular distribution of the scattered photons. It should be pointed out that since the γ-ray beam at HIγS is linearly polarized, it will not be necessary to measure the polarization of the outgoing photon, eliminating the need for costly and inefficient Compton polarimeter (in contrast to the above mentioned bremsstrahlung facilities).

Quite recently, we performed the first NRF experiment at HIγS, in collaboration with N. Pietralla and Z. Berant of Yale University. In this work, the target was ^{138}Ba, and the incident beam was tuned to 5.65 MeV. Four Ge detectors were used to observe the left, right, up and down γ rays. The parity of the state at 5.644 MeV had been previously (tentatively) assigned to be + on the basis of a measurement made using a bremsstrahlung beam and a Compton polarimeter. This was a difficult measurement due to backgrounds and the low-sensitivity of the Compton polarimeter at this energy. On the other hand, the HIγS result was clean and totally unambiguous. In just 8 hours of beam on target, the analyzing power was found to be -0.90 ± 0.05, clearly indicating a negative parity (positive parity would give +1.0, negative gives −1.0 (the −0.90 has not been corrected for finite geometry effects). All told, in just under 20 hours of running at HIγS, unambiguous parity assignments were made to 18 dipole excitations in ^{138}Ba between 5.3 and 6.5 MeV. This result demonstrates the powerful new tool which HIγS represents. The results of this work are being published in Physical Review Letters.[23]

In order to enhance our sensitivity for weak branches even further, we plan to construct Compton suppression shields for at least two of the HPGe detectors mentioned above. These will serve to greatly reduce the background in the measured spectra at γ-ray energies below that of the resonance to ground-state transition. The delineation of such decay pathways will shed light on questions regarding the fragmentation of collective modes and on the degree of harmonicity of presumably multi-phonon vibrational states.

4.4 Studies in Nuclear Astrophysics

As previously discussed, γ-ray fluxes greater than 10^8 photons per second are available below $E_\gamma = 20$ MeV. These fluxes are essentially 100% linearly polarized. Such intense beams of polarized γ rays open the door to new studies which can answer some of the most important questions in the field of nuclear astrophysics. As an example of this, we consider the problem of helium burning and the ^{12}C$(\alpha, \gamma)^{16}$O reaction.[24]

In order to understand the process of helium burning in stars, and in particular the oxygen-to-carbon ratio at the end of the burning, one must understand the ^{12}C$(\alpha, \gamma)^{16}$O reaction at the most effective energy for helium burning, which is 300 keV. At this energy, the cross section is estimated to be about 10^{-8} nbarn, clearly non-measurable in laboratory experiments. The cross section has been measured at various levels of precision down to 1.2 MeV. It must be extrapolated from there, down to 300 keV.[25]

One of the major uncertainties in performing the extrapolation arises from the fact that there are a number of resonances which contribute to the cross section at alpha-particle energies in the vicinity of 1 MeV. Above 1 MeV, the elastic scattering and capture reactions are dominated by a broad 1^- resonance at an excitation energy of 9.59 MeV and a narrow 2^+ state at 9.85 MeV. However, a 1^- state at 7.12 MeV, just 42 keV below threshold, determines the capture cross section in the astrophysically relevant energy region both by itself and by its interference with the higher lying 1^- and 2^+ levels. In addition, broad high-lying states and direct processes produce a coherent background which affects the energy dependence of the cross section and thereby its extrapolation.

Direct measurements of the ^{12}C$(\alpha, \gamma)^{16}$O reaction cross section at energies below 2 MeV have been attempted for over 30 years. The major difficulty encountered in these experiments is the intense neutron background which arises from the $^{13}C(\alpha, n)$ reaction which tends to swamp the γ rays from the capture reaction at these low energies. We have determined that the intense and narrow beam of γ rays which can be produced in the region of 8 to 10 MeV will be able to resolve this problem by studying the inverse reaction: $^{16}O(\gamma, alpha)^{12}C$ with γ rays at the appropriate energies.[26] It is the large flux and the small beam diameter which makes this possible. For example, if we assume that an incident gamma-ray beam having an energy of 9.58 MeV (which is on top of the first 1^- resonance), and an intensity of 5×10^9 γ/sec, then a 100 μg/cm^2 ^{16}O target will produce 6200 α-particles per day having an energy of 3.32 MeV. Likewise, a γ-ray beam at 8.8 MeV will produce α-particles at 1.6 MeV at the rate of 6 counts per day. We are proposing to make a developmental run using parameters such as these and detecting the outgoing α-particles in Silicon Strip detectors which can cover most of the 4π steradians surrounding the target. If such runs are successful, techniques to increase the count rate and extend the measurements to even lower energies can be developed.

Studies of supernovae have suggested that the ejecta of certain supernovae might be the site where the r-process nuclei are synthesized. Critical paths in synthesizing the medium mass nuclei (A < 120) are the bridges across the unstable mass gaps at A = 5 and A = 8. One of the principle bridges across this gap in forming ^{12}C from ^4He is $^4He(\alpha n, \gamma)^9Be(\alpha, n)^{12}C$.

The cross section for the $^9Be(\gamma, n)$ reaction has been measured using γ rays from two sources, either from a standard bremsstrahlung source[27,28] or from radioactive isotopes.[29,30] The shape of the data suggests that there may be a resonance near threshold energies, but the data are inadequate to determine the precise location and the value of the cross section at the resonance energy. We are proposing to measure the $^9Be(\gamma, n)$ cross section in the energy range from 1660 to 2200 keV in order to determine the cross section over the probable resonance in the near threshold region. Preliminary designs of this experiment indicate that a neutron count rate of about 100 Hz can be obtained at the highest proposed energy using a gamma flux of 10^7 γ/s with an energy spread of 1%. While encouraging, further simulations are required before a detailed proposal for this experiment can be constructed. These studies are underway.

Another case we are proposing to study is the depopulation of $^{180}Ta^m$. This odd-odd nucleus is the only isotope occurring naturally in an isomeric state, and is thought to be the rarest "stable" isotope in the universe. The stellar production mechanism of this nucleus is of considerable astrophysical interest. Recently, photodeexcitation of the isomer has been a topic of considerable interest. These studies were performed with bremsstrahlung beams, and failed to find intermediate states at energies low enough to be accessible in stellar photon baths at typical s-process temperatures. We are therefore proposing to repeat these measurements using the high flux, high resolution, gamma-ray beam of HIγS.

This brief description of some possible nuclear physics experiments is meant only to give an idea of the potential power of this beam in nuclear physics studies. At present, linearly polarized beams between 2 and 20 MeV are available with intensities as high as 10^8 γ/s. Beams with energies up to 48 MeV are available with intensities of 10^6 to 10^7 γ/s. Upgrades presently underway are expected to increase these intensities by one-to-two orders of magnitude and to provide both linear and circular polarizations by early 2003. The full-flux at energies as high as 225 MeV is scheduled to be available by mid-2005, following the commissioning of the booster-injector. Proposals from outside users are welcome.

Acknowledgments

I wish to acknowledge all members of the HIγS Collaboration, especially Dr. Vladimir Litvinenko of the Duke Free Electron Laser Laboratory. This work was partially supported by the U.S. Department of Energy under grant number DE-FG02-97ER41033.

References

1. V. Litvinenko *et al.*, "Duke Storage Ring FEL Program," SPIE Vol. 1552 (1991) 2; "UV-UV FEL Program at Duke Storage Ring with OK-4 optical klystron," IEEE PAC 1993, v.2, p. 1442.
2. V. Litvinenko *et al.*, "Commissioning of the Duke Storage Ring," Proc. of 1995 Particle Accelerator Conference, Dallas, TX, May 1995; Y. Wu *et al.*, "The Performance of the Duke FEL Storage Ring," Nucl. Instrum. Methods A 375 (1996) 74.
3. J. Gasser and H. Leutwyler, Phys. Reports C 87 (1982) 77.
4. J. Gasser and H. Leutwyler, Ann. Phys. (N.Y.) 158 (1984) 142, Nucl. Phys. B250, 465 (1985) 539.
5. V. Bernard, N. Kaiser, Ulf-G. Meißner, "Chiral Corrections to the Kroll-Rudermann Theorem," Bonn University preprint TK 96 08, 1996.
6. Ulf-G. Meißner, private communication, May (1995).
7. W.R. Gibbs *et al.*, Phy. Rev. Letts. 74 (1995) 3740; E. Martsinos, preprint ETHZ-IPP, (June 1997).

8. A.M. Bernstein, πN Newsletter 9 (1993) 55.
9. S. Weinberg, Phys. Rev. Lett. 17 (1966) 616.
10. Ulf-G. Meißner, Rep. Prog. Phys. 56 (1993) 903.
11. M. Ahmed and F. W. K. Firk, in "Polarization Phenomena in Nuclear Physics 1980," AIP, N.Y. (1981) p. 389
12. K.W. Rose et al., Nucl. Phys. A 514 (1990) 621.
13. J. Schmiedmayer et al., Phys. Rev. Lett. 66 (1991) 1015.
14. A.I. L'vov, Sov. J. Nucl. Phys. 34 (1981) 597; private communication.
15. S. D. Drell and A. C. Hearn, Phys. Rev. Lett. 16 (1966) 908.
16. S. B. Gerasimov, Sov. J. Nucl. Phys. 2 (1966) 430.
17. M. Hosada and K. Yamamoto, Prog. Theor. Phys. 36 (1966) 425.
18. M. Hosada and K. Yamamoto, Prog. Theor. Phys. 36 (1966) 426.
19. H. Arenhövel and M. Sanzone, "Photodisintegration of the Deuteron: A Review," University of Mainz Report MPKII-T-90-9, 1990.
20. H. Arenhövel et al., Nucl. Phys. A631, 612c (1998)
21. E. Schreiber et al., Phys. Rev. C61 (2000) 061604R.
22. U. Kneissl, H. H. Pitz, and A Zilges, Prog. Part. Nucl. Phys. 37 (1996) 349.
23. N. Pietralla et al., Phys. Rev. Lett. 88, 012502 (2002).
24. C.A. Barnes, Advances in Nuclear Physics, vol.4, ed. M. Baranger and E. Vogt (Plenum Press, New York, 1971), p. 133
25. X. Ji et al., Phys. Rev. C 41 (1990) 1736 and references therein.
26. M. Gai, Univ. of Connecticut, private communication.
27. M. J. Jakobson, Phys. Rev. 123, 229 (1961).
28. B. L. Berman et al., Phys. Rev. 163, 958 (1967).
29. B. Hammermesh and C. Kimball, Phys. Rev. 90, 1063 (1962).
30. M. Fujishiro et al., Can. J. Phys. 60 (1982) 1672.

PARAMETRIZATION OF
PARTON DISTRIBUTION FUNCTIONS IN NUCLEI

S. KUMANO AND M.-A. NAKAMURA

Department of Physics, Saga University, Saga, 840-8502, Japan
Email: kumanos@cc.saga-u.ac.jp, 01sm29@edu.cc.saga-u.ac.jp
URL: http://hs.phys.saga-u.ac.jp

Optimum nuclear parton distributions are investigated by analyzing high-energy nuclear reaction data. Valence-quark distributions at medium x and antiquark distributions at small x are determined by the data of F_2 structure function ratios. However, gluon distributions cannot be determined well. If Drell-Yan data are included in the analysis, the antiquark distributions are restricted in the region $x \sim 0.1$.

1 Introduction

Nuclear structure functions are in general modified from those in the nucleon. Namely, nuclei cannot be described by a simple collection of nucleons. Such nuclear medium effects have been investigated since the discovery of the EMC (European Muon Collaboration) effect in 1983. Now, many theoretical papers are published and they are summarized, for example, in Ref. 1. In order to test the proposed models and to calculate high-energy nuclear reaction cross sections accurately, it is necessary to obtain precise parton distribution functions (PDFs) in nuclei. Although there are strong groups such as CTEQ, GRV, and MRST for the PDF studies in the nucleon, it is rather unfortunate that there are only a few consistent investigations for the nuclear PDFs.[2,3]

Nuclear χ^2 analysis was first reported in Ref. 3. Because it was intended to develop a simple χ^2 analysis technique, only the deep inelastic scattering (DIS) data F_2^A/F_2^D were used. First, we explain χ^2 analysis results in comparison with the data in this paper. Then, although it is still preliminary, we discuss the research in progress by including other data such as Drell-Yan and $F_2^A/F_2^{A'}$ data. Here, A' is a nucleus other than the deuteron. In Sec. 2, a χ^2 analysis method is explained for determining the nuclear PDFs from the DIS data. Analysis results are shown in Sec. 3 in comparison with the DIS data. Computer codes are explained in Sec. 4 so that other users could use obtained nuclear distributions for their studies. The analysis results are summarized in Sec. 5.

2 Analysis method

Nuclear parton distribution functions (NPDFs) are given in an analytical form at a fixed Q^2, which is denoted as Q_0^2. They are evolved to experimental Q^2 points by the ordinary DGLAP evolution equations in order to calculate χ^2. The initial distributions are supplied with a number of parameters, which are then determined by a χ^2 analysis. Because the PDFs in the nucleon are well investigated, it is more practical to use them in the nuclear parametrization. Namely, the NPDFs f_i^A are expressed as the nucleon's PDFs f_i multiplied by weight functions w_i at Q_0^2:

$$f_i^A(x, Q_0^2) = w_i(x, A, Z) f_i(x, Q_0^2), \tag{1}$$

where i indicates u_v, d_v, $\bar{q}$, or g. Because there is no significant data for discriminating among $\bar{u}$, $\bar{d}$, and $\bar{s}$ in nuclei, flavor symmetric antiquark distributions are assumed. In the following discussions, the NPDFs are defined by those per nucleon.

Nuclear medium modification is expressed by the weight functions w_i. In order to obtain the optimum distributions from experimental data, we need to express them in terms of a set of parameters. The mass number dependence is assumed in a simple form for minimizing the parameter number. Namely, the modification part is assumed to be proportional to $1 - 1/A^{1/3}$ by considering that a nuclear cross section per nucleon is given by the volume and surface contributions:[4] $\sigma_A/A = (A\sigma_V + A^{2/3}\sigma_S)/A$. Then, the weight function is expressed by polynomials of the Bjorken variable x and $1 - x$:

$$w_i(x, A, Z) = 1 + \left(1 - \frac{1}{A^{1/3}}\right) \frac{a_i(A, Z) + b_i x + c_i x^2 + d_i x^3}{(1 - x)^{\beta_i}}, \tag{2}$$

where a_i, b_i, c_i, d_i, and β_i are parameters to be determined by a χ^2 analysis. This analysis is called a cubic-polynomial type, and the one without the $d_i x^3$ term is called a quadratic-polynomial type. Using these weight functions, we express nuclear valence u-quark, valence d-quark, antiquark, and gluon distributions as

$$u_v^A(x, Q_0^2) = w_{u_v}(x, A, Z) \frac{Z u_v(x, Q_0^2) + N d_v(x, Q_0^2)}{A},$$

$$d_v^A(x, Q_0^2) = w_{d_v}(x, A, Z) \frac{Z d_v(x, Q_0^2) + N u_v(x, Q_0^2)}{A},$$

$$\bar{q}^A(x, Q_0^2) = w_{\bar{q}}(x, A, Z) \bar{q}(x, Q_0^2),$$

$$g^A(x, Q_0^2) = w_g(x, A, Z) g(x, Q_0^2). \tag{3}$$

The valence-quark expressions are based on the consideration that nuclear distributions are mainly given by proton and neutron contributions and that the distributions in the neutron are related to those in the proton by the isospin symmetry.

There are still many parameters in the NPDFs; however, the total number could be reduced by imposing obvious constraints on nuclear charge, baryon number, and momentum:

$$\text{Charge:} \quad Z = \int dx \, \frac{A}{3} \left(2\, u_v^A - d_v^A \right), \tag{4}$$

$$\text{Baryon Number:} \quad A = \int dx \, \frac{A}{3} \left(u_v^A + d_v^A \right), \tag{5}$$

$$\text{Momentum:} \quad A = \int dx \, A \, x \left(u_v^A + d_v^A + 6\, \bar{q}^A + g^A \right). \tag{6}$$

Then, three parameters could be fixed by these conditions, and a_{u_v}, a_{d_v}, and a_g are chosen for these parameters.

The parameters are determined in Ref. 3 by the DIS experimental data for F_2^A / F_2^D. This first paper is intended to create a possible χ^2 analysis method for the nuclear parametrization, which had not been discussed at all until recently. The x and Q^2 values of the used experimental data are shown in Fig. 1. The SLAC data are located in the large x region with small Q^2. Because their data are taken for various nuclei, they are quite valuable for our nuclear parametrization.

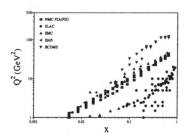

Figure 1. Kinematical range of the used experimental data is shown by x and Q^2 values.

Shadowing data are obtained by NMC and Fermilab-E665 in the small-x region. However, their Q^2 values are confined rather in the small Q^2 region, so that we expect to have difficulty in determining nuclear gluon distributions by scaling violation. In the nucleon case, the gluon distribution is determined by the HERA data at small x. Unless similar facility will become available, such a determination is not possible for nuclei.

Nuclear structure functions F_2^A are calculated in the leading order (LO) of α_s:

$$F_2^A(x, Q^2) = \sum_q e_q^2 \, x \, [q^A(x, Q^2) + \bar{q}^A(x, Q^2)], \tag{7}$$

where e_q is the quark charge, and q^A and $\bar{q}^A$ are the quark and antiquark distributions in the nucleus A, respectively. The same structure function is calculated for the deuteron, and the ratios $R = F_2^A/F_2^D$ are calculated at various experimental x and Q^2 points. Then, these ratios R^{theo} are compared with the data R^{data} to obtain the total χ^2 by

$$\chi^2 = \sum_j \frac{(R_j^{data} - R_j^{theo})^2}{(\sigma_j^{data})^2}. \tag{8}$$

It is minimized by the CERN subroutine MINUIT for obtaining the optimum set of parameters for the initial distributions.

3 Results

We explain conditions for the χ^2 analysis. First, the initial Q^2 point, namely Q_0^2, is chosen as $Q_0^2=1$ GeV^2. Because the DGLAP evolution equations are used, it is desirable to have larger Q_0^2 so that perturbative QCD is valid. On the other hand, smaller Q_0^2 is essential for using the experimental data as many as we can. In order to meet these requirements, we selected $Q_0^2=1$ GeV^2. Second, the PDFs in the nucleon should be given according to Eq. (3). As appropriate LO expressions, we take those from the MRST (Martin, Roberts, Stirling, and Thorne) parametrization in 1998.[5] The MRST distributions are conveniently defined at the initial point, $Q_0^2=1$ GeV^2. The number of flavor is taken three, and the LO analysis is done.

Obtained NPDFs are used for calculating the structure-function ratios F_2^A/F_2^D in Figs. 2–5 in order to compare with the data. Because all the fitted data cannot be shown in this limited paper, typical ratios are selected in these figures. The beryllium-deuteron ratio F_2^{Be}/F_2^D, carbon-deuteron ratio F_2^C/F_2^D, calcium-deuteron ratio F_2^{Ca}/F_2^D, and iron-deuteron ratio F_2^{Fe}/F_2^D are shown. The dashed and solid curves are the results for the quadratic and cubic analyses, respectively. They are calculated at $Q^2=5$ GeV^2, whereas the data are taken at various Q^2 points as shown in Fig. 1. Therefore, the curves cannot be compared directly with the data. However, considering the fact that Q^2 dependence is small and that the agreement between the curves and the data is good, we find that the χ^2 analyses are successful. The obtained χ^2 values are $\chi_{min}^2/d.o.f.=1.93$ and 1.82 for the quadratic and cubic analyses. It implies that the fits are not excellent; however, it is partly due to scattered experimental data as shown by these figures. For example, calcium data are scattered at small x in Fig. 4, and these deviations from the curves produce large χ^2 values.

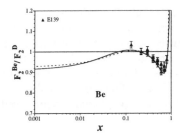

Figure 2. Comparison with the beryllium data.

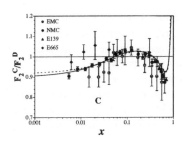

Figure 3. Comparison with the carbon data.

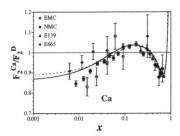

Figure 4. Comparison with calcium data.

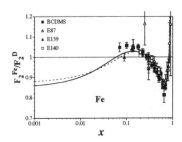

Figure 5. Comparison with iron data.

Obtained weight functions are shown in Fig. 6 for the calcium nucleus. By definition, they indicate nuclear modification effects at $Q^2=1$ GeV^2. The valence-quark distributions have depletion at medium x so as to explain the F_2 modification in this medium x region. Because the valence-quark distributions dominate the F_2 structure functions at medium and large x, they are well determined by the F_2 data in this x region. In

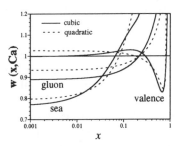

Figure 6. Weight functions for the calcium nucleus.

the similar way, the antiquark distributions dominate F_2 at small x, so that they are also well determined by the shadowing data of F_2 in the small x

region. However, the other parts are not clear at all. The valence-quark distributions at small x and the antiquark distributions at medium x are not determined from the included data. Of course, because of the conservation laws in Eqs. (4), (5), and (6), there are constraints for the x-dependent behavior. For example, the valence modification should be positive at $x \sim 0.1$ so as to cancel the negative contribution at $x \sim 0.5$. The gluon distributions are difficult to be determined at this stage although the χ^2 analyses produced the distributions in Fig. 6.

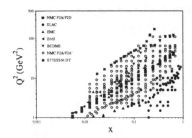

Figure 7. Kinematical range of the experimental data including Drell-Yan and $F_2^A/F_2^{A'}$.

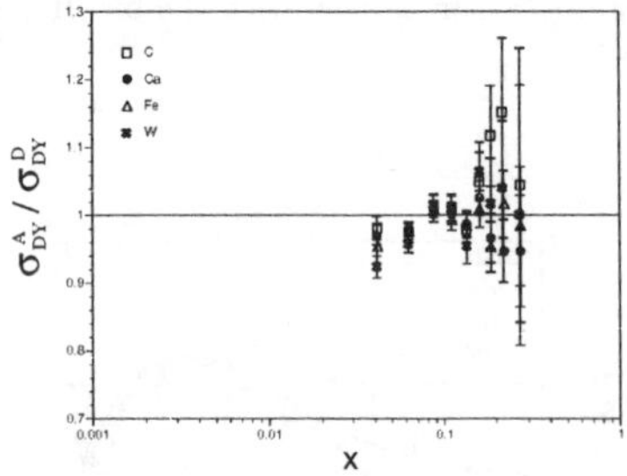

Figure 8. Drell-Yan data σ^A/σ^D.

Recently, we have been working on the determination by including other data such as those for $F_2^A/F_2^{A'}$ ($A' \neq D$) and Drell-Yan processes. Because there are many $F_2^A/F_2^{A'}$ data taken by NMC as shown in Fig. 7, they provide additional information in the χ^2 analysis. The Drell-Yan data are taken at large Q^2 as shown in Fig. 7. Some of the data are shown in Fig. 8, and they are taken mainly in the region $x \sim 0.1$. The Drell-Yan cross section ratios are roughly given by antiquark distribution ratios, so that they are very important for fixing the nuclear antiquark distributions in the region $x \sim 0.1$. If the Drell-Yan data are included in the analysis, the antiquark modification is suppressed in the region $x \sim 0.1$. Then, the bump in the ratios F_2^A/F_2^D in this x region should be explained mainly by the valence-quark modification. If such positive modification exists, the valence-quark distributions show obvious shadowing at small x due to the baryon number conservation in Eq. (5). Including these additional data, we have better determination of antiquark and gluon distributions. However, the gluon distributions are still difficult to be determined well by the χ^2 analysis. This part of the analysis is still in progress, and obtained optimum distributions will be reported in the near future.

4 Computer codes

The obtained distributions are available for other researchers, so that one can use them for one's research project. Typical applications are for heavy-ion reactions[6] and for neutrino-nucleus reactions.[7]

There are two possibilities of using our nuclear distributions. First, analytical expressions are provided at Q^2=1 GeV^2 in the appendix of Ref. 3. They should be evolved to the Q^2 point in one's project by the DGLAP equations. In this case, one should have own Q^2 evolution subroutine or one should get an evolution code from us.[8]

Second, there is an easier way for those who are not familiar with the Q^2 evolution. One could get computer codes at

http://hs.phys.saga-u.ac.jp/nuclp.html ,

for calculating the nuclear parton distributions at any x and Q^2. In principle, requested nucleus should be within the analyzed nuclei with $A = 2 \sim 208$. However, variations are rather small from $A = 208$ to nuclear matter, we expect that the codes could be also used for larger nuclei. The details are explained in the appendix of Ref. 3, and instruction should be found in the program package.

5 Summary

We have explained the results for our χ^2 analysis of the F_2 structure function ratios in order to obtain optimum parton distribution functions in nuclei. Valence-quark distributions at medium x and antiquark distributions at small x are determined from the data. However, the distributions cannot be fixed in other x regions although there are some constraints from nuclear charge, baryon number, and momentum. If the Drell-Yan data are added to the data set, they impose a constraint for the behavior of nuclear antiquark distributions in the range $x \sim 0.1$. The nuclear gluon distributions are still difficult to be determined.

Acknowledgment

S.K. was supported by the Grant-in-Aid for Scientific Research from the Japanese Ministry of Education, Culture, Sports, Science, and Technology.

References

1. For a summary, see D. F. Geesaman, K. Saito, and A. W. Thomas, *Ann. Rev. Nucl. Part. Sci.* **45**, 337 (1995).
2. K. J. Eskola, V. J. Kolhinen, and P. V. Ruuskanen, *Nucl. Phys.* **B535**, 351 (1998). For recent progress, see K. J. Eskola, H. Honkanen, V. J. Kolhinen, P. V. Ruuskanen, and C. A. Salgado, hep-ph/0110348.
3. M. Hirai, S. Kumano, and M. Miyama, *Phys. Rev.* **D64**, 034003 (2001). See http://hs.phys.saga-u.ac.jp/nuclp.html.
4. I. Sick and D. Day, *Phys. Lett.* **B274**, 16 (1992).
5. A. D. Martin, R. G. Roberts, W. J. Stirling, and R. S. Thorne, *Eur. Phys. J.* **C4**, 463 (1998).
6. Shi-yuan Li and Xin-Nian Wang, nucl-th/0110075.
7. E. A. Paschos and J. Y. Yu, hep-ph/0107261.
8. M. Miyama and S. Kumano, *Comput. Phys. Commun.* **94**, 185 (1996). See http://hs.phys.saga-u.ac.jp/program.html for the details.

ROPER ELECTROPRODUCTION AMPLITUDES IN A CHIRAL CONFINEMENT MODEL

M. FIOLHAIS, P. ALBERTO AND J. MARQUES

Departamento de Física and Centro de Física Computacional, Universidade de Coimbra, P-3004-516 Coimbra, Portugal
E-mail: tmanuel@teor.fis.uc.pt

B. GOLLI

Faculty of Education, University of Ljubljana and J. Stefan Institute, Ljubljana, Slovenia

A description of the Roper using the chiral chromodielectric model is presented and the transverse $A_{1/2}$ and the scalar $S_{1/2}$ helicity amplitudes for the electromagnetic Nucleon–Roper transition are obtained for small and moderate Q^2. The sign of the amplitudes is correct but the model predictions underestimate the data at the photon point. Our results do not indicate a change of sign in any amplitudes up to $Q^2 \sim 1$ GeV^2. The contribution of the scalar meson excitations to the Roper electroproduction is taken into account but it turns out to be small in comparison with the quark contribution. However, it is argued that mesonic excitations may play a more prominent role in higher excited states.

1 Introduction

Several properties of the nucleon and its excited states can be successfully explained in the framework of the constituent quark model (CQM), either in its non-relativistic or relativistic version. There are, however, processes where the description in terms of only valence quarks is not adequate suggesting that other degrees of freedom may be important in the description of baryons, in particular the chiral mesons. Typical examples – apart of decay processes – are electromagnetic and weak production amplitudes of the nucleon resonances. Already the production amplitudes for the lowest excited state, the Δ, indicate the important role of the pion cloud in the baryons. The other well known example is the Roper resonance, N(1440), which has been a challenge to any effective model of QCD at low or intermediate energies. Due to the relatively low excitation energy, a simple picture in which one quark populates the 2s level does not work. It has been suggested that the inclusion of explicit excitations of gluons and/or glueballs, or explicit excitations of chiral mesons may be necessary to explain its properties.

The other problem related to the CQM is the difficulty to introduce consistently the electromagnetic and the axial currents as well as the interaction with pions which is necessary to describe the leading decay modes of resonances. Such problems do not exist in relativistic quark models based on effective Lagrangians which incorporate properly the chiral symmetry. Unfortunately, several chiral models for baryons, such as the linear sigma model or various versions of the Nambu–Jona-Lasínio model, though able to describe properly the Δ resonance, are simply not suited to describe higher excited states since they do not confine: for the nucleon, the three valence quarks in the lowest s state are just bound and, for typical parameter sets, the first radial quark excitation already lies in the continuum. In order to resolve this problem, other degrees of freedom have to be introduced in the model to provide binding also at higher excitation energies. The chiral version of the chromodielectric model (CDM) seems to be particularly suitable to describe radial excitations of the nucleon since it contains the chiral mesons as well as a mechanism for confining. The CDM has been used as a model for the nucleon [1] in different approximations. Using the hedgehog coherent state approach supplemented by an angular momentum and isospin projection, several nucleon properties and of the nucleon-delta electromagnetic excitation have been obtained [1,2,3].

In the present work we concentrate on the description of the Roper resonance. Its structure and the electroproduction amplitudes have been considered in several versions of the CQM [4,5,6]. The nature of the Roper resonance has also been considered in a non-chiral version of the CDM using the RPA techniques to describe coupled vibrations of valence quarks and the background chromodielectric field [7]. The energy of the lowest excitation turned out to be 40 % lower than the pure 1s–2s excitations. A similar result was obtained by Guichon [8], using the MIT bag model and considering the Roper as a collective vibration of valence quarks and the bag.

Our description of baryons in the framework of the CDM model provides relatively simple model states which are straightforwardly used to compute the transverse and scalar helicity amplitudes for the nucleon–Roper transition, in dependence of the photon virtuality [9]. The electromagnetic probe (virtual photon) couples to charged particles, pions and quarks. However, in the CDM, baryons have got a weak pion cloud and therefore the main contribution to the electromagnetic nucleon–Roper amplitudes comes from the quarks.

In Section 2 we introduce the electromagnetic transition amplitudes. In Section 3 we briefly describe the model and construct model states representing baryons, using the angular momentum projection technique from coherent states. In Section 4 we present the CDM predictions for the helicity amplitudes for typical model parameters. Finally, in Section 5 we discuss the contribution of scalar meson vibrations.

2 Electroproduction amplitudes in chiral quark models

In chiral quark models the coupling of quarks to chiral fields is written in the form:

$$\mathcal{L}_{q-\text{meson}} = g\,\bar{q}(\hat{\sigma} + i\vec{\tau}\cdot\hat{\vec{\pi}}\gamma_5)\,q\ . \tag{1}$$

Here g is the coupling parameter related to the mass of the constituent quark $M_q = g f_\pi$. In the CDM the parameter g is substituted by the *chromodielectric* field which takes care of the quark confinement as explained in the next section. In the linear σ-model, in the CDM, as well as in different versions of the Cloudy Bag Model, the chiral meson fields, i.e. the isovector triplet of pion fields, $\vec{\pi}$, and the isoscalar σ field (not present in non-linear versions), are introduced as effective fields with their own dynamics described by the meson part of the Lagrangian:

$$\mathcal{L}_{\text{meson}} = \tfrac{1}{2}\partial_\mu\hat{\sigma}\,\partial^\mu\hat{\sigma} + \tfrac{1}{2}\partial_\mu\hat{\vec{\pi}}\cdot\partial^\mu\hat{\vec{\pi}} - \mathcal{U}(\hat{\sigma},\hat{\vec{\pi}}) \tag{2}$$

where $\mathcal{U}$ is the Mexican-hat potential describing the meson self-interaction. In different versions of the Nambu–Jona-Lasínio model [10] the chiral fields are explicitly constructed in terms of quark-antiquark excitations of the vacuum in the presence of the valence quarks.

From (2) and from the part of the Lagrangian corresponding to free quarks,

$$\mathcal{L}_q = i\bar{q}\gamma^\mu\partial_\mu q\,, \tag{3}$$

the electromagnetic current is derived as the conserved Noether current:

$$\hat{J}^\mu_{e.m.}(\boldsymbol{r}) = \bar{q}\,\gamma^\mu\left(\tfrac{1}{6} + \tfrac{1}{2}\tau_3\right)q + (\hat{\vec{\pi}}\times\partial^\mu\hat{\vec{\pi}})_3\ . \tag{4}$$

Note that the operator contains both the standard quark part as well as the pion part. We stress that in all these models the electromagnetic current operator is derived directly from the Lagrangian, hence no additional assumptions have to be introduced in the calculation of the electromagnetic amplitudes.

We can now readily write down the amplitudes for the electroexcitation of nucleon excited states in terms of the EM current (4). Let us denote by $|\tilde{N}_{M,M_T}\rangle$ and $|\tilde{R}_{J,T;M,M_T}\rangle$ the model states representing the nucleon and the resonant state, respectively (the indexes M and M_T stand for the angular momentum and isospin third components). The resonant transverse and scalar helicity amplitudes, A_λ and $S_{1/2}$ respectively, defined in the rest frame of the resonance, are

$$A_\lambda = -\zeta\sqrt{\frac{2\pi\alpha}{k_W}}\int \mathrm{d}^3\boldsymbol{r}\ \langle\tilde{R}_{J,T;\lambda,M_T}|\boldsymbol{J}_{\text{em}}(\boldsymbol{r})\cdot\boldsymbol{\epsilon}_{+1}\,e^{i\boldsymbol{k}\cdot\boldsymbol{r}}|\tilde{N}_{\lambda-1,M_T}\rangle \tag{5}$$

$$S_{1/2} = \zeta \sqrt{\frac{2\pi\alpha}{k_W}} \int d\mathbf{r} \, \langle \tilde{R}_{J,T;+\frac{1}{2},M_T} | J^0_{em}(\mathbf{r}) \, e^{i\mathbf{k}\cdot\mathbf{r}} | \tilde{N}_{+\frac{1}{2},M_T} \rangle, \tag{6}$$

where $\alpha = \frac{e^2}{4\pi} = \frac{1}{137}$ is the fine-structure constant, the unit vector $\boldsymbol{\epsilon}_{+1}$ is the polarization vector of the electromagnetic field, $k_W = (M_R^2 - M_N^2)/2M_R$ is the photon energy at the photon point (introduced rather than ω which vanishes at $Q^2 = M_R^2 - M_N^2$) and ζ is the sign of the $N\pi$ decay amplitude. This sign has to be explicitly calculated within the model; from (1) in our case. In the case of the Δ resonance ($T = J = \frac{3}{2}$), λ takes two values $\lambda = \frac{3}{2}$ and $\frac{1}{2}$, while for the Roper state ($T = J = \frac{1}{2}$), only one transverse amplitude exists ($\lambda = \frac{1}{2}$).

The photon four momentum is $q^\mu(\omega, \mathbf{k})$ and we define $Q^2 = -q_\mu q^\mu$. In the chosen reference frame the following kinematical relations hold:

$$\omega = \frac{M_R^2 - M_N^2 - Q^2}{2M_R}; \qquad k^2 \equiv k^2 = \left[\frac{M_R^2 + M_N^2 + Q^2}{2M_R} \right]^2 - M_N^2. \tag{7}$$

The electroexcitation amplitudes for the Δ resonance have been analyzed in the framework of chiral quark models [3,11,12]. They are dominated by the M1 transition but contain also rather sizable quadrupole contributions E2 and C2. The CQM model predicts here too low values for the M1 piece and almost negligible values for the quadrupole amplitudes. In chiral quark models there is a considerable contribution from the pions (i.e. from the second term in (4)): up to 50 % in the M1 amplitude, and they dominate the E2 and C2 pieces. The absolute values of the amplitudes and their behavior as a function of the photon virtuality Q^2 is well reproduced in the linear σ-model. Though the ratios E2/M1 and C2/M1 are also well reproduced in the CDM, this model gives systematically too low values for the amplitudes, which could be attributed to its rather weak pion cloud. As we shall see in Section 4, this might also explain the small values of the Roper production amplitudes at low Q^2.

In the next section we construct the states $|\tilde{N}\rangle$ and $|\tilde{R}\rangle$ for the Roper in the framework of the CDM and, in Section 4, we present the model predictions for the amplitudes.

3 Baryons in the CDM

The Lagrangian of the CDM contains, apart of the chiral meson fields σ and π, the cromodielectric field χ such that the quark meson-interaction (see (1)) is modified as:

$$\mathcal{L}_{q-meson} = \frac{g}{\chi} \bar{q} (\hat{\sigma} + i\vec{\tau} \cdot \hat{\vec{\pi}}\gamma_5) q. \tag{8}$$

The idea behind the introduction of the χ field is that it acquires a nonzero expectation value inside the baryon but goes to 0 for larger distances from the center of the baryon, thus pushing the effective constituent quark mass to infinity outside the baryon. In addition, the Lagrangian contains kinetic and potential pieces for the χ-field:

$$\mathcal{L}_\chi = \tfrac{1}{2}\partial_\mu\hat{\chi}\,\partial^\mu\hat{\chi} - \frac{1}{2}M_\chi^2\,\hat{\chi}^2\,, \qquad (9)$$

where M_χ is the χ mass. In this work we consider only a simple quadratic potential; other versions of the CDM assume more complicated forms, namely quartic potentials.

The free parameters of the model have been chosen by requiring that the calculated static properties of the nucleon agree best with the experimental values [2]. In the version of the CDM with a quadratic potential, the results are predominantly sensitive to the quantity $G = \sqrt{gM_\chi}$; we take $G = 0.2$ GeV (and $g = 0.03$ GeV). The model contains other parameters: the pion decay constant, $f_\pi = 0.093$ GeV, the pion mass, $m_\pi = 0.14$ GeV, and the sigma mass, which we take in the range $0.7 \leq m_\sigma \leq 1.2$ GeV.

The nucleon is constructed by placing three valence quarks in the lowest s-state, i.e., the quark source can be written as $(1s)^3$. For the Roper the quark source is $(1s)^2(2s)^1$, i.e. one of the three quarks now occupies the first (radially) excited state. The quarks are surrounded by a cloud of pions, sigma mesons and chi field, described by radial profiles $\phi(r)$, $\sigma(r)$ and $\chi(r)$ respectively. The hedgehog ansatz is assumed for the quarks and pions. The quark profiles (described in terms of the upper, u, and the lower component, v) and boson profiles are determined self-consistently.

Because of the hedgehog structure, the solution is neither an angular momentum eigenstate nor an isospin eigenstate, and therefore it cannot be related directly with a physical baryon. However, the physical states can be obtained from the hedgehog by first interpreting the solution as a coherent state of three types of bosons and then performing the Peierls-Yoccoz projection [1,13]:

$$|N_{\frac{1}{2},M_T}\rangle = \mathcal{N}\, P^{\frac{1}{2}}_{\frac{1}{2},-M_T}|Hh\rangle\,, \quad |R'_{\frac{1}{2},M_T}\rangle = \mathcal{N}'\, P^{\frac{1}{2}}_{\frac{1}{2},-M_T}|Hh^*\rangle\,, \qquad (10)$$

where P is the projector and we introduced the symbol $*$ to denote the Roper intrinsic state. Because of their trivial tensor nature, the χ and the σ-fields are not affected by projection. This approach can be considerably improved by determining the radial profiles $\phi(r)$, $\sigma(r)$ and $\chi(r)$, as well as the quark profiles, using the variation after projection method [1], separately for the nucleon and for the Roper.

260

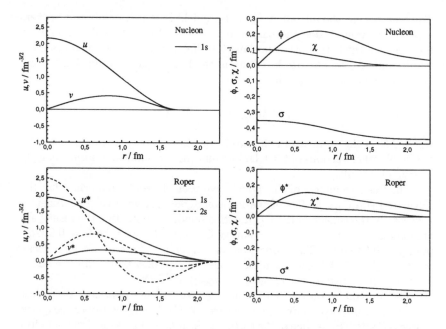

Figure 1. Quark and meson radial wave functions for the $(1s)^3$ (Nucleon) and $(1s)^2(2s)^1$ (Roper) configurations. The vacuum expectation value of the sigma field is $-f_\pi$. Note that the effective quark mass is proportional to the *inverse* of the χ field. We use the symbol * to denote the Roper radial functions. The model parameters are: $M_\chi = 1.4$ GeV, $g = 0.03$ GeV, $m_\pi = 0.14$ GeV, $f_\pi = 0.093$ GeV, $m_\sigma = 0.85$ GeV.

Figure 1 shows the radial profiles for the $(1s)^3$ and $(1s)^2(2s)^1$ configurations. Those corresponding to the Roper extend further. The strength of the chiral mesons is reduced in the Roper in comparison with the nucleon. Another interesting feature is the waving shape acquired by the Roper chromodielectric field, χ^*. A central point in our treatment of the Roper is the freedom of the chromodielectric profile, as well as of the chiral meson profiles, to adapt to a $(1s)^2(2s)^1$ configuration. Therefore, quarks in the Roper experience meson fields which are different from the meson fields felt by the quarks in the nucleon. As a consequence, states (10) are normalized but not mutually orthogonal. They can be orthogonalized taking

$$|R\rangle = \frac{1}{\sqrt{1-c^2}}(|R'\rangle - c|N\rangle), \qquad c = \langle N|R'\rangle. \qquad (11)$$

A better procedure results from a diagonalization of the Hamiltonian in the subspace spanned by (non-orthogonal) $|R'\rangle$ and $|N\rangle$:

$$|\tilde{R}\rangle = c_R^R |R'\rangle + c_N^R |N\rangle, \quad |\tilde{N}\rangle = c_R^N |R'\rangle + c_N^N |N\rangle. \tag{12}$$

In Table 1 the nucleon energies and the nucleon-Roper mass splitting are given. The absolute value of the nucleon energy is above the experimental value but it is known [2] that the removal of the center-of-mass motion will lower those values by some 300 MeV (similar correction applies to the Roper). On the other hand, the nucleon-Roper splitting is small, even in the case of the improved state (12). The smallness of the spitting is probably related with a much too soft way of imposing confinement.

Table 1. Nucleon energies and nucleon-Roper splittings for two sigma masses. E_N is the energy of the nucleon state (10), ΔE was obtained using (11), $\tilde{E}$ and $\Delta \tilde{E}$ are calculated using the states (12). The other model parameters are in the caption of Figure 1. All values are in MeV.

m_σ	E_N	ΔE	$\tilde{E}_N$	$\Delta \tilde{E}$
700	1249	367	1235	396
1200	1269	354	1256	380

4 Amplitudes

Our results for the transverse helicity amplitudes are shown in Figure 2 for the parameter set used for Figure 1. The experimental values at the photon point

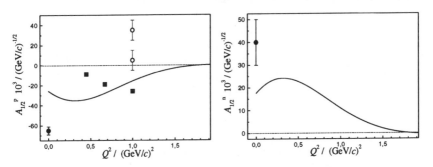

Figure 2. Nucleon-Roper transverse amplitudes. The experimental points at $Q^2 = 0$ GeV^2 are the estimates of the PDG [14]. The solid squares [15] and the open circles [16] result from the analysis of electroproduction data.

are the PDG most recent estimate [14] $A_{1/2}^p = -0.065 \pm 0.004 \ (\text{GeV}/c)^{-1/2}$ and $A_{1/2}^n = 0.040 \pm 0.010 \ (\text{GeV}/c)^{-1/2}$. The pion contribution to the charged states only accounts for a few percent of the total amplitude. The discrepancies at the photon point can be attributed to a too weak pion field, which we already noticed in the calculation of nucleon magnetic moments [2] and of the electroproduction of the Δ [3]. Other chiral models [12] predict a stronger pion contribution which enhances the value of the amplitudes. If we calculate perturbatively the leading pion contribution we also find a strong enhancement at the photon point; however, when we properly orthogonalize the state with respect to the nucleon, this contribution almost disappears.

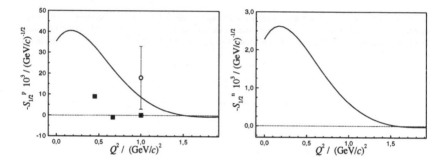

Figure 3. Nucleon–Roper scalar helicity amplitudes (see also caption of Figure 2).

In Figure 3 we present the scalar amplitudes. For the neutron no data are available which prevents any judgment of the quality of our results.

In CQM calculations [4,5,6] which incorporate a consistent relativistic treatment of quark dynamics, the amplitudes change the sign around $Q^2 \sim 0.2$–$0.5 \ (\text{GeV}/c)^2$. The amplitudes with this opposite sign remain large at relatively high Q^2, though, as shown in [5,6], the behavior at high Q^2 can be substantially reduced if either corrections beyond the simple Gaussian-like ansatz or pionic degrees of freedom are included in the model. Other models, in particular those including exotic (gluon) states, do not predict this type of behavior [17]. The present experimental situation is unclear. Our model, similarly as other chiral models [12,18], predicts the correct sign at the photon point, while it does not predict the change of the sign at low Q^2. Let us also note that with the inclusion of a phenomenological three-quark interaction Cano et al. [6] shift the change of the sign to $Q \sim 1 \ (\text{GeV}/c)^2$ beyond which, in our opinion, predictions of low energy models become questionable anyway.

5 Meson excitations

The ansatz (10) for the Roper represents the breathing mode of the three valence quarks with the fields adapting to the change of the source. There is another possible type of excitation in which the quarks remain in the ground state while the χ-field and/or the σ-field oscillate. The eigenmodes of such vibrational states are determined by quantizing small oscillations of the scalar bosons around their expectation values in the ground state [19]. We have found that the effective potential for such modes is *repulsive* for the χ-field and *attractive* for the σ-field. This means that there are no glueball excitations in which the quarks would act as spectators: the χ- field oscillates only together with the quark field. On the other hand, the effective σ-meson potential supports at least one bound state with the energy ε_1 of typically 100 MeV below the σ-meson mass.

We can now extend the ansatz (11) by introducing

$$|R^*\rangle = c_1|R\rangle + c_2\tilde{a}_\sigma^\dagger|N\rangle \,, \tag{13}$$

where $\tilde{a}_\sigma^\dagger$ is the creation operator for this lowest vibrational mode. The coefficients c_i and the energy are determined by solving the (generalized) eigenvalue problems in the 2×2 subspace. The lowest energy solution is the Roper while its orthogonal combination could be attributed to the $N(1710)$, provided the σ-meson mass is sufficiently small. In such a case the latter state is described as predominantly the σ-meson vibrational mode rather than the second radial excitation of quarks. This would manifest in very small production amplitudes since mostly the scalar fields are excited.

The presence of σ-meson vibrations is consistent with the recent phase shift analysis by Krehl at al. [20] who found that the resonant behavior in the P_{11} channel can be explained solely through the coupling to the σ-N channel. In our view, radial excitations of quarks are needed in order to explain relatively large electroproduction amplitudes, which would indicate that the σ-N channel couples to all nucleon $\frac{1}{2}^+$ excitations rather than be concentrated in the Roper resonance alone.

This work was supported by FCT (POCTI/FEDER), Portugal, and by The Ministry of Science and Education of Slovenia. MF acknowledges a grant from GTAE (Lisbon), which made possible his participation in EMI2001.

References

1. M. C. Birse, *Prog. Part. Nucl. Phys.* **25**, 1 (1990); T. Neuber, M. Fiolhais, K. Goeke and J. N. Urbano, *Nucl. Phys.* A **560**, 909 (1993)

2. A. Drago, M. Fiolhais and U. Tambini, *Nucl. Phys. A* **609**, 488 (1996)
3. M. Fiolhais, B. Golli and S. Širca, *Phys. Lett. B* **373**, 229 (1996); L. Amoreira, P. Alberto and M. Fiolhais, *Phys. Rev. C* **62**, 045202 (2000);
4. S. Capstick, *Phys. Rev. D* **46**, 2864 (1992); S. Capstick and B.D. Keister, *Phys. Rev. D* **51**, 3598 (1995)
5. F. Cardarelli, E. Pace, G. Salmè and S. Simula, *Phys. Lett. B* **397**, 13 (1997)
6. F. Cano and P. González, *Phys. Lett. B* **431**, 270 (1998)
7. W. Broniowski, T. D. Cohen and M. K. Banerjee, *Phys. Lett. B* **187**, 229 (1987)
8. P. A. M. Guichon, *Phys. Lett. B* **163**, 221 (1985); *Phys. Lett. B* **164**, 361 (1985)
9. P. Alberto, M. Fiolhais, B. Golli and J. Marques, hep-ph/0103171, *Phys. Lett. B* **523**, 273 (2001)
10. R. Alkofer, H. Reinhardt and H. Weigel, *Phys. Rep.* **265** (1996) 139; C. V. Christov, A. Blotz, H.-C. Kim, P. V. Pobylitsa, T. Watabe, Th. Meissner, E. Ruiz Arriola and K. Goeke, *Prog. Part. Nucl. Phys.* **37** (1996) 1; B. Golli, W. Broniowski and G. Ripka, *Phys. Lett.* **B437** (1998) 24; B. Golli, W. Broniowski and G. Ripka, hep-ph/0107139
11. A. Silva, D. Urbano, T. Watabe, M. Fiolhais and K. Goeke, *Nucl. Phys. A* **675**, 637 (2000) ; D. Urbano, A. Silva, M. Fiolhais, T. Watabe and K. Goeke, *Prog. Part. Nucl. Phys.* **44**, 211 (2000)
12. K. Bermuth, D. Drechsel and L. Tiator, *Phys. Rev. D* **37**, 89 (1988)
13. B. Golli and M. Rosina, *Phys. Lett. B* **165**, 347 (1985); M. C. Birse, *Phys. Rev. D* **33**, 1934 (1986)
14. D. E. Groom et al. (Particle Data Group), *Eur. Phys. J. C* **15**, 1 (2000)
15. C. Gerhardt, *Z. Phys. C* **4**, 311 (1980)
16. B. Boden and G. Krosen, in Proc. of the Conference on Research Program at CEBAF II, eds. V. Burkert et al., CEBAF (USA), 1986
17. Zhenping Li, V. Burkert and Zhujun Li, *Phys. Rev. D* **46**, 70 (1992); E. Carlson and N. C. Mukhopadhyay, *Phys. Rev. Lett.* **67** 3745 (1991)
18. Y. B. Dong, K. Shimizu, A. Faessler and A. J. Buchmann, *Phys. Rev. C* **60**, 035203 (1999)
19. B. Golli, P. Alberto and M. Fiolhais, talk presented at the Mini-Workshop on *Few body problems in hadronic and atomic physics*, Bled, Slovenia, 7-14 July 2001, hep-ph/0111399
20. O. Krehl, C. Hanhart, S. Krewald and J. Speth, *Phys. Rev. C* **62**, 025207 (2000)

S₁₁(1535) RESONANCE IN NUCLEAR MEDIUM OBSERVED WITH THE (γ, η) REACTIONS

H. YAMAZAKI, T. KINOSHITA, K. HIROTA , K. KINO, T. NAKABAYASHI,
T. KATSUYAMA, A. KATOH, T. TERASAWA, H. SHIMIZU, J. KASAGI

Laboratory of Nuclear Science, Tohoku University, Mikamine, Taihaku-ku, Sendai 982-0826, JAPAN

T. TAKAHASHI, H. KANDA, K. MAEDA

Department of Physics, Tohoku University, Aramaki, Aoba-ku, Sendai 980-8578, JAPAN

Y. TAJIMA, H. Y. YOSHIDA, T. NOMA, Y. ARUGA, A. IIJIMA, Y. ITO,
T. FUJINOYA

Department of Physics, Yamagata University, Kojirakawa-machi, Yamagata 990-8560, JAPAN

T. YORITA

Research Center for Nuclear Physics, Osaka University, Mihogaoka, Ibaraki, Osaka 567-0047, JAPAN

O. KONNO

Ichinoseki National College of Technology, Hagisho, Ichinoseki, Iwate 021-8511, JAPAN

In order to study the S₁₁(1535) resonance in the nuclear medium, total cross sections of the (γ, η) reaction on C have been measured for photon energies between 620 and 1100 MeV. This is the first result by using STB Tagger and SCISSORS in LNS, Sendai. Model calculations based on the quantum molecular dynamics (QMD) have been performed. The comparison of the calculation with the total cross section of the η photoproduction suggests that the resonance width of S₁₁(1535) in nuclei is more than 80 MeV broader than the natural width.

1 Introduction

The chiral symmetry with its spontaneous breakdown is one of the most important concepts in the dynamics of strong interaction. It plays a very important role in understanding the properties of hadrons in low energy region. But there has been few theoretical works about the chiral structure of nucleon. DeTar and Kunihiro studied positive and negative parity nucleons under the framework of the linear sigma model [1]. In the chiral restored phase, it will be naively expected that the nucleons becomes massless. However their work showed that the presence of positive and negative parity nucleons made it possible to have the finite nucleon masses without destroying chiral symmetry. Recently Jido and his collaborators studied hard about the chiral structure of nucleons [2,3]. They predicted the transition of the masses and couplings of the negative and positive parity nucleons in nuclear medium. The study of the property of positive and negative parity nucleons in nuclear medium makes it possible to obtain the new information about the chiral structure of nucleons.

The $S_{11}(1535)$ resonance is one of the most considerable candidates of the negative parity nucleons. We have selected the (γ, η) reaction to study the property of the $S_{11}(1535)$ resonance, which strongly couples with $N\eta$ and decays to the $N\eta$ channel with the branching ratio of 45 ∼ 55 %. There is no nucleon resonance which has such a large branching ratio to η-N decay around this energy region. Because of this unique property, one can excite only the S_{11} resonance by using (γ, η) reactions with GeV tagged photons. Thus we have measured the total (γ, η) cross sections on C at KEK-Tanashi using 1.3 GeV electron synchrotron to investigate the property of the $S_{11}(1535)$ resonance in nuclei [4,5]. The conclusion of our experiment was that the (γ, η) reaction cross section can be basically explained by the known effects; Fermi motion, Pauli blocking, η absorption and modest collision broadening. But there was some discrepancy around cross section maximum between the experimental data and the calculation. This discrepancy suggested that the resonance width of S_{11} might be changed in the nuclear medium; more than 60 MeV increased. Thus, more detailed studies were performed at Laboratory of Nuclear Science (LNS), Tohoku University, Sendai.

2   Experiment

The experiment was carried out using a tagged photon beam line in 1.2 GeV electron synchrotron called STretcher-Booster-ring (STB) at Laboratory for Nuclear Science, Tohoku University. Quasi-monochromatic photons were produced from the photon tagging system in the STB-ring (STB Tagger), which consists of the tagging counters and carbon fiber radiator installed in the electron center orbit of the STB ring. The injected electrons from the linac with 200 MeV of an incident energy were accelerated up to 1.2 GeV in about 1.2 second and stored in the STB-ring. After the acceleration the photon production target(radiator), which is a carbon fiber with 11 μm of a diameter, were inserted into electron beam orbit. The position of the radiator was controlled automatically so as to keep the photon flux to be about 3×10^6 tagged photons par one second. The flat top period was about 30 seconds.

Fig. 1 shows the schematic drawing of the experimental setup. The

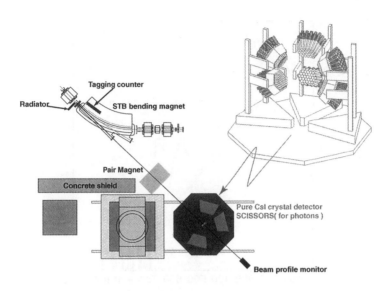

Figure 1. The experimental equipments of the (γ, η) experiment at LNS. Tagged photon beam bombarded targets and produced the η mesons. Two decaying photons from the η mesons were detected the pure CsI calorimeters.

momentum of the electron which is recoiled by the emission of the Bremsstrahlung photon is measured by the analyzer magnet and the photon tagging counters. As shown, a bending magnet of STB-ring is used as an analyzer. Tagging counters consist of the 50 ch finger counters and the 12 ch backup counters made of the plastic scintillator. Scintillation light is transmitted into the photomuliplier tube (HAMAMATSU H6524-01) through the optical fibers of 3.5 m long. The energy range of the photon tagging system is from 0.8 GeV to 1.1 GeV for the 1.2 GeV of the electron energy. The other photon energy range from 0.68 to 0.85 GeV is also obtained with the electron energy of 0.92 GeV. Thus, the STB Tagger is able to provide the tagged photon beam with its energy range from 0.62 GeV to 1.1 GeV by using two electron energies.

The tagged photon beam bombards the nuclear targets and produces η mesons. Two γ-decay of η meson is detected by a calorimeter system called SCISSORS (Sendai CsI Scintillator System On Radiation Search), which is also shown in Fig. 1. Photons were detected by the six sets of pure CsI

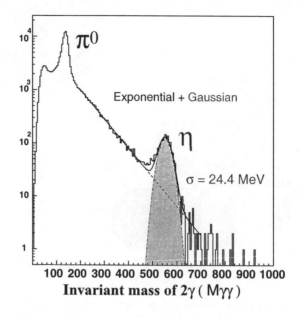

Figure 2. The invariant mass spectrum of two γ events.

calorimeters with the thin plastic scintillators for charged veto, which were placed in front of the calorimeters. Four sets of calorimeters consist of the 37 pure CsI crystals, and two sets of 29 pure CsI crystals. They cover about 1.4 sr. of solid angle. The energy of the incident photon is determined by the sum of the light output in a cluster. The weighted average of the center positions is regarded as the incident position of the photon. The η photoproduction event is identified by using the invariant mass analysis of the two decaying photons.

Fig. 2 shows an invariant mass spectrum measured in the $^{12}C(\gamma, \eta)$ reaction. The yields of the (γ, η) events are deduced from the invariant mass spectrum by subtracting the background as the exponential function as displayed in Fig 2. The invariant mass resolution is about 24 MeV/c^2. Detection efficiency is calculated by the simulation in which the quasi free production process is assumed. The total cross section of η photoproduction on nuclei is deduced by using the yields and detection efficiencies.

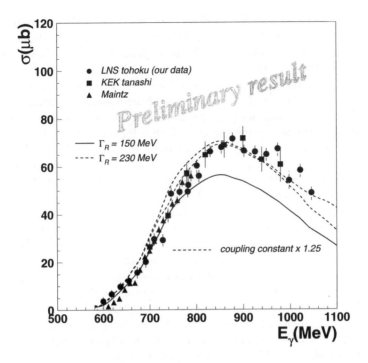

Figure 3. Total cross sections of (γ, η) reactions on C (closed circles) with the QMD calculations (thin dotted line) for the resonance width of 212 MeV, calculation on C for the width of 150 MeV (solid line) and multiplied by 1.25 (thick dotted line). Triangles and squares show the results on C measured in Mainz and KEK(Tanashi), respectively.

3 Results and discussion

In Figure 3, the total η photoproduction cross sections are shown as a function of incident photon energy. Closed circles show the present results of the (γ, η) reaction cross section on C using the STB Tagger. Closed triangles and closed squares correspond to the total cross section of the (γ, η) reaction on C measured in Mainz [6] and KEK(Tanashi) respectively. It should be noticed that present cross sections have the similar dependence on incident photon energy to previous data and better statistics above 900 MeV of incident photon energy. The cross section shows the maximum value at around 900 MeV of photon energy; this value is much higher than naively expected in the quasi free picture. The width of the resonance structure is about 300 MeV, which is also wider than expected. We have performed the calculation based on the Quantum Molecular Dynamics(QMD) to take into account the known nuclear medium effect; Fermi motion, Pauli blocking, η absorption and collision broadening of the N*. QMD was originally developed to describe the high energy heavy ion reactions [7]. Details of the photoreaction in QMD calculation are described in reference [4]. Two sets of the resonance parameter of the $S_{11}(1535)$ resonance were used in the QMD calculations. One parameter set is 1540 MeV for the resonance energy, 150 MeV for the width at the resonance pole and 0.55 for the branching ratio of the η-N channel. These values were determined so as to reproduce the H(γ, η) cross section in references [8,9,10]. The other is 1544 MeV for the resonance energy, 230 MeV for the width at the resonance pole, which were fixed to reproduce our experimental results.

The results of the QMD calculation are plotted in Fig 3. The η absorption cross section, which was incorporated in the calculation, was deduced from the detailed balance analysis of the $\pi^- p \to \eta n$ cross section. The collision broadening of the $S_{11}(1535)$ in a nucleus have been also considered; the cross section given in ref. [11] is used. A solid line is the result on C with the resonance width of 150 MeV. The thin dotted line is the result with 230 MeV. As can be seen in Fig 3, the former parameterization underestimates the measured cross section on C by about 20 μb above 800 MeV of incident photon energy. The result with a 230 MeV resonance width reproduces the experiment well. This suggests that about 80 MeV broadening of the resonance width is needed to reproduce the experiment. In Fig 3, the thick dotted line corresponds to the values of the solid line multiplied by 1.25. These two dotted lines cannot be distinguished each other. Increasing helicity amplitude by 12 % or $S_{11} \to N\eta$ branching ratio by 25 % might cause this 25 % enhancement of the cross section. In conclusion, some properties of the $S_{11}(1535)$ resonance might be changed in nuclear medium; the width, the helicity amplitude or/and decay branch.

References

1. DeTar and T. Kunihiro, *Phys. Rev.* D **39**, 2805 (1989).
2. Hungchong Kim *et al.*, *Nucl. Phys.* A **640**, 77 (1998)
3. D. Jido *et al.*, *Nucl. Phys.* A **671**, 471 (2000).
4. T. Yorita *et al.*, *Phys. Lett.* B **476**, 226 (2000).
5. H. Yamazaki *et al.*, *Nucl. Phys.* A **670**, 202 (2000).
6. M. Robig-Landau *et al.*, *Phys. Lett.* B **373**, 45 (1996).
7. K. Niita *et al.*, *Phys. Rev.* C **52**, 2620 (1995).
8. B. Krusche *et al.*, *Phys. Rev. Lett.* **74**, 3736 (1995).
9. B.H. Schoch, *Prog. Part. Nucl. Phys.* **34**, 43 (1995).
10. S. Homma *et al.*, *J. Phys. Soc. Jpn.* **57**, 828 (1988).
11. M. Effenberger *et al.*, *Nucl. Phys.* A **613**, 353 (1997).

FINAL STATE INTERACTIONS IN ω PHOTOPRODUCTION NEAR THRESHOLD

YONGSEOK OH

Institute of Physics and Applied Physics, Yonsei University, Seoul 120-749, Korea
E-mail: yoh@phya.yonsei.ac.kr

T.-S. H. LEE

Physics Division, Argonne National Laboratory, Argonne, Illinois 60439, U.S.A.
E-mail: lee@theory.phy.anl.gov

Vector meson photoproduction and electroproduction have been suggested as a tool to find or confirm the nucleon resonances. In order to extract more reliable informations on the nucleon resonances, understanding the non-resonant background is indispensable. We consider final state interactions in ω photoproduction as a background production mechanism. For the intermediate states, we consider nucleon–vector-meson and nucleon-pion channels. The role of the final state interactions is discussed in ω meson photoproduction near threshold.

1 Introduction

Vector meson production off nucleons near threshold attracts recent interests in connection with the so-called "missing resonance problem" [1]. By studying various physical quantities of vector meson production one hopes to have information on the nucleon resonances especially which couple rather strongly to the vector-meson–nucleon channel. There have been recent progress to obtain such informations by studying the processes of vector meson photoproduction [2,3,4,5]. Among light vector mesons, ω photoproduction is studied in more detail due to its simple isospin character [3,4].

In order to extract information on the nucleon resonances from vector meson production, it is essential to first understand the background production mechanisms [6]. As the background non-resonant production amplitudes one considers the Pomeron exchange, one-boson exchange, and the nucleon pole terms. Then the gap between the theoretical predictions on the background production and the experimental data, e.g. in total and differential cross sections, are expected to be explained by the terms including nucleon resonances. After adjusting the resonance parameters, other physical quantities, especially polarization asymmetries, are predicted to have more conclusive evidence for the nucleon resonances and it has been shown that some polarization asymmetries are really sensitive to the presence of nucleon resonances because of different helicity structure of the production amplitudes.

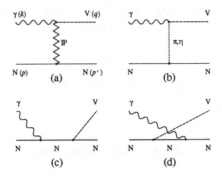

Figure 1. Non-resonant interactions for ω photoproduction at tree level. Here V stands for the ω meson.

However, in order to have more conclusive clues on the nucleon resonances, it is essential to understand the background non-resonant amplitudes in more detail. We can have lessons from the study on the non-resonant part of pion photoproduction and pion-nucleon scattering, which shows that the final state interactions are important to improve meson exchange models [7,8]. Such dynamical studies are important not only in understanding the structure of the nucleon resonances but also for unitarity of the scattering amplitude. Therefore it is legitimate to improve the existing models by imposing unitarity condition.

Such investigations are, however, intricate and many informations are still unavailable to do a reliable model study. In this work, therefore, we study final state interactions in ω photoproduction as our first step to construct a dynamical model for vector meson photoproduction near threshold. Because of its complexity, we only consider some one-loop diagrams in this work that seem to be non-negligible in the production amplitude. In the next Section, we discuss the non-resonant amplitude at tree level and our method to compute the final state interactions for several selected intermediate channels. The preliminary numerical results are given in Sec. III with discussions.

2 Model

As the non-resonant production process for ω photoproduction at tree level, it is widely used to include the Pomeron exchange, π and η exchanges, and the nucleon pole terms as depicted in Fig. 1 [2,3,4,5].

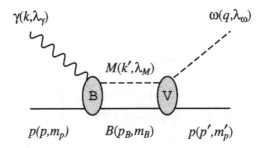

Figure 2. Diagram for final state interactions for ω photoproduction.

However, it is well-known that the amplitudes obtained at tree level such as in Fig. 1 do not satisfy unitarity. For example, the soft Pomeron model of Donnachie and Landshoff [9] that has been used in analyzing ω photoproduction has intercept larger than 1.0 and hence violates the Froissart-Martin bound [10,11] that is a consequence of unitarity and the partial wave expansion. Imposing the unitarity condition to the process has been emphasized in many respects [12,13,14]. Unitarity condition is also crucial in developing dynamical models for pion photoproduction and pion-nucleon interactions [7,8]. Therefore it would be necessary to construct a unitarized model for vector meson photoproduction near threshold to search for the "missing nucleon resonances" by, for example, solving the coupled-channel equations, which would require very complicated calculations. Before tackling to the unitarization of the amplitude directly, we first compute final state interactions by considering several selected intermediate channels. It is the purpose of this work to compute several one loop diagrams in ω photoproduction.

Following Refs. [7,8], what we consider is the diagram shown in Fig. 2, which defines the momenta of the interacting particles and their helicities/spins. Then the amplitude shown in Fig. 2 is written as

$$(\text{FSI})_{BM} = \int d\mathbf{k}' \frac{\langle \mathbf{q}; \lambda_\omega m_p' | V | \mathbf{k}'; \lambda_M m_p'' \rangle \langle \mathbf{k}'; \lambda_M m_p'' | B | \mathbf{k} \lambda_\gamma m_p \rangle}{W - E_B(\mathbf{k}') - E_M(\mathbf{k}') + i\varepsilon}, \qquad (1)$$

after 3-dimension reduction, where $E_{B,M}(\mathbf{k}) = \sqrt{M_{B,M}^2 + \mathbf{k}^2}$. The intermediate baryon and meson masses are denoted by M_B and M_M, respectively, and λ (m) is the helicity (spin) of the particle. Three-dimensional reduction of the full amplitude is not unique [15] and we follow Refs. [7,8] to obtain Eq. (1). It can be further decomposed into the principal integration part and the

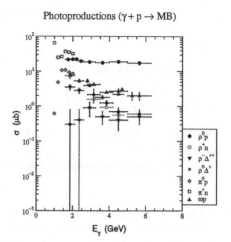

Figure 3. Collected data for the total cross sections for meson photoproduction. The data are from Ref. [16]. The cross sections for $\pi^0 p$ and $\pi^+ n$ photoproductions are from Ref. [17] based on the SAID program.

delta function part as

$$(\text{FSI}) = \mathcal{P} \int dk' k'^2 \frac{\mathsf{V}(q, k'; W)\mathsf{B}(k', k)}{W - E_B(k') - E_M(k')}$$
$$- i\rho_{BM}(k_t)\mathsf{V}(q, k_t; W)\mathsf{B}(k_t, k)\theta(W - M_B - M_M), \quad (2)$$

where

$$\rho_{BM}(k) = \frac{\pi k E_B(k) E_M(k)}{[E_B(k) + E_M(k)]}, \quad (3)$$

and $\theta(x)$ is the step function ($\theta(x) = 1$ for $x > 0$ and 0 otherwise). The delta function part, which contains $\theta(x)$, arises when the intermediate particles are their on mass shell and hence the intermediate state on-shell momentum k_t is defined by

$$W = E_B(k_t) + E_M(k_t). \quad (4)$$

The intermediate states contain various baryon-meson states allowed by symmetries and quantum numbers. Therefore, we start first by selecting the intermediate states that are expected to be rather important. For this purpose, we collect some experimental informations on the cross sections of various meson photoproduction and the available experimental data are shown

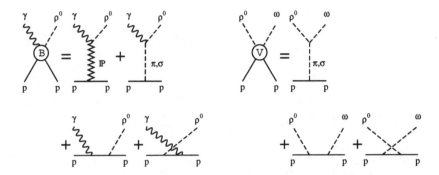

Figure 4. ω photoproduction with intermediate ρ^0-p state.

in Fig. 3. One can see that the cross sections of π and ρ photoproductions are considerably larger than the other reactions. Based on this observation, we consider the intermediate $\pi^+ n$, $\pi^0 p$, and $\rho^0 p$ states.

We first consider the intermediate $\rho^0 p$ state, which is depicted in Fig. 4. Through the studies on ρ photoproduction, we have learned that the σ meson exchange is important at low energies [6,18]. Thus our amplitude for $\gamma p \to \rho^0 p$ contains the Pomeron exchange, π exchange, σ exchange and nucleon pole terms as given in Ref. [6]. The amplitude for $\rho^0 p \to \omega p$ is closely related to the ω photoproduction amplitude at tree level that is given in Fig. 1 via vector meson dominance. In addition to vector meson dominance, what we need is the $\omega\rho\pi$ interaction Lagrangian, which reads

$$\mathcal{L}_{\omega\rho\pi} = \frac{g_{\omega\rho\pi}}{2} \varepsilon^{\mu\nu\alpha\beta} \partial_\mu \omega_\nu \, \text{Tr} \, (\partial_\alpha \rho_\beta \pi), \tag{5}$$

where $\varepsilon^{0123} = +1$, $\pi = \boldsymbol{\pi} \cdot \boldsymbol{\tau}$, $\rho = \boldsymbol{\rho} \cdot \boldsymbol{\tau}$. The coupling constant $g_{\omega\rho\pi}$ was estimated by vector meson dominance, massive Yang-Mills approach, and hidden gauge symmetry approach, etc [19,20,21,22,23] and the estimates are within $10 \sim 15$ GeV^{-1}. In our study, we use

$$g_{\omega\rho\pi} = 12.9 \text{ GeV}^{-1}, \tag{6}$$

where its sign is fixed by SU(3) flavor symmetry. Therefore our amplitude for $\rho^0 p \to \omega p$ contains the pion exchange and the nucleon pole terms.

For intermediate pion-nucleon channel, we need to know the non-resonant part of pion photoproduction. But the most meson-exchange models are constructed to focus on the lower energy region, so its direct extension to our energy region, $W \simeq 2$ GeV, is quite questionable. Because of this reason,

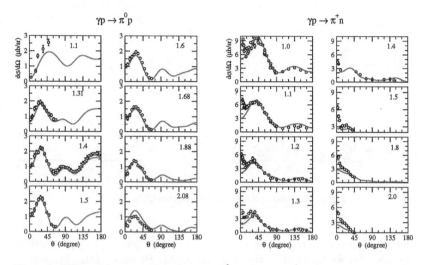

Figure 5. Differential cross sections for $\gamma p \to \pi^0 p$ and $\gamma p \to \pi^+ n$. The numbers in figures represent E_γ. The experimental data are from Ref. [16] and the solid lines are from the calculation of Ref. [17].

we use the SAID program for the pion photoproduction amplitudes. Actually there is no experimental data for the total cross sections for pion photoproduction and the data shown in Fig. 3 are *not* experimental data but are extracted from the SAID program based on Ref. [17]. There can be a few comments on this method. First, the SAID program is established to be valid up to $E_\gamma = 2$ GeV and the extrapolation to the higher energy cannot be guaranteed. Therefore we will use the program only in the limited energy region $W \leq 2$ GeV. Second, the SAID program is believed to represent the experimental data. This means that the extracted amplitude should be assumed to include all nucleon resonance effects. However, since there is no simple meson-exchange model for pion photoproduction within our energy region, we will use this amplitude keeping its limitation in mind. The comparison of the model and the data are given in Fig. 5.

We next need the amplitude for pion induced ω production. This reaction has been discussed in Refs. [24,25,26,27] recently. It was also recently claimed that the final state interactions including nucleon resonances are very crucial to explain the experimental data in Ref. [27]. Following Refs. [26,27], in this study we use the model shown in Fig. 6 concentrating on the low energy

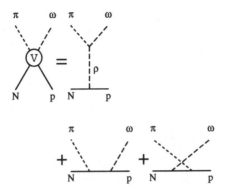

Figure 6. A model for $\pi N \to \omega p$ reaction.

region, which is consistent with those model studies. In addition to the $\omega \rho \pi$ interaction Lagrangian (5), we need the $\omega N N$ and the $\rho N N$ couplings. As in our previous studies [3], we use

$$g_{\omega NN} = 10.35, \qquad \kappa_{\omega} = 0, \qquad g_{\rho NN} = 6.12, \qquad \kappa_{\omega} = 3.1, \qquad (7)$$

where the ρ-nucleon coupling is consistent with the ρ coupling universality. Then the scattering amplitude includes the isospin factor, which reads

$$C_I = \begin{cases} +1 & \text{for } \pi^0 p \to \omega p \\ \sqrt{2} & \text{for } \pi^- p \to \omega n, \, \pi^+ n \to \omega p \\ -1 & \text{for } \pi^0 n \to \omega n \end{cases} \qquad (8)$$

The form factors of the vertices can be found, for example, in Ref. [26]. The calculated total cross section for $\pi^- p \to \omega n$ is shown in Fig. 7.[a]

3 Results and Discussions

With the amplitudes discussed so far, we first compute the differential cross section for ω photoproduction at $E_\gamma = 1.23$ GeV and 1.68 GeV. The preliminary results are shown in Fig. 8. Since the purpose of this calculation is a rough estimate on the role of the final state interactions, we do not try to adjust parameters to fit the experimental data in this calculation. In Fig. 8,

[a]The experimental data for the cross section of this reaction near threshold is controversial [27,28]. In Fig. 7 we follow Ref. [27].

278

$\pi^- p \to \omega n$

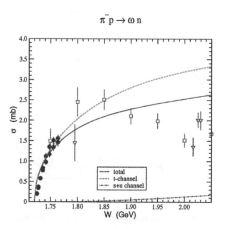

Figure 7. Total cross section for $\pi^- p \to \omega n$.

the solid lines are the results at tree level [3], the dotted lines are from the intermediate $\rho^0 p$ state. The intermediate $\pi^0 p$ and $\pi^+ n$ channels are represented by dashed and dot-dashed lines, respectively.

As we expected from Fig. 3 the contributions from these channels are not suppressed compared to the tree level results. Especially the charged pion channel gives the most important contribution among the intermediate state considered here. The role of the neutral pion intermediate state seems to be smaller than the other channels in the considered energy region. Since the ω photoproduction cross section is at the same order magnitude as the neutral pion photoproduction cross section and smaller than the ρ photoproduction cross section, the contribution from the intermediate ωp channel is expected to be smaller than that from the ρp channel.

We also note that the differential cross section from the intermediate pion channel strongly depends on the momentum transfer t. While the ρp channel differential cross sections do not strongly depend on the scattering angle θ, the pion channel contribution gives rise to strong peaks at backward angles like the u-channel nucleon exchange. Thus, careful analyses are required to distinguish the two mechanisms at backward scattering angles.

In summary, we calculate the one loop contribution to ω photoproduction with selected intermediate states. We found that the final state interactions, especially charged pion intermediate state, are not negligible in the considered energy region and may affect the parameters of the nucleon resonances which

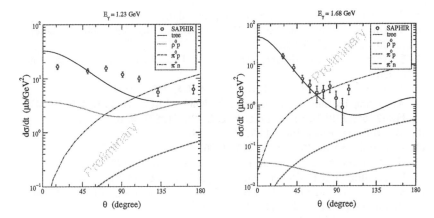

Figure 8. Differential cross sections for $\gamma p \to \omega p$ at $E_\gamma = 1.23$ GeV (left panel) and 1.68 GeV (right panel). θ is the scattering angle in the center-of-mass frame. The experimental data are from SAPHIR [29].

will be extracted from the forthcoming experimental data. Since the polarization asymmetries are suggested to be the most useful tools to investigate nucleon resonances, it would also be important to check the contribution from the final state interactions to the polarization asymmetries.

Acknowledgments

Y.O. is grateful to Prof. M. Fujiwara for the warm hospitality during the symposium. This work was supported in part by the Brain Korea 21 project of Korean Ministry of Education, the International Collaboration Program of KOSEF under Grant No. 20006-111-01-2, and U.S. DOE Nuclear Physics Division Contract No. W-31-109-ENG-38.

References

1. S. Capstick and W. Roberts, JLAB Report (2000), nucl-th/0008028.
2. Q. Zhao, Z. Li, and C. Bennhold, Phys. Rev. C **58**, 2393 (1998).
3. Y. Oh, A. I. Titov, and T.-S. H. Lee, Phys. Rev. C **63**, 025201 (2001); Talk at SPIN 2000 Symposium (2000), nucl-th/0012012; Talk at NSTAR 2001 Workshop (2001), nucl-th/0104046.
4. Q. Zhao, Phys. Rev. C **63**, 025203 (2001); these proceedings.

5. A. I. Titov and T.-S. H. Lee, these proceedings.
6. Y. Oh, A. I. Titov, and T.-S. H. Lee, Talk at NSTAR 2000 Workshop (2000), nucl-th/0004055.
7. S. Nozawa, B. Blankleider, and T.-S. H. Lee, Nucl. Phys. A **513**, 459 (1990).
8. T. Sato and T.-S. H. Lee, Phys. Rev. C **54**, 2660 (1996).
9. A. Donnachie and P. V. Landshoff, Nucl. Phys. B **244**, 322 (1984); **B267**, 690 (1986); Phys. Lett. B **185**, 403 (1987); **296**, 227 (1992).
10. M. Froissart, Phys. Rev. **123**, 1053 (1961).
11. A. Martin, Phys. Rev. **129**, 1432 (1963).
12. U. Maor and P. C. M. Yock, Phys. Rev. **148**, 1542 (1966).
13. K. Schilling and F. Storim, Nucl. Phys. B **7**, 559 (1968).
14. See, e.g., E. Martynov, E. Predazzi, and A. Prokudin, BITP, Ukraine Report (2001), hep-ph/0112242.
15. C.-T. Hung, S. N. Yang, and T.-S. H. Lee, Phys. Rev. C **64**, 034309 (2001) and references therein.
16. The Durham RAL Databases, http://durpdg.dur.ac.uk/HEPDATA/REAC.
17. D. Dutta, H. Gao, and T.-S. H. Lee, MIT Report (2001), nucl-th/0111005.
18. B. Friman and M. Soyeur, Nucl. Phys. A **600**, 477 (1996).
19. F. Klingl, N. Kaiser, and W. Weise, Z. Phys. A **356**, 193 (1996).
20. Ö. Kaymakcalan, S. Rajeev, and J. Schechter, Phys. Rev. D **30**, 594 (1984).
21. P. Jain, R. Johnson, U.-G. Meissner, N. W. Park, and J. Schechter, Phys. Rev. D **37**, 3252 (1988).
22. F. Kleefeld, E. van Beveren, and G. Rupp, Nucl. Phys. A **694**, 470 (2001).
23. T. Fujiwara, T. Kugo, H. Terao, S. Uehara, and K. Yamawaki, Prog. Theor. Phys. **73**, 926 (1985).
24. M. Post and U. Mosel, Nucl. Phys. A **688**, 808 (2001).
25. M. Lutz, G. Wolf, and B. Friman, Nucl. Phys. A **661**, 526c (1999).
26. A. I. Titov, B. Kämpfer, and B. L. Reznik, FZ Rossendorf Report (2001), nucl-th/0102032.
27. G. Penner and U. Mosel, Universität Giessen Report (2001), nucl-th/0111023; nucl-th/0111024.
28. C. Hanhart and A. Kudryavtsev, Eur. Phys. J. A **6**, 325 (1999).
29. F. J. Klein, Ph.D. thesis, Bonn University (1996); πN Newslett. **14**, 141 (1998).

THE Q^2 EVOLUTION OF THE GDH SUM RULE
(ON ³HE AND THE NEUTRON)

GORDON D. CATES

(FOR THE JLAB E94-010 COLLABORATION)

Department of Physics, University of Virginia,
Charlottesville, VA 22903

We discuss the extention of the Gerasimov-Drell-Hearn (GDH) sum rule, which pertains to real photons, to include scattering due to virtual photons. We present data from Jefferson Laboratory experiment E94-010 which measured the inclusive scattering of polarized electrons from a polarized ³He target over the quasielastic and resonance regions. From these data we exctract the transverse-transverse interference cross section σ'_{TT}, and compute the Q^2 depenent extended GDH integral.

1 Introduction

The study of spin structure has proven to be a valuable tool for understanding quantum chromodynamics (QCD). It is natural, therefore, to turn to spin structure to understand the complex transition from parton-like behavior in deep inelastic scattering to hadronic-like behavior that is observed at lower energies. It is thus useful to identify observables that might reveal the relevant dynamics in this transition. One example is the extended GDH integral $I(Q^2)$, a quantity that is intimately connected to both high and low energy sum rules.

In this paper, we begin by discussing the extended GDH sum rule, its connection to other sum rules, and what can be learned through its study. We then go on to present data on the process ³He $(\vec{e}, e')$ taken at Jefferson Laboratory during E94-010[1], the first experiment to use a polarized ³He target at JLab. From this data we extract the transverse-transverse interference cross section σ'_{TT}, and compute the extended GDH integral $I(Q^2)$ for the range $0.1\,\mathrm{GeV}^2 < Q^2 < 1.0\,\mathrm{GeV}^2$.

2 The extended GDH sum rule

An important sum rule is that due to Gerasimov, Drell, and Hearn (GDH) that relates a sum over the total spin-dependent photoabsorption cross sections of the nucleon to the square of the anamoulous magnetic moment κ of the nucleon[2,3]. The sum rule can be written

$$\int_{\nu_0}^{\infty} \frac{d\nu}{\nu} \left[\sigma_{1/2}(\nu) - \sigma_{3/2}(\nu) \right] = -\frac{2\pi^2\alpha}{M^2}\kappa^2, \tag{1}$$

where $\sigma_{1/2}(\nu)$ $(\sigma_{3/2}(\nu))$ is the cross section for absorption when the total spin of the photon and nucleon, projected onto the photon momentum direction, is 1/2 (3/2). The integral is performed over the photon energy ν, beginning at the pion production threshold ν_0. Here α is the fine structure constant, and M is the mass of the nucleon.

The GDH integral can be generalized to include not just the absorption of real photons, but the exchange of virtual photons as occurs in electron scattering. This corresponds to values of the four-momentum transfer squared $Q^2 > 0$, in contrast to the absorption of real photons for which $Q^2 = 0$. Perhaps the most natural way to make this extension is to convert the photoabsorption cross sections of the GDH sum rule into electroproduction cross sections:

$$I(Q^2) = \int_{\nu_0}^{\infty} \frac{d\nu}{\nu} \left[\sigma_{1/2}(\nu, Q^2) - \sigma_{3/2}(\nu, Q^2) \right] = 2 \int_{\nu_0}^{\infty} \frac{d\nu}{\nu} \sigma'_{TT}(\nu, Q^2) \tag{2}$$

where $\sigma_{1/2}(\nu, Q^2)$ and $\sigma_{3/2}(\nu, Q^2)$ are spin-dependent total *virtual* photoabsorption cross sections on the nucleon. The generalized GDH integral is constrained at two limits:

$$\text{As } Q^2 \to 0 \quad I(Q^2) \to -\frac{2\pi^2\alpha}{M^2}\kappa^2 \tag{3}$$

$$\text{As } Q^2 \to \infty \quad I(Q^2) \to \frac{16\pi^2\alpha}{Q^2}\Gamma_1 \tag{4}$$

where $\Gamma_1 = \int_0^1 g_1 dx$ is the first moment of the spin structure function g_1[4]. The first relation follows from the GDH sum rule (for $Q^2 = 0$), and the second relation follows from the definition of $I(Q^2)$, and the asymptotic form of σ'_{TT}.

It is certainly desirable to be able to equate $I(Q^2)$ to some quantity throughout all values of Q^2, and thus have a true sum rule. It has recently been shown by Ji and Osborne that

$$I(Q^2) = 2\pi^2\alpha S_1(0, Q^2) \tag{5}$$

where $S_1(0, Q^2)$ is the forward virtual Compton amplitude[5]. To the extent that we consider $S_1(0, Q^2)$ well defined everywhere, this expression gives us a true extended GDH sum rule. Using chiral perturbation theory, $S_1(0, Q^2)$ can be computed[6,7] for small values of Q^2, probably up to values of around 0.1 GeV^2, or perhaps even larger.

At large values of Q^2, one can also make definite statements about the Compton amplitudes. For instance, beginning with the operator product expansion for the Compton amplitudes for the proton and the neutron, it can be shown that

$$S_1^p(0, Q^2) - S_1^n(0, Q^2) = \frac{4}{3Q^2} g_A \,, \tag{6}$$

where the superscripts p and n refer to the proton and the neutron. This relation, in combination with equation (4), brings us to the equation

$$\Gamma_1^p - \Gamma_1^n = \frac{1}{6} g_A \tag{7}$$

which we recognize as the Bjorken Sum Rule[8]. Similarly, it follows that

$$I^p(Q^2) - I^n(Q^2) = \frac{8}{3} \frac{\pi^2 \alpha}{Q^2} g_A \tag{8}$$

showing that at the very least the extended GDH sum rule for the difference of the proton and the neutron is well defined for $Q^2 \to \infty$. In fact, relations for the proton and neutron taken individually also exist[9]. For the sake of brevity, however, we will stop here.

There are also important experimental constraints on $I(Q^2)$. At $Q^2 = 0$, there are recent measurements of the GDH sum rule on the proton[10], although the neutron has yet to be investigated. At high Q^2, there is also an extensive body of data on both Γ_1^p and Γ_1^n,[11] although most of the data is in the deep inelastic regime where $Q^2 > 1 \, \text{GeV}^2$.

In summary, there are both theoretical and experimental constraints on $I(Q^2)$ over a wide range of values of Q^2, spanning the region where the nucleon is governed by hadronic degrees of freedom and non-perturbative QCD, to the region where the nucleon is governed by quark degrees of freedom and perturbative QCD. At low Q^2 chiral perturbation theory provides theoretical guidance, while at high Q^2 techniques involving the operator product expansion are useful. In the important intermediate transition regime there is at present little theoretical guidance. Since we are concerned with the first moments of the spin structure functions, however, rather than the spin structure functions themselves, it is believed that lattice QCD will be an effective tool[12].

3 Experimental measurement $I^n(Q^2)$

To investigate $I^n(Q^2)$, and more generally the low Q^2 spin structure of the neutron, we investigated the process $^3\text{He} \, (\vec{e}, e')$ in Hall A of Jefferson Laboratory. We took data at six energies: 5.06, 4.24, 3.38, 2.58, 1.72, and 0.86 GeV, all at a nominal scattering angle of 15.5°. The measurements covered values of the invariant mass W from the quasielastic peak through the resonance region, into the deep inelastic regime, as is shown in Fig. 1.

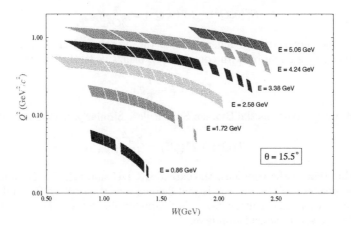

Figure 1. Shown are the regions in the space spanned by Q^2 and W covered by our kinematics. For each energy, each block of space represents a separate spectrometer setting.

Because we were studying inclusive scattering, we had the opportunity to use both of the Hall A high resolution spectrometers for the detection of electrons. Particle identification was accomplished using Čerenkov detectors and Pb-glass shower counters. Momentum analysis was accomplished using drift chambers.

Our experiment was the first at Jefferson Laboratory to utilize the Hall A polarized ^{3}He target[1,13,14]. In fact, the experiment was one of the initial motivations for the target's construction. The target utilizes the technique of spin-exchange optical pumping in which rubidium (Rb) vapor is polarized using lasers, and the ^{3}He is polarized during subsequent collisions[15]. The ^{3}He is contained in sealed glass "target cells" at a pressure of roughly 10 atmosphere. An example of a target cell is shown in Fig. 2. Also contained in the cell is a few droplets of Rb metal and about 70 Torr of nitrogen (N_2). The N_2 is present to "quench" the excited states of Rb without the emission of a photon. Otherwise, the emitted photons, which in general do not have the desired state of polarization, would tend to depolarize the Rb vapor. The target cells are comprised of two chambers: an upper pumping chamber, in which the optical pumping and spin exchange take place, and a lower target chamber, through which the electron beam passes. The temperature of the pumping chamber is regulated in order to control the number density of the Rb vapor. It is also irradiated with a 90 W laser system with a wavelength centered at 795 nm, corresponding to the D_1 line of Rb. The lower chamber is approximately 40 cm in length and has thin "end windows" about 140–150 μm in thickness.

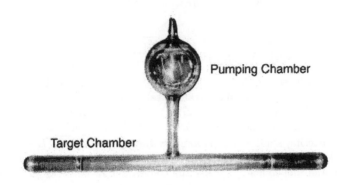

Figure 2. Shown is one of the E94-010 polarized ^{3}He target cells.

The ^{3}He target system included two pairs of Helmholtz coils that produced a static magnetic field of approximately 25 G. The polarization was monitored using the NMR technique of adiabatic fast passage (AFP). The NMR system was calibrated in two ways. In one approach, a glass cell with a geometry closely approaching that of the target cell was filled with water and placed in the apparatus. The small but well defined polarization of the water due to the thermal Boltzman distribution was detected and the resulting water signals were compared with the ^{3}He signals. In another approach, an oscillator was locked to the electron paramagnetic resonance (EPR) frequency corresponding to the Zeeman splitting between two m_F levels of the ^{87}Rb. A small shift in the EPR frequency occured because of an effective magnetic field caused by the presence of the polarized ^{3}He. The relationship between the shift and the absolute polarization of the ^{3}He has been well characterized in separate experiments[16]. By measuring the EPR shift, which yields an absolute polarization measurement, in close proximity in time to measuring ^{3}He AFP signals, a second calibration of the NMR system resulted.

4 Extracting spin observables

We made measurements with the target polarization in both the longitudinal and transverse orientations. The quantities we measured experimentally are related to σ'_{TT} and σ'_{LT} by the relations

$$\frac{d^2\sigma^{\downarrow\Uparrow}}{d\Omega dE'} - \frac{d^2\sigma^{\uparrow\Uparrow}}{d\Omega dE'} = B\left(\sigma'_{TT} + \eta\,\sigma'_{LT}\right) \tag{9}$$

and

$$\frac{d^2\sigma^{\uparrow\Rightarrow}}{d\Omega dE'} - \frac{d^2\sigma^{\downarrow\Rightarrow}}{d\Omega dE'} = B\sqrt{\frac{2\epsilon}{1+\epsilon}}\left(\sigma'_{LT} - \zeta\,\sigma'_{TT}\right) \tag{10}$$

where the left side of eq. (9) (eq. (10)) corresponds to the spin asymmetries measured for parallel (transverse) target orientations. Also $B = -2\,(\alpha/4\pi^2)(K/Q^2)(E'/E)(2/(1-\epsilon))\,(1 - E'\epsilon/E)$, E and E' are the initial and final energies of the electron, $\epsilon^{-1} = 1 + 2[1 + Q^2/4M^2x^2]\tan^2(\theta/2)$, θ is the scattering angle in the laboratory frame, $\eta = \epsilon\sqrt{Q^2}/(E - E'\epsilon)$, and $\zeta = \eta(1 + \epsilon)/2\epsilon$. The quantity K represents the virtual photon flux and is convention dependent. We use the convention $K = \nu - Q^2/2M$, due to Hand[17].

The virtual photoabsorption cross sections σ'_{TT} and σ'_{LT} correspond to the Born Approximation in which scattering is due to a single virtual photon. In practice, the measured cross sections include higher order radiative processes, and it is necessary that these be accounted for when extracting the Born cross sections. The procedure used was that due to Mo and Tsai, which in their first treatment, dealt strictly with unpolarized scattering[18]. For polarized scattering, we used the program POLRAD[19], which we modified in a number of ways. In particular, we incorporated our data for the quasielastic and resonance regions.

5 Preliminary results

The data, with radiative corrections, are shown in Fig. 3a, where σ'_{TT} is plotted as a function of invarient mass W for each of the six energies measured. The error bars indicate the uncertainties due to statistics, and the bands indicate the errors due to systematic effects, which include uncertainty in the beam and target polarizations, uncertainty due to radiative corrections, and uncertainty in the measured cross sections.

Because we are interested in the Q^2 evolution of the GDH integral, we have determined σ'_{TT} for various values of constant Q^2. This was done by

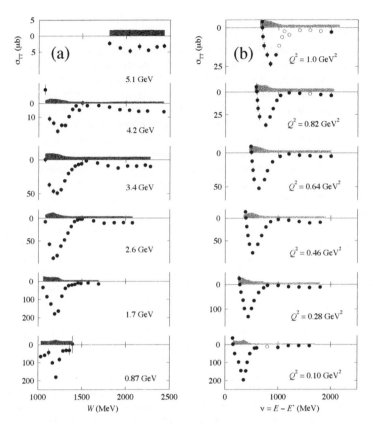

Figure 3. Preliminary data from JLab E94-010. Shown in (a) are measured values of σ'_{TT} plotted as a function of W for each of the six incident beam energies studied. In (b) σ'_{TT} is plotted as a function of energy loss ν for each of six constant values of Q^2. Solid circles indicate interpolated values and open circles indicate extrapolated values.

interpolating between measured values of Q^2, or in some cases, extrapolating to unmeasured regions. The resulting values for σ'_{TT}, plotted as a function of energy loss ν, are shown in Fig. 3b for each of six constant values of Q^2. Considerable care was taken to ensure a smooth interpolation, including accounting as much as possible for the physics determining σ'_{TT} at each point. Several different methods of interpolation/extrapolation were tried to gain a sense of the systematic uncertainty of the process. Typically the error incurred was roughly 5% of σ'_{TT}, although it was somewhat worse in the neighborhood of the delta. The additional error due to interpolation/extrapolation is folded into the systematic errors mentioned earlier in the plotted error bands.

We have computed the extended GDH integral $I(Q^2)$ using equation (2). For each Q^2, we have computed the integral from the pion threshold up to a ν corresponding to a W of 2 GeV. The results are shown in Fig. 4 as open circles, where the error bars represent the uncertainty due to systematics only. The error associated with systmatic effects is shown with a dark band. Because our measurements were made using a ^{3}He target, we must apply a correction to account for the fact that the neutron was imbedded in the nucleus. Here we have used the perscription of C. Ciofi degli Atti and S. Scopetta [20], whose technique is essentially an impulse approximation that takes into account the relative polarizations of the protons and neutrons in the ^{3}He nucleus. The authors estimate that for the extended GDH integral this approach should be accurate at roughly the 5% level. The results for $I(Q^2)$ for the neutron are shown with open squares. The last correction we consider is an estimate off the extended GDH integral for the integrating range of $4\,\text{GeV}^2 < W^2 < 1000\,\text{GeV}^2$. For this we use the parameterization due to Bianchi and Thomas[21], and the results are shown with the closed squares.

Ideally one would like the GDH integral to provide insight into the transition from hadronic behavior at low Q^2 to partonic behavior at high Q^2. Toward this end, it is desirable to compare with theoretical predictions that contain as much physical content as possible. At the $Q^2 = 0$ point, the GDH sum rule provides a solid prediction, which is shown on Fig. 4 with an eight-pointed star. Chiral perturbation theory can also be used to extend this prediction to small values of Q^2. Here, a calculation due to Ji, Kao, and Osborne is shown with a dotted line. At high values of Q^2, operator product expansion (OPE) techniques can be used to expand the Compton amplitudes in a power series of α_S and $1/Q^2$. Our results can either be used to check the results of OPE calculations, or to determine some of the coefficients that appear. For reference, we also show in Fig. 4 a calculation due to Drechsel, Kamalov, and Tiator[22] that is based largely on the phenomenological model MAID. While the basic shape is in agreement with our observations, the overall magnitude is clearly smaller than our data.

In conclusion, we believe that the study of the Q^2 evolution of the GDH integral offers a valuable opportunity to study a single observable over the important transition from hadronic to partonic behavior. Our data is the first such data on the neutron, and provides a valuable glimpse of the current state of agreement between experiment and theory. Further developments,

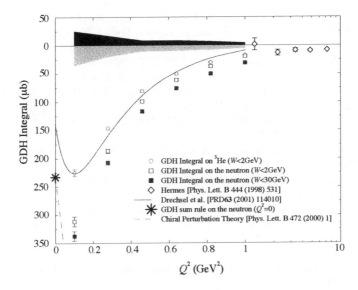

Figure 4. Preliminary results from JLab E94-010 for $I(Q^2)$. The open circles represent our data with no correction for the fact that the neutron is embedded in a ^{3}He nucleus. The open squares have a correction to account for nuclear effects, and the closed squares include an estimate of the unmeasured portion of the integral.

on both the experimental and theoretical fronts, should provide important insights in the future.

We would like to thankfully acknowledge the untiring support of the JLab staff and accelerator division. This work was supported by the U.S. Department of Energy (DOE), the DOE-EPSCoR, the U.S. National Science Foundation, NSERC of Canada, the European INTAS Foundation, the Italian INFN, and the French CEA, CNRS, and Conseil Régional d'Auvergne. The Southeastern Universitites Research Association (SURA) operates the Thomas Jefferson National Accelerator Facility for the DOE under contract DE-AC05-84ER40150.

References

1. For more details on JLab E94-010, see the URL www.jlab.org/e94010.
2. S. B. Gerasimov, Sov. J. of Nucl. Phys. **2**, 430 (1966).

3. S. D. Drell, and A. C. Hearn, Phys. Rev. Lett. **16**, 908 (1966).
4. M. Anselmino, B. L. Ioffe, and E. Leader, Sov. J. Nucl. Phys. **49**, 136 (1989).
5. X. Ji and J. Osborne, J. Phys. G **27**, 127 (2001).
6. X. Ji, C. Kao, and J. Osborne, Phys. Lett. B **472**, 1 (2000).
7. V. Bernard, N. Kaiser, U-G. Meissner, Phys. Rev. D **48**, 3062 (1993); Int. J. Mod. Phys. **E4**, 193 (1995).
8. J. D. Bjorken, Phys. Rev. **148**, 1467 (1966); Phys. Rev. D **1**, 465 (1970); Phys. Rev. D **1**, 1376 (1970).
9. J. Ellis, R. L. Jaffe, Phys. Rev. D **9**, 1444 (1974); Phys. Rev. D **10**, 1669 (1974).
10. J. Ahrens *et al.* (The GDH and A2 Collaborations), Phys. Rev. Lett. **87**, 022003-1 (2001).
11. For recent reviews, see E. W. Hughes and R. Voss, Ann. Rev. Nucl. Part. Sci **49**, 303 (1999) or B. W. Filippone and X. Ji, hep-ph/0101224v1, and references therein.
12. Nathan Isgur and John W. Negele, *Nuclear Theory with Lattice QCD*, a proposal submitted to the U.S. Department of Energy, *unpublished* (2000).
13. J.S. Jensen, Ph.D. Thesis, California Institute of Technology, 2000 (unpublished); I. Kominis, Ph.D. Thesis, Princeton University, 2000 (unpublished).
14. I. Kominis in GDH 2000, ed. by D. Drechsel and L. Tiator, pg. 181, World Scientific, Singapore (2001).
15. A. Ben-Amar Baranga, S. Appelt, M.V. Romalis, C.J. Erickson, A.R. Young, G.D. Cates and W. Happer, Phys. Rev. Lett. **80**, 2801 (1998).
16. M.V. Romalis and G.D. Cates, Phys. Rev. A **58**, 3004 (1998).
17. L.N. Hand, Phys. Rev. **129**, 1834 (1963).
18. L.W. Mo and Y.S. Tsai, Rev. Mod. Phys. **41**, 205 (1969); Y.S. Tsai, Report No. SLAC-PUB-848 (1971).
19. I.V Akushevich and N.M. Shumeiko, J. Phys. G **20**, 513 (1994).
20. C. Ciofi degli Atti and S. Scopetta, Phys. Lett. B **404**, 223 (1997).
21. E. Thomas and N. Bianchi, Nucl. Phys. B **82** (Proc. Suppl.), 256 (2000).
22. D. Drechsel, S.S. Kamalov, and L. Tiator, Phys. Rev. D **63**, 114010 (2001). Note, model I_C is plotted in Fig. 3.

DETAILED STUDY OF THE ^{3}HE NUCLEI THROUGH RESPONSE FUNCTION SEPARATIONS AT HIGH MOMENTUM TRANSFER

D. W. HIGINBOTHAM

Jefferson Lab, Newport News, VA 23606, USA
for the Hall A and E89-044 Collaborations

In order to further our understanding of the few body system, a new series of measurements on the reaction ^{3}He(e,e'p) have been made at Jefferson Lab. Making use of beam energies as high as 4.8 GeV and of the two high resolution in Hall A, kinematics which were previously unattainable have been investigated. In this paper the first preliminary results of this experiment will be presented.

1 Introduction

Coincidence experiments have proven to be very useful tools in studying specific aspects of the nucleus. In particular the (e,e'p) reaction has been used not only to study the single-nucleon structure of nuclei but also to study the behavior of nucleons embedded in the nuclear medium. At JLab, this is being accomplished by extending the domain of momentum transfers towards higher values where short-range effects and possibly the internal structure of the nucleons are manifested, by exploring nuclear structure in its extreme conditions, and by investigating the high momentum part of the wave functions. We also will increase the specificity of the probe by separating the response functions associated with different polarization states of the virtual photon.

The Jefferson Lab E89-044 [1] and E01-108 [2] experiments are designed to exploit these new possibilities by undertaking a series of (e,e'p) measurements on the Helium isotopes. Next to the deuteron, the A=3 and A=4 nuclei are the simplest systems in which all the basic ingredients of a complex nucleus exist. Sophisticated methods to solve the Schrödinger equation almost exactly have been applied to the A=3 nuclei and have been extended to ^{4}He. Microscopic calculations of FSI and MEC contributions have been developed and applied to reactions on few-nucleon systems. The data provided by these experiments test the validity of these models in the high Q^2 and high missing momentum regime. In this paper the preliminary results of the E89-044 experiment will be presented and discussed.

2 Kinematics

The kinematics for the (e,e'p) reaction are shown in Fig. 1. The scattering plane is defined by the incoming electron, $e = (E_e, \mathbf{e})$, and the outgoing electron, $e' = (E'_e, \mathbf{e}')$. The four-momentum of the virtual photon is given by q=(w, $\mathbf{q}$) and the four-momentum of the outgoing proton is given by $p' = (E_p, \mathbf{p}')$. The four-momentum square, $Q^2 = q^2 - \omega^2$, is defined such that for electron scattering Q^2 is always positive. The missing momentum vector is defined as $\mathbf{p}_m = \mathbf{q} - \mathbf{p}'$.

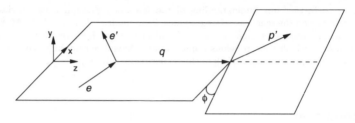

Figure 1. A schematic of the kinematics for the (e,e'p) reaction.

The form of the differential cross section for (e,e'p) reactions in the one-photon exchange approximation without polarization is:

$$\frac{d^6\sigma}{d\Omega_{e'}dE_{e'}d\Omega_{p'}dE_{p'}} = \frac{E_p p_p}{(2\pi)^3}\sigma_{Mott}[v_T R_T + v_L R_L + v_{TL} R_{TL}\cos\phi + $$

$$v_{TT} R_{TT}\cos 2\phi], \qquad (1)$$

with ϕ the angle between the plane defined by $\mathbf{e}$ and $\mathbf{e}'$ and the plane defined by $\mathbf{p}'$ and $\mathbf{q}$, σ_M is the Mott cross section,

$$\sigma_{Mott} = \frac{4\alpha^2 E_{e'}^2}{Q^4}\cos^2\frac{\theta_{e'}}{2}. \qquad (2)$$

The kinematics factors v_L, v_T, v_{TL},and v_{TT} are:

$$v_L = \frac{Q^4}{\mathbf{q}^4}, \qquad (3)$$

$$v_T = \frac{Q^2}{2\mathbf{q}^2} + \tan^2(\theta_e/2), \qquad (4)$$

$$v_{TL} = \frac{Q^2}{\mathbf{q}^2}\left[\frac{Q^2}{\mathbf{q}^2} + \tan^2(\theta_e/2)\right]^{1/2}, and \qquad (5)$$

$$v_{TT} = \frac{Q^2}{2\mathbf{q}^2}. \qquad (6)$$

For the two body break-up channel, ^{4}He(e,e'p)^{3}H, the proton energy and angle with respect to $\mathbf{q}$ are correlated because the missing energy is fixed. In this case the differential cross is written as follows:

$$\frac{d^5\sigma}{d\Omega_{e'}\,d\Omega_{p'}\,dE_{e'}} = \frac{E_p P_p}{(2\pi)^3}\sigma_{Mott}f_{rec}^{-1}\left[v_T R_T + v_L R_L + v_{TL}R_{TL}\cos\phi + v_{TT}R_{TT}\cos2\phi\right],$$

(7)

where f_{rec} is the recoil factor,

$$f_{rec} = \left[1 - \frac{E_{p'}}{E_t}\frac{\mathbf{p_m}\cdot\mathbf{p'}}{p'^2}\right].$$

(8)

The response functions, R_L, R_T, R_{TL}, R_{TT} can be separated by a suitable choice of the kinematic parameters. In perpendicular in-plane kinematics, i.e. constant $\mathbf{q}$ and ω kinematics, one can separate R_T, R_{TL}, and a combination of the R_L and R_{TT} response functions, denoted in this proposal as R_{L+TT}. One can also measure the cross section asymmetry A_{TL} for a given $\mathbf{q}$ and ω. This asymmetry is defined as:

$$A_{TL} = \frac{\sigma(\phi=0) - \sigma(\phi=180)}{\sigma(\phi=0) + \sigma(\phi=180)}.$$

(9)

In parallel and anti-parallel kinematics, i.e. when the out-going proton is in the direction of $\mathbf{q}$, one can separate the R_L and R_T response functions. In parallel kinematics p_m points in the opposite direction as $\mathbf{q}$ with $x_B < 1$ while in anti-parallel kinematics p_m points in the same direction as $\mathbf{q}$ with $x_B > 1$ where

$$x_B = \frac{Q^2}{2M\omega}$$

(10)

is the Bjorken scaling variable. For $x_B > 1$, the region in ω between the quasi-elastic peak and the elastic peak is being probed; while for $x_B < 1$, the region ω towards the delta peak is being probed. The region in ω between the quasi-elastic peak and delta peak is often referred to as the dip region.

3 Measurements

In perpendicular kinematics, with a constant $\mathbf{q} = 1.5$ GeV/c and $\omega = 0.845$ GeV, the single nucleon structure of ^{3}He was studied with special emphasis on high momenta, up to 1 GeV/c in missing momentum. We also did a complete in-plane separation of the response functions R_{TL}, R_T, and the combination of R_{L+TT} up to missing momenta of 0.55 GeV/c. In parallel kinematics, the $\mathbf{q}$ dependence of the reaction was determined by performing an R_L/R_T (longitudinal/transverse) Rosenbluth separation for protons emitted along $\mathbf{q}$ (in parallel kinematics), up to $\mathbf{q} = 3$ GeV/c. This was performed in both quasifree kinematics ($p_m = 0$) and for q = 1 and 2 GeV/c at $p_m \pm$ 0.3 GeV/c. Also, the continuum region was studied in order to search for correlated nucleon pairs. This was done in both parallel and perpendicular kinematics with full in-plane separation of the response functions.

Perpendicular Kinematics	$\vec{q}$ [GeV/c]	E_0 [GeV]	ω [GeV]	ϵ	P_m [GeV/c]
Kin01	1.50	4.80	0.837	0.943	0.00
Kin02	1.50	4.80	0.837	0.934	0.00
Kin03	1.50	1.25	0.837	0.108	0.00
Kin04	1.50	4.80	0.837	0.943	0.150
Kin05	1.50	4.80	0.837	0.943	0.150
Kin06	1.50	1.25	0.837	0.108	0.150
Kin07	1.50	4.80	0.837	0.943	0.300
Kin08	1.50	4.80	0.837	0.943	0.300
Kin09	1.50	1.25	0.837	0.108	0.300
Kin10	1.50	4.80	0.837	0.943	0.425
Kin11	1.50	4.80	0.837	0.943	0.425
Kin12	1.50	1.25	0.837	0.108	0.425
Kin13	1.50	4.80	0.837	0.943	0.550
Kin14	1.50	4.80	0.837	0.943	0.550
Kin15	1.50	1.25	0.837	0.108	0.550
Kin28	1.50	4.80	0.837	0.943	0.750
Kin29	1.50	4.80	0.837	0.943	1.00
Kin31	1.50	1.25	0.880	0.069	0.401
Kin32	1.50	1.25	0.880	0.069	0.509
Kin33	1.50	1.95	0.837	0.615	0.000
Kin34	1.50	1.95	0.837	0.615	0.150
Kin35	1.50	1.95	0.837	0.615	0.150

Table 1. *The perpendicular kinematics measured during the E89-044 experiment. The kinematics cover a range in missing momentum from 0 to 1000 GeV/c and span a range in ϵ from 0.108 to 0.934, all with a constant $\vec{q}$ and ω. From these cross section measurements, the response functions R_{TL}, R_T and a combination of R_L and R_{TT} can be extracted. Due to the large momentum acceptance of the Hall A spectrometers, data includes both the two-body break-up, $^3He(e,e'p)d$ channel, and the continuum $^3He(e,e'p)pn$ channel out to missing energies of approximately 150 MeV/c.*

Parallel Kinematics	$\vec{q}$ [GeV/c]	E_0 [GeV]	ω [GeV]	ϵ	P_m [GeV/c]
Kin16	1.00	4.04	0.438	0.966	0.00
Kin17	1.00	0.845	0.438	0.221	0.00
Kin18	1.94	4.80	1.22	0.898	0.00
Kin19	1.94	1.95	1.22	0.314	0.00
Kin20	3.00	4.80	2.21	0.718	0.00
Kin21	3.00	2.90	2.21	0.180	0.00
Kin22	1.00	2.90	0.700	0.925	-0.300
Kin23	1.00	1.25	0.700	0.527	-0.300
Kin24	1.94	4.80	1.52	0.891	-0.300
Kin25	1.94	1.95	1.52	0.204	-0.300
Kin26	1.00	4.05	0.262	0.968	+0.300
Kin27	1.00	0.845	0.262	0.342	+0.300
Kin36	1.00	1.95	0.439	0.847	0.000

Table 2. *Shown here are the parallel kinematics measured during the E89-044 experiment. The kinematics include measurements made at 0 and 300 MeV/c missing momentum. The 300 MeV/c missing momentum data is taken with Bjorken x both greater and less than one. Due to the large momentum acceptance of the Hall A spectrometers, data includes both the two-body break-up ^{3}He(e,e'p)d channel and the continuum ^{3}He(e,e'p)pn channel out to missing energies of approximately 150 MeV/c.*

4   Results

The preliminary cross section results of the in ^{3}He(e,e'p)d reaction in perpendicular kinematics are shown in Figures 2,3, and 4 have generated considerable theoretical interest. The curves show the most recent calculations of Jean-Marc Laget. The agreement of the theory to missing momentum of 750 MeV/c is striking. There is as yet no clear indication as to what is causing the cross section at the largest missing momentum to be much greater than predicted, though Jean-Marc Laget is investigating the possibility that it is due to multiple rescattering which has not yet been included in his calculations.

Jose Udias is also working on making theoretical calculations. He has already provide the collaboration with a rough A_{TL} calculation, shown in Figure 5 and is now working on making a full calculation of the cross sections and A_{TL} using a realistic potential. Rocco Schiavilla and Sabine Jeschonnek are also planning to provide calculations for the collaboration.

With the cross section analysis of the perpendicular kinematics now nearly finished, the process of making the response function separations is now un-

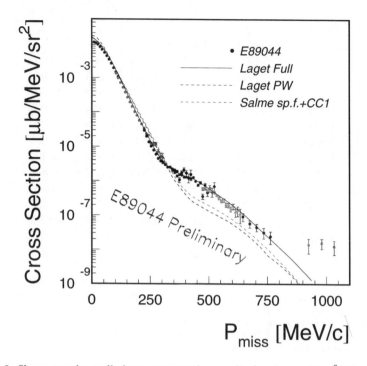

Figure 2. Shown are the preliminary cross section results for the reaction ^{3}He(e,e'p)d as a function of missing momentum with a beam energy of 4807 MeV and with a fixed $\mathbf{q} = 1500$ MeV/c and $\omega = 840$ MeV with $\phi = 180$ degrees. The theory curves show a PWIA calculation using a spectral function a Salme along with the latest calculations of Jean-Marc Laget. The enhancement in the cross section near 300 MeV/c and continuing to larger missing momentum is predominately due to final state interactions. There is not any clear indication from theory what causing the enhancement of the cross section near 1000 MeV/c, though Jean-Marc Laget is investigating the possibility that it is due to multiple rescattering which has not yet been included in his calculations.

derway and in early 2002 preliminary response function separations results should be available. Also next year the cross section and response function separations for the parallel kinematics and for the continuum data should be available.

5 Conclusion

The Jefferson Lab E89-044 ^{3}He(e,e'p) experiment has been successfully completed and preliminary results are starting to become available From the pre-

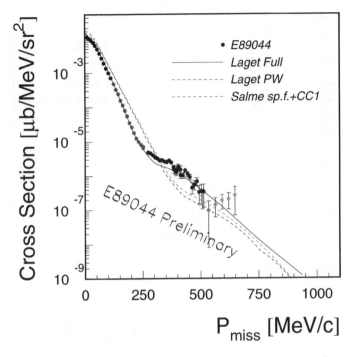

Figure 3. Shown are the preliminary cross section results for the reaction ^{3}He(e,e'p)d as a function of missing momentum with a beam energy of 4807 MeV and with a fixed $q = 1500$ MeV/c and $\omega = 840$ MeV with $\phi = 0$ degrees. When combined with the results shown in Figure 2 this data can be used to extract the A_{TL} asymmetry and the R_{TL} response function.

liminary cross section results along with the A_{TL} asymmetry we can already see regions where the theoretical models clearly predicted the experimental results, such as the rise in the cross section at 300 MeV/c in Figure 2. Also seen are the deficiencies in the theories, as indicated by much larger then expected cross section at extremely large missing momentum in Figure 2 and by the A_{TL}. As the analysis now continue with the response function separations, it should become clear what effects are causing the discrepancies between theory and experiment, and allow a significant improvement in our understand of the three-body system. In the coming years, the E01-108 ^{4}He(e,e'p) experiment will be run in Hall-A and will provide a valuable investigation of the four-body system.

298

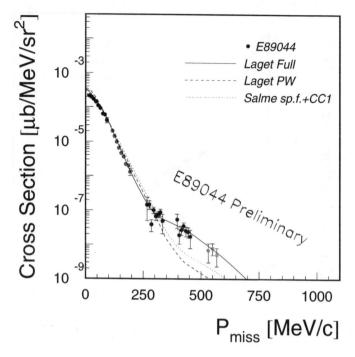

Figure 4. Shown are the preliminary cross section results for the reaction ^{3}He(e,e'p)d as a function of missing momentum with a beam energy of 1250 MeV and with a fixed $q = 1500$ MeV/c and $\omega = 840$ MeV with $\phi = 180$ degrees. This data, which has an ϵ of 0.108, will be combined with the results shown in Figure 2, which has an ϵ of 0.943 to extract R_T and a combination of R_L and R_{TT}.

Acknowledgments

I would like to acknowledge the hard work and dedication of the doctoral students working on the experiment. Marat Rvachev with M.I.T. has done the perpendicular analysis and is responsible for the analysis shown in the work. Emilie Penel-Nottaris with Grenoble is working on the parallel analysis and plans to have results available next year and Fatiha Benmokhtar with Rutgers University is analyzing the continuum data for her doctoral research. Also, I would also like to thank the experimental spokespersons, Marty Epstein, Arun Saha, and Eric Voutier, for their hard work and dedication along with both the Hall A collaboration and the E89044 collaboration.

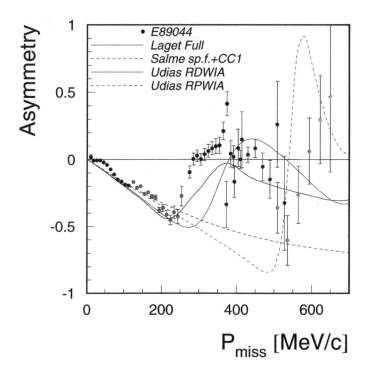

Figure 5. Shown is the preliminary A_{TL} data. This result is obtained by combining the taking the sum over the difference of the results shown in Figures 2 and 3 (see Equation 9). The curves show the latest calculation of Jean-Marc Laget along with a preliminary results of Jose Udias. At lower missing momentum the theories are most sensitive to relativistic effects, while at the larger missing momentum the theories become sensitive to final state interactions effects.

References

1. M. Epstein, A. Saha, and E. Voutier, Selected Studies of the ^{3}He and ^{4}He Nuclei through Electrodisintegration at High Momentum Transfer, Jefferson Lab Experiment 89-044.
2. K. Aniol, D.W. Higinbotham, S. Gilad, and A. Saha, Detailed Study of the ^{4}He Nuclei Through Response Function Separations at High Momentum Transfers, Jefferson Lab Experiment 01-108.
3. J.M. Laget, *Phys. Lett.* B **199**, 493 (1987).
4. J.M. Laget, *Phys. Rev.* D **579**, 333 (1994).

FINAL STATE INTERACTION IN ^{4}He(e,e'p)^{3}H REACTION - STUDY OF FINITE FORMATION TIME EFFECTS -

H. MORITA

Sapporo Gakuin University, Bunkyo-dai 11, Ebetsu, Hokkaido 069-8555, Japan
E-mail: hiko@earth.sgu.ac.jp

C. CIOFI DEGLI ATTI

Department of Physics, University of Perugia, and Istituto Nazionale di Fisica Nucleare, Sezione di Perugia, Via A. Pascoli, I-06100 Perugia, Italy
E-mail: Claudio.Ciofi@pg.infn.it

D. TRELEANI

Department of Theoretical Physics, University of Trieste, Strada Costiera 11, Istituto Nazionale di Fisica Nucleare, Sezione di Trieste, and ICTP, I-34014 Trieste, Italy
E-mail: daniel@ts.infn.it

M.A. BRAUN

Department of High-Energy Physics, S.Petersburg University, 198904 S.Petersburg, Russia
E-mail: braun@ts.infn.it

Finite Formation Time (FFT) effects in the exclusive reaction $^4He(e,e'p)^3$H at high values of Q^2 are introduced and discussed. From its mechanism, the FFT effects attenuate the Glauber-type Final State Interaction(FSI) as the value of Q^2 grows. We investigate how and to what extent this effect becomes apparent. As a result it is shown that a dip in the distorted momentum distributions $n_D(k)$ predicted by the Plane Wave Impulse Approximation (PWIA), which is filled by the Glauber-type Final State Interaction (FSI), is completely recovered at $Q^2 \sim 20(GeV/c)^2$.

1 Introduction

Recently [1] the Glauber approach to Final State Interaction (FSI) in the inclusive quasi-elastic $A(e, e'p)X$ process, has been extended by taking into account the virtuality of the hit nucleon after $\gamma*$ absorption. It has been found that at large Q^2, due to the fact that the hit nucleon needs a Finite Formation Time (FFT) to reach its asymptotic form, the interaction with the remainder of the target nucleus becomes very weak and vanishes in the asymptotic limit. Therefore once FFT effects are taken into account, the FSI in $(e, e'p)$ reaction will be reduced at large Q^2 region. In fact Braun et al.[1] show such tendency in the case of $^2H(e,e')X$ reaction.

However for the deuteron the FSI is small in any case, therefore the FFT effects do not lead to any qualitative change in the cross section. One expects that these effects would be much more pronounced for more compact nuclei, such as ^{4}He. Therefore we apply the FFT approach [1] to the exclusive $^4He(e,e'p)^3H$ reaction. In this paper we present the results of distorted momentum distributions $n_D(k)$ for the $^4He(e,e'p)^3H$ reaction at various Q^2 and discuss how and to what extent the FFT effects make an influence on them.

In Sec.2, the expression of the distorted momentum distributions is given and the formulation of FSI is explained in Sec.3. After the formulation of the wave function used in the present calculations is briefly shown in Sec.4, we show the results of our calculations in Sec.5. Finally our findings are summarized in Sec.6.

2 The Distorted Momentum Distributions

The cross section for the process $^4He(e,e'p)^3H$, can be shown to have the following form

$$\frac{d^5\sigma}{d\vec{k}_{e'}d\Omega_{\vec{k}_p}} = \mathcal{K}n_D(\vec{k}_m), \qquad \vec{k}_m = \vec{q} - \vec{k}_p, \qquad (1)$$

where $\mathcal{K}$ is a kinematical factor and $n_D(\vec{k}_m)$ mean the proton distorted momentum distributions, which are defined as follows

$$n_D(\vec{k}_m) = |w(\vec{k}_m)|^2 \qquad (2)$$

where

$$w(\vec{k}_m) = (2\pi)^{-3/2}\int d\vec{r}exp(-i\vec{k}_m \cdot \vec{r})A(\vec{r}) \qquad (3)$$

and

$$A(\vec{r}) = \sqrt{4}\int d\vec{R}_1 d\vec{R}_2 d\vec{R}_2 \psi_t^*(\vec{R}_1,\vec{R}_2)S_G^\dagger\psi_\alpha(\vec{R}_1,\vec{R}_2,\vec{R}_3 = \vec{r}). \qquad (4)$$

In the above equation ψ_t and ψ_α denote the ^{3}H and ^{4}He wave function respectively. To get these wave functions we adopt the ATMS method[3] whose formulation is briefly explained in Sec.4. $\vec{R}_i's$ are usual Jacobi coordinates $\vec{R}_1 = \vec{r}_2 - \vec{r}_1$, $\vec{R}_2 = \vec{r}_3 - (\vec{r}_1 + \vec{r}_2)/2$, $\vec{R}_3 = \vec{r}_4 - (\vec{r}_1 + \vec{r}_2 + \vec{r}_3)/3$, and S_G is the Glauber operator, whose explicit form is given in the next section.

3   The Glauber Operator and the Final State Interaction

The Glauber operator S_G is explicitly given by[2]

$$S_G = \prod_{i=1}^{3} G(4i), \qquad G(4i) = 1 - \theta(z_4 - z_i)\Gamma(\vec{b}_4 - \vec{b}_i), \qquad (5)$$

where the knocked out nucleon is denoted by "4". Here the z-axis is oriented along the direction of the motion of the knocked out proton, $\vec{b}$ is the component of the nucleon coordinate in the xy plane, and Γ stands for the usual Glauber profile function

$$\Gamma(\vec{b}) = \frac{\sigma_{tot}(1 - i\alpha)}{4\pi b_0^2} e^{-\vec{b}^2/2b_0^2}, \qquad (6)$$

with σ_{tot} denoting the total proton-nucleon cross section, and α the ratio of the real to imaginary parts of the forward elastic pN scattering amplitude. The value of σ_{tot} and α were chosen at the proper values of the invariant mass of the process, and the numerical values were taken from Ref.4, whereas the value of b_0 has been determined by using the relation $\sigma_{el} = \sigma_{tot}^2(1+\alpha^2)/16\pi b_0^2$.

When FFT effects are considered, the $G(4i)$ in eq. (5) is replaced by [1]

$$G(4i) = 1 - \mathcal{J}(z_i - z_4)\Gamma(\vec{b}_4 - \vec{b}_i), \qquad (7)$$

where $\mathcal{J}$ is given by

$$\mathcal{J}(z) = \theta(z)(1 - exp(\frac{z}{l(Q^2)})), \qquad l(Q^2) = \frac{Q^2}{xmM^2}, \qquad Q^2 = \vec{q}^2 - \nu^2. \quad (8)$$

Where $\vec{q}$ and ν are momentum and energy transfer of the incident electron. The quantity l can be interpreted as a formation length growing linearly with Q^2. x is the Bjorken scaling variable, m the nucleon mass, and M represents the average virtuality defined by $M^2 = (m_{Av}^*)^2 - m^2$. In our calculations the value of the average excitation mass m_{Av}^* was taken to be $1.8(GeV/c)$ [1].

Eq.(8) shows that for large value of Q^2 one obtains a large formation length and, correspondingly, a small FSI. When the ejected nucleon is on shell, that is, the formation length is zero, $\mathcal{J}$ is reduced to $\theta(z)$ and the standard Glauber picture is recovered.

4   ATMS wave function

The ATMS wave function has the following form:

$$\Psi_{ATMS} = F \cdot \Phi_0, \qquad (9)$$

where F represents a proper correlation function, and Φ_0 is the mean field (uncorrelated) wave function. The correlation function F has the following form [3]:

$$F = D^{-1} \sum_{ij}(w(ij) - \frac{(n_p - 1)}{n_p}u(ij)) \prod_{kl \neq ij} u(kl), \qquad (10)$$

$$D = \sum_{ij}(1 - \frac{(n_p - 1)}{n_p}u(ij)) \prod_{kl \neq ij} u(kl), \qquad (11)$$

where $n_p = A(A-1)/2$ is the number of pair, $w(ij)$ and $u(ij)$ are on-shell and off-shell two-body correlation functions, respectively. Here the realistic NN interaction generates a state dependence of NN correlations, which is taken into account by introducing the following state dependence for the on-shell correlation function:

$$w(ij) =^1 w_S(ij)\hat{P}^{1E}(ij) +^3 w_S(ij)\hat{P}^{3E}(ij) +^3 w_D(ij)S_{ij}\hat{P}^{3E}(ij), \qquad (12)$$

where $\hat{P}^{1E}(\hat{P}^{3E})$ is a projection operator to the singlet-even (triplet-even) state and S_{ij} is the usual tensor operator. We also includes the tensor-type off-shell correlation function and the explicit form of wave function is given in Ref. 5. The best set of correlation functions $\{u\} = \{w's, u's\}$ are determined by the Euler-Lagrange equation.

In the following calculation the Reid soft core V_8 model potential is used as the realistic NN force. The calculated binding energies for ^{4}He and ^{3}H are 21.2 MeV and -7.20 MeV, respectively.

5 Results and Discussions

The distorted momentum distributions $n_D(\vec{k}_m)$ for the $^4He(e, e'p)^3H$ reaction are shown in Fig.1 at $Q^2 = 2, 5, 10, 20(GeV/c)^2$. Here we take the parallel kinematics, i.e. when the missing momentum $\vec{k}_m$ is oriented along the virtual photon momentum $\vec{q}$.

From these figures one can see that the "NO FSI" result, or in other word the PWIA result, represented by the dotted curve shows a dip around $k \sim 2.2(fm^{-1})$, which is totally masked by the Glauber-type FSI as is shown by the solid curves. However once the FFT effects are taken into account, the results approach to the PWIA one as Q^2 becomes larger, that is because FFT effects weaken the Glauber-type FSI by the way represented by eq.(8). As a result at around $Q^2 = 10(GeV/c)^2$ a dip behaviour starts to appear

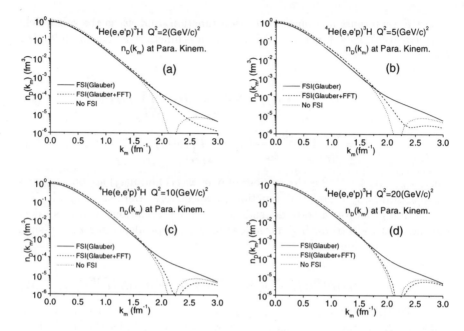

Figure 1. Results of $n_D(\vec{k}_m)$ at parallel kinematics. Here the dotted curve means the undistorted(No FSI) $n(k_m)$, and the solid and dashed curves denote the "Glauber-type FSI" and "Glauber-type FSI+FFT" $n(k_m)$, respectively. (a) $\sim$ (d) correspond to $Q^2 = 2, 5, 10, 20$ $(GeV/c)^2$, respectively.

and finally at $Q^2 = 20(GeV/c)^2$ it is totally recovered, which means that the Glauber-type FSI is almost diminished by FFT effects.

One can also recognize from figures that though the Glauber-type FSI has little Q^2 dependence FFT effects induce a large Q^2 dependence at around dip region. This is more clearly seen in Fig.2. Fig.2(a) shows results with only Glauber-type FSI, which reveal little differences for $Q^2 = 2 \sim 20(GeV/c)^2$. On the other hand as is seen from Fig.2(b), if the FFT effects are taken into account $n_D(k)$ has a large Q^2 dependence at around dip region.

The reason of little Q^2 dependence of the Glauber-type FSI is ascribed to the fact that the scattering parameter used in $\Gamma(\vec{b})$ defined by eq.(6), which are σ, b_0 and α, are almost constant in the present kinematics. For instance, σ_{tot} versus the incident proton momentum is shown in Fig.3. Figure shows a flat region at $P_{Lab} \geq 4(GeV/c)$, which corresponds to the present kinematics.

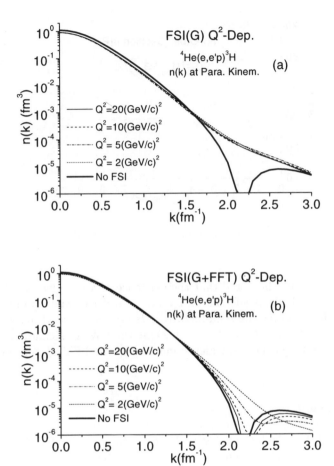

Figure 2. Q^2 dependence of $n_D(\vec{k}_m)$ at parallel kinematics. In the case of (a), only the Glauber-type FSI is taken into account, on the other hand (b) is the results with also FFT effects. Here the bold curve represents the "No FSI" results, whereas the other thin lines denote results with FSI at various Q^2 value.

Thus at large Q^2 region the Glauber-type FSI predict less Q^2 dependence, which means that the FFT mechanism can be studied by observing the Q^2 dependence of $^4He(e, e'p)^3H$ reaction cross section.

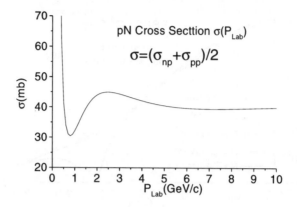

Figure 3. σ_{tot} versus incident proton momentum(P_{Lab}).

Finally it should be noted that the existence of a dip gives a benefit for the study of small effects such as FFT ones. In order to understand this, let us consider for example the semi-inclusive $^4He(e, e'p)^3H$ reaction, which are shown in Fig.4. Figure shows that in this case the PWIA calculation predicts no dip and due to this Q^2 dependence induced by the FFT effects can not be significant.

6 Summary

The results of the calculations that we have exhibited, show that the exclusive process $^4He(e, e'p)^3H$ at high values of Q^2 could become a promising probe to detect the hadronic mechanism which goes beyond the treatment of FSI in terms of Glauber-type rescattering. In particular, the FFT approach of Ref.1 predicts a clean and regular Q^2 behaviour leading to the vanishing of FSI effects at moderately large values of Q^2; such a prediction would be validated by the experimental observation of a dip in the cross section at $k_m \simeq 2.2 fm^{-1}$. Recently, Benhar et al [6] have also analyzed the $^4He(e,e'p)^3H$ reaction, using a colour transparency model. However their model predict somewhat different results from ours and does not lead to the vanishing of FSI at $Q^2 \simeq 20(GeV/c)^2$. We conclude, therefore, that exclusive electron scattering off ^4He would really provide a powerful tool to discriminate various models of hadronic final state rescattering.

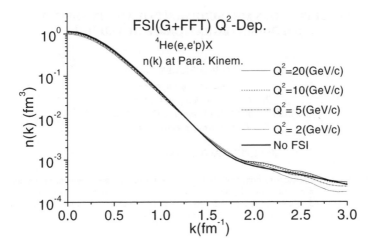

Figure 4. Same as Fig.2(b), but for the $^4He(e,e'p)X$ reaction.

Acknowledgments

H.M. thanks the INFN, Sezione di Perugia for hospitality.

References

1. M.A. Braun, C. Ciofi degli Atti and D. Treleani, Phys. Rev. C62 (2000) 034606.
2. H. Morita, C. Ciofi degli Atti and D. Treleani, Phys. Rev. C60 (1999) 034603.
3. M. Sakai, I. Shimodaya, Y. Akaishi, J. Hiura and H. Tanaka, Prog. Theor. Phys. Suppl.56 (1974) 32; Y. Akaishi, *Cluster and Other Topics(World Scientific, 1987)*, p.335.
4. http://pdg.lbl.gov/ and http://said.phys.vt.edu/.
5. H. Morita, Y.Akaishi, O. Endo and H. Tanaka, Prog. Theor. Phys. 78 (1987) 1117.
6. O. Benhar, N.N. Nikolaev, J. Speth, A.A. Usmani and B.G. Zakharov, Nucl. Phys. A673 (2000) 241.

SIMULTANEOUS MEASUREMENT OF THE TWO-BODY PHOTODISINTEGRATION OF ^{3}H AND ^{3}HE

G.V. O'RIELLY

Department of Physics
The George Washington University
725 21st Street NW, Washington DC, 20052, USA
E-mail: orielly@gwu.edu

FOR THE SAL071[†] EXPERIMENT

The 90° differential cross sections have been measured for the two-body photodisintegration of ^{3}H and ^{3}He in the reactions ^{3}H$(\gamma, d)n$ and ^{3}He$(\gamma, d)p$ using tagged photons with energies between 18 and 50 MeV. The differential cross sections have been compared with recent predictions using different models of the nucleon-nucleon potential.

1 Introduction

The A=3 systems ^{3}H and ^{3}He can be considered to be the simplest real nuclei; consequently, they are fundamental testing grounds for exact calculations employing realistic nucleon-nucleon potentials. The photodisintegration channels are a particularly useful place to test the predictions of these calculations since the electromagnetic interaction of the photon probe is well understood, simplifying the interpretation of the results.

Given this, it is surprising that the existing data sets for the two-body photodisintegration of ^{3}H [2,3,4] and ^{3}He [5,7,6,8] in the energy range from threshold to 50 MeV are rather sparse, and that these data sets are generally not consistent with one another, particularly in the case of ^{3}H, making reliable comparison between theory and experiment difficult.

We have addressed this issue by performing the first simultaneous measurement of the absolute cross sections at $\theta_{lab} = 90°$ for the two-body photodisintegration of ^{3}H and ^{3}He using tagged photons. The previous measurements were all made more than 20 years ago and so lack the precision possible with modern experimental techniques. The availability of CW electron beams together with tagged photon facilities, in which the photon energy is measured for each detected event and an accurate determination of the photon flux is obtained, provides the opportunity for new measurements free of the potential uncertainties associated with bremsstrahlung photon beams.

2 Experimental Details

The present measurement was performed using the Tagged Photon Facility at the Saskatchewan Accelerator Laboratory (SAL). Bremsstrahlung photons were tagged from 18 to 50 MeV using a magnetic spectrometer and a 31-channel focal-plane array[1] to detect the recoil electron, giving a photon energy resolution of approximately 1 MeV. The spectrometer magnet also served to deflect the unconverted primary electron beam into a well-shielded beam dump. The photon beam was incident on two identical gas cells, filled with tritium and ^{3}He respectively, located in series along the photon beam axis such that the photon flux and detector geometry were the same for both cells, thus minimizing systematic uncertainties in the comparison between the two cross sections. Charged recoil particles were detected in one of four silicon detector range telescopes mounted at 90° on each of the gas cells, with thin exit windows to minimize energy loss of the recoil particles. A coincidence between a recoil event and tagger focal-plane hit (within a 100 ns coincidence resolving time) was required to generate a tagged-event trigger for the data acquisition system.

Each target gas cell was machined from a single block of stainless steel, with 0.1 mm thick Havar beam port windows at each end, and 15 μm thick Havar foil exit windows in front of each detector telescope. The total cell length was 100 mm, with each detector viewing a shorter length of the target volume in order to ensure that any background produced by interactions in the beam-port windows was not incident on the detectors.

Each target cell had a total volume of about 100 cm^3 and was at ambient temperature and approximately 440 kPa pressure. The absolute pressure and temperature of the gas were monitored continuously during the data acquisition in order to accurately determine the target density. With these target parameters, there was approximately 1 kCi of tritium in the system during the production running period.

The tritium was supplied by Ontario Hydro Corporation of Canada, produced as a by-product of their heavy water production. It was stored and transported with the tritium gas absorbed onto a bed of natural uranium metal pellets contained in a sealed container with a valved gas line and equipped with an electrical heater. When heated to above 190°C, the tritium outgasses from the uranium to fill the target system. This method had the advantage that when the valve to the cold uranium bed was opened, the tritium was reabsorbed quickly, emptying the target system in less than one second. Due to the safety consideration necessary with the use of a gaseous tritium target, the two target cells were located within an evacuated vacuum box which served as a secondary-containment vessel in the event that the tritium cell failed in some fashion, and there was additional radiological monitoring of the environment immediately around the experimental set-up during the measurement. A schematic of of the experimental apparatus is shown in Figure 1.

310

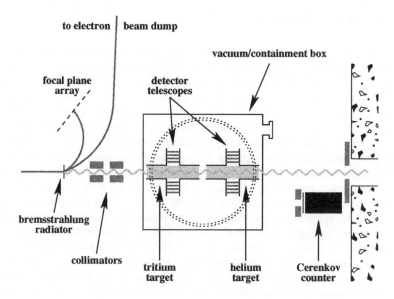

Figure 1. Schematic of the experimental apparatus

The two-body photodisintegration channels were unambiguously identified by detecting the recoil deuteron from each target in one of four silicon surface barrier detector telescopes arranged symmetrically about the beam axis at 90°. Silicon detectors were chosen because they have good energy and timing resolution, excellent linear response with energy, and high detection efficiency for charged particles in the energy range of this measurement. Each telescope consisted of 4 detectors in a ΔE–E arrangement, with thicknesses of 50, 150, 1000 and 1000 μm respectively. This configuration optimized the particle species separation and energy resolution for deuterons with energies from 3 to 20 MeV, covering to the full range of deuteron energies produced by photons in the tagged energy range of 18 – 50 MeV.

The event trigger was a signal above threshold (set to 0.3 MeV) in either of the first two elements in any of the detector telescopes. This threshold was below the minimum energy deposited by a proton or deuteron in these detector elements. For each event trigger, data were acquired in event-by-event mode, recording the energy and timing information from each detector telescope element, together with the tagger focal-plane hit pattern and timing information from each of the focal-plane counters that fired. The total event rate was approximately 25 Hz, giving a data acquisition live-time of over 98%.

3 Analysis Details

To determine the deuteron yield at each photon energy, events were selected using particle identification to select deuterons, requiring a true coincidence between a tagged photon and the detected deuteron, and finally using the reaction kinematics to reconstruct the photon energy and comparing this with the photon energy known from the tagger.

3.1 Particle Identification

Deuterons from the two-body photodisintegration reactions were identified using the energy deposited in the silicon detector telescope, comparing the full energy of the particle with the energy deposited in the detectors before the stopping detector. The resulting stopping-power plot of ΔE vs. E_{total}, shown in Figure 2a, clearly illustrates the separation of proton and deuteron events.

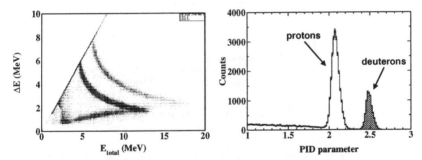

Figure 2. The left plot shows ΔE vs. E_{total} for a single detector telescope. The right plot show the linearized PID for the same detector telescope, together with the cut used to select deuterons.

These stopping-power bands were then linearized and projected onto a PID axis perpendicular to the linearized bands, and a cut applied to this linearized PID parameter to select deuteron events (see Figure 2b). The use of this linearized PID simplifies the analysis and allows a reliable estimation of the misidentification fractions (excluded deuterons and misidentified protons) necessary to correct the final deuteron yields. For the very lowest energy deuterons, *i.e.* those which stop in the first ΔE element of a telescope, only deuterons with a total energy greater than the maximum energy loss of a proton can be uniquely identified as a deuteron. For the 50 μm detectors used in this measurement, this energy is 2.5 MeV, matching the deuteron energy, after energy losses in the target and exit window, expected from the lowest energy tagged photon ($E_\gamma = 19$ MeV).

3.2   Event Timing

The timing information from the detector telescopes and tagger focal-plane counters was used to identify those deuteron events which formed a true coincidence with an electron hit in a tagger focal-plane counter. These appear in the tagger TDC as a true coincidence peak sitting on a background due to accidental coincidences caused by recoil electrons from photons uncorrelated with the detected event arriving within the 100 ns wide coincidence resolving time used in the data acquisition. Because of the very high tagged photon rate used in this measurement (an integrated rate of approximately 1.6×10^8 photons/sec), there were usually multiple hits (typically 16) in the tagger focal-plane counters during this coincidence resolving time, resulting in a large accidental background as seen in Figure 3, with a signal-to-noise ratio of 0.29.

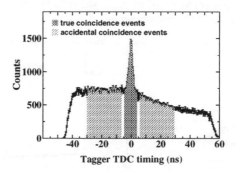

Figure 3. Tagger focal-plane TDC for a single detector telescope showing the true coincidence peak (arbitrarily positioned at 0 ns) on a background due to uncorrelated (accidental) coincidence events.

A cut between -4.4 ns to +4.4 ns was applied to the tagger timing spectrum to select the true coincidence events. Most of the accidental coincidences were eliminated by this timing cut; however, some of these accidental coincidence events will have the same timing as the true events, appearing under the true coincidence peak. This accidental background must be subtracted in order to determine the true deuteron yield.

3.3   Kinematics

For the two-body photodisintegration reaction, the measured deuteron energy (corrected for energy loss in the target and exit-window foils) and angle (90°) can be used to uniquely reconstruct the energy of the incident photon. This can be compared with the photon energy known from the tagger, kinematically over-determining the system. The missing energy between the kinematically reconstructed and measured photon energy was determined (see Figure 4a).

Even with all the analysis cuts applied, there was still a contribution from accidental coincidence events. The correction for these accidental coincidences was made by selecting events outside the prompt coincidence timing region (from 6 to 30 ns both before and after the true coincidence peak) and generating the missing energy for these events in the same way as was done for events from the true coincidence region. The resulting missing-energy spectrum was empirically scaled to the missing energy reconstructed from the prompt events to match the area in the region outside of $\Delta E_\gamma = \pm 5$ MeV. This scaled random spectrum reproduced the shape of the prompt spectrum, with the final missing energy spectrum showing a flat background consistent with zero away from the kinematic peak, as seen in Figure 4b. This gave confidence that the random contribution in the good event region has been correctly accounted for. A final cut to select those events within ± 3.5 MeV from zero (the expected energy difference for good events) was applied.

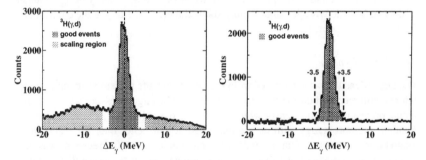

Figure 4. The left plot shows the missing energy $[E_\gamma]_{reconstructed} - [E_\gamma]_{tagger}$ for deuteron events from the tritium target with true timing cut (summed over all photon energies). The right plot shows the missing energy after subtraction of the scaled accidental missing-energy spectrum. The range used to determine the deuteron yields is shown by the two arrows.

3.4   Yield Corrections

Since the TDCs recorded only a single hit for each event, it is possible for a true coincidence to be lost (or stolen) if an earlier (uncorrelated) recoil electron hits the same tagger focal-plane counter. The correction for these lost events, referred to as the stolen coincidence correction, is the probability that a second electron arrived before the true event time but still within the coincidence resolving time. This depends on the rate R of electron hits in the tagger focal-plane counter and the time t from the TDC start to true peak position. This correction factor was calculated separately for each tagger channel and detector to account for the different focal-plane counter rates and for timing differences between the silicon detectors and focal plane counters. The individual stolen coincidence corrections varied from 8 to 30% (for the fastest counting tagger channels), with an overall uncertainty of $\pm 3\%$.

3.5 Cross Sections

The extraction of the absolute 90° cross sections at each photon energy required knowledge of the deuteron yield, target density, detector acceptance and the flux of tagged photons incident on the target.

The final deuteron yield was determined after using particle identification to select deuterons, applying a tagger TDC timing cut to select true coincidence events and using the reaction kinematics, as described above. Corrections were made to subtract the accidental coincidence background which survived the analysis cuts, and for stolen coincidences due to the single-hit operation of the TDCs. Analysis of data runs made with the target empty showed there was a negligible contribution from non-target sources.

3.6 Target Density

The target density was determined from the target absolute pressure and temperature information recorded during the data acquisition. The target pressure was measured with a precision pressure sensor which has been cross-calibrated against an absolute standard pressure gauge with an absolute accuracy of better than 1%. The target temperature, measured with a standard thermocouple to $\pm 2°C$ precision, was also known to an accuracy of less than 1%. Combining these gives an overall systematic uncertainty in the target density of $\pm 1.1\%$ for each target.

3.7 Detector Acceptance

The detector acceptance was determined using a Monte-Carlo simulation to account for the extended target geometry, finite detector size and position, and the collimation. Based on realistic estimates of the possible uncertainties with each of the simulation inputs, the acceptance has an estimated uncertainty of $\pm 3\%$.

3.8 Photon Flux

The tagger focal-plane counter scalers provided the number of tagged electrons, which has a one-to-one correspondence with the number of photons produced. However, due to collimation of the photon beam, not all of these photons are incident on the target. In order to determine the absolute photon flux incident on the target, the fraction of tagged electrons which correspond to a photon after the collimation was determined by direct counting of the photons, in measurements of the tagging efficiency. These were frequent high-statistics measurements performed at reduced electron beam currents with the photons counted in a large lead-glass Čerenkov detector placed directly in the photon beam. The tagging efficiency is simply the ratio of electron-photon coincidences to electrons detected in the tagger focal-plane counters. The tagging efficiency determined in a single measurement was typically 0.24, determined with a statistical accuracy of better than 1% for each tagger channel. These measurements were made repeatedly throughout the production data-taking period. Analysis of the full set of tagging-efficiency measurements indicated a systematic uncertainty of 1% leading to an overall uncertainty (statistical and systematic combined in quadrature) of less than 1.5% in the final photon flux determination.

3.9 Final Uncertainties

The final statistical uncertainties for the deuteron yields at each photon energy (in 1 MeV bins) ranged from 2 to 8% for the tritium photodisintegration (with the largest uncertainties at the higher photon energies where the cross section is smallest). For the ^{3}He photodisintegration, the final statistical uncertainties for each photon energy bin were from 3 to 15%, the ^{3}He deuteron yields being smaller because of the lower target density since helium is monoatomic.

The total systematic uncertainty, with contributions from the target density ($\pm 1.1\%$), photon flux ($\pm 1.5\%$), detector acceptance ($\pm 3\%$) and the stolen coincidence correction ($\pm 3\%$) combined in quadrature, was $\pm 4.6\%$, comparable with the statistical uncertainty on each of the data points.

4 Results

The results for the two-body photodisintegration of tritium are shown in Figure 5. The present results, which have significantly better statistical uncertainties and extend to a higher energy than any previous measurement, are clearly inconsistent with these previous data sets. The data have also been compared with the results of theoretical predictions by Schadow [9] for three different realistic nucleon-nucleon potentials (using the Bonn, Paris and Nijmegen potentials). Good agreement is observed between the calculations using the Bonn and Paris potential and the results of the present measurement.

The results for the two-body photodisintegration of ^{3}He are shown in Figure 6. These results have uncertainties comparable with the best previous data sets. This

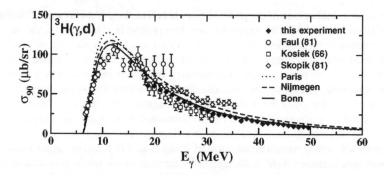

Figure 5. The ^{3}H$(\gamma,d)n$ 90° cross section from 18 to 50 MeV. The error bars are statistical only. Results from previous measurements, Faul[3], Kosiek[2] and Skopik[4] are shown. The lines show theoretical predictions by Schadow[9] for different NN potentials.

measurement is consistent with the previous data sets for photon energies above about 25 MeV. Below this energy, the results from the present measurement are inconsistent with the previous data sets, lying between the higher data set of Kundu[6] and the lower band of data sets by Berman[5], Stewart[7] and Ticcioni[8]. The data have been compared with the results of theoretical predictions by Schadow[9] using the the Bonn, Paris and Nijmegen nucleon-nucleon potentials. As in the ^{3}H(γ,d) reaction, good agreement is observed between the calculations using the Bonn and Paris potential and the results of the present measurement.

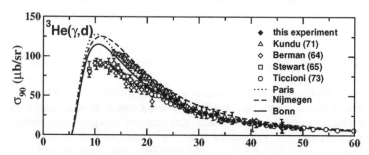

Figure 6. The ^{3}He$(\gamma,d)n$ 90° cross section from 18 to 50 MeV. The error bars are statistical only. Results from previous measurements Kundu[6], Berman[5], Stewart[7] and Ticcioni[8]. are shown. The lines show theoretical predictions by Schadow[9] for different nucleon-nucleon potential theoretical predictions for different NN potentials.

5 Conclusion

We have completed a measurement of the two-body photodisintegration of ^{3}H and ^{3}He in the energy range from 18 to 50 MeV. The results of the present measurements have better systematic uncertainties than the previous photodisintegration measurements on the A=3 system in this energy range, and comparable or better statistical uncertainties than the previous data. The present results are inconsistent with the previous data sets for the ^{3}H(γ, d) reaction, and agree with the previous data sets for the ^{3}He(γ, d) only for photon energies above 25 MeV. Comparison with the recent theoretical calculations by Schadow[9] show good agreement between the results of this measurement and the calculations made using the Bonn and Paris nucleon-nucleon potentials for both the ^{3}H(γ, d) and ^{3}He(γ, d) reactions.

Acknowledgments

The author gratefully acknowledge the support of NSF grant PHY-9703049 and DOE grant DE-FG02-95ER

† SAL071

G.V. O'RIELLY[a], G. FELDMAN[a], J.R. CALARCO[b], N.M. KARLSSON[c], B.L. BERMAN[a], W.J. BRISCOE[a], R. IGARASHI[d], N.R. KOLB[d], R.E. PYWELL[d] AND D.M. SKOPIK[d]

[a] *Physics Department, The George Washington University, 725 21st Street NW, Washington DC, 20052, USA*

[b] *Physics Department, University of New Hampshire, Durham, NH 03824, USA*

[c] *Physics Department, Lund University, Box 117, SE-221 00 Lund, SWEDEN*

[e] *Department of Physics and Engineering Physics, University of Saskatchewan 116 Science Place, Saskatoon, SK S7N 5E2 CANADA*

References

1. J.M. Vogt et al., *Nucl. Instrum. Methods* **A324**, 198 (1983).
2. R. Kosiek et al., Phys. Lett. **21**, 199 (1966).
3. D.D. Faul et al., *Phys. Rev.* **C24**, 849 (1981).
4. D.M. Skopik et al., *Phys. Rev.* **C24**, 1791 (1981).
5. B.L. Berman et al., Phys. Rev. **133B**, 117 (1964).
6. S.K. Kundu et al., *Nucl. Phys.* **A171**, 384 (1971).
7. J.R. Stewart et al., *Phys. Rev.* **138**, B372 (1965).
8. G. Ticcioni et al., *Phys. Lett.* **46B**, 369 (1973).
9. W.Schadow et al., *Phys. Rev.* **C63**, 044006 (2001).

NUCLEAR MEDIUM EFFECTS IN HADRON LEPTOPRODUCTION

N. BIANCHI

I.N.F.N. Laboratori Nazionali di Frascati via E. Fermi 40 I-00044 Frascati, Italy
E-mail: nicola.bianchi@lnf.infn.it

The production of charged hadrons in semi-inclusive deep inelastic scattering off nuclei has been studied by the HERMES experiment at DESY using 27.5 GeV positrons. A reduction of the multiplicity of charged hadrons and identified charged pions from nuclei relative to that from deuterium has been measured as a function of the relevant kinematic variables ν, z and p_t^2. A larger reduction of the multiplicity ratio R_M^h has been found for the krypton with respect to the one previously measured on nitrogen in agreement with a $\sim A^{2/3}$ power law. Both the krypton and nitrogen data show that the multiplicity ratio is the same for positive and negative pions, while a significant difference is observed between R_M^h for positive and negative hadrons. This result can be interpreted in terms of a difference between the formation time of protons and pions. It has been also suggested that the observed differences between positive and negative hadrons can be attributed to a different modification of the quark and antiquark fragmentation functions in nuclei.

1 Introduction

Inclusive Deep-Inelastic-Scattering has been widely used to extract information of the nuclear medium modification of the Parton Distribution functions. A large effect (the EMC effect) was found. Despite much experimental and theoretical progress, no unique and universally accepted explanation has emerged so far. Clearly the conventional mucleon-meson dynamics seems not adequate to reproduce the experimental data.

On other hand, Semi-Inclusive DIS can be used to have new information of the effect of the nuclear medium on the quark fragmentation process. The understanding of quark propagation through the nuclear environment is crucial for the interpretation of ultra-relativistic heavy ion collisions and high energy proton-nucleus and lepton-nucleus interactions. Quark propagation in the nuclear medium involves competing processes like gluon radiation, the quark energy loss through multiple scattering, and the hadronization of quarks. The multiple scattering process is directly associated with multiparton correlation functions in a nucleus. The hadronization is linked to quark confinement, and a characterization of the hadronization process is likely to provide important insights into the nature and the origin of confinement. Semi-inclusive deep inelastic lepton-nucleus collisions provide a unique opportunity to study these effects. In the simplest scenario, the nucleus, which has the size of a few fermi, acts as an ensemble of targets with which the struck quark or the formed hadron may interact. In contrast to proton-nucleus scattering, in SIDIS no deconvolution of the distributions of the projectile and target fragmentation particles has to be made, so that hadron distributions and multiplicities from different nuclei can be directly related to nuclear effects in quark propagation and hadronization.

2   Experiment

The measurement has been performed at HERMES using D, ^{14}N and ^{84}Kr gas targets internal to the 27.5 GeV HERA positron storage ring, by identifying both the scattered positron and the produced hadrons in the HERMES spectrometer [1]. Kinematic cuts have been considered on the scattered lepton to ensure the hard scattering regime: $Q^2 > 1$ GeV^2, $W^2 > 4$ GeV^2. Among the hadrons, pions were identified with a Cerenkov detector in a wide momentum range.

The experimental results are presented in terms of the multiplicity ratio R_M^h, which represents the ratio of the number of hadrons of type h produced per DIS event for a nuclear target of mass A to that from a deuterium target (D):

$$R_M^h(z, \nu) = \frac{\left.\frac{N_h(z,\nu)}{N_e(\nu)}\right|_A}{\left.\frac{N_h(z,\nu)}{N_e(\nu)}\right|_D} = \frac{\left.\frac{\sum e_f^2 q_f(x) D_f^h(z)}{\sum e_f^2 q_f(x)}\right|_A}{\left.\frac{\sum e_f^2 q_f(x) D_f^h(z)}{\sum e_f^2 q_f(x)}\right|_D}, \tag{1}$$

where the expression on the right-hand-side represents the interpretation of R_M^h in the framework of the quark-parton model. The multiplicity ratio is seen to depend on the the virtual photon energy ν and on the fraction z of this energy transferred to the hadron; $N_h(z, \nu)$ is the number of semi-inclusive hadrons in a given (z, ν)-bin and $N_e(\nu)$ the number of inclusive DIS leptons in the same ν-bin. This ratio can be expressed in term of the fragmentation functions $D_f^h(z)$ of a quark of flavor f, and the quark distribution functions $q_f(x)$, with x the Bjorken scaling variable. Additional cuts on the detected hadron energy were considered to minimize the contamination from the target fragmentation region : $z > 2$.

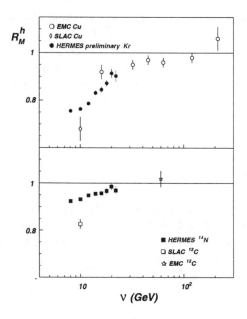

Figure 1. Charged hadron multiplicity ratio R_M^h as a function of ν for values of z larger than 0.2. The error bars represent the statistical uncertainty only.

3 Results

3.1 ν- and z- dependence

The HERMES multiplicity ratios for [2] all charged hadrons with $z > 0.2$ are presented as a function of ν in Fig. 1 together with data of previous experiments on nuclei of similar size [3,4]. The HERMES data for R_M^h are observed to increase with increasing ν and are consistent with the high-energy EMC data. The apparent discrepancy between HERMES and SLAC data can be partially attributed to the reduction of the inclusive nuclear cross section which was not included in the SLAC results. Clearly the HERMES kinematic region is well suited for the study of the quark propagation and hadronization.

In Fig. 2, the multiplicity ratio for charged hadrons with $\nu > 7$ GeV is given as a function of z. A stronger attenuation of factor of ~ 3.7 is observed for ^{84}Kr with respect to ^{14}N. This value is in good agreement with the predicted mass number-dependence of the modifications of the quark fragmentation functions in DIS [5], which are expected to depend quadratically on the nuclear size $\sim A^{2/3}$.

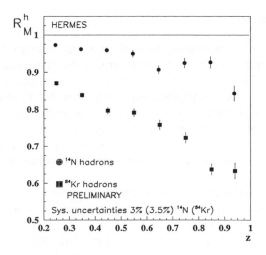

Figure 2. The multiplicity ratio for charged hadrons versus z.

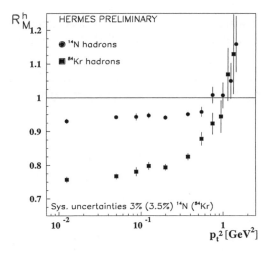

Figure 3. The multiplicity ratio for charged hadrons versus P_t^2.

3.2 P_t^2- dependence

Experimental results on hadron production in high-energy hadron-nucleus collisions show an increase at high hadron transverse momentum P_t over what would be expected based on a simple scaling of the appropriate proton-proton cross section. The standard physical explanation[6] is that the hadron traveling through the nucleus gains extra transverse momentum due to random soft collisions increasing the parton transverse momentum (Cronin efftect). New heavy-ion data from the SPS [7] and RHIC [8] show, for the central collisions, a weaker P_t enhancement in contradiction with the expectations based on the Cronin effect.

In lepton-nucleus collisions neither multiple scattering of the incident particle nor the interaction of its constituents complicate the interpretation of the data. Hence, the residual nuclear effect can be studied directly. A nuclear enhancement at high P_t^2 is observed from the HERMES data shown in Fig. 3. This enhancement is similar to the one reported for proton-nucleus and nucleus-nucleus collisions. Within the Glauber formalism [9] for the multiple parton scattering, the transition between soft and hard processes is predicted to occur at a P_t-scale of about 1-2 GeV, which is in agreement with HERMES results.

The multiple scattering process is directly associated with multiparton correlation functions. It has been shown [10] that the transverse momentum

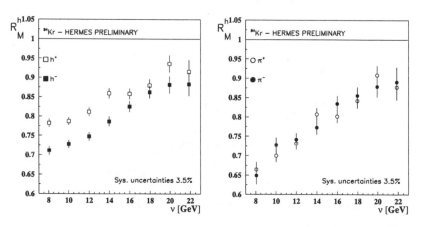

Figure 4. Multiplicity ratios for all charged hadrons including pions , and those for identified pions as a function of ν. The open (closed) symbols represent the positive (negative) charge states.

broadening of leading pions in deep inelastic lepton-nucleus scattering is an excellent observable to probe the parton correlation functions in the nucleus, and that a measurement of the mass-number dependence of the P_t-enhancement provides information on the functional form of the parton correlation functions. The preliminary HERMES results for charged hadrons shown in Fig. 3 indicate that the high P_t-enhancement (with respect to the R_M^h data at low P_t^2) is significantly larger in ^{84}Kr relative to ^{14}N.

3.3 Hadron-charge dependence

For both hadrons and pions the multiplicity ratios have been determined separately for the two charge states. Pions were identified in the momentum range between 4 and 13.5 GeV using a Cerenkov detector.

In Fig. 4 the multiplicity ratios for positive and negative hadrons and pions are displayed as a function of ν for ^{84}Kr. The data show that the multiplicity ratio is the same for positive and negative pions while a significant difference is observed between positive and negative hadrons. This result, which agrees with the one reported for the ^{14}N data [2] can be interpreted in terms of a difference between the formation time of protons and pions. In particular the HERMES data seem to suggest that a proton has a larger formation time than a pion [2]. Alternatively, it has been suggested that the observed differences between positive and negative hadrons can be attributed to a different modification of the quark and antiquark fragmentation functions in nuclei [12]. In order to clarify this issue an analysis has been started of multiplicity ratios for identified kaons, protons and anti-protons on various nuclei using the RICH detector at HERMES.

3.4 Model calculations

Recent QCD inspired models which account for the space-time development of the process, have provided theoretical descriptions of both the hadronization process and the observables in relativistic heavy-ion collisions and in Drell-Yan reactions on nuclear targets.

In the gluon-bremsstrahlung model of Ref. [11] radiative energy loss of a highly virtual quark originating from a DIS electron, plays a crucial role in the production of leading hadrons off nuclei. The struck quark dissipates energy via the emission of gluons until, in the case of meson formation, a $q\bar{q}$ configuration is formed consisting of the struck quark and an antiquark originating from the last emitted gluon. The interactions of the formed hadron with the nuclear environment causes a nuclear suppression of the hadron multiplicity. Moreover, the interaction of the initial quark with the nuclear medium

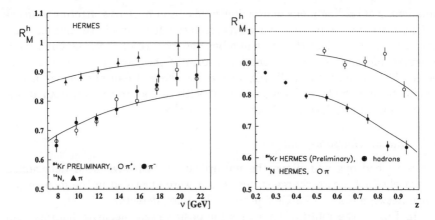

Figure 5. The multiplicity ratio for charged pions and charged hadrons compared with the gluon-bremsstrahlung model predictions.

causes the emission of additional soft gluons. The model predictions for ^{84}Kr are shown in Fig. 5. Both the ν and z dependence of the data are well described by the model expectations, which is also in agreement with the results obtained for ^{14}N.

Alternatively, in Ref. [12] the nuclear modification of the quark fragmentation in DIS has been evaluated taking into account multiple parton scattering and induced energy loss in the medium. In the framework of perturbative QCD, the modified quark fragmentation functions and their QCD evolution have been calculated including the next-to-leading twist contribution. It has been shown that the modification of the fragmentation functions depends on twist-four parton matrix elements in nuclei, and its magnitude depends quadratically on the nuclear size. The calculated ratio of the fragmentation function in nuclei $D_A(z, Q^2)$ and in deuterium $D_d(z, Q^2)$, is shown in Fig. 6 where it is compared to the measured multiplicity ratio for both ^{84}Kr and ^{14}N nuclei. A good description of the z and ν dependence for both nuclei is obtained. No subsequent interaction of the produced hadrons in the nuclear medium has been included in this calculation [12], which is at variance with the description of the hadronization process in terms of both hard-parton and soft-hadron interactions.

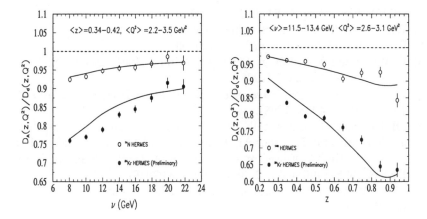

Figure 6. The multiplicity ratio for charged hadrons compared with the prediction of the nuclear modification of the fragmentation function.

4 Conclusion

New data on the nuclear medium effect of hadron multiplicity in leptoproduction have been presented. The data, which were collected by the HERMES experiment, show a strong attenuation of the multiplicity ratios which is significantly larger for ^{84}Kr as compared to ^{14}N. The ν and z-dependence of the data are well described by two independent model calculations. While the nuclear attenuation is mainly attributed to hadron-nuclear interactions in the one case [11], nuclear higher-twist contributions to the fragmentation functions are introduced in the other case [12] to explain the data. A large enhancement of nuclear hadron multiplicity has been found at large P_t^2, similar to the one measured in heavy-ion collision. Further theoretical and experimental work is needed in order to investigate the interplay between hadronic and partonic degrees of freedom.

References

1. HERMES, K. Ackerstaff et al., *Nucl. Instr. Meth.* A **417**, 230 (1998).
2. HERMES, A. Airapetian et al., *Eur. Phys. J.* C **20**, 479 (2001).
3. L.S. Osborne et al., *Phys. Rev. Lett.* **40**, 1624 (1978).
4. EMC, J. Ashman et al., *Zeit. Phys.* C **52**, 1 (1991).
5. X. Guo and X.N. Wang, *Phys. Rev. Lett.* **85**, 3591 (2000).
6. X.N. Wang, *Phys. Rep.* **280**, 287 (1997), *Phys. Rev. Lett.* **81**, 2655

(1998), *Phys. Rev.* C **61**, 064910 (2000).

7. WA98 Collaboration nucl-ex/0108006.

8. PHENIX Collaboration nucl-ex/0109003.

9. E. Wang and X.N. Wang, *Phys.Rev.* C **64** 034901 (2001) ; X.N. Wang *Phys. Rev.* C **58**, 2321 (1998).

10. X. Guo and J. Qiu, *Phys. Rev.* D **61**, 096003 (2000); X. Guo *Phys. Rev.* D **58**, 114033 (1998).

11. B. Kopeliovich, J. Nemchik, private communication and B. Kopeliovich, J. Nemchik and E. Predazzi, Proceedings of the workshop on Future Physics at HERA, Edited by G. Ingelman, A. De Roeck and R. Klanner, DESY, 1995/1996, vol 2, 1038 (nucl-th/9607036).

12. X.N. Wang private communication and X. Guo and X.N. Wang hep-ph/0102230.

QUASIFREE PROCESSES FROM NUCLEI: MESON PHOTOPRODUCTION AND ELECTRON SCATTERING

L.J. ABU-RADDAD[1] *AND J. PIEKAREWICZ[2] †

[1] *Theory Group, Research Center for Nuclear Physics, Osaka University, 10-1 Mihogaoka, Ibaraki City, Osaka 567-0047, Japan*

[2] *Department of Physics, Florida State University, Tallahassee, FL 32306, USA*

We have developed a relativistic formalism for studying quasi-free processes from nuclei. The formalism can be applied with ease to a variety of processes and renders transparent analytical expressions for all observables. We have applied it to kaon photoproduction and to electron scattering. For the case of the kaon, we compute the recoil polarization of the lambda-hyperon and the photon asymmetry. Our results indicate that polarization observables are insensitive to relativistic, nuclear target, and distortion effects. Yet, they are sensitive to the reactive content, making them ideal tools for the study of modifications to the elementary amplitude — such as in the production, propagation, and decay of nucleon resonances — in the nuclear medium. For the case of the electron, we have calculated the spectral function of ^{4}He. An observable is identified for the clean and model-independent extraction of the spectral function. Our calculations provide baseline predictions for the recently measured, but not yet fully analyzed, momentum distribution of ^{4}He by the A1-collaboration from Mainz. Our approach predicts momentum distributions for ^{4}He that rival some of the best non-relativistic calculations to date.

1 Introduction

Faced by an increasing demand for studying quasifree processes from nuclei, we have developed a general fully relativistic treatment for studying such interactions [1,2,3]. The power of the theoretical approach employed here lies in its simplicity. Analytic expressions for the response of a mean-field ground state may be provided in the plane-wave limit. The added computational demands placed on such a formalism, relative to that from a free on-shell proton, are minimal. The formalism owes its simplicity to an algebraic trick, first introduced by Gardner and Piekarewicz [4], that enables one to define a "bound" (in direct analogy to the free) nucleon propagator. Indeed, the Dirac structure of the bound nucleon propagator is identical to that of the free Feynman propagator. As a consequence, the power of Feynman's trace

---

*ELECTRONIC ADDRESS: LAITH@RCNP.OSAKA-U.AC.JP
†ELECTRONIC ADDRESS: JORGEP@CSIT.FSU.EDU

techniques may be employed throughout the formalism.

We have applied this formalism to two kinds of processes: kaon photoproduction [1,2] and electron scattering [3]. Further, there is a promising potential of applying it to many processes being studied experimentally at various laboratories. We will give here a brief introduction to this formalism and we will discuss some of the results of using it.

The investigation of the quasifree kaon photoproduction process is impelled by recent experimental advances and the increasing interest in the study of strangeness-production reactions from nuclei. These reactions form our gate to the relatively unexplored territory of hypernuclear physics. Moreover, these reactions constitute the basis for studying novel physical phenomena, such as the existence of a kaon condensate in the interior of neutron stars[5].

As for electron scattering, the appeal of this reaction is due to the perceived sensitivity of the process to the nucleon momentum distribution. Interest in this reaction has stimulated a tremendous amount of experimental work at electron facilities such as NIKHEF, MIT/Bates, and Saclay, who have championed this effort for several decades. Our motivation for studying this process is twofold: First, we use this formalism to compute the spectral function of ^{4}He in anticipation of the recently measured, but not yet fully analyzed, A1-collaboration data from Mainz [6,7,8,9,10]. Second, we take advantage of the L/T separation at Mainz to introduce what we regard as the cleanest physical observable from which to extract the nucleon spectral function.

2 Formalism

We provide here a brief discussion of our formalism. We use a plane-wave formalism and incorporate no distortions. Our rationale for this is that we concentrate on polarization observable which are typically insensitive to distortions. Moreover, in some occasions the effect of distortions is determined from other treatments and thus we are able to concentrate on the fundamental physics with no diversions. Notably, there is a definite appeal in terms of practicality: we can use now the Gardner's and Piekarewicz's [4] trick which renders transparent analytical results for all observables.

The Gardner and Piekarewicz trick enables us to introduce the concept of a "bound-state propagator":

$$S_\alpha(\mathbf{p}) = \frac{1}{2j+1} \sum_m U_{\alpha,m}(\mathbf{p}) \overline{U}_{\alpha,m}(\mathbf{p})$$
$$= (\not{p}_\alpha + M_\alpha), \quad \left(\alpha = \{E, \kappa\}\right). \tag{1}$$

The mass-, energy-, and momentum-like quantities in this expression are defined in terms of the upper component of the Dirac spinor $g_\alpha(p)$ and the lower component of the Dirac spinor $f_\alpha(p)$ [1,2,3].

The evident similarity in structure between the free and bound propagators for the direct product of spinors results in an enormous simplification; we can now employ the powerful trace techniques developed by Feynman to evaluate all observables — irrespective if the nucleon is free or bound to a nucleus. It is important to note, however, that this enormous simplification would have been lost if distortion effects would have been incorporated.

In order to automate the straightforward but lengthy procedure of calculating these Feynman traces, we rely on the *FeynCalc 1.0*[11] package with *Mathematica 2.0* to calculate all traces involving γ-matrices.

3   Results

3.1   kaon quasifree process

We start the discussion of our results by examining the nuclear dependence of the polarization observables. Fig. 1 displays the recoil polarization ($\mathcal{P}$) of the

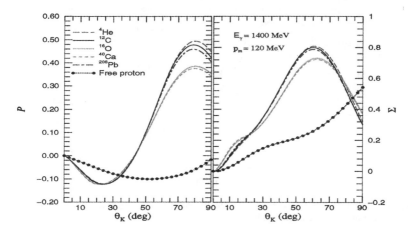

Figure 1. The polarization observables for the knockout of a valence proton from a variety of nuclei and for a free proton.

Λ—hyperon and the photon asymmetry (Σ) as a function of the kaon scattering angle for the knockout of a valence proton for a variety of nuclei, ranging from ^{4}He all the way to ^{208}Pb. These observables were evaluated at a photon energy of $E_\gamma = 1400$ MeV and at a missing momentum of $p_m = 120$ MeV. We have used the Saclay-Lyon model for the elementary amplitude [12]. We have included also polarization observables from a single proton to establish a baseline for comparison against our bound–nucleon calculations. The sensitivity of the polarization observables to the nuclear target is rather small. Moreover, the deviations from the free value are significant. This indicates important modifications to the elementary process in the nuclear medium. Although not shown, we have studied the importance of relativity and found that these observables are insensitive to relativistic dynamics.

Having established the independence of polarization observables to relativistic effects and to a large extent to the nuclear target we are now in a good position to discuss the sensitivity of these observables to the elementary amplitude (note that an insensitivity of polarization observables to final-state interaction has been shown in Ref. [13]). We display in Fig. 2 the differential cross section as a function of the kaon scattering angle for the knockout of a

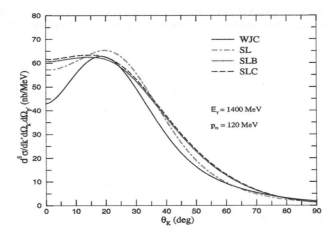

Figure 2. The differential cross section for the knockout of a proton from ^{12}C using various models for the elementary amplitude.

proton from the $p^{3/2}$ orbital in ^{12}C using four different models for the elementary amplitude [12,14,15]. Although there are noticeable differences between the models, primarily at small angles, these differences are relatively small. Much more significant, however, are the differences between the various sets for the case of the polarization observables displayed in Fig. 3. The added sensitivity

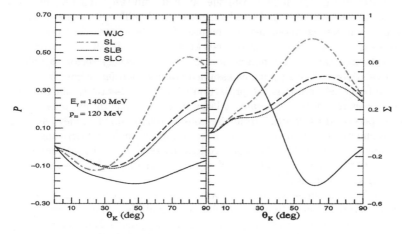

Figure 3. The polarization observables for the knockout of a proton from ^{12}C using various models for the elementary amplitude.

to the choice of amplitude exhibited by the polarization observables should not come as a surprise; unraveling subtle details about the dynamics is the hallmark of polarization observables. In particular, polarization observables show a strong sensitivity to the inclusion of the off-shell treatment for the various high-spin resonances, as suggested in Ref. [15]

3.2 electron quasifree process

There is a vast amount of literature on $(e, e'p)$ reaction in the quasifree region. Most relevant to our present discussion is the one pertaining to fully relativistic calculations such as the extensive set of studies conducted by the *"Spanish"* group of Udias and collaborators [16,17,18,19,20,21,22]. These studies have shown that the many subtleties intrinsic to the relativistic approach challenge much of the "conventional wisdom" developed within the non-relativistic framework and that, as a result, a radical revision of ideas may be required.

The experimental extraction of the spectral function is based on a non-relativistic plane-wave result that is typically referred to as the factorization [23]:

$$S(E, \mathbf{p}) = \frac{1}{p' E'_p \sigma_{eN}} \frac{d^6\sigma}{dE'_e d\Omega_{\mathbf{k}'} dE'_p d\Omega_{\mathbf{p}'}} . \tag{2}$$

However, this procedure is problematic. First, the quasifree cross section [the numerator in Eq. (2)] suffers from the off-shell ambiguity; different on-shell equivalent forms for the single-nucleon current yield different results. Second, the problem gets compounded by the use of an elementary electron-proton cross section (σ_{eN}) evaluated at off-shell kinematics [24]. Finally, the projection of the bound-state wave-function into the negative-energy sector as well as other relativistic effects spoil this assumed cross section factorization [19].

To be noted here that the projection of the bound-state spinor into the negative-energy states dominate at large missing momenta and may mimic effects perceived as "exotic" from the non-relativistic point of view, such as an asymmetry in the missing-momentum distribution [4] or short-range correlations [25]. Indeed, Caballero and collaborators have confirmed that these contributions can have significant effect on various observables, especially at large missing momenta [19].

While a consistent relativistic treatment seems to have spoiled the factorization picture obtained from a non-relativistic analysis, and with it the simple relation between the cross-section ratio and the spectral function [Eq. (2)], the situation is not without remedy. Having evaluated all matrix elements analytically in the plane-wave limit, the source of the problem can be readily identified in the form of several ambiguous kinematical factors when evaluated off-shell. Thus we search for an observable that exhibits a weak dependence on these quantities and we find, perhaps not surprisingly, that the longitudinal component of the hadronic tensor could be such an observable which is given (in parallel kinematics) by

$$R_{\rm L} \equiv W^{00} \simeq F_1^2 (E_p' + M)\, \rho(p), \tag{3}$$

where $\rho(p)$ is nothing but the momentum distribution of the bound nucleon.

This expression depends on unambiguous kinematics quantities and is valid up to small (1-3 %) second-order corrections. It is also independent of the small components of the Dirac spinors and of the negative-energy states. Moreover, it is free from of off-shell ambiguities.

The momentum distribution for ^{4}He is displayed in Fig. 4 using various methods for its extraction. The solid line gives the "canonical" momentum distribution, obtained from the Fourier transform of the $1S^{1/2}$ proton wavefunction. The momentum distribution extracted from the longitudinal response (dot-dashed line) is practically indistinguishable from the canonical momentum distribution. To be noted here that the contribution from the anomalous form factor F_2 to the longitudinal response (see the dotted line in the figure) is small because it appears multiplied by two out of three "small" quantities in the problem.

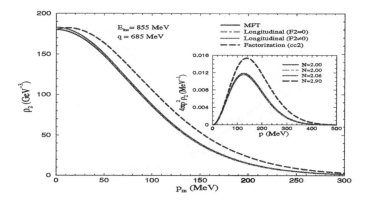

Figure 4. The proton momentum distribution ρ_2 for ^{4}He as a function of the missing momentum extracted using various methods. The inset shows the corresponding integrand from which the shell occupancy may be extracted.

The last calculation displayed in Fig. 4 corresponds to a momentum distribution extracted from the factorization approximation (long dashed line). The momentum distribution extracted in this manner overestimates the canonical momentum distribution over the whole range of missing momenta and integrates to 2.9 rather than 2; this represents a discrepancy of 45 percent.

In Fig. 5 a comparison is made between our results and non-relativistic state-of-the-art calculations of the momentum distribution of ^{4}He. The solid line displays, exactly as in Fig. 4, the canonical momentum distribution. We see no need to include the momentum distribution extracted from the longitudinal response as it has been shown to give identical results.

In addition to our own calculation, we have also included the variational results of Schiavilla and collaborators [26], for both the Urbana [27] (dashed line) and the Argonne [28] (long-dashed line) potentials, with both of them using Model VII for the three-nucleon interaction. The variational calculation of Wiringa and collaborators [29,30,31] (dashed-dotted) has also been included; this uses the Argonne v18 potential [32] supplemented with the Urbana IX three-nucleon interaction [33]. Figure 5 also shows NIKHEF data by van den Brand and collaborators [34,35] as well as preliminary data from MAINZ by Florizone and collaborators [6,7]. Comparisons to the preliminary Mainz data of Kozlov and collaborators [8,9,10] have also been made (although the data

334

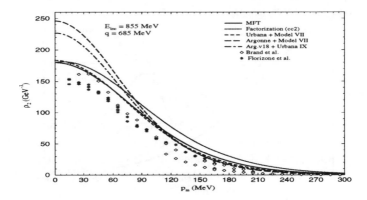

Figure 5. A comparison between our relativistic calculations, non-relativistic calculations reported elsewhere, and experimental data for the proton momentum distribution in ^{4}He.

is not shown). Thus, high-quality data for the momentum distribution of ^{4}He is now available up to a missing momentum of about 200 MeV. We find the results of Fig. 5 quite remarkable. It appears that a simple relativistic mean-field calculation of the momentum distribution rivals — and in some cases surpasses — some of the most sophisticated non-relativistic predictions. Still, theoretical predictions of the momentum distribution overestimate the experimental data by up to 50-60%. Part of the discrepancy is attributed to distortion effects which are estimated at about 12% [6,36]. However, distortions are not able to account for the full discrepancy. We have argued earlier that an additional source of error may arise from the factorization approximation used to extract the spectral function from the experimental cross section. We are confident that the approach suggested here, based on the extraction of the spectral function from the longitudinal response, is robust. While the method adds further experimental demands, as a Rosenbluth separation of the cross section is now required, the extracted spectral function appears to be weakly dependent on off-shell extrapolations and relativistic effects.

4 Conclusions

We have developed a relativistic formalism for studying quasi-free processes from nuclei. The formalism can be applied with ease to a variety of processes and renders transparent analytical expressions for all observables. We have applied it to the processes of kaon photoproduction and electron scattering.

For the kaon quasifree process, we have found that the polarization observables are very sensitive to the fundamental physics in this process, but at the same time mostly insensitive to distortion effects, relativistic effects, and nuclear target effects. We conclude that the polarization observables are one of the cleanest tools for probing both the elementary amplitude ($\gamma p \to K^+\Lambda$) and nuclear medium modifications.

For the electron quasifree process, we have derived a robust procedure for extracting the momentum distribution using the longitudinal response. Furthermore, we found that the relativistic mean-field calculation of the momentum distribution in ^{4}He rivals — and in some cases surpasses — some of the most sophisticated non-relativistic predictions to date.

Acknowledgments

This work was supported in part by the United States Department of Energy under Contract No. DE-FG05-92ER40750 and in part by a joint fellowship from the Japan Society for the Promotion of Science and the United States National Science Foundation.

References

1. L.J. Abu-Raddad and J. Piekarewicz, Phys. Rev. C **61**, 014604 (2000).
2. L.J. Abu-Raddad, *"Photoproduction of pseudoscalar mesons from nuclei"*, Ph.D. thesis, Florida State University (unpublished). Available at: nucl-th/0005068 (2000).
3. L.J. Abu-Raddad and J. Piekarewicz, Phys. Rev. C **64**, 064902 (2001).
4. S. Gardner, and J. Piekarewicz, Phys. Rev. C **50**, 2822 (1994).
5. D.B. Kaplan and A.E. Nelson, Phys. Lett. B **175**, (1986), 57; B **179**, (1986), 409(E).
6. R.E.J. Florizone, Ph.D. thesis, Massachusetts Institute of Technology (unpublished). Available at: (http://wwwa1.kph.uni-mainz.de/A1/publications/doctor/).
7. R.E.J. Florizone *et al.*, to be published.
8. A. Kozlov, Ph.D. thesis, University of Melbourne (unpublished). Available at: (http://wwwa1.kph.uni-mainz.de/A1/publications/doctor/).
9. A. Kozlov *et al.*, Nucl. Phys. **A684**, 460 (2001).
10. A. Kozlov *et al.*, to be published.
11. R. Mertig and A. Hubland, *Guide to FeynCalc 1.0*, downloaded from

the internet, 1992; R. Mertig, Comp. Phys. Comm. **60**, 165 (1991); http://www.feyncalc.org/.

12. J.C. David, C. Fayard, G.H. Lamot, and B. Saghai, Phys. Rev. C **53**, 2613 (1996).

13. C. Bennhold, F.X. Lee, T. Mart, and L.E. Wright, Nucl. Phys. **A639**, 227c (1998).

14. R. Williams, C.R. Ji, and S. R. Cotanch, Phys. Rev. D **41**, 1449 (1990).

15. T. Mizutani, C. Fayard, G,-H. Lamot, and B. Saghai, Phys. Rev. C **58**, 75 (1998).

16. J.M. Udias, P. Sarriguren, E. Moya de Guerra, E. Garrido, and J.A. Caballero, Phys. Rev. C **48**, 2731 (1993).

17. J.M. Udias, P. Sarriguren, E. Moya de Guerra, and J.A. Caballero, Phys. Rev. C **53**, 1488 (1996).

18. K. Amir-Azimi-Nili, J.M. Udias, H. Muther, L.D. Skouras, and A. Polls, Nucl. Phys. **A625**, 633 (1997).

19. J.A. Caballero, T.W. Donnelly, E. Moya de Guerra, and J.M. Udias, Nucl. Phys. **A632**, 323 (1998).

20. J.A. Caballero, T.W. Donnelly, E. Moya de Guerra, and Nucl. Phys. **A643**, 189 (1998).

21. J.M. Udias, J.A. Caballero, E. Moya de Guerra, J.E. Amaro, and T.W. Donnelly, Phys. Rev. Lett. **83**, 5451 (1999).

22. J.M. Udias, J.A. Caballero, E. Moya de Guerra, Javier R. Vignote, and A. Escuderos, nucl-th/0101038.

23. S. Frullani and J. Mougey, Adv. Nucl. Phys. **14**, 1 (1984).

24. T. de Forest Jr., Nucl. Phys. **A392**, 232 (1983).

25. J. Piekarewicz and R.A. Rego Phys. Rev. C **45**, 1654 (1992).

26. R. Schiavilla, V.R. Pandharipande, and R.B. Wiringa, Nucl. Phys. **A449**, 219, (1986).

27. I.E. Lagaris and V.R. Pandharipande, Nucl. Phys. **A359**, 331, (1981).

28. R.B. Wiringa *et al.*, Phys. Rev. C **29**, 1207, (1984).

29. R. B. Wiringa, Phys. Rev. C **43**, 1585 (1991).

30. R. B. Wiringa, private communication.

31. J.L. Forest *et al.*, Phys. Rev. C **54** 646 (1996).

32. R.B. Wiringa *et al.*, Phys. Rev. C **51**, 38, (1995).

33. B.S. Publiner *et al.*, Phys. Rev. Lett. **74**, 4396, (1995).

34. J.F.J. van den Brand *et al.*, Nucl. Phys. **A534**, 637, (1991).

35. J.F.J. van den Brand *et al.*, Phys. Rev. Lett. **60**, 2006, (1988).

36. R. Schiavilla *et al.*, Phys. Rev. Lett. **65**, 835 (1990).

QUASIELASTIC AND Δ EXCITATION IN ELECTRON SCATTERING

K. S. KIM, S. W. HONG, B. T. KIM

BK21 Physics Research Division and Institute of Basic Science
Sungkyunkwan University, Suwon, 440-746, Korea

T. UDAGAWA

Department of Physics, University of Texas, Austin, Texas 78712

We present theoretical investigation of quasielastic and Δ excitations for the ^{12}C target induced by virtual photon in (e, e') reaction. We attempt to describe the quasielastic and Δ regions in one formalism. We use the Tamm-Dancoff Approximation to deal with the nuclear many-body effects. The longitudinal and transverse structure functions are separately studied with and without including the particle-hole correlation. We investigate the contributions from different physical processes by decomposing the inclusive cross sections.

1 Introduction

Electron scattering has been proved to be one of the most effective tools for probing nuclear properties and structure over a long time. The reason is that the leptonic interaction is well understood and thus the hadronic electromagnetic currents and the nuclear properties can be investigated without so much distortion effects as in hadronic probes.

Although experimental data[1,2,3,4] have been available in both quasielastic and Δ regions, theoretical calculations have been done separately for the quasielastic[5,6,7,8] or Δ excitation region.[9,10] In the present work, we consider both regions with one formalism so that the dip region can be treated consistently. The dip region is particularly interesting because it has not been yet explained satisfactorily.

Our calculation is based on the nuclear response function theory. In order to include the particle-hole correlation effects in the nuclear response, we use the Tamm-Dancoff Approximation. To construct the external field operator for inclusive (e, e') reaction we use the plane wave impulse approximation, ignoring Coulomb distortion of electron wave function by target nucleus.

In Sec. 2 we review briefly the formalism for the inclusive (e, e') reaction in the quasielastic and the Δ excitation regions. In Sec. 3 we discuss the non-relativistic transition operator for the virtual photon exchange in the quasielastic scattering and the Δ-resonance in the nucleus. In Sec. 4 we set up a set of coupled-channel equations for the excited nucleon and Δ to include the particle-hole correlation effects due to $NN^{-1} - NN^{-1}$, $NN^{-1} - \Delta N^{-1}$, and $\Delta N^{-1} - \Delta N^{-1}$ couplings. The coupled-channel equations is transformed into equations for localized functions so that numerical calculations can be done. We then apply the Lanczos method for solving the equations.[11] The calculated cross section is separated into its components corresponding to each physical process: the coherent pion production (CPP), the quasifree decay (QF), the spreading (SP), the nucleon knockout (KO), and the nucleon spreading (or nucleon knockout-fusion, KF) cross sections. In Sec. 5 we show the calculation results for ^{12}C target and discuss each decomposed cross section and the effect of correlation on the cross sections. Finally, we summarize the work in Sec. 6.

2 Inclusive Electron Cross Section

The cross section for the inclusive (e, e') reaction can be written as

$$\frac{d^2\sigma}{d\Omega d\omega} = \Gamma_v [\sigma_T(\omega, \mathbf{k}) + \epsilon \sigma_L(\omega, \mathbf{k})], \tag{1}$$

where the kinematical factors are given by[9]

$$\Gamma_v = \frac{\alpha}{2\pi^2} \frac{K}{k_\mu^2} \frac{E_f}{E_i} \frac{1}{1 - \epsilon} \qquad \epsilon = [1 + 2\frac{\mathbf{k}^2}{k_\mu^2} \tan^2\frac{\theta}{2}]^{-1}$$

$$k_\mu^2 = \mathbf{k}^2 - \omega^2 \qquad K = \frac{E^2 - M^2}{2M}.$$

E and M denote the total energy and the nucleon mass, respectively. The transverse and longitudinal cross sections are given by

$$\sigma_T = \frac{2\pi^2}{K} \alpha R_T(\omega, \mathbf{k}) = \frac{8\pi^3 \alpha}{K} S_T(\omega),$$

$$\sigma_L = \frac{4\pi^2}{K} \alpha \frac{k_\mu^2}{\mathbf{k}^2} R_L(\omega, \mathbf{k}) = \frac{16\pi^3 \alpha}{K} \frac{k_\mu^2}{\mathbf{k}^2} S_L(\omega). \tag{2}$$

The one-body external field operator $\hat{\rho}$ exciting the target is defined as

$$\hat{\rho} = (\phi_b|\hat{O}|\phi_a), \tag{3}$$

where $\hat{O}$ is the transition operator and $\phi_a(\phi_b)$ is the incident (outgoing) electron wave function. The strength functions in Eq. (2) are then written as

$$S(\omega) = \sum_f |\langle f|\hat{\rho}|\Phi_A\rangle|^2 \delta(\omega - E_f)$$

$$= \frac{1}{\pi} Im[-\langle \Phi_A|\hat{\rho}^+ G\hat{\rho}|\Phi_A\rangle]$$

$$= \frac{1}{\pi} Im[-\langle \rho | \Psi \rangle], \tag{4}$$

where

$$|\Psi\rangle = G|\rho\rangle. \tag{5}$$

$|\rho\rangle$ is the doorway state excited by the external field operator $\hat{\rho}$; $|\rho\rangle = \hat{\rho}|\Phi_A\rangle$. It is propagated to the final state by the propagator G having many-body nuclear Hamiltonian, which will be discussed in Sec. 4. We calculate the final state $|\Psi\rangle$ by solving the coupled-channel equations in Sec. 4.

3 Transition Operators through Virtual Photon Exchange

The non-relativistic transition operators for quasielastic scattering are given by

$$\hat{\rho}_N(\mathbf{r}) = \sum_j e_j \delta(\mathbf{r} - \mathbf{r}_j),$$

$$\hat{\mathbf{J}} = \hat{\mathbf{J}}_N(\mathbf{r}) + \nabla \times \hat{\mu}_N(\mathbf{r}), \tag{6}$$

where

$$\hat{\mathbf{J}}_N(\mathbf{r}) = \sum_j e_j \delta(\mathbf{r} - \mathbf{r}_j) \frac{1}{iM} \nabla$$

$$\hat{\mu}_N(\mathbf{r}) = \sum_j \mu_j \frac{1}{2M} \sigma_j \delta(\mathbf{r} - \mathbf{r}_j)$$

$$e_j = \frac{1}{2}(1 + \tau_3(j))$$

$$\mu_j = \frac{\mu_p + \mu_n}{2} + \frac{-\mu_p + \mu_n}{2} \tau_3(j). \tag{7}$$

The transition operator for $\gamma N \to \Delta$ in free space may be given by

$$F_{\gamma N \Delta} = \frac{g_{\gamma N \Delta}}{M_\Delta} f_\Delta(k_\mu^2) \boldsymbol{\varepsilon} \cdot \mathbf{k} \times \mathbf{S}^+ T_3^+. \tag{8}$$

In the nuclear medium the transition operator needs to be modified. We use the medium modified transition operator as given by Ref. (9):

$$\tilde{F}_{\gamma N \Delta} = \frac{\tilde{g}_{\gamma N \Delta}(E, k_\mu^2)}{M_\Delta} e^{i\phi(E, k_\mu^2)} \boldsymbol{\varepsilon} \cdot \mathbf{k} \times \mathbf{S}^+ T_3^+, \tag{9}$$

where $\tilde{g}_{\gamma N\Delta}$ and ϕ are parametrized as

$$\tilde{g}_{\gamma N\Delta}(E, k_\mu^2) = \tilde{g}_{\gamma N\Delta} f(k_\mu^2), \qquad \tilde{g}_{\gamma N\Delta} = 1.03,$$

$$\phi(E, k_\mu^2) = \phi_0(E) e^{-\frac{k_\mu^2}{\Lambda^2}} \tag{10}$$

$$\phi_0 = \frac{q^3}{(a_1 + a_2 q^2)},$$

$$f(k_\mu^2) = (1 + \frac{k_\mu^2}{a^2})^{-2}(1 + \frac{k_\mu^2}{\xi a^2})^{-1}, \tag{11}$$

with $a_1 = 1.27$ fm^{-3}, $a_2 = 4.46$ fm^{-1}, $\Lambda^2 = 0.2$ GeV^2, $a^2 = 0.71$ GeV^2 and $\xi = 5$. Here, q is the pion momentum.

4 Inhomogeneous Coupled Channel Equations

The propagator G in Eqs. (4) and (5) can be expressed as

$$G = \frac{1}{\omega - H_h - T_p - U_p - V_{ph} + i\Gamma_p/2 + i\epsilon}, \tag{12}$$

where H_h stands for the nuclear shell model Hamiltonian for the hole-nucleus, which satisfies

$$H_h|\Phi_h\rangle = \epsilon_h|\Phi_h\rangle. \tag{13}$$

The complex single particle potential U_p for the excited particles is given by

$$U_p = V_\Delta + iW_\Delta \quad \text{for} \quad p = \Delta$$
$$= V_N + iW_N \quad \text{for} \quad p = N. \tag{14}$$

For the present calculations we take $V_\Delta = -25$ MeV and $W_\Delta = -45$ MeV. V_N and W_N are taken as energy-dependent. Also,

$$\Gamma_p = 0 \quad \text{for} \quad p = N$$
$$= \Gamma_\Delta \quad \text{for} \quad p = \Delta. \tag{15}$$

The decay width of Δ is given by[9]

$$\Gamma_\Delta(E) = (\frac{q}{q_R})^3 \frac{M_\Delta}{E} [\frac{\nu(q)}{\nu(q_R)}]^2 \Gamma_R, \tag{16}$$

$$\nu(q) = (1 + q^2/\beta^2)^{-1}, \tag{17}$$

with $\Gamma_R = 110$ MeV and $\beta = 300$ MeV. Here, q_R denotes the pion momentum at the resonance.

The residual ph interaction V_{ph} in Eq. (12) is treated in the Tamm-Dancoff approximation within the $(\pi + \rho + g_0')$ model: $V_{ph} = V_\pi + V_\rho + g_0'$ (see Ref. (12) and the references therein). In the ρ-exchange we keep only the tensor interaction and drop the central part assuming that the latter can be effectively included in the short-range interaction. The matrix elements $V_{ph,p'h'}^{j_t}$ consist then of four couplings $V_{NN^{-1},NN^{-1}}$, $V_{NN^{-1},\Delta N^{-1}}$, $V_{\Delta N^{-1},NN^{-1}}$, and $V_{\Delta N^{-1},\Delta N^{-1}}$. In the present study, we use $g_{NN}' = 0.60$[13] for $NN^{-1} - NN^{-1}$ and $g_{N\Delta}' = 0.45$ for $NN^{-1} - \Delta N^{-1}$, and treat $g_{\Delta\Delta}'$ as a free parameter which have to be determined by fitting the calculated cross section to the data as in Ref. (11). Later we shall show the results for $g_{\Delta\Delta}' = 0.33$ or 0.60.

The calculation of $|\Psi\rangle$ is done as following. The equation for $|\Psi\rangle$

$$|\Psi\rangle = G|\rho\rangle \tag{18}$$

is first integrated over all coordinates except the radial coordinates of particle p ($= N$ or Δ). Then Eq. (18) is reduced to a set of coupled channel (CC) equations for the radial wave functions of particle p. In order to achieve this we expand $|\rho\rangle$ in terms of the channel wave functions

$$|[y_p \Phi_h]_{jm}\rangle = \Sigma_{m_p m_h} \langle j_p m_p j_h m_h | jm \rangle | y_{j_p m_p} \Phi_{j_h m_h}\rangle, \tag{19}$$

where $y_{j_p m_p}$ is the spin-angle wave function of p and $\Phi_{j_h m_h}$ is the hole nucleus wave function. The channel expansion is then given by

$$|\rho\rangle = \Sigma_{j_t \ell_t m_{j_t} m_{\ell_t} m_{s_t}} \sqrt{2} \langle s_b m_b s_a - m_a | s_t m_{s_t} \rangle (-1)^{\ell_t + m_{\ell_t}}$$
$$\langle j_t m_{j_t} s_t - m_{s_t} | \ell_t - m_{\ell_t} \rangle \Sigma_{ph}^{N_c} \frac{\rho_{ph}(r)}{r} | [y_p \Phi_h]_{j_t m_{j_t}}\rangle, \tag{20}$$

where j_t, ℓ_t, and s_t are the total, orbital, and spin angular momenta transferred in the reaction, respectively; m_{j_t}, m_{ℓ_t}, and m_{s_t} are the corresponding angular momentum projections. Once $|\rho\rangle$ is known, it is easy to calculate $|\Psi\rangle$. In partial wave expansion $|\Psi\rangle$ is given by

$$|\Psi\rangle = \Sigma_{j_t \ell_t m_{j_t} m_{\ell_t} m_{s_t}} \sqrt{2} \langle s_b m_b s_a - m_a | s_t m_{s_t} \rangle (-1)^{\ell_t + m_{\ell_t}}$$
$$\langle j_t m_{j_t} s_t - m_{s_t} | \ell_t - m_{\ell_t} \rangle \sum_{ph}^{N_c} \frac{\psi_{ph}(r)}{r} | [y_p \Phi_h]_{j_t m_{j_t}}\rangle. \tag{21}$$

Then we obtain coupled channel equations for $\psi_{ph}(r)$

$$(\varepsilon_{ph} - T_p - U_p + i\Gamma_p/2)\psi_{ph}(r) = \rho_{ph}(r)$$
$$+ \sum_{p'h'} \langle [ph]_{JM} |V_{ph,p'h'}| [p'h']_{JM} \ \psi_{p'h'}(r')\rangle, \tag{22}$$

which are solved by using the Lanczos method.[11]

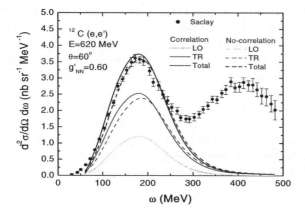

Figure 1. Comparison of the cross sections in the quasielastic region. The solid curves represent the cross section with correlation, and the dashed curves without correlation. The data are from Ref. (4).

5 Results

We present the calculations on the inclusive (e, e') scattering with the incident energy of 620 MeV and the scattering angle of 60^0 for ^{12}C target. We compare the calculated cross section with the experimental data[4] in the entire energy range from the quasielastic to Δ regions.

In Fig. 1 we show the calculated cross section in the quasielastic region. The solid (dashed) curves denote the cross sections calculated with (without) including $NN^{-1} - NN^{-1}$ correlation. g'_{NN} is taken as 0.60.[13] The cross section is decomposed into transverse and longitudinal components. It is shown that the transverse cross section is shifted towards the low excitation energies by about 10 MeV due to the correlations. However, the longitudinal cross section remains the same regardless of whether or not we include the correlation. It is because the correlation effects we considered in this work have to do with the spin-isospin excitations whereas the longitudinal external field operator probes only the charge response of the target nucleus and is blind to the spin-isospin excitation. After summing the longitudinal and the transverse cross sections, the agreement between theoretical cross section and the experimental data[4] is quite good.

In Fig. 2 we present a similar comparison for the Δ region. $g'_{\Delta\Delta}$ is taken to

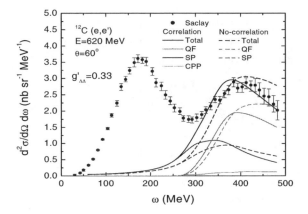

Figure 2. Comparison of the cross sections in the Δ region. The solid curves represent the cross sections with correlation, and the dashed curves without correlation. The data are from Ref. (4).

be 0.33 as in Ref. (11). Again the cross sections including the correlation are plotted by the solid curves and the ones without the correlation are plotted by the dashed curves. The summed correlated cross section (black, solid curve) is shifted to the left by about 30 MeV relative to the summed uncorrelated cross section (black, dashed curve). The cross section is a little overestimated in magnitude and the peak position is not well reproduced. The reason for this poor fit to the data can be better seen if we decompose the summed cross section into its components. We can see that both the QF and the SP components are shifted to the low excitation region due to correlation. The CPP cross section is almost negligible. This may be understandable because the external probe excites the Δ in the spin-transverse direction only but the CPP, coming from the pole of the pion propagator, is closely related to the spin-longitudinal correlation.

Figure 3 shows the $g'_{\Delta\Delta}$ dependence of each decomposed cross section. As is well known,[12] g'_0 term produces repulsion, and thus the cross section with larger $g'_{\Delta\Delta}$ is peaked at higher excitation energies. The results with $g'_{\Delta\Delta}=0.60$ fit the data better than those with $g'_{\Delta\Delta}=0.33$.

Figure 4 shows the cross section calculated by including both nucleon and Δ excitations. In Figs. 2 and 3 it was shown that the calculated cross sections for the Δ peak did not fit the experimental Δ peak position. However, when

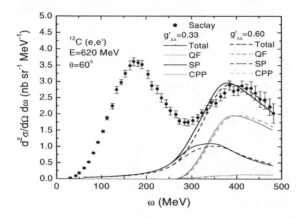

Figure 3. Comparison of the cross sections in the Δ region. The solid curves are obtained with $g'_{\Delta\Delta} = 0.33$ and the dashed lines are wth $g'_{\Delta\Delta} = 0.60$. The data are from Ref. (4).

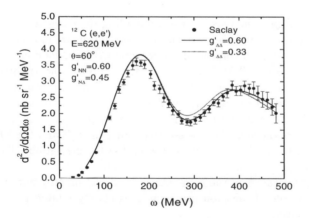

Figure 4. Comparison of the cross sections in the entire energy region with $g'_{\Delta\Delta} = 0.33$ and $g'_{\Delta\Delta} = 0.60$. The data are from Ref. (4).

we take into account the $NN^{-1} - \Delta N^{-1}$ correlation by including both the nucleon and Δ excitation, the Δ peak is rather well reproduced as shown in

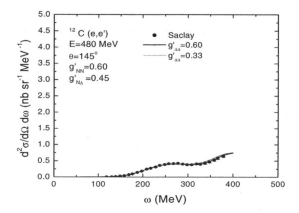

Figure 5. The same as in Fig. 4 but with the incident electron energy of 480 MeV and the scattering angle of 145°. The data are from Ref. (4).

Fig. 4. It is because some of the strength in the Δ region is shifted to the quasielastic and the dip regions due to the attractive correlation. We can also see in Fig. 4 the high energy side of the quasielastic peak is somewhat overestimated. This is due to the long tail of the SP component of the Δ cross section as seen in Figs. 2 and 3. We also show in Fig. 4 the dependence of the cross section on the values of $g'_{\Delta\Delta}$. The curve calculated with $g'_{\Delta\Delta} = 0.60$ reproduces the experimental data quite well.

We then considered a backward angle of $\theta = 145°$ at the incident energy of 480 MeV. As shown in Fig. 5 the data are well reproduced. Note that the magnitude of the quasielastic cross section at this large angle is quite small, and our calculation reproduces the data very well with the same model parameters.

6 Summary

We have considered both quasielastic and Δ excitation regions with one formalism in order to treat the dip region consistently. Our formalism can separate the cross section into 5 components: CPP, QF, Δ-SP, nucleon KO, and nucleon SP. When the residual ph interaction is included, the equations for the nucleon and Δ wave functions become coupled. We have solved these equations by using the Lanczos method. Our calculations seem to reproduce the cross sections in the entire energy region including the dip region quite well. The effect of the correlation is small in the quasielastic region but causes about 30 MeV shift in the Δ excitation region. Further investigation of the Δ-dynamics is needed, particularly for the SP and QF processes. The electron Coulomb distortion needs to be included for medium and heavy nuclei.

Acknowledgement

This work was partially supported by KOSEF (Grant No. 2000-6-111-01-2 and No. 2000-2-11100-004-4).

References

1. L. Chinitz et al., Phys. Rev. Lett. **67**, 568 (1991).
2. Z. E. Meziani, Nucl. Phys. **A446**, 113 (1985); Z. E. Meziani et al., Phys. Rev. Lett. **52**, 2130 (1984); **54**, 1233 (1985).
3. M. Deady et al., Phys. Rev. C **33**, 1897 (1986); **28**, 631 (1983).
4. P. Barreau, et al., Nucl. Phys. **A402**, 515 (1983).
5. Yanhe Jin, D. S. Onley, and L. E. Wright, Phys. Rev. C **45**, 1311 (1992).
6. K. S. Kim, L. E. Wright, Yanhe Jin, and D. W. Kosik, Phys. Rev. C **54**, 2415 (1996): K. S. Kim, L. E. Wright, and D. A. Resler, Phys. Rev. C **64**, 04607 (2001).
7. C. Giusti and F. D. Pacati, Nucl. Phys. **A473**, 717 (1987).
8. M. Trani, S. Turck-Chieze, and A. Zghiche, Phys. Rev. C **38**, 2799 (1988).
9. J. H. Koch and N. Ohtsuka, Nucl. Phys. **A435**, 765, (1984); J. H. Koch, E. J. Moniz and N. Ohtsuka, Ann. of Phys. **154**, 99, (1984).
10. T. W. Donnelly, J. W. van Orden, T. de Forest Jr., and W. C. Hermans, Phys. Lett. B **76**, 249 (1978).
11. T. Udagawa, P. Oltmanns, F. Osterfeld and S. W. Hong, Phys. Rev. C**49**, 3162 (1994): T. Udagawa, S. W. Hong, and F. Osterfeld, Phys. Lett. B **245**, 1 (1990).
12. F. Osterfeld, Rev. Mod. Phys. **64**, 491 (1992).
13. Kimiaki Nishida and Munetake Ichimura, Phys. Rev. C **51**, 269 (1995).

KAON PHOTO- AND ELECTROPRODUCTION ON THE DEUTERON WITH BEAM AND RECOIL POLARIZATIONS

K. MIYAGAWA

Department of Applied Physics, Okayama University of Science, 1-1 Ridai-cho, Okayama 700, Japan

T. MART

Jurusan Fisika, FMIPA, Universitas Indonesia, Depok 16424, Indonesia

C. BENNHOLD

Center for Nuclear Studies, Department of Physics, The George Washington University, Washington, DC 20052, U.S.A.

W. GLÖCKLE

Institut für Theoretische Physik II, Ruhr-Universität Bochum, D-4478 Bochum, Germany

Kaon photo- and electroproduction processes on the deuteron are investigated theoretically. Modern hyperon-nucleon forces as well as an updated kaon production operator on the nucleon are used. Sizable effects of the hyperon-nucleon final state interaction are seen in various observables. Especially the photoproduction double polarization observables are shown to provide a handle to distinguish different hyperon-nucleon force models.

1 Introduction

Recent precise calculations[1,2,3] of light hypernuclei have revealed important features of the $\Lambda N - \Sigma N$ interaction. The $\Lambda N - \Sigma N$ coupling has been found to be crucial for the binding of hypernucear states. Hypertriton calculations[1] have yielded significant insight into the S-wave force components, showing that the two YN soft-core interactions of the Nijmegen group[4] NSC89 and NSC97f reproduce a reasonable binding energy. However, recent calculations[3] of $^4_\Lambda$H and $^4_\Lambda$He have demonstrated that those interactions can not reproduce the 4-body hypernuclear binding energies. Thus, even the strength of the S-wave forces has not yet been well determined.

Photo- and electroproduction processes of kaons on light nuclei open a unique possibility for studying the $\Lambda N - \Sigma N$ interaction in the continuum, especially around the Σ threshold where the YN tensor force is expected to play an important role. An inclusive $d(e, e'K^+)YN$ experiment[5] has already been performed, and the data for $d(\gamma, K^+Y)N$ and $^3\text{He}(\gamma, K^+Y)N$ are being

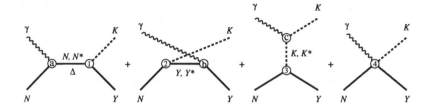

Figure 1. Feynman diagram for the $\gamma + N \to K^+Y$ process. Contributions from the Δ are only possible in Σ production. The contact diagram (4) is required in both PS and PV couplings in order to restore gauge invariance after introducing hadronic form factors. The Born terms contain the N, Y, K intermediate states and the contact term.

analyzed at Jlab. Kaon photoproduction on the deuteron is also important since it allows access to the elementary cross sections on the neutron, such as $\gamma + n \to K^+ + \Sigma^-$, in kinematic regions where final-state interaction effects are small.

We have analyzed the inclusive $d(\gamma, K^+)$ and exclusive $d(\gamma, K^+\Lambda(\Sigma))$ processes, and present preliminary results of the electroproduction process $d(e, e' K^+)$. The coupled $\Lambda N - \Sigma N$ interaction in the final state are precisely incorporated and the modern YN interactions of the Nijmegen group NSC97f and NSC89 are used. Furthermore, we have employed a recently updated elementary operator[6] for the $\gamma + N \to K^+Y$ process.

2 Elementary operator

For the elementary operator, we choose an isobar model without final-state interactions which provides a simple tool to parameterize meson photoproduction off the nucleon. Without rescattering contributions the T-matrix is simply approximated by the driving term alone which is assumed to be given by a series of tree-level diagrams. The selected Feynman diagrams for the s-, u-, and t-channel shown in Fig. 1 contain some unknown coupling parameters to be adjusted in order to reproduce experimental data. Final state interaction is effectively absorbed in these coupling constants which then cannot easily be compared to couplings from other reactions. Guided by recent coupled-channel results[7] we have therefore constructed a tree-level amplitude that reproduces all available $K^+\Lambda$, $K^+\Sigma^0$ and $K^0\Sigma^+$ data and thus provides an effective parameterization of the process. In this model, we include the three resonances that have been found to decay into the $K\Lambda$ channel, the $S_{11}(1650)$, $P_{11}(1710)$, and $P_{13}(1720)$. For $K\Sigma$ production we also allow con-

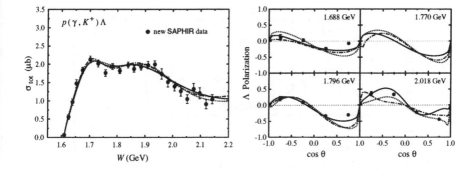

Figure 2. Total cross section (left) and recoil polarization (right) for the $p(\gamma, K^+)\Lambda$ process for different hadronic form factors $\widehat{F}$. Dotted lines are the result obtained by using $\widehat{F}$ given in Ref. [11], while solid and dash-dotted lines show the results of $\widehat{F}$ proposed by Ref. [13]. See Ref. [10] for a detailed discussion.

tributions from the $S_{31}(1900)$ and $P_{31}(1910)$ Δ resonances. Furthermore, we include not only the usual 1^- vector meson $K^*(892)$, but also the 1^+ pseudovector meson $K_1(1270)$ in the t-channel since a number of studies have found this resonance to give a significant contribution.

The $K^+\Lambda$ total cross section data reveal an interesting structure around $W = 1900$ MeV. A constituent quark model by Capstick and Roberts[8] predicts a missing D_{13} at 1960 MeV that has a large branching ratio both into the γN and the $K\Lambda$ channel. In order to study this structure more closely, we [9] have included a D_{13} resonance but allowed the mass and the width of the state to vary as free parameters. A significant reduction in χ^2/N for a mass of 1895 MeV and a total width of 372 MeV was achieved.

To account for the structure in hadronic vertices we include hadronic form factors in the operator. It has been well known that this procedure can lead to the violation of gauge invariance in the Born amplitude. To overcome this problem, we use the method proposed by Haberzettl[11,12] to restore gauge invariance. Recently, Ref. [13] noticed that an indiscriminate choice of hadronic form factor $\widehat{F}$ could create a pole in the Born amplitude. A simple solution is suggested by taking special form factors which can also simultaneously satisfy the crossing symmetry of Born terms. A comparison of several form factors $\widehat{F}$ for photoproduction is shown in Fig. 2, while the result for electroproduction is given in Fig. 3. A more detailed discussion can be found in Ref. [10].

Since the use of different $\widehat{F}$ in the elementary operator does not change

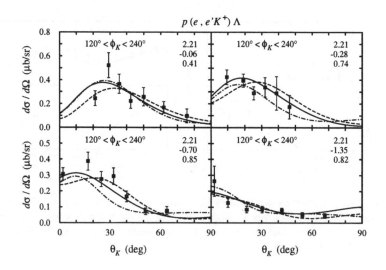

Figure 3. Same as in Fig.2 but for electroproduction.

our calculation on the deuteron dramatically, we performed our calculation by using the Haberzettl's method.

3 Photoproduction

Numerical results of the inclusive $d(\gamma, K^+)$ cross sections are shown as a function of lab momentum P_K in Fig. 4. The incident photon energy is 1.3 GeV, and the outgoing kaon angle is fixed to $\theta_K = 1°$. The two pronounced peaks around $p_K = 945$ and 809 MeV/c are mainly due to quasifree scattering between the photon and one of the nucleons with the outgoing Λ and Σ, respectively. Results including the YN final-state interaction NSC97f are compared to PWIA results. Sizable FSI effects are seen around both Λ and Σ thresholds, and to a lesser degree at the two quasifree peak positions. The enhancement around the Σ threshold is not a simple threshold effect but is caused by a YN t-matrix pole in the complex momentum plane[14].

We also calculated the inclusive cross sections for various kaon angles to find the region where the cross sections are large. The results are depicted on the θ_K-P_K plane in Fig. 5. The two peaks in Fig. 4 continue into the region of larger θ_K values and form the ridges accordingly. We confirm that these ridges lie along the quasifree scattering conditions.

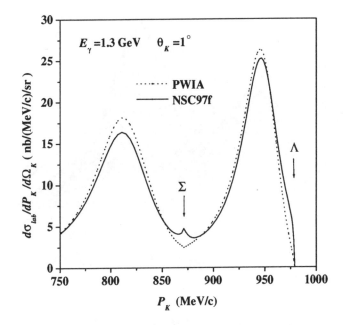

Figure 4. Inclusive $d(\gamma, K^+)$ cross section as a function of kaon lab momentum P_K. The incident photon energy is 1.3 Gev, and the outgoing kaon angle is fixed to $\theta_K = 1°$. The results with the final state YN interaction NSC97f are compared to the PWIA results. The $K^+\Lambda N$ and $K^+\Sigma N$ thresholds are indicated by the arrows.

For the exclusive process, we have calculated various observables changing θ_K and P_K with the same photon energy 1.3 GeV as in the inclusive processes. The five observables, cross section, hyperon recoil polarization P_y, beam polarization asymmetry Σ, double polarization C_z, and C_x are predicted in Figs. 6 and 7. We show only two cases, the $d(\gamma, K^+\Lambda)n$ process at $\theta_K = 1°$ and $P_K = 870$ MeV/c in Fig. 6 and $d(\gamma, K^+\Sigma^-)p$ at $\theta_K = 1°$ and $P_K = 870$ MeV/c in Fig. 7. In the both cases, the results with the NSC97f and NSC89 YN interactions are compared with the PWIA results.

The θ_K and P_K values in Fig. 6 correspond to the final states very close to the $K^+\Sigma N$ threshold as indicated in Fig. 4. The cross sections in Fig. 6 show a rather gentle falloff for increasing Λ lab angle, θ_Λ^{lab}. Above $\theta_\Lambda^{lab} = 20°$, large FSI effects are seen especially in the double polarization observables C_z and C_x. Furthermore, the two YN forces of NSC97f and NSC89 become clearly distinguishable in C_z.

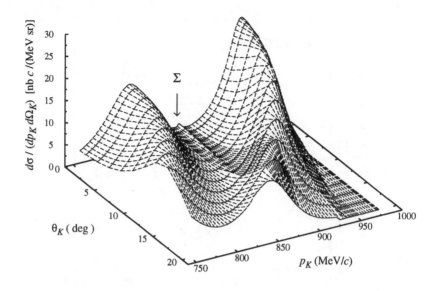

Figure 5. Inclusive $d(\gamma, K^+)$ cross section as a function of kaon lab momentum P_K and angle θ_K. The incident photon energy is the same as in Fig. 1.

In Fig. 7, shown are the observables for another exclusive process which corresponds to one of the peak position above the Σ threshold in Fig. 4. In this case, the cross sections exhibit a sharp forward peak, but above $\theta_\Lambda^{lab} = 15°$, considerable FSI effects can be seen in P_y, C_z and C_x.

4 Electroproduction

Here we present preliminary results for the electroproduction process $d(e, e'K^+)$. Two sets of the results are shown in Figs. 8 and 9. The incident electron energy E_e and momentum transfer Q^2 in Fig. 8 is set to reproduce the conditions of the recent Hall C experiment at Jlab [5], while the Q^2 value in Fig. 9 is searched for to find sizeable FSI effects. As in the case of the photoproduction $d(\gamma, K^+)$, YN, FSI effects are seen near both Λ and Σ

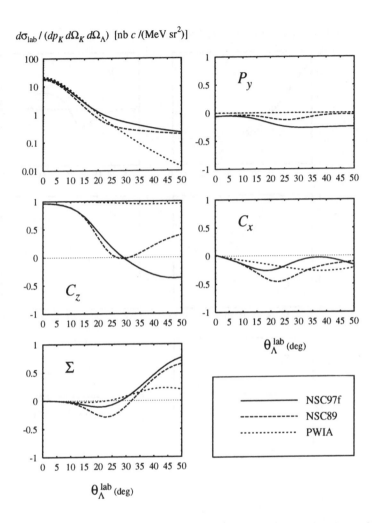

Figure 6. The exclusive $d(\gamma, K^+\Lambda)$ cross section, Λ recoil polarization P_y, photon beam asymmetry Σ, double polarization C_z and C_x, as the function of Λ lab momentum. The kaon lab momentum and angle are $P_K = 870$ MeV/c and $\theta_K = 1°$, respectively.

threshold. However, in Fig. 8, the FSI has effects in a wide range above the Σ threshold. Figure 9 shows a prominent enhancement around the Σ threshold.

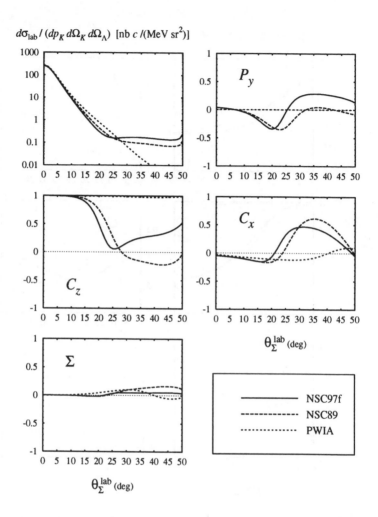

Figure 7. The exclusive $d(\gamma, K^+\Sigma^-)$ cross section, Σ^- recoil polarization P_y, photon beam asymmetry Σ, double polarization C_z and C_x, as the function of Σ^- lab momentum. The kaon lab momentum and angle are $P_K = 810$ MeV/c and $\theta_K = 1°$, respectively.

5 Summary and Outlook

We investigate cross sections and hyperon recoil polarization for K^+ photo-production on the deuteron, and find large hyperon-nucleon FSI effects in the

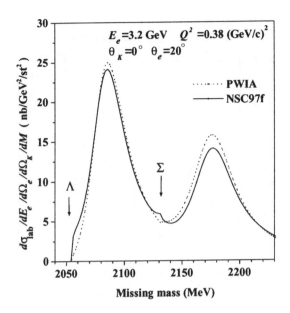

Figure 8. Missing mass spectrum for the reaction $d(e, e' K^+)$. Results with the YN final state interaction NSC97f are compared to PWIA results.

double polarization observable C_z and C_x. Also, in electroproduction, for suitable Q^2 values, cross sections show a prominent enhancement around the Σ threshold. A systematic analysis for a wide range of kinematics for both photo- and electroproduction processes is planed. Future studies will investigate final-state interaction effects in kaon photo- and electroproduction on the $A = 3$ system.

References

1. K. Miyagawa, H. Kamada, W. Glockle, H.Yamamura, T. Mart, and C.Bennhold, Few-Body Systems Suppl. **12**, 324 (2000); nucl-th/0002035, and references therein.
2. E. Hiyama, M. Kamimura, T. Motoba, T. Yamada and Y. Yamamoto, Nucl. Phys. **A691**, 107c (2001).
3. A. Nogga, Ph. D Thesis, Ruhr-Universität Bochum, 2001 (unpublished).
4. Th. A. Rijken, V.G.J. Stoks, and Y. Yamamoto, Phys. Rev. **C59**, 21 (1999), and references therein.

356

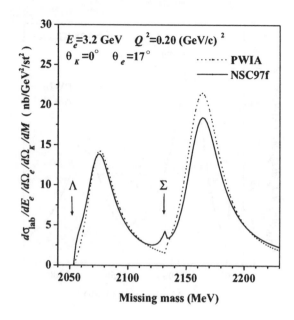

Figure 9. Same as Fig. 8, but for $Q^2 = 0.20$ $(\mathrm{GeV/c})^2$, $\theta_e = 17°$.

5. B. Zeidman et al., Nucl. Phys. **A691**, 37c (2001).
6. F.X. Lee, T. Mart, C. Bennhold, H. Haberzettl, and L.E. Wright, Nucl. Phys. **A695**, 237 (2001); C. Bennhold, H. Haberzettl, T. Mart, nucl-th/9909022.
7. T. Feuster and U. Mosel, Phys. Rev. C **59**, 460 (1999).
8. S. Capstick and W. Roberts, Phys. Rev. D **58**, 074011 (1998).
9. T. Mart and C. Bennhold, Phys. Rev. C **61**, 012201(R) (2000).
10. T. Mart, C. Bennhold, and H. Haberzettl, πN Newslett. **16**, 86 (2001).
11. H. Haberzettl, C. Bennhold, T. Mart and T. Feuster, Phys. Rev. C **58**, 40 (1998).
12. H. Haberzettl, Phys. Rev. C **56**, 2041 (1997)
13. R. M. Davidson and R. Workman, Phys. Rev. C **63**, 025210 (2001).
14. K. Miyagawa and H.Yamamura, Phys. Rev. **C60**, 024003 (1999).

ELECTROPRODUCTION OF STRANGE NUCLEI*

E. V. HUNGERFORD[†]

Department of Physics, University of Houston,4800 Calhoun Rd,Houston TX 77204, USA
E-mail: Hunger@uh.edu

The advent of high-energy, CW-beams of electrons now allows electro-production and precision studies of nuclei containing hyperons. Previously, the injection of strangeness into a nucleus was accomplished using secondary beams of mesons, where beam quality and target thickness limited the missing mass resolution. We review here the theoretical description of the $(e, e'K^+)$ reaction mechanism, and discuss the first experiment demonstrating that this reaction can be used to precisely study the spectra of light hypernuclei. Future experiments based on similar techniques, are expected to attain even better resolutions and rates.

1 Introduction

It is generally understood that high energy electron scattering can illuminate the substructure of elementary hadronic particles, and even the more complicated quark/gluon substructure of nuclei. However it is less known, that electromagnetic reactions can also implant strange hadrons into nuclear matter, providing a new tool to precisely study the hadronic many-body problem.

Of course it is not possible, nor is it reasonable, to reproduce nuclear physics with strange hadrons. However the addition of strangeness into a nucleus adds a new degree of freedom which emphasizes questions impossible or difficult to address in non-strange nuclear systems. The discussion here will be developed in terms of quantum hadro-dynamics, where the interacting particles are the fundamental baryons and mesons having strangeness 0 or -1, although at some point in the future the quark substructure of nuclei may be required to adequately describe these systems[1].

Previously it has been established that a Lambda within a nucleus can be described as a distinguishable particle, moving in the nuclear mean-field[2]. In this representation a Lambda can reside deeply within the nuclear interior, interacting with the other hadrons near this position[3]. However one-pion-exchange with nucleons cannot occur due to isospin conservation, so the shorter range components of the hadronic interaction, including multi-pion exchange and $\Lambda - \Sigma$ coupling, are enhanced[4]. In addition, multi-strange nu-

---
*SUPPORTED IN PART BY THE US DOE
[†]FOR THE HNSS COLLABORATION AT THE JEFFERSON LABORATORY

clei are predicted to be particle stable, and neutron stars may contain equal numbers of strange and non-strange hadrons[5]. However as electromagnetic reactions effectively restrict strangeness changing reactions to Δ—S— = 1, we concentrate here only on the production and analysis of light Λ hypernuclei.

2 Electroproduction of Hypernuclei

The cross section for electroproduction can be written in a very intuitive form by separating out a factor, which under certain conditions, may be identified as the virtual photon flux produced by (e, e') scattering[6]. Using this notation the electroproduction cross section can be written;

$$\frac{\partial^3 \sigma}{\partial E'_e \partial \Omega'_e \partial \Omega_k} = \Gamma \left[\frac{\partial \sigma_T}{\partial \Omega_k} + \epsilon \frac{\partial \sigma_L}{\partial \Omega_k} + \epsilon \cos(2\phi) + \cos(\phi_k) \sqrt{2\epsilon(1 + \epsilon)} \frac{\partial \sigma_I}{\partial \Omega_k} \right].$$

The factor, Γ, is the virtual flux factor evaluated for electron kinematics in the lab frame. It has the form;

$$\Gamma = \frac{\alpha}{2\pi^2 Q^2} \frac{E_\gamma}{1 - \epsilon} \frac{E'_e}{E_e}.$$

In the above equation, ϵ is the virtual photon polarization factor;

$$\epsilon = \left[1 + \frac{2|\mathbf{k}|^2}{Q^2} \tan^2(\Theta_e/2) \right]^{-1};$$

which vanishes when the kaon is produced along the direction of the virtual gamma. The differential cross sections in the above expression are separated into transverse, longitudinal, polarized, and interference terms. For real photons of course, the 4-momentum (Q^2) vanishes, and only the transverse cross section is non-zero. In the production of strange nuclear systems, the kaon angle must be small in order to maintain a reasonable reaction rate, and for the experimental geometry described in this paper, $Q^2 \to \frac{m_e^2 \omega^2}{E_e E_{e'}}$. Therefore only the transverse cross section significantly contributes to the rate, and we approximate the electroproduction cross section by its on-shell value, $i.e.$ by the (γ, K^+) cross section, so that the electro-production cross section is given by the virtual flux factor multiplied by the photo-production cross section[6].

2.1 The photo-production cross section

In order to describe the strangeness-producing reaction mechanism in a nucleus, we use the elementary process, $p(\gamma, K^+)\Lambda$, as was discussed above. The total cross section for this reaction was studied over a 1 GeV range above threshold[7]. The cross section rises from threshold at 0.91 GeV (photon energy) to a peak of 2.1 μb at 1.07 GeV. It remains nearly constant for about 0.5

GeV before gradually falling to about 1 μb at 1.9 GeV. Attempts have been made to describe this reaction in terms of an effective field theory, however there are no dominant Feynman diagrams as exist for pion photo-production, and many s,t, and u channel contributions must be included[8]. There are at least 10-15 undetermined coupling constants, and also a number of nucleon and hyperon resonances which must be included in any theory. Presently the data sample is inadequate to constrain the model, but quality data is now becoming available, and the situation may improve in the near future[9]. In any event, the forward-angle, unpolarized nuclear photo-production is insensitive to the details of the elementary amplitude.

In the laboratory frame the photo-production amplitude can be written in the general form[11];

$$\langle (\vec{k} - \vec{p}), \vec{p}|t|\vec{k}, 0 \rangle = \epsilon_0 (f_0 + g_0 \sigma_0) + \epsilon_x (g_{-1} \sigma_1 + g_+ \sigma_{-1}).$$

Here ϵ is the photon polarization such that ϵ_z is perpendicular to the scattering plane and ϵ_x is along the incident photon direction. Then as the kaon production angle $\to 0$ we obtain;

$$f_0 \to 0$$
$$g_0 \to a_1$$
$$g_\pm \to \pm \frac{a_1}{\sqrt{2}}$$

This means that;

$$\langle (\vec{k} - \vec{p}), \vec{p}|t|\vec{k}, 0 \rangle \to a_1 (\vec{\sigma} \cdot \vec{\epsilon}).$$

2.2   The virtual Flux

The virtual flux factor rises very steeply to a maximum at an electron scattering angle of $\frac{m_e^2 \omega^2}{E_e E_{e'}}$, Figure 1. For the energies involved here, this is very small, ≈ 2 milliradians, and falls to less than one-tenth its maximum value in 10 mr. Therefore if the electron spectrometer is placed at zero degrees, it need only have a small acceptance to assure that essentially all the scattered electrons are transported to the focal plane. This also means that aberrations in the optics of the magnetic transport system will be small, and easily corrected. The disadvantage of using the forward virtual flux is that bremsstrahlung also peaks in the forward direction and dominates the electron scattering process[6].

Therefore any attempt to observe a coincident reaction with zero degree virtual photons must cope with a large background of bremsstrahlung electrons.

Figure 1. The Virtual Flux Factor as a Function of the Electron Scattering angle

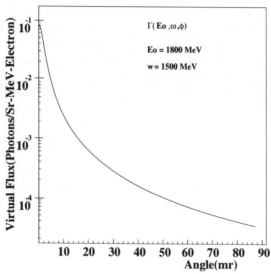

3   Photo-production of Strange Nuclei

The photo-production cross section on a nucleus can be written as[11];

$$\frac{d\sigma}{d\Omega} = \kappa |T_{ij}|^2$$

where the transition matrix, T_{ij} is;

$$T_{ij} = \frac{1}{2J_i + 1} \sum |\langle J_f m_f | O_{\gamma k} | J_i m_i \rangle|^2$$

and;

$$O_{\gamma k} = \int d^3x \, \chi_k^{-*} \chi_\gamma^+ \sum V \, \delta(r - \nu_v) \langle (\vec{k} - \vec{p}), \vec{p} | t | \vec{k}, 0 \rangle.$$

In this expression χ are the nuclear distorted waves, V is the transition matrix, $\delta(r - \nu v)$ is a recoil correction and the remaining term is the one-body photo-production amplitude.

It has been demonstrated that the (γ, K^+) reaction mainly contains only the $\vec{\sigma} \cdot \vec{\epsilon}$ amplitude in the forward direction and that wave distortion decreases the cross section by perhaps 20% for light hypernuclei, leaving the angular distribution essentially unchanged[10].

4 Theoretical Expectations

Because of the spin dependence of the photo-production process, spin-flip transitions leading to non-natural parity hypernuclear states are favored. In addition because of the high momentum transfer of the (γ, K^+) reaction, high angular momentum transfer is also expected. This contrasts to mesonic production of hypernuclei, where natural parity states are preferentially populated. However the structure of the hypernuclear spectrum produced by all reaction processes is similar, owing to the weak spin dependence of the Λ-N interaction[11]. However, even though the shape of the spectra is similar, the spin structure of the unresolved states is completely different, and the strength of the transitions to the various hypernuclear states varies due to changes in the momentum transfer of the reactions.

Finally, based on the effective Λ-Nucleus interaction we expect that the $^{12}\text{C}(\gamma, K^+)_{\Lambda}^{12}\text{B}$ reaction will have s and p shell structure separated by approximately 11 MeV, and that there will be some small excitation of core excited states (nuclear excitations with the Λ in the s shell)[12].

5 Experimental Considerations

It is important to keep the incident electron beam energy below $\approx$ 1.8 GeV. A low beam momentum increases the experimental resolution, given a fixed $\Delta p/p$ in the spectrometer magnets, and minimizes the size of these magnets. In addition, it limits kaon production to elementary processes involving either Λs or Σs, as opposed to unwanted hyperon and meson resonances. Since the maximum hypernuclear photoproduction rate occurs at photon energies of about 1.5 GeV, the energy of the scattered electron would be 300 MeV in this case.

Strangeness production using either real or virtual photons, requires the associated production of a strange quark/anti-quark pair, and therefore one must detect multiple particles in the final state in order to undertake a kinematically complete experiment. Although strangeness production by mesons may also be associated, meson-induced reactions can be arranged so that only one particle need be measured in the final state, assuming of course, that the

momentum of the incident meson is determined. In addition, the rather high
3-momentum transfer, $\vec{q}$, for associated production is $\geq$ to the nuclear Fermi
momentum. While this may be advantageous for the study of quark behavior,
it decreases the probability of leaving a bound nuclear system in the final state.
Thus the nuclear form factor, which falls rapidly with momentum transfer,
essentially cuts off production of reaction particles at all but the very forward
angles.

Thus, the greatest disadvantage of electroproduction is the requirement
that both the electron and the K^+ must be detected in the restricted phase
space region about the forward beam direction. To do this, the detection
apparatus must consist of two momentum sensitive instruments placed at
essentially the same forward-angle location, and having sufficient solid angle
to capture a substantial fraction of the reaction particles.

The electron beam at the Jefferson National laboratory, Jlab, is ≈ 40 μm
in diameter with a momentum dispersion of $\Delta p/p \approx 10^{-5}$. At a beam energy
of ~ 2 GeV, this introduces negligible error in the missing mass resolution.
Beam stability is a much more important issue, i.e. the requirement that
whatever the beam momentum, it must be stable and reproducible for days
(weeks) during the course of an experimental run. In the experiment reported
here, this was accomplished to a level of 10^{-4} by implementing momentum
measurements in the beam arcs of the accelerator which were fed-back to the
accelerator controls. In addition changes in beam momentum were recorded as
each experimental run progressed and the data corrected for the momentum
shift.

The layout of the HNSS (E89-009) experiment is shown in Figure 2. An
electron beam of primary energy ≈ 1.8 GeV and ~ 1 μA current strikes a
thin target placed just before a small zero-degree dipole magnet. This magnet splits the scattered particles; the e^- of about 300 MeV/c into a split-pole
spectrometer and the K^+ of about 1.2 GeV/c into the Short Orbit Spectrometer, SOS, fixed to the Hall C pivot at the Jefferson Laboratory. The magnetic
optics of the splitting magnet/SOS system must be designed so that there
is a virtual focus along the optic axis extension of the SOS magnet. In this
case the image occurs somewhat further away than for a system without the
splitting magnet, reducing slightly the acceptance, but preserving the optics.
The splitting magnet is necessary in order to capture the scattered electrons
and reaction kaons at very forward angles ($\leq 3°$). The resolution for the
HNSS experimental arrangement is expected to approach 600 keV. Various
contributions to this resolution are shown in Table 1, and are combined as if
they were statistically independent. The system resolution is dominated by
the SOS magnet, which is not designed for high resolution spectroscopy as

it has low dispersion and large momentum bite. However, it is a short-orbit magnet (important as the decay length for a 1.2 GeV/c kaon is about 9m), is already mounted and tested at the Hall C pivot, and has the sophisticated particle identification package, PID, required to extract kaons from the large background of pions and positrons.

Figure 2. The Experimental Geometry Showing the Zero Degree Splitting Magnet, the SOS Kaon Spectrometer, and the Enge Split-Pole Spectrometer.

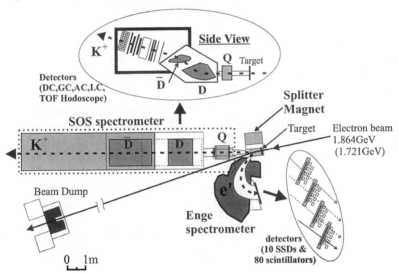

Nuclear targets, 22 mg/cm^2 of natural C and 10 mg/cm^2 of CH$_2$ were used. The CH$_2$ target was also used to observe Λ and Σ production from hydrogen and calibrate the missing mass spectrum. The reaction of interest, $(e,e'K^+)$, resulted in the production of the $^{12}_{\Lambda}$B hypernucleus when C was the target. Beam intensities were tuned to produce an acceptable signal to accidental ratio, which in the unprocessed coincident-time spectrum was $\geq$ 0.2. For the C target this resulted in a current of approximately 0.6 μA, or an experimental luminosity of $\sim 3.5 \times 10^{33}$.

To satisfy the somewhat conflicting beam requirements of the different experimental Halls, the experiment was required to use two different beam energies, 1721 and 1864 MeV. Data analysis has shown that differences in the $^{12}_{\Lambda}$B spectra produced by these two beam energies were small. It was found that the reaction proton and pion flux dominated the kaons by several orders of magnitude.

Table 1. Anticipated Missing Mass Resolution in E89-009.

Component	Resolution Contribution (keV)
Splitter and Split-Pole Magnets	120
Beam	160
Splitter and SOS Magnets	550
Target	20
Scattering Angle Kinematics	200
Total	600

The electrons were detected near the focal plane of the Split-pole spectrometer by a set of silicon strip detectors (SSD). This detector package was specifically designed for the HNSS experiment to meet the spatial resolution and count-rate requirements. The SSD was backed by a scintillation hodoscope to provide timing signals. The coincident time resolution of this system was about 800 ps FWHM, and was sufficient to resolve the real and accidental time peaks in the 2 ns micro-structure of the Jlab beam. A set of 10 SSD boards covering a spatial range of 78 cm were used for the focal plane detector.

A spectrum with accidental background is shown in Figure 3. Clearly evident in the spectrum is $^{12}_{\Lambda}$B hypernuclear structure in which a Λ is inserted in the s and p shells at a B_Λ of -11 and 0, respectively. There also appears to be a core excited states between these peaks. The broad peak at about 38 MeV is due to Λ production from H in the CH_2 target, spread by the incorrect kinematics.

This spectrum is similar to that predicted by Motoba, et al[11] who indicated that the gs doublet is split by only 100 keV with the 2^- excited state receiving most of the strength. The resolution of the spectrum shown in Figure 3 is about 850 keV, and the spin structure of the major shells is not resolved. A discussion of the structure will require additional analysis, including the differential cross section scale.

6 Conclusions

The first HNSS experiment has successfully observed the electroproduction of $^{12}_{\Lambda}$B and collected data on the production of $^{7}_{\Lambda}$He. The missing mass resolution is consistent with the predicted value of 600 to 1000 keV, and is a factor of ~ 3 better that existing measurements. The spectrum is also reasonably consistent with theory. Further analysis is expected to increase the statistics, and improve the backgrounds and resolution.

Figure 3. The Preliminary Hypernuclear Spectrum with Accidental Background. Clearly seen are the s and p shell excitations.

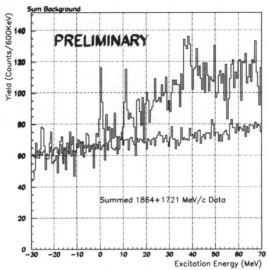

Targets for future experiments could consist of almost any separated-isotope of light mass since thicknesses of only $\approx$ 10mg/cm^2 are required. However as the target Z increases, electrons produced by bremsstrahlung processes increase as Z^2/A, but hyperon production increases as only as Z/A. Since bremsstrahlung limits the luminosity, the hypernuclear production rate should fall as $1/Z$. Of course nuclear effects must be applied, but from this analysis the HNSS geometry is suitable for lighter, p-shell targets. A newly designed geometry[13] with a specially constructed kaon spectrometer is planned which will allow extension of these studies well beyond the p-shell. This new system could improve the resolution by a factor of 2, with a increase in rate by a factor of as much as 50.

References

1. K. Maltman and M. Shmatikov, Nucl. Phys. **A585**(1995)343c; E. V. Hungerford and L. C. Biedenharn, Phys. Lett. **142B**(1984)232.
2. D. J. Millener, A. Gal, and C. B. Dover, Phys. Rev. **C38**(1988)2700.
3. H. Feshbach, *Proceedings of the Summer Study Meeting on Nuclear and Hypernuclear Physics with Kaon Beams*, H. Palevsky, ed. BNL Report

18335, July 1973.

4. B. F. Gibson and E. V. Hungerford, Phys. Rep., **257**(1995)350.

5. H. Bethe, G. E. Brown, and J. Cooperstein, Nucl. Phys. **A462**(1987)791; M. Prakish, *et. al.*, Phys. Rev. **D52**(1994)661; N. K. Glendenning and S. A. Moszkowski, Phys. Rev. Lett. **67**(1991)2414, N. Glendenning, Phys. Rev. **64C**(2001)025801.

6. Jlab experiment 89-009 and related documents, E. Hungerford, spokesperson, E. V. Hungerford, Prog. Theor. Phys. Sup. **117**(1994)135; Proceedings of the APCTP Workshop(SNP'99), IL-T. Cheon, S. W. Hong, and T. Motoba, eds, World Scientific, Singapore, 2000.

7. B. Saghai, Jlab Workshop on Electroproduction of Strangeness in Nuclei, December, 1999, Hampton University; B. Saghai, Euro. Jou. Phys. **C15**(2001).

8. M. Mart and C. Bennhold, Phys. Rev. **C61**(1999); C. Bennhold, Jlab Workshop on Electroproduction of Strangeness in Nuclei, December, 1999, Hampton University.

9. see the large numbers of papers presented at this conference.

10. R. A. Adelseck, *et al*, Ann. of Phys. **184**(1988)33; O. Richter, M. Sotona, and J. Zofka, Phys. Rev. **C43**(1991)2753

11. T. Motoba, M. Sotona, and K. Itonaga, Prog. Theor. Phys. **117**(1994)123; M. Motoba, *et al*, Prog. Theor. Phys. Sup. **117**(1994)135; M. Sotona and S. Frulani, *et al*, Prog. Theor. Phys. Sup. **117**(1994)135.

12. D. J. Millener, Nucl. Phys. **A691**(2001)93c.

13. O. Hashimoto, this conference, private communication, Jlab experiment 97-008.

PHOTOPRODUCTION OF THE φ(1020) NEAR THRESHOLD IN CLAS

D. J. TEDESCHI FOR THE CLAS COLLABORATION

Department of Physics and Astronomy, University of South Carolina Columbia,
SC 20208, USA
E-mail: tedeschi@sc.edu

The differential cross section for the photoproduction of the φ (1020) near threshold (E_γ = 1.57GeV) is predicted to be sensitive to production mechanisms other than diffraction. However, the existing low energy data is of limited statistics and kinematical coverage. Complete measurements of φ meson production on the proton have been performed at The Thomas Jefferson National Accelerator Facility using a liquid hydrogen target and the CEBAF Large Acceptance Spectrometer (CLAS). The φ was identified by missing mass using a proton and positive kaon detected by CLAS in coincidence with an electron in the photon tagger. The energy of the tagged, bremsstrahlung photons ranged from φ-threshold to 2.4 GeV. A description of the data set and the differential cross section for (E_γ = 2.0 GeV) will be presented and compared with present theoretical calculations.

1 Introduction

Early photoproduction measurements of φ-mesons are consistent with diffractive production via pomeron exchange as the dominant mechanism.[1,2,3] The differential cross sections for this reaction are forward peaked and vary slowly with energy.[4] In addition, the polarization data agree with the predictions of the vector dominance model in which an $s\bar{s}$ pair is produced by the transformation of the incoming photon.[5,6] However, recent measurements of OZI violations in $p\bar{p}$ collisions at CERN have drawn speculation about pre-existing strangeness in the proton.[7] Furthermore, the spin crisis in deep inelastic scattering, and the large value of the nucleon sigma term have led to the discussion of a strange quark component to the nucleon wavefunction, although analyses do exist that interpret the data without the need for an $s\bar{s}$ component. [8,9,10] Presented here are new CLAS data using unpolarized photons from threshold up to E_γ = 2.4 GeV and covering a range in momentum transfer up to $|t|$ = 1.6 (GeV/c)^2. While an initial, low-statistics analysis of this data set did not support interpretation beyond diffraction;[11] as the statistics have become available, a finer binned analysis (discussed here) is beginning to show interesting features that may support some of the new interpretations.[12,13,14,15]

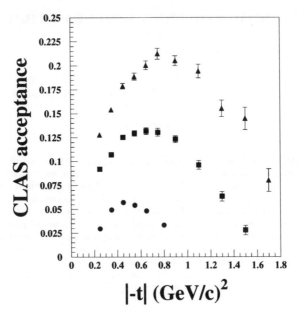

Figure 1. CLAS acceptance for pK^+ events used in the analysis. The circles are for the energy range $1.6 \leq E_\gamma \leq 1.8$ GeV, the squares for $1.9 \leq E_\gamma \leq 2.1$ GeV, and the triangles for $2.1 \leq E_\gamma \leq 2.3$ GeV.

2   Data Analysis

A preliminary analysis of ϕ-meson photoproduction, $\gamma + p \rightarrow p + \phi$, using bremsstrahlung photons of energy from $E_\gamma = 1.6$ GeV up to $E_\gamma^{max} = 2.45$ GeV, has be done using data taken with CLAS during the g1c running period (Fall 1999). Approximately 1.3 billion triggers containing at least one charged particle have been analyzed. The data have been separated in three photon energy bins: $1.6 \leq E_\gamma \leq 1.8$ GeV, $1.9 \leq E_\gamma \leq 2.1$, and $2.1 \leq E_\gamma \leq 2.3$. The bin size in photon energy is limited by the statistics of the data set analyzed which accounts for about 40% of the triggers on tape from the entire g1 running period. Finer bins in energy will be used as the statistics of the data set increases.

The large kinematic coverage of the CLAS detector is evident in Figure

1. The vertical scale is the acceptance for ϕ-meson photoproduction events (pK^+) determined from Monte Carlo simulation. It is clear that CLAS has good acceptance ($5 - 22\%$) over the entire kinematic range from threshold to the highest photon energies studied. The drop off at small values of four-momentum (t) is due to the forward opening in CLAS.

The ϕ photoproduction channel was identified through the decay of the ϕ-meson into two charged kaons ($\phi \to K^+ + K^-$) with the final state proton and positive kaon detected by CLAS in coincidence with a valid photon tagger signal. The missing mass (m_x) of the final state was calculated from the four vectors of the incident photon (p_γ), target proton (p_t), detected proton (p_p), and detected positive kaon (p_{K^+}). Events that fell within the range $0.48 \leq m_x \leq 0.50$ GeV/c^2 were considered to contain an undetected negative kaon ($m_{K^-} = 0.493$ GeV/c^2). The missing kaon four vector was then constructed using the missing energy and momentum, and constrained to have the kaon mass. Pions that were mis-identified as kaons were removed from the data sample on a bin-by-bin basis.

To isolate the K^+K^- events resulting from ϕ-meson production, a cut ($1.48 \leq m_{pK^-} \leq 1.55$ GeV/c^2) has been made to exclude events resulting from the decay of the $\Lambda^*(1520)$. The K^+K^- continuum was modeled with a Monte-Carlo simulation, fit to the K^+K^- mass spectrum and subtracted from underneath the ϕ-meson peak. Additionally, the contribution from the $a0/f0(980)$ meson was accounted for in the K^+K^- mass spectrum.

To achieve the differential cross section as a function of the four-momentum transfer (t), the above procedure was performed for data selected in each t-bin and photon energy bin. The ϕ meson yield was determined by a fit to the data of a peak shape that included the intrinsic meson width folded with the CLAS detector resolution via a GEANT simulation. The nominal CLAS photon normalization was then applied resulting in the preliminary cross section shown in Figure 2.

The CLAS data are the solid circles and the open squares are the data from Bonn.[3] The CLAS data are in good agreement with the entire range of Bonn data and extend the measurement to $|t| = 1.6$ $(GeV/c)^2$ where the low statistics begin to affect the cross section determination procedure. The error bars on the CLAS data include statistics from the data and Monte Carlo. The solid line is an exponential fit with a slope of -3.733 ± 0.2 that is consistent with the Bonn result.

The dotted curve in Figure 2 is from a regge model[14] that combines pomeron, f_2, and π meson exchange. In this model, the pomeron and pion couplings are fixed; the only free parameter is the f_2 coupling. At this photon energy, the only contributing mechanism is diffractive scattering via pomeron

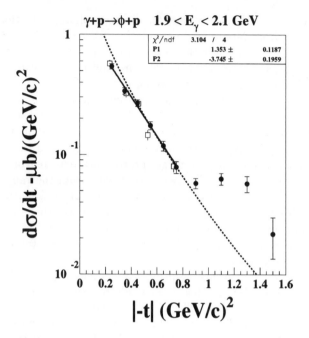

Figure 2. Differential cross section for the photon energy range $1.9 \leq E_\gamma \leq 2.1$ GeV. The circles are CLAS data, the open squares are the $E_\gamma = 2.0$ GeV data from Bonn.[3] The solid line is an exponential fit and the dashed line is a Regge model prediction with no intrinsic strangeness. [14]

exchange. The f_2 and pion amplitudes are negligible. The data and theory begin to disagree around a momentum transfer of $t = 1.0$ $(GeV/c)^2$. This discrepancy can accounted for by other photoproduction models that include s-channel (and u-channel) mechanisms such as phi-knock out [13] or resonance coupling. [15] However, further analysis, such as the evolution of the density matrix elements with momentum transfer and photon energy is needed to support or refute these alternate interpretations. This analysis will proceed as the photoproduction data set improves.

3 Summary

Reported here is a preliminary differential cross section of ϕ-meson photoproduction off the proton near threshold. Using CLAS, we have extended the measurement to the region in the variable t where production mechanisms beyond that of diffraction are expected to become significant. Recent theoretical calculations treat the diffractive component of the production due to the pomeron in sufficient detail, and all agree on the small size of the pion exchange contribution. Thus any additional strength in the cross section at high-t may be due to new mechanisms such as s and u channel production. The present analysis of the data allows the presence of these effects but further analysis of the angular distributions are needed to improve our understanding of the photoproduction process.

References

1. D.P. Barber et al., Z. Phys. **C12**, 1 (1982).
2. H.-J. Behrend et al., Nucl. Phys. **B144**,22 (1978).
3. H.J. Betsch,et al., Nuc. Phys. **B70**, 257 (1974);
4. T.H. Bauer, R.D. Spital, D.R. Yennie, and F.M. Pipkin, Rev. Mod. Phys. **50**, 261 (1978).
5. J. Ballam et al., Phys. Rev. **D7**, 3150 (1973).
6. H.J. Halpern , Phys. Rev. Lett. **29**, 1425 (1972).
7. A.M. Cooper et al., Nucl. Phys. **B146**, 10 (1978).
8. J. Ellis, E. Gabathuler, M. Karliner, Phys. Lett. **B217**, 173 (1989).
9. M. Anselmino, M.D. Scadron, Phys. Lett. **B229**, 117 (1989).
10. M.A. Nowak, J.J.M. Verbaarschot, I. Zahed, Phys. Lett, **B217**, 157 (1989).
11. The Photoproduction of the phi(1020) Near Threshold, Strange Quarks in Nucleons, Nuclei and Nuclear matter, ed. K. Hicks, World Scientific, 2001.
12. E.M. Henley, G. Krein, S.J. Pollock, A.G. Williams, Phys. Lett. **B269**, 31 (1991).
   E.M. Henley,G. Krein, S.J. Pollock, A.G. Williams , Phys. Lett. **B281**, 178 (1992).
13. A.I. Titov,T.-S.H. Lee, H. Toki, O. Streltsova, Phys. Rev.**C60**, 035205-1 (1999).
14. J.-M. Laget, hep-ph/0003213(2000).
   J.-M. Laget, R. Mendez-Galain, Nucl. Phys.**A581**, 397 (1995). Private communication.
15. Q. Zhao, J.-P. Didelez, M. Guidal, B. Saghai, Nucl. Phys.**A660**, 323 (1999).

K+ PHOTOPRODUCTION AT LEPS/SPRING-8

R.G.T. ZEGERS*, Y. ASANO, N. MURAMATSU

Advanced Science Research Center, JAERI, SPring-8, 1-1-1 Kouto, Mikazuki, Sayo, Hyogo 679-5198, Japan
present address: RCNP, Osaka University, 10-1 Mihogaoka, Ibaraki, Osaka 567-0047, Japan

M. FUJIWARA, H. FUJIMURA, T. HOTTA, H. KOHRI, T. MATSUMURA, N. MATSUOKA, T. MIBE, M. MORITA, T. NAKANO, K. YONEHARA, T. YORITA

RCNP, Osaka University, 10-1 Mihogaoka, Ibaraki, Osaka 567-0047, Japan

S. DATÉ, N. KUMAGAI, Y. OHASHI, H. OOKUMA

JASRI, SPring-8, 1-1-1 Kouto, Mikazuki, Sayo, Hyogo 679-5198, Japan

W.C. CHANG, D.S. OSHUEV, C.W. WANG, S.C. WANG

Institute of physics, Academia Sinica, Nankang, Taipei 11529, Taiwan

K. IMAI, T. ISHIKAWA, M. MIYABE, N. NIIYAMA, M. YOSOI

Department of Physics, Kyoto University, Kitashirakawa Oiwake-cho, Sakyo-ku, Kyoto 606-8502, Japan

H. KAWAI, T.OOBA, Y. SHIINO

Department of Physics, Chiba University, 1-33 Yayoi-cho, Inage, Chiba 263-8522, Japan

D.S. AHN, J.K. AHN

Department of Physics, Pusan National University, Pusan 609-735, Korea

S. MAKINO

Wakayama Medical College, Kimiidera 811-1, Wakayama, 641-0012, Japan

T. IWATA, Y. MIYACHI, A. WAKAI

Department of Physics, Nagoya University, Chikusa, Nagoya 464-8602, Japan

M. NOMACHI, A. SAKAGUCHI, Y. SUGAYA, M. SUMIHAMA

Department of Physics, Osaka University, 1-1 Machikaneyama, Toyonaka, Osaka 560-0043, Japan

H. AKIMUNE

Department of Physics, Konan University, 8-9-1 Okamoto, Higasginada-ku, Kobe,

658-8501, Japan

C. RANGACHARYULU

Department of Physics and Engineering Physics, University of Saskatchewan, 116 Science Place, Saskatoon, SK S7N 5E2, Canada

K. HICKS

Department of Physics, Ohio University, Athens, Ohio, 45701, USA

H. SHIMIZU

Department of Physics, Yamagata University, Kojirakawa 1-4-12, Yamagata 990-8560, Japan

P. SHAGIN

Institute for high-energy physics (IHEP), RU-142284, Protvino, Moscow region, Russia

H.C. BHANG, Z.Y. KIM

Department of Physics, Seoul National University, Sinlim-dong, Kwanak-gu, Seoul 151-742, Korea

K^+ photoproduction is expected to provide new insight into the structure of nucleon resonances. However, due to the lack of data and ambiguities in the theoretical descriptions, many questions remain unanswered. At the LEPS beam line at SPring-8, data for the K^+ photoproduction of the proton have been taken between December 2000 and June 2001. Both differential cross section and photon-polarization asymmetries have been measured for the $p(\gamma, K^+)\Lambda$ and $p(\gamma, K^+)\Sigma^0$ reactions. The photon polarization asymmetries in particular will provide valuable information to further the understanding of these reaction channels and the presence of missing resonances. The center-of-mass energy range covered is 1.9 GeV $< W < 2.3$ GeV. The angular acceptance for detection of the K^+ is between 0deg and 60deg in the center-of-mass frame. Here, we discuss the status of the analysis and the relevance of the observables that are accessible.

1 Introduction

Knowledge about the excitation spectrum of the nucleon has come mainly from the study of electromagnetic induced π-production and π induced reactions. The study of strangeness photoproduction on the nucleon is a potentially powerful tool to deepen the insight into baryon structures, because of the additional degree of freedom involved. The $p(\gamma, K^+)\Lambda$ and $p(\gamma, K^+)\Sigma^0$ reactions are good candidates for such studies, since they provide a way to access a rather large part of the rich excitation spectrum of the nucleon. For

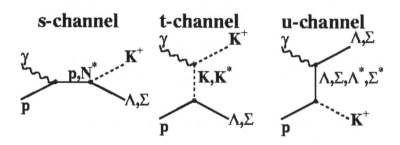

Figure 1. Contributions to the $p(\gamma, K^+)\Lambda$ and $p(\gamma, K^+)\Sigma^0$ reactions at the tree level; Born terms (involving p, K or Λ, Σ) and exchanges of excited particles or resonances (involving N^*, K^* or Λ^*, Σ^*) in the s,t and u channels.

a number of these nucleon resonances, the branching into strange channels is well known. On the other hand, a large number of resonances predicted in quark models [1,2] have so far been undiscovered and are hence referred to as 'missing resonances'. It has been prepositioned that some of these missing resonances could couple strongly to the $K\Lambda$ and $K\Sigma$ channels.

Although the above-mentioned reaction channels had been under investigation for an extensive period of time, data taken at SAPHIR [3] has resulted in renewed interest. This is largely due to the fact that good statistics revealed structure in the total cross section spectrum of the $p(\gamma, K^+)\Lambda$ reaction near $W=1900$ MeV. Mart and Bennhold [4] showed that this structure can be explained if an additional $D_{13}(1895)$ resonance is included in the calculations. An appreciable decay of this resonance is indeed expected from constituent quark-model calculations [2]. Alternatively, if from the set of possible candidates in this mass range a $P_{13}(1950)$ resonance is chosen, the data is equally well described [5].

Unfortunately, it is difficult to draw strong conclusions based on cross section measurements only because of ambiguities in the theoretical descriptions. In the region $W < 2.5$ GeV, most calculations are performed in a tree-level effective-Lagrangian approach, in which the Feynman diagrams as depicted in figure 1 are taken into account. Intermediate particles are treated as effective fields using their specific masses, couplings and decay widths. Furthermore, hadronic form factors are introduced at each strong vertex in order to describe the finite extension of the hadrons. This causes Gauge-invariance breaking and contact terms are required for restoration.

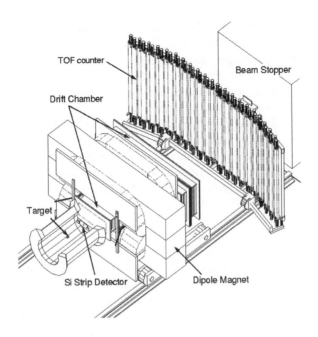

Figure 2. The LEPS detection system. The silicon-strip vertex detector, 3 driftchambers (1 up-stream and 2 down-stream from the dipole magnet) and time-of-flight scintillator array are indicated. For details, see text.

Besides the problem of which resonances should be included in the calculations, the inclusion of any resonance requires the introduction of freely chosen coupling constants whose values must be obtained from fitting to the data. Although the result can be compared to quark-model predictions, ambiguity remains because SU(3) symmetry is broken due to the mass difference between the strange and up/down quarks. Moreover, final-state interactions, which are essentially integrated with the effective couplings complicate the comparison further.

Essential in the theoretical descriptions is to find a proper way to suppress the strength from the Born terms, since these by themselves give rise to cross sections much larger than seen in the data. As shown by Janssen et al. various procedures lead to good descriptions of the data [5]. The choice for the functional shape of the form factor is another source of ambiguity [6,7] and leads to the introduction of another free parameter, namely the cut-off mass. Further problems arise from the fact that in principle the calculations should be performed in a coupled-channels scheme. Chiang et al. [8] showed that coupling to the πN channels gives a sizeable (20%) contribution to the Kaon photoproduction cross section. Saghai [9] showed that off-shell effects, inherent to fermions with spin larger than 3/2 can give rise to structure in the cross sections similar to that produced by missing resonances and, hence, that it is dangerous to draw any conclusion about the existence of such resonances.

It can, therefore, be concluded that at present we are left with an unacceptable level of model dependence in the theoretical description of experimental results. Improvement in this situation is possible in two ways. On the theoretical side, approaches should be changed as to reduce the number of parameters and limit the model dependencies. Saghai [9] suggested to use a constituent quark approach, which would have the benefits that all known resonances can be included, the number of fit parameters are reduced and their ranges more constrained.

On the experimental side, one strives to increase the amount of available data, especially by measuring previously unprobed observables. Inclusion or exclusion of certain resonances [4] as well as the use of different model ingredients [5] have large effects on the predictions for (double) polarization observables. Therefore, at various places in the world (GRAAL, JLAB, ELSA, MAMI, LEGS and SPring-8) experimental programs are in progress to measure these observables, in various kinematical ranges. It has been shown that the photon-polarization asymmetry (the asymmetry measured in the comparison between data taken with different polarizations for the incoming photon) in particular contains large sensitivities. Mart and Bennhold [4] predict a sign change for the photon polarization asymmetry if the missing $D_{13}(1895)$ resonance is present. Janssen et al. [5] showed that various methods to suppress the strength of the Born terms, as well as the choice of form factor also changes the photon polarization asymmetry significantly.

In the present paper we wish to describe the efforts to measure the K^+ photoproduction and its photon-polarization asymmetry (Σ) at the Laser-Electron-Photon beam line at SPring-8, Japan.

2 The LEPS beam line at SPring-8 and K^+ photoproduction

At the Laser-Electron-Photon beam line at SPring-8, (LEPS [10,12]) GeV photons are produced by Compton backscattering of laser photons from 8 GeV electrons in the storage ring. Depending on the wavelength of the laser photons, different photon-energy regions can be investigated. So far, measurements have been performed using a 351-nm Ar laser, which corresponds to a maximum photon energy of 2.4 GeV. Electrons that are participants in the backscattering process are detected in a tagger system on the inside of

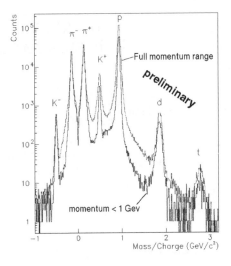

Figure 3. Reconstructed mass spectrum measured at LEPS from a liquid Hydrogen target. Events stemming from the plastic-scintillator start counter are include and hence deuterons and tritons are also detected. Increased resolution for particles with low momentum can be seen.

the SPring-8 storage ring and provide information about the energy of the photons. Because of limitations on how close the tagger can be positioned to the circulating electron beam, only photons whose energies are above 1.5 GeV can be tagged. One of the most important advantages of using Compton backscattering is that the photon polarization can be determined by polarizing the laser light. At the maximum photon energy the degree of linear polarization of the laser photons is almost fully transferred to the backscattered photon [13]. At 1.5 GeV still more than 50% of the laser polarization is transferred. These high transfer percentages, combined with the a-priori knowledge of the degree of polarization as a function of photon energy, make Compton backscattering ideally suitable for polarization studies. At LEPS, tagger rates of 600 kHz are routinely obtained and changing between different states of linear polarization, in order to minimize systematical errors, is a matter of minutes. The backscattered photons transverse a 70-m long beam line before hitting a liquid Hydrogen target with a thickness of 5 cm. Reaction products are detected in an elaborate detection system. In figure 2 this system is displayed. Particles are momentum-analyzed in a dipole magnet. Three sets of drift chambers, one positioned up-stream and two positioned

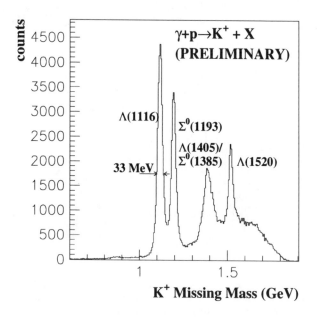

Figure 4. Preliminary K^+ missing mass spectrum taken at LEPS. The $\Lambda(1116)$, $\Sigma^0(1193)$, $\Lambda(1520)$ and combined $\Lambda(1405)/\Sigma^0(1385)$ are indicated.

down-stream of the dipole magnet, are used to trace the particle trajectories. Each set consists of 5 wire planes. A silicon-strip detector, put close to the LH$_2$ target cell, also provides tracking information and is used to determine the vertex position accurately. An array of scintillators provides time of flight information for particle identification. The data-taking system of the present experiment is described in detail by Sugaya et al. [11]. The detector system is optimized for ϕ-photoproduction studies [12]. Acceptance for single K^+ mesons is limited to center-of-mass scattering angles of 60deg and full azimuthal coverage is only obtained up to about 30deg. The hadronic trigger rate for the LH$_2$ target is about 20 Hz.

In figure 3, the reconstructed mass spectrum is shown. The mass resolution is better for particles with low momentum, as can clearly be seen when only particles with a momentum lower than 1 GeV are selected. π's, K's and proton peaks can clearly be distinguished and cleanly selected. Contamination of protons and π^+'s into the K^+ mass range amounts to a few percent maximally, event at the highest momenta. Narrowing the mass cut for the

K^+ will lower this number even more and the small fraction of lost events can easily be corrected for. Since figure 3 includes events coming from the plastic scintillator start counter, deuterons and tritons are also detected, due to the presence of Carbon.

After selecting the K^+'s from the mass spectrum, one can reconstruct the missing mass. Figure 4 shows the missing mass in the K^+ channel for about 50% of the data taken between December 2000 and June 2001. Peaks for the $\Lambda(1116)$, $\Sigma^0(1193)$, $\Lambda(1520)$ and a combined peak for the $\Lambda(1405)/\Sigma^0(1385)$ can be identified. The (preliminary) missing-mass resolution is approximately 33 MeV and the latter two resonances can therefore not be separated. A small tail below a missing mass of 1 GeV corresponds to small contaminations from π's and protons. The $\Lambda(1116)$ can adequately, albeit not completely, be seperated from the $\Sigma^0(1193)$ and small contaminations from one into the other must be taken into account. In total about 50k $K^+\Lambda$ and 30k $K^+\Sigma^0$ events are detected. The broad structure underlying the sharp states are due to the opening of the three-body phase space.

After selecting the $\Lambda(1116)$ or $\Sigma^0(1193)$ in the missing mass spectrum, one can determine the photon-polarization asymmetries by comparing the azimuthal distributions of the K^+'s for data taken with horizontal polarized photons with data taken with vertical polarized photons:

$$\Sigma P cos(2\phi) = \frac{N_v(\phi) - N_h(\phi)}{N_v(\phi) + N_h(\phi)}, \qquad (1)$$

where Σ is the photon-polarization asymmetry, P the averaged degree of polarization for the horizontal and vertical polarized data, ϕ the azimuthal angle of the K^+ and $N_h(N_v)$ the measured azimuthal distributions for horizontal (vertical) polarized data.

In figure 5, the right-hand side of the above equation is plotted for the $\Lambda(1116)$ (left) and $\Sigma^0(1193)$ (right). No selections for scattering angle or photon energy has been made. A clear $cos(2\phi)$ structure can be seen and the full lines show fits to the data with a function of the form $A cos(2\phi)$. After dividing A by P one would find Σ.

To demonstrate the dependence on photon energy, figure 6 displays similar curves as in figure 5, but events with $1.6 < E_\gamma < 1.8$ GeV (open data points, dashed fit curves) and $2.1 < E_\gamma < 2.3$ GeV(full data points, solid fit curves) have been separated. Of course, part of the difference in asymmetry is caused by the lower degree of photon polarization (P) for the lower photon energies.

From figures 5 and 6 it is clear that at LEPS the photon polarization asymmetry for the K^+, Λ and K^+, Σ^0 can be accurately determined. Since the photon polarization asymmetry is expected to fluctuate rather quickly

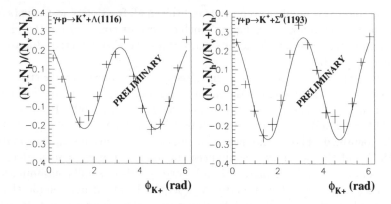

Figure 5. Scattering-angle and photon-energy integrated asymmetry curves for the $\Lambda(1116)$ (left) and $\Sigma^0(1193)$ (right).

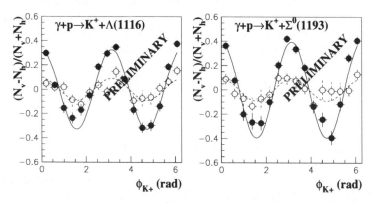

Figure 6. Scattering-angle and photon-energy integrated asymmetry curves for the $\Lambda(1116)$ (left) and $\Sigma^0(1193)$ (right). Open data points/dashed fits are for $1.6 < E_\gamma < 1.8$ GeV and full data points/solid fits are for $2.1 < E_\gamma < 2.3$ GeV.

as a function of photon energy in the resonance region, it is important to determine asymmetries for as narrow energy bins as possible. It is envisioned that with the current statistics a binning of 100 MeV in E_γ (corresponding to approximately 50 MeV in W) can be achieved for 4 different values of t between 0 and -0.6 $(\text{GeV/c})^2$.

3 Status and outlook

At present, acceptances and efficiencies for the LEPS detector are being studies. These include Monte-Carlo simulations to determine the geometrical acceptance accurately as well as detailed studies of the separate parts of the detector system, photon-beam properties and tracking errors. With the various details understood, results for the photon polarization asymmetry and the differential cross section are expected to come out in the course of 2002.

Acknowledgments

Support by the Japan International Science and Technology Exchange Center (JISTEC) and the Science and Technology Agency (STA) of Japan for the fellowship of R.G.T. Zegers hosted by the Japan Atomic Energy Reseach Institute (JAERI) is gratefully acknowledged.

References

1. S. Capstick and N. Isgur, Phys. Rev. D **34**, 2809 (1986).
2. S. Capstick and W. Roberts, Phys. Rev. D. **49**, 4570 (1994).
3. M.Q. Tran et al., Phys. Lett. B **445**, 20 (1998).
4. T. Mart and C. Bennhold, Phys. Rev. C **61** (R) 012201 (2000).
5. S. Janssen et al., nucl-th/0107028, 2001.
6. H. Haberzettl et al., Phys. Rev. C **58** (R) 40 (1998).
7. R. Davidson and R. Workman, Phys. Rev. C **63**, 058201 (2001).
8. Wen-Tai Chiang it et al., nucl-th/0104052, 2001.
9. B. Saghai, nucl-th/0105001, 2001.
10. T. Nakano et al., Nucl. Phys. **A684**, 71c (2001).
11. Y. Sugaya et al., IEEE Trans. Nucl. Phys. 48, 1282 (2001).
12. T. Nakano et al., these proceedings.
13. A. D'Angelo, Nucl. Instr. & Meth. **A455**, 1 (2000).

POLARIZATION OBSERVABLES
IN KAON ELECTROPRODUCTION WITH CLAS
AT JEFFERSON LABORATORY

DANIEL S. CARMAN

Department of Physics, Ohio University, Athens, OH 45701, USA
E-mail: carman@ohio.edu

An extensive program of strange particle production off the proton is currently underway with the CEBAF Large Acceptance Spectrometer (CLAS) in Hall B at Jefferson Laboratory. Precision measurements of ground-state and low-lying excited-state hyperons are being carried out with both electron and real photon beams, both of which are available with high polarization at energies up to 6 GeV. This talk will focus on selected aspects of our strangeness physics program regarding electroproduction measurements of single and double-polarization observables.

1 Introduction

During the last decade there has been considerable effort to develop theoretical models for the kaon electroproduction process. However, the present state of understanding is still limited by a sparsity of data. Model fits to the existing cross section data are generally obtained at the expense of many free parameters, which leads to difficulties in constraining existing theoretical descriptions. Moreover, cross section data alone are not sufficiently sensitive to fully understand the reaction mechanism, as they probe only a small portion of the full response. In this regard, measurements of spin observables are essential for continued theoretical development in this field, as they allow for improved understanding of the dynamics of this process and provide for strong tests of QCD-inspired theoretical models.

The strange particle electroproduction program with CLAS at Jefferson Laboratory focusses on the associated production reaction $ep \to e'K^+Y$. Individual experiments have measured cross sections and spin observables for $K^+\Lambda$ and $K^+\Sigma^0$ final states at beam energies from 2.5 to 4.8 GeV. With the existing approved program, the present lack of data should soon be remedied with a wealth of very high quality measurements spanning a broad range in invariant energy W and momentum transfer Q^2.

The large acceptance of the CLAS spectrometer has enabled us to detect the scattered electron and the electroproduced kaon, as well as the proton from the mesonic decay of the final-state Λ hyperon, over a range of Q^2 from 0.4 to 2.7 $(\text{GeV/c})^2$ and W from threshold to 2.4 GeV, while providing full angular coverage in the center-of-mass of the kaon. The measured angular

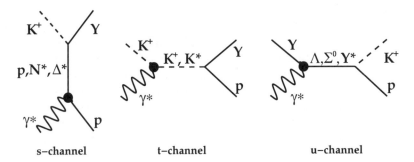

s-channel   t-channel   u-channel

Figure 1. Feynman diagrams for the process $\gamma^* p \to K^+ Y$ for the s, t, and u channel reactions listing the Born and non-Born terms in the intermediate state.

correlation of the decay proton allows for the determination of the final-state Λ or Σ^0 polarization. The CLAS detector enables simultaneous study of the reactions over kinematical regions where the contributing s, t, and u reaction channels processes (Fig. 1) have varying strengths. By emphasizing specific channel processes we can effectively limit the intermediate hadronic resonances involved in the reaction.

In order to better understand the reaction mechanism of open-strangeness production, it is important to better understand which baryon resonances contribute to the intermediate state, and to allow for a better determination of their associated coupling strengths and form factors. Presently our knowledge on the baryon resonances comes mainly from pion channels[1,2]. Studies of strange final states could uncover baryonic resonances that do not couple or couple only weakly to the πN channel due to the different hadronic vertices. This allows for insight into the so-called "missing" quark model baryon states. Recent quark model calculations[3] have shown that there are several N^* resonances with appreciable photocouplings that also provide for sizeable strength to decay to KY. These resonances, including the $S_{11}(1650)$, $P_{11}(1710)$, $P_{13}(1720)$, and most recently the $D_{13}(1895)$, also appear to be the main s-channel resonances required to fit the data within the framework of some hadrodynamic effective Lagrangian models[4,5]. However, this phenomenological framework is not without inherent ambiguities in fitting the existing data. Higher quality data, including polarization observables, are important to make progress. Ultimately the data must be analyzed with full partial wave analysis in a coupled-channels framework[2].

This leads to an important question, namely, in the absence of direct QCD predictions, can effective theories and models allow us to understand the reaction mechanism for open-strangeness production by providing a framework that describes the full set of available cross section and polarization data? In this regard, polarization observables have been demonstrated to be extremely sensitive to different assumptions about baryonic structure. Polarization effects arise due to interference of amplitudes in the intermediate state, and thus can provide for high sensitivity to resonance-resonance interference, non-resonant components, and small amplitude contributions. In this talk, I will focus on the CLAS beam asymmetry and hyperon single and double-polarization observables, and discuss the preliminary data within the framework of available hadrodynamic models.

2 Experimental Setup

The Continuous Electron Beam Accelerator Facility (CEBAF) at Jefferson Laboratory is based on a recirculating multi-GeV electron linear accelerator. Located in Experimental Hall B of this facility is the CLAS spectrometer, a detector for use with electron and tagged-photon beams. CLAS is constructed around six iron-free superconducting coils that generate a toroidal magnetic field. The particle detection system (Fig. 2) consists of drift chambers to determine charged-particle trajectories, Čerenkov detectors for electron/pion separation, scintillation counters for flight-time measurements, and calorimeters to identify electrons and high-energy neutral particles.

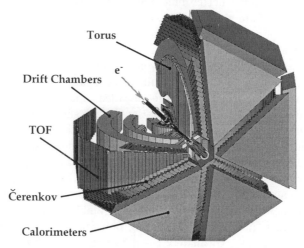

Figure 2. Three dimensional representation of CLAS with a portion of the system cut away to highlight the elements of the detector system.

CLAS was designed to track charged particles emerging from the target with momenta greater than 200 MeV/c over polar angles from 8° to 142°, while covering up to 80% of the azimuth. The average acceptance of CLAS for detecting the final state e' and K^+ is about 30%, and drops to 5% when also requiring detection of the decay proton. The electron beam longitudinal polarization is roughly 70%, with typical beam-target luminosities of $6\text{-}7\times10^{33}$ cm^{-2} sec^{-1}.

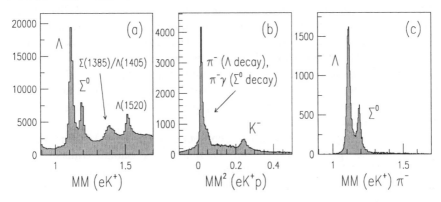

Figure 3. Missing-mass spectra (GeV) for the reactions (a) $p(e, e'K^+)X$ and (b) $p(e, e'K^+p)X$. (c) The hyperon distribution after cutting on the low-mass peak in (b). Reconstructions from 4.247 GeV CLAS data.

Hyperon final-state identification with CLAS relies on missing-mass reconstructions to identify neutral particles in exclusive reactions. Shown in Fig. 3a is the missing-mass distribution at 4.247 GeV for $p(e, e'K^+)X$ after reconstructing the scattered electron and kaon. This spectrum shows substantial, well-separated peaks for the low-lying hyperon states. Fig. 3b shows the missing mass for the reaction $p(e, e'K^+p)X$. The final-state proton in this case can come from the decay of the $\Lambda(1115)$ (missing π^-), the $\Sigma^0(1192)$ (missing $\gamma\pi^-$), or the $\Lambda(1520)$ (missing K^-). Fig. 3c shows the resulting spectrum from a cut on the low-mass peak in Fig. 3b. The width of the hyperon peaks in this spectrum, summed over all Q^2 and W, is about 14 MeV.

3 Physics Results

3.1 Formalism

The most general form for the virtual photo-absorption cross section of the kaon from an unpolarized-proton target, allowing for both a polarized-electron beam and recoil hyperon is given by[6]:

$$\frac{d^2\sigma_v}{d\Omega_K^*} = \sigma_0 \left[1 + hA_{TL'} + \sum_{i=t,n,l} \left(P_i^0 + hP_i' \right) \right], \tag{1}$$

$$\sigma_0 = K_f \left[R_T^{00} + \epsilon_L R_L^{00} + \sqrt{2\epsilon_L(1+\epsilon)} R_{TL}^{00} \cos\Phi + \epsilon R_{TT}^{00} \cos 2\Phi \right]. \tag{2}$$

Here σ_0 represents the unpolarized cross section, $A_{TL'}$ the polarized-beam asymmetry, and P_i refers to the hyperon polarization. The R_i terms represent the longitudinal, transverse, and interference response functions that account for the dynamics of the hadronic system. Here ϵ (ϵ_L) is the degree of transverse (longitudinal) polarization of the virtual photon, h is the electron beam helicity, K_f is a kinematic factor given by the ratio of CM momenta of the outgoing kaon and the virtual photon, and Φ is the relative angle between the electron and hadron planes.

Each of the hyperon polarization components (P_t, P_n, P_l) is further split into the induced polarization P^0, and the helicity-dependent transferred polarization P', defined with respect to a particular set of spin-quantization axes. In the (t, n, l) coordinate system defined in Fig. 4, the three induced and transferred polarization components are given by[6]:

$$P_t^0 = -K_f \left(\sqrt{2\epsilon_L(1+\epsilon)} R_{TL}^{x'0} \sin\Phi + \epsilon\, R_{TT}^{x'0} \sin 2\Phi \right) / \sigma_0$$

$$P_n^0 = K_f \left(R_T^{y'0} + \epsilon_L R_L^{y'0} + \sqrt{2\epsilon_L(1+\epsilon)} R_{TL}^{y'0} \cos\Phi + \epsilon R_{TT}^{y'0} \cos 2\Phi \right) / \sigma_0$$

$$P_l^0 = -K_f \left(\sqrt{2\epsilon_L(1+\epsilon)} R_{TL}^{z'0} \sin\Phi + \epsilon R_{TT}^{z'0} \sin 2\Phi \right) / \sigma_0, \tag{3}$$

$$P_t' = -K_f \left(\sqrt{2\epsilon_L(1-\epsilon)} R_{TL'}^{x'0} \cos\Phi + \sqrt{1-\epsilon^2} R_{TT'}^{x'0} \right) / \sigma_0$$

$$P_n' = K_f \sqrt{2\epsilon_L(1-\epsilon)} R_{TL'}^{y'0} \sin\Phi / \sigma_0$$

$$P_l' = -K_f \left(\sqrt{2\epsilon_L(1-\epsilon)} R_{TL'}^{z'0} \cos\Phi + \sqrt{1-\epsilon^2} R_{TT'}^{z'0} \right) / \sigma_0. \tag{4}$$

3.2 Hyperon Polarization

An attractive feature of the mesonic decay $\Lambda \to p\pi^-$ comes from its self-analyzing nature. From the decay-proton angular distribution, the average hyperon polarization about each of the three spin axes can be determined. The decay distribution in the Λ rest frame is of the form:

$$dN/d\Omega_p^{RF} \propto 1 + \alpha P_\Lambda \cos\theta_p^{RF}. \tag{5}$$

In this expression, $\alpha=0.642$ is the weak-decay asymmetry parameter and θ_p^{RF} is the decay-proton polar angle relative to the spin-quantization axes. To improve the statistical precision, our analysis has proceeded by summing over all relative Φ angles. In this case, the polarization components P_t^0, P_l^0, and P_n' vanish and our definitions (with $K_I = (R_T^{00} + \epsilon_L R_L^{00})^{-1}$) become:

$$P_n^0 = K_I(R_T^{y'0} + \epsilon_L R_L^{y'0}) \tag{6}$$

$$P_t' = -K_I\sqrt{1 - \epsilon^2}R_{TT'}^{x'0} \qquad P_l' = -K_I\sqrt{1 - \epsilon^2}R_{TT'}^{z'0}. \tag{7}$$

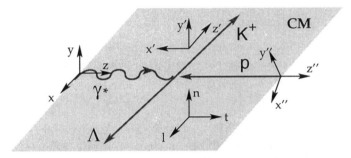

Figure 4. Center-of-mass (CM) coordinate systems used in the polarization analysis.

For polarization observables, the (t, n, l) coordinate system is not a unique choice. Other choices for the spin-quantization axes, shown in Fig. 4, could include an axis along the virtual photon direction (z, z'') or perpendicular to the electron scattering plane (y''). The Φ-integrated polarization observables will then be sensitive to a different subset of response functions.

3.3 Theoretical Models

A comprehensive theoretical framework has been developed for the electromagnetic production of low-lying Λ and Σ^0 hyperons[4,5]. Model parameters are determined by a simultaneous fit to the low-energy γp and $\gamma^* p \to K^+ Y$ photo- and electroproduction data, along with the $K^- p \to \gamma Y$ radiative capture data. In this hadrodynamic effective Lagrangian approach, several different elementary models have been developed from fits to the data by adding Born terms with a number of resonances and leaving their coupling constants as free parameters. Different models have markedly different ingredients and coupling constants. The different resonances in the various models include:

t chan.: $K^*(893)$, $K_1(1270)$ | s chan. : $P_{11}(1440)$, $S_{11}(1650)$,
u chan.: $\Lambda(1405)$, $\Lambda(1670)$, $\Lambda(1800)$ | $P_{11}(1710)$, $P_{13}(1720)$, $D_{13}(1895)$.

Beyond selecting different elementary models of the production process
that incorporate different intermediate resonant states, the theoretical framework also allows for the selection of different forms for the Q^2 dependence of
the electromagnetic form factors of the kaon and the hyperon. In the results
that follow in this paper, simple dipole form factors have been employed and
the results for different elementary models have been compared. These models were developed by Bennhold and Mart (BM)[4], Williams, Ji, and Cotanch
(WJC)[8], and Adelseck and Wright (AW)[9].

3.4 Beam Asymmetry

The polarized-beam asymmetry $A_{TL'}$ for the $K^+\Lambda$ final state provides direct
access to the fifth response function $R^{00}_{TL'}$. This asymmetry is given by:

$$A_{TL'} = \frac{1}{P_e} \frac{N^+ - N^-}{N^+ + N^-} = \frac{K_f}{\sigma_0} \sqrt{2\epsilon_L(1-\epsilon)} \, R^{00}_{TL'} \sin\Phi. \tag{8}$$

Here the N^+ and N^- helicity-gated yields are extracted by selecting events
in the $e'K^+$ missing-mass spectrum consistent with a final-state Λ. This
term allows for separation of the reaction mechanisms that contribute to the
intermediate-state excitations from non-resonant processes. This observable
accesses imaginary parts of interfering longitudinal and transverse amplitudes.
These imaginary parts vanish identically if the resonant state is determined
by a single complex phase, which is the case for an isolated resonance.

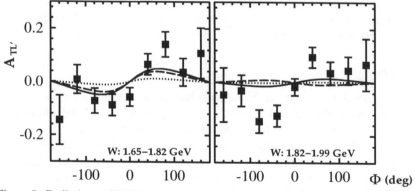

Figure 5. Preliminary CLAS beam asymmetry for $\vec{e}p \rightarrow e'K^+\Lambda$ as a function of Φ for
$W = 1.7$ GeV (left) and $W = 1.9$ GeV (right) summed over Q^2 at 4.247 GeV. The curves
correspond to different hadrodynamic models: AW-solid, WJC - dashed, BM - dotted.

The preliminary CLAS results at 4.247 GeV are shown in Fig. 5 for two different W bins. These asymmetries have been acceptance corrected and had the dominant background subtracted. This dominant background arises from misidentification of final-state pions as kaons. The results are compared with three different hadrodynamic models. The models are in reasonable agreement with the low-W data, but underpredict the data for the higher-W bin. One of the models (WJC) even comes in at higher W with the wrong phase.

3.5 Induced Polarization

For the induced Λ polarization measurements, the acceptance-corrected decay-proton yields have been summed over both electron helicity states and fit with the form of eq(5) to extract P_n^0. P_n^0 represents the only Φ-integrated induced polarization component that is not required to vanish. Results have been summed together from data sets at 2.567 and 4.247 GeV as the data show essentially no ϵ dependence.

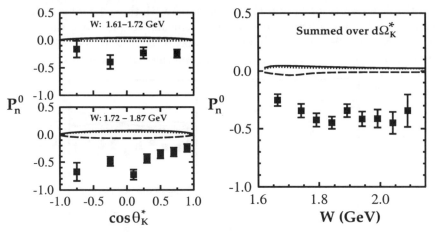

Figure 6. Preliminary CLAS induced polarization for $ep \rightarrow e'K^+\Lambda$ as a function of $\cos\theta_K^*$ for two different W bins summed over Q^2 (left) and as a function of W summed over Q^2 and $d\Omega_K^*$ (right) from the combined 2.567 and 4.247 GeV CLAS data. The curves correspond to different hadrodynamic models: AW-solid, WJC - dashed, BM - dotted.

The preliminary CLAS results are shown vs. $\cos\theta_K^*$ and W in Fig. 6. The dependence of P_n^0 with respect to W is very similar to what has been found from analysis of photoproduction data at SAPHIR[7]. However the angular dependence is quite different, indicating an important Q^2 dependence. The results are compared with three different hadrodynamic model calculations to show how well the models are doing relative to the data, as well as the spread in their predictions. Clearly there is not much spread between the calculations, indicating a possible commonality in their disagreement.

3.6 Transferred Polarization

The transferred Λ polarization is determined from the acceptance-corrected yield asymmetries:

$$A_i = \frac{N^+ - N^-}{N^+ + N^-} = \frac{\alpha P_e \cos\theta_p^{RF} P_i'}{1 + \alpha \cos\theta_p^{RF} P_i^0}, \qquad i = t, n, l. \qquad (9)$$

This asymmetry is formed separately from the decay-proton helicity-gated yields with respect to the (t, n, l) axes. With this method, we are quite insensitive to the CLAS acceptance function. In forming these asymmetries and summing over Φ, A_n must vanish, but A_l and A_t can be non-zero. Fig. 7 shows results of our preliminary analysis at 4 GeV as a function of both $\cos\theta_K^*$ and W. The results for P_l' and P_t' are compared with three different hadrodynamic models. The accuracy of the measurements, coupled with the spread in the theory predictions, clearly indicates that these data are sensitive to the resonant and non-resonant structure of the intermediate state.

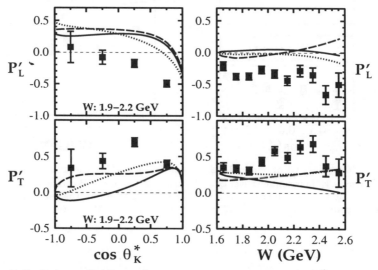

Figure 7. Preliminary CLAS transferred polarization from $\vec{e}p \rightarrow e'K^+\vec{\Lambda}$ vs. $\cos\theta_K^*$ (left) summed over Φ and Q^2 and vs. W (right) summed over Q^2 and $d\Omega_K^*$ at 4 GeV. The curves correspond to different hadrodynamic models: AW-solid, WJC - dashed, BM - dotted.

4 Summary and Conclusions

In this talk I have reviewed some of the key reasons why electroproduction processes of open-strangeness production are important for the investigation of baryonic structure and "missing" quark model states. I have highlighted several aspects of the CLAS strangeness physics program regarding $ep \to e'K^+Y$ focussing on single and double-polarization observables with polarized beam and/or polarized hyperon recoils. Detailed analysis of data sets at 2.5 and 4 GeV have shown sizeable Λ polarization signatures with very little W or Q^2 dependence. The polarization data have been studied within the framework of hadrodynamic models. The results indicate that the data are highly sensitive to the ingredients of the models, including the specific baryonic resonances included, along with their associated form factors and coupling constants.

The CLAS program is presently ongoing with a polarized-electron beam at 6 GeV. This will extend the W range of the polarization data to nearly 3.5 GeV. This will allow for study of the dynamics of open-strangeness production beyond the resonance region, where the effective degrees of freedom are expected to evolve from mesons and baryons to quarks and gluons.

Acknowledgments

Some of the physics results presented in this talk were obtained through the work of Brian Raue and Rakhsha Nasseripour from Florida International University and Simeon McAleer from Florida State University. The author also wishes to thank Cornelius Bennhold and Terry Mart for the use of their theory code. This work has been supported by the U.S. Department of Energy and the National Science Foundation.

References

1. Particle Data Group, Eur. Phys. J **15**, 1 (2000).
2. T.P. Vrana, S.A. Dytman, and T.S.H. Lee, Phys. Rep. **328**, 181 (2000).
3. S. Capstick and W. Roberts, Phys. Rev. D **58**, 74011 (1998).
4. T. Mart and C. Bennhold, Phys. Rev. C **61**, 012201 (2000).
5. B. Saghai, Preprint nucl-th/0105001, (2001).
6. G. Knöchlein, D. Drechsel, L. Tiator, Z. Phys. **A 352**, 327 (1995).
7. M.Q. Tran et al., Phys. Lett B **445**, 20 (1998).
8. R.A. Williams, C. Ji, and S.R. Cotanch, Phys. Rev. C **46**, 1617 (1992).
9. R.A. Adelseck and L.E. Wright, Phys. Rev. C **38**, 1965 (1988).

CAN THE SCALAR MESONS $a_0/f_0(980)$ BE DESCRIBED BY $K + \overline{K}$?

R. T. JONES

University of Connecticut, Storrs CT 06269-3049, USA
E-mail: richard.t.jones@uconn.edu

The scalar mesons $f_0(980)$ and $a_0(985)$ are two of the best-verified states in the excited meson spectrum. They appear as narrow resonances in the 2π and $\eta\pi$ mass spectra, respectively, observed in hadroproduction reactions and in the decays of heavier particles such as the J/Ψ and the Z^0. Their relatively low mass and narrow width do not match well with quark-model expectations for the lightest scalar multiplet, fueling speculation that these states might be different in nature from conventional $q\bar{q}$ mesons. A variety of theoretical models have been put forward to explain these states, most notably the so-called "2-kaon molecule" model in which they appear as kaon-antikaon bound states. Within this picture, the radiative decays of the $\phi(1020)$ meson two two-pseudoscalar final states are expected to be dominated by f_0 and a_0 intermediate states, in spite of their appearance very near the edge of the available phase space. Recent experimental results from e^+e^- colliders have recently provided striking verification of this effect. Meanwhile, new work within a more general theoretical approach to the two-meson scattering problem has shown that the appearance of narrow scalar resonances near the two-kaon threshold is not a unique prediction of the molecule model, but are a consequence of the general principles of unitarity and chiral symmetry. The relative merits of these different viewpoints are examined in light of the ϕ decay data.

1 Introduction

Within the context of the quark model, the lightest L=1 mesons belong to the fundamental scalar nonet. The scalars appear as the first rung of a ladder of three nonets closely spaced in mass, with J^{PC} assignments 0^{++}, 1^{++} and 2^{++}. A tidy correspondence between the quark model and the observed spectrum of mesons has been achieved for the axial and tensor nonets, with central masses around 1.3 GeV/c^2. The widths of these states are governed by decay selection rules which determine the minimum number and angular momentum of light mesons in the final state, with the broadest states having allowed s-wave decays into 2 L=0 mesons. For example, the a1(1260) decays predominantly into $\rho\pi$ s-wave, with a partial width of several hundred MeV. Based upon this pattern, one should expect to find the fundamental scalar nonet with non-strange members in the mass range 1.2 - 1.3 GeV/c^2 and with 2-pseudoscalar decay widths of several hundred MeV. When one looks at the resonance structure in $\pi\pi$ scattering, however, what one sees instead is a scalar state just below 1.0 GeV/c^2 that is less than 100 MeV/c^2 in width observed through its interference with a broad non-resonant s-wave. This state has isoscalar quantum numbers and is called the $f_0(980)$. A similar narrow scalar state around $1.0 \text{ GeV}/c^2$ is seen in $\pi\eta$ scattering, called the $a_0(980)$.

Several reasons have been suggested for this apparent failure of the quark model in the scalar sector. One idea arises within the context of a coupled-channel analysis of 2-pseudoscalar s-wave scattering [1]. In this analysis, the $q\bar{q}$ scalars are put in with their conventional quark-model masses and couplings. It is shown that, as the couplings are increased from zero, the renormalized scalar propagator starts off looking like an ordinary Breit-Wigner peak centered at the quark-model mass with a width increasing with the coupling. For larger couplings, however, the shape deviates from a Breit-Wigner and large distortions appear in the tails at the locations of thresholds in coupled channels. These distortions are understood as rescattering effects in the final state, but they act very much like physical particles because they are associated with poles in the scattering amplitude that happen to lie near the real axis. Naively interpreting the observed masses and widths of such dynamical poles as the parameters of quark-model states, it is argued, has given rise to the problem of the scalars.

Another common explanation for the problem of the scalars is the glaring absence in the quark model of any account being taken of the $q\bar{q}$ condensate in the QCD vacuum. The condensate arises within an analysis of low-energy QCD that is built upon the light quarks of the QCD Lagrangian as the basic degrees of freedom instead of the constituent quarks of the quark model. One consequence of making the quarks very light is that the spectrum of physical states should exhibit chiral symmetry, where every positive parity mode has a degenerate partner with negative parity and vice versa. The fact that the pion has no scalar partner nearby in mass is explained within the context of sigma models by the mechanism of spontaneous symmetry breaking, whereby the pion emerges as a massless Goldstone boson and the lightest scalar (sigma meson) is pushed up in mass through its interaction with the $q\bar{q}$ background. This picture of a fundamental pion and sigma can be augmented by the explicit inclusion of light quarks upon which the vectors and the rest of the meson spectrum is built. The fundamental role played by the scalars and pseudoscalars with respect to the QCD vacuum sets them apart from the rest of the mesons in this picture, suggesting that the masses and structure of the scalars and pseudoscalars cannot be understood within the same framework that accounts for the rest of the $q\bar{q}$ meson spectrum, as the quark model attempts to do.

As a description of the dynamics of low-energy QCD, the sigma model is no doubt correct. But a clear statement of what this has to do with the failure of the quark model for the light scalars cannot be made until it is understood how the constituent quark is related to the fundamental fermion. The contrast between these two approaches can be stated as follows. The quark model has a clear relation to the meson spectrum, but inasmuch as it invokes the ill-defined constituent quark, its relation to QCD is an open question. The sigma model has a clear relation to QCD, but its relation to the meson spectrum is an open question (apart from its prediction of a light pion). This latter point is illustrated by the fact that no obvious candidate stands out in the meson spectrum to fit the role of the sigma meson.

The $a_0(980)$ and $f_0(980)$ mesons present a challenge to both models. Within the quark model, the naive assignment of these states to the fundamental scalar nonet is problematic. The explanation that these states are dynamical resonances arising from rescattering between coupled channels appears sound, but needs to be tested by experiment. Within the sigma model, the naive identification of the f_0 as the sigma begs the question of the identity of its twin, the a_0. An alternative explanation for these states within the context of the sigma model has recently been proposed [2]. Again the context of the calculation is a coupled-channel analysis of 2-pseudoscalar scattering, but instead of beginning with the quark model one starts with the Lagrangian of Chiral Perturbation Theory. Including only the leading-order interactions in the Lagrangian, but taking into account unitarity by summing over diagrams with any number of rescattering loops, the isoscalar and isovector scattering amplitudes exhibit narrow resonance structures just below 1 GeV. Their appearance there is not accidental, but coincides with the opening of the kaon-antikaon threshold.

Thus an unexpected convergence has appeared between two very different ways of looking at low-energy QCD regarding the nature of the f_0 and a_0, with the hypothesis that they are dynamical resonance phenomena associated with the opening of the $K\overline{K}$ threshold. They are dynamical in the sense that they are not present in the calculation at leading order, but appear when the infinite chain of rescattering loops is summed, as required by unitarity.

2 Structure

Besides the fact that the light a_0 and the f_0 lie very close to the $K\overline{K}$ threshold, they have another feature that supports their interpretation as dynamical resonances. Both states have quite large couplings to $K\overline{K}$ decay modes. For the f_0 the ratio $g_{2K}^2/g_{2\pi}^2$ is about 4 and for the a_0 the analogous ratio is about 1. For quark model states, by contrast (eg. the vectors), the lightest isoscalar and isovector states are expected to have suppressed couplings to strange decay modes. However in a rescattering picture, these states are not $q\bar{q}$ mesons but enhancements in the $\pi\pi$ and $\eta\pi$ scattering amplitude near the $K\overline{K}$ threshold that come about through intermediate kaon-antikaon loops. Thus any process that has kaons in the final state will be at least as strong or stronger than the corresponding non-strange process.

In this way it is seen that the somewhat obscure mathematical notion that these states arise from the unitarization of a one-loop scattering calculation actually has distinct consequences for their structure: they should behave like a low-energy kaon-antikaon pair. This is reminiscent of an old idea suggested by Weinstein and Isgur [3] that the $a_0(980)$ and $f_0(980)$ are weakly-bound kaon-antikaon states. Within the context of their potential model they found that the short-range interaction between the kaon and antikaon was sufficient to give a bound state with a binding energy of several MeV. Such a state has been called a $K\overline{K}$ molecule.

While there are structural similarities between the molecule and dynamical resonance models, it is obvious from the complexity of a coupled-channel analysis that the latter cannot be reduced to the former. This is why the two notions are not identified in this work. Nevertheless the molecule hypothesis is not in opposition to dynamic resonance pictures. Regardless of what mechanism gives rise to the binding, it is an intriguing question whether the *essential physics* involving the a_0/f_0 resonances can be subsumed into the properties of a single hypothetical bound $K\overline{K}$ state. In the limit where the contribution of infrared kaon loops dominates all other phenomena in the region around the $2K$ threshold one expects that the two viewpoints should be complementary. If however the interference with other 2-meson loops is important then nothing short of a full coupled-channels analysis will be adequate.

Other hypotheses for the structure of the light scalars are also found in the literature. The most noteworthy among these are the conventional $q\bar{q}$ hypothesis and the 4-quark exotic state. The $q\bar{q}$ hypothesis asserts that the structure of the light scalars is that of an L=1 meson with an unusual flavor signature, masses and widths. The 4-quark exotic state is distinguished from the molecule in that the former is confined within a typical bag radius of 1fm whereas the latter is not.

3 Phi radiative decays

It has long been known that the decays of the $\phi(1020)$ meson to γa_0 and γf_0 afford a unique opportunity to test models of the light scalars. In a landmark paper [4], Close, Isgur and Kumano calculated the partial width for this decay within the framework of a $K\overline{K}$ molecule. In particular they found

Table 1. Predictions for the ratio R of branching ratios for the reactions $(\phi \to \gamma a_0)$: $(\phi \to \gamma f_0)$

$q\bar{q}$ state	0:1
$K\overline{K}$ molecule	1:1
$qq\bar{q}\bar{q}$ state	9:1

that the ratio R of branching ratios to the two scalar states was sensitive to the hypothesized structure. Their results are summarized in Table 1. The reason for the suppression of the a_0 relative to the f_0 in the $q\bar{q}$ model is the OZI rule: the f_0 has a large $s\bar{s}$ admixture as required by its strong $K\overline{K}$ coupling, whereas the I=1 a_0 has none. In the case of the molecule, the reason for the equality of the a_0 and f_0 branching ratios is the fact that the photon only couples to the K^+K^- piece of the $K\overline{K}$ wave function, which piece has equal projections onto the I=0 and I=1 partial waves. The assumption in that prediction is that the f_0 is pure isoscalar and the a_0 isovector, but it can be generalized to allow for isospin mixing. In the case of the 4-quark state, almost any value for R can be obtained by a suitable choice of flavor couplings. The value of 9 in the table emerges for one possible choice which illustrates that the 4-quark exotic offers a unique explanation for the case where the f_0 branch is strongly suppressed relative to the a_0. More generally, the couplings of the exotic a_0 and f_0 can be adjusted to give any value for R.

The predictions for the dynamic resonance models are difficult to interpret in terms of R because in such models the part of the amplitude coming from the kaon loops associated with the a_0 and f_0 peak is part of a 2K continuum and produces a non-Breit Wigner tail in the $\pi\eta$ and $\pi\pi$ invariant mass spectra, respectively, whose integral depends on how the a_0 and f_0 regions are defined (see Fig. 1). Nevertheless one would expect in the absence of strong interference with non-resonant background that the I=0 and I=1 yields should be comparable.

Experimental results for these decays have recently been published by SND [5] and KLOE [6]. A third experiment to measure these decays called RADPHI is underway at Jefferson Lab [7], with no published result at present. The SND results are shown in Fig. 1. Corresponding results from KLOE are shown in Fig. 2. The branching ratios to the respective final states are obtained by integrating over the peak and correcting for unseen decay modes of the a_0 and f_0. These are given in Table 2. The agreement between the two measurements is better than appears from the table because the SND result for the f_0 channel includes the entire $\pi^0\pi^0$ phase space, whereas KLOE

Table 2. Experimental results for the branching ratios of phi radiative decays in I=0 and I=1 partial waves. The statistical error is given first, followed by the systematic error.

channel	SND	KLOE
$\phi \to \gamma a_0$	$(0.88 \pm 0.14 \pm 0.09) \times 10^{-4}$	$(0.58 \pm 0.05 \pm 0.04) \times 10^{-4}$
$\phi \to \gamma f_0$	$(3.5 \pm 0.3 \pm 0.9) \times 10^{-4}$	$(2.4 \pm 0.6 \pm 0.2) \times 10^{-4}$
R	0.25 ± 0.04	0.24 ± 0.03

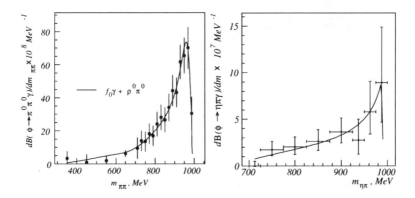

Figure 1. Results from VEPP-2M experiment SND for the decay $\phi \to \gamma\pi^0\pi^0$ (left) and $\phi \to \gamma\eta\pi^0$ (right). The prominent peak at high mass in the figures is attributed to the scalar resonances. The $\pi^0\pi^0$ spectrum also contains a small contribution from the decay $\phi \to \rho^0\pi^0$ where the ρ^0 is seen via its small decay branch to $\gamma\pi^0$.

reports only the integrated yield above 700 MeV/c^2 in $\pi\pi$ invariant mass. To a large extent this difference drops out in the ratio R.

4 Interpretation

These new data exhibit two surprising features: a large departure of the ratio R from the expected value of unity and a larger magnitude for the individual branching ratios than was anticipated.

The prediction that $R = 1$ is a general result that applies to any model in which the ϕ couples to the photon through intermediate *kaon − antikaon* loops. This picture applies both to the coupled-channel analyses where the a_0 and f_0 are dynamical rescattering effects and the molecule model where they are $K\overline{K}$ bound states. Since the photon does not couple to the neutral kaon, only the K^+K^- loop contributes, whose projection onto final states of good

398

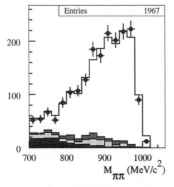 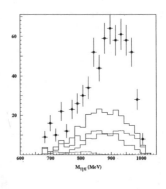

Figure 2. Results from DAΦNE experiment KLOE for the decay $\phi \to \gamma\pi^0\pi^0$ (left) and $\phi \to \gamma\eta\pi^0$ (right). The prominent peak at high mass in the figures is attributed to the scalar resonances. The histograms under the data in the plots indicate the expected contribution from backgrounds that are present in the final sample.

isospin yields a 1:1 ratio of I=0 to I=1.

In the molecule model in particular it is difficult to explain the observed ratio of 1:4. A very large isospin violation in the interactions is required, beyond what can be accounted for by the neutral-charged kaon mass difference. Within this model the fact that the kaon mass difference is comparable to the binding energy leads naturally to strong mixing between I=0 and I=1 in the mass eigenstates a_0 and f_0. But as long as these eigenstates remain more or less degenerate, that mixing is unable to explain a significant deviation of R from 1. The reasoning that leads to this conclusion is as follows.

We assume that the interaction vertices themselves respect isospin, and that ϕ radiative decays proceed through an intermediate charged-kaon loop. The scalar state formed when the two charged kaons annihilate in the s-wave is the superposition $\sin(\theta+\pi/4)a_0 + \cos(\theta+\pi/4)f_0$ where θ is the scalar mixing angle. For example, if the a_0 were pure I=1 and the f_0 pure I=0 then $\theta = 0$ and the 1:1 ratio for the isospin-symmetric limit is obtained. However no matter what the value of the mixing angle, the intermediate scalar superposition state can always be written $\sin(\pi/4)|I = I > + \cos(\pi/4)|I = 0 >$ which is just a statement of isospin conservation at the scalar production vertex. In the case that the physical a_0 and f_0 are degenerate in mass, this isospin superposition propagates unchanged to the decay vertex, where the it couples to final states of good isospin $\pi\eta$ and $\pi\pi$ with relative amplitudes $\sin(\pi/4)$ and $\cos(\pi/4)$, respectively. A key to this argument is the observation that the physical

a_0 and f_0 states are degenerate in mass, or at least their splitting is small compared to their widths, so that significant oscillations between I=0 and I=1 does not take place during their lifetimes.

Close and Kirk have recently argued [8] based upon data taken in central pp collisions that there is a large mixing angle between the I=0 and I=1 components in the a_0/f_0 system. General arguments indicate that the source of particles formed at mid-rapidity in such high-energy collisions is isoscalar in character, which would lead to a ratio of 0 for centrally produced a_0/f_0 in the absence of mixing. From the experimental value of this ratio they deduce a mixing angle of $\tan^2(\theta) = 0.30 \pm 0.05$. The authors point out the remarkable similarity of their value of $\tan^2(\theta)$ to the measured value of 0.25 ± 0.05 for R in radiative ϕ decays. One issue not addressed by the authors is how they can identify the physical a_0 with the $\pi\eta$ channel and the physical f_0 with $\pi\pi$ even though the physical states contain strong admixtures of both I=0 and I=1 components. In addition to mixing, it would appear that a large isospin violation in the decay couplings are required.

The same argument described above for ϕ radiative decays would also prevent the large isospin-violating effect in these data from being arising simply from a_0/f_0 mixing. Repeating the argument in another way, if the a_0 and f_0 were really identical apart from isospin then the mixing angle would simply be a matter of definition and could have no observable consequences. Allowing for a small mass splitting does introduce mixing effects, but inasmuch as the physical a_0-f_0 mass difference is small compared to the widths, the interference between the two scalars cannot give rise to large isospin-violating amplitudes. This argument applies equally to the case of central production and to radiative ϕ decays.

The way that this problem can be resolved is to take into account interference between the resonant amplitude coming from the scalar intermediate states and the nonresonant background. It is well known that there is a strong nonresonant background in the I=0 channel that must be taken into account to understand the structure of the $\pi\pi$ s-wave below 1.2 GeV. Trying to understand this interference involves a comprehensive analysis of data from several 2-pseudoscalar channels and leads inevitably to the kinds of coupled-channel analyses presented in Refs. [1],[2]. In such a picture, the value of R is ill-defined because the resonance-background interference makes the extraction of a branching ratio for the resonant part model dependent. Instead it is simpler to directly compare the fit with the measured spectral decay rate, as shown in Fig. 3 for the analysis of Ref. [2]. Apart from sign ambiguities, the prediction has zero parameters. From this figures it appears that these data may be described quite well within such an approach.

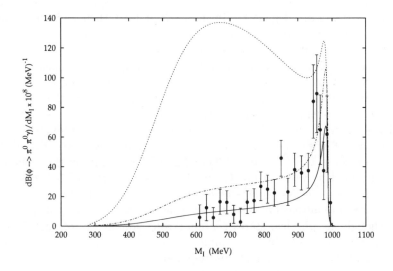

Figure 3. Comparison between coupled-channel calculation of Oset *et.al.* and early data from SND and CMD-2 for the reaction $\phi \to \gamma\pi^0\pi^0$. The different curves represent different signs for terms in the analysis that are not constrained by fits to other $\pi\pi$ scattering data.

Referring to Table 1 it may be asked why a small value for R may not be taken as evidence for a conventional $q\bar{q}$ for the scalars. However this is effectively ruled out by the sheer magnitude of the individual $\phi \to f_0\gamma$ and $\phi \to a_0\gamma$ decay rates. For a conventional $q\bar{q}$ state the individual branching ratios are expected on the order of 10^{-5} [4] whereas the measured values in Table 2 are of order 10^{-4} or larger.

The SND collaboration interprets their results in the light of a $qq\bar{q}\bar{q}$ model [5]. The curves in Fig. 1 show excellent agreement with the data, although this is not surprising in view of the fact that such a model is essentially unconstrained in its choice of couplings. The mainstay of the argument put forward for the $qq\bar{q}\bar{q}$ interpretation is that the magnitude of the decays is too large to be explained in any other way. This conclusion would appear to be supported by Ref. [4] which predicts branching ratios on the order of 5×10^{-5}. However they arrive at that prediction based on an estimate for the effective range of the $K\overline{K}$ potential which is an unknown parameter in the molecule model. The effective range can be decreased without changing the scattering length, and doing so increases their prediction up to a limiting value of about 3×10^{-4}. The measured value falls just at the edge of the range allowed for the molecule, and so the $K\overline{K}$ bound-state model should not be excluded on this basis.

5 Conclusions

The $K\overline{K}$ bound-state model of the $a_0(980)$ and $f_0(980)$ is appealing because of how it explains a number of unusual properties of these states under a conceptually and calculationally simple hypothesis. This simplicity has enabled a straightforward prediction for the ratio R in radiative ϕ decays that is relatively robust within the model. Two independent experiments have now published results indicating that $R = 0.25 \pm 0.05$ rather than the expected $R = 1$, which can only be explained within the molecule model by anomalously large isospin-breaking effects in the scalar-pseudoscalar couplings.

Data taken in central production have been used to estimate the isospin mixing angle between the scalars, with the result $\tan^2(\theta) = 0.30 \pm 0.05$. It has been suggested that such mixing might also be responsible for the value of R in ϕ decays. The similarity of the values for θ obtained from these two very different experiments is interesting, but in order to understand the *phi* decay result in this way a large isospin violation is needed at an interaction vertex; mere mixing of the nearly-degenerate a_0/f_0 states is not sufficient.

Acknowledgments

This work is supported by National Science Foundation grant 0072416.

References

1. N.A. Tornqvist, Z.Phys.C 68 (1995) 647 [hep-ph/9504372].
2. E. Marco, S. Hirenzaki, E. Oset and H. Toki, Phys.Lett.B 470 (1999) 20 [hep-ph/9903217].
3. J. Weinstein and N. Isgur, Phys.Rev.D 27 (1983) 588.
4. F. Close, N. Isgur and S. Kumano, Nucl.Phys.B 389 (1993) 513 [hep-ph/9301253].
5. M.N. Achasov *et.al.*, Phys.Lett.B 485 (2000) 349 [hep-ex/0001048], Phys.Lett.B 479 (2000) 53 [hep-ex/0003031].
6. A. Aloisio *et.al*, to be published in Proceedings of Lepton Photon Conference 2001, Rome, July 23-28 2001 [hep-ex/0107024].
7. R.T. Jones, AIP Conference Proceedings 432, Eds. S.U. Chung and H.J. Willutzki (1998) 635.
8. F. Close and A. Kirk, Phys.Lett.B 515 (2001) 13 [hep-ph/0106108].

MESON PHOTOPRODUCTION AT GRAAL

O. BARTALINI[1,2,3],V. BELLINI[4,5], J.P. BOCQUET[6], M. CAPOGNI[1,7], M.
CASTOLDI[8], A. D'ANGELO[1,7], ANNELISA D'ANGELO[1,3], J.P. DIDELEZ[9], R.
DI SALVO[1,7], A. FANTINI[1,7], G. GERVINO[10], F. GHIO[11,12], B.
GIROLAMI[11,12], M. GUIDAL[9], E. HOURANY[9], I. KILVINGTON[13], R.KUNNE[9],
V. KUZNETSOV[1,7,14], A. LAPIK[14], P. LEVI SANDRI[3], A. LLERES[6], D.
MORICCIANI[7], V. NEDOREZOV[14], L. NICOLETTI[6,4,5], D. REBREYEND[6], F.
RENARD[6], N.V. RUDNEV[15], C. SCHAERF[1,7], M.L. SPERDUTO[4,5], M.C.
SUTERA[16,4], A. TURINGE[17], A. ZUCCHIATTI[8]

INVITED PAPER PRESENTED BY R. DI SALVO[1,7]

(1) Università di Roma "Tor Vergata", I-00133 Roma, Italy
(2) Università di Trento, I-38100 Trento, Italy
(3) INFN, Laboratori Nazionali di Frascati, I-00044 Frascati, Italy
(4) Università di Catania, I-95123 Catania, Italy
(5) INFN, Laboratori Nazionali del Sud, I-95123 Catania, Italy
(6) Institut des Sciences Nucléaires de Grenoble, 38026 Grenoble, France
(7) INFN, Sezione Roma II, I-00133 Roma, Italy
(8) INFN, Sezione di Genova, I-16146 Genova, Italy
(9) Institut de Physique Nucléaire, 91406 Orsay Cedex, France
(10) INFN, Sezione di Torino and Università di Torino, I-10125 Torino, Italy
(11) Istituto Superiore di Sanità, I-00161 Roma, Italy
(12) INFN, Sezione Roma1, 00185 Roma, Italy
(13) ESRF, Polygone Scientifique, F-38043 Grenoble, France
(14) Institute for Nuclear Research, 117312 Moscow, Russia
(15) Institute of Theoretical and Experimental Physics, 117259 Moscow, Russia
(16) INFN, Sezione di Catania, I-95123 Catania, Italy
(17) Kurchatov Institute of Atomic Energy, 123182 Moscow, Russia

At the GRAAL facility, a polarised and tagged $\vec{\gamma}$ ray beam is produced in the
energy range from 500 MeV up to 1500 MeV. Results of beam polarisation asymmetries and cross sections for the photoproduction of η and π^0 on the proton are
presented. These very precise measurements cover the angular range $30°$-$150°$,
providing stringent constraints to theoretical models.

1 Introduction

The GRAAL $\vec{\gamma}$-ray beam is produced [1] by the Compton backscattering of
laser light against the 6 GeV electrons circulating in the ESRF storage ring
in Grenoble. The kinematics of the reaction is shown in figure 1.

The interaction takes place in one of the straight sections of the ring and

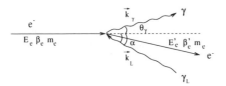

Figure 1. Kinematics of the reaction $\vec{\gamma}+e \to \vec{\gamma}+e$

the laser light strikes the electron at a relative angle very close to 180°. In this situation the energy of the outgoing γ is given by the formula:

$$E_\gamma = \frac{4\gamma^2 E_L}{1 + (\gamma\theta_\gamma)^2 + 4\gamma E_L/m_e}$$

where: E_e and m_e are the energy of the incident electron and its mass, $\gamma = E_e/m_e$, E_L is the energy of the laser light, θ_γ is the angle of the backscattered photon with respect to the incident electron direction.

Two independent sets of data were collected using the green laser line (2.41 eV) and a UV laser line (3.53 eV), allowing to reach a maximum γ energy of 1100 MeV and 1500 MeV respectively.

2 Tagging and polarisation of the beam

The energy of the outgoing γ is calculated as the difference between the incoming and the scattered electrons energies. The diffused electron traverses the first bending magnet after the interaction region, deviating from the main orbit. The measurement of this deviation allows to determine the electron energy. The tagging detector consists of 128 silicium micro-strips (with a 300 μm pitch) in coincidence with a set of 10 scintillator bars. The spatial resolution of the strips corresponds to an energy resolution of 1.1% (FWHM) for the scattered photon. The signals from the plastic scintillators are in coincidence with the radio frequency of the ring and are used to start the trigger of the whole experimental acquisition.

The light emitted from the laser is linearly polarized by two Brewster windows. Moreover it can be circularly polarized using a quarter-wave-length plate. The $\vec{\gamma}$-ray retains almost entirely the laser polarization for the following two reasons: 1) the helicity is a good quantum number for the ultrarelativistic electrons circulating inside the ring, so that the spin-flip amplitude in the Compton scattering with the laser light is negligible; 2) if the γ is diffused at an angle $\theta_\gamma = 0°$ with respect to the incident electron direction, the relative angular momentum between the two particles is zero and no depolarisation may occur due to angular momentum exchange. For angles different from zero the transfer of polarisation is in any case higher than 60% for a wide energy range.

3 The GRAAL experimental set-up

The GRAAL apparatus is a large solid angle detector, as shown in figure 2, for the detection of neutral and charged particles [13].

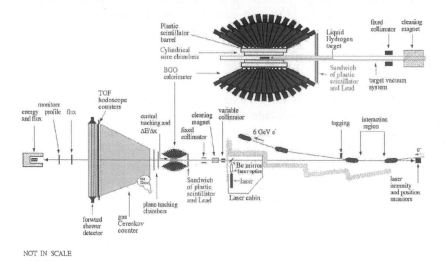

Figure 2. The GRAAL experimental apparatus.

In the central region, for polar angles between 25° and 155°, the apparatus consists of: 1) an electromagnetic calorimeter, made of 480 BGO crystals, each of thickness corresponding to 21 r.l., providing a high energy resolution for γ, $\simeq$ 3% (FWHM) at 1 GeV, and a good response to protons with energies up to 400 MeV [3]; 2) a cylindrical barrel of 32 plastic scintillators, for charged particle discrimination and identification; 3) two cylindrical MWPC's for vertex reconstruction, providing a spatial resolution of $\simeq$1 mm.

In the forward direction, for polar angles smaller than 25°, the apparatus consists of: 1) two plane chambers, providing a spatial resolution on track reconstruction of $\simeq$1 mm; 2) a double layer scintillator wall of 3×3 m^2 size, providing ΔE and t.o.f. measurements for charged particles with a time resolution of $\simeq$600 ps (FWHM); 3) a shower detector consisting of 16 vertical modules made of layers of lead and plastic scintillators, with a 3 cm thick iron sheet in front of it, providing ΔE and t.o.f. measurement for charged and neutral particles, with a good efficiency also for photons [4].

Two disks of plastic scintillators are located at backward angles and are separated by a layer of 1 cm of lead.

The feasibility of a gas Čerenkov detector was studied [5], the construction was started and it will be installed between the plane chambers and the double wall for electrons identification.

In the forward direction, a beam monitor, composed of three scintillators separated by a gamma converter, constantly measures the incident flux. The efficiency in the detection of photons is measured by a spaghetti calorimeter working at low beam intensities [2]. The measured flux over the entire energy spectrum is about 2×10^6 γ/s with a laser power of 2.8 W.

4 Physics motivations

Meson photoproduction on the proton is a powerful tool for probing the internal structure of the proton itself. Mesons are produced via the excitation of hadronic resonances, whose parameters can be extracted from the data, providing stringent inputs to theoretical models.

Asymmetries are much more sensitive than cross sections to the contribution of the less dominating resonances. In facts, while the cross sections are proportional to the sum of the square of the helicity amplitudes, the asymmetries contain the interference between them, thus amplifying the effect of the less contributing resonances. For this reason there is a strong demand for very precise measurements particularly for polarization asymmetry data.

These requests are well matched by the GRAAL 4π apparatus and highly polarised and tagged $\vec{\gamma}$ beam. An extended program of meson photoproduction with linearly polarised photons was performed since 1997, including η, π^0, $n\pi^+$, $2\pi^0$, ω, $K\Lambda$ and Compton scattering channels. High quality results on differential cross sections and beam asymmetries for the photoproduction of pions [9,17] and etas [13,8,9] have already been produced. In the near future the experiment will also be enriched by the insertion of a $\vec{H}\vec{D}$ polarised target, that will allow double polarisation measurements and the test of the Gerasimov-Drell-Hearn sum rule on the neutron and the proton.

5 Physics results

We will present in the following some results of beam asymmetries and cross sections for the two channels of η and π^0 with some theoretical interpretation. They cover the $\vec{\gamma}$ energy range 500 MeV - 1500 MeV and a very wide angular interval, between 30° and 150°. The GRAAL data often extend to unexplored regions, both in energy and in angles, while for the regions where previous results already existed, they considerably reduce the existing errors bars.

5.1 η photoproduction

The η channel is attractive among the others because it naturally selects $N^*(I=1/2)$ resonances as intermediate states. The η isospin being 0, the nucleonic excited state must carry the same isospin as the proton (I=1/2), forbidding the contribution of the Δ resonances that strongly dominate the total photoproduction cross section.

The $S_{11}(1535)$ resonance has been shown to be the dominant contribution to the η photoproduction [7]. This result is strongly confirmed by the GRAAL data [8,9], that show a nearly isotropic behaviour of the differential cross section for energies near threshold (E_γ up to 900 MeV). At the higher energies a deviation from the isotropy can be interpreted as an arising contribution of the $D_{13}(1520)$ resonance.

Figure 3 shows the preliminary results for the total cross section of the reaction $\vec{\gamma} + p \to \eta + p$. Full circles represent the Graal data, that cover for the first time all the energy range of the $S_{11}(1535)$ resonance, allowing the extraction of its total width $\Gamma_{S_{11}}$:

$$\Gamma_{S_{11}(1535)} = 160 \pm 20 \text{ MeV}$$

The year 2000 edition of the Review of Particle Physics (RPP) [?] quotes values between 100 and 250 and suggests a value of 150 MeV. More recently a similar result has been obtained at TJNAF [6]. Open circles and triangles represent the results from Mainz [7] and Bonn [10] (electroproduction at low Q^2) respectively. The dashed line shows the latest Partial Wave Analysis of the GW - SAID group [11], including GRAAL data. The solid and the dashed-dotted curves show the analysis based on the Li and Saghai quark model [12], with and without the insertion of a third "missing" $S_{11}(1730)$ resonance, that can be interpreted as a *quasi-bound* K-Λ state. The bold solid line shows the latest results from the isobar model ETA-MAID [24] for the study of η photo- and electroproduction, based on experimental data from Mainz [7], GRAAL [8] and JLab [22,23].

Figure 4 shows the GRAAL results for the beam polarisation asymmetry

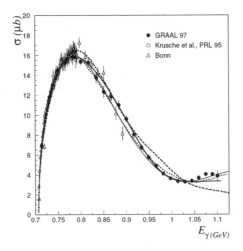

Figure 3. Preliminary results for the total cross section for the reaction: $\vec{\gamma} + p \rightarrow \eta + p$. See text for explanations.

Σ. No previous data existed for η photoproduction asymmetry. The results represented by open circles and triangles have already been published [13] and have been obtained with the green laser line and a maximum γ-ray energy of 1100 MeV: circles correspond to events in which the two or six γ's coming from the η decay are detected in the central region; triangles correspond to data in which one of the two γ's is detected in the forward direction. These results are compared with preliminary results (open squares) obtained with the UV laser line and a maximum γ-ray energy of 1487 MeV. The two sets of data are completely independent and are in excellent agreement between themselves, proving the good quality and the high precision of our results. The analysis confirms the dominant role played by the $S_{11}(1535)$ resonance. This resonance gives a negligible (slightly negative) contribution to the asymmetry through its interference with the Born term. The insertion of the $D_{13}(1520)$ resonance produces a positive clear contribution to Σ, which is symmetric in the forward and backward directions. At the highest energies, a marked forward peak was initially seen in the higher bin of green line data and was then confirmed by the UV line data. This effect may be explained by the contribution of the higher angular momentum resonance $F_{15}(1680)$. The signification of the curves is the same as in figure 3.

Other parameters extracted by theoretical analyses from GRAAL data are:

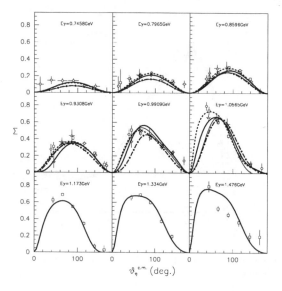

Figure 4. Published and preliminary GRAAL results for beam asymmetry in η photoproduction. See the text for details.

- the branching ratio of the $D_{13}(1520)$ and $F_{15}(1680)$ resonances in the η-nucleon channel by [14,?]:

$$\Gamma(D_{13}(1520)) = (0.08 \pm 0.01)\% \qquad \Gamma(F_{15}(1680)) = (0.15^{+0.35}_{-0.10})\%$$

- the mixing angles θ_S, between the $S_{11}(1535)$ and $S_{11}(1650)$ resonances, and θ_D, between the $D_{13}(1520)$ and $D_{13}(1700)$ resonances by [16]:

$$\theta_S = -31° \pm 2° \qquad \theta_D = +10° \pm 2°$$

5.2 π^0 photoproduction

Figure 5 shows the beam asymmetry Σ in the $\vec{\gamma} + p \to \pi^0 + p$ channel as a function of the π^0 angle in the c.m. system for nine energy bins [9]. Open circles and squares correspond to the GRAAL data obtained using the green and UV laser lines, respectively. The two independent sets of data are in excellent agreement between themselves. Full lines show the SM00 solution of the GW-SAID group [11], obtained by a fit of all existing world data, including the GRAAL green line ones. Results have been obtained also for cross sections. The comparison of GRAAL data with the existing results of cross sections shows a good agreement in the common regions, proving the reliability of our knowledge of the apparatus efficiency and of the whole data analysis procedure.

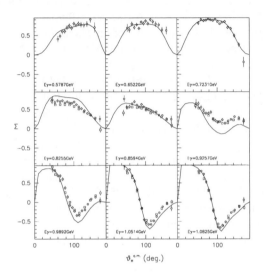

Figure 5. Preliminary GRAAL results for beam asymmetry in π^0 photoproduction. See the text for details.

6 Conclusions

The GRAAL apparatus and polarised and tagged $\vec{\gamma}$ ray beam has become operational since several years, allowing the collection of a very high statistics of data. Beam asymmetries and cross sections have been extracted for several channels and other reactions are presently under analysis.

Acknowledgments

We are grateful to our theoretical colleagues who collaborated to the interpretation of our data. We thank the ESRF staff for the stable and reliable operation of the storage ring.

References

1. Babusci et al., *La Rivista del Nuovo Cimento* **19**,5 (1996)
2. Bellini et al., Nucl. Instrum. Methods **A386**, 254 (1997)
3. P. Levi Sandri et al., *Nucl. Instrum. Methods* **A370**, 396 (1996); F. Ghio et al., *Nucl. Instrum. Methods* **A404**, 71 (1998)
4. V. Kouznetsov et al., to be published in *Nucl. Instrum. Methods* **A**
5. M. Castoldi et al., Technical Report INFN/TC-99/17
6. Review of Particle Physics, Year 2000
7. Armstrong et al., *Phys. Rev.* **D60**, 052004 (1999)
8. B. Krusche et al., *Phys. Rev. Lett.* **74**, 3736 (1995)
9. F. Renard et al., hep-ex/0011098, accepted by *Phys. Lett.* **B**
10. F. Renard, PhD Thesis, Univ. J. Fourier (Grenoble 1999)
11. B. Schoch et al., *Prog. Part. Nucl. Phys.* **34**, 43 (1995)
12. R.Arndt, I.Strakovsky, and R.Workman,*Bull. Am. Phys.Soc.* 45, 34(2000);R. Arndt, I. Strakovsky, R. Workman, *Phys. Rev.* C53 430 (1996). The SAID SP01, WI98, WI00, and SM95 solutions and the single pion photoproduction data base are available via telnet gwdac.phys.gwu.edu, user said .
13. B. Saghai and Z. Li, DAPNIA/SPhN-01-04 Submitted to *Eur. Phys. J. A*
14. J. Ajaka et al., *Phys. Rev. Lett.* **81**, 1797 (1998)
15. L. Tiator et al., *Phys. Rev.* **C60**, 35210 (1999)
16. A. Bock et al., *Phys. Rev. Lett.* **81**, 534 (1995)
17. B. Saghai and Z. Li, Proceeding of the Conference N°2000
18. J. Ajaka et al., *Phys. Lett.* **B475**, 372 (2000)
19. P.J. Bussey et al., *Nucl. Phys.* **B154**, 205 (1979)
20. G. Knies et al., *Phys. Rev* **D10**, 2778 (1974)
21. R. Zdarko and E. Dolly, *Nuovo Cimento* **10A**, 10 (1972)
22. D. Drechsel, O. Hanstein, S.S. Kamalov, L. Tiator, *Nucl. Phys* **A645**, 145 (1999)
23. C.S. Armstrong et al., *Phys. Rev.* **D60**, 052004 (1999)
24. R. Thompson et al., *Phys. Rev. Lett.* **86**, 1702 (2001)
25. W.T. Chiang, S.N. Yang, L. Tiator, D. Drechsel, nucl-th/0110034. The MAID program is available on the webpage: http://www.kph.uni-mainz.de/MAID/maid.html

GIANT RESONANCES IN NUCLEI NEAR AND FAR FROM β-STABILITY LINE

H. SAGAWA

Center for Mathematical Sciences, University of Aizu,
Aizu-Wakamatsu, Fukushima 965-8580, Japan
E-mail: sagawa@u-aizu.ac.jp

The self-consistent HF+RPA (TDA) model is applied to describe the excitation energies and the strength distributions of Gamow-Teller and spin-dipole resonances for charge exchange reactions on ^{208}Pb. We found that the model predictions show reasonable agreement to recent experimental data. Isoscalar monopole resonances were also studied in relation to the nuclear compression modulus by using the same model. It is shown in drip line nuclei that strong peaks of the monopole resonances appear at the excitation energies not only in the giant resonance region but also near the threshold of the RPA response. The asymmetry term of nuclear compression modulus is discussed with giant monopole resonances in Ca-isotopes.

1 Introduction

Giant resonances (GR) in nuclei are characterized by the multipolarity, the spin and the isospin as typical collective modes. Similar collective excitations have been found also in other quantum many-body systems like microclusters. Two different types of GR are studied in the present manuscript. The first one is spin dependent excitation modes such as Gamow-Teller (GT) and spin-dipole (SD) excitations. Recently, the sum rule strength of GT transitions of ^{90}Nb was obtained experimentally by charge exchange (p, n) reactions [1], and the importance of the coupling to many-particle and many-hole states was pointed out on the quenching of the transition strength. Experimental investigations of GT and SD strengths in ^{208}Bi have been done also at RCNP by ^{208}Pb(^{3}He,t) ^{208}Bi reactions [2,3]. Pigmy strength of GT transitions and a large width of SD peak are found in these experiments. These empirical information might be important to give the cross sections of energetic neutrino scattering from supernova [4]. The second problem of GR is the compression mode in nuclei, which is often discussed in relation to the nuclear compression modulus. We study the isoscalar (IS) giant monopole states (GMR) in nuclei near and far from the stability line to obtain the information of the nuclear matter compressibility and the symmetry term of the compression modulus.

As a theoretical model to study GR, the self-consistent Hartree-Fock (HF) + random phase approximation (RPA) or Tamm-Dancoff approximation (TDA) is applied for both the compression modes [5,6] and the spin dependent excitations [7,8,9,10] with the use of Skyrme interactions. It was shown that the model predicts successfully GT and SD states in ^{48}Sc and ^{90}Nb [8,9]. In order to obtain the effect of the continuum on the width, we solve the Green's function in the coordinate space to take into account properly the continuum coupling. In section 2, we discuss the results of GT and SD strength in ^{208}Bi. The self-consistent calculations of ISGMR are presented in section 3 and compared with available experimental data. Summary is given in section 4.

2 Spin excitations in ^{208}Pb

The operators for GT and SD transitions are defined by

$$\hat{G}_{\pm} = \sum_{i=1}^{A} \sum_{m} \tau_{\pm}^{i} \sigma_{m}^{i} \qquad \text{for GT strength} \tag{1}$$

$$\hat{S}_{\pm}^{\lambda} = \sum_{i=1}^{A} \sum_{\mu} \tau_{\pm}^{i} r_{i} [\sigma \times Y_{1}(\hat{r}_{i})]_{\mu}^{\lambda} \qquad \text{for SD strength,} \tag{2}$$

respectively, with the isospin operators $\tau_3 = \tau_z, \tau_{\pm} = \frac{1}{2}(\tau_x \pm i\tau_y)$. The model independent sum rule for the GT transition is expressed as

$$G_- - G_+ \equiv \sum_n |\langle n | \hat{G}_- | 0 \rangle|^2 - \sum_n |\langle n | \hat{G}_+ | 0 \rangle|^2$$
$$= \langle 0 | [\hat{G}_-, \hat{G}_+] | 0 \rangle = 3(N - Z) \tag{3}$$

where the state n denotes the final state in a daughter nucleus. The model independent sum rule is also derived for the SD λ–pole operator (2),

$$S_-^{\lambda} - S_+^{\lambda} \equiv \sum_n |\langle n | \hat{S}_-^{\lambda} | 0 \rangle|^2 - \sum_n |\langle n | \hat{S}_+^{\lambda} | 0 \rangle|^2$$
$$= \langle 0 | [\hat{S}_-^{\lambda}, \hat{S}_+^{\lambda}] | 0 \rangle = \frac{(2\lambda + 1)}{4\pi}(N\langle r^2 \rangle_n - Z\langle r^2 \rangle_p). \tag{4}$$

Calculated results of GT strength in ^{208}Bi obtained by HF + TDA with the use of the SIII interaction are shown in Fig. 1. The calculated GT (SD) results are averaged by using a weighting factor ρ,

$$\frac{dB(GT/SD)_{ave}}{dE_x} = \int \frac{dB(GT/SD)}{dE_x'} \rho(E_x' - E_x) dE_x' \tag{5}$$

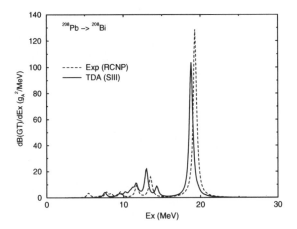

Figure 1. GT transition strength from the parent ground state ^{208}Pb to GT states in the daughter nucleus ^{208}Bi calculated by continuum HF+TDA model with SIII interaction. The excitation energy is measured from the ground state of ^{208}Pb . The transition strength (5) in ^{208}Bi calculated with a weighting factor (6) with Δ =1MeV. Experimental data are taken from ref. [3]. The absolute magnitude of the experimental data is in an arbitrary unit.

where

$$\rho(E'_x - E_x) = \frac{1}{\pi} \frac{\Delta/2}{(E'_x - E_x)^2 + (\Delta/2)^2} \tag{6}$$

with Δ=1MeV for GT states (2MeV for SD states). To check the effect of the ground state correlations on the charge-exchange spin excitations (GT and SD transitions), we performed both TDA and RPA calculations with discrete basis and found no appreciable difference between two calculations. This is due to the large excess neutrons in ^{208}Pb, which block the backward amplitudes of RPA for almost all GT and SD configurations. The calculated main GT peak is located at energy of E_x = 18.8 MeV with respect to the parent, and exhausts 63.6 % of the sum rule strength (3). There are also substantial strengths below 15MeV, which exhaust 33.3% of the sum rule. The energy obtained for the main GT peak shows good agreement with the experimental one, E_x = 19.2 ±0.2 MeV [2]. The experimental data show also similar amount of the low energy GT strength to those of the calculations.

We have done the continuum HF+TDA calculations with the same SIII interaction for SD states with J^π= 0$^-$, 1$^-$ and 2$^-$. Calculated results for SD strength are shown in Fig. 2 averaged by a weighting factor (6) with Δ =2MeV. The integrated sum of the calculated SD strengths are 160 fm^2,

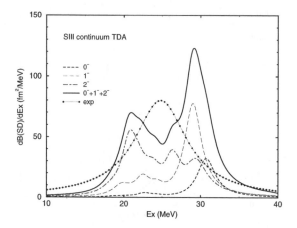

Figure 2. Averaged SD transition strength (5) in ^{208}Bi calculated from continuum HF+TDA results and a weighting factor (6) with Δ =2MeV. Calculated results are taken from ref. [11]. Experimental data are taken from [2]. The absolute magnitude of the experimental data is in an arbitrary unit. The excitation energy is measured from the ground state of ^{208}Pb.

427 fm^2 and 649 fm^2 for J= 0$^-$, 1$^-$ and 2$^-$ states, respectively. The ratio is roughly proportional to 2J+1. The total sum of the strength gives $\langle 0 \mid \hat{S}^\dagger \hat{S} \mid 0 \rangle$ = 1236.4 fm^2. The model independent sum rule for the SD transitions (4) is evaluated for ^{208}Pb to be

$$S_-^\lambda - S_+^\lambda = \begin{cases} 121\text{fm}^2 & \lambda = 0^- \\ 363\text{fm}^2 & \lambda = 1^- \\ 604\text{fm}^2 & \lambda = 2^- \end{cases} \qquad (7)$$

using HF radii of SIII interaction. The S_+^λ contributions to the sum rule (7) are found to be 39fm^2 for 0$^-$, 64fm^2 for 1$^-$ and 45fm^2 for 2$^-$, respectively.

The strength of J$^\pi$= 2$^-$ spreads over three peaks at around E$_x$=20, 27 and 30MeV, while the single peaks for J$^\pi$= 0$^-$ and 1$^-$ are found at around E$_x$=30MeV. Integrated transition strengths of the peaks at around E$_x$=30MeV are 135 fm^2 for 0$^-$, 309 fm^2 for 1$^-$ and 130 fm^2 for 2$^-$, which exhausts 84%, 72 % and 20% of the corresponding total strength, respectively. The calculated summed strength of all multipoles shows two peaks, in which the lower (higher) one is dominated by 2$^-$ (1$^-$) states. The height of 0$^-$ peak is about a factor two smaller than that of 1$^-$ centered at E$_x$ $\sim$30MeV. Experimentally, a broad SD bump in ^{208}Bi is found by ^{208}Pb(^{3}He, t)^{208}Bi reaction centered at E$_x$=24.8±0.8MeV with a very large total width of 8.4±1.7MeV [2] as is shown in Fig. 2. Because of relatively poor statistics of the ^{208}Pb(^{3}He,

$t)^{208}$Bi reactions, it is rather difficult to see detailed structure of SD strength in the experimental spectra. On the average, however, the calculated results show good agreement with the experimental observations as far as the excitation energy and the width are concerned, while the two peak structure obtained by the calculation is not seen in the experiment. Notice that main part of the calculated width is due to the Landau damping effect together with the coupling to the continuum. The latter effect is at most 1MeV in the higher energy region than $E_x=27$MeV and plays a minor role on the width of SD resonance. The effect of the coupling to the many-particle many-hole states on the width might be simulated by the weighting factor (6) with the width of $\Delta=2$MeV. A quantitative agreement between the calculated and the empirical width suggests that the width $\Gamma^\downarrow$ due to the coupling to many-particle many-hole states will be a magnitude of 2MeV in the case of SD resonances in ^{208}Bi.

It is pointed out that the GT and SD strength distributions are crucial to study energetic neutrinos from super nova when Pb target is used to observe it. In this respect, more quantitative study of these resonances are desperately wanted experimentally and theoretically especially below 15MeV region measured from the parent nucleus [4].

3 ISGMR in drip line nuclei

The transition operator for ISGMR is given by

$$\hat{O}^{\lambda=0,\tau=0} = \sum_{i=1}^{A} r_i^2 \frac{1}{\sqrt{4\pi}} \qquad \text{for isoscalar monopole strength} \qquad (8)$$

The excitation energy of GMR is often calculated by using the energy moment of the transition strength ,

$$m_k = \sum_n (E_n)^k |< n | \hat{O}^{\lambda=0,\tau} | 0 >|^2 \qquad (9)$$

where n denotes the RPA state. The IS GMR energy, referred to as the scaling energy, is defined by

$$E_0^s = \sqrt{\frac{m_3}{m_1}} \qquad (10)$$

The energy (10) is particularly interesting for the discussion of the compression modulus since the scaling hypothesis for the isoscalar monopole vibration gives a relation between E_0^s and the nuclear compression modulus K_A^s [13],

$$K_A^s = \frac{m(E_0^s)^2 < r^2 >_m}{\hbar^2} \tag{11}$$

where m is the nucleon mass and $< r^2 >_m$ is the mean squared mass radius. There are other ways to define the energy of GMR by using the moments m_k. The average energy $\overline{E}$ is defined by

$$\overline{E} = m_1/m_0 \tag{12}$$

which is close to the peak energy of GMR, if it is a single peak. The IS GMR energy may also be defined by

$$E_0^c = \sqrt{\frac{m_1}{m_{-1}}} \tag{13}$$

The energy (13) has a direct relation with the compression modulus K_A^c which is derived from the constrained HF calculation with the nuclear radius parameter [12]

$$K_A^c = \left[R^2 \frac{d^2(E/A)}{dR^2} \right]_{R=R_0} = \frac{2A < r^2 >_m^2}{m_{-1}} = \frac{m(E_0^c)^2 < r^2 >_m}{\hbar^2} \tag{14}$$

where $R_0^2 = <r^2>_m$ and E/A is the HF energy.

The relation between K_A and K_{nm} is often discussed using an analogy with the expression of the semi-empirical mass formula. The nuclear compression modulus may be expressed in terms of the volume, the surface, the symmetry and the Coulomb contributions,

$$K_A = K_{vol} + K_{surf} A^{-1/3} + K_{sym} \left(\frac{N-Z}{A} \right)^2 + K_{Coul} \frac{Z^2}{A^{4/3}} + \ldots \tag{15}$$

Other terms like a curvature term or a surface symmetry term are also considered in the literatures [12,13]. The various terms in Eq.(15) are determined by fitting experimental IS GMR energies. Then, the nuclear matter compression modulus K_{nm} may be identified as the volume term K_{vol} in the limit of $A \to \infty$, which is the case when the giant monopole vibration is described by the scaling model of the ground state density [13,5]. However it is pointed out in ref. [14] that the value of K_{vol} is mal determined from the comparison between experimental data and Eq.(15). On the contrary, a promising procedure to determine K_{nm} is to calculate the response of IS GMR by self-consistent RPA with several effective interactions of different K_{nm} values, and the results will be compared directly with well established experiments to obtain empirical information of K_{nm}.

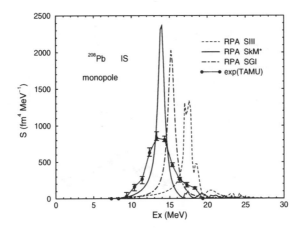

Figure 3. RPA IS monopole strength function of ^{208}Pb as a function of excitation energy. The results with SkM*, SIII and SGI interactions are shown by solid, dotted and dashed curves, respectively. Calculated results are taken from ref. [6]. The experimental data are taken from ref. [15].

The IS monopole strength in ^{208}Pb are given in Fig. 3 for SkM*, SGI and SIII interactions. The three interactions have different nuclear compression moduli in nuclear matter ; $K_{nm} = 217$, 256 and 355 MeV for SkM*, SGI and SIII, respectively. Among realistic effective interactions which give reasonable binding energies and mean square radii for the ground states of stable nuclei, the SkM* and the SIII interactions are two extremes which have a small and a large compression modulus, while the SGI interaction has a medium value of the compression modulus. The IS response for all three interactions shows, in a good approximation, a single peak with the width of about 1 MeV. The peaks for the SkM* and the SGI interaction exhaust about 91% of the EWSR. The IS peak of the SIII interaction around E_x=17 MeV exhausts only 71% of the EWSR. The IS GMR in ^{208}Pb is experimentally best explored among IS GMR in various nuclei. The (α, α') data were reported to show a peak at E_x=(14.18±0.11) MeV exhausting (90±20) % of the EWSR value [15]. Another (α, α') data give the peak energy at E_x=(13.90±0.30) MeV [16]. The (α, α') data agree well with the calculated peak at E_x=13.9 MeV using the SkM* interaction.

We discuss hereafter the numerical results of ISGMR in Ca-isotopes obtained by using the SkM* interaction, since the comparison of the calculated GMR properties of β-stable nuclei with experimental data favors the SkM*

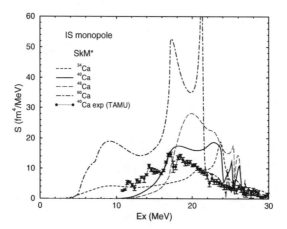

Figure 4. RPA IS monopole strength of Ca-isotopes. The interaction SkM* is used for the response calculations. Calculated results are taken from ref. [6]. Experimental data is taken from ref. [17].

interaction which has the low compression modulus K_{nm}=217 MeV. The RPA IS responses of four Ca-isotopes are shown in Fig. 4. A considerable amount of the proton strength in ^{34}Ca and that of the neutron strength in ^{60}Ca , which lie in the energy region of E_x=(4−12) MeV, express the threshold strength that is a characteristic feature of the response in drip line nuclei. The calculated RPA IS strength extends over a large energy range especially in drip line nuclei, ^{34}Ca and ^{60}Ca , but also in β-stable nuclei, ^{40}Ca and ^{48}Ca . The IS monopole response in all Ca-isotopes is far away from a single peak. Thus, the collectivity is considered to be rather weak in IS GMR of Ca-isotopes compared with that of ^{208}Pb. The integrated area under each curve is equal to the IS EWSR, $m_1 = 2\hbar^2 A < r^2 >_m /m$, and is an increasing function of the mass number.

The nuclear compression moduli of Ca-isotopes are shown in Fig. 5 calculated by using Eqs. (11) and (14). The IS GMR strength above E_x=14MeV is adopted to obtain K_A values since the lower energy region than E_x=14MeV is very much influenced by the neutron and proton skin effects in the drip line nuclei and does not have the IS nature. It is clear that the asymmetry term K_{sym} is large and negative although the quantitative argument is rather difficult. This might be due to the effect of the large surface diffuseness on the K_{surf} due to the existence of the thick proton and neutron skins in drip line nuclei ^{34}Ca and ^{60}Ca, respectively. Because of a large contribution of the sur-

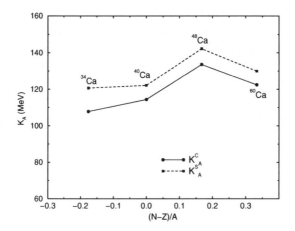

Figure 5. Nuclear compression moduli of Ca-isotopes calculated from Eqs. (11) and (14). The Skyrme interaction SkM* is used for HF+RPA calculations.

face term K_{surf}, the nuclear compression moduli do not show the maximum at the $(N-Z)/A = 0$.

4 Summary

We have studied the GT and SD states in ^{208}Bi using the self-consistent HF + TDA model within the 1p-1h configuration space taking into account the coupling to the continuum. The calculated main GT peak is found at E_x=18.8MeV which agrees with the experimental peak at E_x=19.2 ±0.2MeV. We showed also that the calculated SD strength gives good agreement with the observed data by ^{208}Pb(^{3}He, t)^{208}Bi as far as the excitation energy and the width are concerned. The Landau damping effect is shown to be highly responsible for the large observed width of SD resonance, while the couplings to the continuum and to the many-particle and many-hole states play minor roles.

The GMR in nuclei near and far from the stability line is studied by using the same HF+RPA model with the three Skyrme interactions and found the SkM* interaction, having a small nuclear compression modulus K_{nm}=217 MeV, gives better agreement with available experimental data in ^{40}Ca and ^{208}Pb than those of other interactions SGI and SIII, which have larger compression moduli K_{nm}=256 and 355 MeV. The IS GMR of Ca-isotopes towards drip lines are studied in comparison with those of the stable nuclei, using the

Skyrme interaction SkM*. The distribution of the monopole strength is much affected by the presence of the low-energy threshold strength in both the proton and neutron drip line nuclei. We pointed out that the negative curvature of the K_{sym} term of the compression modulus from the analysis of calculated IS GMR in Ca-isotopes.

Acknowledgments

The authors would like express his thanks to M. Fujiwara and Y. -W. Lui for informing him the current experimental results. He also thanks I. Hamamoto and Toshio Suzuki for fruitful collaborations. This work was supported in part by Grant-in-Aid for Scientific Research (C) (No. 12640284 and 13640257) from Ministry of Education, Culture, Sports, Science, and Technology.

References

1. T. Wakasa et al., Phys. Rev. **C55**, 2909 (1997);
   H. Sakai, H. Okamura, H. Otsu, T. Wakasa, S. Ishida, N. Sakamoto, T. Uesaka, Y. Satou, S. Fujita and K. Hatanaka, Nucl. Instrum. Methods Phys. Res. **A369**, 120 (1996).
2. H. Akimune et al., Phys. Rev. **C52**, 604 (1995);
   H. Akimune, I. Daito, Y. Fujita, M. Fujiwara, M. N. Harakeh, J. Janecke and M. Yosoi, Phys. Rev. **C61**, 011304(R) (2000).
3. A. Kraznahorkay et al., Phys. Rev. **C64** 067302 (2001).
4. G. M. Fuller, W. C. Haxton and G. C. McLaughlin, Phys. Rev. **D59**, 085005 (1999).
   E. Kolbe and K. Langanke, Phys. Rev. **C63**, 025802 (2001).
5. J. P. Blaizot, J. F. Berger, D. Decharge and M. Girod, Nucl. Phys. **A591**, 435 (1995).
6. I. Hamamoto, H. Sagawa and X. Z. Zhang, Phys. Rev. **C56**, 3121 (1997).
7. Nguyen Van Giai and H. Sagawa, Phys. Lett. **B106**, 379 (1981).
8. G. Colò, Nguyen Van Giai, P. F. Bortignon and R. A. Broglia, Phys. Rev. **C50**, 1496 (1992).
9. T. Suzuki, H. Sagawa and Nguyen Van Giai, Phys. Rev. **C57**, 139 (1998).
10. I. Hamamoto and H. Sagawa, Phys. Rev. **C60**, 064314 (2000).
11. T. Suzuki and H. Sagawa , Eur. Phys. **A9**, 49 (2000).
12. J. P. Blaizot, Phys. Report **64**, 171 (1980).

13. J. Treiner, H. Krivine, O. Bohigas and J. Martorell, Nucl. Phys. **A371**, 253 (1981).
14. S. Shlomo and D. H. Youngblood , Phys. Rev. **C47**, 529 (1993).
15. D. H. Youngblood, H. L. Clark and Y. -W. Lui, Phys. Rev. Lett. **82**, 691(1999).
16. S. Brandenburg et al., Nucl. Phys. **A466**, 29 (1987).
17. D. H. Youngblood, Y. -W. Lui and H. L. Clark, Phys. Rev. **C63**, 067301 (2001).

INDIRECT MEASUREMENTS OF THE ^{7}Be(p,γ)^{8}B REACTION

T. MOTOBAYASHI

Department of Physics, Rikkyo University, Nishi-Ikebukuro, Toshima, Tokyo
171-8501, Japan
E-mail: motobaya@rikkyo.ne.jp

The high-energy solar neutrino flux is determined by the low-energy cross sections of the ^{7}Be(p,γ)^{8}B reaction. As well as many direct measurements, Coulomb dissociation with intermediate-energy ^{8}B beams and low-energy proton-transfer reactions with ^{7}Be beams have been investigated to determine indirectly the astrophysical S_{17}-factors of the ^{7}Be(p,γ)^{8}B reaction. The results of these studies are generally in good agreement, though recent demand, S-factor in 5% accuracy, requires more detailed understanding of the reaction mechanism.

1   Solar fusion reactions

For a long time, the low-energy ^{7}Be(p,γ)^{8}B reaction has attracted much attention, because it is the source of high-energy neutrinos created in the solar center. The flux of these neutrinos measured on the earth is systematically lower than expected. The first terrestrial neutrino measurement was performed by Davis *et al.* at Homestake mine in 1960's. [1] It used ^{37}Cl as material detecting neutrinos through the weak process ^{37}Cl(ν,e$^-$)^{37}Ar, and is sensitive mainly to the neutrinos from ^{8}B through its β^+ decay ^{8}B$\rightarrow^8$Be(2^+)+e$^+$+ν with a high end-point energy of about 14 MeV. The measurement is still continuing, and the most recent result quoted the flux of 2.56±0.23 SNU, [2] where SNU abbreviates Solar Neutrino Unit (10^{-36} interactions per atom per sec in the detector material). This is only 34±6% of the prediction by the standard solar model of Bahcall, Pinsonneault and Basu. [3]

Another solar neutrino measurement at Kamioka mine detects water Cherenkov radiation from the neutrino-electron scattering. It measures almost only the ^{8}B neutrinos. The measured flux in their latest report is 48±9 % [4] of the prediction of the solar model. [3] Since the flux of the solar neutrino originating from ^{8}B depends directly on the ^{7}Be(p,γ)^{8}B cross section at around 20 keV, the Gamow energy or the effective burning energy in the sun, its experimental information is essential for predicting the flux corresponding to the Homestake and Kamioka experiments.

New solar neutrino measurements used the ^{71}Ga(ν,e$^-$)^{71}Ge reaction that has a low threshold energy of neutrino detection, so that the measured flux contains about 50% contribution from the p+p$\rightarrow$D+e$^+$+ν and p+p+e$^-$ $\rightarrow$D+ν processes. All the experiments called SAGE, [5] Gallex [6] and

GNO [7] observed the flux of about 75 SNU, which is lower than the prediction 128^{+9}_{-7} SNU. [3] This suggests that the solar neutrino puzzle is not entirely due to ^{8}B.

Much effort has been devoted to the determination of the astrophysical S-factor of the ^{7}Be(p,γ)^{8}B reaction, S_{17} . The S-factor is related to the cross section $\sigma(E)$ as,

$$\sigma(E) = SE\exp[-2\pi\eta]. \tag{1}$$

The term $E\exp[-2\pi\eta]$ accounts for the steep energy dependence of $\sigma(E)$ due to the Coulomb penetration in S-wave, where η denotes the Sommerfeld parameter $e^2 Z_1 Z_2/\hbar v$. Since ^{7}Be is an unstable isotope with 53 day half life, direct (p,γ) experiments employ a radioactive target. This requires special attention in determination of the beam-target luminosity. Since the first experimental study by Kavanagh was reported in 1960, [8] many experiments have been performed for accurate measurements of direct capture. [9,10,11,12,13,14,15,16,17,18] The most recent recommendation for the S factor at zero energy is given as $S_{17}(0)=19^{+4}_{-2}$ eV-b, which is obtained by evaluating the direct capture data up to 1998. [19]

Attempts to employ indirect methods have also been made for determination of S_{17} hoping that the measurements are independent of the difficulties associated with the direct measurements. So far two methods have been investigated, Coulomb dissociation with intermediate-energy ^{8}B beams and low-energy transfer reactions with ^{7}Be beams to extract the asymptotic normalization coefficient (ANC) for the ^{8}B ground state.

2 Coulomb dissociation of ^{8}B

2.1 Coulomb dissociation method

In the Coulomb dissociation method, the residual nucleus B of the A(x,γ)B process bombards a high-Z target, and is Coulomb excited to an unbound state that decays to the A+x channel. Since the process is regarded as absorption of a virtual photon, i.e. B(γ,x)A, the radiative capture (the inverse of the photoabsorption) cross section can be extracted from the dissociation yield. This idea was first proposed by Baur, Bertulani and Rebel. [20] In addition to the advantage of using thick targets, the Coulomb dissociation enhances the original capture cross section by a large factor. This is due to the large virtual-photon number and the phase space factor. The two factors can be in the order of 100 to 1000 in the case of ^{8}B dissociation.

After pioneering studies for the stable Li isotopes, the first Coulomb dissociation experiments with radioactive beams were made for the ^{208}Pb(^{14}O,^{13}N p)^{208}Pb reaction at higher incident energies of 87.5 MeV/nucleon [21] and 70 MeV/nucleon. [22] The results demonstrated the usefulness of the method, and stimulated further studies such as ^{12}N→^{11}C+p. [23]

2.2 Experimental studies

The first ^{8}B dissociation experiment was performed at RIKEN. [24,25] Radioactive ^{8}B nuclei were produced by the ^{12}C+^{9}Be interaction at 92 MeV/nucleon, and were analyzed by the RIPS fragment separator. [26] The ^{8}B beam energy in the center of the target, 50 mg/cm^2 ^{208}Pb, was 46.5 MeV/nucleon. A ΔE-E plastic scintillator hodoscope detected the outgoing particles of the Coulomb dissociation, ^{7}Be and p, in coincidence. The energy-dependent coincidence yield was then converted to the ^{7}Be(p,γ)^{8}B cross section.

The second RIKEN experiment at 51.9 MeV/nucleon [27,28] used also a stack of sixty NaI(Tl) scintillators called DALI, which measured the deexcitation γ rays from the first excited state of ^{7}Be at 429 keV populated in the dissociation process. The contribution from this process was measured to be about 5% of the Coulomb dissociation yield as an average for 500 keV< $E_{\rm rel}$ <3 MeV.

In Fig. 1 the astrophysical S_{17}-factors obtained in the first and second experiments are shown together with the results from direct (p,γ) measurements. The Coulomb dissociation data are consistent within errors with the direct-capture results by Vaughn et al., [11] Filippone et al. [13] and Hammache et al. [14] which gave lower S_{17} factors than the ones by Parker [9] and Kavanagh. [10] The S factor extrapolated to zero energy, $S_{17}(0)$, for the first experiment is 16.7±3.2 eV-b, and the second experiment with better accuracy gives 18.9±1.8 eV-b. [28] The latter is within the range of the latest recommendation 19^{+4}_{-2} eV-b. [19]

An experiment at a higher incident energy of 254 MeV/nucleon was performed at GSI. [29] The reaction products were momentum-analyzed by the spectrometer KaoS, [30] and detected in coincidence at the focal plane. The scattering angles of the fragments were determined by silicon microstrip detectors set between the target and the spectrometer. The result $S_{17}(0)=20.6\pm1.2\pm1.0$ eV-b agrees with the RIKEN data and hence the direct capture data with lower S-factors.

Recently another p-^{7}Be coincidence experiment was performed at MSU with 83 MeV/nucleon ^{8}B beams. [31] The reaction products were momentum-analyzed by a dipole magnet. Though the experimental results are essentially consistent with those of the previous studies at RIKEN and GSI, the authors propose a slightly lower $S_{17}(0)$ value of $17.8^{+1.4}_{-1.2}$ eV-b by subtracting

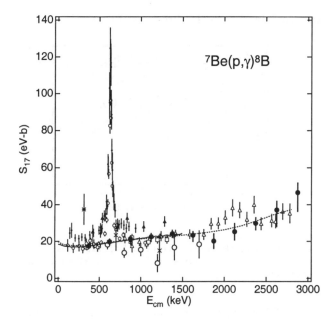

Figure 1. Astrophysical S-factors for the ^{7}Be(p,γ)^{8}B reaction extracted from the first (large open circles) and second (large solid circles) experiments at RIKEN. Direct (p,γ) data are also shown. The solid and dashed curves represent the fits to the data from the second Coulomb dissociation experiment with the theoretical energy dependence of Barker and Spear [33] and that of Descouvemont and Baye, [34] respectively.

the E2 component evaluated from their parallel-momentum distribution measurements [32] discussed in the next subsection.

2.3 Reaction mechanism

These results indicate that the S_{17} factors extracted from the Coulomb dissociation studies are consistent with the one recommended and used in solar model calculations. However, detailed investigation on the reaction mechanism is required for more accurate determination of the S_{17}-factor.

It was pointed out that a possible E2 contribution may affect the Coulomb dissociation results. [35] The ^{7}Be(p,γ)^{8}B reaction is dominated by E1 γ-emission through continuum states. The E2 amplitude is very small, but it is enhanced in the Coulomb dissociation as demonstrated in Fig. 2. On the other hand, the M1 transition is suppressed in the Coulomb dissociation, while it might have a certain contribution in the direct (p,γ) process.

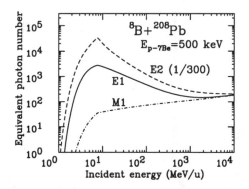

Figure 2. Number of virtual photons plotted as a function of the incident energy of the ^{8}B+^{208}Pb interaction. The p-^{7}Be relative energy in the final state is assumed to be 500 keV. The solid, dashed and dash-dotted curves are for the E1, E2 (scaled by 1/300) and M1 transitions, respectively.

The second RIKEN experiment [27] measured a precise angular distribution of θ_8, the scattering angle of the p-^{7}Be center-of-mass. In the vicinity of the grazing angle, the E1 angular distribution decreases, whereas the E2 cross section stays almost constant. Therefore, the E2 contribution may be observed at large angles even if it is small. The results shown in Fig. 3 suggests that the mixture of the E2 component is negligibly small. It should be pointed out that the analysis shown in Fig. 3 includes the nuclear component with the angular momentum transfer $\ell=2$, which will be discussed later.

For the same purpose, the extraction of E2 component, the parallel-momentum distribution of the ^{7}Be fragment from the Coulomb dissociation of ^{8}B has been measured at MSU. [32] They observed asymmetric peak shapes that were interpreted as an interference between E1 and E2 amplitudes. The extracted E2 component is in the same order as those predicted by nuclear structure models for ^{8}B. This conclusion is in contradiction to the above angular distribution measurement, requiring further studies.

Post acceleration is one of the effects of higher order processes. Various theoretical investigations have been performed. Most of their results point to only a few % correction to S_{17} for the conditions of the RIKEN experiments.

The higher order effects are reduced as the collision time gets shorter or the incident energy gets higher. In this respect, the agreement between the results obtained by the first-order analysis at the different energies (50 MeV/nucleon at RIKEN, 80 MeV/nucleon at MSU and 250 MeV/nucleon

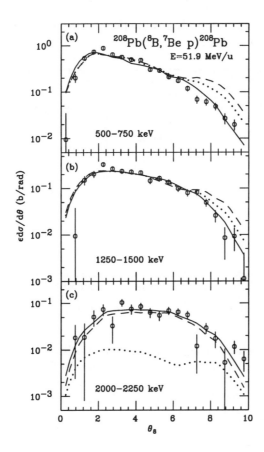

Figure 3. Angular distributions for the Coulomb dissociation of ^{8}B measured at 51.9 MeV/nucleon for three relative energy bins. The best fits result in pure E1 transitions (solid curves in (a) and (b)). Dashed and dotted curves in (a) and (b) correspond respectively to two different models for the ^{7}Be(p,γ)^{8}B process which are normalized to the data by fixing the ratios between the $\ell=1$ and $\ell=2$ components to the theoretical predictions. For $\ell=2$, the nuclear excitation amplitudes are included together with the E2 Coulomb amplitude, whereas the $\ell=1$ nuclear amplitudes are not included, because they are expected to be negligibly small.

at GSI) might indicate the minor contribution of the higher order process. A group of Notre Dame University measured ^{7}Be fragments from the ^{8}B+^{58}Ni interaction at a low energy of 3.2 MeV/nucleon. [36,37] The angular distribution taken in the range from 20° to 70° is quite inconsistent with first-order theories, whereas calculations including higher-order effects [38,39] well reproduced the data. This demonstrates the importance of the higher-order effects at sub-Coulomb energies.

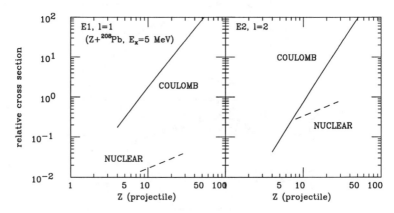

Figure 4. Atomic number dependence of the Coulomb and nuclear excitation cross sections for the process, $Z+{}^{208}\text{Pb}\rightarrow Z^*+{}^{208}\text{Pb}$, where a nucleus Z with atomic number Z is excited to its 5 MeV state. The incident energy is set to be 70 MeV/nucleon. The left panel corresponds to $\ell=1$ (E1) and the right panel to $\ell=2$ (E2) transition.

For nuclear-breakup contribution in the ${}^8\text{B}$ dissociation, Bertulani made a calculation based on a one-body potential model for the nuclear structure of ${}^8\text{B}$, [40] and found that the $\ell=1$ nuclear-excitation cross section is only in the order of 1% of the E1 cross section. However, for E2 transitions the $\ell=2$ nuclear and E2 amplitudes are expected to be in the same order as shown in Fig. 4. Therefore, the nuclear excitation with $\ell=2$ affects the Coulomb dissociation results, if the E2 mixture is sizable. For that case, full microscopic calculations are desirable.

3 ANC measurements by proton transfer reactions

The Asymptotic Normalization Coefficient (ANC) is a normalization factor for the single particle component of a wave function (overlap function) in its tail part. Since the low-energy radiative capture is only sensitive to the wave function outside the nuclear radius, the ANC can be related to the capture cross section with a good accuracy. [41] The ANC method employs a particle transfer reaction to determine the coefficient.

Experiments for the S_{17} determination have been performed with radioactive ${}^7\text{Be}$ beams in reversed kinematics. The ${}^2\text{H}({}^7\text{Be},{}^8\text{B})\text{n}$ reaction measurement led to a large $S_{17}(0)$ factor of 27.4 ± 4.4 eV-b. [42] . On the other hand, a Texas A&M group employed the targets of ^{10}B [43] and ^{14}N. [44] Radioactive ^{7}Be beams of 12.1 MeV/nucleon was produced by the ^{1}H(^{7}Li,^{7}Be)n reaction, and delivered to the secondary target by the MARS recoil spectrometer. The measured angular distributions of the (^{7}Be,^{8}B) reaction were analyzed with the DWBA theory, and the ANC's were deduced. The combined result of the two Texas A&M experiments is $S_{17}(0)=17.3\pm1.8$ eV-b, [45] which is consistent with the recommended value 19^{+4}_{-2} eV-b, [19] and hence with direct capture and Coulomb dissociation results.

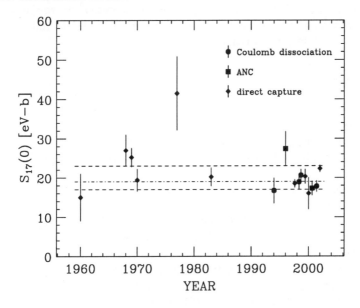

Figure 5. The $S_{17}(0)$ values extracted by the Coulomb dissociation method (solid circles) and the ANC method (solid squares). The ones obtained by direct capture measurements are also shown by solid diamonds. The latest recommendation $S_{17}(0)=19^{+4}_{-2}$ eV-b [19] is indicated by the dashed and dot-dashed lines.

4   Summary

Intermediate-energy Coulomb dissociation and low-energy proton transfer (ANC determination) have been studied to extract the astrophysical S-factor of the ^{7}Be(p,γ)^{8}B reaction. In Fig. 5, the $S_{17}(0)$ values from Coulomb dissociation, ANC and direct capture measurements are plotted. The Coulomb-dissociation and ANC results are generally in good agreement with the recent

430

evaluation from direct capture measurements. This gives a support to the recommended reaction rate of the ^{7}Be(p,γ)^{8}B reaction used in solar models. For more accurate determination, however, further studies are necessary to evaluate possible corrections due to the mixture of E2 and nuclear components and higher-order processes for the Coulomb dissociation method. More detailed confirmation of the ANC method should also be made. It should be also pointed out that either the direct or Coulomb dissociation measurements could not reach the Gamow energy of the solar burning. Extrapolation is necessary to estimate S_{17} at lower energies with a theoretical energy-dependence. Efforts to obtain lower energy cross sections are desirable to reduce the ambiguity of the ^{7}Be(p,γ)^{8}B reaction rate.

References

1. R. Davis Jr., D.S. Harmer and K.C. Hoffman, *Phys. Rev. Lett.* **20**, 1205 (1968).
2. K. Lande *et al.*, *Nucl. Phys.* B **91**, 50 (2001).
3. J.N. Bahcall, M.H. Pinsonneault and S. Basu, *Astrophys. J.* **555**, 990 (2001).
4. Y. Suzuki *et al.* (Super-Kamiokande Collaboration), *Nucl. Phys.* B **91**, 29 (2001).
5. V. Gavrin *et al.* (SAGE Collaboration), *Nucl. Phys.* B **91**, 36 (2001).
6. W. Hampel *et al.* (GALLEX Collaboration), *Phys. Lett.* B **447**, 127 (1999).
7. E. Bellotti, *Nucl. Phys.* B **91**, 44 (2001).
8. R.W. Kavanagh, *Nucl. Phys.* **15**, 411 (1960).
9. P.D. Parker, *Phys. Rev.* **150**, 851 (1966).
10. R.W. Kavanagh *et al.*, *Bull Am. Phys. Soc.* **14**, 1209 (1969); Cosmology, Fusion and other Matters, Colorado Assoc. Univ. Press, Boulder, 1972, P. 169.
11. F.J. Vaughn, R.A. Chalmers, D. Kohler, and L.F. Chase, Jr., *Phys. Rev.* C **2**, 1657 (1970).
12. C. Wiezorek, H. Krawinkel, R. Santo and L. Wallek, *Z. Phys.* A **282**, 121 (1977).
13. B. Filippone, S.J. Elwyn, C.N. Davids, and D.D. Koetke, *Phys. Rev. Lett.* **50**, 412 (1983); *Phys. Rev.* C **28**, 2222 (1983).
14. F. Hammache *et al.*, *Phys. Rev. Lett.* **80**, 928 (1998).
15. M. Hass *et al.*, *Phys. Lett.* B **462**, 237 (1999)).
16. L. Gialanella *et al.*, *Eur. Phys. J.* A **7**, 303 (2000).
17. F. Hammache *et al.*, *Phys. Rev. Lett.* **86**, 3985 (2001).

18. A.R. Junghans *et al.*, *Phys. Rev. Lett.* **88**, 041101 (2002).
19. E.G. Adelberger *et al.*, *Rev. Mod. Phys.* **70**, 1265 (1998).
20. G. Baur, C.A. Bertulani, and H. Rebel, *Nucl. Phys.* A **458**, 188 (1986).
21. T. Motobayashi *et al.*, *Phys. Lett.* B **264**, 259 (1991).
22. J. Kiener *et al.*, *Nucl. Phys.* A **552**, 66 (1993).
23. A. Lefebvre *et al.*, *Nucl. Phys.* A **592**, 69 (1995).
24. T. Motobayashi *et al.*, *Phys. Rev. Lett.* **73**, 2680 (1994).
25. N. Iwasa *et al.*, *J. Phys. Soc. Jpn.* **65**, 1256 (1996).
26. T. Kubo *et al.*, *Nucl. Instr. Meth.* B **70**, 309 (1992).
27. T. Kikuchi *et al.*, *Phys. Lett.* B **391**, 261 (1997).
28. T. Kikuchi *et al.*, *Eur. Phys. J.* A **3**, 209 (1998).
29. N. Iwasa *et al.*, *Phys. Rev. Lett.* **83**, 2910 (1999).
30. P. Senger *et al.*, *Nucl. Instrum. Methods* A **327**, 393 (1993).
31. B. Davids *et al.*, *Phys. Rev. Lett.* **86**, 2750 (2001).
32. B. Davids *et al.*, *Phys. Rev. Lett.* **81**, 2209 (1998).
33. F.C. Barker and R.H. Spear, *Astrophys. J.* **307**, 847 (1986).
34. P. Descouvemont and D. Baye, *Nucl. Phys.* A **567**, 341 (1994).
35. K. Langanke and T.D. Shoppa, *Phys. Rev.* C **49**, R1771 (1994); Erratum, *Phys. Rev.* C **51**, 2844 (1995); *Phys. Rev.* C **52**, 1709 (1995).
36. V. Guimarães *et al.*, *Phys. Rev. Lett.* **84**, 186 (2000).
37. J. Kolata *et al.*, *Phys. Rev.* C **63**, 024616 (2001).
38. F.M. Nunes and I.J. Thompson, *Phys. Rev.* C **59**, 2652 (1999).
39. H. Esbensen and G.F. Bertsch, *Phys. Rev.* C **59**, 3240 (1999).
40. C.A. Bertulani, *Phys. Rev.* C **49**, 2688 (1994).
41. H.M. Xu *et al.*, *Phys. Rev. Lett.* **73**, 2027 (1994); A.M. Mukhamedzhanov *et al.*, *Phys. Rev.* C **56**, 1302 (1995).
42. W. Liu *et al.*, *Nucl. Phys.* A **616**, 131c (1997).
43. A. Azhari *et al.*, *Phys. Rev. Lett.* **82**, 3960 (1999).
44. A. Azhari *et al.*, *Phys. Rev.* C **60**, 055803 (1999).
45. C.A. Gagliardi, *Nucl. Phys.* A **682**, 369c (2001).

SEARCH FOR AN ORBITAL MAGNETIC QUADRUPOLE TWIST MODE IN NUCLEI WITH ELECTRON SCATTERING AT 180°

P. VON NEUMANN-COSEL

Institut für Kernphysik, Technische Universität Darmstadt,
64289 Darmstadt, Germany

The nuclei ^{48}Ca, ^{58}Ni and ^{90}Zr were investigated with high-resolution inelastic electron scattering at 180° in a momentum transfer range $q \simeq 0.35 - 0.8$ fm^{-1}. Complete M2 strength distributions could be extracted up to excitation energies of about 15 MeV utilizing a fluctuation analysis technique. Microscopic calculations including the coupling to 2p-2h states successfully describe the experimentally observed strong fragmentation of the M2 mode. A quantitative reproduction of the data requires a quenching similar to the M1 case and the presence of appreciable orbital strength which can be interpreted as a torsional elastic vibration (the so-called twist mode). A comparison to proton scattering in kinematics which favor spin-isospin excitation of $J^{\pi} = 2^{-}$ states reveals direct evidence for dominantly orbital excitations in ^{58}Ni in line with microscopic predictions.

1 Introduction

The possible existence of an orbital magnetic quadrupole (M2) resonance in finite Fermi systems is of fundamental quantum statistical signifance. This mode was originally suggested in nuclei within a fluid-dynamic approach for finite Fermi systems[1] but it should also exist in metal clusters[2] and ultracold atomic Fermi gases[3]. For systems with small particle number its bulk properties are well described [4-7] within the random phase approximation (RPA) which is the microscopic analog of the fluid-dynamical picture. Macrosopically, the orbital M2 mode in nuclei can be viewed as a vibrational counterrotation of different layers of fluid in the upper and lower hemisphere where the rotational angle is proportional to the distance along the axis of rotation, hence the name 'twist mode'[1]. Its experimental observation invalidates the hydrodynamical picture of collective modes in finite Fermi systems because in an ideal liquid there is no restoring force for a twist mode[8]. Rather, a zero sound character is suggested where the restoring force is provided by the quantum-kinetic energy and the bulk behavior is that of an elastic medium[9].

Backward electron scattering presents the most promising tool to search for the twist mode in nuclei. However, the experimental observation is complicated by a significant fragmentation of the strength and, in particular, by the mixing with spin-flip M2 excitations which are microscopically generated by the same particle-hole transitions and thus centered at about the same excitation energies. Recently, extensive investigations on the M2 strength in nuclei have been performed at the new system [10] for 180° electron scattering at the superconducting Darmstadt electron linear accelerator S-DALINAC. Here, the status of experiments aiming at an experimental verification of the twist mode is reviewed.

2 Experiments

Measurements have been performed at the 180° scattering facility of the S-DALINAC coupled to a large solid-angle, large momentum-acceptance spectrometer [11]. Isotopically enriched targets of ^{48}Ca, ^{58}Ni and ^{90}Zr have been investigated in a momentum transfer range $q \simeq 0.35 - 0.8$ fm^{-1}. The energy resolution, mainly due to the target thickness, ranged from 50 to 90 keV. The spectrometer settings covered an excitation energy range $E_x = 4 - 15$ MeV with additional measurments up to very high energies ($E_x \approx 25$ MeV) for ^{58}Ni.

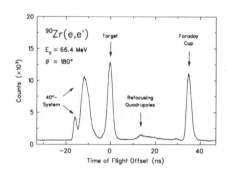
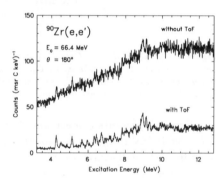

Figure 1: TOF spectrum of electrons with respect to a 10 MHz pulsed beam detected in the magnetic spectrometer. Different scattering sources can be clearly distinguished.

Figure 2: Spectra of the ^{90}Zr(e,e') reaction at 180° and $E_0 = 66.4$ MeV with and without a gate on the target-related TOF peak in Fig. 1.

An important improvement over previous 180° experiments was achieved by using a 10 MHz pulsed beam originally developed for the free-electron laser. It permits a distinction of target-related events from other beam-induced radiation sources by time-of-flight (TOF) techniques [12]. As an example, Fig. 1 displays the measurement for the ^{90}Zr target and an energy $E_0 = 66.4$ MeV. Several peaks are visible in the TOF spectrum which can be related to definite scattering sources. Besides target-related events, indicated by a relative zero TOF offset, backscattering from the Faraday cup and contributions from an energy-defining slit system in front of the target area (40° system) can be clearly identified as major background sources. Thus, the spectra are considerably improved by gating on the target TOF peak. This is demonstrated in Fig. 2, where spectra with and without correction on the TOF are shown. An improvement of the peak-to-background ratio up to an order of magnitude could be achieved with this method.

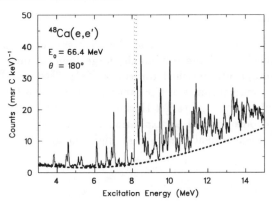

Figure 3: Spectrum of the ^{48}Ca(e,e') reaction at $E_0 = 66.4$ MeV and $\Theta_e = 180°$. The dotted peak is due to a ^{1}H contamination. The energy dependence of the background (dashed line) follows theoretical predictions [13] at lower energies and can be fixed quantitatively above $E_x = 11$ MeV by a fluctuation analysis technique.

Figure 3 presents a spectrum of the ^{48}Ca(e,e') reaction taken at an energy $E_0 = 66.4$ MeV roughly corresponding to the maximum of the M2 form factor. Two important features emerge. Firstly, the spectrum shows a rich amount of spectroscopic information. Above 8 MeV it is dominated by transitions to 2^- states. The spin information and reduced transition probabilities can be derived from fits of RPA form factors to the experimental data. Secondly, above $E_x \simeq 11$ MeV the level density of 2^- states in ^{48}Ca becomes very high leading to a considerable fragmentation of the transition strength. Thus, the usual unfolding procedure of the spectra as a superposition of discrete lines is no longer possible, and parts of the magnetic quadrupole strength might be hidden in the background of the spectra. Quantitative knowledge of the background is prerequisite for an extraction of the complete M2 strength.

A solution to this problem is provided by a fluctuation analysis technique [14,15] based upon a statistical treatment, i.e. assuming Wigner-type level spacings and Porter-Thomas intensity distributions (for details see [16]). The dashed line shows the background whose energy dependence (but not its magnitude) can be described by theoretical predictions [13] at lower energies. Above $E_x = 11$

MeV, it was fixed by the method described above.

3 Magnetic quadrupole response

The resulting M2 strength distribution, presented in the top part of Fig. 4, is discussed for the case of ^{90}Zr but similar findings are observed for ^{48}Ca (see [15,17]). Attempts to describe the experimental results by RPA calculations shown in the midle part fail (independent of details of the residual interaction) to describe the complexity and fragmentation of the extracted M2 strength distribution. One has to invoke the second-RPA (SRPA) displayed in the bottom part which extends the model space to include 2p-2h excitations on the correlated ground state. Because mean-field and collisional damping are included, SRPA should provide the fine structure of nuclear modes [8,18].

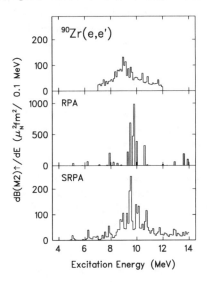

Figure 4: Upper part: B(M2) strength distribution in ^{90}Zr from the (e,e′) experiment [15]. Middle part: RPA calculation, see text. Lower part: SRPA calculation, see text.

Single-particle energies and wavefunctions were obtained from a static Woods-Saxon potential with parameters which optimally reproduce the ground-state properties [19]. Alternatively, the energies were taken from experiment when available. All 2p-2h states up to 28 MeV (^{48}Ca) and 21 MeV (^{90}Zr), respectively, were included. As residual interaction we choose the M3Y interaction [20] which is a finite-range parameterization of the G-matrix.

The RPA results predict a concentration of the strength in essentially 5 states clustered around 9.5 MeV (note the scale difference in comparison to the data) in contrast to the strong fragmentation visible in the experimental results. However, if the coupling to 2p-2h excitations is taken into account in the SRPA calculation, the description is considerably improved. The main structure of the experimental strength distribution with a centroid close to 9 MeV and a width of about 2 MeV can be quite well reproduced.

The role of spin and orbital components in forming the M2 resonance can be deduced from the SRPA predictions. This is of course model dependent but the very good agreement with the data should make such an analysis meaningful. One finds sizable orbital parts of about equal magnitude with respect to the spin matrix elements in the resonance region. The interference pattern leads to a suppression of the total strength at low excitation energies and an enhancement above approximately 7 MeV which is particularly pronounced in the main bump of the M2 resonance in ^{90}Zr around 9 MeV [15]. A quenching of the spin M2 strength similar to the M1 case in medium-heavy nuclei [21-23] is needed to obtain quantitative agreement with the data.

4 Direct evidence for the twist mode

A more direct proof for the existence of the twist mode is attained by a comparison of 180° electron scattering with proton scattering under forward angles. This method has been shown to be a useful tool for the decomposition of spin and orbital parts of the M1 strength in light [24,25] and heavy [26-28] nuclei and also to extract the role of meson-exchange current contributions [29-31] in sd-shell nuclei. Proton scattering or charge-exchange reactions on a target with zero ground-state spin in the appropriate kinematics (bombarding energies $E_0 \simeq 150 - 300$ MeV/nucleon, forward angles) selectively excite isovector spin-dipole resonances (IVSDR) with quantum numbers $\Delta L = 1, \Delta S = 1, \Delta T = 1, J^{\pi} = 0^-, 1^-, 2^-$. The $J^{\pi} = 2^-$ component is excited by the same operator $O \propto r[\sigma \otimes Y_1]^{2^-}$ as the spin-flip part of the M2 strength observed in electron scattering. While a spin decomposition of the IVSDR has not yet been achieved experimentally, calculations predict the excitation of the 2^- states to dominate the reaction cross sections [32].

Here, a comparison of the (e,e') and (p,p') reactions is presented [33] for ^{58}Ni which provides compelling evidence for the existence of the twist mode. The proton scattering data were measured at KVI Groningen for an incident proton energy $E_0 = 172$ MeV. Details of the experiment are described elsewhere [33]. Figure 5 presents spectra of both reactions for $E_x = 7 - 16$ MeV and kinematics which favor the excitation of $J^{\pi} = 2^-$ states. The comparison exhibits a good correspondence between the peaks observed in the (e,e') and (p,p') spectra (as shown by the dashed lines) for excitation energies up to 11 MeV indicating that the 2^- component of the IVSDR is significantly evident in the proton scattering cross section. At $E_x > 11$ MeV some resemblance of the peaks observed with both probes is found, but possible contributions of M3 strength in the (e,e') scattering and the excitation of the other spin components $0^-, 1^-$ of the IVSDR and $\Delta L = 2$ excitations in proton scattering complicate the picture. A detailed comparison has to await a full analysis of both experiments.

A striking difference between the two spectra is observed for the peak at $E_x = 9.87$ MeV, strongly populated in electron scattering. As indicated by the arrow in Fig. 5, no comparable excitation is visible in the proton scattering data. The same is true for the adjacent transition at $E_x = 10.04$ MeV which actually corresponds to a local minimum in the (p,p') cross section. At both energies $J^\pi = 2^-$ states were identified from the form factor behaviour in previous high-resolution (e,e') experiments [34] and in the present work.

The comparison of (e,e') and (p,p') spectra provides a clear signature for a dominantly orbital character of the observed transitions. Orbital contributions to the proton scattering cross sections, which might arise from exchange effects, the spin-orbit potential, or the tensor part of the projectile-target interaction, are expected to be very weak [35].

The observation of rather pure orbital transitions is surprising because microscopically the same 1p-1h transitions generate both, spin and orbital currents. However, the above conclusion is corroborated by calculations of the ^{58}Ni(e,e') cross sections of the present experiment within the microscopic quasiparticle-phonon model (QPM) which is similar to the SRPA. The basics of the model are described in [36]. The calculations have been performed with wavefunctions for the excited states which include one- and two-phonon configurations, i.e. 2p-2h coupling. The resulting 2^--strength functions are presented in Fig. 6 using an energy averaging parameter of 90 keV which corresponds to the experimental resolution. The total cross sections in the kinematics of the data shown in Fig 5 as well as a decomposition into spin and orbital parts are shown. Note that a quenching factor $g_s^{eff} = 0.8 g_s^{free}$ is included for the spin part of the M2 operator which was determined by a global fit to M1 and M2 transitions in spherical nuclei. The magnitude of quenching is in reasonable agreement with shell-model [21,22] and SRPA [15,23] predictions in this mass region.

The conclusions on the nature of the observed excitations resulting from the comparison of electron and proton scattering data are supported by the theoretical results shown in Fig. 6. A strong M2 resonance is obtained between 8 and about 11 MeV with comparable contributions from spin and orbital parts and a mostly constructive interference. These features qualitatively agree with

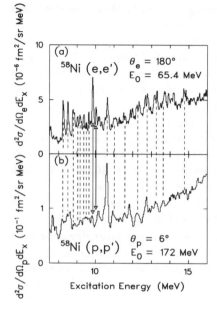

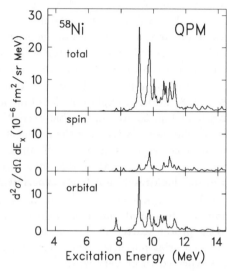

Figure 5: (a) Spectrum of the ^{58}Ni(e,e') reaction at $E_0 = 65.4$ MeV and $\Theta_e = 180°$. (b) Spectrum of the ^{58}Ni(p,p') reaction at $E_0 = 172$ MeV and $\Theta_p = 6°$. The arrows indicate the energies of transitions strongly excited in electron scattering, but not in proton scattering. The dashed lines connect possible candidates of $J^\pi = 2^-$ states excited in both experiments (from [33]).

Figure 6: Quasiparticle-phonon model calculation of the ^{58}Ni(e,e') cross sections due to M2 excitations for the kinematics of the spectrum shown in Fig. 5(a). Upper part: total cross section. Middle part: spin contribution. Lower part: orbital contribution. (From [33]).

the SRPA calculations of the M2 strength in ^{48}Ca and ^{90}Zr [15]. However, the most prominent peak in the QPM calculation exhibits a dominantly orbital character as suggested to be present in the experimental data.

As pointed out in Refs. [4,6], the (e,e') formfactor of the twist mode has a maximum at larger q value than the one of the spin mode. The conditions of the present (e,e') experiment are in fact close to the maximum of the twist mode form factor. Thus, the contribution of the spin mode in Fig. 6 is somewhat suppressed as compared to its role in the B(M2) strength distribution.

Finally, it is interesting to see to what extent the microscopic results reflect a twist-like motion as predicted by the fluid-dynamical model [1]. In Fig. 7 the orbital transition current density $j_{22}^l(r)$ is shown for the 2^- state with the largest orbital M2 cross section in Fig. 6. Figure 8 presents a three-dimensional plot of the current. Cuts are shown in the xy-plane for fixed values $z = z_0$ and $z = -z_0$. In the xy-plane the current vectors are perpendicular to the radius vector $\vec{r}_{xy}$. This is a direct consequence of the properties of the vector spherical harmonics $\vec{Y}_{221}^{\mu}(\hat{r})$. The angle of rotation is proportional to the z value. The current vanishes at $z = 0$ and the rotation has opposite signs in the upper and lower semispheres. Thus, the properties of the twist mode as predicted in the fluid-dynamical model appear in a natural way.

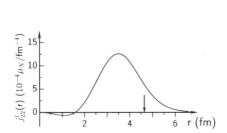

Figure 7: Orbital transition current density, $j_{22}^{l}(r)$, for the excitation of the 2^- state in Fig. 6 with the largest orbital cross section part. The vertical arrow at 4.6 fm indicates the radius defined as $R_0 = 1.2 \cdot A^{1/3}$.

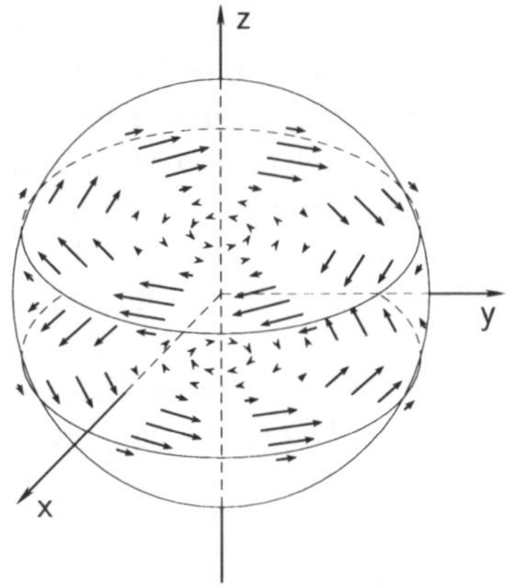

Figure 8: Three-dimensional plot of the nuclear current (Fig. 7) for cuts in the xy-plane, see text. From [33].

The current velocity as a function of $\vec{r}_{xy}$ – indicated by the length of the arrows in Fig. 8 – is proportional to the total $j_{22}(r)$ current density. In the microscopic calculation the current density results from an interference of various 1p-1h configurations. Thus, the current distribution of any 2^- state complies with the qualitative picture suggested by the fluid-dynamical model, but the magnitude of the velocity varies from state to state. Another difference is observed for the present example in the region of the nuclear interior. At radii below 1.5 fm the transition current density changes its sign and correspondingly the direction of flow is reversed.

5 Conclusions

Considerable progress in our understanding of the scarcely explored M2 response of nuclei has been achieved in recent $180°$ electron scattering studies of

^{48}Ca, ^{58}Ni and ^{90}Zr. With the combination of efficient background suppression using TOF methods and the application of a fluctuation analysis technique it is possible to extract complete strength distributions in the investigated energy ranges. Microscopic calculations are only capable to describe the fine structure of the highly fragmented mode if the coupling to 2p-2h configurations is fully included. The total observed B(M2) strengths can be reproduced assuming a magnitude of the quenching of the spin part comparable to the M1 case (not discussed here in detail, but see [15,17]). Large orbital contributions and their interference effects are necessary to obtain a quantitative agreement with experiment, indicating the presence of the twist mode.

More direct evidence for the twist mode is obtained from a comparison of electron and proton scattering on ^{58}Ni measured at kinematics where the excitation of isovector spin-flip transitions with $\Delta L = 1$ is favored. For excitation energies below 11 MeV a rather good correspondence of the spectra is found indicating the dominance of the excitation of $J^\pi = 2^-$ states. Two transitions near $E_x = 10$ MeV are strongly excited in the (e,e') reaction, but not in proton scattering. This comparison provides a clear signature for the predominantly orbital character of the transitions, thereby providing direct evidence for the excitation of the twist mode in nuclei. This interpretation is corroborated by QPM calculations of the M2 strength function in ^{58}Ni which predict dominantly orbital transitions carrying a considerable fraction of the full twist mode strength. One may hope to back this result further by a detailed study of the (e,e') form factors of the twist mode candidates because differences between the spin-flip and orbital components are predicted by the QPM and SRPA results which should be measurable for transitions with not too strong mixing of spin and orbital contributions.

Acknowledgements

This work represents the efforts of a larger collaboration and I would like to thank F. Hofmann, F. Neumeyer, Y. Kalmykov, V.Yu. Ponomarev, C. Rangacharyulu, B. Reitz, A. Richter, G. Schrieder, and J. Wambach for their contributions and support. This work has been supported by the DFG under contracts FOR 272/2-1 and 446 JAP 113/267/3-1.

References

1. G. Holzwarth and G. Eckart, *Z. Phys.* A **283**, 219 (1977).
2. V.O. Nestorenko *et al.*, *Phys. Rev. Lett.* **85**, 3141 (2000).
3. X. Vinas *et al.*, *Phys. Rev.* A **64**, 055601 (2001).

4. B. Schwesinger, K. Pingel and G. Holzwarth, *Nucl. Phys.* A **341**, 1 (1982).
5. B. Schwesinger, *Phys. Rev.* C **29**, 1475 (1984).
6. V.Yu. Ponomarev, *J. Phys.* G **10**, L177 (1984).
7. J. Kvasil *et al.*, *Phys. Rev.* C **63**, 054305 (2001).
8. J. Speth and J. Wambach, in: *Electric and Magnetic Giant Resonances in Nuclei*, Ed. J. Speth (World Scientific, Singapore, 1991) 1.
9. G.F. Bertsch, *Ann. Phys. (N.Y.)* **86**, 138 (1974); G.F. Bertsch and R.A. Broglia, *Oscillations in Finite Quantum Systems* (Cambridge University Press, 1994).
10. C. Lüttge *et al.*, *Nucl. Instrum. Methods* A **366**, 325 (1995).
11. H. Diesener *et al.*, *Phys. Rev. Lett.* **72**, 1994 (1994).
12. F. Hofmann *et al.*, *Phys. Rev.* C , in press (2002).
13. L.C. Maximon, *Rev. Mod. Phys.* **47**, 183 (1969).
14. J. Enders *et al.*, *Phys. Rev. Lett.* **79**, 2010 (1997).
15. P. von Neumann-Cosel *et al.*, *Phys. Rev. Lett.* **82**, 1105 (1999).
16. P.G. Hansen, B. Jonson and A. Richter, *Nucl. Phys.* A **518**, 13 (1990).
17. P. von Neumann-Cosel, *Nucl. Phys.* A **649**, 77c (1999).
18. J. Wambach, *Rep. Prog. Phys.* **51**, 989 (1988).
19. F. Neumeier, *Dissertation D17*, Technische Universität Darmstadt (1998).
20. G. F. Bertsch *et al.*, *Nucl. Phys.* A **284**, 399 (1977).
21. G. Martinez Pinedo *et al.*, *Phys. Rev.* C **53**, R2602 (1996).
22. P. von Neumann-Cosel *et al.*, *Phys. Lett.* B **443**, 1 (1998).
23. P. von Neumann-Cosel *et al.*, *Phys. Rev.* C **62**, 034307 (2000).
24. Y. Fujita *et al.*, *Phys. Rev.* C **55**, 1137 (1997).
25. Y. Fujita et al., *Phys. Rev.* C **62**, 044314 (2000).
26. D. Frekers et al., *Phys. Lett.* B **244**, 178 (1990).
27. A. Richter, *Prog. Part. Nucl. Phys.* **34**, 261 (1995).
28. P. von Neumann-Cosel, *Prog. Part. Nucl. Phys.* **38**, 213 (1997).
29. A. Richter *et al.*, *Phys. Rev. Lett.* **65**, 2519 (1990).
30. C. Lüttge *et al.*, *Phys. Rev.* C **53**, 127 (1996).
31. P. von Neumann-Cosel *et al.*, *Phys. Rev.* C **55**, 532 (1997).
32. S. Drożdż *et al.*, *Phys. Lett.* B **189**, 271 (1987).
33. B. Reitz *et al.*, *Phys. Lett.* B , to be published (2002).
34. W. Mettner *et al.*, *Nucl. Phys.* A **473**, 160 (1987).
35. W.G. Love and M.A. Franey, *Phys. Rev.* C **24**, 1073 (1981); M.A. Franey and W.G. Love, *Phys. Rev.* C **31**, 488 (1985).
36. V. G. Soloviev, *Theory of Atomic Nuclei: Quasiparticles and Phonons* (Institute of Physics, Bristol, 1992).

SPIN-ISOSPIN INTERACTION AND PROPERTIES IN STABLE AND EXOTIC NUCLEI

TAKAHARU OTSUKA

Department of Physics, University of Tokyo, Hongo, Tokyo 113-0033, Japan
RIKEN, Hirosawa, Wako-shi, Saitama 351-0198, Japan
E-mail: otsuka@phys.s.u-tokyo.ac.jp

RINTARO FUJIMOTO

Department of Physics, University of Tokyo, Hongo, Tokyo 113-0033, Japan

YUTAKA UTSUNO

Japan Atomic Energy Research Institute, Tokai, Ibaraki 319-1195, Japan

B.ALEX BROWN

National Superconducting Cyclotron Laboratory, Michigan State University, East
Lansing, MI 48824

MICHIO HONMA

Center for Mathematical Sciences, University of Aizu, Tsuruga, Ikki-machi,
Aizu-Wakamatsu, Fukushima 965-8580, Japan

TAKAHIRO MIZUSAKI

Department of Law, Senshu University, Higashimita, Tama, Kawasaki, Kanagawa,
214-8580, Japan

The magic numbers are the key concept of the shell model, and are different in
exotic nuclei from those of stable nuclei. Its novel origin and robustness will be
discussed in relation to the spin-isospin part of the nucleon-nucleon interaction.

I talk about new magic numbers in exotic nuclei. In exotic nuclei far
from the β-stability line, some usual magic numbers disappear while new
ones arise. This is a very intriguing problem, and its mechanism is related to
basic properties of nucleon-nucleon interaction in a very robust way.

The magic number is the most fundamental quantity governing the nuclear structure. The nuclear shell model has been started by Mayer and Jensen
by identifying the magic numbers and their origin [1]. The study of nuclear
structure has been advanced on the basis of the shell structure associated
with the magic numbers. This study, on the other hand, has been made predominantly for stable nuclei, which are on or near the β-stability line in the
nuclear chart. This is basically because only those nuclei have been accessible

experimentally. In such stable nuclei, the magic numbers suggested by Mayer and Jensen remain valid, and the shell structure can be understood well in terms of the harmonic oscillator potential with a spin-orbit splitting.

Recently, studies on exotic nuclei far from the β-stability line have started owing to development of radioactive nuclear beams. The magic numbers in such exotic nuclei can be a quite intriguing issue. We shall show [2] that new magic numbers appear and some others disappear in moving from stable to exotic nuclei in a rather novel manner due to a particular part of the nucleon-nucleon interaction.

In order to understand underlying single-particle properties of a nucleus, we can make use of *effective (spherical) single-particle energies (ESPE's)*, which represent mean effects from the other nucleons on a nucleon in a specified single-particle orbit. The two-body matrix element of the interaction depends on the angular momentum J, coupled by two interacting nucleons in orbits j_1 and j_2. Since we are investigating a mean effect, this J-dependence is averaged out with a weight factor $(2J+1)$, and only diagonal matrix elements are taken. Keeping the isospin dependence, $T=0$ or 1, the so-called monopole Hamiltonian is thus obtained with a matrix element [3,4]:

$$V_{j_1 j_2}^T = \frac{\sum_J (2J+1) <j_1 j_2 |V| j_1 j_2 >_{JT}}{\sum_J (2J+1)}, \quad \text{for } T = 0, 1, \tag{1}$$

where $<j_1 j_2 |V| j_1' j_2' >_{JT}$ stands for the matrix element of a two-body interaction, V.

The ESPE is evaluated from this monopole Hamiltonian as a measure of mean effects from the other nucleons. The normal filling configuration is used. Note that, because the J dependence is taken away, only the number of nucleons in each orbit matters. As a natural assumption, the possible lowest isospin coupling is assumed for protons and neutrons in the same orbit. The ESPE of an *occupied* orbit is defined to be the separation energy of this orbit with the opposite sign. Note that the separation energy implies the minimum energy needed to take a nucleon out of this orbit. The ESPE of an *unoccupied* orbit is defined to be the binding energy gain by putting a proton or neutron into this orbit with the opposite sign.

In Fig. 1 (a), ESPE's are shown for O isotopes. The Hamiltonian and the single-particle model space are the same as those used in Utsuno *et al.*[4], where the structure of exotic nuclei with $N \sim 20$ has been successfully described within a single framework.

A significant gap is found at $N=16$ with the energy gap between the $0d_{3/2}$ and $1s_{1/2}$ orbits equal to about 6 MeV. This is a quite large gap comparable

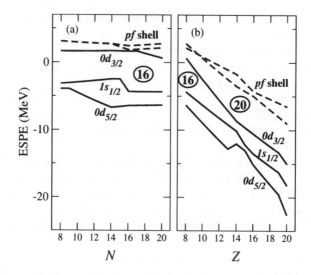

Figure 1. Effective single-particle energies of neutrons of (a) O isotopes for $N=8$ to 20 and (b) $N=20$ isotones for $Z=8$ to 20.

to the gap between the sd and pf shells in ^{40}Ca. The neutron number $N=16$ should show features characteristic of magic numbers as pointed out by Ozawa et al.[5] for observed binding energy systematics. A figure similar to Fig. 1 (a) was shown by Brown[6] for the USD interaction [7], while only nuclei with subshell closures were taken. Basically because the $0d_{3/2}$ orbit has positive energy as seen in Fig. 1 (a), O isotopes heavier than ^{24}O are unbound for the present Hamiltonian in agreement with experiments [8,9], whereas the $0d_{3/2}$ orbit has negative energy for the USD interaction [7,6].

One finds that the gap between the $0d_{3/2}$ and $1s_{1/2}$ orbits is basically constant within a variation of $\sim \pm 1$ MeV. In lighter O isotopes, valence neutrons occupy predominantly $0d_{5/2}$ and this gap does not make much sense to the ground or low-lying states. The gap becomes relevant to those states only for $N>14$. Thus, the large $0d_{3/2}$-$1s_{1/2}$ gap exists for O isotopes in general, while it can have major effects on the ground state for heavy O isotopes, providing us with a magic nucleus ^{24}O at $N=16$.

Figure 2 shows the effective $0d_{3/2}$-$1s_{1/2}$ gap, i.e., the difference between ESPE's of these orbits, in $N=16$ isotones with $Z=8\sim 20$ for three interactions: "Kuo" means a G-matrix interaction for the sd shell calculated by Kuo [10], and USD was obtained by adding empirical modifications to "Kuo" [7]. The present shell-model interaction is denoted SDPF hereafter, and its sd-shell

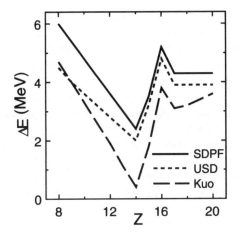

Figure 2. Effective $1s_{1/2}$-$0d_{3/2}$ gap in N=16 isotones as a function of Z. Shell model Hamiltonians, SDPF, USD and "Kuo" are used. See the text.

part is nothing but USD with small changes [4]. Steep decrease of this gap is found in all cases, as Z departs from 8 to 14. In other words, a magic structure can be formed around Z=8, but it should disappear quickly as Z deviates from 8 because the gap decreases very fast. The slope of this sharp drop is determined by $V^{T=0,1}_{0d_{5/2}0d_{3/2}}$ in eq. (1), where the dominant contribution is from T=0.

The gap can be calculated from the Woods-Saxon potential. The resultant gap is rather flat, and is about half of the SDPF value for Z=8.

The occupation number of the neutron $1s_{1/2}$ is calculated by Monte Carlo Shell Model [11,12,13,14] with full configuration mixing for the nuclei shown in Fig. 2. It is nearly two for ^{24}O as expected for a magic nucleus, but decreases sharply as Z increases. It remains smaller (< 1.5) in the middle region around Z=14, and finally goes up again for $Z{\sim}20$. This means that the N=16 magic structure is broken in the middle region of the proton sd shell, where deformation effects also contribute to the breaking. The N=16 magic number is thus quite valid at both ends. It is of interest that the gap becomes large again for larger Z, due to other monopole components.

We now discuss, in more detail, the sharp drop of the gap indicated in Fig. 2 for Z moving away from 8. This drop is primarily due to the rapid decrease of the $0d_{3/2}$ ESPE for neutrons. Figure 3 shows ESPE's for ^{30}Si and ^{24}O, both of which have N=16. Note that ^{30}Si has six valence protons in the

sd shell on top of the $Z=8$ core and is indeed a stable nucleus, while ^{24}O has no valence proton in the usual shell-model. In Fig. 3 (a), the neutron $0d_{3/2}$ and $1s_{1/2}$ are rather close to each other, while keeping certain gaps from the other orbits. Thus, the $0d_{3/2}$-$1s_{1/2}$ gap becomes smaller as seen in Fig. 2.

In Fig. 3 (b), shown are ESPE's for an exotic nucleus, ^{24}O. The $0d_{3/2}$ is lying much higher, very close to the pf shell. A considerable gap (~ 4 MeV) is between the $0d_{3/2}$ and the pf shell for the stable nucleus ^{30}Si, whereas an even larger gap (~ 6 MeV) is found between $0d_{3/2}$ and $1s_{1/2}$ for ^{24}O. The basic mechanism of this dramatic change is the strongly attractive interaction shown schematically in Fig. 3 (c), where $j_> = l + 1/2$ and $j_< = l - 1/2$ with l being the orbital angular momentum. In the present case, $l=2$. One now should remember that valence protons are added into the $0d_{5/2}$ orbit as Z increases from 8 to 14. Due to a strong attraction between a proton in $0d_{5/2}$ and a neutron in $0d_{3/2}$, as more protons are put into $0d_{5/2}$, a neutron in $0d_{3/2}$ is more strongly bound. Thus, the $0d_{3/2}$ ESPE for neutrons is so low in ^{30}Si as compared to that in ^{24}O.

The process illustrated in Fig. 3 (d) produces the attractive interaction in Fig. 3 (c). The NN interaction in this process is written as

$$V_{\tau\sigma} = \tau \cdot \tau \, \sigma \cdot \sigma \, f_{\tau\sigma}(r). \tag{2}$$

Here, the symbol "$\cdot$" denotes a scalar product, τ and σ stand for isospin and spin operators, respectively, r implies the distance between two interacting nucleons, and $f_{\tau\sigma}$ is a function of r. In the long range (or no r-dependence) limit of $f_{\tau\sigma}(r)$, the interaction in eq.(2) can couple only a pair of orbits with the same orbital angular momentum l, which are nothing but $j_>$ and $j_<$.

The σ operator couples $j_>$ to $j_<$ (and vice versa) much more strongly than $j_>$ to $j_>$ or $j_<$ to $j_<$. Therefore, the spin flip process is more favored in the vertexes in Fig. 3 (d). The same mathematical mechanism works for isospin: the τ operator favors charge exchange processes. Combining these two properties, $V_{\tau\sigma}$ produces large matrix elements for the spin-flip isospin-flip processes: proton in $j_> \rightarrow$ neutron in $j_<$ and vice versa. This gives rise to the interaction in Fig. 3 (c) with a strongly attractive monopole term in the $T=0$ channel for the appropriate (of course, usual) sign of $V_{\tau\sigma}$. This feature is a general one and is maintained with $f_{\tau\sigma}(r)$ in eq.(2) with reasonable r dependences.

Although $V_{\tau\sigma}$ yields sizable attraction between a proton in $j_>$ and a neutron also in $j_>$, the effect is weaker than in the case of Fig. 3 (c).

In stable nuclei with $N \sim Z$ with ample occupancy of the $j_>$ orbit in the valence shell, the proton (neutron) $j_<$ orbit is lowered by neutrons (protons) in the $j_>$ orbit. In exotic nuclei, this lowering can be absent, and then the

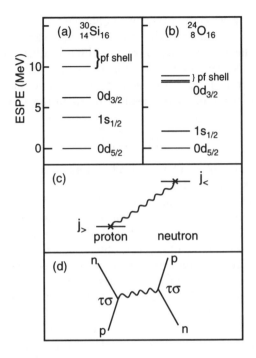

Figure 3. ESPE's for (a) ^{30}Si and (b) ^{24}O, relative to $0d_{5/2}$. (c) The major interaction producing the basic change between (a) and (b). The process relevant to the intearction in (c).

$j_<$ orbit is located rather high, not far from the upper shell. In this sense, the proton-neutron $j_>$-$j_<$ interaction enlarges a gap between major shells for stable nuclei with proper occupancy of relevant orbits. This interaction has been known, for instance [15,16], to play important roles also in other issues, *e.g.*, the onset of the deformation.

The origin of the strong $V_{\tau\sigma}$ is quite clear. The One-Boson-Exchange-Potentials (OBEP) for π and ρ mesons have this type of terms as major contributions. While the OBEP is one of major parts of the effective NN interaction, the effective NN interaction in nuclei can be provided more quantitatively by the G-matrix calculation with core polarization corrections and other various effects. Such effective NN interaction will be called simply G-matrix interaction for brevity. The G-matrix interaction should maintain the basic features of meson exchange processes, and, in fact, existing G-matrix interactions generally have quite large matrix elements for the cases shown in

Fig. 3 (c), including strongly attractive monopole terms [17].

We would like to point out that the $1/N_c$ expansion of QCD by Kaplan and Manohar indicates that $V_{\tau\sigma}$ is one of three leading terms of the long-range part of the NN interaction [18]. Since the next order of this expansion is smaller by a factor $(1/N_c)^2$, the leading terms should have rather distinct significance. One of the other two leading terms in the $1/N_c$ expansion [18] is a central force,

$$V_0 = f_0\,(r), \tag{3}$$

where f_0 is a function of r. The last leading term is a tensor force. We shall come back to these forces later.

We now turn to exotic nuclei with $N \sim 20$. The ESPE has been evaluated for them by Utsuno et al.[4]. The small effective gap between $0d_{3/2}$ and the pf shell for neutrons is obtained, and is found to play essential roles for various anomalous features. This small gap is nothing but what we have seen for ^{24}O in Fig. 3 (b). Figure 3 (b) indicates that the $N=20$ gap becomes wider as Z increases within the $N=20$ isotones, ending up with a normal gap for ^{40}Ca. Thus, the disappearance of $N=20$ magic structure in $Z=9\sim14$ exotic nuclei and the appearance of the new magic structure in ^{24}O have the same origin.

A very similar mechanism works for p-shell nuclei. We consider the structure of a stable nucleus ^{13}C, and exotic nuclei ^{11}Be and ^{9}He, all of which have $N=7$. Shell model calculations are performed with the PSDMK2 Hamiltonian [19] in the $p+sd$ shell model space on top of the ^{4}He core. Figure 4 indicates that the experimental levels are reproduced well for ^{13}C [20], whereas notable deviations are found in ^{11}Be and ^{9}He [20,21].

The p-shell part of PSDMK2 is the Cohen-Kurath (CK) Hamiltonian [22], while the other parts are from [10,23]. The single-particle energies of $0p_{3/2}$ and $0p_{1/2}$ are 1.38 and 1.68 MeV, respectively, in PSDMK2, and are basically the same as those used in CK. These energies correspond to the observed spectra of ^{5}He: 1.15 for $0p_{3/2}$ and ~ 5 MeV for $0p_{1/2}$. We use these observed values as single particle energies, while $0p_{1/2}$ is quite low in PSDMK2. As compared to the G-matrix interaction [17], the interaction in Fig. 3 (c) for the p-shell is too weak in CK, and is enlarged so that its $T=0$ monopole component becomes more attractive by 2 MeV. Thus, the Hamiltonian is modified only for three parameters. Figure 4 (a) indicates that the levels of ^{13}C calculated from the present Hamiltonian are similar to the ones obtained from PSDMK2, because the higher-lying $0p_{1/2}$ is pulled down by a stronger $T=0$ $0p_{3/2}$-$0p_{1/2}$ monopole interaction and the $0p_{3/2}$ occupancy.

The ground state of ^{11}Be and ^{9}He is known for the *inversion* between $1/2^+$ and $1/2^-$ [24]. The present calculation reproduces this inversion for both nuclei,

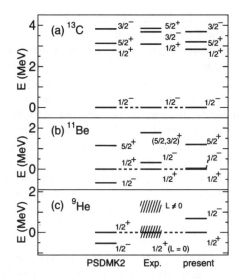

Figure 4. Energy levels for (a) ^{13}C, (a) ^{11}Be and (c) ^{9}He, relative to the state corresponding to the experimental ground state. The results by the PSDMK2 and present Hamiltonians are compared to experiments. Hatched area indicate continuum states with resonance(-like) structures.

whereas the PSDMK2 fails in either case. With the present Hamiltonian, neutron effective $1s_{1/2}$-$0p_{1/2}$ gap decreases from 8.6 MeV for ^{13}C down to 0.8 MeV for ^{9}He. Many-body correlations including pairing and deformation finally invert $1/2^+$ and $1/2^-$ eigenstates, partly due to more holes in the p shell for $1/2^+$ [25]. Thus, we achieve a reasonable description of stable and exotic nuclei with $N=7$.

The neutron $0p_{1/2}$ orbit becomes higher as the nucleus loses protons in its spin-flip partner $0p_{3/2}$. In nuclei such as He, Li and Be, the $N=8$ magic structure then disappears. In some cases, $N=6$ becomes magic: for instance, bound ^{8}He and unbound ^{9}He are obtained, similarly to bound ^{24}O and unbound ^{25}O.

The spin-isospin interaction should affect spin-isospin properties. Suzuki [26] presents a systematic survey on p-shell nuclei, suggesting that a stronger spin-isospin interaction improves spin-isospin properties.

Moving back to heavier nuclei, from the strong interaction in Fig. 3 (c), we can predict other magic numbers, for instance, $N=34$ associated with the $0f_{7/2}$-$0f_{5/2}$ interaction. In heavier nuclei, $0g_{7/2}$, $0h_{9/2}$, $etc.$ are shifted upward in neutron-rich exotic nuclei, disturbing the magic numbers $N=82$, 126, *etc.* It is of interest how the r-process of nucleosynthesis is affected by it.

In conclusion, we showed how magic numbers are changed in nuclei far from the β-stability line: $N=6$, 16, 34, *etc.* can become magic numbers in neutron-rich exotic nuclei, while usual magic numbers, $N=8$, 20, 38/40, *etc.*, may disappear. Since such changes occur as results of the nuclear force, there is isospin symmetry that similar changes occur for the same Z values in mirror nuclei. The mechanism of this change can be explained by the strong $V_{\tau\sigma}$ interaction which has robust and more fundamental origins in OBEP, G-matrix and QCD. In fact, simple structure such as magic numbers should have a simple and sound basis. Since it is unlikely that a mean central potential can simulate most effects of $V_{\tau\sigma}$, we should treat $V_{\tau\sigma}$ rather explicitly. It is nice to build a bridge between very basic feature of exotic nuclei and the basic theory of hadrons, QCD. In existing Skyrme HF calculations except for those with Gogny force, effects of $V_{\tau\sigma}$ may not be well enough included, because the interaction is truncated to be of δ-function type. The Relativistic Mean Field calculations must include pion degrees of freedom to be consistent with $V_{\tau\sigma}$. Thus, the importance of $V_{\tau\sigma}$ opens new directions for mean field theories of nuclei. In addition, $f_0(r)$ in eq.(3) and $f_{\tau\sigma}(r)$ in eq.(2) should have different ranges. Since the latter has smaller contributions in exotic nuclei, this difference should produce very interesting effects on nuclear size. Loose-binding or continuum effects are important in some exotic nuclei. By combining such effects with those discussed in this Letter one may draw a more complete picture for the structure of exotic nuclei. Finally, we would like to mention once more that the $V_{\tau\sigma}$ interaction should produce large, simple and robust effects on various properties, and may change the landscape of nuclei far from the β-stability line in the nuclear chart.

Acknowledgments

This work was supported in part by Grant-in-Aid for Scientific Research (A)(2) (10304019) from the Ministry of Education, Science and Culture.

References

1. M.G. Mayer, Phys. Rev. **75** 1969 (1949); O. Haxel, J.H.D. Jensen and H.E. Suess, Phys. Rev. **75** 1766 (1949).
2. T. Otsuka, *et al.*, Phys. Rev. Lett. **87**, 082502 (2001).
3. A. Poves and A. Zuker, Phys. Rep. **70**, 235 (1981).

4. Y. Utsuno, T. Otsuka, T. Mizusaki, and M. Honma, Phys. Rev. C **60**, 054315 (1999).
5. A. Ozawa *et al.*, Phys. Rev. Lett. **84**, 5493 (2000).
6. B. A. Brown, Revista Mexicana de Fisica **39**, Suppl. 2, 21 (1983).
7. B. A. Brown and B. H. Wildenthal, Annu. Rev. Nucl. Part. Sci. **38**, 29 (1988).
8. D. Guillemaud Mueller *et al.*, Phys. Rev. C **41**, 937 (1990); M. Fauerbach *et al.*, Phys. Rev. C **53**, 647 (1996); O. Tarasov *et al.*, Phys. Lett **409**, 64 (1997).
9. H. Sakurai *et al.*, Phys. Lett B **448**, 180 (1999); references therein.
10. T.T.S. Kuo, Nucl. Phys. **A103**, 71 (1967).
11. M. Honma, T. Mizusaki, and T. Otsuka, Phys. Rev. Lett. **75**, 1284 (1995).
12. M. Honma, T. Mizusaki, and T. Otsuka, Phys. Rev. Lett. **77**, 3315 (1996).
13. T. Otsuka, M. Honma, and T. Mizusaki, Phys. Rev. Lett. **81**, 1588 (1998).
14. T. Otsuka, M. Honma, T. Mizusaki, N. Shimizu, and Y. Utsuno, Prog. Part. Nucl. Phys. **47**, 319 (2001).
15. P. Federman, S. Pittel and R. Campos, Phys. Lett. B **82**, 9 (1979).
16. K. Heyde *et al.*, Phys. Lett. B **155**, 303 (1985).
17. M. Hjorth-Jensen, T.T.S. Kuo and E. Osnes, Phys. Rep. **261**, 125 (1995); M. Hjorth-Jensen, private com.
18. D.B. Kaplan and A.V. Manohar, Phys. Rev. C **56**, 76 (1997).
19. T. Suzuki and T. Otsuka, Phys. Rev. C **56**, 847 (1997).
20. *Table of Isotopes*, ed. by R. B. Firestone *et al.* (Wiley, New York, 1996).
21. L. Chen *et al.*, submitted; references therein.
22. S. Cohen and D. Kurath, Nucl. Phys. **73**, 1 (1965).
23. D.J. Millener and D. Kurath, Nucl. Phys. **A255**, 315 (1975).
24. I. Talmi and I. Unna, Phys. Rev. Lett. **4**, 469 (1960).
25. For instance, see H. Sagawa, B.A. Brown and H. Esbensen, Phys. Lett. B **309**, 1 (1993).
26. T. Suzuki *et al.*, talk presented in this conference.

PHOTONUCLEAR REACTIONS OF LIGHT NUCLEI AND FEW BODY PROBLEMS

T. SHIMA, Y. NAGAI, S. NAITO, K. TAMURA

Research Center for Nuclear Physics, Osaka University, Mihogaoka 10-1, Ibaraki, Osaka 567-0047, Japan
E-mail: shima@rcnp.osaka-u.ac.jp

H. OHGAKI, T. KII

Institute for Advanced Energy, Kyoto University, Gokasho, Uji, Kyoto 611-0011, Japan

H. TOYOKAWA

Photonics Research Institute, National Institute of Advanced Industrial Science and Technology (AIST), Tsukuba Central 2, 1-1-1 Umezono, Tsukuba, Ibaraki 305-8568, Japan

Absolute cross sections of the photonuclear reaction of ^{3}He and ^{4}He are measured near the peak energies of the excitation functions. The results are compared with the recent calculations based on Faddeev-AGS formalism and Lorentz-kernel transform method.

1 Introduction

Nuclear reactions of light nuclei have attracted considerable interests in the field of nuclear physics and astro-nuclear physics. They provide a nice tool to investigate basic properties of the few-nucleon dynamics as well as the nuclear force. Also they play crucial roles in nucleosynthesis occurring at the beginning of the universe and the stellar interior. For such studies great efforts have been devoted to construct a framework of few-nucleon reactions. For the deuteron, the measurements are in good agreement with each other, and they are successfully reproduced by the theories, as shown in Figure 1.

The dependence on the NN potentials has been studied using modern realistic potentials such as Paris, Reid-soft core, AV14, etc., and the results with those potentials are consistent with each other[8]. From those studies, it was found that the contribution of the meson-exchange current (MEC) amounts to about 20 % of the transition strength. and can be fully included by using Siegert's theorem. And the effect of the final-state interaction (FSI) is found to be practically negligible. Therefore, in summary, the photodisintegration reaction of the deuteron is well understood on the basis of the present theories with realistic NN potentials.

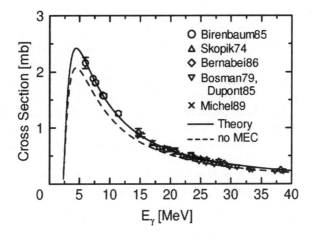

Figure 1. Photodisintegration cross sections of deuteron. Circles, triangles and diamonds denote experimental data of the deuteron photodisintegration from Refs. 1, 2
and 3, respectively. Reverse-triangles and crosses are the data of the neutron capture on proton from Refs. 4, 5 and 6, respectively. Solid curve and dashed curve are
theoretical values with and without MEC, respectively[7].

On the other hand, the situation for the nuclei beyond $A = 2$ is much
more complicated. As it is well known, there is no exact general solution for
an equation of motion for a system beyond two particles. Therefore, many
theoretical approaches have been proposed to have approximate description
of few-nucleon reactions. And in order to test those theories, accurate experimental data of nuclear reactions on light nuclei are highly demanded. For
this purpose, we have recently made new precise measurements of the photonuclear reactions of ^{3}He and ^{4}He in the energy region below 30 MeV. In
this paper the results of the experiments will be presented and discussed.

2 Photodisintegration of ^{3}He

Recent development of the theories for treating three-nucleon systems provides
methods to carry out precise calculations for the photodisintegration cross sections of ^{3}He and ^{3}H. In these calculations it has been pointed out that the photodisintegration cross sections in the peak energy region ($E_\gamma = 10{\sim}20$ MeV)
are sensitive to the NN potential. For example, Schadow et al.[9] calculated
the ^{3}He(γ,p)d cross section at $E_\gamma{\sim}10$ MeV using the Alt-Grassberger-Sandhas

(AGS) theory, and found that the cross section varies by ~20 % depending on the choice of the NN potential. In the AGS formalism, the equation of motion for the N-body system can be reduced to the one for the (N-1)-body system, if the NN potential can be approximated by a non-local and separable one (the Yamaguchi-type potential), and the equation of motion can be solved exactly. For the total photoabsorption cross section of ^{3}He, Efros et al.[10 11] applied the Lorentz-kernel transform method. The most important point of their method is that the method gives a transition probability without requiring the explicit wave function for the final scattering state, and can fully include the effect of FSI. Using the method they found that the maximum cross section of the total photoabsorption is reduced by ~10 % if the 3N forces like the Urbana-VIII and Tucson-Melbourne potentials are incorporated. Therefore it is quite important to determine the experimental cross sections in the peak region. However, there has been a large inconsistency between the previous data, and the present situation is quite unsatisfactory for the above studies. For the above reason we made a new precise measurement of the photodisintegration cross sections of ^{3}He using a new experimental method as shown below.

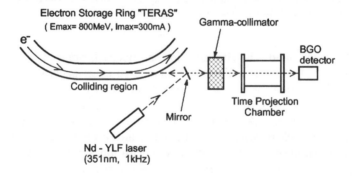

Figure 2. Experimental set up of the photodisintegration measurement at AIST.

The experimental setup is schematically shown in Figure 2. To determine the cross sections accurately, we used the time projection chamber (TPC) and Laser Compton-Scattered (LCS) γ beam at AIST[12]. Since the TPC contains an active gas target made of ^{3}He (640 Torr) together with the quenching gas of methane (160 Torr), charged particles from the photodisintegration of ^{3}He can be detected with almost 100 % detection efficiency and 4π acceptance. TPC detects the tracks of the charged particles from the reactions to distinguish precisely the reaction events of ^{3}He(γ,p)d and ^{3}He(γ,pp)n from the background ones. Furthermore, since the LCS γ beam is quasi-monochromatic and well-collimated, the background events can be rejected efficiently. The above characteristic features of the present method are advantages for precise measurement of the cross sections. The measurement was carried out at $E_\gamma =$ 10.5 MeV and 16.5 MeV.

Figure 3, 4 and 5 show the cross sections obtained for ^{3}He$(\gamma,$p$)$d, ^{3}He$(\gamma,$pp$)$n and the total photoabsorption, respectively. Thanks to the advantages of the present method, the systematic error of the measurement was less than 3% [13]. The present results are consistent with the cross sections from the previous experiments for ^{3}He$(\gamma,$p$)$d[16 20], ^{3}He$(\gamma,$pp$)$n[15 24], and total photoabsorption[26]. On the other hand, the presently obtained cross sections are smaller than the theoretical calculations beyond the ambiguities due to the choice of the NN potential and the effect of the 3N force. Therefore, it may be still needed to improve the theoretical method for treating three-nucleon systems. Further investigations for the effects of the E1-E2 interference and the contribution of the meson-exchange currents are also important.

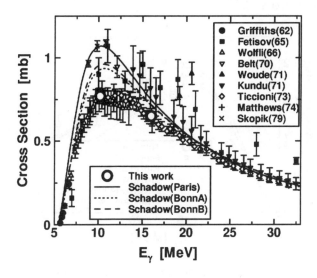

Figure 3. Cross sections of the ^{3}He$(\gamma,$p$)$d reaction. Circle; present, closed circle; Ref. 14, closed square; Ref. 15, open triangle; Ref. 16, open-reverse triangle; Ref. 17, closed triangle; Ref. 18, closed-reverse triangle; Ref. 19, diamond; Ref. 20, cross; Ref. 21, diagonal cross; Ref. 22. The solid, dotted and dashed lines are the calculations from Ref. 9.

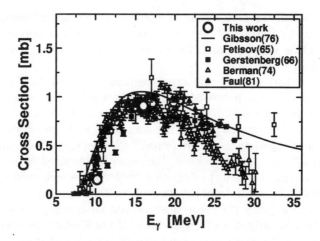

Figure 4. Cross sections of the ^{3}He(γ,pp)n reaction. Circle; present, open square; Ref. 15, closed square; Ref. 24, open triangle; Ref. 25, closed triangle; Ref. 26. The solid curve is the calculations by Gibson et al.[23].

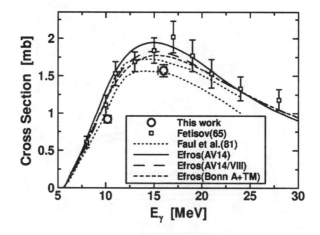

Figure 5. Cross sections of the ^{3}He total photoabsorption. Circles; present, open squares; Ref. 15. The dotted curves are the upper and lower limits of the experimental cross sections given by Faul et al.[26]. The solid, long-dashed and short-dashed curves are the calculations by Efros et al. with the NN potentials of AV14, AV14+Urbana/VIII and BonnA+TM, respectively[27].

3 Photodisintegration of ^{4}He

The photodisintegration of ^{4}He has attracted special interests, because the ratio R_γ of the (γ,p) cross section $\sigma_{\gamma p}$ to the (γ,n) cross section $\sigma_{\gamma n}$ in the peak energy region is sensitive to the isospin mixing in the ground state of ^{4}He, and it can be used to search for possible charge-symmetry breaking (CSB) in nuclear force[28]. Many experimental and theoretical efforts have been made to determine the accurate value of R_γ. The previous experimental data are, however, suffered from severe inconsistency (Refs. 29~41); The value of R_γ is still uncertain between 1 and 2. In addition, Efros et al. pointed out that some of previous data and calculations[42] provide rather smaller (about 60%) values of the energy-integrated transition strength compared to the expected one from the E1 sum rule[43].

In order to solve the above problems, reliable data of the cross sections are demanded. We performed a new measurement using the same method as used in the ^{3}He experiment. We used an active gas target made of natural helium and CH$_4$. The measurement was carried out at E_γ =24.4 MeV, 27.2 MeV and 31.1 MeV. In addition, we made measurement with a gas made of helium and CD$_4$ at E_γ =21.4 MeV in order to check the overall sensitivity of our experiment. Here it should be noted the d(γ,p)n reaction cross section has been well known as discussed in Section 1. Actually the present value of the d(γ,p)n cross section is 0.5±0.05 mb in good agreement with the previous data. The present result on the total cross section of ^{4}He is shown in Figure 6. The obtained cross sections increase to 2.8 mb at E_γ =32 MeV. This result is completely different from most of the previous data and calculations. The shape of the obtained excitation function is similar to the one from the calculation by Efros et al.[43], but the absolute value is lower by ~30 % than the calculation.

For the cross section ratio R_γ, we obtained the values consistent with the theoretical ones with no anomalous CSB effect as shown in Figure 7.

4 Summary

The photonuclear reaction cross sections of ^{3}He and ^{4}He have been measured accurately. The absolute sensitivity of the present method has been confirmed by measuring the d(γ,p)n cross section. The obtained cross sections of ^{3}He and ^{4}He are 10~30 % smaller than the recent theoretical calculations. Since the Lorentz-kernel transform method gives systematically larger cross sections for ^{3}He and ^{4}He, it will be meaningful to search for a reason common for the calculations for both nuclei. Also it is interesting to compare the present data

458

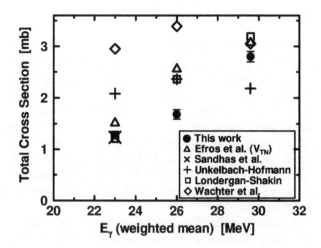

Figure 6. ^{4}He total cross section. Closed circle; present, open triangle; Ref. 43, diagonal cross; Ref. 7, cross; Ref. 44, open square; Ref. 45, open diamond; Ref. 42.

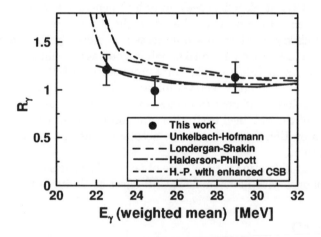

Figure 7. Ratio R_γ of the ^{4}He(γ,p)^{3}H to ^{4}He(γ,n)^{3}He cross section. Open circle; present, solid curve; Ref. 44, long-dashed curve; Ref. 45, dashed-dotted curve; Ref. 46, short-dashed curve; calculation with enhanced CSB effect[46].

with the calculation based on AGS formalism including the effect of 3N force.

Another important point is that the present cross section is smaller than the one expected from the E1 sum rule below E_γ =32 MeV. It is interesting to measure the reaction cross section in the energy range from 30 MeV to 100 MeV in order to search for the missing γ-transition strength.

Finally, the anomalous CSB effect on the ^{4}He photodisintegration is ruled out on base of the present result on R_γ.

Acknowledgment

This work was supported in part by Grant-in-Aid for Scientific Research of the Japan Society for the Promotion of Science (JSPS).

References

1. Y. Birenbaum, S. Kahane and R. Moreh, Phys. Rev. **C32** (1985) 1825.
2. D.M. Skopik, Y.M. Shin, M.C. Phenneger and J.J. Murphy, Phys. Rev. **C9** (1974) 531.
3. R. Bernabei, A. Incicchitti, M. Mattioli, P. Picozza, D. Prosperi, L. Casano, S. d'Angelo, M.P. De Pascale, C. Schaerf, G. Giordano, G. Matone, S. Frullani and B. Girolami, Phys. Rev. Lett. **57** (1986) 1542.
4. M. Bosman, A. Bol, J.F. Gilot, P. Leleux, P. Lipnik and P. Macq, Phys. Lett. **82B** (1979) 212.
5. C. Dupont, P. Leleux, P. Lipnik, P. Macq and A. Ninane, Nucl. Phys. **A445** (1985) 13.
6. P. Michel, K. Möller, J. Mösner and G. Schmidt, J. Phys. G. **15** (1989) 1025.
7. W. Sandhas, W. Schadow, G. Ellerkmann, L.L. Howell and S.A. Sofianos, Nucl. Phys. **A631** (1998) 210c.
8. T. Sato, M. Niwa and H. Ohtsubo, in *Proc. of the Int. Symp. on Weak and Electromagnetic Interactions in Nuclei*, edited by H. Ejiri, T. Kishimoto and T. Sato (World Scientific, Singapore, 1995), p.488.
9. W. Schadow, O, Nohadani and W. Sandhas, Phys. Rev. **C63** (2001) 044006.
10. V.D. Efros, W.Leidemann and G.Orlandini, Phys. Lett. **408B** (1997) 1.
11. V.D. Efros, W. Leidemann and G. Orlandini, Nucl. Phys. **A631** (1998) 658c.
12. H. Ohgaki, S. Sugiyama, T. Yamazaki, T. Mikado, M. Chiwaki, K. Yamada, R. Suzuki, T. Noguchi and T. Tomimasu, IEEE Trans. Nucl. Sci. **38**, 386 (1991).
13. S. Naito, Ph.D thesis, Osaka University, 2001.
14. G.M. Griffiths, E.A. Larson and L.P. Robertson, Can. J. Phys. **40** (1962) 402.

15. V.N. Fetisov, A.N. Gorbunov and A.T. Varfolomeev, Nucl. Phys. **71** (1965) 305.
16. W. Wölfli, R. Bösch, J. Lang, R. Müller and P. Marmier, Phys. Lett. **22** (1966) 75.
17. B.D. Belt, C.R. Bingham, M.L. Halbert and A. van der Woude, Phys. Rev. Lett. **24** (1970) 20.
18. A. van der Woude, M.L. Halbert, C.R. Bingham, and B.D. Belt, Phys. Rev. Lett. **26** (1971) 909.
19. S.K. Kundu, Y.M. Shin and G.D. Wait, Nucl. Phys. **A171** (1971) 384.
20. G. Ticcioni, S.N. Gardiner, J.L. Matthews and R.O. Owens, Phys. Lett. **46B** (1973) 369.
21. J.L. Matthews, T. Kruse, M.E. Williams, R.O. Owens and W. Savin, Nucl. Phys. **A223** (1974) 221.
22. D.M. Skopik, H.R. Weller, N.R. Roberson and S.A. Wender, Phys. Rev. **C19** (1979) 601.
23. B.F. Gibson and D.R. Lehman, Phys. Rev. **C13** (1976) 477.
24. H.M. Gerstenberg and J.S. O'Connell, Phys. Rev. **144** (1966) 834.
25. B.L. Berman, S.C. Fultz and P.F. Yergin, Phys. Rev. **C10** (1974) 2221.
26. D.D. Faul, B.L. Berman, P. Meyer and D.L. Olson, Phys. Rev. **C24** (1981) 849.
27. V.D. Efros, W. Leidemann, G. Orlandini and E.L. Tomusiak, Phys. Lett. **484B** (2000) 223.
28. F.C. Barker and A.K. Mann, Philos. Mag. **2**, 5 (1957).
29. C.C. Gardner and J.D. Anderson, Phys. Rev. **15** (1962) 626.
30. D.S. Gemmell and G.A. Jones, Nucl. Phys. **33** (1962) 102.
31. W.E. Meyerhof, M. Suffert and W. Feldman, Nucl. Phys. **A148** (1970) 211.
32. J.D. Irish, Phys. Rev. **C8** (1973) 1211.
33. C.K. Malcom, D.V. Webb, Y.M. Shin and D.M. Skopik, Phys. Lett. **47B** (1973) 433.
34. B.L. Berman, D.D. Faul, P. Meyer and D.L. Olson, Phys. Rev. **C22** (1980) 2273.
35. L. Ward, D.R. Tilley, D.M. Skopik, N.R. Roberson and H.R. Weller Phys. Rev. **C24** (1981) 317.
36. R. Bernabei, A. Chisholm, S. d'Angelo, M.P. De Pascale, P. Picozza, C. Schaerf, P. Belli, L. Casano, A. Incicchitti, D. Prosperi and B. Girolami, Phys. Rev. **C38** (1988) 1990.
37. G. Feldman, M.J. Balbes, L.H. Kramer, J.Z. Williams, H.R. Weller and D.R. Tilley, Phys. Rev. **C42** (1990) 1167.
38. L. Van Hoorebeke, R. Van de Vyver, V. Fiermans, D. Ryckbosch, C. Van

den Abeele and J. Dias, Phys. Rev. **C48** (1993) 2510.

39. K.I. Hahn, C.R. Brune and R.W. Kavanagh, Phys. Rev. **C51** (1995) 1624.
40. D.P. Wells, D.S. Dale, R.A. Eisenstein, F.J. Federspiel, M.A. Lucas, K.E. Mellendorf, A.M. Nathan and A.E. O'Neill, Phys. Rev. **C46** (1992) 449.
41. R.E.J. Florizone, J. Asai, G. Feldman, E.L. Hallin, D.M. Skopik, J.M. Vogt, R.C. Haight and S.M. Sterbenz, Phys. Rev. Lett. **72** (1994) 3476.
42. B. Wachter, T. Mertelmeier and H.M. Hofmann, Phys. Rev. **C38** (1988) 1139.
43. V.D. Efros, W. Leidemann and G. Orlandini, Phys. Rev. Lett. **78** (1997) 4015.
44. M. Unkelbach and H.M. Hofmann, Nucl. Phys. **A549** (1992) 550.
45. J.T. Londergan and C.M. Shakin, Phys. Rev. Lett. **78** (1972) 1729.
46. D. Halderson and R.J. Philpott, Phys. Rev. **C28** (1983) 1000.

DETERMINATION OF S_{17} BASED ON CDCC ANALYSES FOR $^7\text{Be}(d,n)^8\text{B}$

KAZUYUKI OGATA

Research Center for Nuclear Physics (RCNP), Osaka University, Ibaraki, Osaka 567-0047, Japan
E-mail: kazuyuki@rcnp.osaka-u.ac.jp

MASANOBU YAHIRO

Department of Physics and Earth Sciences, University of the Ryukyus, Nishihara-cho, Okinawa 903-0213, Japan
E-mail: yahiro@sci.u-ryukyu.ac.jp

YASUNORI ISERI

Department of Physics, Chiba-Keizai College, Todoroki-cho 4-3-30, Inage, Chiba 263-0021, Japan
E-mail: iseri@chiba-kc.ac.jp

MASAYASU KAMIMURA

Department of Physics, Kyushu University, Fukuoka 812-8581, Japan
E-mail: kami2scp@mbox.nc.kyushu-u.ac.jp

The astrophysical factor S_{17} for $^7\text{Be}(p,\gamma)^8\text{B}$ reaction is reliably extracted from the transfer reaction $^7\text{Be}(d,n)^8\text{B}$ at $E = 7.5$ MeV with the asymptotic normalization coefficient method. The transfer reaction is analyzed with CDCC based on the three-body model. This analysis is free from uncertainties of the optical potentials in the previous DWBA analyses.

The solar neutrino problem is a central issue of the neutrino physics [1]. The major source of the high-energy neutrinos observed by solar neutrino detectors is ^8B produced in the $^7\text{Be}(p,\gamma)^8\text{B}$ reaction. The astrophysical factor S_{17} for the reaction, nevertheless, is one of the most poorly determined reaction rates in the standard solar model. The latest recommendation for the factor $S_{17}(0)$ at zero energy, based on recent direct measurements, is 19^{+4}_{-2} eV b [2]. However, this is still far from our goal of determining $S_{17}(0)$ within several percent. The main difficulty in the direct measurement is from ambiguities of determining the effective target thickness of the radioactive ^7Be beam. Indirect measurements are thus highly expected.

The transfer reaction $^7\text{Be}(d,n)^8\text{B}$ is important as an indirect measurement. Once the asymptotic normalization coefficient (ANC) of the overlap function between the proton, ^7Be, and ^8B ground states is determined from the data of the transfer reaction, $S_{17}(0)$ can be accurately derived from the ANC, as long as the reaction is peripheral [3]. The reaction has been measured at $E = 7.5$ MeV and the ANC is extracted with the distorted wave Born approximation (DWBA) [4]. The $S_{17}(0)$ obtained with the ANC is 27.4±4.4 eV b, leading to an inconsistency with the recommended value. The reaction is found with DWBA to be peripheral [5,6]. For lack of data on the corresponding elastic scattering, however, distorted potentials used in DWBA are quite ambiguous [5,6], since these are derived from proton and deuteron optical potentials for different targets and/or energies. In particular, uncertainties of the d-^{7}Be optical potential bring about large errors for S_{17}, typically 30 % in magnitude.

In the present paper, we analyze ^{7}Be$(d, n)^8$B at 7.5 MeV with the three-body model $(p + n + ^7$Be$)$, assuming ^{7}Be to be an inert core. This model is reliable, since effects of core excitations of ^{7}Be are shown to be negligible [7,8]. An advantage of this analysis is that we do not need ambiguous d-^{7}Be and n-^{8}B optical potentials. The three-body dynamics is explicitly treated by means of continuum-discretized coupled-channels (CDCC) method [9,10]. The theoretical foundation of CDCC is shown in Ref. 11. This theory is then established as a method of solving the three-body system with good accuracy. The CDCC provides a precise description of the wave function in the entrance channel. The wave function in the exit channel is consistently calculated on the basis of the three-body model. We can then avoid using the ambiguous n-^{8}B optical potential extrapolated from different targets.

The effective Hamiltonian based on the three-body model contains the optical potential between n and ^{7}Be. Data [12] on the neutron elastic scattering are available for target ^{7}Li, the mirror nucleus of ^{7}Be, at 4 MeV, approximately half the deuteron energy considered here. The potential is then determined accurately from the data. The potential is used as an input in CDCC calculation for deuteron elastic scattering on ^{7}Li at 8 MeV, and the numerical result is compared with the experimental data [13]. This is a good test for the neutron optical potential determined here. The neutron optical potential thus obtained is only an input in CDCC calculation for ^{7}Be$(d, n)^8$B. The ANC is then obtained with no ambiguity of distorted wave functions in both the entrance and the exit channel.

The transition amplitude for the transfer reaction, based on the three-body model $(p + n + ^7$Be$)$, is

$$T_{\text{fi}} = S_{\text{exp}}^{1/2} < \Psi_{\text{i}}^{(-)}|V_{\text{np}}|\Psi_{\text{f}}^{(+)} > . \tag{1}$$

The three-body wave function $\Psi_{\text{i}}^{(+)}$ in the initial channel is governed by the three-body Hamiltonian

$$H_i = K_{np} + K_{dBe} + V_{np}(r_{np}) + U_{nBe}(r_{nBe}) + U_{pBe}(r_{pBe}), \qquad (2)$$

where r_{XY} is the coordinate of nucleus X relative to nucleus Y. The potential V_{np} is the interaction between n and p, U_{pBe} (U_{nBe}) is the proton (neutron) optical potential for the target ^{7}Be at half the deuteron incident energy, and K_{np} and K_{dBe} show kinetic energy operators for two-body systems denoted by the subscripts. The nuclear part of U_{pBe} is assumed to have the same as U_{nBe}. The wave function $\Psi_i^{(+)}$ is solved with CDCC. In CDCC, deuteron breakup states are truncated into a model space. The present model space is $k_{max} = 1.0$ fm^{-1}, $\ell = 0, 2$ and $\Delta = 1/16$ fm^{-1} in the notation of Ref. 10. The CDCC solution converges at these values, when the model space is enlarged.

The three-body wave function $\Psi_f^{(-)}$ in the exit channel is determined by the three-body Hamiltonian

$$H_f = K_{pBe} + K_{nB} + V_{pBe}(r_{pBe}) + U_{nBe}(r_{nBe}). \qquad (3)$$

In the three-body model, ^{8}B is treated by the two-body (p+^{7}Be) model with the potential V_{pBe}. The spectroscopic factor S_{exp} is introduced by taking into account the incompleteness of the model. Note that H_f does not contain V_{np} since the interaction is already treated as an operator in T.

In principle, H_f allows transitions between different states in the p+^{7}Be system. However, effects of the transitions to the transfer reaction may be estimated with the adiabatic approximation, since the ground state of ^{8}B has a binding energy (0.137 MeV) considerably smaller than an energy (3.713 MeV) of outgoing neutron. Following Johnson and Soper [14], we replace the sub-Hamiltonian of the p+^{7}Be system by the binding energy and take the zero-range approximation to V_{np}. This application of the Johnson-Soper approximation leads to a simple form $\Psi_f^{(-)} = \chi_f^{(-)}\phi_{nBe}$, where ϕ_{nBe} is the wave function of ^{8}B in its grand state and $\chi_f^{(-)}$ is the wave function of outgoing neutron distorted by the potential $U_{nBe}(8r_{nB}/7)$. The factor 8/7 in the argument of the potential shows the ^{8}B breakup effect. In H_f, actually, when the argument r_{nBe} of U_{nBe} is simply replaced by r_{nB}, the latter never induces the effect, since r_{nB} is the center-of-mass coordinate of ^{8}B relative to neutron.

Figure 1 shows results of the optical potential search for neutron scattering at 4.08 MeV from ^{7}Li, the mirror nucleus of ^{7}Be. The resultant potential shows a good agreement with data [12]. The optical potential is applied for deuteron scattering at 8.0 MeV from ^{7}Li. The CDCC calculation with the potential gives a good agreement with data [13] at angles $\theta < 70°$, as shown in Figure 2. It is then concluded that the potential is reliable especially at forward angles. This potential is used in H_i.

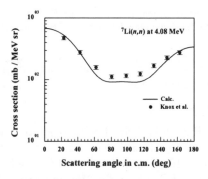
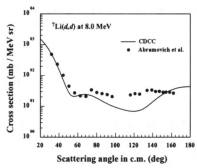

Figure 1. Results of the optical potential search for neutron elastic scattering at 4.08 MeV from ^{7}Li. Experimental data are taken from Ref. 12.

Figure 2. Comparison between the CDCC calculation and the experimental data [13] for ^{7}Li$(d,d)^7$Li at 8.0 MeV.

The transfer reaction ^{7}Be$(d,n)^8$B is analyzed with the three-body model. In H_f, we take the model of Kim $et\ al.$ [15] for the p+^{7}Be system, and the neutron optical potential U_{nBe} which reproduces data [16] on ^{7}Li$(n,n)^7$Li at 3.83 MeV, as shown in Figure 3. All the potential parameter sets are listed in Table 1. At forward angles $\theta < 55°$, as shown in Figure 4, the calculated cross

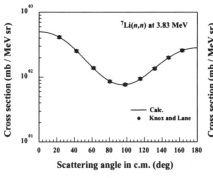
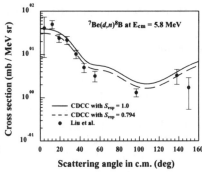

Figure 3. Same as in Figure 1 but at 3.83 MeV. Experimental data are taken from Ref. 16.

Figure 4. The calculated cross sections for ^{7}Be$(d,n)^8$B at 7.5 MeV with $S_{exp} = 1.0$ (solid line) and 0.794 (dashed line), compared with the experimental data [4].

section well reproduces data [4] with the spectroscopic factor $S_{exp} = 0.794$. The astrophysical factor S_{17} is obtained from the S_{exp} with the relation $S_{17}(0) =$

Table 1. Parameters for the optical potential between neutron and ^{7}Li at E_n= 4.08 and 3.83 MeV corresponding to the initial and the final channels for ^{7}Be$(d,n)^8$B at E_d= 7.5 MeV, respectively.

E_n	V_0	r_0	a_0	W_d	r_i	a_i	V_{so}	r_{so}	a_{so}
4.08 MeV	32.19	1.628	0.781	2.352	1.917	0.153	6.090	1.623	1.408
3.83 MeV	47.40	1.030	1.005	1.128	2.571	0.234	5.855	0.500	1.434

$S_{exp}b^2/0.026$ in the ANC method [3]. Here b is defined with ϕ_{pBe} and the Whittaker function W as $\phi_{pBe} = bW$ at r_{pBe} larger than the range of the nuclear part of V_{pBe}. Our result is $S_{17}(0) = 19.99$, since $b = 0.8091$ for the Kim model. After investigating the dependence of $S_{17}(0)$ on the ^{8}Be single particle wave function and taking account of the experimental error [4] (13%) of the ^{7}Be$(d,n)^8$B cross section, we obtain the following value: $S_{17}(0) = 19.2 \pm 2.5$.

In conclusion, the present CDCC analyses for the transfer reaction ^{7}Be$(d,n)^8$B provide a precise value $S_{17}(0) = 19.2 \pm 2.5$, where the almost all of the error comes from that of the experimental data, being consistent with the recent recommendation 19^{+4}_{-2} eV b. The present analyses are free from uncertainties of the optical potentials in both initial and exit channels.

Acknowledgments

The authors would like to thank T. Motobayashi, T. Kajino, Y. Watanabe, and M. Kawai for helpful discussions.

References

1. J. N. Bahcall, M. H. Pinsonneault, and S. Basu, Astrophys. J. 555, 990 (2001) and references therein.
2. E. G. Adelberger et al., Rev. Mod. Phys. 70, 1265 (1998).
3. H. M. Xu, C. A. Gagliardi, R. E. Tribble, A. M. Mukhamedzhanov, and N. K. Timofeyuk, Phys. Rev. Lett. 73, 2027 (1994).
4. W. Liu et al., Phys. Rev. Lett. 77, 611 (1996).
5. C. A. Gagliardi et al., Phys. Rev. Lett. 80, 421 (1998).
6. J. C. Fernandes, R. Crespo, F. M. Nunes, and I. J. Thompson, Phys. Rev. C59, 2865 (1999).
7. F. M. Nunes, I. J. Thompson, and R. C. Johnson, Nucl. Phys. A596, 171 (1996).
8. F. M. Nunes, R. Crespo, and I. J. Thompson, Nucl. Phys. A615, 69 (1997); A627, 747 (1997).

9. M. Kamimura, M. Yahiro, Y. Iseri, Y. Sakuragi, H. Kameyama and M. Kawai, Prog. Theor. Phys. Suppl. **89** (1986); N. Austern, Y. Iseri, M. Kamimura, M. Kawai, G.H. Rawitscher and M. Yahiro Phys. Reports. **154**, 125 (1987).

10. R. A. D. Piyadasa, M. Kawai, M. Kamimura and M. Yahiro, Phys. Rev. **C60**, 044611 (1999).

11. N. Austern, M. Yahiro and M. Kawai, Phys. Rev. Lett. **63**, 2649 (1989); N. Austern, M. Kawai and M. Yahiro, Phys. Rev. **C53**, 314 (1996).

12. H. D. Knox *et al.*, Nuclear Science and Engineering **69**, 223 (1979).

13. S. N. Abramovich *et al.*, Izv. Rossiiskoi Akademii Nauk, Ser. Fiz. **40**, 842 (1974).

14. R. C. Johnson and P. R. J. Soper, Phys. Rev. **C1**, 976(1970).

15. K. H. Kim, M. H. Park, and B. T. Kim, Phys. Rev. **C35**, 363 (1987).

16. H. D. Knox and R. O. Lane, Bulletin of the American Physical Society, **23**, 942 (1978).

E2 AND M1 TRANSITIONS AMONG TRIAXIALLY
SUPERDEFORMED BANDS
IN LU ISOTOPES

KAZUKO SUGAWARA-TANABE

Otsuma Women's University, Tama,
Tokyo 206-8540, Japan
E-mail: tanabe@otsuma.ac.jp

KOSAI TANABE

Department of Physics, Saitama University,
Saitama 338-8570, Japan
E-mail: tanabe@riron.ged.saitama-u.ac.jp

Triaxial superdeformed bands in ^{163}Lu is analysed by applying the Holstein-Primakoff transformation both to the total angular momentum and to the single-particle angular momentum. Quite good agreements with the experimental values are reproduced in the energy difference between two superdeformed bands as a function of angular momentum, and also in the ratio of E2 transitions among these bands. The results coming from the exact diagonalization of the particle plus rotor Hamiltonian are compared with the approximate calculation.

1 Introduction

Recently, triaxial superdeformed bands in ^{163}Lu nucleus[1] have been observed, which are interpreted as the wobbling motion[2]. We have already proposed the theory for the triaxially deformed bands nearly thirty years ago[3]. The difference between our treatment and Bohr-Mottelson's text book is in the quantization axis and in the order of approximation. We chose z-axis as the quantization axis, while Bohr-Mottelson chose x-axis. We include the whole effect coming from $1/I$ where I is the total angular momentum. These two differences help us to get the exact energy eigenvalue at the axially symmetric limit. Moreover, our paper is the first work which apply the Holstein-Primakoff transformation to the nuclear physics.

In this paper we extend our old theory to the odd nuclei by introducing two kinds of bosons for the total angular momentum I and the single-particle angular momentum j. We get quite good approximation both for the energy difference and the transition rates. Our approximation is confirmed by the exact diagonalization of the total Hamiltonian. The formulation is in Sec. 2 and the numerical analysis in Sec. 3. The discussion is in the final Sec. 4.

2 Formulation

We apply the Holstein-Primakoff transformation to the total angular momentum **I** and to the single-particle angular momentum **j**.

$$I_+ = I_x + iI_y = -a^\dagger (2I - \hat{n})^{1/2},$$
$$I_- = I_x - iI_y = (2I - \hat{n})^{1/2} a,$$

$$I_z = I - \hat{n},$$
$$j_+ = j_x + ij_y = b^\dagger (2j - \hat{k})^{1/2},$$
$$j_- = j_x - ij_y = (2j - \hat{k})^{1/2} b,$$
$$j_z = j - \hat{k}, \tag{1}$$

where a and b are two kinds of boson operators with $\hat{n} = a^\dagger a$, and $\hat{k} = b^\dagger b$. The commutation relations among I_i $(i = x, y, z)$ become $[I_i, I_j] = -iI_{i \times j}$ with $-$ sign, while those among j_i $(i = x, y, z)$ become $[j_i, j_j] = ij_{i \times j}$ with $+$ sign. The operator I_i commutes with j_j as $[I_i, j_j] = 0$. Here, x, y and z correspond to the principal axes of the rotor. We diagonalize the rotor Hamiltonian H_{rot} using Eq. (1) in two steps, where H_{rot} is given by,

$$H_{rot} = \sum_i \frac{(I_i - j_i)^2}{2\mathcal{J}_i} = \sum_i A_i (I_i - j_i)^2. \tag{2}$$

At first, bosons a and b are transformed into another bosons γ and β, and then γ and β into the final bosons ρ and σ.

$$\gamma = \eta_+ \frac{\sqrt{I}a + \sqrt{j}b^\dagger}{\sqrt{I - j}} - \eta_- \frac{\sqrt{I}a^\dagger + \sqrt{j}b}{\sqrt{I - j}},$$
$$\beta = \frac{\sqrt{j}a^\dagger + \sqrt{I}b}{\sqrt{I - j}}, \tag{3}$$

with

$$\eta_\pm^2 = \frac{1}{2}(\frac{\zeta}{\sqrt{\zeta^2 - \eta^2}} \pm 1), \tag{4}$$

where $\zeta = A_z - \xi$, $\xi = (A_x + A_y)/2$, and $\eta = (A_y - A_x)/2$. At the symmetric limit of $A_x = A_y$, η_+^2 becomes 1 and η_-^2 to 0. On the other hand, Ref. 2 uses the following equation

$$I_+ = I_y + iI_z = (2I)^{1/2}c^\dagger, \quad I_- = I_y - iI_z = (2I)^{1/2}c, \quad I_x = I. \tag{5}$$

It defines the quantization axis as x axis, and the rotation axis is also x axis. They transform c into new boson operator $\hat{c}$,

$$c^\dagger = \eta_+ \hat{c}^\dagger + \eta_- \hat{c},$$
$$\eta_\pm = \frac{1}{2}(\frac{A_y + A_z - 2A_x}{2\sqrt{(A_y - A_x)(A_z - A_x)}} \pm 1), \tag{6}$$

which diverges at the symmetric limit of $A_x = A_y$.

In the second step, γ and β are transformed into the final boson operators ρ and σ through the relations

$$\rho = \gamma + x\gamma^\dagger - z\beta - t\beta^\dagger,$$
$$\sigma = \beta + y\beta^\dagger + z\gamma - t\gamma^\dagger. \tag{7}$$

With the use of the bosons ρ and σ, Eq. (2) is diagonalized as

$$E_{I\Omega\kappa} = \zeta(I - \Omega + \frac{1}{2} - \kappa)^2 + A_z(I - \Omega + \frac{1}{2})^2$$
$$- 2\theta(I - \Omega + \frac{1}{2})(I - \Omega + \frac{1}{2} - \kappa) - \frac{\xi}{4}, \quad (8)$$

where $\theta = \sqrt{\zeta^2 - \eta^2}$. In Eq. (8), $\kappa = K - \Omega$ equals the expectation value of $I_z - j_z$, with $n_\rho = <\rho^\dagger\rho>$, $n_\sigma = <\sigma^\dagger\sigma>$, $I - n_\rho = K$ and $j - n_\sigma = \Omega$. Minimum energy of $E_{I\Omega\kappa}$ in Eq. (8) is near $\kappa \sim 0$, and $n_\sigma \sim j$. The Bohr symmetry[4], which requires the invariance under the rotation about z axis by π, limits the values of κ as even $(0, \pm 2 \pm 4 \cdots)$. At the axially symmetric limit $(\eta = 0, \xi = A_\perp)$, Eq. (8) becomes

$$E_{I\Omega\kappa} = (A_z - A_\perp)\kappa^2 + A_\perp(I - j + n_\sigma)(I - j + n_\sigma + 1). \quad (9)$$

Now we consider of the single-particle Hamiltonian H_{sp}.

$$H_{sp} \propto \cos\gamma Y_{20} + \sin\gamma(Y_{22} + Y_{2-2})/\sqrt{2}$$
$$= \frac{V}{j(j+1)}[\cos\gamma(3j_z^2 - \vec{j}^2) + \sqrt{3}\sin\gamma(j_x^2 - j_y^2)]. \quad (10)$$

We can apply Eq. (1) to $\vec{j}$, and the result is easily calculated from our old paper[3].

$$\frac{E_{sp}j(j+1)}{V} = 2\cos\gamma j(j+1) - [\sqrt{3(4\cos^2\gamma - 1)}(2j + 1) - 3\cos\gamma]$$
$$\times (n_c + 1/2) + 3\cos\gamma n_\alpha^2, \quad (11)$$

where $n_\alpha = <\alpha^\dagger\alpha>$ and $\alpha = u_+b - u_-b^\dagger$, with

$$u_\pm = \frac{1}{2}(f \pm \frac{1}{f})(1 + \frac{3}{16j}(f \mp \frac{1}{f})^2),$$
$$f = \left(\frac{3\cos\gamma + \sqrt{(3)}\sin\gamma}{3\cos\gamma - \sqrt{3}\sin\gamma}\right)^{1/4}. \quad (12)$$

Minimum energy of E_{sp} is around $n_\alpha \sim j$. We assume α in Eq. (11) as σ in Eq. (8). Both from the conditions of $\partial E_{sp} = 0$ and $\partial E_{I\Omega\kappa} = 0$, the yrast superdeformed band may have $n_\sigma \sim j$ and $\kappa \sim 0$.

As is used in the experimental analysis[1], we choose $E'_{I\Omega\kappa} = E_{I\Omega\kappa} - AI(I + 1)$ with $A = 0.0075$ MeV. Then the derivative of $E'_{I\Omega\kappa}$ by I can be negative depending on the factor of $(A_z + \zeta - 2\theta - A)$.

$$\frac{\partial E'_{I\Omega\kappa}}{\partial I} = (A_z + \zeta - 2\theta - A)(2I + 1) - (A_z + \zeta - 2\theta)2\Omega - (\zeta - \theta)2\kappa \quad (13)$$

Here, we compare the energy difference between two bands with κ and $\kappa + 2$.

$$E'_{I\Omega\kappa+2} - E'_{I\Omega\kappa} = 4(\theta - \zeta)(I - \Omega + \frac{1}{2}) + 4\zeta(\kappa + 1) \quad (14)$$

This difference decreases with I, as $\theta - \zeta < 0$. On the other hand, Ref. 2 gives $E(n) = A_x I(I+1) + \hbar\omega(n + \frac{1}{2})$ with $\hbar\omega = 2I\sqrt{(A_y - A_x)(A_z - A_x)}$. In this case the energy difference $E(n+1) - E(n) = \hbar\omega$ increases with I.

Here we pay attention to the important nature of the nucleus. The rotor itself is composed of many nucleons and the last odd nucleon is coupled to this rotor. There is the D_2-symmetry, i.e. the rotor Hamiltonian is invariant with respect to rotations through the angle π about each of the three principle axes, $R_k(\pi) = \exp(-i\pi I_k)$ $(k =, x, y, z)$. The symmetry group has the four representations. The intrinsic Hamiltonian has also D_2 symmetry as $[I_i, j_j] = 0$. The D_2-symmetry is based on the time-reversal invariance of the deformation. The intrinsic Hamiltonian is also the function of deformations β and γ, and there appears another symmetry in 5-dimentional space, i.e. three Eulerian angles, β and γ. The symmetry is called Bohr's symmetry [4], which restricts $\kappa =$ even values in our case. Then among four kinds of D_2-symmetry, only one symmetry appears in the nucleus. It is different from the classical precession around rigid body given in the text book by Landav-Lifshits [5]. In nuclear physics, there exist the intrinsic states irrespective of the even or odd mass nucleus, which are influenced by the rotor and also give influence to the rotor.

We transform the body-fixed frame to the space-fixed frame through the relation,

$$|IM> = \sum_{\kappa,\Omega} D^I_{MK}|I\Omega\kappa > < I\Omega\kappa|IK > . \tag{15}$$

It is very dificult to obtain the analytic formula for $< I\Omega\kappa|IK >$ in the odd nucleus, and so we assume $< I\Omega\kappa|IK > \sim \delta_{\kappa, K-\Omega}$. More precise wave function is obtained for the case of even nuclei as $G_{nk}(x) = < I\kappa|IK >$, whose details are in our old paper [3]. Thus, we adopt an approximate wave function as following,

$$\Psi_{IM} = \sqrt{\frac{5}{16\pi^2}}[D^I_{MK}\phi_{\Omega\kappa} + (-)^{I-j-K+\Omega}D^I_{M-K}\phi_{-\Omega-\kappa}]. \tag{16}$$

Then the energy eigenvalue becomes

$$\bar{E}_{I\Omega\kappa} = (E'_{I\Omega\kappa} + E'_{I-\Omega-\kappa})/2. \tag{17}$$

And the energy difference between $\kappa + 2$ and κ becomes

$$\bar{E}_{I\Omega\kappa+2} - \bar{E}_{I\Omega\kappa} = 4(\zeta - \theta)\Omega + 4\zeta(\kappa + 1). \tag{18}$$

which is independent from I.

3 Numerical Analysis

The quadrupole moments are defined as proportional to the moments of inertia.

$$Q_0 \propto \mathcal{J}_x + \mathcal{J}_y - 2\mathcal{J}_z,$$

$$Q_2 \propto \sqrt{\frac{3}{2}}(\mathcal{J}_y - \mathcal{J}_x). \tag{19}$$

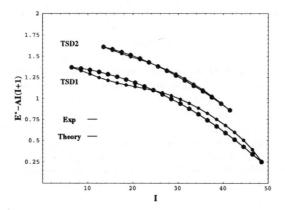

Figure 1: The excited energy E^*. $AI(I+1)$ versus I for TSD1 and TSD2 bands. The experimental values are shown by smaller circles connected by dashed lines, and the theoretical values by larger circles connected by solid lines.

There are two models to define the functional dependence of the moments of inertia on β and γ.

(1) irrotational flow

$$\mathcal{J}_i \propto \beta^2 \sin^2(\gamma - \frac{2\pi}{3}i),$$
$$\frac{Q_2}{Q_0} = -\frac{\tan(2\gamma)}{\sqrt{2}}, \tag{20}$$

(2) rigid rotor

$$\mathcal{J}_i \propto 1 - \sqrt{\frac{5}{4\pi}}\beta \cos(\gamma - \frac{2\pi}{3}i),$$
$$\frac{Q_2}{Q_0} = \frac{\tan(\gamma)}{\sqrt{2}}, \tag{21}$$

for $i = x, y, z$.

In the numerical analysis for ^{163}Lu, we adopt the state for the triaxially deformed superdeformed band 1 (TSD1) as $K = \frac{1}{2}$ $\Omega = \frac{1}{2}$ ($\kappa = 0$), and for the triaxially deformed superdeformed band 2 (TSD2) as $K = -\frac{3}{2}$ $\Omega = \frac{1}{2}$ ($\kappa = 2$). We adopt $\beta = 0.38$ and $\gamma = 19°$ given by Ref. [1], and unique-parity level i13/2 for the last nucleon. We compared two kinds of moments of inertia, i.e., irrotational moments of inertia in Eq. (20) and rigid moments of inertia in Eq. (21), and found the irrotational moments of inertia is preferable. It suggests that the nucleus is not

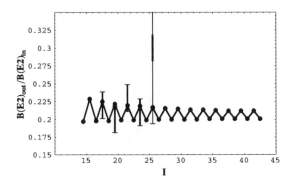

Figure 2: The transition rate among the TSD1 and TSD2 bands as a function of I. The experimental values are shown by error bars, and the theoretical values by circles connected by solid lines.

rigid body but have the nature of the superfluidity. The moments of inertia has the parameter of C through the formula of $\mathcal{J}_i^{irr} = C\beta^2 \sin^2(\gamma - \frac{2\pi}{3}i)$ for $i = x, y, z$. The parameter β is included in the parameter C. The band head energies of TSD1 and TSD2 are adjusted to the experimental energies. Then, parameters are γ, C and two band head energies. Three moments of inertia becomes $\mathcal{J}_x^{irr} = 118.98$, $\mathcal{J}_y^{irr} = 53.15$ and $\mathcal{J}_z^{irr} = 13.09$ in unit of $(\text{MeV})^{-1}$. We show the numerical results of $E^* - AI(I+1)$ as a function of I in Fig. 1. In the figure E^* is the excited energy measured from the ground state. The experimental values [1] are shown by smaller circles connected by dashed lines, and the theoretical values by larger circles connected by solid lines. The agreement is quite good. The energy difference between TSD1 and TSD2 is given by

$$\bar{E}_{I\Omega=-1/2\kappa=2} - \bar{E}_{I\Omega=1/2\kappa=0} = 4\sqrt{\zeta^2 - \eta^2} = 4\theta, \qquad (22)$$

which is positive and independent from I. The experimental energy difference between TSD1 and TSD2 shows the gradual decreace with increasing I.

Next we consider of the E2 transitions among TSD1 and TSD2 bands. The E2 transition operator in the space-fixed frame is defined as $M(E2\mu) \propto Q_0 D_{\mu 0}^2 + Q_2(D_{\mu 2}^2 + D_{\mu -2}^2)$ Then, by using the wave function given by Eq. (16), the transitions within the same bands for $I \rightarrow I - 2$, i.e. $B(E2)_{in}$, and the transition between the different bands for $I \rightarrow I - 1$, i.e. $B(E2)_{out}$, are given by the following equations,

$$B(E2)_{in}^{TSD1} \propto Q_0^2 (I\frac{1}{2}2\,0\,|\,I-2\frac{1}{2})^2,$$

$$B(E2)_{in}^{TSD2} \propto Q_0^2 (I\frac{3}{2}2\,0\,|\,I-2\frac{3}{2})^2,$$

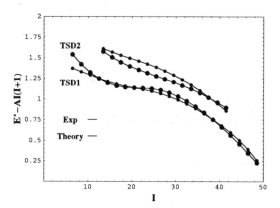

Figure 3: The results by the exact diagonalization of $H_{rot} + H_{sp}$. The longitudinal axes is for the excited energy E^* $AI(I+1)$, and the horizontal axes is for I. The experimental values are shown by smaller circles connected by dashed lines, and the theoretical values by larger circles connected by solid lines.

$$B(E2)_{out}^{TSD1 \to TSD2} \propto Q_2^2 (I - \frac{1}{2} \, 2 \, 2| \, I - 1 \, \frac{3}{2})^2,$$

$$B(E2)_{out}^{TSD2 \to TSD1} \propto Q_2^2 (I \, \frac{3}{2} \, 2 \, -2| \, I - 1 \, - \frac{1}{2})^2. \tag{23}$$

At the limit of large I, the relative ratio of the transitions converges to the constant.

$$B(E2)_{out}^{TSD1 \to TSD2} / B(E2)_{in}^{TSD1} \longrightarrow \frac{2}{3} Q_2/Q_0,$$

$$B(E2)_{out}^{TSD2 \to TSD1} / B(E2)_{in}^{TSD2} \longrightarrow \frac{2}{3} Q_2/Q_0. \tag{24}$$

On the other hand, Ref. 2 gives the following formula.

$$\frac{B(E2)_{out}}{B(E2)_{in}} = (\frac{\sqrt{3}Q_0\eta_+ - \sqrt{2}Q_2\eta_-}{Q_2})^2 \frac{n}{I}, \tag{25}$$

for $n \to n - 1$ transition, and

$$\frac{B(E2)_{out}}{B(E2)_{in}} = (\frac{\sqrt{3}Q_0\eta_- - \sqrt{2}Q_2\eta_+}{Q_2})^2 \frac{n + 1}{I}, \tag{26}$$

for $n \to n + 1$ transition. Both decreases with increasing I.

In Fig. 2, we compare our theoretical results with experimental data [1]. The theoretical values are shown by circles connected by solid lines, and the experimental

Table 1: The magnitudes of the amplitudes in the eigenfunction for I.

			$TSD1$		
I	$[\Omega, K]$	$[\Omega, K]$	$[\Omega, K]$	$[\Omega, K]$	$[\Omega, K]$
	$[1/2, 1/2]$	$[1/2, 3/2]$	$[3/2, 1/2]$	$[3/2, 3/2]$	$[5/2, 1/2]$
25/2	0.750	-0.298	-0.392	0.340	-0.234
41/2	0.791	-0.181	-0.372	0.334	-0.255
61/2	0.755	0.140	-0.225	0.238	-0.227
81/2	0.672	0.305	-0.077	0.211	-0.167
			$TSD2$		
	$[1/2, 1/2]$	$[1/2, 3/2]$	$[3/2, 1/2]$	$[3/2, 3/2]$	$[1/2, 5/2]$
27/2	0.331	0.581	0.281	0.484	-0.383
43/2	0.311	0.590	0.331	0.482	-0.355
63/2	0.313	0.572	0.394	0.492	-0.239
83/2	0.304	0.514	0.424	0.486	-0.115

values are by rectangles with error bars. The agreement is quite good within error bars.

Now, we consider of the M1 transitions. The transition operator in the space-fixed frame $M_{1\mu}$ is related with that in the body fixed frame $M'_{1\nu}$ through $M_{1\mu} = \sum_\nu M'_{1\nu} D^1_{\mu\nu}$, where

$$M'_{1\nu} = \sqrt{\frac{3}{4\pi}} \mu_N [g_R(\vec{I} - \vec{j})_\nu + g_s \vec{s} + g_\ell \vec{\ell})_\nu].$$ (27)

We see M1 transitions from TSD2 → TSD1 are forbidden in the approximate wave function given by Eq. (16), because of the selection rule by both K and Ω. However, if the states with different κ values are mixed, $B(M1)_{out}$ becomes finite. For example if the band TSD2 has small mixture of the component $K = \frac{1}{2}, \Omega = \frac{1}{2}$ ($\kappa = 0$), then there is the M1 transition from TSD2 to TSD1. In this case we need another mixing parameter. Without mixing parameter, the value of $B(M1)^{TSD2 \to TSD1}_{out}/B(E2)_{in} \sim 0.185$, while the experimental data gives 0.005. Thus, the mixing parameter may be around 0.164.

Now, we give a result from the exact diagonalization of the total Hamiltonian, $H_{rot} + H_{sp}$. The parameters are γ, C, V and the attenuation parameter. In this paper, we choose $\gamma = 12°$, $V = 6$ MeV and attenuation parameter as 0.85. The three moments of inertia are $\mathcal{J}^{irr}_x = 70.15$, $\mathcal{J}^{irr}_y = 42.82$ and $\mathcal{J}^{irr}_z = 3.35$ in unit of $(\text{MeV})^{-1}$. We show the numerical results of $E^* - AI(I+1)$ as a function of I in Fig. 3. In this case we have not adjusted the band head energies to the experimental data. Comparing with Fig. 1, the agreement with the experimental values is not good, but still the numerical results simulate the experimental data quite well. Now we see the relative magnitudes of the components of the wave function, which is an eigenfunction of the energy at fixed I. As is seen in the table, The most dominant component in TSD1 is $\Omega = 1/2$ and $K = 1/2$ ($\kappa = 0$), and the most dominant component in TSD2 is $\Omega = 1/2$ and $K = -3/2$ ($\kappa = 2$). This agrees with our assumption in the approximate calculation shown in Figs. 1 and 2. As for $B(M1)_{out}/B(E2)_{in}$ values, the components of $[1/2, 1/2]$ in $I = 27/2$ is 0.331 and $[1/2, 1/2]$ in $I = 25/2$ is 0.750. On the other hand $[1/2,-3/2]$ in $I = 27/2$ is 0.58.

Thus the ratio of $(0.33/0.58)^2 \sim 0.32$ is comparable with the ratio of $(0.005/0.185) \sim 0.27$. The preliminary calculation of $B(M1)_{out}/B(E2)_{in}$ values between $I = 15/2$ and $I = 13/2$ gives 0.0045, which agrees with the experimental data.

4 Conclusion

We derived an algebraic expression for the triaxially deformed rotor plus single-particle model in odd mass nucleus by applying the Holstein-Primakoff transformation to the angular momenta $\vec{I}$ and $\vec{j}$. We have chosen the quantization axis as z axis, which confirms to agree with the axially symmetric limit. Our formula is free from the divergence at the axially symmetric limit, and applicable for the small γ case.

We have carefully paid attension to the D_2-symmetry together with the Bohr symmetry. Then the numerical analysis reproduces both the excitation energies and B(E2) ratios among the triaxially superdeformed bands in ^{163}Lu very well. The moments of inertia derived from the irrotational flow is more favourable than that from the rigid rotor.

In order to describe $B(M1)_{out}/B(E2)_{in}$ we need more precise wave function. We are now carrying the exact more precise wave function. We are now carrying the exact diagonalization of the total Hamiltonian numerically. Our preliminary calculation of the exact diagonalization gives quite reasonable fit to the experimental data in the excited energy difference and $B(M1)_{out}/B(E2)_{in}$. Moreover, it confirms our assumption, i.e. the dominant component in TSD1 is $\Omega = 1/2$ and $K = 1/2$ ($\kappa = 0$), while the dominant component in TSD2 is $\Omega = 1/2$ and $K = -3/2$ ($\kappa = 2$).

References

1. S. W. Ødegard et al, *Nucl. Phys.* A **682**, 427C (2001): PRL **86**, 5866 (2001).
2. A. Bohr and B. R. Mottelson, *Nuclear Structure* **vol. II** (Benjamin, 1975).
3. K. Tanabe and K. Sugawara-Tanabe, *Phys. Lett.* B **34**, 575 (1971): K. Sugawara-Tanabe and K. Tanabe, *Nucl. Phys.* A **208**, 317 (1973).
4. A. Bohr, *Mat. Fys. Medd. Dan. Vid. Selsk.* **26**, no.14 (1952): A. Bohr and B. R. Mottelson, *Mat. Fys. Medd. Dan. Vid. Selsk.* **27**, no.16 (1953).
5. L. D. Landau and E. M. Lifshitz, *Mechanics*, **vol.1** (Pergamon Press, 1960).

ON DESCRIPTION OF PHOTONUCLEAR REACTIONS ACCOMPANIED BY EXCITATION OF THE GIANT DIPOLE RESONANCE

V.A. RODIN AND M.H. URIN

Moscow State Engineering and Physics Institute,
115409 Moscow, Russia

A semimicroscopical approach is applied to describe the photoabsorption and partial photonucleon reactions accompanied by excitation of the giant dipole resonance (GDR). The approach is based on both continuum-RPA (CRPA) and a phenomenological description for doorway-state coupling to many-quasiparticle configurations. Apart from a phenomenological mean field, the separable isovector momentum-dependent forces and momentum-independent Landau-Migdal particle-hole interaction are used as the input quantities for CRPA calculations. The photoabsorption and partial (n, γ)-reaction cross sections in the vicinity of the GDR are satisfactorily described for ^{89}Y, ^{140}Ce and ^{208}Pb target nuclei.

1 Introduction

Excitations and decays of giant resonances (GRs) in medium-heavy mass nuclei have been the subjects of intensive experimental and theoretical studies during many years. The permanent interest to the subjects is quite understandable because the collective, single- and many- quasiparticle aspects of nuclear dynamics are involved in the giant-resonance phenomenon. A rather full theoretical description of GR properties is a serious test for nuclear structure models and needs (semi)microscopical approaches for realization. In particular, the semimicroscopical approach was developed recently to describe properties of a number of charge-exchange and isoscalar GRs (Refs. [1,2] and [3], respectively). The ingredients of the approach are the following: (i) the continuum RPA (CRPA); (ii) the phenomenological isoscalar part (including the spin-orbit term) of the nuclear mean field, the momentum-independent Landau-Migdal particle-hole interaction and some partial self-consistency conditions; (iii) the phenomenological description (in terms of a smearing parameter) for the coupling of collective particle-hole-type doorway states to many-quasiparticle configurations (i.e., for the spreading effect).

However, it was found that the experimental energies of the familiar (isovector) giant dipole resonance (GDR) [4] are systematically underestimated in calculations within the semimicroscopical approach [5]. At the same time, the observed exceeding of the sum rule σ_{GDR}^{int}, exhausted by the GDR in photoabsorption cross section, over the TRK sum rule $\sigma_{TRK}^{int} = 15\,A$ MeV·mb can only be reproduced in selfconsistent calculations provided the momentum-dependent forces are taken into account. An attempt to do that was undertaken in Ref. [6]. Taking the isovector separable momentum-dependent forces into account allowed the authors of Ref. [6] to reproduce the GDR energy for ^{208}Pb (if the experimental exceeding was used to normalize the strength of the forces) as well as to describe satisfactorily the experimental partial ^{208}Pb(n, γ)-reaction cross sections [7].

In the present work we extend the approach of Ref. [6] along the following baselines. First, we use a more simple and much easier in the practical realization way for the phenomenological description of the spreading effect by introducing the energy-dependent smearing parameter directly into the CRPA equations. Second, we take into consideration also the isoscalar part of the momentum-dependent forces in terms of the effective nucleon mass. Third, the photonucleon-reaction cross sections are calculated for a number of nuclei and the results are compared with available experimental data. We make also predictions for several partial (γ, n)-reaction cross sections in the vicinity of the GDR in ^{208}Pb and, therefore, for the partial direct-neutron decay branching ratios for this GR.

2 Calculation scheme and results

The basic CRPA equations for calculations of the photoabsorption cross section σ_a and partial photonucleon-reaction cross section σ_μ within the semimicroscopical approach are given explicitly in Ref. [6]. For brevity sake, we use here the notations of this paper and refer to the equations from the paper.

To calculate directly the energy-averaged cross sections accounting for the spreading effect, we solve the respective CRPA equations (Eqs. (4)-(10) of Ref. [6]) with the replacement of the excitation energy ω by $\omega + \frac{i}{2}I(\omega)$. The smearing parameter $I(\omega)$ (the mean doorway spreading width) is taken in the form of the function having a saturation-like behaviour (Eq. (16a) of Ref. [6]). A reasonable description of the GR total width was obtained in Refs. [1,3,6] making use of the mentioned parametrization.

The choice of the mean field parameters is described in Refs. [3,6]. In addition, the isoscalar mean field amplitude U_0 along with the dimensionless strength $f' = 1.0$ of the isovector Landau-Migdal particle-hole interaction is chosen to reproduce in calculations the experimental nucleon separation energies for the nuclei in question. The dimensionless strength κ' of the isovector (separable) momentum-dependent forces and the intensity $\alpha = 0.06$ MeV^{-1} of the smearing parameter are taken to reproduce the experimental GDR energy and width, respectively. The parameters of U_0 and κ' are listed in Table 1 and vary just slightly for the nuclei in question confirming the stability of the model parameters. The quality of description of the experimental data can be seen from Fig. 1, where the energy-averaged photoabsorption cross section $\bar{\sigma}_a(\omega) = \sigma_a(\omega + \frac{i}{2}I(\omega))$, calculated according to Eq. (1) of Ref. [6], is shown for ^{89}Y, ^{140}Ce and ^{208}Pb target nuclei. It is worth to mention that the use of two adjustable parameters κ' and α has allowed us to describe satisfactorily three GDR parameters: the energy, the total width and the sum rule σ_{GDR}^{int}.

Table 1. Model parameters U_0 and κ' used in calculations.

Nucleus	m^*/m	U_0, MeV	κ'
^{89}Y	1.0	53.3	0.53
	0.9	58.0	0.38
^{140}Ce	1.0	53.7	0.56
	0.9	57.9	0.39
^{208}Pb	1.0	53.9	0.56
	0.9	58.9	0.42

The differential energy-averaged partial cross sections of the neutron radiative capture $d\bar{\sigma}_\mu/d\Omega$ are calculated according to Eqs. (9),(10),(14),(15) of Ref. [6] without introducing any new adjustable parameters. Index μ denotes quantum numbers of product-nucleus single-particle states populated in the capture. Each calculated cross section has been multiplied by the experimental spectroscopic factor S_μ of the final single-particle state, which are listed in Table 2. In distinction to the strength function, the energy-averaged reaction-amplitude $\bar{M}_\mu(\omega) = M_\mu(\omega + \frac{i}{2}I(\omega))$ is determined by not only the effective dipole operator, but also the continuum wave function (see Eq. (10) of Ref. [6]). We calculate the wave function as an optical-model one with the imaginary part of the optical potential being $\frac{1}{2}I(\omega)$ multiplied by Saxon-Woods radial dependence with the radius to be the nuclear one. Such a choice of the continuum wave function makes the model more consistent in the region of the respective single-particle resonance, as it can be seen from a schematical solution of the RPA equations with the replacement ω by $\omega + \frac{i}{2}I(\omega)$. It should be stressed that the optical model in our approach plays an compensative role and is not a realistic one.

The calculated cross sections $d\bar{\sigma}_\mu/d\Omega$ at $90°$ and the anysotropy parameters a_μ are shown in Figs. 2-4 for the nuclei in question in comparison with the respective experimental data. It is noteworthy, that the difference between the cross sections calculated for ^{208}Pb in the vicinity of the GDR by the methods of this work and Ref. [6] does not exceed 10 % provided the parameter α

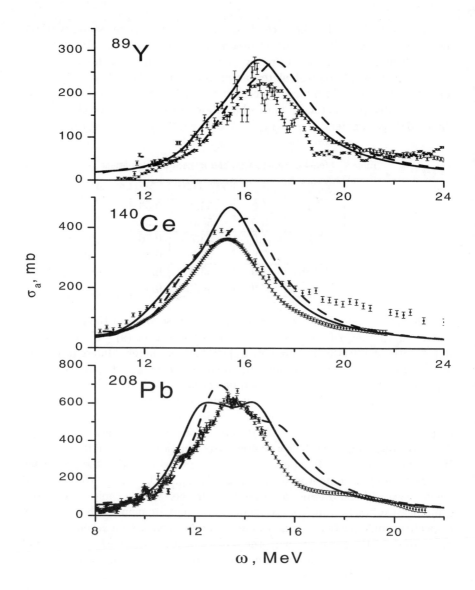

Figure 1. The calculated photoabsorption cross sections $\bar{\sigma}_a$ for ^{89}Y, ^{140}Ce and ^{208}Pb (here-after the solid and dashed lines correspond to calculations with $m^*/m = 1$ and 0.9, respectively). The experimental data (black squares and circles) are taken from Refs. [4].

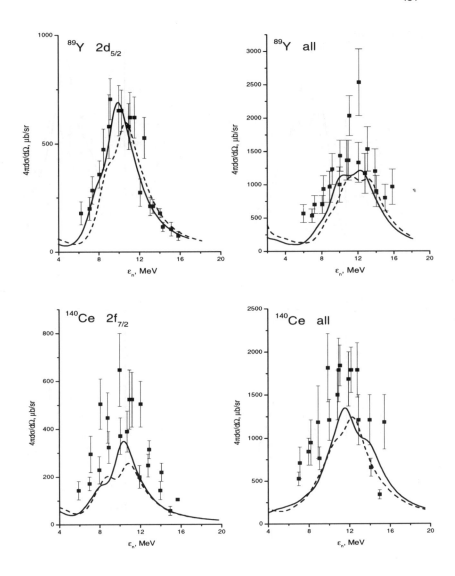

Figure 2. The calculated partial cross sections at 90^0 multiplied by 4π as functions of neutron energy for the neutron radiative capture to the ground state and to the all single-particle states in ^{90}Y and ^{141}Ce. The experimental data are taken from Ref. [8].

482

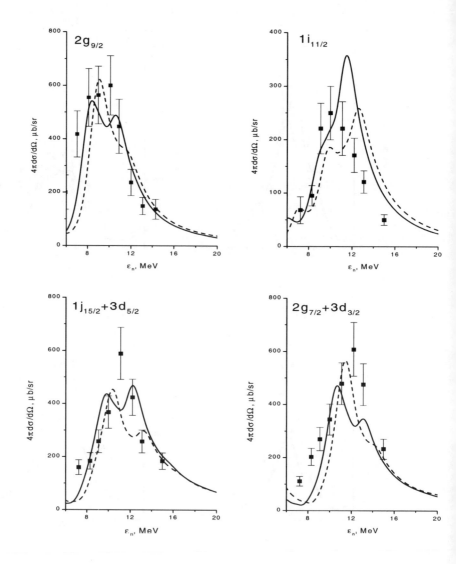

Figure 3. The calculated partial cross sections at 90^0 multiplied by 4π as functions of neutron energy for the neutron radiative capture to some single-particle states in ^{209}Pb. The experimental data are taken from Ref. [7].

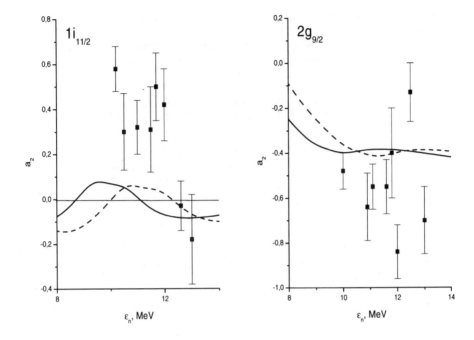

Figure 4. Calculated anisotropy parameter a_2 for some partial ^{208}Pb(n, γ)-reactions. The experimental data are taken from Ref. [7].

is chosen to be the same as in Ref. [6].

We calculate also some partial ^{208}Pb(γ, n)-reaction cross sections $d\bar{\sigma}_\mu/d\Omega$ at 90° with population of the single-hole states in ^{207}Pb (Fig. 5). Cross sections $\bar{\sigma}_\mu(\omega)$ determine the partial direct-neutron-decay branching ratios b_μ for the GDR in ^{208}Pb according to the equation:

$$b_\mu = \int_{GDR} \bar{\sigma}_\mu(\omega)d\omega / \int_{GDR} \bar{\sigma}_a(\omega)d\omega. \qquad (1)$$

The values of b_μ, calculated with the use of the unit spectroscopic factor for the final one-hole states, are listed in Table 3, so that the upper limit for total

Table 2. Experimental spectroscopic factors of the valence-neutron states in ^{90}Y,^{141}Ce and ^{209}Pb

^{90}Y		^{141}Ce		^{209}Pb	
μ	S_μ	μ	S_μ	μ	S_μ
$1g_{7/2}$	0.6	$2f_{5/2}$	0.8	$3d_{3/2}$	0.9
$2d_{3/2}$	0.7	$1h_{9/2}$	1.0	$2g_{7/2}$	0.8
$1h_{11/2}$	0.4	$1i_{13/2}$	0.6	$4s_{1/2}$	0.9
$3s_{1/2}$	1.0	$3p_{1/2}$	0.4	$3d_{5/2}$	0.9
$3d_{5/2}$	1.0	$3p_{3/2}$	0.4	$1j_{15/2}$	0.5
		$2f_{5/2}$	0.8	$1i_{11/2}$	1.0
				$2g_{9/2}$	0.8

branching ratio $b = \sum_\mu b_\mu$ is found to be equal about 14 %.

As applied to the description of the isovector 1^- excitations in the limit $(N - Z)/A \ll 1$, the Galileo-invariant separable momentum-dependent forces can be approximated as follows:

$$\hat{F}_{m-d} = -\frac{1}{4mA} \sum_{1,2} (\kappa + \kappa'(\vec{\tau}_1 \vec{\tau}_2))(\vec{p}_1 - \vec{p}_2)^2 \simeq -\frac{k}{2m} \sum_1 \vec{p}_1^2 + \frac{k'}{2mA} \sum_{1,2} (\vec{\tau}_1 \vec{\tau}_2)(\vec{p}_1 \vec{p}_2)$$

(2)

As in Ref. [6], the above-described calculation results are obtained supposing $\kappa = 0$. According to Eq. (2), nonzero κ leads to the effective mass $\frac{m}{m^*} = 1 - \kappa$ and, therefore, to the change in the sum rule: $\sigma^{int} = (1 - \kappa)(1 + \kappa')\sigma_{TRK}^{int}$. As an example, we use in calculations of the photonuclear reactions with ^{208}Pb a "realistic" value $m^* = 0.9\, m$. Such a choice leads to using different values of the mean field amplitude U_0 and the intensities of κ' (allowing us to get the GDR at the right position), also listed in Table 1. The respective calculation results for the partial cross sections are also shown in Figs. 2-5.

In principle, an appropriate value of m^* could be chosen as giving the best description of the partial cross sections of the neutron radiative capture

Table 3. Calculated branching ratios for direct neutron decay of the GDR in ^{208}Pb.

	m^*/m	Final single-hole states					
		$3p_{\frac{1}{2}}$	$2f_{\frac{5}{2}}$	$3p_{\frac{3}{2}}$	$1i_{\frac{13}{2}}$	$2f_{\frac{7}{2}}$	$1h_{\frac{9}{2}}$
b_μ,	1.0	2.0	4.4	3.4	1.3	2.5	0.7
%	0.9	1.8	3.6	3.1	1.0	1.8	0.4

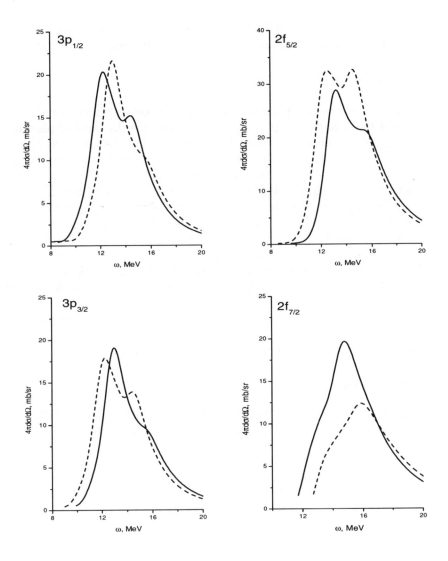

Figure 5. The calculated partial cross sections at 90^0 multiplied by 4π of the (γ, n) reaction with population of some single-hole states in ^{207}Pb as functions of photon energy.

in the vicinity of the GDR. It can be seen from the Figures that the choice $m^* = m$ gives better describtion of the respective experimental data.

3 Conclusion

The obtained calculation results allow us to make several comments on abilities of the semimicroscopical approach to describe the simplest photonuclear reactions accompanied by the GDR excitation. (1) The use of the alternative way for taking the spreading effect into account allows us to simplify the calculations of the energy-averaged reaction cross sections. This way can be only used for description of highly excited giant resonances within the approach. (2) We take also the isoscalar part of the separable momentum-dependent forces into consideration. That results in difference of the nucleon effective mass from the free value. It is found, however, that the use of the "realistic" effective mass does not improve description of the data. (3) We get reasonable description of two sets of the rather old experimental data on the partial (n, γ)-reactions for medium-heavy mass nuclei [7,8]. (4) We make predictions for partial ^{208}Pb(γ, n)-reaction cross sections bearing in mind opportunities being open now at new facilities (e.g., SPring-8) to measure the photonuclear reaction cross sections with high accuracy. Such measurements allow to make electromagnetic probes more efficient to study the nuclear structure.

References

1. E.A. Moukhai, V.A. Rodin, and M.H. Urin, *Phys. Lett.* B **447**, 8 (1999); V.A. Rodin and M.H. Urin, *Nucl. Phys.* A **687**, 276c (2001).
2. M.L. Gorelik, and M.H. Urin, *Phys. Rev.* C **63**, 064312 (2001).
3. M.L. Gorelik, S. Shlomo, and M.H. Urin, *Phys. Rev.* C **62**, 044301 (2000); M.L. Gorelik and M.H. Urin, *Phys. Rev.* C **64**, 047301 (2001).
4. B.L. Berman, S.C. Fultz, *Rev. Mod. Phys.* **47**, 713 (1975); I.N. Boboshin, A.V. Varlamov, V.V. Varlamov, D.S. Rudenko, and M.E. Stepanov, The Centre for Photonuclear Experiment Data, CDFE nuclear data bases, *http://depni.npi.msu.su/cdfe*. INP preprint 99-26/584, Moscow, 1999.
5. G.A. Chekomazov and M.H. Urin, *Phys. Lett.* B **354**, 7 (1995).
6. V.A. Rodin, and M.H. Urin, *Phys. Lett.* B **480**, 45 (2000).
7. I. Bergqvist, D.M. Drake and D.K. McDaniels, *Nucl. Phys.* A **191**, 641 (1972).
8. I.Bergqvist, B.Palson, L.Nilsson, A.Lindholm, D.M.Drake, E.Arthur, D.K.McDaniels and P.Varghese, *Nucl. Phys.* A **295**, 256 (1978).

ISOSCALAR AND ISOVECTOR, SPIN AND ORBITAL CONTRIBUTIONS IN $M1$ TRANSITIONS

Y. FUJITA,^{A} Y. SHIMBARA,^{A} T. ADACHI,^{A} G.P.A. BERG,^{B} H. FUJIMURA,^{B}
H. FUJITA,^{A} I. HAMAMOTO,^{C} K. HARA,^{B} K. HATANAKA,^{B}
F. HOFMANN,^{D} J. KAMIYA,^{B} T. KAWABATA,^{B} Y. KITAMURA,^{B}
P. VON NEUMANN-COSEL,^{D} A. RICHTER,^{D} Y. SHIMIZU,^{B} M. UCHIDA,^{E}
K. YAMASAKI,^{F} M. YOSHIFUKU,^{A} M. YOSOI^E

^{A} Department of Physics, Osaka University, Toyonaka, Osaka 560-0043, Japan
^{B} RCNP, Osaka University, Ibaraki, Osaka 567-0047, Japan
^{C} Division of Mathematical Physics, LTH, University of Lund, P.O. Box 118,
S-22100 Lund, Sweden
^{D} Institut für Kernphysik, TU Darmstadt, D-64289 Darmstadt, Germany
^{E} Department of Physics, Kyoto University, Sakyo, Kyoto 606-8224, Japan
^{F} Department of Physics, Konan University, Higashinada, Kobe 658-8501, Japan

The electromagnetic $M1$ operator contains not only the usually dominant isovector (IV) spin ($\sigma\tau$) term, but also IV orbital ($\ell\tau$), isoscalar (IS) spin (σ) and IS orbital (ℓ) terms. On the other hand the Gamow-Teller (GT) operator contains only the $\sigma\tau$ term. Under the assumption that the isospin is a good quantum number, isobaric analog structure is expected for the nuclei with the same mass number A, and thus analogous transitions are found. For $M1$ transitions in the $T = 1/2$ mirror nuclei pairs ^{23}Na-^{23}Mg and ^{27}Al-^{27}Si, contributions of these terms are studied by comparing the strengths of the $M1$ γ and GT transitions obtained in high-resolution $(^3\mathrm{He},t)$ charge-exchange reactions. For $M1$ transitions in the $T = 0$ even-even nucleus ^{24}Mg, IV orbital and IV spin contributions were studied for the $\Delta T = 1$, IV $M1$ transitions by comparing the $M1$ strengths of (e,e') and GT strengths of $(^3\mathrm{He},t)$ charge-exchange reactions.

1 Introduction

The magnetic dipole operator $\boldsymbol{\mu}$ for the $M1$ γ-transition and the Gamow-Teller (GT) operator for the GT β-decay are similar in the sense that they have the same major components of the isovector (IV) spin ($\sigma\tau$) term, although transitions originate from electromagnetic and weak interactions, respectively.[1,2,3] The difference is that the electromagnetic $M1$ operator contains not only the $\sigma\tau$ term, but also IV orbital ($\ell\tau$), and isoscalar (IS) (σ and ℓ) terms.

Assuming that isospin is a good quantum number, isospin multiplet states are found in nuclei with the same mass A but different isospin z component $T_z = (N - Z)/2$. Because of the analogous nature of multiplet states, $M1$ and GT transitions from the ground state (or its analog state) to an excited state (or its analog state) are all analogous. Analogous $M1$ and GT transitions are expected to have corresponding energies and strengths. Through the study of the similarity and/or difference of strengths of analogous transitions, we can study orbital and spin contributions for individual $M1$ transitions in the $T = 1/2$ nuclei ^{27}Al and ^{23}Na and in the $T = 0$ nucleus ^{24}Mg.

The $M1$ transition strengths $B(M1)$ are obtained from the study of γ-decay and also from backward-angle (e, e') measurements. The most direct values of $B(GT)$ are obtained from studies of GT β-decay, but they are Q-value limited. On the other hand, CE reactions, like (p, n) or $(^3\text{He}, t)$, performed at intermediate energies (> 100 MeV/nucleon) can be used as a means to map the GT strengths over a wide range of excitation energies.[4] For this purpose, one relies on the approximate proportionality between the reaction cross sections measured at $\theta = 0°$ and $B(GT)$ values. We found the $(^3\text{He}, t)$ reaction to be very suitable for this purpose because of its high-resolution capability by using magnetic spectrometers.

2 Characteristics of $M1$ and GT Transitions

2.1 Analogous $M1$ and GT Transitions

After applying the Wigner-Eckart theorem in the isospin space, the reduced $M1$ transition strength $B(M1)$ is expressed as [5,6]

$$B(M1) = \frac{1}{2J_i+1} \frac{3}{4\pi} \mu_N^2 \left[\left(g_\ell^{\text{IS}} M_{M1}(\ell) + \tfrac{1}{2} g_s^{\text{IS}} M_{M1}(\sigma) \right) \right.$$
$$\left. - \frac{C_{M1}}{\sqrt{2T_f+1}} \left(g_\ell^{\text{IV}} M_{M1}(\ell\tau) + \tfrac{1}{2} g_s^{\text{IV}} M_{M1}(\sigma\tau) \right) \right]^2 \qquad (1)$$

$$= \frac{1}{2J_i+1} \frac{3}{4\pi} \mu_N^2 \left[M_{M1}^{\text{IS}} - \frac{C_{M1}}{\sqrt{2T_f+1}} M_{M1}^{\text{IV}} \right]^2 , \qquad (2)$$

with μ_N the nuclear magneton, and C_{M1} the isospin Clebsch-Gordan (CG) coefficient $(T_i T_{zi} 10 | T_f T_{zf})$, where $T_{zf} = T_{zi}$ holds. The matrix elements are $M_{M1}(\ell) = \langle J_f T_f ||| \sum_{j=1}^{A} \ell_j ||| J_i T_i \rangle$ and $M_{M1}(\sigma) = \langle J_f T_f ||| \sum_{j=1}^{A} \sigma_j ||| J_i T_i \rangle$ for the IS part, and $M_{M1}(\ell\tau) = \langle J_f T_f ||| \sum_{j=1}^{A} \ell_j \tau_j ||| J_i T_i \rangle$ and $M_{M1}(\sigma\tau) = \langle J_f T_f ||| \sum_{j=1}^{A} \sigma_j \tau_j ||| J_i T_i \rangle$ for the IV part. The coefficients for these matrix elements are the IS and IV combinations of gyromagnetic factors (g factors): $g_\ell^{\text{IS}} = \tfrac{1}{2}(g_\ell^\pi + g_\ell^\nu)$, $g_s^{\text{IS}} = \tfrac{1}{2}(g_s^\pi + g_s^\nu)$, $g_\ell^{\text{IV}} = \tfrac{1}{2}(g_\ell^\pi - g_\ell^\nu)$ and $g_s^{\text{IV}} = \tfrac{1}{2}(g_s^\pi - g_s^\nu)$. For g factors of bare protons and neutrons, we get $g_\ell^{\text{IS}} = 0.5$, $g_s^{\text{IS}} = 0.880$, $g_\ell^{\text{IV}} = 0.5$ and $g_s^{\text{IV}} = 4.706$, respectively. Because of the large value of the coefficient g_s^{IV}, the term $(1/2)g_s^{\text{IV}} M_{M1}(\sigma\tau)$ is usually the largest.[7] The IS term M_{M1}^{IS} is much smaller than the IV term M_{M1}^{IV}. The IS term interferes destructively or constructively with the IV term. In addition, the

IV orbital term interferes with the IV spin term.

A GT transition is caused by the $\sigma\tau$ operator. Its reduced strength in isospin is given by [6]

$$B(\text{GT}) = \frac{1}{2J_i+1} \frac{1}{2} \frac{C_{\text{GT}}^2}{2T_f+1} \left[M_{\text{GT}}(\sigma\tau) \right]^2, \tag{3}$$

with $M_{\text{GT}}(\sigma\tau)$ the (IV and spin type) GT transition matrix element and C_{GT} the isospin CG coefficient $(T_i T_{zi} 1 \pm 1 | T_f T_{zf})$, where $T_{zf} = T_{zi} \pm 1$.

The difference of the so-called meson-exchange currents (MEC) contributions to the $M1$ and GT operators has been studied theoretically [8,9] and experimentally.[10,11,12,13,14,15] These effects are expressed by the ratio $R_{\text{MEC}} = [M_{M1}(\sigma\tau)]^2 / [M_{\text{GT}}(\sigma\tau)]^2$. The most probable value of 1.25 is deduced for the nuclei in middle of sd shell.[6]

Let us compare the analogous $M1$ and GT transitions. If the IV spin term in the $M1$ transition is dominant, a similar transition strengths are expected for corresponding $M1$ and GT transitions. The "quasi" proportionality between $B(\text{GT})$ and $B(M1)$ is expressed by

$$B(M1) \approx \frac{3}{8\pi}g_s^{\text{IV}})^2 \mu_N^2 \frac{C_{M1}^2}{C_{\text{GT}}^2} R_{\text{MEC}} B(\text{GT}) = 2.644 \mu_N^2 \frac{C_{M1}^2}{C_{\text{GT}}^2} R_{\text{MEC}} B(\text{GT}) . \tag{4}$$

In order to compare the $M1$ and the GT strengths directly, the different coupling constants should be renormalized. From Eq. (4), we find that a renormalized $B(M1)$ defined by

$$B^R(M1) = \frac{1}{2.644 \mu_N^2} \frac{C_{\text{GT}}^2}{C_{M1}^2} B(M1) , \tag{5}$$

can be compared directly with the values of $B(\text{GT})$.

2.2 IS and Orbital Contributions to $M1$ Transitions

In order to see the interference of IS and IV orbital terms with the IV spin term in a $M1$ transition, we define the ratio [15]

$$R_{\text{ISO}} = \frac{1}{R_{\text{MEC}}} \frac{B^R(M1)}{B(\text{GT})} . \tag{6}$$

By comparing Eq. (1) and Eq. (3), it can be seen that $R_{\text{ISO}} > 1$ and < 1 show the constructive and destructive contributions, respectively. The contribution of the IS term is usually minor.[6]

If the $M1$ transition is from the ground state of a $T = 0$ nucleus to a $T = 1$ excited state, the transition is of pure IV nature. In this case the ratio given by Eq. (6) shows only the contribution of the IV orbital term. We then call the ratio R_{OC}.

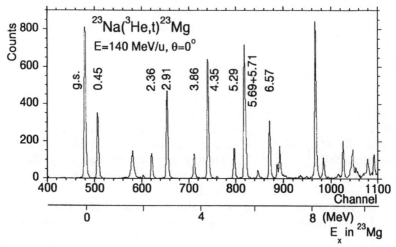

Figure 1. High-resolution ^{23}Na(^{3}He, t) spectrum measured at 0°. Major states with $L = 0$ character are indicated by their excitation energies.

3 Experiment

In intermediate-energy CE reactions, such as (p, n) or $(^3$He, $t)$, the GT states become prominent at forward angles near $\theta = 0°$ because of their $L = 0$ nature and the dominance of the IV spin part of the effective interaction $V_{\sigma\tau}$ at small momentum transfer q.[16] It was found that the cross sections at 0° are proportional to the B(GT) values, if the transitions are not too weak.[4] The proportionality is given by [4,17,18]

$$\frac{d\sigma^{\mathrm{CE}}}{d\Omega}(0°) \simeq K^{\mathrm{CE}} N^{\mathrm{CE}}_{\sigma\tau} |J_{\sigma\tau}(0)|^2 B(\mathrm{GT}), \qquad (7)$$

where $J_{\sigma\tau}(0)$ is the volume integral of $V_{\sigma\tau}$ at $q = 0$, K^{CE} the kinematic factor and $N^{\mathrm{CE}}_{\sigma\tau}$ the distortion factor.

The limited energy resolution in (p, n) reactions ($\Delta E \geq 200 - 400$ keV) could be dramatically improved by using the $(^3$He,$t)$ reaction.[19] The momenta of the outgoing tritons are analyzed by a high-resolution magnetic spectrometer. At the QQDD-type Grand Raiden spectrometer [20] at RCNP, Osaka, up to 150 MeV/nucleon tritons can be analyzed. At this beam energy it has been shown for mirror GT transitions in ^{27}Al(^{3}He, $t)^{27}$Si and ^{27}Si → ^{27}Al β decay,[15] that the proportionality given by Eq. (7) is valid if the B(GT) values are larger than 0.04.

The (^{3}He, t) experiments were performed by using 140-150 MeV/nucleon ^{3}He beams from RCNP Ring Cyclotron. The ^{3}He^{2+} beam with 5 nA was stopped in a Faraday cup inside the first dipole magnet of the spectrometer set at 0°. The ejectile tritons were momentum analyzed and detected at the focal plane by a multi-wire drift-chamber system capable of determining horizontal and vertical (x and y) positions and angles of each ray.[21] Track reconstruction made it possible to subdivide the acceptance angle of the spectrometer.

Figure 1 shows the spectrum obtained around $\theta = 0°$ in the ^{23}Na(^{3}He, t) experiment. Precision *dispersion matching* and *angular dispersion matching* were realized by using the newly commissioned WS beam line.[22,23] Owing to the development of a new "faint beam method" for diagnosing the matching conditions,[24,25] a resolution of 50 keV (FWHM) can be realized routinely. A good angular resolution in y direction was achieved by applying the over-focus mode for the spectrometer.[26] The ^{27}Al(^{3}He, t)^{27}Si spectrum is shown in Ref. [15].

4   Comparison of Analogous GT and $M1$ Transitions

4.1   GT and $M1$ Transition Strengths

In order to obtain B(GT) values by using Eq. (7), a standard B(GT) value is needed. In a mirror-nuclei pair, very similar B(GT) values are expected for the corresponding GT transitions studied in a (^{3}He, t) reaction on the ground state of $T_z = +1/2$ nucleus and the β^+ decay from the ground state of $T_z = -1/2$ nucleus. The B(GT) value obtained from the β^+ decay of ^{23}Mg to the 0.440 MeV state of ^{23}Na was used as calibration standard for the B(GT) values in the ^{23}Na(^{3}He, t)^{23}Mg experiment.

The $M1$ γ-transition strength $B(M1)$ (in units of μ_N^2) from an excited state to the ground state of ^{23}Na is calculated (see *e.g.* Ref. [3]) using the measured lifetime (mean life) τ_m, gamma-ray branching ratio b_γ (in %) to the ground state, $M1$ and $E2$ mixing ratio δ and the γ-ray energy E_γ. By using the data compiled in Ref. [27], the $B(M1)$ values were obtained for states up to $E_x = 6$ MeV.

In order to determine the $B(M1)\uparrow$ value, which would be obtained in an (e, e')-type transition from the ground state with the spin-value $J_{\text{g.s.}}$ to the jth excited state with the spin-value J_j, the $B(M1)$ value obtained in the γ decay are modified by the $2J + 1$ factor of the jth state and the ground state. From the obtained $B(M1)\uparrow$ values, the $B^R(M1)$ values were calculated by using Eq. (5) to directly compare them with B(GT) values from the CE reactions.

4.2 Combined IS and Orbital Contributions

The R_{ISO} values calculated by using Eq. (6) from the $B(M1)$ and $B(\mathrm{GT})$ values of the analogous $M1$ and GT transitions are shown in Fig. 2 for the $^{27}\mathrm{Al}$-$^{27}\mathrm{Si}$ mirror nuclei as a function of $B(M1)\uparrow$. It is seen that the R_{ISO} value tends to deviate from unity by more than a factor of two when the $B(M1)\uparrow$ is less than approximately 0.1. This shows that the combined IS-orbital contribution is rather large in weaker transitions and the "quasi" proportionality between $B(M1)$ values and the analogous $B(\mathrm{GT})$ values is lost. This finding is interpreted as follows. Since the IS and the IV orbital terms are usually small, the dominance of the IV spin term of the $M1$ operator is expected if the transitions are at least of average strength. The IV spin term, however, can also be small. Then the relative contribution of the IS and the IV orbital terms becomes significant although the transition itself is weak.

In the $^{23}\mathrm{Na}$-$^{23}\mathrm{Mg}$ mirror nuclei, corresponding $M1$ and GT transitions were identified for six pairs of excited states. The calculated R_{ISO} values assuming a R_{MEC} value of 1.25 [6] are again plotted as a function of $B(M1)\uparrow$ of the $M1$ γ-transition in $^{23}\mathrm{Na}$ (see Fig. 3).

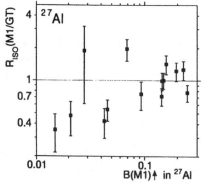
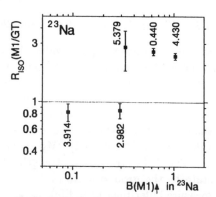

Figure 2. The ratio R_{ISO} for the $M1$ transitions in $^{27}\mathrm{Al}$. Values of $R_{\mathrm{ISO}} > 1(<1)$ show constructive (destructive) interference of these terms with the IV spin term.

Figure 3. The ratio R_{ISO} for the $M1$ transitions in $^{23}\mathrm{Na}$. The excitation energies of the states in $^{23}\mathrm{Na}$ are given in unit of MeV.

Compared to the $^{27}\mathrm{Al}$-$^{27}\mathrm{Si}$ system, we notice three major differences for the $^{23}\mathrm{Na}$-$^{23}\mathrm{Mg}$ system: 1) $B(M1)\uparrow$ values are much larger than those in $^{27}\mathrm{Al}$, 2) R_{ISO} values are large even for some of the strong transitions, 3) there exists two groups having $R_{\mathrm{ISO}} \sim 1$ and $R_{\mathrm{ISO}} \sim 3$. These observations will be interpreted in the following.

5   Discussion Based on the Particle-Rotor Model

The static quadrupole moment of ^{23}Na is $Q_0 = +10.1$ fm^2,[28] which suggests that it has a large prolate deformation with $\delta \approx 0.5$. It has been discussed that the orbital contribution in the $M1$ transition becomes large in deformed nuclei.[29,30] The contribution, however, is largely dependent on the configurations involved in the transitions.

5.1   Spin and Orbital Contributions for $M1$ Transitions between Rotational Bands

Let us consider an odd A deformed nucleus with the even-even core. We assume that a band is formed on a pure Nilsson orbit, which is identified by using asymptotic quantum numbers $[Nn_z\Lambda\Omega]$,[31] where, by using the spin projection along the z axis $\Sigma(= s_z)$, Ω is expressed as $\Omega = \Lambda + \Sigma$. On top of the single-particle orbits, rotational bands with spin values J are formed.

The $M1$ transitions in deformed nuclei are categorized into three different groups: 1) intra-band transitions, 2) inter-band transitions with $\Delta\Lambda = 1$, and 3) inter-band transitions with $\Delta\Sigma = 1$. The contributions of spin and orbital operators are different in these three cases. Based on selection rules for the transitions between members of various rotational bands,[32] we get the following results. 1) For the intra-band transitions, both spin and orbital operators can contribute. 2) In the transition in which Λ changes, only orbital operator can contribute. 3) In the transition in which Σ changes, only spin operator can contribute.

5.2   Interpretation of the Experimental Results

The low-lying states in ^{23}Na are found to form rotational bands, and the structure is accounted for in particle-rotor calculations including the effects of pair correlations.[33] It is suggested that the ground state is the band head of the $[2\,1\,1\,3/2]$ band. Starting from the $J^\pi = 3/2^+$ ground state, transitions to the $J^\pi = 1/2^+, 3/2^+$, and $5/2^+$ states are allowed.

The transition from the ground state to the 0.44 MeV, $3/2^+$ state is an intra-band transition. Since the origin of the ground state band $[2\,1\,1\,3/2]$ is $d_{5/2}$, a constructive interference is expected between the spin and orbital contributions for this transition. This explains the relatively large R_{OC} value of this transition (see Fig. 3).

The 4.43 and 5.38 MeV states are assigned to be members of $[2\,2\,0\,1/2]$ band. In the transitions to these states, the asymptotic quantum number Λ decreases (and n_z increases) by one unit. In principle, such transitions are caused by the orbital (ℓ_-) operator. The R_{ISO} values are large for these transitions as shown in Fig. 3.

The $J = 1/2, 3/2$, and $5/2$ members of the $[2\,1\,1\,1/2]$ band are assigned to be the states at 2.39, 2.98, and 3.91 MeV, respectively.[33] In principle, the transitions to these states are caused only by the spin (σ) operator. For the transitions to the 3/2 and 5/2 states, the obtained R_{ISO} values are nearly unity, in agreement with the expectation that the transitions are caused by the σ operator.

5.3 Orbital Contributions to M1 Transitions in T = 0 Even-Even Nuclei

The $B(M1)$ derived from ^{24}Mg(e, e') measured at S-DALINAC and $B(\text{GT})$ from ^{24}Mg$(^3\text{He}, t)^{24}$Al have been compared for the analogous transitions.[34] Since the GT states $(J^\pi = 1^+)$ in ^{24}Al have $T = 1$, the corresponding $M1$ states in ^{24}Mg are also of $T = 1$ nature. Due to the IV nature of the transitions to these $T = 1$ states, the R_{OC} values without the IS influence are deduced for the states in the $E_x = 10 - 14$ MeV region. The R_{OC} values, shown as a function of excitation energy in Fig. 4, indicate that the orbital contribution is large and constructive $(R_{\text{OC}} > 1)$ in the lowest excited state, destructive in the 12.8 MeV state, and small $(R_{\text{OC}} \sim 1)$ in the higher excited state. A similar E_x dependence was also observed for the $M1$ excitations in ^{28}Si.

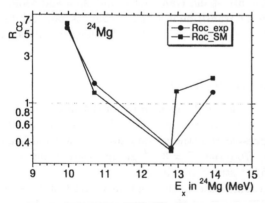

Figure 4. The ratio R_{OC} as a function of E_x for each $T = 1$, $M1$ state in ^{24}Mg. The experimental values (circles) are compared with those from shell model calculations using the USD interaction (squares).[35] Values larger (smaller) than unity show constructive (destructive) interference. The lines are drawn to guide the eye.

In a naive shell model picture, it is suggested that the lower energy states are excited by $d_{5/2} \to d_{5/2}$ type transitions, while the higher energy states by $d_{5/2} \to d_{3/2}$ type transitions. For the former transitions, the orbital and spin operators work constructively, which makes the R_{OC} value large. In addition, for some transitions of this type, the orbital contribution can be favored, because of the selection rules for the deformed nucleus ^{24}Mg. For the $d_{5/2} \to d_{3/2}$ type transitions, however, mainly the spin operator is active, which makes the R_{OC} value near unity.

6 SUMMARY

A GT transition has the nature of pure IV spin transition, while an $M1$ transition is caused by the IS and IV orbital operators as well. By comparing the strengths of analogous $M1$ and GT transitions, which are caused by the electro-magnetic and the weak interactions, respectively, the contributions from IS and/or IV orbital operators in the $M1$ transition can be studied in terms of the ratios R_{ISO} and R_{OC}. In order to make such a comparison in a wider excitation energy range, we used $(^3He, t)$ reactions in which the cross sections at $0°$ and at intermediate incident energies were proportional to the $B(GT)$ values. The $B(M1)$ values are obtained from the studies of γ decay and (e, e') reaction. In each reaction, a good resolution was important to make a transition-by-transition comparison.

For the $M1$ transitions in ^{27}Al, it was observed that the R_{ISO} values tend to deviate from unity in weaker transitions, suggesting larger contributions from IS and IV orbital terms are found only in weaker $M1$ transitions. On the other hand for the $M1$ transitions in ^{23}Na, large R_{ISO} values were observed even for stronger transitions. It was found that the observation can be interpreted by the large deformation of $A = 23$ nuclei and the selection rules for the transitions between bands with different asymptotic quantum numbers. The R_{OC} values obtained for $M1$ transitions in ^{24}Mg showed a characteristic dependence on excitation energy.

The $^{23}Na(^3He, t)^{23}Mg$ experiment was performed at RCNP, Osaka University under the Experimental Program E158. The authors are grateful to the accelerator group of RCNP for providing a high-quality ^3He beam.

References

1. E. K. Warburton and J. Weneser, *Isospin in Nuclear Physics*, ed. D. H. Wilkinson, (North-Holland, Amsterdam, 1969) Chap. 5.
2. S. S. Hanna, *Isospin in Nuclear Physics*, ed. D. H. Wilkinson, (North-Holland, Amsterdam, 1969) Chap. 12, and references therein.
3. H. Morinaga and T. Yamazaki, *In Beam Gamma-Ray Spectroscopy* (North-Holland, Amsterdam, 1976), and references therein.
4. T. N. Taddeucci et al., Nucl. Phys. **A469**, 125 (1987).
5. L. Zamick and D. C. Zheng, Phys. Rev. C **37**, 1675 (1988).

496

6. Y. Fujita *et al.*, Phys. Rev. C 62, 044314 (2000), and refs. therein.
7. *Isospin in Nuclear Physics*, Chap. 5 and 12, ed. D.H. Wilkinson, (North-Holland, Amsterdam, 1969), and refs. therein.
8. A. Arima, K. Shimizu, and W. Bentz, *Advances in Nuclear Physics*, ed. J. W. Negele and E. Vogt, (Plenum, New York, 1987), Vol.18, p. 1, and references therein.
9. I. S. Towner, Phys. Rep. 155, 263 (1987).
10. A. Richter et al., Phys. Rev. Lett. 65, 2519 (1990).
11. C. Lüttge et al., Phys. Rev. C 53, 127 (1996).
12. P. von Neumann-Cosel et al., Phys. Rev. C 55, 532 (1997).
13. Y. Fujita et al., Phys. Rev. C 55, 1137 (1997).
14. Y. Fujita et al., Nucl. Instrum. Meth. Phys. Res. A 402, 371 (1998).
15. Y. Fujita et al., Phys. Rev. C 59, 90 (1999).
16. W. G. Love and M. A. Franey, Phys. Rev. C 24, 1073 (1981).
17. C.D. Goodman et al., Phys. Rev. Lett. 44, 1755 (1980).
18. W.G. Love, K. Nakayama and M.A. Franey, Phys. Rev. Lett. 59, 1401 (1987).
19. Y. Fujita *et al.*, Nucl. Phys. A687, 311c (2001).
20. M. Fujiwara et al., Nucl. Instr. Meth. Phys. Res. A 422, 484 (1999).
21. T. Noro et al., RCNP Annual Report 1991, RCNP, Osaka Univ., p. 177.
22. Y. Fujita et al., Nucl. Instr. Meth. Phys. Res. B 126, 274 (1997); and references therein.
23. T. Wakasa et al., RCNP Annual Report 1999, RCNP, Osaka University, p. 95; Nucl. Instr. Meth. Phys. Res. A, to be published.
24. H. Fujita et al., RCNP Annual Report 1999, RCNP, Osaka University, p. 87; Nucl. Instr. Meth. Phys. Res. A, to be published.
25. Y. Fujita et al., J. Mass Spectrom. Soc. Jpn. 48(5), 306 (2000).
26. H. Fujita et al., Nucl. Instr. Meth. Phys. Res. A, A 469, 55 (2001).
27. P. M. Endt, Nucl. Phys. A521, 1 (1990), and references therein.
28. P. Raghavan, At. Data Nucl. Data Tables 42, 189 (1989).
29. D. Bes and R. Broglia Phys. Lett. B 137, 141 (1984).
30. I. Hamamoto and W. Nazarewicz Phys. Lett. B 297, 25 (1992).
31. A. Bohr and B. Mottelson, in: Nuclear Structure II (Benjamin, New York, 1969), and references therein.
32. J.P. Boisson, R. Piepenbring, Nucl. Phys. A 168, 385 (1971).
33. M. Guttormsen, T. Pedersen, J. Rekstad, T. Engeland, E. Osnes, and F. Ingebretsen, Nucl. Phys. A 338, 141 (1980).
34. Y. Fujita *et al.*, Nucl. Phys. A690, 243c (2001).
35. B. H. Wildenthal, Prog. Part. Nucl. Phys. 11, 5 (1984).

ELECTROMAGNETIC PROCESSES IN DRIP-LINE NUCLEI

Toshio SUZUKI

Department of Physics, College of Humanities and Sciences, Nihon University,
Sakurajosui 3-25-40, Setagaya-ku, Tokyo 156-8550, Japan

Takaharu OTSUKA[1,2] and Rintaro FUJIMOTO[1]

[1] *Department of Physics, University of Tokyo, Hongo, Bunkyo-ku, Tokyo 113-0033*
[2] *RIKEN, Hirosawa, Wako-shi,Saitama 351-0198, Japan*

Change of the shell structure near drip-lines is pointed out and enhancement of low-energy dipole strength is shown in light drip-line nuclei. Change of the shell magicity can be explained by taking into account an important role of spin-isospin interaction. New aspects of quenching effects in Gamow-Teller (GT) transitions and magnetic moments are investigated for p-shell nuclei using improved shell-model interactions with enhanced spin-flip neutron-proton interactions and modified single-particle energies. Manifestation of less quenching due to a weakening of intermediate coupling nature of the Cohen-Kurath interction is shown in GT transitions; $^{12}C \rightarrow {}^{12}N$, $^{11}B \rightarrow {}^{11}Be$ and $^{9}Li \rightarrow {}^{9}Be$. Better agreement with experimental values is obtained by using the present interactions in most of GT transitions and magnetic moments in p-shell nuclei.

1 Introduction

Magic numbers are found to be modified in nuclei near drip-lines owing to recent intensive studies of neutron and proton drip line nuclei by radioactive beam facilities[1]. Melting of the shell magicity at N=8 and Z=8 has been found in light drip-line nuclei such as ^{11}Be, ^{11}Li, ^{12}Be and ^{13}O [2,3,4,5]. A possibility of a new magic number at N=16 is pointed out by measurement of interaction cross sections as well as separation energies [6]. It has also been shown [7] that hole-vibration coupling to ^{24}O-core leads to a lowering of $1s_{1/2}$ shell and a new magic number at N=16.

The importance of the role of spin-isospin dependent interaction is pointed out to explain the change of the shell magicity near drip lines [8]. Spin-flip neutron-proton interactions become weak for proton deficient, that is, neutron rich nuclei, leading to larger energy gap between $j_{\rangle} = \ell + \frac{1}{2}$ and $j_{\langle} = \ell - \frac{1}{2}$ shells. In case of $\ell = 2$, this mechanism explains the vanishing of the shell magicity at N = 20 and the appearance of the shell magicity at N = 16. It leads to a vanishing of the shell magicity at N = 8 and a new magic number at N = 6 for the case of $\ell = 1$.

In ref. [8], spin-flip n-p interactions are enhanced in T = 0 channel and single-particle energies of p-shell are changed to have as large as 3.85 MeV shell gap between $0p_{1/2}$ and $0p_{3/2}$ orbits in the Millener-Kurath interaction [9] (PSDMK2 [10]). Parity inversion in the low-lying energy levels in ^{9}He and ^{11}Be are reproduced, while energy levels in ^{13}C are little modified in the shell model calculations with the use of the modified interaction within $0-1$ $\hbar\omega$ shell model space.

One of the new aspects of the modified interaction is a possible different effect of quenching in spin-dependent transitions. As the shell gap between $0p_{1/2}$ and $0p_{3/2}$ shells becomes larger than that in the original interaction, the intermediate coupling nature characteristic to the Cohen-Kurath interaction [11] in the p-shell may be weakened and the spin-dependent transitions can become more single-particle like with larger strength.

In the present contribution, we study Gamow-Teller (GT) transitions and magnetic moments in p-shell nuclei using modified shell model interactions. Here, we take configuration space up to $2\hbar\omega$ or $3\hbar\omega$ excitations in the shell model calculations to properly take into account the quenching effects in the spin-isospin modes. We thus modify the shell model interaction for use in the $2-3$ $\hbar\omega$ configuration space. The shell gap between $0p_{1/2}$ and $0p_{3/2}$ orbits is increased up to 3.92 MeV, and spin-flip n-p interaction in the T $= 0$ channel is enhanced as in the original modified interaction for use in $0-1$ $\hbar\omega$ space [8] (this interaction wil be referred as 'present 1' hereafter). Furthermore, some of the two-body interactions, $\langle\, 0p^2 \mid V \mid 0p^2\rangle$ and $\langle\, 0p^2 \mid V \mid (1s0d)^2\rangle$, are renormalized. This renormalization effect is found to be important to reproduce energy spectra of oxygen isotopes, $^{14\sim18}$O [12]. We will refer this interaction as 'present 2'.

When the 'present 2' interaction is used, level inversions in ^{9}He and ^{11}Be are reproduced and, moreover, the ordering of $1/2^+$, $3/2^-$ and $5/2^+$ states in ^{13}C is also reproduced, which was not the case for the 'present 1' interaction; ordering of $3/2^-$ and $5/2^+$ states was reversed compared to the observation. Energy levels in other p-shell nuclei such as ^{9}Li, ^{10}Be, ^{10}B, ^{11}B, ^{12}B and ^{12}C are also rather well reproduced.

In the next section, we will study the effects of the modified interactions in a GT transition in ^{12}C and magnetic moment of ^{13}C. GT transitions in ^{11}B and ^{9}Li are investigated in sect.3, where we will find large effects of the present new interactions. In sect. 4, log ft values of GT transitions and magnetic moments are studied in p-shell nuclei for which experimental data are available. Summary is given in sect. 5.

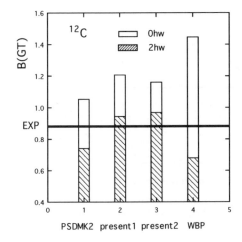

Figure 1: B(GT) values for the transition ^{12}C → ^{12}N ($1^+_{g.s.}$). They are obtained by the shell model calculations within 0p-1s0d shells including up to $0\hbar\omega$ (white bars) and $2\hbar\omega$ (dashed bars) excitations with the use of the PSDMK2, present 1, present 2 and WBP interactions. Experimental value is taken from ref. [13].

2   Gamow-Teller Transition in ^{12}C

Now, we first investigate a GT transition in ^{12}C. Calculated results for B(GT) are shown in Fig. 1 for the GT transition ^{12}C → ^{12}N (1^+, T=1) as well as the experimental value[13]. Results of the shell model calculations obtained by using the PSDMK2, WBP[14], present 1 and present 2 interactions are given for both the cases of up to $0\hbar\omega$ and $2\hbar\omega$ excitations. The B(GT) values are enhanced by about 30% compared to those obtained by the PSDMK2 interaction. Here, we see clearly the effects of the weakening of the intermediate coupling nature of the Cohen-Kurath interaction. The occupation number of $0p_{1/2}$ ($0p_{3/2}$) orbit, which is 1.570 (6.220) for the PSDMK2 interaction, changes to 1.442 (6.415) for the present 2 interaction in the ground state of ^{12}C. We notice less mixing of the $0p_{1/2}$ and $0p_{3/2}$ shells in the modified interaction. The results obtained by the present 2 and present 1 interactions with the space up to $2\hbar\omega$ excitations are close to the observed value and show a fair improvement over those obtained by the standard interactions, PSDMK2 and WBP. Possible reductions induced by including higher configurations as well as the Δ_{33}-isobar exchange current lead to further improvement of the agreement with the experiment.

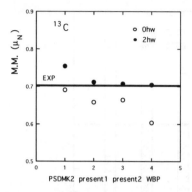

Figure 2: Magnetic moment of ^{13}C. Calculated values are obtained by the shell model calculations within 0p-1s0d shells including up to $0\hbar\omega$ (open circles) and $2\hbar\omega$ (filled circles) excitations with the use of the PSDMK2, present 1, present 2 and WBP interactions. Experimental value is taken from ref. [15].

We here also show the results of the study of the magnetic moment of ^{13}C. Calculated and experimental values are shown in Fig. 2. Calculated values of the magnetic moment obtained by our interactions within $2\hbar\omega$ space are close to the observed value. We thus find that spin degrees of freedom are well described and improved by the use of our new interactions for ^{12}C and ^{13}C.

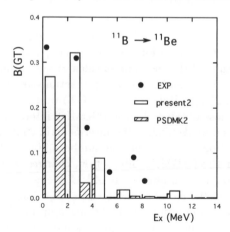

Figure 3: B(GT) values for ^{11}B $\rightarrow$ ^{11}Be ($1/2^-$, $3/2^-$, $5/2^-$). They are obtained by the shell model calculations with the use of the PSDMK2 (white) and present 2 (dashed) interactions including up to $2\hbar\omega$ excitations. Experimental data are taken from ref. [16].

3 Gamow-Teller Transitions in ^{11}B and ^{9}Li

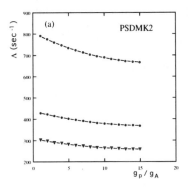 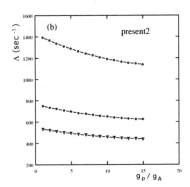

Figure 4: Averaged transition rates for ^{11}B (μ^-, ν_μ) ^{11}Be $(1/2^-, 0.32$ MeV) from the two hyperfine states, F=1 and F=2. Open circles are obtained by using harmonic oscillator wave functions while black circles and triangles denote those obtained by using halo wave functions for $\nu 1s_{1/2}$ and $\nu 0p_{1/2}$ orbits by ref. [18] and ref. [19], respectively.

In this section, we investigate GT transitions from ^{11}B and ^{9}Li and magnetic moment of ^{11}Be. We show that fairly large effects of the modification are found in the GT transitions and calculated results become more consistent with the observations.

Calculated B(GT) values are shown in Fig. 3 for ^{11}B $\rightarrow$ ^{11}Be $(1/2^-, 3/2^-,$ $5/2^-)$ transitions obtained by using the PSDMK2 and present 2 interactions with the space up to $2\hbar\omega$ excitations. Contributions from the Δ_{33}-isobar exchange current are also included. The B(GT) values with the present 2 interaction show a large enhancement of 70±10% over those with the PSDMK2 interaction at $E_x \leq 4$ MeV. This enhancement is considered to come from the weakening of the intermediate coupling nature of the Cohen-Kurath interaction. A recent observation[16] supports this enhancement of the B(GT) values.

A similar enhancement is obtained in the muon capture reaction ^{11}B (μ, ν_μ) ^{11}Be $(1/2^-, 0.32$ MeV). Transition rates are calculated by using harmonic oscillator wave functions as well as halo wave functions[18,19] for $1s_{1/2}$ and $0p_{1/2}$ orbits. Calculated averaged transition rates from the two hyperfine states, F=1 and F=2, are shown in Fig. 4 as a function of g_p/g_A, where g_p and g_A are pseudoscalar and axial-vector coupling constants, respectively. The value of g_A is taken to be -1.26. The rates obtained by the present 2 interaction are enhanced by about 70~80 % compared to those by the PSDMK2 interaction[17], which were smaller than an observed value[20] and remained as a puzzle since we needed enhancement instead of quenching to explain the experiment in the spin-isospin dependent transition. The experimental value of the rate, $\Lambda \sim 1000 \pm 100$ sec^{-1} [20], is more consistent with the results of our new interaction as the new results support the necessity of the quenching effects.

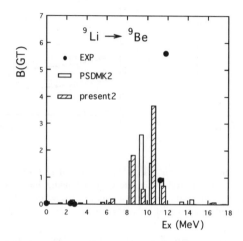

Figure 5: B(GT) values for ^{9}Li $\rightarrow$ ^{9}Be ($1/2^-$, $3/2^-$, $5/2^-$). They are obtained by the shell model calculations with the use of the PSDMK2 (white) and present 2 (dashed) interactions including up to $2\hbar\omega$ excitations. Experimental values are taken from ref. [21].

A remarkable improvement is also seen in GT transitions ^{9}Li $\rightarrow$ ^{9}Be ($1/2^-$, $3/2^-$, $5/2^-$). Results are shown in Fig. 5. B(GT) obtained for the present 2 interaction is enhanced and the strength is shifted toward higher excitation energy region compared to those for the PSDMK2 interaction. Agreement with the experimental values[21] is considerably improved for the new interaction.

We here comment on the magnetic moment of ^{11}Be. Calculated values are shown in Fig. 6. Observed value[22] is consistent with those obtained by the PSDMK2, present 1 and present 2 interactions. The probabilities for the component of the $1s_{1/2}$ orbit coupled to ^{10}Be (0^+) state in the ^{11}Be wave function are $73 \sim 74\%$ for the interactions used here. These values are consistent with the experimental values $0.73 \sim 0.77\%$ [23].

4   Magnetic Moments and Log ft Values in p-Shell Nuclei

Calculated log ft values for GT transitions and those of magnetic moments are summarized in Table 1 and Table 2, respectively, for p-shell nuclei whose

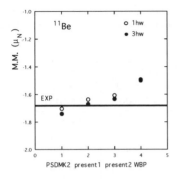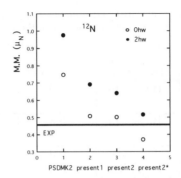

Figure 6: The same as in Fig. 2 for the magnetic moment of ^{11}Be (left side) and ^{12}N (right side). Experimental value is taken from refs. [22] and [15,25]. For ^{12}N, present2* includes the effect of modified orbital g-factor.

experimental data are available. As we see from the tables, in most cases in which the PSDMK2 interaction gives good account of the experimental values, our modified interactions give generally better agreement with the observations though the changes of the calculated values are rather small. In cases where there are some deviations in the calculated values by the PSDMK2 interaction from the experimental values, the modified interactions give remarkable improvements and calculated values come closer to the observations.

For example, in the magnetic moment of ^{9}C, the deviation of the calculated value from the experimental one[24] becomes small. Especially in the GT transition ^{14}C $\rightarrow$ ^{14}N $(1^+_{g.s.})$, the log ft value is found to be increased by 2.2 when the present 2 interaction is used instead of the PSDMK2 interaction, and gets closer to the observed value[5].

Remaining deviations from experimental values in the magnetic moments of ^{8}B, ^{11}C, ^{12}B and ^{12}N can be diminished by including an effective orbital g-factor, $\delta g_\ell^{\nu(\pi)} = +0.1\ (-0.1)$. Calculated values of the magnetic moments are 0.8846, -0.9405, 0.8665, and 0.5151 μ_N for ^{8}B, ^{11}C, ^{12}B and ^{12}N, respectively, for the present 2 interaction. An example of a remarkable improvement is shown in Fig. 6 for the magnetic moment of ^{12}N. The effect of modified orbital g-factor is included in the present 2* case. Magnetic moments of unstable nuclei recently measured, ^{14}B, ^{15}B, ^{17}B, ^{16}N, ^{17}N and ^{19}O [26], are also well described by the present interactions (see Table 2).

Table 1: Calculated and observed log ft values in p-shell nuclei. Calculated values are obtained within 2~3 $\hbar\omega$ space while those in the parentheses are obtained within 0~1 $\hbar\omega$ space.

Transition	experiment	PSDMK2	present 1	present 2
$^6\text{He} \rightarrow {}^6\text{Li}$	2.9	2.8505	2.8386	2.8403
		(2.8515)	(2.8386)	(2.8403)
$^7\text{Be} \rightarrow {}^7\text{Li}$	3.3	3.2370	3.2556	3.2578
		(3.2372)	(3.2557)	(3.2578)
$^9\text{C} \rightarrow {}^9\text{B}$	5.3	4.7280	4.6967	4.7089
$^9\text{Li} \rightarrow {}^9\text{Be}$	5.3	(4.6625)	(4.6407)	(4.6709)
$^{11}\text{C} \rightarrow {}^{11}\text{B}$	3.6	3.5201	3.5120	3.5024
		(3.5241)	(3.5090)	(3.5006)
$^{12}\text{C} \rightarrow {}^{12}\text{B}$	3.6	3.7187	3.6130	3.6019
$^{12}\text{C} \rightarrow {}^{12}\text{N}$	3.64	(3.5655)	(3.5062)	(3.5235)
$^{12}\text{Be} \rightarrow {}^{12}\text{B}$	3.8	3.4723	3.3484	3.4581
		(3.2700)	(3.2769)	(3.2700)
$^{13}\text{B} \rightarrow {}^{13}\text{C}$	4.0	3.9474	3.8887	3.8890
$^{13}\text{O} \rightarrow {}^{13}\text{N}$	4.1	(3.8678)	(3.8287)	(3.8358)
$^{13}\text{N} \rightarrow {}^{13}\text{C}$	3.7	3.6165	3.6247	3.6256
$^{14}\text{C} \rightarrow {}^{14}\text{N}$	9.0	5.6683	6.3984	7.8609
$^{14}\text{O} \rightarrow {}^{14}\text{N}$	7.3	(5.2340)	(5.6232)	(6.0707)

5 Summary

In summary, we have shown that the present shell-model interactions can explain strength of GT transitions in $^{11}\text{B} \rightarrow {}^{11}\text{Be}$ and $^9\text{Li} \rightarrow {}^9\text{Be}$, where the quenching effects are small. This is due to the weakening of the intermediate coupling nature of the Cohen-Kurath interaction in the present interactions. We find that agreement of calculated log ft values and magnetic moments with experimental values are generally improved in most of the p-shell nuclei investigated here. Recently, the present interactions are found to explain branching ratios for proton emission better than the Cohen-Kurath interaction in GT transitions in $^{13}\text{C} \rightarrow {}^{13}\text{N}$[27].

Acknowledgments

The author would like to thank H. Sagawa for collaborations on low-energy dipole transitions.

Table 2: Calculated and observed values of magnetic moments in p-shell nuclei in units of μ_N. Calculated values are obtained within 2~3 $\hbar\omega$ space.

Nucleus	experiment	PSDMK2	present 1	present 2
^{7}Li	3.2564286	3.23547	3.23635	3.23236
^{8}B	1.0355	1.2817	0.8653	0.8394
^{9}Be	−1.1778	−1.2772	−1.0877	−1.0809
^{9}Li	3.4391	3.4455	3.4157	3.4225
^{9}C	−1.3914	−1.5657	−1.5114	−1.5139
^{10}B	1.80064	1.81315	1.81321	1.81429
^{1}Be	−1.6816	−−1.7400	−1.6697	−1.6317
^{11}B	2.68865	2.47982	2.56627	2.61386
^{11}C	−0.964	−0.75345	−0.82976	−0.87414
^{12}B	1.00306	0.39851	0.68942	0.74090
^{12}N	0.4573	0.97613	0.69204	0.64068
^{13}B	3.1778	3.0083	3.0185	3.0315
^{13}C	0.7024	0.7553	0.7122	0.7071
^{15}C	−1.720	−1.7853	−1.7750	−1.7694
^{15}N	−0.2831884	−0.27892	−0.27719	−0.27756
^{14}B	1.185	0.9889	1.0031	1.0137
^{15}B	2.659	2.5963	2.5736	2.5793
^{17}B	2.545	2.3743	2.3479	2.3499
^{16}N	−1.9859	−2.0604	−1.9756	−2.0085
^{17}N	−0.352	−0.3159	−0.3004	−0.3047
^{19}O	−1.53195	−1.4772	−1.4771	−1.4689

References

1. I. Tanihata et al., *Phys. Lett.* B **287**, 307 (1992), and references therein; I. Tanihata, *J. Phys.* G **22**, 157 (1996).
2. T. Suzuki and T. Otsuka, *Phys. Rev.* C **56**, 847 (1997); T. Suzuki and T. Otsuka, *Nucl. Phys.* A **635**, 86 (1998).
3. T. Suzuki, H. Sagawa and P. F. Bortignon, *Nucl. Phys.* A **662**, 282 (2000).
4. H. Iwasaki et al., *Phys. Lett.* B **491**, 8 (2000).
5. H. Sagawa, T. Suzuki, H. Iwasaki and M. Ishihara, *Phys. Rev.* C **63**, 034310 (2001).
6. A. Ozawa et al., *Phys. Rev. Lett.* **84**, 5493 (2000).

7. G. Colò, T. Suzuki and H. Sagawa, *Nucl. Phys.* A **695**, 167 (2001).
8. T. Otsuka, R. Fujimoto, Y. Utsuno, B. A. Brown, M. Honma and T. Mizusaki, *Phys. Rev. Lett.* **87**, 082502 (2001).
9. D. J. Millener and D. Kurath, *Nucl. Phys.* A **255**, 315 (1975).
10. OXBASH, The Buenos-Aires, Michigan State, Shell Model Program, B. A. Brown, A. Etchegoyen and W. D. M. Rae, MSU Cyclotron Laboratory Repoers No. 524, 1986.
11. S. Cohen and D. Kurath, *Nucl. Phys.* **73**, 1 (1965).
12. T. Sebe et al., to be published.
13. R. E. McDonald, J. A. Becker, R. A. Chalmers and D. H. Wilkinson, *Phys. Rev.* C **10**, 333 (1974).
14. E. K. Warburton and B. A. Brown, *Phys. Rev.* C **46**, 923 (1992).
15. F. Ajzenberg-Selove, *Nucl. Phys.* A **523**, 1 (1991); A **449**, 1 (1986).
16. T. Ohnishi et al., *Nucl. Phys.* A **687**, 38c (2000).
17. T. Suzuki, in *Proc. of the Int. Symp. on "Non-Nucleonic Degrees of Freedom Detected in Nuclei"*, Osaka, 1996, eds. by T. Minamisono et al., (World Scientific, 1997), p.349.
18. T. Otsuka, A. Muta, M. Yokoyama, N. Fukunishi and T. Suzuki, *Nucl. Phys.* A **588**, 113c (1995).
19. D. Ridikas and J. S. Vaagen, ECT*-96-006 (1996).
20. J. P. Deutsch et al., *Phys. Lett.* B **28**, 178 (1968).
21. G. Nyman et al., *Nucl. Phys.* A **510**, 189 (1990).
22. W. Geithner et al., *Phys. Rev. Lett.* **83**, 3792 (1999).
23. D. L. Auton, *Nucl. Phys.* A **157**, 305 (1970);
    B. Zweiglinski, W. Benenson and R. G. H. Robertson, *Nucl. Phys.* A **315**, 124 (1979).
24. K. Matsuta et al., *Nucl. Phys.* A **588**, 153c (1995).
25. *Table of Isotopes*, eds. by R. B. Firestone et al., (Wiley, New York, 1996).
26. H. Okuno et al., *Phys. Lett.* B **354**, 41 (1995);
    H. Ueno et al., *Phys. Rev.* C **53**, 2142 (1996);
    T. Minamisono et al., *Phys. Lett.* B **457**, 9 (1999);
    K. Matsuta et al., *Phys. Rev. Lett.* **86**, 3735 (2001).
27. M. Fujimura et al., private communication.

TOTAL PHOTO-ABSORPTION OF ACTINIDE NUCLEI AT INTERMEDIATE ENERGIES

V.G.NEDOREZOV

Institute for Nuclear Research, Moscow, 117312

A brief review on photo-absorption and photo-fission of actinide nuclei at intermediate energies is given. To explain the experimental data for ^{237}Np which contradict to universal behavior of the photo-absorption process vs the nuclear mass, we extrapolate the data on electromagnetic dissociation of relativistic heavy ions to the photo-fission cross section.

1 Introduction

It is well known that intermediate energy photons (above the giant resonance region) interact with a nucleus through excitation of quasi-free deuterons (below pion production threshold) or quasi-free nucleons (above 150 MeV), where the total nuclear photo-absorption cross section, $\sigma_{\gamma A}$, is proportional to the quasi-free deuteron photo-disintegration cross section or quasi-free nucleon photo-absorption cross section, respectively. "Quasi-free" means that nuclear media effects (Fermi motion, Pauli blocking, Intra-nuclear cascade) cause a modification in the cross section shape for the bound nucleon. However, this modification does not depend on the nuclear mass for middle heavy and heavy nuclei. Such universal behavior was especially measured in the Δ-region for various nuclei from ^{7}Li up to ^{238}U (see Ref.[1]).

$\sigma_{\gamma A}$ was measured in different ways. In the case of heavy actinide nuclei, the total photo-fission cross section, $\sigma_{\gamma F}$, has been thought to be a good approximation for $\sigma_{\gamma A}$. Specifically, for the case of ^{238}U, experimental measurements and theoretical calculations suggested that the photo-fission probability is consistent with unity for photon energies larger than about 40 MeV [2,3].

However, recent experimental results [4] on photo-fission of actinide nuclei discussed below provoke a more detail investigation of photo-absorption mechanisms at intermediate energies. In particular, we analyze the fission cross sections, mass distributions of fission fragments and other data which can be considered as an evidence for complementary mechanism of nuclear excitation at intermediate energies. Therefore, we evaluate the probability of collective excitations of nuclei in the photo-absorption process using the data on electromagnetic dissociation of relativistic heavy ions.

2   Photo-absorption and Photo-fission

Experimental result [4] on relative fission probabilities for some actinide nuclei is shown in fig. 2. One can see that the total fission cross section for ^{237}Np exceeds that of ^{238}U (which fits to the "universal curve" in accordance with generally accepted knowledge) by 20% in the energy region 50 - 250 MeV. Traditionally the fissilities of the actinide nuclei have been presented relative to ^{238}U by dividing $\sigma_{\gamma F}$ (X) by $\sigma_{\gamma F}$ (^{238}U). However, since the $\sigma_{\gamma F}$ (^{237}Np) is higher, it makes more sense to present the fissilities relative to this isotope.

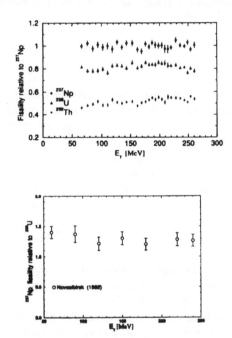

Figure 1. Fissility of ^{232}Th and ^{238}U nuclei relatively to ^{237}Np from [4] (left figure) and [5] (right figure), respectively.

Photo-fission cross sections per nucleon for ^{237}Np and ^{238}U [4] in comparison with photo-absorption cross sections for Be [6] and C [7] are shown in fig. 2 and 3.

It is seen that the total photo-fission cross section per nucleon for ^{237}Np exceeds the photo-absorption cross section for Be and C by 10%, about. This

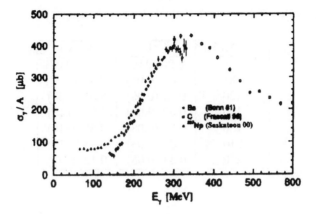

Figure 2. Photo-fission cross sections per nucleon for ^{237}Np in comparison with photo-absorption cross sections for Be and C.

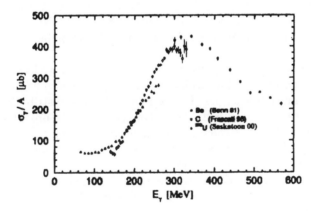

Figure 3. Photo-fission cross sections per nucleon for ^{238}U in comparison with photo-absorption cross sections for Be and C.

difference is less than the fissility difference between ^{237}Np and ^{238}U. So, the current conclusion is rather limited: the width and strength of Δ - resonance for the actinide nuclei measured by the fission method does not agree with

Table 1. Probability of symmetric fission R = S/(S+A) [9]; systematic error bars are shown in brackets.

E_γ (MeV)	^{235}U	^{237}Np
60	.0.25 (0.09)	.0.22 (0.08)
90	0.32 (0.09)	.0.21 (0.07)
120	0.29 (0.09)	0.30 (0.08)
150	0.39 (0.09)	0.25 (0.10)
180	0.35 (0.08)	0.35 (0.10)
210	0.37 (0.08)	0.38 (0.11)
240	0.34 (0.08)	0.12 (0.07)

the "universal curve" but this dependence is not sufficiently studied yet.

We would take into account that the results for uranium isotopes [4] are generally in good agreement with previous results, especially those from Mainz [8], Therefore, we can conclude that minimal excess of the ^{237}Np fission cross section above the "universal curve" is equal to 10%, at least. Also, we have to emphasize that fissility of ^{237}Np can be less than unit also (up to 0.8) in accordance with theoretical evaluations [4]. So, the ^{237}Np case is a better approximation to the total photo-absorption cross section but may be not the perfect one.

To clarify the result on relative ^{238}U / ^{237}Np fissility we look on the complementary experimental data which are available for photo-fission of actinide nuclei at intermediate energies. For example, the mass distribution of fission fragments [9] are presented in table 1. One can see that probability for the symmetric fission mode (which is the signature of high excitation energy of fissile nucleus) does not exceed 40%. Here S means symmetric fission, A - asymmetric fission, respectively ; (S + A) = 1.

A large contribution of the asymmetric fission mode means that the average excitation energy is not so high as compared with the INC (Intra-nuclear Cascade) model predictions or the neutron induced fission data [10]. Therefore, one can assume that about 20% of the total yield is caused by the complementary low energy excitation mechanism in addition to the universal curve. This can explain qualitatively the data presented in fig. 1-3. We have to remind the principal difference between ^{237}Np and ^{238}U nuclei in fission widths: at low excitation energies (near the barrier) the fission probability for ^{237}Np exceeds the ^{238}U one in factor 3 , approximately [11].

Another signature for direct collective excitation mechanism can be seen in photo-production of spontaneously fissioning isomers (delayed fission) [12].

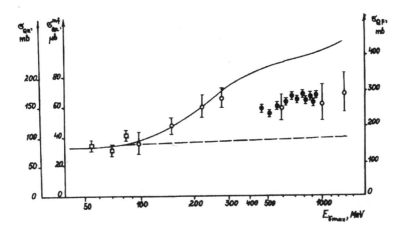

Figure 4. The delayed reaction cross section, σ^{mf}, for the ^{243}Am $(\gamma, n)^{242mf}$Am reaction (the second from the left scale). Points are the experimental data [12]. The solid curve is the averaged cross section per an equivalent photon, σ_{qn}, for ^{243}Am, (the right scale). The dashed line shows the contribution from the giant resonance, calculated with the realistic bremsstrahlung gamma spectrum. The extreme left scale corresponds to the photo-neutron ^{243}Am(γ,n) cross section obtained by dividing of σ_{qn} by the isomer ratio $\alpha = 4 \times 10^{-4}$.

Spontaneously fissioning isomers are interpreted as shape isomers associated with a second minimum in the fission barrier (see, for example, Ref. [13]). Therefore, the isomeric ratio is very sensitive to the excitation energy, indicating isomers as low-lying nuclear states. One can see in fig. 4 the noticeable contribution to the isomer production cross section from the intermediate energy region.

This experimental result confirms an assumption about a complementary low energy excitation mechanism. Therefore, we assume a mechanism of nuclear excitations associated with inelastic e^+e^- pair production. Theoretically the probability of nuclear excitations in atomic processes like inelastic pair production or inelastic bremsstrahlung was studied at first in [14].

However, theoretical attempt [15] to evaluate quantitatively the probability of fission by a complementary mechanism using the assumption of inelastic e^+e^- pair production (see fig. 5) gave a very low upper limit (less than 1%) for such a process whereas the experimental value is about 20%. But the theoretical result depends strongly on a model, especially on the nuclear form factor. Therefore, we have tried to evaluate this effect qualitatively basing on the data of electromagnetic dissociation (EMD) of relativistic heavy ions,

512

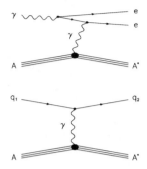

Figure 5. Diagrams for inelastic e^+e^- pair production (above) and electromagnetic dissociation (below).

which mechanism is very similar with the inelastic e^+e^- pair production, as seen in fig. 5.

3   Electromagnetic dissociation of relativistic heavy ions

It was shown experimentally [16] that photo-nuclear mechanism dominates in the total cross section for interaction of relativistic heavy ions. Therefore, we can extrapolate the cross section in the scales of nuclear charge and interaction energy. Extrapolation of the EMD cross section (taken from [17,18] and shown in fig. 6), to the respective photo-nuclear one, gives a value about 10 mb per nucleon in assumption that low energy nuclear excitations (near fission threshold) can take place due to inelastic e^+e^- pair production. This value is in reasonable agreement with the photo-nuclear results presented above.

Indeed, the accuracy of such evaluations is not good enough for final conclusions. Therefore, new correlation experiments are needed with simultaneous measurements of fission fragments in coincidence with fast charged particles. The fission fragment can be easily identified and provide a reliable trigger in the coincidence measurements. Moreover, this can provide a reliable normalization of measured probabilities for different processes which are preceded with the fission decay because the fission cross section is very close to the total photo-absorption one.

4   Conclusions

The experimental data on photo-fission of actinide nuclei at intermediate energies allow to assume the existence of a complementary mechanism for nuclear

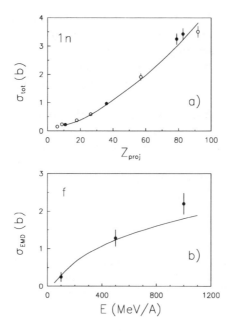

Figure 6. The partial neutron emission EMD cross section as a function of Z of the target for the 1 GeV ^{197}Au + X reaction [17] and the fission cross section for the ^{238}U + Pb reaction vs the projectile energy [18]. Points and curves represent the experimental data and calculations within the virtual photon method, respectively.

excitation near the fission barrier, in addition to the well known "universal curve". The contribution of such a mechanism to the total photo-absorption yield (about 20% for fissile nuclei) does not depend on Eγ in the region above the giant resonance. This can be related to direct collective nuclear excitations.

To explain this effect we assume the inelastic e^+e^- pair production mechanism which can describe qualitatively the experimental results on total photo-absorption cross sections and mass distributions of fission fragments. Nevertheless, one can not exclude other possibilities, for example, low energy nuclear excitations when a pion or nucleon emitted from a nucleus takes all the initial energy of the photon. Therefore, new correlation experiments will be very desirable.

Moreover, a coincidence measurement between fast nuclear products and slow fission fragments can provide new valuable information on collective, hadron and non-hadron degrees of freedom of nuclear matter and energy dissipation processes. The availability of new generation facilities as SPring-8 makes such experiments possible due to the high quality of the beam and low backgrounds. Therefore, this presentation can be considered as proposal for experiments at SPring-8. Especially, measurements of charged particles (e^+e^- pairs, pions, protons) in coincidence with fission fragments would be very interesting.

The important feature of experiments with fissile nuclei is the high fissility value (closed to unite). Therefore, the measurement of fission fragments will practically not decrease the statistics in spite of the coincidence type of the experiment. In fact, this means that any kind of photo-absorption and nuclear excitation (first stage of reaction) will lead to the fission mode (second stage) of decay. So, different mechanisms of reaction can be studied very thoroughly.

Acknowledgments

I would like to thank Prof. B. Berman and Dr. J. Ahrens for discussions during the preparation of the manuscript, Prof. M. Fujiwara for hospitality at the conference EMI2001. This work is supported by RFBR; grant 17235.

References

1. J. Ahrens, Nucl. Phys. A446 (1985) 229.
2. H. Ries, U. Kneissl, G. Mank *et al.*, Phys. Lett. B139 (1984) 254.
3. A. Lepretre, R. Bergere, P. Bourgeois *et al.*, Nucl.Phys. A472 (1987) 533.
4. J. Sanabria, B.L. Berman, C. Getina *et al.*, Phys. Rev. C61 (2000) 134604.
5. A.S. Iljinov *et al.*, Nucl. Phys. A539 (1992) 263.
6. J. Arends *et al.*, Phys. Lett 146B (1981) 423.
7. N. Bianchi *et al.*, Phys.Rev C54 (1996) 1688.
8. T. Frommhold *et al.*, Z. Phys. A350 (1994) 249.
9. D.I. Ivanov *et al.*, Sov. J. Nucl. Phys. 35 (1992) 907.
10. M. Mebel. private communication.
11. B.M. Alexandrov *et al.*, Sov. J. Nucl. Phys. 2(1986) 290.
12. D.I. Ivanov *et al.*, Nucl. Phys. A485 (1988) 668.
13. V.G. Nedorezov and S.M. Polikanov. Elementary Particles and Atomic Muclei 8 (1977), 374.
14. R. Horvat *et al.*, Phys. Rev C29 (1981) 5, 1614.
15. A.I. Lvov *et al.*, Sov. J. Nucl. Phys. 55 (1986) 1, 3.
16. D.L. Olson, B.L. Berman *et al.*, Phys. Rev. C24 (1981) 849.
17. T. Aumann *et al.*, Nucl. Phys. A569 (1994) 157.
18. S.M. Polikanov *et al.*, Z. Phys. A350 (1994) 221.

COMPTON SCATTERING ON THE PROTON AND LIGHT
NUCLEI IN THE Δ-RESONANCE REGION

O. SCHOLTEN

Kernfysisch Versneller Instituut, University of Groningen,
9747 AA Groningen, The Netherlands

S. KONDRATYUK

TRIUMF, 4004 Wesbrook Mall, Vancouver, British Columbia,
Canada V6T 2A3

A microscopic coupled-channels model for Compton and pion scattering off the nucleon is introduced which is applicable at the lowest energies (polarizabilities) as well as at GeV energies. To introduce the model first the conventional K-matrix approach is discussed to extend this in a following chapter to the "Dressed K-Matrix" model. The latter approach restores causality, or analyticity, of the amplitude to a large extent. In particular, crossing symmetry, gauge invariance and unitarity are satisfied. The extent of violation of analyticity (causality) is used as an expansion parameter.

1 Introduction

In a K-matrix approach an infinite series of rescattering loops is taken into account with the approximation of incorporating only the pole contributions of the loop diagrams. In this approximation lies both the strong and the weak sides of this approach. By including only the pole contributions, which correspond to rescattering via physical states, unitarity in obeyed and the infinite series can be expressed as a geometric series and summed as given in Eq. (1). Such an approach can be formulated in a co-variant approach, where electromagnetic-gauge invariance is obeyed through the addition of appropriate contact terms. It can also be shown that crossing symmetry is obeyed. In the structure of the kernel a large number of resonances can be accounted for. The weak point of including only the pole contributions of the loop corrections (i.e. only the imaginary part of the loop integrals) is that the resulting amplitude violates analyticity constraints and thus causality.

In Section 4 an extension of the K-matrix model is discussed where, without violating the other symmetries, an additional constraint, that of analyticity (or causality), is incorporated approximately. In this approach, called the "Dressed K-Matrix Model", dressed self-energies and form factors are included in the K-matrix. These functions are calculated self-consistently in an iteration procedure where dispersion relations are used at each recursion step to relate real and imaginary parts.

2 The K-Matrix approach

In K-matrix models the T-matrix is written in the form,

$$\mathcal{T} = (1 - Ki\delta)^{-1} K \,, \tag{1}$$

where δ indicates that the intermediate particles have to be taken on the mass shell and all physics is put in the kernel, the K-matrix. This amounts to re-sum an infinite series of pole contributions ofloop corrections. It is straightforward to check that $S = 1 + 2i\mathcal{T}$ is unitary provided that the kernel K is Hermitian. Since Eq. (1) involves integrals only over on-shell intermediate particles, it reduces to a set of algebraic equations when one is working in a partial wave basis. When both the $\pi - N$ and $\gamma - N$ channels are open, the coupled-channel K-matrix becomes a 2×2 matrix in the channel space, i.e.

$$K = \begin{bmatrix} K_{\gamma\gamma} & K_{\gamma\pi} \\ K_{\pi\gamma} & K_{\pi\pi} \end{bmatrix}. \tag{2}$$

It should be noted that due to the coupled channels nature of this approach the widths of resonances are generated dynamically.

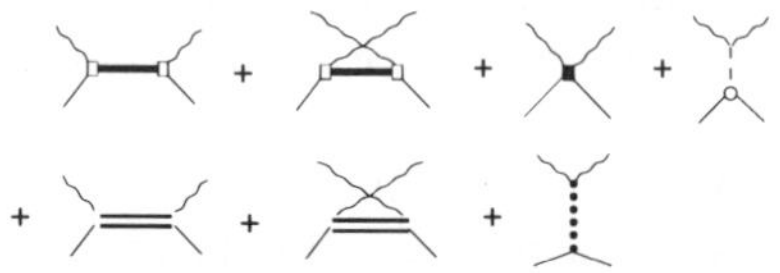

Figure 1. The sum of diagrams included in the K matrix for Compton scattering. A full spectrum of baryon resonances has been included.

In traditional K-matrix models the kernel, the K-matrix, is built from tree-level diagrams[1,2,3,4]. In the present investigation the type of diagrams included in $K_{\gamma\gamma}$ are similar to that of ref. [4] except that the Δ is treated now as a genuine spin-3/2 resonance [5] in order to be compatible with the later treatment of the in-medium Δ resonance. This K-matrix is indicated in Fig. (1). Most of the (non-strange) resonances below 1.7 GeV have been included. The different coupling constants were fitted to reproduce pion scattering, pion photoproduction and Compton scattering on the nucleon. A comparable fit to the data as in ref. [4] could be obtained. In Fig. (2) the results for Compton scattering are compared to data.

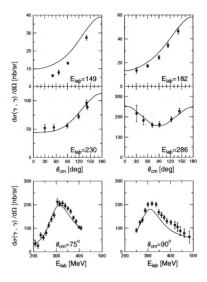

Figure 2. The calculated cross section for Compton scattering off the proton as a function of angle at fixed photon energy, and as a function of photon energy at fixed angle. Data are taken from ref. [11].

3 Basic symmetries

A realistic scattering amplitude for a particular process should obey certain symmetry relations, such as Unitarity, Covariance, Gauge invariance, Crossing and Causality. In the following each of these symmetries will be shortly addressed, stating its physical significance. It is also indicated which of these is obeyed by the K-matrix approach discussed in the previous section. The comparative success of the K-matrix formalism can be regarded as due to the large number of symmetries which are being obeyed. A violation of any-one of these symmetries will directly imply some problems in applications. Improvements are thus important.

3.1 Unitarity

The unitarity condition for the scattering matrix S reads $S^\dagger S = 1$. Usually one works with the T-matrix operator which can be defined as $S = 1 + 2\,i\,T$, and the unitarity condition is rewritten as $2i(TT^\dagger)_{fi} = T_{fi} - T_{if}^*$. If the T-matrix is symmetric (which is related to time-reversal symmetry), the last formula becomes $Im\ T_{fi} = \sum_n T_{fn}T_{in}^*$, where the sum runs over physical intermediate states. The latter relation is the generalization of the well-known optical theorem for the scattering amplitude. Unitarity can only be obeyed in a coupled channel formulation; the imaginary part of the amplitude "knows" about the flux that is lost in other channels.

In the K-matrix formalism the T-matrix is expressed as $T = \frac{K}{1-iK}$ which implies that $S = \frac{1+iK}{1-iK}$ is clearly unitary provided the the kernel K is Hermitian. This kernel is a matrix, where the different rows and columns correspond to different physical outgoing (incoming) channels. The coupled channels nature is thus inherent in such an approach. As explained earlier the kernel is usually written as the sum of all possible tree-level diagrams. In a partial-wave basis K is a matrix of relatively low dimensionality and the inverse, implied in the calculation of the T matrix, can readily be calculated.

3.2 Covariance

The scattering amplitude is said to be covariant if it transforms properly under Lorentz transformations. As a consequence the description of the reaction observables is independent of the particular reference frame chosen for the calculations. It naturally implies that relativistic kinematics is used.

Since the appropriate four-vector notation and γ-matrix algebra are used throughout our calculation, the condition of Lorentz covariance is fulfilled.

3.3 Gauge invariance

Gauge invariance means that there is certain freedom in the choice of the electromagnetic field, not affecting the observables. Its implication is current conservation, $\nabla \vec{J} = \frac{\partial \rho}{\partial t}$, or in four-vector notation, $\partial_\mu J^\mu = 0$. Using the well known correspondence between momenta and derivatives, current conservation can be re expressed as $k_\mu J^\mu = 0$. If the electromagnetic current obeys this relation it can easily be shown that observables, such as a photo-production cross section, are independent of the particular gauge used for constructing the photon polarization vectors.

One of the sources for violation of gauge invariance is the form factors used in the vertices. A form factor implies that at a certain (short range) scale a particle appears 'fuzzy'. At distances smaller than this scale deviations from a point-like structure are important; however in the formulation the dynamics at this short scale is not sufficiently accurate. For one thing, the flow of charge at this scale is not properly accounted for, implying violation of charge conservation. To correct for this, so-called contact terms are usually included in the K-matrix kernel. In the present model these contact terms are constructed using the minimal substitution rules. The corresponding T-matrix, as well as the observables, are independent of the photon gauge.

3.4 Crossing Symmetry

Physical consequences of the crossing symmetry are more difficult to explain. It basically means that in a proper field-theoretical framework the scattering amplitudes of processes in the so-called crossed channels can be obtained from each other by appropriate replacements of kinematics. This assumes that the amplitude can be analytically continued from the physical region of one channel to the physical regions of other channels. An example of the crossed channels is $\gamma N \to \pi N$, $\pi N \to \gamma N$ and $N\bar{N} \to \gamma\pi$.

Crossing symmetry puts a direct constraint on the amplitude for the case that direct and crossed channels are identical, as for example for the processes $\pi N \to \pi N$ or $\gamma N \to \gamma N$. In these reactions crossing symmetry leads to important symmetry properties of the amplitudes under interchange of s and u variables. Due to the fact that in the K-matrix formalism the rescattering diagrams which are taken into account have only on-shell intermediate particles, it can be shown that the s-u crossing symmetry is obeyed provided that the kernel itself is crossing symmetric. Since the latter is the case, crossing symmetry is obeyed.

3.5 Analyticity

Analyticity of the scattering matrix is not really a symmetry. Rather, it requires that the amplitude be an analytic function of the energy variable and in particular that it obeys dispersion relations. The physics of analyticity is closely related to causality of the amplitude as is illustrated in the following example.

Assume that a signal is emitted by an antenna at time $t = 0$. At all subsequent times the signal is given by a function $F(t)$ while causality requires that at earlier times there was no signal, $F(t < 0) = 0$. This signal can be Fourier-transformed, $f(\omega) = \int_0^{+\infty} dt \, e^{i\omega t} F(t)$ to explicitly show its energy or frequency dependence. Note that the integration region from $t = -\infty$ to $t = 0$ gives zero contribution due to the causality requirement. This transformation can also be considered for complex values of ω. Since the integration interval runs only over positive values for t the Fourier integral exists and is a well behaving function for all complex values of energy ω for which $Im(\omega) > 0$ i.e. it is an analytic function. For such a function contour integrals in the complex ω plane can be performed and the function obeys the Cauchy theorem which

520

in this context is usually formulated as a dispersion relation,

$$\mathrm{Re} f(\omega) = \frac{\mathcal{P}}{\pi} \int_{-\infty}^{+\infty} d\omega' \frac{\mathrm{Im} f(\omega')}{\omega' - \omega}$$

showing that for an analytic function the real and imaginary parts are closely related. For example, if the imaginary part of an analytic function is given by the curve on the left-hand side of Fig. (3) the real part of this function is given by the right-hand side.

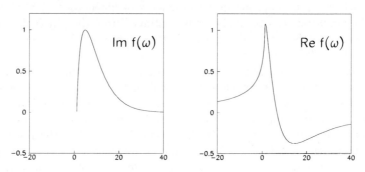

Figure 3. An example of the real and imaginary parts of an analytic function which are related through a dispersion relation.

In the traditionally used K-matrix approaches the analyticity constraint is badly violated. The origin of this is explained in the following.

In a field-theoretical calculation of a scattering amplitude one includes rescattering contributions of intermediate particles which are expressed as loop integrals. In Fig. (4) a typical loop contribution to the self energy is shown. Ignoring terms in the numerator which are irrelevant for the analyticity properties, the corresponding integral can be expressed as

$$J(p^2) = \int d^4 k \, \frac{1}{[k^2 - \mu^2 + i\epsilon] \, [(p-k)^2 - m^2 + i\epsilon]} = \mathrm{Re} J(p^2) + i \mathrm{Im} J(p^2) \quad (3)$$

where the right hand side in this equation and in Fig. (4) expresses the fact this integral has a real and an imaginary part, each of which corresponds to some particular physics. The imaginary part of the integral arises from the integration region where the denominators vanish, corresponding to four-momenta k where the intermediate particles in the loop are on the mass shell -or equivalently- are physical particles with $k^2 = \mu^2$ and $(p-k)^2 = m^2$. Conventionally this is indicated by placing a slash through the loop (see Fig. (4)) to indicate that the loop can be cut at this place since it corresponds to a physical state. The other parts of the integration region contribute to the real part of the integral. In the latter case the particles in the loop are off the mass shell.

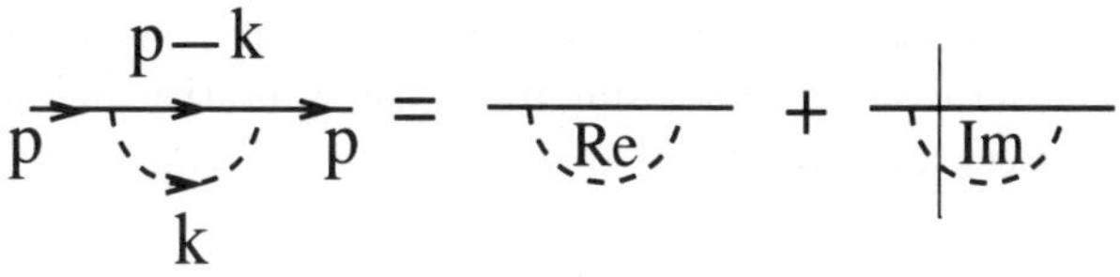

Figure 4. Loop integral contributing to the self energy.

It can be shown that the K-matrix formulation for the T-matrix corresponds to including only the imaginary (or cut-loop) contributions of a certain class of loop diagrams. This guarantees (as was shown before) that unitarity is obeyed. Analyticity of the scattering amplitude is however violated due to ignoring the real contributions of these loop integrals. As a consequence causality will be violated!

To (partially) recover analyticity of the scattering amplitude the so-called "Dressed K-matrix approach"[9] has been developed. It is described in the following section.

4 The Dressed K-matrix Model

As discussed in the previous section, the coupled channels K-matrix approach is quite successful in reproducing Compton scattering. However it fails in predicting nucleon polarizabilities. The reason is that, in spite of the many symmetry properties that are satisfied, analyticity or causality of the amplitude is badly violated. In the "Dressed K-matrix" approach the constraint of analyticity is incorporated in an approximate manner without spoiling the other symmetries. In fact analyticity is used as a kind of expansion parameter where presently only the leading contributions are included. The ingredients of the Dressed K-Matrix Model were described in Refs. [6,7,8] and the main results were presented in Ref. [9]. The essence of this approach lies in the use of *dressed* vertices and propagators in the kernel K.

The objective of dressing the vertices and propagators is solely to improve on the analytic properties of the amplitude. The imaginary parts of the amplitude are generated through the K-matrix formalism (as imposed by unitarity) and correspond to cut loop corrections where the intermediate particles are taken on their mass shell. The real parts have to follow from applying dispersion relations to the imaginary parts. We incorporate these real parts as real vertex and self-energy functions. Investigating this in detail (for a more extensive discussion we refer to [6]) shows that the dressing can be formulated in terms of coupled equations, schematically shown in Fig. (5), which generate multiple overlapping loop corrections. The coupled nature of the equations is necessary to obey simultaneously unitarity and analyticity at the level of vertices and propagators.

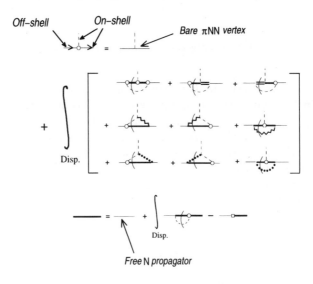

Figure 5. Graphical representation of the equation for the dressed irreducible πNN vertex, denoted by an open circle, and the dressed nucleon propagator, denoted by a solid line. The dashed lines denote pions, the double lines denote Δs and the zigzag and dotted lines are ρ and σ mesons, respectively. The resonance propagators are dressed. The last term in the second equation denotes the counter-term contribution to the nucleon propagator, necessary for the renormalization.

The equations presented in Fig. (5) are solved by iteration where every iteration step proceeds as follows. The imaginary – or pole – contributions of the loop integrals for both the propagators and the vertices are obtained by applying cutting rules. Since the outgoing nucleon and the pion are on-shell, the only kinematically allowed cuts are those shown in Fig. (5). The principal-value part of the vertex (i.e. the real parts of the form factors) and self-energy functions are calculated at every iteration step by applying dispersion relations to the imaginary parts just calculated, where only the physical one-pion–one-nucleon cut on the real axis in the complex p^2-plane is considered. These real functions are used to calculate the pole contribution for the next iteration step. This procedure is repeated to obtain a converged solution. We consider irreducible vertices, which means that the external propagators are not included in the dressing of the vertices.

Bare πNN form factors have been introduced in the dressing procedure to regularize the dispersion integrals. The bare form factor reflects physics at energy scales beyond those of the included mesons and which has been left out of the dressing procedure. One thus expects a large width for this factor, as is indeed the case.

The dressed nucleon propagator is renormalized (through a wave function renormalization factor Z and a bare mass m_0) to have a pole with a unit residue at the physical mass. The nucleon self-energy is expressed in terms of self-energy functions $A(p^2)$ and $B(p^2)$ as $\Sigma_N(p) = A_N(p^2)\not{p} + B_N(p^2)\,m$.

The procedure of obtaining the γNN vertex [7] is in principle the same as for the πNN vertex. Contact $\gamma\pi NN$ and $\gamma\gamma NN$ vertices, necessary for gauge invariance of the model, are constructed by minimal substitution in the dressed πNN vertex and nucleon propagator, as was explained in [7].

The present procedure restores analyticity at the level of one-particle reducible diagrams in the T-matrix. In general, violation due to two- and more-particle reducible diagrams can be regarded as higher order corrections. An important exception to this general rule is formed by, for example, diagrams where both photons couple to the same intermediate pion in a loop (so-called "handbag" diagrams). This term is exceptional since at the pion threshold the S-wave contribution is large, due to the non-zero value of the $E_{0+}^{1/2}$ multipole in pion-photoproduction, leading to a sharp near-threshold energy dependence of the related f_{EE}^{1-} Compton amplitude[13]. In the K-matrix formalism, the imaginary (pole) contribution of this type of diagrams is taken into account. Not including the real part of such a large contribution would entail a significant violation of analyticity. To correct for this, the $\gamma\gamma NN$ vertex also contains the (purely transverse) "cusp" contact term whose construction is described in Section 4 of Ref. [7]. Since, due to chiral symmetry, the S-wave pion scattering amplitude vanishes at threshold, the mechanism that gives rise to the important "cusp" term in compton scattering does not contribute to $\pi\pi NN$ or $\pi\gamma NN$ contact terms. The analogons to the "cusp" $\gamma\gamma NN$ term will thus be negligible and have therefore not been considered.

4.1 Results

Results for pion-nucleon scattering and pion-photoproduction obtained in the dressed K-matrix model and in the traditional K-matrix approach are of comparable quality. One should, however, expect the two approaches to have significant differences for Compton scattering since for this case constraints imposed by analyticity will be most important [12,13].

The effect of the dressing on the f_{EE}^{1-} amplitude can be seen in Fig. (6), where also the results of dispersion analyses are quoted for comparison. Note that the imaginary parts of f_{EE}^{1-} from calculations B (Bare, corresponding to the usual K-matrix approach) and D (Dressed, the full Dressed K-matrix results) are rather similar in the vicinity of threshold.

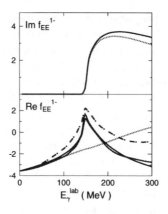

Figure 6. The f_{EE}^{1-} partial amplitude of Compton scattering on the proton in units $10^{-4}/m_\pi$. Solid line: dressed K-matrix, D; dotted line: bare K-matrix, B. Also shown are the results of the dispersion analyses of Ref. [12] (dash-dots) and Ref. [13] (dashed).

The polarizabilities characterize response of the nucleon to an externally applied electromagnetic field [14,15]. They can be defined as coefficients in a low-energy expansion of the cross section or partial amplitudes of Compton scattering. Since gauge invariance, unitarity, crossing and CPT symmetries are fulfilled in both models the Thompson limit at vanishing photon energy is reproduced. Our results for the electric, magnetic and spin polarizabilities of the proton are given in Table 1, where they are compared with the results given in refs. [14,16] and with the values extracted from recent experiments. The contribution from the the t-channel π^0-exchange diagram has been subtracted. The effect of the dressing on the polarizabilities can be seen by comparing the values given in columns D (dressed) and B (bare). In particular, the dressing tends to decrease α while increasing β. Among the spin polarizabilities, γ_{E1} is affected much more than the other γ's. The effect of the additional "cusp" $\gamma\gamma NN$ contact term[7], strongly influences the electric polarizabilities rather than the magnetic ones. This is because the "cusp" contact term affects primarily the electric partial amplitude f_{EE}^{1-} (corresponding to the the total angular momentum and parity of the intermediate state $J^\pi = 1/2^-$) rather than the magnetic amplitude f_{MM}^{1-} ($J^\pi = 1/2^+$).

Table 1. Polarizabilities of the proton. The units are $10^{-4} fm^3$ for α and β and $10^{-4} fm^4$ for the γ's (the anomalous π^0 contribution is subtracted). The first two columns contain the polarizabilities obtained from the present calculation; D (full, dressed) and B (bare K-matrix). The two columns named χPT contain the polarizabilities calculated in the chiral perturbation theory [14,16]. Results of recent dispersion analyses are given in the last column (Ref. [17] for α and β and Ref. [18] for the γ's).

	D	B	χPT		DA
			Gel00	Hem98	
α	12.1	15.5	10.5	16.4	11.9
β	2.4	1.7	3.5	9.1	1.9
γ_{E1}	-5.0	-1.7	-1.9	-5.4	-4.3
γ_{M1}	3.4	3.8	0.4	1.4	2.9
γ_{E2}	1.1	1.0	1.9	1.0	2.2
γ_{M2}	-1.8	-2.3	0.7	1.0	0.0
γ_0	2.4	-0.9	-1.1	2.0	-0.8
γ_π	11.4	8.9	3.5	6.8	9.4

Of special interest is to check whether polarizabilities as extracted form the low energy behavior of the amplitude are in agreement with the values as extracted from energy weighted sumrules. The derivation of sumrules is based on the fact that the amplitudes obey certain symmetries where analyticity is of particular importance. This comparison is still in progess and results will be published in a forthcoming paper [10]. Preliminary results indicate that the different sum rules are obeyed, with the exception of the sum rule for the spin polarizability γ_0. This may be due to an incomplete dressing of the Δ-resonance in the present calculational scheme.

5 Summary and Conclusions

Results are presented for Compton scattering on a free proton as well as on a nucleus. For processes on the proton the usual K-matrix model as well as the recently developed dressed K-matrix model are discussed. In the latter approach the real self energies and vertex functions are obtained from the imaginary parts using dispersion relations imposing self-consistency conditions. It is indicated that such an approach is essential to understand features seen in the data, in particular at energies around and below the pion production threshold.

Acknowledgments

O. S. thanks the Stichting voor Fundamenteel Onderzoek der Materie (FOM) for their financial support. S. K. thanks the National Sciences and Engeneering Research Council of Canada for their financial support.

References

1. P. F. A. Goudsmit, H. J. Leisi, E. Matsinos, B. L. Birbrair, and A. B. Gridnev, Nucl. Phys. A **575**, 673 (1994).
2. O. Scholten, A.Yu. Korchin, V. Pascalutsa, and D. Van Neck, Phys. Lett. B **384**, 13 (1996).
3. T. Feuster and U. Mosel, Phys. Rev. C **58**, 457 (1998); ibid, Phys. Rev. C **59**, 460 (1999).
4. A. Yu. Korchin, O. Scholten, and R.G.E. Timmermans, Phys. Lett. B **438**, 1 (1998).
5. V. Pascalutsa, Phys. Lett. B **503**, 85 (2001).
6. S. Kondratyuk and O. Scholten, Phys. Rev. C **59**, 1070 (1999); ibid, Phys. Rev. C **62**, 025203 (2000).
7. S. Kondratyuk and O. Scholten, Nucl. Phys. A **677**, 396 (2000).
8. S. Kondratyuk and O. Scholten, Nucl. Phys. A **680**, 175c (2001).
9. S. Kondratyuk and O. Scholten, Phys. Rev. C **64**, 024005 (2001).
10. S. Kondratyuk and O. Scholten, nucl-th/0109038.
11. E.L. Hallin et al., Phys. Rev. C **48**, 1497 (1993).
12. W. Pfeil, H. Rollnik, and S. Stankowski, Nucl. Phys. B **73**, 166 (1974).
13. J.C. Bergstrom and E.L. Hallin, Phys. Rev. C **48**, 1508 (1993).
14. T.R. Hemmert, B.R. Holstein, and J. Kambor, Phys. Rev. D **57**, 5746 (1998).
15. B.R. Holstein, hep-ph/0010129.
16. G.C. Gellas, T.R. Hemmert, and Ulf-G. Meissner, Phys. Rev. Lett. **85**, 14 (2000).
17. A.I. L'vov, V.A. Petrun'kin, and M. Schumacher, Phys. Rev. C **55**, 359 (1997).
18. D. Drechsel, M. Gorchtein, B. Pasquini, and M. Vanderhaeghen, Phys. Rev. C **61**, 015204 (2000).

SEMI-INCLUSIVE LARGE–P_T LIGHT HADRON PAIR PRODUCTION AS A PROBE OF POLARIZED GLUONS

T. MORII

Faculty of Human Development and Graduate School of Science and Technology,
Kobe University, Nada, Kobe 657-8501, Japan
E-mail: morii@kobe-u.ac.jp

Y. B. DONG

Institute of High Energy Physics, Academia Sinica, Beijing 100039, P. R. China
E-mail: dongyb@mail.ihep.ac.cn

T. YAMANISHI

Department of Management Science, Fukui University of Technology, Gakuen,
Fukui 910-8505, Japan
E-mail: yamanisi@ccmails.fukui-ut.ac.jp

We propose a new formula for extracting the polarized gluon distribution Δg in the proton from the large–p_T light hadron pair production in deep inelastic semi-inclusive reactions. Though the process dominantly occurs via photon–gluon fusion(PGF) and QCD Compton, we can remove an effect of QCD Compton from the combined cross section of light hadron pair production by using symmetry relation among fragmentation functions and thus, rather clearly extract information of Δg.

1 Introduction

Since the measurement of the polarized structure function of the proton g_1^p for wide kinematic range of Bjorken x and Q^2 by the EMC in 1988[1], the so-called proton spin puzzle has been still one of the most challenging topics in particle and nuclear physics. After the EMC experiment, much progress has been attained experimentally and theoretically.[2] A large amount of data on polarized structure functions of proton, neutron and deuteron were accumulated. The progress in the data precision is also remarkable. Furthermore, with much theoretical effort based on the next–to–leading order QCD analyses of the longitudinally polarized structure function g_1, a number of excellent parameterization models of the polarized parton distribution functions (pol–PDFs) were obtained from fitting to many/precise data with various targets for polarized deep inelastic scattering (pol–DIS) [3,4,5,6]. All of these parameterization models tell us that quarks carry a little of the proton spin, i.e. only 30% or smaller. Due to an intense experimental and theoretical effort, a rather good knowledge of the polarized valence quark distribution has been obtained so far. However, knowledge of Δg is still very limited. As is well-known, the next–to–leading order QCD analyses on g_1 bring about information on the first moment of the polarized gluon Δg.[7] However, there are large uncertainties in Δg extracted from g_1 alone. To solve the so-called spin puzzle, it is very important to precisely know how the gluon polarizes in the nucleon.

So far, a number of interesting processes such as direct prompt photon production in polarized proton–polarized proton collisions[8], open charm[9] and J/ψ[10] productions in polarized lepton scattering off polarized nucleon targets, were proposed for studying longitudinally polarized gluon distributions. Recently, HERMES group at DESY reported the first measurement of the polarized gluon distribution from di–jet analysis of semi–inclusive processes in pol–DIS, though only one data point was given as a function of Bjorken x.[11] In general, a large–p_T hadron pair is produced via photon–gluon fusion (PGF) and QCD Compton at the lowest order of QCD. The PGF gives us a direct information on the polarized gluon distribution in the nucleon, whereas QCD Compton is background to the signal process for extracting Δg.[12,13] For the case of heavy hadron pair($D\bar{D}$, $D^*\bar{D}^*$, etc.) production, we could safely neglect the contribution of QCD Compton because the heavy quark content in the proton is extremely small and also the fragmenting probability of light quarks to heavy hadrons is very small. However, in this case the cross section itself is small at the energy in the running(HERMES) or forthcoming(COMPASS) experiments and thus, we could not have enough data for carrying out detailed analysis. For the case of light hadron pair production, we will have a rather large number of events because of large cross sections. However, in this case QCD Compton can no longer be neglected and hence, it looks rather difficult to unambiguously extract the behavior of Δg from those processes.

In this work, to overcome these difficulties we propose a new formula for clearly extracting the polarized gluon distribution from the light hadron pair production of pol–DIS by removing the QCD Compton component from the cross section. As is well-known, the cross section of the hadron pair production being semi-inclusive process, can be calculated based on the parton model with various fragmentation functions. Then, by using symmetry relations among fragmentation functions and taking an appropriate combination of various hadron pair production processes, we can remove the contribution of QCD Compton from the cross section and thus, get clear information of Δg from those processes. Here, to show how to do this practically, we consider the light hadron pair production with large transverse momentum.

2 A new formula for extracting $\Delta g(x)$ from large–p_T pion pair production

Let us consider the process of $\ell + N \to \ell' + h_1 + h_2 + X$ in polarized lepton scattering off polarized nucleon targets, where h_1 and h_2 denote light hadrons in a pair. As mentioned above, the spin–dependent differential cross section at the leading order of QCD can be given by the sum of the PGF process and QCD Compton as follows,

$$d\Delta\sigma = d\Delta\sigma_{PGF} + d\Delta\sigma_{QCD} , \tag{1}$$

with

$$d\Delta\sigma = d\sigma_{-+} - d\sigma_{++} + d\sigma_{+-} - d\sigma_{++} ,$$

where, for example, $d\sigma_{-+}$ denote that the helicity of an initial lepton and the one of a target proton is negative and positive, respectively. Each term on the right hand side of eq.(1) is given by

$$d\Delta\sigma_{PGF}$$
$$\sim \Delta g(\eta, Q^2) d\Delta\hat{\sigma}_{PGF} \sum_{i=u,d,s,\bar{u},\bar{d},\bar{s}} e_i^2 \{ D_i^{h_1}(z_1, Q^2) D_{\bar{i}}^{h_2}(z_2, Q^2) + (1 \leftrightarrow 2) \}, \tag{2}$$

$$d\Delta\sigma_{QCD}$$
$$\sim \sum_{q=u,d,s,\bar{u},\bar{d},\bar{s}} e_q^2 \Delta f_q(\eta, Q^2) d\Delta\hat{\sigma}_{QCD} \{ D_q^{h_1}(z_1, Q^2) D_g^{h_2}(z_2, Q^2) + (1 \leftrightarrow 2) \} \tag{3}$$

where $\Delta g(\eta, Q^2)$, $\Delta f_q(\eta, Q^2)$ and $D_i^h(z, Q^2)$ denote the polarized gluon and the quark distribution function of q with momentum fraction η and the fragmentation function of a hadron h_i with momentum fraction z_i emitted from a parton i, respectively. $d\Delta\hat{\sigma}_{PGF}$ and $d\Delta\hat{\sigma}_{QCD}$ are the polarized differential cross sections of hard scattering subprocesses for $\ell g \to \ell' q\bar{q}$ and $\ell \overset{(-)}{q} \to \ell' g \overset{(-)}{q}$ at the leading order QCD, respectively.

Here we consider the following 4 pairs of the produced hadrons h_1 with z_1 and h_2 with z_2,

(i) (π^+, π^-) ,   (ii) (π^-, π^+) ,   (iii) (π^+, π^+) ,   (iv) (π^-, π^-) ,

where (particle 1, particle 2) corresponds to (h_1 with z_1, h_2 with z_2). Then, the differential cross section of eq.(1) for each pair can be written as

(i)
$$d\Delta\sigma^{\pi^+\pi^-}$$

$$\sim \Delta g(\eta, Q^2) d\Delta\hat\sigma_{PGF} \left\{ \frac{4}{9} D_u^{\pi^+}(z_1, Q^2) D_{\bar u}^{\pi^-}(z_2, Q^2) + \frac{1}{9} D_d^{\pi^+}(z_1, Q^2) D_{\bar d}^{\pi^-}(z_2, Q^2) \right.$$

$$\left. + \frac{1}{9} D_s^{\pi^+}(z_1, Q^2) D_{\bar s}^{\pi^-}(z_2, Q^2) + (\pi^+(z_1) \leftrightarrow \pi^-(z_2)) \right\}$$

$$+ \frac{4}{9} \Delta u(\eta, Q^2) d\Delta\hat\sigma_{QCD} \left\{ D_u^{\pi^+}(z_1, Q^2) D_g^{\pi^-}(z_2, Q^2) + D_u^{\pi^-}(z_2, Q^2) D_g^{\pi^+}(z_1, Q^2) \right\}$$

$$+ (\text{contributions from } \Delta d, \Delta s, \Delta \bar u, \Delta \bar d \text{ and } \Delta \bar s) , \tag{4}$$

(ii)

$$d\Delta\sigma^{\pi^-\pi^+}$$

$$\sim \Delta g(\eta, Q^2) d\Delta\hat\sigma_{PGF} \left\{ \frac{4}{9} D_u^{\pi^-}(z_1, Q^2) D_{\bar u}^{\pi^+}(z_2, Q^2) + \frac{1}{9} D_d^{\pi^-}(z_1, Q^2) D_{\bar d}^{\pi^+}(z_2, Q^2) \right.$$

$$\left. + \frac{1}{9} D_s^{\pi^-}(z_1, Q^2) D_{\bar s}^{\pi^+}(z_2, Q^2) + (\pi^-(z_1) \leftrightarrow \pi^+(z_2)) \right\}$$

$$+ \frac{4}{9} \Delta u(\eta, Q^2) d\Delta\hat\sigma_{QCD} \left\{ D_u^{\pi^-}(z_1, Q^2) D_g^{\pi^+}(z_2, Q^2) + D_u^{\pi^+}(z_2, Q^2) D_g^{\pi^-}(z_1, Q^2) \right\}$$

$$+ (\text{contributions from } \Delta d, \Delta s, \Delta \bar u, \Delta \bar d \text{ and } \Delta \bar s) , \tag{5}$$

(iii)

$$d\Delta\sigma^{\pi^+\pi^+}$$

$$\sim \Delta g(\eta, Q^2) d\Delta\hat\sigma_{PGF} \left\{ \frac{4}{9} D_u^{\pi^+}(z_1, Q^2) D_{\bar u}^{\pi^+}(z_2, Q^2) + \frac{1}{9} D_d^{\pi^+}(z_1, Q^2) D_{\bar d}^{\pi^+}(z_2, Q^2) \right.$$

$$\left. + \frac{1}{9} D_s^{\pi^+}(z_1, Q^2) D_{\bar s}^{\pi^+}(z_2, Q^2) + (\pi^+(z_1) \leftrightarrow \pi^+(z_2)) \right\}$$

$$+ \frac{4}{9} \Delta u(\eta, Q^2) d\Delta\hat\sigma_{QCD} \left\{ D_u^{\pi^+}(z_1, Q^2) D_g^{\pi^+}(z_2, Q^2) + D_u^{\pi^+}(z_2, Q^2) D_g^{\pi^+}(z_1, Q^2) \right\}$$

$$+ (\text{contributions from } \Delta d, \Delta s, \Delta \bar u, \Delta \bar d \text{ and } \Delta \bar s) , \tag{6}$$

(iv)

$$d\Delta\sigma^{\pi^-\pi^-}$$

$$\sim \Delta g(\eta, Q^2) d\Delta\hat\sigma_{PGF} \left\{ \frac{4}{9} D_u^{\pi^-}(z_1, Q^2) D_{\bar u}^{\pi^-}(z_2, Q^2) + \frac{1}{9} D_d^{\pi^-}(z_1, Q^2) D_{\bar d}^{\pi^-}(z_2, Q^2) \right.$$

$$\left. + \frac{1}{9} D_s^{\pi^-}(z_1, Q^2) D_{\bar s}^{\pi^-}(z_2, Q^2) + (\pi^-(z_1) \leftrightarrow \pi^-(z_2)) \right\}$$

$$+ \frac{4}{9} \Delta u(\eta, Q^2) d\Delta\hat\sigma_{QCD} \left\{ D_u^{\pi^-}(z_1, Q^2) D_g^{\pi^-}(z_2, Q^2) + D_u^{\pi^-}(z_2, Q^2) D_g^{\pi^-}(z_1, Q^2) \right\}$$

$$+ (\text{contributions from } \Delta d, \Delta s, \Delta \bar u, \Delta \bar d \text{ and } \Delta \bar s) . \tag{7}$$

Due to the isospin symmetry and charge conjugation invariance of the fragmentation functions, various fragmentation functions in eqs.(4)–(7) can be classified into the following 4 functions,[14] $D \equiv D_u^{\pi^+} = D_d^{\pi^+} = D_d^{\pi^-} = D_{\bar u}^{\pi^-}$,

$\widetilde{D} \equiv D_d^{\pi^+} = D_{\bar{u}}^{\pi^+} = D_u^{\pi^-} = D_{\bar{d}}^{\pi^-}$, $D_s \equiv D_s^{\pi^+} = D_{\bar{s}}^{\pi^+} = D_s^{\pi^-} = D_{\bar{s}}^{\pi^-}$ and $D_g \equiv D_g^{\pi^+} = D_g^{\pi^-}$, where D and $\widetilde{D}$ are called favored and unfavored fragmentation functions, respectively. Considering the suppression of the s quark contribution to the pion production compared with the u and d quark contribution, we do not identify D_s with $\widetilde{D}$.[15,16] By using these 4 kinds of pion fragmentation functions, we can make an interesting combination of cross sections which contains only the PGF contribution as follows;

$$d\Delta\sigma^{\pi^+\pi^-} + d\Delta\sigma^{\pi^-\pi^+} - d\Delta\sigma^{\pi^+\pi^+} - d\Delta\sigma^{\pi^-\pi^-} \tag{8}$$

$$\sim \frac{10}{9}\Delta g(\eta, Q^2)d\Delta\hat{\sigma}_{PGF}\left\{D(z_1, Q^2) - \widetilde{D}(z_1, Q^2)\right\}\left\{D(z_2, Q^2) - \widetilde{D}(z_2, Q^2)\right\}.$$

From this combination, we can calculate the double spin asymmetry A_{LL} defined by

$$A_{LL} = \frac{d\Delta\sigma^{\pi^+\pi^-} + d\Delta\sigma^{\pi^-\pi^+} - d\Delta\sigma^{\pi^+\pi^+} - d\Delta\sigma^{\pi^-\pi^-}}{d\sigma^{\pi^+\pi^-} + d\sigma^{\pi^-\pi^+} - d\sigma^{\pi^+\pi^+} - d\sigma^{\pi^-\pi^-}}$$

$$= \frac{\Delta g(\eta, Q^2)}{g(\eta, Q^2)} \cdot \frac{d\Delta\hat{\sigma}_{PGF}}{d\hat{\sigma}_{PGF}}, \tag{9}$$

where the factor of the fragmentation function in eq.(8) is dropped out from the numerator and the denominator of A_{LL}.[a] Therefore, from the measured A_{LL}, one can get clear information of $\Delta g/g$ with reliable calculation of $d\Delta\hat{\sigma}_{PGF}/d\hat{\sigma}_{PGF}$.

3 Numerical calculation of cross section and double spin asymmetry

Now, let us calculate numerically the cross section and double spin asymmetry A_{LL} for the large-p_T pion pair production of pol–DIS. The spin–independent (spin–dependent) differential cross sections for producing hadrons h_1 and h_2 are given by[17]

$$\frac{d(\Delta)\sigma^{h_1 h_2}}{dz_1 d\cos\theta_1 dz_2 d\cos\theta_2 dxdy} = \frac{d(\Delta)\sigma_{PGF}^{h_1 h_2}}{dz_1 d\cos\theta_1 dz_2 d\cos\theta_2 dxdy}$$

$$+ \frac{d(\Delta)\sigma_{QCD}^{h_1 h_2}}{dz_1 d\cos\theta_1 dz_2 d\cos\theta_2 dxdy}. \tag{10}$$

---

[a] For large Q^2 regions, heavy quarks might contribute to the PGF and QCD Compton. Even so, the final form of eq.(9) for A_{LL} remains unchanged if $D_Q^{\pi^+} = D_{\bar{Q}}^{\pi^+} = D_Q^{\pi^-} = D_{\bar{Q}}^{\pi^-}$.

Each term in the right hand side of eq.(10) is written as

$$\frac{d(\Delta)\sigma_{PGF}^{h_1 h_2}}{dz_1 d\cos\theta_1 dz_2 d\cos\theta_2 dxdy}$$

$$= (\Delta)g(\eta, Q^2)\, C(\theta_1, \theta_2)\frac{d(\Delta)\widehat{\sigma}_{PGF}^{h_1 h_2}}{dz_i d\cos\theta_1 dz_{\bar{i}} d\cos\theta_2 dx_g dy}$$

$$\times \sum_{i=u,d,s,\bar{u},\bar{d},\bar{s}} e_i^2\{D_i^{h_1}(\xi_1, Q^2)D_{\bar{i}}^{h_2}(\xi_2, Q^2) + (1 \leftrightarrow 2)\}, \tag{11}$$

$$\frac{d(\Delta)\sigma_{QCD}^{h_1 h_2}}{dz_1 d\cos\theta_1 dz_2 d\cos\theta_2 dxdy}$$

$$= \sum_{q=u,d,s,\bar{u},\bar{d},\bar{s}} e_q^2(\Delta)f_q(\eta, Q^2)\, C(\theta_1, \theta_2)\frac{d(\Delta)\widehat{\sigma}_{QCD}^{h_1 h_2}}{dz_q d\cos\theta_1 dz_g d\cos\theta_2 dx_q dy}$$

$$\times\{D_q^{h_1}(\xi_1, Q^2)D_g^{h_2}(\xi_2, Q^2) + (1 \leftrightarrow 2)\}, \tag{12}$$

where $\eta = x + (1-x)\tau_1\tau_2$, $Q^2 = xys$, $\xi_1 = \left(\frac{\tau_1+\tau_2}{\tau_2}\right)z_1$, $\xi_2 = \left(\frac{\tau_1+\tau_2}{\tau_1}\right)z_2$ and $C(\theta_1, \theta_2) = \frac{(\tau_1+\tau_2)^2}{\eta\tau_1\tau_2}\frac{1-x}{8\cos^2\frac{1}{2}\theta_1\cos^2\frac{1}{2}\theta_2\sin^2\frac{1}{2}(\theta_1+\theta_2)}$, with $\tau_{1,2} = \tan\frac{1}{2}\theta_{1,2}$.

Here we simply assume the scattering angle of outgoing hadrons $\theta_{1,2}$ to be the same with the one of scattered partons in the virtual photon–nucleon c.m. frame. This assumption might not be unreasonable if observed particles are light hadrons with high energy. s is the total energy square of the lepton scattering off the nucleon. x, y and $z_{1,2}$ are familiar kinematic variables for semi–inclusive processes in DIS, defined as $x = \frac{Q^2}{2P \cdot q}$, $y = \frac{P \cdot q}{P \cdot \ell}$ and $z_{1,2} = \frac{P \cdot P_{1,2}}{P \cdot q}$, where ℓ, q, P and $P_{1,2}$ are the momentum of the incident lepton, virtual photon, target nucleon and outgoing hadrons, respectively. At the leading order of QCD, differential cross sections of hard scattering subprocesses, $\ell g \to \ell' q\bar{q}$ and $\ell \overset{(-)}{q} \to \ell' g \overset{(-)}{q}$, with two outgoing partons in an opposite azimuth angle are given as[18]

$$\frac{d(\Delta)\widehat{\sigma}_{PGF(QCD)}}{dz_i d\cos\theta_1 d\phi_1 dz_{\bar{i}(g)} d\cos\theta_2 dx_{g(q)} dyd\phi_\ell} = \frac{1}{128\pi^2}\frac{\alpha^2\alpha_s}{(p_0 \cdot \ell)Q^2}\frac{y(\eta-x)(1-\eta)^2}{x}$$

$$\times B(\theta_1, \theta_2)\, e_\ell^2\, e_i^2|(\Delta)M|_{PGF(QCD)}^2, \tag{13}$$

with

$$\frac{1}{B(\theta_1, \theta_2)} = \sin(\theta_1 + \theta_2)$$

$$\times \left[\frac{\{z_i(1-\eta) + (\eta-x)\}\sin\theta_1 + \{z_{\bar{i}(g)}(1-\eta) + (\eta-x)\}\sin\theta_2}{\sin\theta_1\sin\theta_2}\right], \tag{14}$$

where z_i, $z_{\bar{i}}$ and z_g are the momentum fraction of the outgoing parton i, $\bar{i}$ and g, respectively, to the incoming parton, and are given as $z_i = \frac{\tau_2}{\tau_1+\tau_2}$, $z_{\bar{i}(g)} = \frac{\tau_1}{\tau_1+\tau_2}$.[17] The amplitude $|(\Delta)M|^2_{PGF(QCD)}$ in eq.(13) is presented elsewhere[19].

By using these formulas and pion fragmentation functions[15], we have calculated the spin–dependent and spin–independent cross sections of the large–p_T pion pair production and estimated the double spin asymmetry at the energy of COMPASS experiments.[b] Here, we have taken the AAC[6] and GS96[4] parameterizations at LO QCD as polarized parton distribution functions and GRV98[20] and MRST98[21] as unpolarized ones. At $\sqrt{s} = 13.7$GeV, $y = 0.75$, $Q^2 \geq 1$GeV^2 and $W^2 \geq 10$GeV^2 with kinematical values of $\theta_{1,2}$ and $z_{1,2}$ for the produced pion pair, the calculated results of the spin–independent (spin–dependent) differential cross sections and A_{LL} are shown as a function of η in Figs.1 and 2, respectively. In Fig.2, error was estimated by[22] $\delta A_{LL} = \frac{1}{A_c P \sqrt{L}} \sqrt{\frac{1}{d\sigma_1} + \frac{1}{d\sigma_2}}$, where $A_c = \frac{d\sigma_1 - d\sigma_2}{d\sigma_1 + d\sigma_2}$ with $d\sigma_1 = d\sigma(\pi^+\pi^-) + d\sigma(\pi^-\pi^+)$ and $d\sigma_2 = d\sigma(\pi^+\pi^+) + d\sigma(\pi^-\pi^-)$. L is the integrated luminosity and $P = P_B P_T f$, where P_B, P_T are the polarization of beam and target, respectively, and f is the dilution factor. Here we used $L = 2$ fb^{-1}(150 days running), $P_B = 0.80$ and $P_T \cdot f = 0.25$. From Fig.2, one can see a big difference of the behavior of A_{LL} depending on the models of $\Delta g/g$ and hence, we can extract the behavior of Δg rather clearly from this analysis.

4 Summary and discussion

We proposed a new formula for extracting the polarized gluon distribution from the large–p_T light hadron pair production in pol–DIS by taking an appropriate combination of hadron pair productions. Since the double spin asymmetry A_{LL} for this combination is directly proportional to $\Delta g/g$, the measurement of this quantity is quite promising for getting rather clear information on the polarized gluon distribution in the nucleon.

The analysis can be applied also for the kaon pair or the proton pair production by considering the reflection symmetry along the V–spin axis, the isospin symmetry and charge conjugation invariance of the fragmentation functions.

The same combinations of cross sections for light hadron pair production were discussed for photoproduction by other people[23], while we studied the lepton-proton processes in deep inelastic region in this work. Both of those processes are useful for extracting Δg.

---

[b]Calculation for HERMES energy was done[18] without error estimation.

534

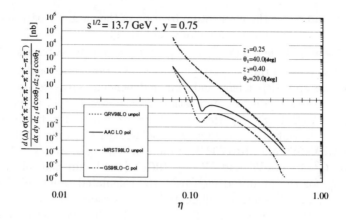

Figure 1. η dependence of combined spin-independent and spin-dependent differential cross sections defined at the denominator and numerator, respectively, of eq.(9) as a function of η at $\sqrt{s} = 13.7$ GeV, $y = 0.75$ for the deep inelastic regions ($Q^2 \geq 1\text{GeV}^2$ and $W^2 \geq 10$ GeV^2) with kinematical values of $\theta_{1,2}, z_{1,2}$ for the produced pion pair.

This work is supported by the Grant–in–Aid for Science Research, Ministry of Education, Science and Culture, Japan (No.11694081).

References

1. J. Ashman et al.(EMC), Phys. Lett. **B206** (1988) 364; Nucl. Phys. **B328** (1989) 1.
2. For recent reviews, see B. Lampe and E. Reya, Phys. Rep. **332** (2000) 1; B. W. Filippone and X.-D. Ji, hep-ph/0101224.
3. M. Glück, E. Reya, M. Stratmann and W. Vogelsang, Phys. Rev. **D53** (1996) 4775; hep-ph/0011215.
4. T. Gehrmann and W. J. Stirling, Phys. Rev. **D53** (1996) 6100.
5. L. E. Gordon, M. Goshtasbpour and G. P. Ramsey, Phys. Rev. **D58** (1998) 094017; G. P. Ramsey, Prog. Part. Nucl. Phys. **39** (1997) 599; E. Leader, A. V. Sidrov and D. B. Stamenov, Phys. Rev. **D58** (1998) 114028; Phys. Lett. **B445** (1998) 232; *ibid.* **B462** (1999) 189.

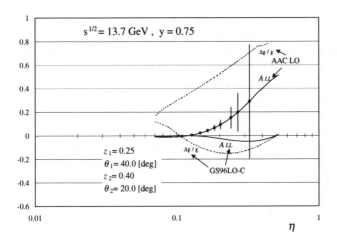

Figure 2. η dependence of A_{LL} at $\sqrt{s}$ = 13.7 GeV, y = 0.75 for kinematical values of $\theta_{1,2}, z_{1,2}$ for the produced pion pair. Solid line and dotted line are for AAC LO and GS96LO-C parameterizaton models, respectively. $\Delta g/g$ itself is also presented for both parameterization models.

6. Y. Goto et al., Asymmetry Analysis Collaboration, Phys. Rev. **D62** (2000) 034017.
7. R. Mertig and W. L. van Neerven, Z. Phys. **C70** (1996) 637; W. Vogelsang, Phys. Rev. **D54** (1996) 2023.
8. N. S. Craigie, K. Hidaka, M. Jacob and F. M. Renard, Phys. Rep. **99** (1983) 69.
9. A. D. Watson, Z. Phys. **C12** (1982) 123; M. Glück and E. Reya, Z. Phys. **C39** (1988) 569; G. Altarelli and W. J. Stirling, Particle World 1 (1989) 40; M. Glück, E. Reya and W. Vogelsang, Nucl. Phys. **B351** (1991) 579.
10. T. Morii, S. Tanaka and T. Yamanishi, Phys. Lett. **B322** (1994) 253.
11. A. Airapetian, et al., HERMES Collaboration, Phys. Rev. Lett. **84** (2000) 2584.
12. A. Bravar, D. von Harrach and A. Kotzinian, Phys. Lett. **B421** (1998) 349.
13. A. De Roeck, A. Deshpande, V. W. Hughes, J. Lichtenstadt and G. Rädel, Eur. Phys. J. **C6** (1999) 121.

14. S. Kumano, Phys. Rep. **303** (1998) 183.
15. S. Kretzer, Phys. Rev. **D62** (2000) 054001.
16. K. Abe, et al., SLD Collaboration, Phys. Rev. Lett. **78** (1997) 3442; *ibid.* **79** (1997) 959; G. Abbiendi, et al., OPAL Collabration, hep–ex/0001054.
17. R. D. Peccei and R. Rückl, Phys. Lett. **B84** (1979) 95; Nucl. Phys. **B162** (1980) 125.
18. Yu-Bing Dong, T. Morii and T. Yamanishi, Phys. Lett. **B520** (2001) 99.
19. E. Mirkes, hep–ph/9711224.
20. M. Glück, E. Reya and A. Vogt, Eur. Phys. J. **C5** (1998) 461.
21. A. D. Martin, R. G. Roberts, W. J. Stirling and R. S. Thorne, Eur. Phys. J. **C4** (1998) 463.
22. T. Iwata, private communication.
23. G. Grispos, A. P. Contogouris and G. Veropoulos, Phys. Rev. **62** (2000) 014023.

STRANGE-QUARK CURRENT IN THE NUCLEON FROM LATTICE QCD

RANDY LEWIS

Department of Physics, University of Regina, Regina SK, Canada S4S 0A2

W. WILCOX

Department of Physics, Baylor University, Waco TX 76798-7316, U.S.A.

R.M. WOLOSHYN

TRIUMF, 4004 Wesbrook Mall, Vancouver BC, Canada V6T 2A3

The contribution of the strange-quark current to the electromagnetic form factors of the nucleon is studied using lattice QCD. The strange current matrix elements from our lattice calculation are analyzed in two different ways, the differential method used in an earlier work by Wilcox and a cumulative method which sums over all current insertion times. The preliminary results of our simulation indicate the importance of high statistics, and that consistent results between the varying analysis methods can be achieved. Although this simulation does not yet yield a number that can be compared to experiment, several criteria useful in assessing the robustness of a signal extracted from a noisy background are presented.

1   Introduction

An important theme of contemporary Hadron Physics is the role of nonvalence degrees of freedom. In particular, the contribution of strange quarks to a variety of nucleon properties has been studied.[1] For nucleon form factors the contribution of the strange-quark current can be extracted using information obtained from parity violation in polarized electron scattering.[2] A number of experimental results have been reported[3,4] and new measurements are planned. As well, there are numerous calculations using a number of different approaches.[2]

In this work we focus on the calculation of the strange-quark current loop using lattice QCD. Calculations of these so-called disconnected current insertions have been reported previously.[5,6,7] However, these calculations differ in the method used to analyze the results of their lattice simulations and also differ in their conclusions. Wilcox[5] used a differential method to analyze the time dependence of the three-point function and no signal for the strange current was found. On the other hand, Dong and co-workers [6,7] used a summation of the current insertion over lattice times which requires a further identification and fitting of a linear time dependence. They claim to see a

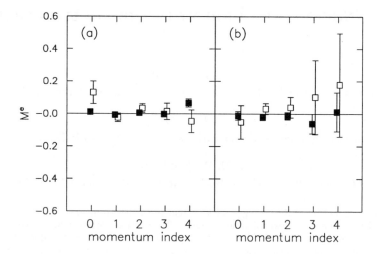

Figure 1. Comparison of the electric matrix element calculated with the differential method for samples of 100 configurations (open symbols) and 1050 configurations (filled symbols). (a) Average of times 10 to 12. (b) Average of times 15 to 17.

definite signal for the strangeness form factors.

We present preliminary results from a lattice QCD simulation of the strange-quark current loop using a Monte-Carlo sample of gauge field configurations about ten times larger than in previous studies. Both differential and cumulative time analyses are carried out on the simulation data. Comparisons of results obtained from a 100 configuration subsample (the same size used in previous work[5,6,7]) with results from the full data set show quite clearly how large fluctuations, that could be interpreted as a signal in low-statistics data, disappear with improved statistics. Using our complete data set consistency between different analysis methods is achieved. By comparing different analysis methods and results using different size gauge field samples, criteria for assessing the robustness of a strange quark current signal are presented. Using these criteria, no compelling signal for the strange quark current is observed.

2   Lattice Calculations

The standard methods of the path integral formulation of lattice QCD in Euclidean space are used. The Wilson action is used for both quark and gluon fields. We need two and three-point functions which describe, respectively, the

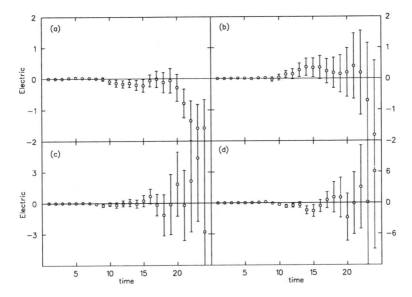

Figure 2. Electric matrix element using the cumulative method for a sample of 100 configurations. The plots correspond to momentum transfer (a) (1,0,0), (b) (1,1,0), (c) (1,1,1), (d) (2,0,0).

propagation of nucleon states as a function of Euclidean lattice time and the strange quark current in the presence of a nucleon. The two-point function $G^{(2)}(t; \vec{q})$ correlates the excitation of a nucleon state with momentum $\vec{q}$ at some initial time (called 0) and its annihilation at time t. For large Euclidean time t this quantity decreases exponentially like $e^{-E_q t}$.

To calculate the three-point function $G^{(3)}(t, t'; \vec{q})$ an insertion of the strange-quark vector current is made at time t'. Since there are no strange valence quarks in the nucleon this insertion amounts to a correlation of a strange-quark current loop with the nucleon two-point function. The strange-quark current loop is calculated using a so-called noisy estimator with Z_2 noise[8,9]. The noise and perturbative subtraction methods employed here are the same as those of Wilcox[5] expect that 60 noises are used instead of 30.

The three-point function also has an exponential time dependence but is more complicated due to the presence of two times. It can be shown that the exponential time factors can be cancelled by taking the ratio

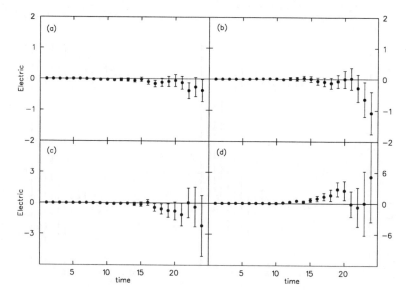

Figure 3. Electric matrix element using the cumulative method for a sample of 1050 configurations. The plots correspond to momentum transfer (a) (1,0,0), (b) (1,1,0), (c) (1,1,1), (d) (2,0,0).

$$R(t, t'; \vec{q}) = \frac{G^{(3)}(t, t'; \vec{q})G^{(2)}(t'; 0)}{G^{(2)}(t; 0)G^{(2)}(t'; \vec{q})}. \tag{1}$$

It is these ratios R that are analyzed to get the final results.

Note that in writing the two and three-point functions, labels associated with the Dirac indices of the nucleon fields and the Lorentz index of the current have been suppressed. Also in writing Eq. (1) the necessary spin sums and projections are not shown, the expression is given schematically to show the time dependence of the various factors. The detailed calculations follow previous work.[5,6,7]

The ratio R is usually summed in to order to try to improve the signal. The differential method uses a difference of $R(t, t'; \vec{q})$ on neighbouring time slices. It can be shown[5] that the quantity $M(\bar{t}, \vec{q})$ given by

$$M(\bar{t}, \vec{q}) = \sum_{t'=1}^{t+1} (R(t, t'; \vec{q}) - R(t - 1, t'; \vec{q})) \tag{2}$$

is the current matrix element of interest.

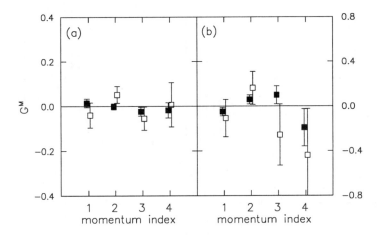

Figure 4. Comparison of the magnetic matrix element calculated with the differential method for samples of 100 configurations (open symbols) and 1050 configurations (filled symbols) using $\kappa = 0.152$ for the valence quark. (a) Average of times 10 to 12. (b) Average of times 15 to 17.

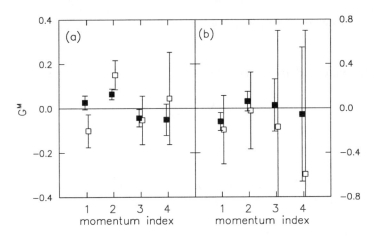

Figure 5. Comparison of the magnetic matrix element calculated with the differential method for samples of 100 configurations (open symbols) and 1050 configurations (filled symbols) using $\kappa = 0.154$ for the valence quark. (a) Average of times 10 to 12. (b) Average of times 15 to 17.

An alternative is to simply sum R and then fit to a linear time dependence

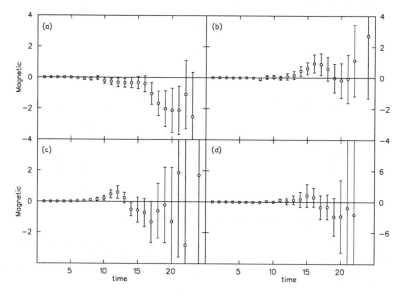

Figure 6. Magnetic matrix element using the cumulative method for a sample of 100 configurations with $\kappa = 0.152$ valence quark. The plots correspond to momentum transfer (a) (1,0,0), (b) (1,1,0), (c) (1,1,1), (d) (2,0,0).

to get the matrix element. We use

$$S(t, \vec{q}) = \sum_{t'=1}^{t'=t} R(t, t'; \vec{q}),$$ (3)

$$\rightarrow constant + tM(t, \vec{q}).$$ (4)

Summing current insertions up to $t' = t$ follows the suggestion of Viehoff et al.[9] and helps to reduce the statistical noise. Dong and co-workers[6,7] actually use a different upper limit $(t' = t_{fixed}, t_{fixed} > t)$.

3 Results

Calculations were carried out in quenched approximation at gauge field coupling of $\beta = 6.0$. The lattice size was $20^3 \times 32$. A total of 1050 gauge field configurations were generated using a pseudo-heat bath algorithm. Two thousand sweeps were used between saved configurations.

Since quenched lattice QCD does not provide a perfect description of hadrons, there is some ambiguity in fixing the parameters (overall scale and

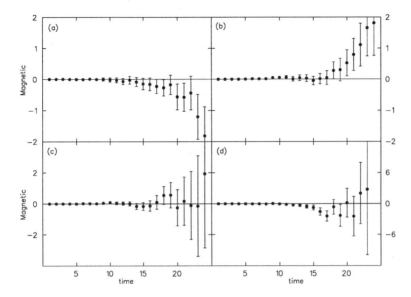

Figure 7. Magnetic matrix element using the cumulative method for a sample of 1050 configurations with $\kappa = 0.152$ valence quark. The plots correspond to momentum transfer (a) (1,0,0), (b) (1,1,0), (c) (1,1,1), (d) (2,0,0).

quark masses) in the calculations. In this work we use 0.152 as the hopping parameter for the strange quark. Using $a^{-1} = 2GeV$ and the ϕ-meson to fix the strange quark mass would suggest a hopping parameter closer to κ_s =0.153. On the other hand, using the scale of Dong and co-workers[6,7] $a^{-1} = 1.74GeV$ gives κ_s smaller than 0.152.

Results for electric matrix element with valence quark κ_v =0.152 calculated using the differential method Eq. (2) are shown in Fig. (1), averaging over two different time windows. The 100 configuration sample results are consistent with those obtained by Wilcox[5]. No signal is seen when the gauge field sample size is increased to 1050. The summed ratio Eq. (3) is shown in Fig. (2) and (3) as function of the nucleon sink time for different momentum transfers. It shows oscillations characteristic of lattice correlation function ratios[10]. The magnitude of these oscillations decreases slowly as the configuration sample size is increased.

Results for magnetic matrix element for different valence quark masses calculated using the differential method are plotted in Fig. (4) and (5). As in the electric case the results are consistent with zero.

Finally, the summed ratios for the magnetic current are given in Fig. (6)

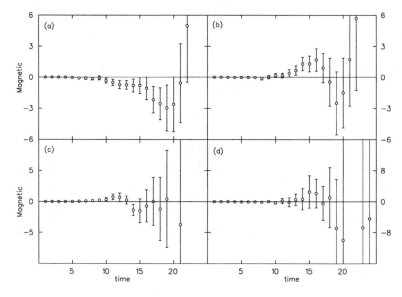

Figure 8. Magnetic matrix element using the cumulative method for a sample of 100 configurations with $\kappa = 0.154$ valence quark. The plots correspond to momentum transfer (a) (1,0,0), (b) (1,1,0), (c) (1,1,1), (d) (2,0,0).

to Fig. (9). Note that a kinematic factor (see Eq. (3) in Ref.[6]) of $q/(E+M)$ has been removed. The results for 100 configurations, Fig. (8) and (8), should be compared to Fig. (1) of Dong et al.[6] and Fig. (2) of Mathur and Dong[7] where calculations with the same statistics are reported. Then, comparing with Fig. (7) and (9), one sees that the kind of oscillations in the time range 10 to 15 which Dong and co-workers[6,7] took to be their signal, have largely disappeared in the higher statistics results. Of course, fluctuations still persist at larger times but even higher statistics simulations will be necessary to establish if they go away or if indeed a strange quark current signal is hiding under them.

4 Conclusions

To get an estimate of the strange-quark current matrix elements requires the extraction of a small signal in the presence of large statistical fluctuations. The results presented here suggest a number of useful criteria that should be met before one can claim a credible signal:

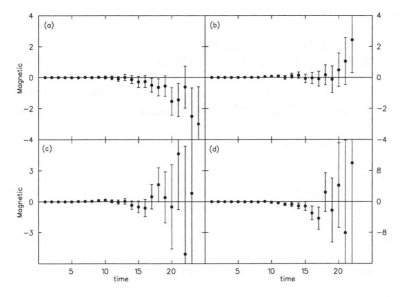

Figure 9. Magnetic matrix element using the cumulative method for a sample of 1050 configurations with $\kappa = 0.154$ valence quark. The plots correspond to momentum transfer (a) (1,0,0), (b) (1,1,0), (c) (1,1,1), (d) (2,0,0).

- There should be consistency between different analysis methods.

- The signal should appear in the same lattice time region and its statistical significance should increase as the size of the Monte-Carlo sample of gauge fields is increased.

- The signal should appear in the same time window for different masses.

Not all of these criteria have been met at our present level of statistics. Work is continuing and it is hoped to have final results with an increased configuration sample size in the not too distant future.

Acknowledgments

This is work is supported in part by the National Science Foundation under grant 0070836 and in part by the Natural Sciences and Engineering Research Council of Canada. We thank N. Mathur for helpful discussions.

References

1. K. F. Liu, *J. Phys. G* **27**, 511 (2001).
2. D. H. Beck and R. D. McKeown, arXiv:hep-ph/0102334 contains an extensive review.
3. R. Hasty *et al.* [SAMPLE Collaboration], *Science* **290**, 2117 (2000).
4. K. A. Aniol *et al.* [HAPPEX Collaboration], *Phys. Lett.* B **509**, 211 (2001).
5. W. Wilcox, *Nucl. Phys. Proc. Suppl.* **94**, 319 (2001).
6. S. J. Dong, K. F. Liu and A. G. Williams, *Phys. Rev.* D **58**, 074504 (1998).
7. N. Mathur and S. J. Dong, *Nucl. Phys. Proc. Suppl.* **94**, 311 (2001).
8. S. J. Dong and K. F. Liu, *Nucl. Phys. Proc. Suppl.* **26**, 353 (1992).
9. S. J. Dong and K. F. Liu, *Phys. Lett.* B **328**, 130 (1994)
10. J. Viehoff *et al.* [SESAM Collaboration], *Nucl. Phys. Proc. Suppl.* **63**, 269 (1998).
11. S. Aoki *et al.* [JLQCD Collaboration], *Nucl. Phys. Proc. Suppl.* **47**, 354 (1996).

QUANTUM ENTANGLEMENT MEASUREMENTS OF TWO SPIN 1/2 HADRONS

C.RANGACHARYULU

Department of Physics, University of Saskatchewan,
Saskatoon, SK., Canada, S7N 5E2
E-mail: chary@sask.usask.ca

In recent years, the meaning of quantum entanglement has received renewed interest due to possible implications to the emerging fields of quantum computing and information theory. We have recently carried out feasibility studies of these measurements on the $^1S_0$ pair of protons produced in the $12C(d,^2He)^{12}B$ reaction. This talk will present the motivation of these studies, present status and future prospects.

1 Introduction

It is well known that Einstein, Podolsky and Rosen [1] published a paper questioning the completeness of physical reality as described by quantum mechanics. During the last 65 years, this question has been refined by several authors, notably by Bohm [2], and Bell [3], which made it possible to test the predictions of quantum mechanics against those of hidden variable theories. These topics have been treated in an exhaustive manner in the book of Duck and Sudarshan [4] and review article by Laloë [5]. However, a few recent developments are worth mentioning. The first is the contextuality theorem due to Bell, Kochen and Specker. It refers to the constraints to an observable due to the values that orthogonal components may take. It is best illustrated by the spin projections of a pair of spin 1/2 particles in a triplet state. The spin eigen value equations are

$$S^2\psi = (S_x^2 + S_y^2 + S_z^2)\psi = S(S+1)\psi \tag{1}$$

For a particle of spin one, the eigen values are limited to $(1,1,0),(1,0,1)$ and $(0,1,1)$. It thus implies that if a projection is measured to be zero, then the other two each have a magnitude one. If a projection is measured to be of magnitude one, then there is another projection of zero value.

Wigner [6] gave another interesting example where one can look for contradictions with quantum mechanics(QM) or Bell's inequalities, which can be illustrated as below: Consider a spin-singlet state of the pair: $\sigma_+(1)\sigma_-(2) -$

$\sigma_-(1)\sigma_+(2)$. According to QM, the probability that both components are either positive or negative, along any directions is

$$P_{++} = P_{--} = \frac{1}{2}sin^2\left[\frac{\theta_{12}}{2}\right] \qquad (2)$$

And the probability that they have opposite signs for the components is

$$P_{-+} = P_{+-} = \frac{1}{2}cos^2\left[\frac{\theta_{12}}{2}\right] \qquad (3)$$

Wigner considers a much general case. Take three directions in quite an arbitrary manner. Denote $(+ - +; - + +)$, the probability-weighted integral over the values of hidden variables.

Of all the possibilities, only eight combinations mentioned above exist, since there is a constraint that sum of the spin projections along any direction is zero.

Write down the probabilities for those outcomes of measurements which yield a positive spin component of both partilces if these are measured in two different directions and equate them to the above equations. If the spin components are measured in e_1 and e_2 directions, we have for the probability the positive components is

$$(+, -, +; - + -) + (+ - -; - + +) = \frac{1}{2}sin^2\left(\frac{\theta_{12}}{2}\right) \qquad (4)$$

For the measurements in e_2 and e_3 directions, we have the probabilities

$$(+ + -; - - +) + (- + -; + - +) = \frac{1}{2}sin^2\left(\frac{\theta_{23}}{2}\right) \qquad (5)$$

And for the

$$(+ + -; - - +) + (+ - -; - + +) = \frac{1}{2}sin^2\left(\frac{\theta_{31}}{2}\right) \qquad (6)$$

From these three equations, we deduce

$$\frac{1}{2}sin^2\left(\frac{\theta_{12}}{2}\right) + \frac{1}{2}sin^2\left(\frac{\theta_{23}}{2}\right) = \frac{1}{2}sin^2\left(\frac{\theta_{13}}{2}\right) + 2(+-+;-+-) + 2(-+-;+-+)$$

$$\geq \tfrac{1}{2}sin^2\left(\tfrac{\theta_{13}}{2}\right) \tag{7}$$

This is Bell's inequality. It is easy to find where the inequality does not hold. Eg:

$$\theta_{12} = \theta_{23} = \frac{\pi}{3} \quad \& \quad \theta_{13} = \frac{2\pi}{3} \tag{8}$$

does violate this inequality.

More recently, several theoretical attempts have been made to offer Bell's theorem without inequalities. Note worthy among them are the GHZ theorem [7] and refinements due to Peres [8] and Mermin [9]. These authors show that the contextuality is relevant for observables corresponding to some commuting operators. Hardy [10] presents a Gedanken experiment from which he deduces that if realism is retained, quantum mechanics implies both nonlocality and violation of Lorentz invariance.

In the recent years, the question of entanglement has become increasingly important due to interest in quantum computing and quantum information theory [11].

It is worth mentionining that, despite the fact that the works of both Bell, Bohm and Wigner refer to the entanglement of pairs of spin $1/2$ particles, most of the experimental work has been carried out using pairs of photons, especially atomic cascades and laser down conversion photons [12,13]. A notable exception, of particular relevance to the present work was due to Lamehi-Rachti and Mittig [14].

They measured polarization correlations of proton pairs, a result of elastic scattering of 13 MeV protons off a hydrogen target. Two carbon scatterers, one in the path of each of the outgoing protons served as the polarization analyzers. Four solid state detectors were employed to measure the left-right asymmetries. As the correlations betweeen the two protons were of interest, they defined a correlation function as

$$P_{measured}(\theta) = \frac{N_{LL} + N_{RR} - N_{LR} - N_{RL}}{N_{LL} + N_{RR} + N_{LR} + N_{RL}} \tag{9}$$

where N_{LL} are the number of counts with protons in both arms detected in the detectors on the left side of their flight direction. Similar terminology applies for the other terms. The analyzer detectors for one arm were kept in the scattering plane, while the other pair was set at a few orientations off the scattering plane denoted by the angle θ. This pioneering experiment, nevertheless, had some deficiencies. For one thing, the final result was a record of coincidences only. Due to the limited geometrical acceptance of the solid state detectors, the system was blind to most of the phase space. This, what we nuclear physicists term as trigger bias, lends to the criticism of observer dependent reality. The choice of reference axes was a part of the experimental design, which also contributes to the concerns about the information exchange between the two protons in the final states.

Secondly, the limited acceptance also rendered the predictions of Bell's inequalities and that of quantum mechanics not much different, thus the experiment was not very discriminating between the two results. From the experimental result, one can confidently conclude that the qunatum mechanical predictions are proved correct, but at the same time they do not prove Bell's inequalities to be invalid.

2 The present experiment

The current project was undertaken to improve on this experiment in several ways to provide a stringent test of Bell's inequalities and later developments.

2.1 $^1S_0$ State Preparation

A prerequisite for the success of this experiment is to ensure that pairs of protons in pure $^1S_0$ state are prepared. To this end, unlike the scattering state of the previous experiment, we resort to a resonant state formation in the ^2He configuration. Of particular interest is the single charge exchange reaction of the type $^AX_Z(d,^2He)^AY_{Z-1}$ leaving the recoil nucleus in a bound state. Due to Pauli exclusion principle, the ^2He states are limited to $^1S_0$, $^1D_2$, ... configurations for the positive parity pairs, with $^1S_0$ as the lowest energy state. From the earlier nuclear spectroscopy studies [15], it is known that a constraint on the internal energy of ^2He defined as the $M(^2He^*) - M(^2He) \leq 0.5MeV$ ensures that the intermediate $^2He^*$ is in the $^1S_0$ state with a probability of more than 99.9%. It is to be remarked that the analogue nucleus ^2d has only one stable state, viz., the ground state $J^\pi = 1^+, T = 0$ and one quasi-bound state at 2.3 MeV excitation $(J^\pi, T = 1^+, 0)$ and it is the analogue of the ^2He lowest energy state. The absence of other excited levels in the ^2d allows one to conclude that a well defined resonant state ensures that the ^2He is indeed of $^1S_0$ configuration. The reconstruction of the $M(^2He^*)$ can be done with invariant mass and missing mass techniques very common to nuclear and particle physics experiments.

With the availability of intermediate energy high resolution facilities such as AGOR cyclotron at KVI, Groningen, Ring cyclotron at RCNP, Osaka and RIKEN facility in Tokyo, these experiments are possible. A resonant state thus formed has a life time of less than 10^{-20} seconds. As the final state protons are flying at speeds greater than 0.5 c, space-like separation in the context of strong interactions is assured. Several nuclear targets are promising candidates for this measurement. The self-conjugate ^{12}C nucleus and hydrogen target are of particular interest. The reaction $^{12}C(d,^2He)$ can be used to populate the isobaric analogue states in the residual ^{12}B nucleus with well defined angular momentum and parities. Among them, the ground state of ^{12}B, analogue of the 15.11 MeV excited level in ^{12}C of quantum numbers $J^\pi, T = 1^+, 1$ is well suited for these studies. The reaction process can be characterized as

$$^{12}C \; + \; d \Rightarrow ^{12}B +^2 He$$
$$J^\pi, T \quad 0^+, 0 \quad + \quad 1^+, 0 \Rightarrow 1^+, 1 \quad + \quad 0^+, 1 \;(10)$$

which amounts to spin and isospin-flip process. Thus, the $^1S_0$ configuration is favoured from energy and angular momentum considerations. The facilities at KVI and RIKEN with resolutions of about 200 keV are well suited for this purpose. The process $^1H + d \to^2 He + n$ is also a charge exchange reactions with a large production cross section [16]. This channel has the additional advantage that the recoil neutron is a well defined state with no phase space spreading ambiguities once the two protons in the final state are detected.

2.2 Preliminary Measurements

In summer 2001, we [17] ran a test experiment at the AGOR facility of KVI, Groningen. We used the Focal plane polarimeter system of the Big Byte Spectrometer at the KVI. The experimental layout is shown in the figure 1.

A 170 MeV deuteron beam of 1 nA beam current was incident on a carbon target of $9mg/cm^2$ thickness. The emerging protons were detected in the BBS set up at zero degrees in the large momentum acceptance (19%). Two drift chambers VDC1 and VDC2 serve to determine the tracks of the two protons for measuring the momentum vectors. A carbon target of $40mg/cm^2$ thickness serves as the second scatterer for polarization analyses. The wire

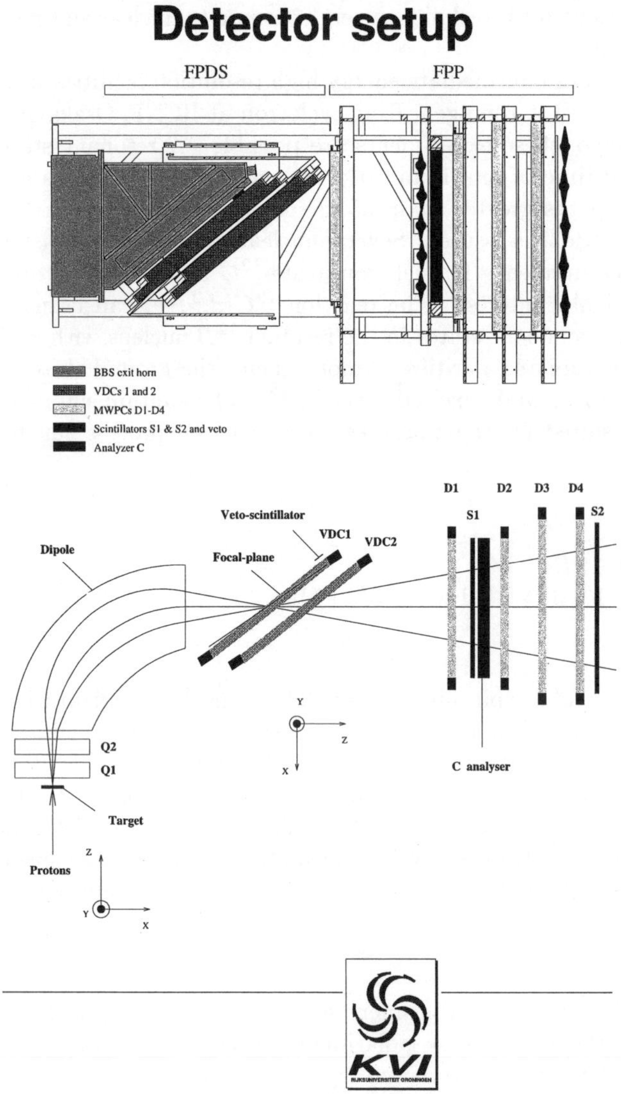

Detector setup

Figure 1. The layout of the Focal plane detector system (FPDS) and Focal plane polarimeter(FPP) set up at the Big Byte Spectrometer of AGOR facility, KVI, Groningen,NL.

chambers D1, D2, D3 and D4 are used to the determine the directions of protons after the scatter in the analyzer. The plastic scintillator paddles S1 and S2 measure the time of flight, which serve the purpose of particle identification. The segmentation of S1 and S2 provides a moderate position resolution and more importantly, it enables the system to handle higher count rates. The trigger logic simply requires that there is a proton passing through the focal plane, making all the way up to S2.

In the off-line analyses, two proton events are selected and a further constraint is set on them that they correspond to ^2He like events with the internal energy of the $^2He \leq 0.5 MeV$. Since the AGOR facility has a repetition rate of 100 MHz, while the coincidence resolving time is set for 40 ns., one sees the accidentals corresponding to the coincidences between protons from different beam bursts. As to be expected, this background contributes to the overall countrate, but does not add to the events under the peak of the ground state transition. It is easily handled by subtracting the normalized accidental spectrum from that of the main spectrum. A spectrum thus obtained is shown in the figure 2.

It is quite clear that we were able to produce a clean spectrum of $^{12}C(d,^2He)$ in which we see not just the ground state transition but also events feeding a few excited states in the ^{12}B. The analyses of the data for the estimates of the figure of merit of the setup are in progress.

3 Merits of the Current Experiment

While the final analyses of the current feasibility study are still awaited, we may point to the strengths of this arrangement to test the quantum mechanics against other hidden variable theories.

- We employ a resonant state with a life time of $\leq 10^{-20}$ sec, for the formation of singlet state, the entanglement times are much shorter than the transit time of protons in the focal plane system (a few ns). The detection system itself has transit time of about 100 ns. ie., the particles leave the detection region much before an event is registered.

- The triggering system requires that there is at least one proton in the focal plane satisfying the momentum selection. Thus, both single-proton and two-proton events are recorded. The carbon analyzer is down stream of the focal plane. All protons leaving the focal plane, whether or not they undergo secondary scattering in the carbon analyzer, are recorded. Thus, this experiment is free from the trigger bias.

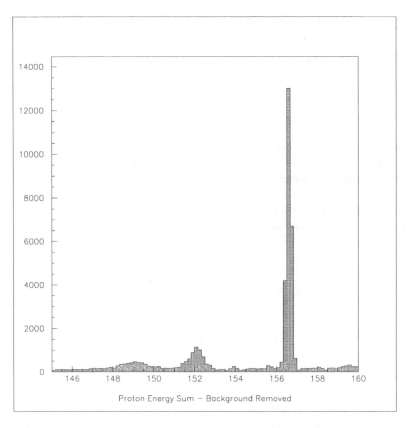

Figure 2. The sum energy of the two protons in the $^{12}C(d, 2p)$ reaction, corresponding to the internal energy of intermediate ^2He state to be $\leq 0.5MeV$. The sharp peak at 156 MeV is the process feeding the ^{12}B ground state. Also seen are transitions to levels at about 4 and 6 MeV excitations in ^{12}B.

- The requirement that the internal energy of the $^2He \leq 0.5MeV$, results in a difference of the momenta of two protons such that they are well separated from each other as they reach focal plane detectors. This, in turn, satisfies the space-like separation at least as a first approximation.

- We have access to four inertial frames of reference, viz., the ^2He rest frame, frames of the two protons separately and also the laboratory frame.

- The reference axes for the polarization asymmetries can be chosen at will during the off-line analyses. The system allows one to make an arbitrary choice of axes in a plane perpendicular to the momentum direction of individual protons. This flexibility allows us to test Wigner type correlations also.

4 Future prospects

We are apprising the figure of merit of the system and plan for a full fledged experiment in the year 2002. Meanwhile, it is very likely that an experiment with hydrogen target will be carried out at RIKEN [16]. While the $(d, ^2He)$ reaction has the advantage of clean $^1S_0$ state, it lacks the flexibility where one can manipulate the polarization states of intermediate state. Also, one may never be able to resolve the criticisms of information exchange between the final state protons, as long as we have both protons detected in the same spectrometer. I am of the personal view point that this latter criticism is not going to be laid to rest unless the experimental arrangement is made to satisfy space-like separation for electromagnetic interactions.

In this context, the elastic scattering of protons on protons are again of interest. The ultimate goal of these measurements should be to employ the polarized beams and polarized targets. The polarization states of the outgoing protons are to be measured. This arrangement will allow one to control the polarization states of the protons in the initial state to either triplet or singlet configurations and the large acceptance polarimeter will permit one to measure the polarization correlations along axes chosen at the will of the person analyzing the data. One can thus avoid all the loopholes, topics of debate in these types of experiments. Needless to say, these experiments are more difficult but they are certainly feasible. It should, then, be possible to extend these tests beyond Bell's inequalities in terms of BKS theorem and Hardy's impossibilities.

Acknowledgements

It is a real pleasure to acknowledge the discussions with many of my collaborators on this project. Special thanks are owed to Heinrich Wörtche of KVI, Groningen.

References

1. A. Einstein, B. Podolsky and N. Rosen Phys. Rev. **47**, 777 (1935).
2. D. Bohm Phys. Rev. **85**, 166 (1957).
3. J.S. Bell Physics **1**, 195 (1965).

4. I. Duck and E.C.G. Sudarshan *100 years of Planck's Quantum* (world Scientific, Singapore, 2001).
5. F. Laloë Am. J. Phys. **69**, 655 (2001).
6. E. Wigner lectures reprinted in *Quantum theory and Measurement* , ed. J.A. Wheeler and W.H. Zurek (Princeton University Press, 1983).
7. D. Greenberger, M.A. Horne, A. Shimony and A. Zeilinger, Am. J. Phys. **58**, 1131 (1990).
8. A. Peres Phys. Lett. **A151**, 107 (1990).
9. N. David Mermin Phys. Rev. Lett. **65**, 3373 (1990).
10. L. Hardy Phys. Rev. Lett. **68**, 2981 (1992)
11. see for example, http://www.qubit.org
12. A. Aspect, P. Grangier and G. Roger Phys. Rev. Lett. **49**, 91 (1982);
13. T. Jennewein, G. Weihs, J-W. Pan, and A. Zeilinger, Phys. Rev. Lett. **88**, 017903 (2002).
14. M. Lamehi-Rachti and W. Mittig Phys.Rev. D **14**, 2543 (1976).
15. H. J. Wörtche, private communication.
16. H. Sakai, spokesperson for RIKEN proposal(Dec. 2001)
17. C. Rangacharyulu and H. J. Wörtche,*Co-spokespersons, S-24 proposal,* KVI, Groningen (October 2000).

SINGLE AND DOUBLE SPIN AZIMUTHAL ASYMMETRIES IN SEMI-INCLUSIVE PION ELECTROPRODUCTION

K. A. OGANESSYAN

INFN-Laboratori Nazionali di Frascati I-00044 Frascati, via Enrico Fermi 40, Italy
DESY, Deutsches Elektronen Synchrotron Notkestrasse 85, 22603 Hamburg,
Germany
E-mail: kogan@mail.desy.de

The leading and sub-leading order results for pion electroproduction in polarized and unpolarized semi-inclusive deep-inelastic scattering, are considered putting emphasis on transverse momentum dependent effects appearing in azimuthal asymmetries. In particular, single-, and double-spin asymmetries of the distributions in the azimuthal angle ϕ of the pion related to the lepton scattering plane are discussed. A possibility to measure spin-independent and double-spin analyzing powers, simultaneously, is also discussed.

Introduction

Semi-inclusive deep inelastic scattering (SIDIS) of leptons off a nucleon is an important process to study the internal structure of the nucleon and its spin properties. In particular, measurements of azimuthal distributions of the detected hadron provide valuable information on hadron structure functions, quark-gluon correlations and parton fragmentation functions. Defining a coordinate system in the laboratory frame with the z axis along the momentum transfer q and x axis in the leptonic plane, the component of the detected hadron momentum transverse to q, $P_{h\perp}$, and its azimuthal orientation, ϕ, (see Fig.1) provide interesting variables to study non-perturbative and perturbative effects.

In SIDIS one assumes the factorization of the cross section, schematically

$$d\sigma^{lN \to l'hX} = \sum_q f^{H \to q} \otimes d\sigma^{eq \to eq} \otimes D^{q \to h}, \tag{1}$$

where the soft parts, the distribution function f and the fragmentation function D depend not only on x and z, respectively, but also on quark's transverse momenta; $d\sigma^{eq \to eq}$ describes the scattering among elementary constituents and can be calculated perturbatively in the framework of quantum chromodynamics (QCD). Due to the non-zero parton intrinsic transverse momentum, in the SIDIS cross section besides the conventional twist-2 non-perturbative blocks, there are different combinations of twist-two and twist-three structures, which could be probed in azimuthal asymmetries of hard scattering

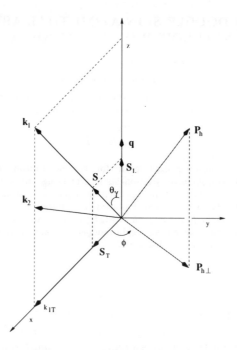

Figure 1. The kinematics of semi-inclusive deep-inelastic processes. For the specific case in which the target is polarized parallel (anti-parallel) to the beam a transverse spin in the virtual photon frame arises which only can have azimuthal angle 0 (π). The value of this transverse spin component is [1] $S_T = S \sin \theta_\gamma$, where θ_γ is the virtual photon emission angle and S is target polarization parallel/antiparallel to the incoming lepton momentum.

processes. The complete tree-level description expression containing contributions from twist-two and twist-three distribution and fragmentation functions in SIDIS has been given in Ref. [2]. The full result for the SIDIS cross section contains a large number of terms. In this respect it is more practical to split up it in the parts involving the lepton polarizations, unpolarized (U) or longitudinally (L) polarized and target polarizations, unpolarized (U), longitudinal (L), and transverse (T) polarized keeping only the ϕ-independent and ϕ-dependent terms, relevant in following. Then it can be presented in the following way[a]:

$$\frac{d\sigma^{eN \to ehX}}{dx\,dy\,dz\,dP_{h\perp}^2} \propto d\sigma_{UU}^{(0)} + \frac{1}{Q}\cos\phi \, d\sigma_{UU}^{(1)}$$

[a] Up to the $1/Q$ order.

$$+ \frac{1}{Q} \sin \phi \lambda d\sigma_{LU}^{(2)} + \frac{1}{Q} \sin \phi S_L d\sigma_{UL}^{(3)}$$
$$+ \sin(\phi + \phi_S) S_T d\sigma_{UT}^{(4)}$$
$$+ \lambda S_L d\sigma_{LL}^{(5)} + \frac{1}{Q} \lambda S_L \cos \phi d\sigma_{LL}^{(6)}$$
$$+ \lambda S_T \cos(\phi - \phi_S) d\sigma_{LT}^{(7)}, \tag{2}$$

where the first subscript corresponds to beam polarization and the second one to the target polarization. Here the terms proportional to $1/Q$ indicate the "kinematical" or dynamical twist-3 contributions. Here the variables x, y, and z, are the standard leptoproduction variables defined by

$$x = \frac{Q^2}{2(P \cdot q)}, \quad y = \frac{(P \cdot q)}{(P \cdot k_1)}, \quad z = \frac{(P \cdot P_h)}{(P \cdot q)}.$$

In inclusive processes, at leading $1/Q$ order, besides the well-known parton distribution $f_1(x)$, often denoted as $q(x)$, the longitudinal spin distribution $g_1(x)$, often denoted as $\Delta q(x)$, there is a third twist-two distribution function, the transversity distribution function $h_1(x)$, also often denoted as $\delta q(x)$. It was first discussed by Ralston and Soper[3] in double transverse polarized Drell-Yan scattering. The transversity distribution $h_1(x)$ measures the probability to find a transversely polarized quark in a transversely polarized nucleon. It is equally important for the description of the spin structure of nucleons as the more familiar function $g_1(x)$; their information being complementary. In the non-relativistic limit, where boosts and rotations commute, $h_1(x) = g_1(x)$; then difference between these two functions may turn out to be a measure for the relativistic effects within nucleons. On the other hand, there is no gluon analog on $h_1(x)$. This may have interesting consequences for ratios of transverse to longitudinal asymmetries in polarized hard scattering processes (see e.g. Ref. [4]).

The transversity remains still unmeasured, contrary to the case for spin-average and helicity structure functions, which are known over large ranges of Q^2 and x. The reason is that it is a chiral odd function, and consequently it is suppressed in inclusive deep inelastic scattering (DIS) [5,6]. Since electroweak and strong interactions conserve chirality, $h_1(x)$ cannot occur alone, but has to be accompanied by a second chiral odd quantity.

In principle, transversity distributions can be extracted from cross section asymmetries in polarized processes involving a transversely polarized nucleon. In the case of hadron-hadron scattering these asymmetries can be expressed through a flavor sum involving a product of two chiral-odd transversity distributions. This is one of the main goals of the spin program at RHIC [7].

In the case of SIDIS off transversely polarized nucleons there exist several methods to access transversity distributions. One of them, the twist-3 pion production [8], uses longitudinally polarized leptons and measures a double spin asymmetry. Other methods do not require a polarized beam, and rely on the *polarimetry* of the scattered transversely polarized quark. They consist on:

- the measurement of the transverse polarization of Λ's in the current fragmentation region [9],

- the observation of a correlation between the transverse spin vector of the target nucleon and the normal to the two-meson plane [10,11],

- the observation of the "Collins effect" in quark fragmentation through the measurement of pion single target-spin asymmetries [12,13,2].

In Sec. 2 I will focus on the last method – Collins fragmentation function, $H_1^\perp$ which can be simply interpreted as the production probability of an unpolarized hadron from a transversely polarized quark [12]. A first indication of a nonzero $H_1^\perp$ comes from analysis of the 91-95 LEP1 data (DELPHI) [14] [b]

Due to non-zero quark transverse momentum in semi-inclusive processes, at leading $1/Q$ order, in addition to the above discussed distribution functions, there are three non-vanishing distribution functions, $g_{1T}, h_{1L}^\perp, h_{1T}^\perp$. These functions relate the transverse (longitudinally) polarization of the quark to the longitudinally (transverse) polarization of the nucleon.

Single-spin azimuthal asymmetries

In recent years significant SSAA have been observed in experiments with transversely polarized proton and anti-proton beams, respectively[16].

Very recently a significant target-spin asymmetry of the distributions in the azimuthal angle ϕ of the pion related to the lepton scattering plane for pion electroproduction in a *longitudinally* polarized hydrogen target has been observed by the HERMES collaboration [17]. At the same time the SMC collaboration has studied the azimuthal distributions of pions produced in deep inelastic scattering off *transversely* polarized protons and deuterons [18].

These non-zero asymmetries may originate from multi-parton correlations in initial or final states and non-zero parton transverse momentum. They have initiated a number of phenomenological approaches to evaluate these asymmetries using different input distribution and fragmentation functions. An

---

[b]Possibilities of measuring $H_1^\perp$ at BELLE [15] are currently being examined.

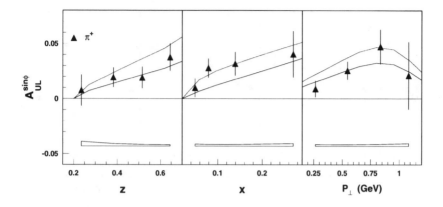

Figure 2. $A_{UL}^{\sin\phi}$ for π^+ electroproduction as a function of z, the Bjorken-x, and of the $P_{h\perp}$ from Ref. [17]. Error bars include the statistical uncertainties only. The open bands at the bottom of the panel represent the systematic uncertainty. The curves show the range of predictions of a model calculation [21].

analysis of different approximations, which aim at explaining the experimental data, have been provided in Ref. [19]. The approximation where the twist-2 *transverse* quark spin distribution in the *longitudinally* polarized nucleon, $h_{1L}^{\perp(1)}(x)$, is considered small enough to be neglected [20,21] with the assumption of the u-quark dominance are in good agreement with the Bjorken-x, z, and $P_{h\perp}$ behaviors of the *sinϕ* asymmetry for charged and neutral pion production observed at HERMES (see Figs. 2,3) [c].

Results on SSAA provide evidence in support of the existence of non-zero chiral-odd structures that describe the transverse polarization of quarks. New data are expected from future HERMES, COMPASS measurements on a transversely polarized target, which will give direct access to the transversity [22].

Double-spin azimuthal asymmetries

Here I will focus my attention on the *cosϕ* moment of the double-spin azimuthal asymmetry (DSAA) for pion electroproduction in semi-inclusive deep inelastic scattering of longitudinally polarized leptons off longitudinally polar-

---

[c]For π^- production the data are consistent with zero in agreement with the result of the approximation.

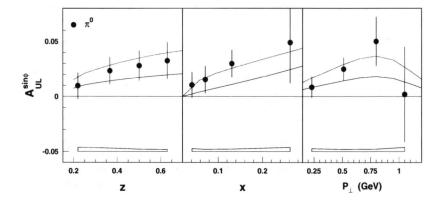

Figure 3. The same as in Fig.2 for neutral pion electroproduction. Data are from Ref. [17]. The curves show the range of predictions of a model calculation [21].

ized protons recently considered in Ref. [23]. The contribution to double-spin $cos\,\phi$ asymmetry with the combinations of different leading and sub-leading distribution and fragmentation function can be symbolically presented in the following way

$$d\sigma \propto \quad \lambda S_L d\sigma_{LL}^{(5)}$$
$$+ \frac{1}{Q}\lambda S_L \cos\phi d\sigma_{LL}^{(6)}$$
$$+ \lambda S_T \cos\phi \sin\theta_\gamma d\sigma_{LT}^{(7)}, \tag{3}$$

where

$$d\sigma_{LL}^{(5)} \propto \sum_a e_a^2\, g_1^a(x)\, D_1^a(z) \tag{4}$$

$$d\sigma_{LL}^{(6)} \propto \sum_a e_a^2 \left(g_1^a(x)\, \tilde{D}^{\perp a(1)}(z) - g_L^{\perp a(1)}(x) D_1^a(z) - h_{1L}^{\perp(1)a}(x)\tilde{E}^a(z) \right) \tag{5}$$

$$d\sigma_{LT}^{(7)} \propto \sum_a e_a^2 g_{1T}^{a(1)}(x)\, D_1^a(z) \tag{6}$$

The $cos\,\phi$ DSAA in the SIDIS cross-section can be defined as appropriately

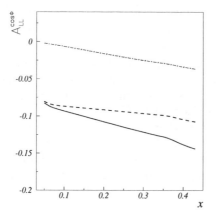

Figure 4. $A_{LL}^{\cos\phi}$ for π^+ production as a function of Bjorken x. The dashed line corresponds to contribution of the $d\sigma_{LL}^{(6)}$, dot-dashed one to $\sin\theta_\gamma d\sigma_{LT}^{(7)}$ and the solid line is their sum.

weighted integral of the cross section asymmetry:

$$< |P_{h\perp}| \cos\phi >_{LL} = \frac{\int d^2 P_{h\perp} |P_{h\perp}| \cos\phi \, (d\sigma^{++} + d\sigma^{--} - d\sigma^{+-} - d\sigma^{-+})}{\frac{1}{4} \int d^2 P_{h\perp} \, (d\sigma^{++} + d\sigma^{--} + d\sigma^{+-} + d\sigma^{-+})}.$$
(7)

Here $++, --$ $(+-, -+)$ denote the antiparallel (parallel) polarization of the beam and target [d] and M_h is the mass of the final hadron. The above defined weighted asymmetry is related to the experimentally observable asymmetry through the following relation

$$A_{LL}^{\cos\phi} \approx \frac{1}{\langle P_{h\perp} \rangle} \langle |P_{h\perp}| \cos\phi \rangle_{LL}$$
(8)

Using the Eqs.(2), (3) one can get

$$A_{LL}^{\cos\phi} = 4 \frac{d\sigma_{LL}^{(6)} + \sin\theta_\gamma d\sigma_{LT}^{(7)}}{d\sigma_{UU}^{(0)}}.$$
(9)

To estimate that asymmetry we take into account only the $1/Q$ order contribution to the DSAA which arises from intrinsic p_T similar to the $\cos\phi$ asymmetry in *unpolarized* SIDIS [24], i.e. all twist-3 interaction dependent

---

[d]this leads to positive $g_1(x)$.

distribution and fragmentation functions are set to zero [23]. To estimate the transverse asymmetry contribution $(d\sigma_{LT}^{(7)})$ into the $A_{LL}^{\cos\phi}$, one can act as in the Ref. [25].

In Fig.4, the asymmetry $A_{LL}^{\cos\phi}(x)$ of Eq.(9) for π^+ production on a proton target is presented as a function of x-Bjorken. The curves are calculated by integrating over the HERMES kinematic ranges taking $\langle P_{h\perp} \rangle = 0.365$ GeV as input. For more details see Ref. [23].

From Fig.4 one can see that the approximation where all twist-3 DF's and FF's are set to zero gives the large negative double-spin $cos\phi$ asymmetry at HERMES energies. Note that the 'kinematic' contribution to $A_{LL}^{\cos\phi}(x)$ coming from the transverse component of the target polarization, is negligible.

Now, instead of $\cos\phi$ weighted asymmetry (Eq.7), lets consider the asymmetry defined as [26]

$$A = \frac{d\sigma^{++} + d\sigma^{--} - d\sigma^{+-} - d\sigma^{-+}}{d\sigma^{++} + d\sigma^{--} + d\sigma^{+-} + d\sigma^{-+}}, \qquad (10)$$

which, obviously, free of acceptance effects.

Then at leading order one obtains

$$A = \frac{d\sigma_{LL}^{(5)}/2d\sigma_{UU}^{(0)} + \langle\cos\phi\rangle_{LL}\ \cos\phi}{1/2 + \langle\cos\phi\rangle_{UU}\ \cos\phi}, \qquad (11)$$

where $\langle\cos\phi\rangle_{LL}$ and $\langle\cos\phi\rangle_{UU}$ are unpolarized and double polarized $\cos\phi$ moments, respectively:

$$\langle\cos\phi\rangle_{LL} = \frac{d\sigma_{LL}^{(6)}}{2d\sigma_{UU}^{(0)}}, \quad \langle\cos\phi\rangle_{UU} = \frac{d\sigma_{LL}^{(1)}}{2d\sigma_{UU}^{(0)}}. \qquad (12)$$

In Fig.5 the $A(\phi)$ asymmetry of Eq.(11) for π^+ production on a proton target is presented as a function of ϕ at different values of x-Bjorken. By squires denoted the values of $A(\phi)$ at $\phi = \pi/2$ and $\phi = 3\pi/2$. From Fig.5 one can conclude that there is a large (up to $10 - 15\%$) deviation from those values, which makes it available experimentally. Thus, the consideration of such asymmetry will provide the measurement of unpolarized and double polarized $\cos\phi$ moments, simultaneously. For more details see Ref. [26].

Conclusions

I have considered the leading and sub-leading order results for pion electroproduction in polarized and unpolarized semi-inclusive deep-inelastic scattering,

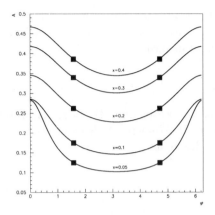

Figure 5. Azimuthal dependence of the asymmetry A for π^+ production at different values of x-Bjorken.

putting emphasis on transverse momentum dependent effects appearing in azimuthal asymmetries. In particular, I have discussed the single-, and double-spin azimuthal asymmetries, which contain valuable information of hadron spin structure.

I have discussed the results on single-spin azimuthal asymmetries which provide evidence in support of the existence of non-zero chiral-odd structures that describe the transverse polarization of quarks.

At HERMES kinematics a sizable negative $cos\phi$ double-spin asymmetry for π^+ electroproduction in SIDIS is predicted taking into account only the $1/Q$ order contribution to the DSAA which arises from intrinsic p_T effects similar to the $cos\phi$ asymmetry in unpolarized SIDIS: all twist-3 interaction dependent distribution and fragmentation functions are set to zero. The "kinematical" contribution from target transverse component (S_T) is well defined. It is shown that its contribution to $A_{LL}^{\cos\phi}$ is negligible. Double-spin $\cos\phi$ DSAA is a good observable to investigate the weights of twist-2 and twist-3 contributions at moderate Q^2. It may give a possibility to estimate $< p_T^2 >$ in the nucleon. The complete behavior of azimuthal distributions may be predicted only after inclusion of higher-twist and pQCD contributions, nevertheless, if one consider the kinematics with $P_{h\perp} < 1$ GeV and $z < 0.8$, one can isolate the non-perturbative effects conditioned by the intrinsic transverse momentum of the partons in the nucleon.

Finally, an acceptance-independent azimuthal asymmetry has been considered, which leads to measurements of unpolarized and double-polarized $\cos\phi$ asymmetries, simultaneously.

Acknowledgments

I would like to thank M. Anselmino, D. Boer, A. Efremov, R. L. Jaffe, R. Jakob, A. Kotzinian, P. Mulders, A. Schäfer, and O. Teryaev for many useful and stimulating discussions.

References

1. K.A. Oganessyan, et al., hep-ph/9808368; Proceedings of the workshop Baryons'98, Bonn, Sept. 22-26, 1998.
2. P.J. Mulders and R.D. Tangerman, Nucl Phys. B461 (1996) 197.
3. J. Ralston, D.E. Soper, Nucl. Phys. B 152 (1979) 109.
4. R.L. Jaffe, hep-ph/9602236; hep-ph/9603422.
5. R.L. Jaffe, X. Ji, Phys. Rev. Lett. 67 (1991) 552
   and Nucl. Phys. B 375 (1992) 527;
6. X. Artru, M. Mekhfi, Z. Phys. C45, 669 (1990).
7. D. Hill et al., RHIC Spin Collaboration: Letter of Intent, BNL, May 1991.
8. R.L. Jaffe, X.J. Ji, Phys. Rev. Lett. 71, 2547 (1993).
9. R.L. Jaffe, Phys. Rev., D54, 6581 (1996).
10. R.L. Jaffe, hep-ph/9710465.
11. R.L. Jaffe, X. Jin, J. Tang, Phys. Rev. Lett, 80, 1166 (1998).
12. J. Collins, Nucl. Phys. B 396 (1993) 161.
13. A. Kotzinian, Nucl. Phys. B 441 (1995) 234.
14. A.V. Efremov, O.G. Smirnova, L.G. Tkachev, Nucl.Phys.Proc.Suppl. 74 (1999) 49.
15. A. Ogawa, Proceedings of the Transverse Spin Physics workshop, Zeuthen, July 9 – 11, 2001.
16. D. Adams et al., Phys. Lett. B 264 (1991) 462; Phys. Rev. Lett. 77 (1996) 2626; B.E. Bonner et al., Phys. Rev. D 41 (1990) 13.
17. HERMES Collaboration, A. Airapetian, et.al, Phys. Rev. Lett. 84 (2000) 4047; Rev. D 64 (2001) 097101.
18. A. Bravar, Nucl. Phys. (Proc. Suppl.) B79 (1999) 520.
19. K.A. Oganessyan, et al., Nucl. Phys. A 689 (2001) 784.
20. D. Boer, hep-ph/9912311; Boglione and P.J. Mulders, Phys. Lett. B478 (2000) 114; A.V. Efremov, hep-ph/0001214.

21. E. De Sanctis, W.-D. Nowak, and K.A. Oganessyan, Phys. Lett. **B483** (2000) 69.
22. V.A. Korotkov, W.-D. Nowak, and K.A. Oganessyan, Eur. Phys. J. **C** **18** (2001) 639.
23. K. A. Oganessyan, P. J. Mulders, E. De Sanctis, hep-ph/0201061.
24. R.N. Cahn, Phys. Lett. B **78** (1978) 269; Phys. Rev. D **40** (1989) 3107.
25. A.M. Kotzinian, and P.J. Mulders, Phys.Rev. D54 (1996) 1229.
26. K. A. Oganessyan, et.al., in preparation.

ON USE OF QUARK-HADRON DUALITY IN PHOTOABSORPTION SUM RULES

S.B. GERASIMOV

Bogoliubov Laboratory of Theoretical Physics,
Joint Institute for Nuclear Research, 141980 Dubna, Russia
E-mail: gerasb@thsun1.jinr.ru

The idea of the quark-hadron duality is developed and applied to specific integral sum rules for the photoexcitation of hadron resonances. Within the constituent quark model approach, the relations between different bremsstrahlung weihgted integrals of the nucleon resonance amplitudes and/or cross sections and correlation functions of the quark dipole moments in the nucleon ground state are obtained. These functions are of interest for checking the detailed quark-configuration structure of the nucleon state vector. Some applications of the presented approach in the meson sector are made, and the role of meson and scalar diquark cluster degrees of freedom in the electromagnetic hadron observables is discussed.

1 Introduction

The connection of the quark-gluon and hadronic pictures without explicit solution of the hadronization phase of the quark-gluon degrees of freedom is of intrinsic and practical interest while new examples of (in)applicability of this idea may lead to a better understanding of its benefits and limits. In the present report, we follow, to some extent, our old work[1,2] where an attempt was made to demonstrate that, under certain conditions, conventional field-theoretical constructions and lowest order graphs can be used in the description of real photon-hadron processes and such hadron properties that are connected with essentially nonperturbative effects of binding and confinement. An underlying idea is the global duality of two complete sets of state vectors, saturating certain integral sum rules, one of the sets being the solution of the bound state problem with colour-confining interaction, while the other describes free particles. The choice of sum rules satisfying the assumed duality condition is suggested by correspondence with the well-known results in the nonrelativistic theory of interaction of the radiation with matter. The message is that sum rules connected with the dipole moment fluctuation seem to be singled both in the nonrelativistic[3] and relativistic regions. Following this idea, that will firstly be illustrated and tested in the nonrelativistic theory of the nuclear photoeffect and in the models of quantum field theory, we present, within the constituent quark model approach, the relations between different bremsstrahlung-weighted integrals of the nucleon resonance photoexcitation cross sections and correlation functions of the quark dipole moments in the nucleon ground state. These functions are of interest for checking the detailed quark-configuration structure of the nucleon state vector. Some applications of this approach for polarized gamma-gamma cross sections are made, and the role of meson and scalar diquark cluster degrees of freedom in the electromagnetic hadron observables will then be discussed.

Sum rules equating integrals over photoabsorption cross-sections to static parameters of the ground state have long been in the use in nonrelativistic atomic and nuclear physics (e.g.[5]). To illustrate the essence of specific meaning of the "duality" to be used later on, we find it convenient and relevant to turn to the fact derived from application of the Cabibbo-Radicati sum rule to the photoabsorption processes on the isodublet $^3He - ^3H$-nuclei[6,7]. In the context of the nonrelativistic version of this sum rule and using a fully symmetric orbital wave function for the ground state of these nuclei, the bremsstrahlung-weighted integrals of the electric dipole photodisintegration of ^3He leading to final states with isospins $I = 1/2$ and $I = 3/2$ turn out be equal to each other, and, moreover, the isodublet (or isoquartet) part exhausts the integral of cross section, respectively, of the two-body (i.e. dp) or three-body $(n2p)$ channel. The same equality takes place for the integrals of the "model" cross sections with the final state interactions switched-off and when we deal the three-nucleon state described by plane waves. So we have here the integral duality (i.e the equality of integrals not only over total dipole cross sections, but also over partial cross sections with fixed different isospins in the final state) of two sets of final wave functions, one - with nucleon interactions inluded, and the other - with no interaction at all for both possible values of the isospin. This means that for the isodublet channel the duality, in the indicated sense, takes place between physically quite different states having, e.g., differing thresholds due to the deuteron binding energy, and under adopted assumptions, this duality does not depend on the strength of FSI. We propose just this type of duality to exploit in the relativistic dynamics of two- and three- particle systems including the genuine colour-confining interaction between quarks composing mesons and baryons.

In the relativistic domain, one should meet the problem of the convergence of corresponding integrals and the necessity to treat all radiative transitions relativistically, i.e. to include into consideration all higher multipoles in contrast with the atomic and nuclear photoeffect where one can confine oneself by the inclusion of the electric dipole amplitudes in a reasonably large energy interval. Here, our main idea is to try to merge the duality concept of the Regge-resonance type and the concept of the parton-hadron duality within a definite class of sum rules for the photoabsorption cross sections which, at the same time, demonstrate a qualitatively justifiable common base with the known and much simpler sum rules in the nonrelativistic domain. The popular idea of the sixties is that of semi-local duality between the description of hadronic scattering behaviour at high energies via the leading non-vacuum Regge trajectory exchanges in the t-channel and the sum of the s-channel resonances[8]. The vacuum Pomeron exchanges should then be associated with the non-resonance s-channel amplitudes[9]. Accepting that kind of duality between the resonance photoexcitation cross-sections and contributions of the positive charge-parity Regge exchanges, the tensor f_2- and a_2 - trajectories, we conclude that the integrals over bremsstrahlug-weighted total resonance photoproduction cross-sections should be convergent.

2 Quark-hadron duality for the bremsstrahlung-weighted sum rules

The constituent quark model is a useful descriptive tool to systematize the phenomenology of the resonance physics. Despite many successes, the fundamental question of the connection between the three (for baryons) (or the $q\bar{q}$-pair, for mesons) "spectroscopic" quarks, with quantum numbers of a given hadron and an infinite number of the "current" (or fundamental/Lagrangian) quarks required, e.g., by deep-inelastic lepton-hadron scattering, is still a problem. The sum rules to be discussed may contribute to investigation of this problem.

2.1 Mesons

We first remind that applying the adopted duality principle of two complete sets of final state vectors (one - with the confined $q\bar{q}$-states representing the sum over all resonances, the other- with the "gedanken" free $q\bar{q}$ -pair) to the Cabibbo-Radicati sum rule for the pion, we get[2]

$$< r^2 >_{\pi^\pm} = \frac{3}{4\pi^2 F_\pi^2} \tag{1}$$

The same result, in the approximation $m_\pi \to 0$, follows from the relativistic dipole-moment-fluctuation sum rule

$$\frac{4}{3}\pi^2\alpha < r^2 >_{\pi^+} = J_{tot}(\gamma\pi^+) + \frac{4}{5}J_{tot}(\gamma\pi^o), \tag{2}$$

where $J_{tot}(\gamma\pi)$ is the bremsstrahlung-weighted integral of $\sigma_{tot}(\gamma+\pi \to \bar{q}+q)$. The sum rule for the charged kaon radius

$$\frac{4}{3}\pi^2\alpha < r^2 >_{K+} = J_{tot}(\gamma K^+) + 2J_{tot}(\gamma K^o). \tag{3}$$

where the radius is given by the standard triangle diagram has been analytically checked against the value computed via the integrals of corresponding cross sections of processes represented by the Born diagrams. However, the numerical value of $< r^2 >_{K\pm}$ turns out to be larger than the experimental value, demonstrating worse applicability of the local approximation for the Kqs-vertices in comparison with the pion case. Further useful relations were obtained[2] from the assumed approximate equality of the derivatives as $q^2 \to 0$

$$F'_{\gamma^* \gamma\pi^o}(q^2) \simeq F'_{\gamma^* \pi^\pm \pi^\pm}(q^2)$$

of form factors normalized to unity at zero momentum transfers:

$$m_q \simeq \sqrt{\frac{2}{3}}\pi F_\pi, (q = u, d) \tag{4}$$

$$\frac{g^2_{\pi^o qq}}{4\pi} \simeq \frac{\pi}{6} \tag{5}$$

where the numerial value of the pseudoscalar πqq-coupling constant is following from an analogue of the Goldberger-Treiman relation for quarks.

We turn now to sum rules for meson resonances in photon-photon collisions. Varying the polarizations of colliding photons, one can show that the linear combination of certain $\gamma\gamma \to q\bar{q}$ cross-sections will dominantly collect, at low and medium energies most important for saturation of the integral sum rules considered, the $q\bar{q}$- states with definite spin-parity, hence, by the adopted quark-hadron duality, the meson resonances with the same quantum numbers. In particular, the combinations of the integrals over the bremsstrahlung-weighted and polarized $\gamma\gamma \to q\bar{q}$ cross-sections, $I_\perp - (1/2)I_p, I_\parallel - (1/2)I_p, I_p$ will be related with low mass meson resonances having spatial quantum numbers $J^{PC} = 0^{-+}$ and 2^{-+}, 0^{++} and $2^{++}(\lambda = 0)$, $2^{++}(\lambda = 2)$, respectively (λ being the z-projection of the total angular momentum, and we confine ourselves to the meson spins $J \leq 2$ for further discussion).

The $\gamma\gamma$ - cross-sections $\sigma_{\perp(\parallel)}$ (and the integrals thereof) refer to plane-polarized photons with the perpendicular (parallel) polarizations, and σ_p corresponds to circularly polarized photons with parallel spins.

Evaluating cross-sections and elementary integrals we get the sum rules for radiative widths of resonances with different values of J^{PC}

$$\sum_{i(PS)} \frac{\Gamma(i \to 2\gamma)}{m_i^3} + 5 \sum_{j(PT)} \frac{\Gamma(j \to 2\gamma)}{m_j^3} \simeq \frac{3}{16\pi^2} \sum_q \langle Q(q)^2 \rangle^2 \frac{\pi\alpha^2}{m_q^2} \qquad (6)$$

$$\sum_{i(S)} \frac{\Gamma(i \to 2\gamma)}{m_i^3} + 5 \sum_{j(T;\lambda=0)} \frac{\Gamma(j \to 2\gamma)}{m_j^3} \simeq$$

$$\simeq \frac{3}{16\pi^2} [\sum_q \langle Q(q)^2 \rangle^2 \frac{5\pi\alpha^2}{9m_q^2} + \sum_{qq} \langle Q(qq)^2 \rangle^2 \frac{2\pi\alpha^2}{9m_{qq}^2}] \qquad (7)$$

$$5 \sum_{i(T;\lambda=2)} \frac{\Gamma(i \to 2\gamma)}{m_i^3} + (2J+1) \sum_{R(J\geq 2;\lambda=2)} \frac{\Gamma(R \to 2\gamma)}{m_R^3} \simeq$$

$$\simeq \frac{3}{16\pi^2} \sum_q \langle Q(q)^2 \rangle^2 \frac{14\pi\alpha^2}{9m_q^2} \qquad (8)$$

All the integrals over the resonance cross sections are taken in the narrow width approximation and, further, the contributions of the states with $J \geq 3$ are dropped in sum rules (6) and (7). For generality of the consideration, we included the (last) term in (7) that corresponds to a possible role of scalar diquarks as constituent "partons" composing, at least in part, either one or two scalar meson nonets (for discussion of this acute problem in hadron spectroscopy see, e.g.,minireview in [14] and references therein). Assuming no-mixing between light and heavy quark sectors, one can read every sum rule separately for mesons constructed of the u-,d-, s-, and c-quarks. Moreover, one can split sum rules into isovector and isoscalar resonance parts in the light quark sector, which enables one to carry a more detailed comparison with experimental data. For numerical estimation of radiative widths we use the mass values $m_c = 1.5$ GeV, $m_u = m_d \simeq 240$ MeV, m_s is taken from [2] as expressed via the strangeness-changing GT-relation, $m_s \simeq 350$ MeV, and for the diquark masses we take the values following from the extended chiral model including scalar diquarks into the low-energy effective Lagrangian approach [16]: $m_{qq} \simeq 310 \div 330$ MeV, $m_{qs} \simeq 545 \div 570$ MeV, where $q = u, d$.

In the charm sector, we have $\Gamma_{\gamma\gamma}(\eta_c) \leq 7.4$ (6.9 ± 1.7 ± 0.8), $\Gamma_{\gamma\gamma}(\chi_{c0}) \leq 6.1$ (3.76±0.65±1.81), $\Gamma_{\gamma\gamma}(\chi_{c2}) \leq 6.1$ [(0.53±0.15±0.23)]÷[(1.76±0.47±0.37)] where the most recent results on the two-photon widths of resonances in the units of keV, taken from [17], are given in parentheses and the least (largest) value for the width $\Gamma_{\gamma\gamma}(\chi_{c2})$ refers to the registered χ_{c2} cascade-decay final states $2\pi^+2\pi^-$ $(l^+l^-\gamma)$. The closeness of $\Gamma_{\gamma\gamma}^{ex}(\eta_c)$ to the upper value corresponding in all cases to the neglect of radial excited and higher orbital excited resonances to the considered sum rules tells about relative smallness of the higher resonance part compared with the ground state contribution. The cross section of the transition $\gamma\gamma \to q\bar{q}(J^{PC} = 2^{++})$ is known [18] to fall much slower with rise of photon energies as compared with the transitions to the states with $J^{PC} = 0^{\pm+}$, and this is reflected in rather a large contribution from the higher resonances with the larger masses and spins, while their decay widths are unknown. The situation with the light tensor mesons is largely the same, and we shall instead focus on the pseudoscalar and scalar meson sum rules in the light quark sector. The specific property of the "pseudoscalar" sum rule is that it includes also the contribution of the pseudotensor ($J^{PC} = 2^{-+}$) mesons which may be quite sizable. If we take for granted $\Gamma_{\gamma\gamma}(\pi_2(1670)) = 1.35 \pm .26$ keV, following from measurements of the CB and CELLO Collaborations [14], we can estimate the missing contribution of $\eta_2(1645)$ and $\eta_2(1870)$ assuming the ratio of the respective terms in the sum rule as $< Q^2 >_{isovec}:< Q^2 >_{isosc} = 1 : 3$ which fulfills approximately also for the ratio of pseudoscalar meson contributions. After that one obtains a somewhat marginal correspondence between $(16.5 \pm 1.2) \cdot 10^{-6}$ GeV^{-2} in the l.h.s. of Eq.(6) and $12.3 \cdot 10^{-6}$ GeV^{-2} in the r.h.s. thereof (much less upper values of the radiative width of $\pi_2(1670)$, reported by the L3 and ARGUS Collab.[14] would correspond much better to our sum rule).

We apply sum rule (7) for light scalar mesons to test the exploratory suggestion (see, e.g.[19]) that the low-lying scalar states can be classified as (at least) two (in principle, mixed) nonets formed by the scalar triplets of light diquarks and anti-diquarks and usual fermionic triplets of light (u-,d-,s-) quarks. For the isovector mesons $a(S = 1) \equiv a_0(980)$ and $a(S = 2) \equiv a_0(1450)$ we have

$$\sum_{S=1,2} \frac{\Gamma_{\gamma\gamma}(S)}{m(S)^3} \simeq \frac{\alpha^2}{96\pi}\left(\frac{5}{9m_q{}^2} + \frac{2}{9m_{qs}{}^2}\right) \tag{9}$$

With $\Gamma_{\gamma\gamma}(a_0(980)) \simeq 0.3$ keV, and $m_{q,(qs)} = 240(550)$ MeV, one gets $\Gamma_{\gamma\gamma}(a_0(1450)) \simeq 3.5$ keV which cannot be compared with experiment yet.

Assuming the orthogonal mixing of the isovector $|q\bar{q} >$ and $|(qs)(\bar{q}\bar{s}) >$ basis states and, further,

$$\frac{\Gamma_{\gamma\gamma}(S = 1)}{m(S = 1)^3} : \frac{\Gamma_{\gamma\gamma}(S = 2)}{m(S = 2)^3} =$$
$$[\frac{\sqrt{2}}{3m_{qs}}\cos\theta - \frac{\sqrt{5}}{3m_q}\sin\theta]^2 : [\frac{\sqrt{2}}{3m_{qs}}\sin\theta + \frac{\sqrt{5}}{3m_q}\cos\theta]^2 , \tag{10}$$

we get for the mixing angle $\theta \simeq 39°$ showing the considerable mixing between the quark-antiquark and diquark-antidiquark configurations in the two scalar nonets.

2.2 Baryons

In this section, we discuss sum rules giving a connection between the valence quark part of the Dirac (charge) radii of nucleons $< r^2 >$ and the photoexcitation cross sections of the nucleon resonances. We assume that both the ground and excited states of hadrons are the bound states of three (or two) constituent quarks and, tentatively, that the electroweak coupling constants of "constituent" and "current" quarks are the same (i.e. obtained via the minimal coupling principle). It follows then that the usual current algebra relations are valid for the currents constructed of the constituent quark field operators and, at the same time, one can replace the sum over a complete set of the hadron states in the dispersion integrals by the sum over all resonance states constructed, for example, of three constituent quarks and having the needed quantum numbers.

Our basic idea consists in relating the electric dipole moment correlator and the squared radii operators sandwiched by the nucleon state vectors in the "infinite - momentum frame"

$$2 < \hat{D}^2 >_N - < \hat{D}^2 >_P + 8 < \hat{D}_S^2 >_{P(N)} = 2 < \hat{r}_1^2 >_N + < \hat{r}_1^2 >_P \quad (11)$$

where

$$\hat{D} = \int \vec{x}\rho(\vec{x})d^3x = \sum_{j=1}^{3} Q_q(j)\vec{\xi}_j \quad (12)$$

$$\hat{r}_1^2 = \int \vec{x}^2\rho(\vec{x})d^3x = \sum_{j=1}^{3} Q_q(j)\vec{\xi}_j^{\,2} \quad (13)$$

$Q_q(j)$ and $\vec{\xi}_j$ are the electric charges and configuration variables of presumably point-like constituent quarks in the infinite-momentum frame, while the electric charge density operator $\hat{\rho} = \hat{\rho}^S + \hat{\rho}^V$ is a sum of the isoscalar and isovector parts. We introduce the following parametrization of the matrix elements

$$< r_1^2 >_P = \frac{4}{3}\alpha - \frac{1}{3}\beta \quad (14)$$

$$< r_1^2 >_N = -\frac{2}{3}\alpha + \frac{2}{3}\beta \tag{15}$$

$$< \hat{D}^2 >_P = \frac{8}{9}\alpha + \frac{1}{9}\beta + \frac{8}{9}\gamma - \frac{8}{9}\delta \tag{16}$$

$$< \hat{D}^2 >_N = \frac{2}{9}\alpha + \frac{4}{9}\beta + \frac{2}{9}\gamma - \frac{8}{9}\delta \tag{17}$$

$$< \hat{D}_S^2 >_{P,N} = \frac{1}{36}(2\alpha + \beta + 2\gamma + 4\delta), \tag{18}$$

where $< \vec{\xi}_1^2 >=< \vec{\xi}_2^2 >= \alpha, < \vec{\xi}_3^2 >= \beta$, $< \vec{\xi}_1 \cdot \vec{\xi}_2 >= \gamma, < \vec{\xi}_1 \cdot \vec{\xi}_3 >=< \vec{\xi}_2 \cdot \vec{\xi}_2 >$ $= \delta$, indices "1" and "2" refer to the like quarks (i.e. to the $u(d)$- quarks inside the proton (neutron)), and the "3", to the odd quark. The matrix elements over the proton and neutron wave function are only assumed to obey the charge symmetry relations.

A subsequent procedure is a standard technique of the sum rule derivation within the framework of the dipole moment algebra in the "$p_z \to \infty$" - frame [10,11,12].

However, we attribute a new meaning to all appearing quantities, namely, all the cross sections are understood as the nucleon resonance excitation cross sections; all radii $< r_1^2 >_{p,n}$, as the constituent quark distribution radii $< r_1^2 >_{p,n}^b$, not including the sea quark and/or meson current effects.Further, we will replace the intermediate one-nucleon contributions proportional to the anomalous magnetic moments of nucleons by the corresponding integrals entering into the anomalous magnetic (*i.e.*, GDH) sum rules.

One can then get

$$\frac{4}{3}\pi^2\alpha < \hat{D}^2 >_{P(N)} = J_p^{\gamma P(N)} = J_p^{\gamma P(N)}(\frac{1}{2}) + J_p^{\gamma P(N)}(\frac{3}{2}) \tag{19}$$

$$\frac{4}{3}\pi^2\alpha < r_1^2 >_N^b = J_a^{\gamma P}(\frac{3}{2}) - J_p^{\gamma P}(\frac{1}{2}) + 4J_p^S(\frac{1}{2}) \tag{20}$$

where

$$J_{p,a}^{V(S)}(I) = \int_{\nu_{thr}}^{\infty} \frac{d\nu}{\nu}\sigma_{p,a}(\gamma^{V(S)}N \to N^\star(I)) \tag{21}$$

and $\sigma_{p,a}$ refers to the interaction cross section of polarized "isovector"("isoscalar") ($\gamma^{V,S}$) photons and polarized nucleons with parallel (or antiparallel) spins.

The relativistic dipole-moment fluctuation sum rule,Eq.(19), was checked[4] in a number of the field-theoretical models, while the limiting case of Eq.(20) with the assumption of fully symmetric quark distributions in nucleons, *i.e.*, when $\alpha = \beta$ and $\gamma = \delta$, giving $< r_1^2 >_n^b = 0$, was first discussed in [1] .

The calculations show that the Dirac charge radius of the neutron is indeed a small quantity, especially, if the model amplitudes are used from the fit taking into account relativistic corrections and effects of the mixing of the basis $SU(6) \otimes O(3)$- configurations in the ground and excited nucleon wave functions [13].

However, if the use is made of the phenomenological amplitudes (especially of enhanced, as compared with the quark model calculations, the $\Delta(1232)$ excitation amplitude), the right-hand side of Eq.(20) acquires a larger positive value

$$\frac{4}{3}\pi^2\alpha < \hat{r_1^2} >_N^b \simeq +4, (+63)\mu b, \tag{22}$$

where the first value corresponds to the model [13]; and the second one, to phenomenological $\gamma N \to N^\star (I = 1/2, 3/2)$-transition amplitudes given in [14].

The most precise determination of the neutron charge radius follows from the neutron-electron scattering lengths measured in the thermal neutron scattering off the inert gases (Ne,Ar,Kr,Xe), or metal (W,Pb,Bi) atoms. The dominant contribution to the (Sachs) charge radius of the neutron

$$< \hat{r_{ch}^2} >_N = < \hat{r_1^2} >_N + \frac{3\kappa_N}{2m_N^2} \tag{23}$$

is due to the second (Foldy) term proportional to $\kappa_N = -1.913$ n.m.

The experimental uncertainties ascribed to measured values of $< r_{ch}^2 >_N$ enable one to extract a very small contribution of the Dirac (charge) radius $< r_1^2 >_N$ from Eq.(23). The latest experimental value of $< r_1^2 >_N$ [15] allows one to obtain :

$$\frac{4}{3}\pi^2\alpha < r_1^2 >_N^{exp} \simeq +10\mu b \tag{24}$$

with the uncertainty of $\pm 30\%$. This is markedly less than the second value in Eq.(22), and it may signal about a more complicated set of the involved dynamical degrees of freedom.

3 Concluding remarks

For mesons, the above-mentioned application of the developed approach to derive the $\gamma\gamma$- sum rule looks very encouraging. As far as it is the most clean test of our approach, the measurements of radiative decays of higher spin meson resonances would be very desirable and interesting.

For the nucleon and nucleon resonances, the situation looks so that the $SU(6)$- value (remaining also in the $N_c \to \infty$ approach) of the ΔN - transition magnetic moment is lower by about 20% than the experimental one and it turns out intact to the estimation of pionic corrections via the exchange current contributions to magnetic moments expressed in terms of the nucleon variables. Therefore the resolution of this problem can be found, probably, by adopting the significant "live" meson component in the Fock-state vector of this resonance. The pertinent generalization of sum rules taking into account these additional degrees of freedom is still to be done.

In conclusion, the author wishes to thank the organizers of the EMI-2001 Conference for invitation, warm hospitality, and support.

References

1. S.B. Gerasimov, JINR E2-4295, Dubna, 1969.
2. S.B.Gerasimov, Yad.Fiz. **29** (1979) 513 (Sov.J.Nucl.Phys. **29** (1979) 259; Erratum-ibid. **32** (1980) 156).
3. Yu.K. Khokhlov, Doklady AN SSSR **97** (1954) 239; ZhETP **32** (1957) 124;
   L.L. Foldy, Phys.Rev. **107** (1957) 1303.
4. S.B.Gerasimov and J.Moulin, Nucl. Phys. **B 98** (1975) 349.
5. J.S. Levinger, *Nuclear Photodisintegration*, Oxford Univ.Press, New York, 1960.
6. S.B.Gerasimov, ZhETP Pis'ma, **5** (1967) 412; (Soviet Phys.- JETP Letters, **5** (1967) 337).
7. G.Barton, Nucl.Phys. **A104** (1967) 289.
8. R.Dolen, D.Horn and C.Schmidt, Phys. Rev. **166** (1968) 1768.
9. H. Harari, Phys. Rev. Lett. **22** (1968) 1395. P.G.O. Freund, Phys. Rev. Lett. **20** (1968) 235.
10. N.Cabibbo and L.Radicati, Phys.Lett. **19** (1966) 697.
11. K.Gottfried, Phys. Rev. Lett. **18** (1967) 1174.
12. M.Hosoda and K.Yamamoto, Prog.Theor.Phys. **36** (1966) 425.
13. Z.Li and F.E.Close, Phys. Rev. **D42** (1990) 2194,2207.
14. Particle Data Group, Eur.Phys.J. **C15** (2000) 1.
15. S.Kopecky et al., Phys.Rev.Lett. **74** (1995) 2427.
16. A.S.Korotkov and A.G. Pronko, Yad.Fiz.**62** (1999) 1844.
17. S.Braccini, Acta Phys.Polon.**31** (2000) 2143; hep-ex/0111039.
18. W.Robson, Can.J.Phys. **64** (1986) 1359.
19. D.Black, A.H.Fariborz, F.Sannino, and J.Schechter, Phys.Rev. **D58** (1998) 054012

EXPERIMENTAL STUDIES ON GERASIMOV-DRELL-HEARN SUM RULE

T. IWATA

Department of Physics, Nagoya University, Furo-cho, Chikus a-ku, Nagoya, 464-8602, Japan
E-mail: iwata@kiso.phys.nagoya-u.ac.jp

The Gerasimov-Drell-Hearn(GDH) sum rule relates the helicity dependent total photo-absorption cross sections of circularly polarized photons on longitudinally polarized nucleons to the statistic properties of the nucleon. Recently, a direct measurement for the proton has been carried out at Mainz in the the energy range 200< ν < 800 MeV. The contribution to the GDH sum rule in this energy range was found to be $226 \pm 5(stat) \pm 12(sys)$ μb. The running GDH integral from 200 MeV is still increasing even at 800 MeV. For the measurements over 800 MeV, an experiment has been carried out by the same collaboration at Bonn. The preliminary results on the helicity dependent total photoabsorption cross-section difference in the energy range of 800< ν < 1400 MeV show a structure in the 2nd resonance region around 1 GeV. And the data give positive contribution to the sum rule. In the higher energy regions, an experiment is planned using the Laser-Electron-Photon facility at SPring-8 (LEPS). The circularly polarized photon beam produced by the backward-Compton scattering of laser photons from 8 GeV electrons will allow measurements up to the maximum energy of 2.8 GeV. The experimental plan along with the setup will be shown.

1   Introduction

The Gerasimov-Drell-Hearn(GDH) sum rule relates the total absorption cross section of circularly polarized photons on longitudinally polarized nucleons to the statistic properties of the nucleon. The two relative spin configurations, parallel or antiparallel, determine the two absorption cross sections $\sigma_{3/2}$ and $\sigma_{1/2}$. The integral over the photon energy ν of the difference of these cross sections, weighted by the inverse of ν, is related to the mass m and anomalous magnetic moment κ of the nucleon as:

$$\int_{\nu_0}^{\infty} (\sigma_{3/2} - \sigma_{1/2}) \frac{d\nu}{\nu} = \frac{2\pi^2 \alpha}{m^2} \kappa^2 \tag{1}$$

where ν_0 is the pion threshold and α the fine-structure constant. The GDH values for the proton and the neutron are calculated to be 205 μb and 233 μb, respectively. The GDH sum rule was derived in the 1960's by Gerasimov [1] and independently by Drell and Hearn [2]. They derived it on the bases of important principles such as Lorentz-invariance, gauge invariance, unitarity, causality

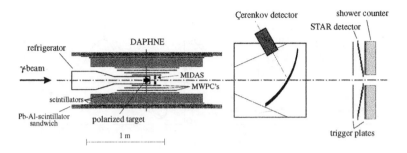

Figure 1. Schematic view of the Mainz setup.

and the unsubtracted dispersion relation applied to the forward Compton amplitude. Another independent derivation using a current algebra technique was made almost at the same time by Hosoda and Yamamoto [3] of Osaka University, although it is less well known.

2 The GDH Experiment at Mainz

The first experiment to study the helicity dependent total photo-absorption cross sections was carried out at the tagged photon facility of Mainz MAMI accelerator. Circularly polarized photons are produced by bremsstrahlung of longitudinally polarized electrons. The polarized electrons are produced by a strained GaAs source giving the polarization of about 80 % [4]. The produced photons were tagged by a tagging spectrometer with a hodoscope with 352 channels giving an energy resolution of about 2 MeV. The tagging region is from 50 to 800 MeV.

Polarized nucleons were available with the frozen spin target [5] which was constructed by Bonn, Bochum and Nagoya groups. The target material is butanol. It is polarized at 2.5 T provided by a movable polarizing coil to give the maximum polarization of about 90 %. Once it is polarized, the coil is moved out and the detectors are installed around the target. Then, data taking was carried out in the frozen spin mode at 0.4 T provided by the internal holding coil mounted inside the cryostat.

The reactions were registered in the setup as shown in Fig. 1 with the central detector DAPHNE [6] which consists of MWPC and plastic scintillation counters. It has capability to detect also gamma-rays although with a moderate efficiency of about 20 %. It can separate protons, pions and electrons by dE/dx and total energy measurement. These performances allow

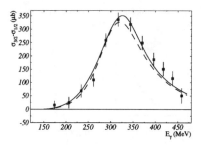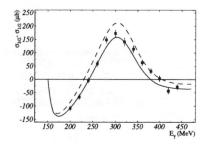

Figure 2. Difference of the total cross sections for $\gamma p \to p\pi^0$ and $\gamma p \to n\pi^+$ for the helicity states 3/2 and 1/2 as a function of photon energy in comparison to HDT (solid line) and SAID (dashed line) multipole analyses.

to make studies not only for the inclusive channel but also partial channels. Electromagnetic showers were removed in the trigger level by a Cerenkov counter. The forward detector system consists of the annular ring detector STAR [7] sensitive to charged particles and a forward sandwich counter for gamma-rays.

2.1 Results in Partial Channels

The difference $(\sigma_{3/2} - \sigma_{1/2})$ was obtained for $p\pi^0$ and $n\pi^+$ channels [8] as shown in Fig. 2. The lines are predictions from multipole analyses of HDT [9] and SAID [10]. The data and the predictions are consistent with each other in $p\pi^0$ channel. On the other hand, in $n\pi^+$ channel, the SAID prediction shows systematic disagreement to the data in particular in the lower energy region around 200 MeV. The contributions to the GDH integral in the energy region from 200 to 450 MeV were found to be 144 ± 7 μb in $p\pi^0$ channel and 32 ± 3 μb in $n\pi^+$ channel, respectively.

Preliminary data have been obtained for the difference $(\sigma_{3/2} - \sigma_{1/2})$ in $p\pi^+\pi^-$ and $n\pi^+\pi^0$ channels. These data show structures in the energy region $E_\gamma = 0.6 \sim 0.7$ GeV which are caused by the contributions from the $\sigma_{3/2}$ components.

Another interesting channel is ηp channel. S_{11} resonance located at $E_\gamma = 0.78$ GeV strongly couples to the channel. It dominantly contributes to $\sigma_{1/2}$. The ηp channel can be identified in the missing mass distribution against a proton detected in DAPHNE acceptance. Fig. 3 shows the missing mass distribution preliminary obtained for a hydrogen target. In the analy-

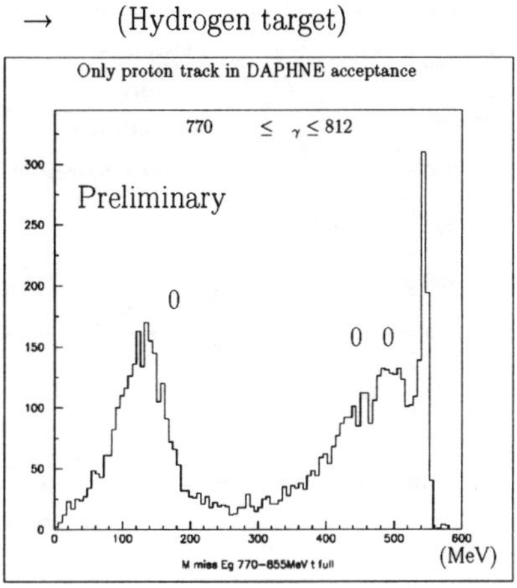

$\longrightarrow$ (Hydrogen target)

Figure 3. Missing mass distribution against a proton detected in DAPHNE acceptance for hydrogen target. The photon energy region is $770\ MeV \leq E_\gamma \leq 812\ MeV$.

sis, only a proton track was required in DAPHNE acceptance. A clear peak corresponding to η is seen along with π^0 and $\pi^0\pi^0$. Further analysis for a polarized butanol target is in progress.

2.2 Results in Inclusive Measurement

The total cross-section can be directly accessed by measuring the numbers of the events with charged hadrons and π^0s. The number of the π^0 events can be corrected by using the π^0 detection efficiency evaluated with simulation. The details on the analysis procedure can be found in elsewhere[11]. The data for the cross-section difference $(\sigma_{3/2} - \sigma_{1/2})$ obtained in the inclusive measurement are shown in fig 4 along with the one for single pion channels. In the Δ region, the single pion channel gives a large contribution. And the data set is well described by theories. On the other hand, in the second resonance region above 500 MeV, the theories give lower values than the data points obtained in the inclusive measurement. The unitary isobar model, taking account of also double pion channels and eta channel, even can not describe the data.

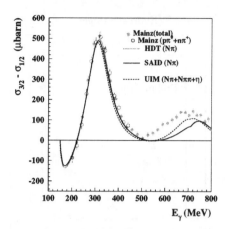

Figure 4. The cross-section difference $(\sigma_{3/2} - \sigma_{1/2})$ on the proton is compared to theories.

Table 1. Theoretical evaluations for the contribution to the GDH integral in the un-measured energy regions.

$\nu \leq 200$ MeV	UIM [13]	$-30\ \mu b$
$800 \leq \nu \leq 1650$ MeV	UIM [13]	$+40\ \mu b$
1650 MeV $\leq \nu$	Bianchi and Thomas [14]	$-26\ \mu b$
	total from un-measured regions	$-16\mu b$
$\nu_\pi \leq \nu \leq \infty$	Exp + evaluations	210μ b
	GDH prediction	$205\ \mu b$

2.3 GDH Integral

From the the data for the cross-section difference $(\sigma_{3/2} - \sigma_{1/2})$ obtained in the inclusive measurement, one can calculate the running GDH integral from 200 MeV as shown in fig. 5. It is not saturated even at 800 MeV. The GDH integral from 200 MeV to 800 MeV was found to be $226 \pm 5 \pm 12\ \mu b$ [12]. The theoretical predictions give lower values: disagreement from the data is about 20 %.

The GDH integral for full energy region can be evaluated with theoretical evaluations as shown in Table 1 for the un-measured regions. The total contribution from un-measured regions is evaluated to be $-16\ \mu b$. This leads to

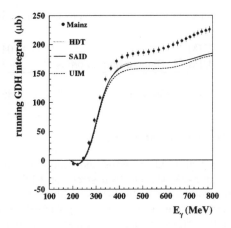

Figure 5. The running GDH integral from 200 MeV.

210 μb for the integral from the pion threshold to infinity which is compared with the GDH sum rule prediction, 205 μb. These values are in good agreement with each other. However, the theoretical evaluations should be verified in future experiments.

2.4 Forward Spin Polarizability

From the same data for the cross-section difference, one can extract another important quantity, namely, forward spin polarizability γ_0 which reflects the internal structure of the nucleon. γ_0 is expressed with the cross-section difference as

$$\gamma_0 = -\frac{1}{4\pi^2} \int_{\nu_\pi}^{\infty} (\sigma_{3/2} - \sigma_{1/2}) \frac{d\nu}{\nu^3}. \tag{2}$$

Since the cross section difference is divided by the 3rd power of the photon energy, contribution in low energy regions is significant. We obtained $-187 \pm 8(stat)\pm10(sys)\times10^{-6}$ fm^4 for the integral corresponding to γ_0 from 200 MeV to 800 MeV. With evaluated values as shown in Table 2 for the contributions from un-measured regions, -86×10^{-6} fm^4 was obtained for full energy region. This value can be compared with theoretical calculations [15].

Table 2. Theoretical evaluations for the contribution to γ_0 in the un-measured energy regions.

$\nu \leq 200$ MeV	UIM [13]	$+104 \times 10^{-6} \ fm^4$
800 MeV $\leq \nu$	UIM [13], Bianchi and Thomas [14]	$-3 \times 10^{-6} \ fm^4$
γ_0	Exp + evaluations	$-86 \times 10^{-6} \ fm^4$

3   The GDH Experiment at Bonn

The GDH experiment at Bonn covering the energy region above 800 MeV has been carried out using the electron stretcher ELSA. Circularly polarized photons were produced also by bremsstrahlung of polarized electrons. The polarized electrons were available by the polarized electron source with a super-lattice photo-cathode giving a polarization of 80 % [16]. The energy setting of the electron beam is between 1.0 GeV and 3.2 GeV. A main difference from the Mainz experiment is that ELAS is a storage type accelerator, so that it has depolarization resonances during acceleration of the polarized electrons. To overcome these resonances, closed orbit correction and fast tune jumps have been adopted. As a result, reasonable polarizations remained in accelerator tests [16]:$P_e = 72$ %($E_e \leq 2 \ GeV$), 50 %($E_e = 2.4 \ GeV$) and $P_e = 30$ %($E_e = 3.2 \ GeV$). The same target system as in Mainz was used. On the other hand, the Bonn detector system has different concept. The Mainz setup is capable of identifying partial channels. However, the Bonn system is dedicated only for inclusive measurement. At high energies multiplicity of the events becomes higher, so that to identify partial channels is not meaningful. The detector system consists of the central components, the forward component. The central components are made of plastic scintillator and lead plates. It is sensitive to both charged particles and gamma-rays. The Cerenkov counter and the forward components are the same as in Mainz.

3.1   Preliminary Data

Preliminary data for the cross-section difference ($\sigma_{3/2} - \sigma_{1/2}$) have been obtained in the energy region up to about 1.4 GeV as shown in Fig. 6 along with the data set of Mainz. Two data sets are in good agreement with each other in the overlapping energy region. The Bonn data set show a clear structure in the 3rd resonance region around 1 GeV. As for the higher energy regions, additional data are expected in future analysis. However, due to depolarization problems of the beam in particular in the highest energy region, the data

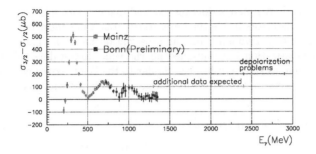

Figure 6. Preliminary data obtained in the Bonn experiment for the cross-section difference $(\sigma_{3/2} - \sigma_{1/2})$ on the proton.

may be limited up to 2.4 GeV.

Preliminary value for the running GDH integral from 200 MeV to the energies up to 1.35GeV was derived by using the data of Mainz and Bonn. It was found to be about 250 μb at 1.35 GeV. One can evaluate the running GDH integral based on the GDH sum rule prediction with following assumptions: (1)no additional contribution in the higher energy region above 1.35 GeV, (2)contribution from the energy region below 200 MeV is evaluated to be -30 μb from UIM model [13]. The evaluated running GDH integral is 235 μb which can be compared with the experimental data. At this stage, the disagreement between them is only about 15 μb level.

4 The GDH Experiment at SPring-8

An experiment to study the GDH sum rule for proton has been proposed at SPring-8. It aims at measurements in the energy range from 1.8 GeV to 2.8 GeV. Circularly polarized photons are produced by laser backward Compton scattering at the Laser-Electron-Photon(LEP) beam line of SPring-8. The LEP beam gives high polarization, e.g. 100 % at the maximum energy point. The measurements are planned with two energy settings, up to 2.4 GeV and up to 2.8 GeV. A polyethylene polarized target developed at Naogya University will be adopted. It is suitable to reduce systematical uncertainty originating from determination of the thickness of the target. A different feature from the Mainz and Bonn experiments is the target is continuously polarized. In other words, the frozen spin mode which requires a movable detector system,

586

will not be used. A superconducting solenoid magnet with a large bore allows to install the detectors inside the bore, so that one can continuously enhance the polarization in Dynamic Nuclear Polarization method. This means the maximum polarization is maintained all the time during data taking. The detector system as shown in fig. 7 is similar to the Bonn experiment with the geometrical acceptance of 97 %. A charged particle detector and a gamma-

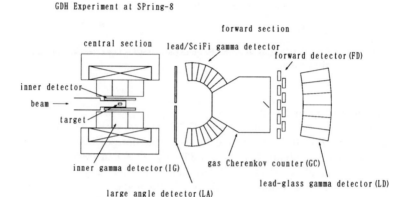

Figure 7. Schematic view of the SPring-8 GDH setup.

ray detector will be installed around the target inside the magnet bore. A Cherenkov counter will be adopted to veto the electromagnetic showers.

Recently, the proposal of the experiment has been approved. The international collaboration for the experiment has been organized with the following institute: Nagoya Univ., Miyazaki Univ., UCLA, KEK, Royal Melbourne Institute for Technology (RMIT), Bonn Univ., Catholic University of America (CUA) and Melbourne Univ. R & D work for detectors and the polarized target is in progress. The data taking is expected in the middle of 2003.

5 Summary

The GDH sum rule is one of the fundamental theoretical predictions to be verified in experiments. The GDH integral for proton from 200 MeV to 800 MeV was obtained for the first time in Mainz-GDH experiment. The Bonn-GDH experiment to cover higher energy range is providing preliminary data . The

SPring-8 GDH experiment is aiming at measurements in higher energy range up to 2.8 GeV with the LEP beam.

References

1. S.B. Gerasimov Yad.Fiz. 2(1965)598, Sov.J.Nucl. Phys. 2 (1996) 430.
2. S.D. Drell and A.C. Hearn, *Phys. Rev. Lett.* **16**, 908 (1966)
3. M. Hosoda and K. Yamamoto, Prog. Theo. Phys., 16, 908 (1966).
4. K. Aulenbacher, Proceedings of the Symposium on the Gerasimov-Drell-Hearn Sum Rule and the Nucleon Spin Structure in the Resonance Region, Mainz, Germany, 14-17 June 2000, eds. D.Drechsel and L.Tiator.
5. Ch. Bradtke et al., *Nucl. Instrum. Methods* A **436**, 430 (1999)
6. G. Audit et al., *Nucl. Instrum. Methods* A **301**, 473 (1991)
7. M. Sauer et al., *Nucl. Instrum. Methods* A **378**, 143 (1996)
8. J. Ahrens et al., *Phys. Rev. Lett.* **84**, 5950 (2000)
9. O. Hanstein et al., *Nucl. Phys.* A **632**, 561 (1998)
10. R.A. Arndt et al., *Phys. Rev.* C **53**, 430 (1996)
11. M. MacCormick et al., *Phys. Rev.* C **53**, 41 (1998)
12. J. Ahrens et al., *Phys. Rev. Lett.* **87**, 022003 (2001)
13. D. Drechsel et al., *Nucl. Phys.* A **645**, 145 (1999)
14. N. Bianchi and T. Thomas, *Phys. Lett.* B **450**, 439 (1999)
15. Theoretical predictions of the forward spin polarizability γ_0 of the proton(in units of 10^{-6}).
    D. Babusci et al., *Phys. Rev.* C **58**, 1013 (1998), D. Drechsel et al., *Phys. Rev.* C **61**, 015204 (2000), V. Bernard et al., *Int. J. Mod. Phys.* E **4**, 193 (1995), T.R. Hemmert et al., *Phys. Rev.* D **57**, 5746 (1998), X. Ji et al., *Phys. Lett.* B **472**, 1 (2000), K.B. Vijaya Kumar et al., *Phys. Lett.* B **479**, 167 (2000), G.C. Gellas et al., *Phys. Rev. Lett.* **85**, 14 (2000) and Y. Tanushi et al., *Phys. Rev.* C **60**, 065213 (1999)
16. Proceedings of the Symposium on the Gerasimov-Drell-Hearn Sum Rule and the Nucleon Spin Structure in the Resonance Region, Mainz, Germany, 14-17 June 2000, eds. D. Drechsel and L. Tiator.

FORWARD COMPTON SCATTERING: SUM RULES FOR PARITY VIOLATING SPIN POLARIZABILITIES

LESZEK ŁUKASZUK

The Andrzej Sołtan Institute for Nuclear Studies, Hoża 69, 00-689 Warsaw, Poland
E-mail: lukaszuk@fuw.edu.pl

Sum rules connecting parity violating(p.v.)threshold terms of the Compton amplitudes with the suitable integrals over p.v. parts of the total cross sections for photoproduction are given and discussed. They hold both for hadron and nuclear targets of arbitrary spin. Theoretical legitimacy of sum rules is satisfactory; they are valid in the lowest contributing order of the electroweak theory and are exact in strong interactions. Model dependent discussion presented in this talk will be concerned with application of p.v. sum rules to the cases of proton and deuteron targets.

1 Introduction

The recent revival [1,2,3,4] of interest in the weak part of gamma - hadron interactions seems to be closely connected with the advent of intense polarized beams of photons [5,6,7]. Generally it is sound to expect that future experiments would be a good source of information on theoretically difficult domain of low energy hadronic structure. For example the threshold production asymmetries in pion photoproduction on proton might establish strength of weak $NN\pi$ coupling [1], $h^{(1)}_{\pi NN}$. Similiar expectations are connected with low energy Compton scattering [3] on proton. The $h^{(1)}_{\pi NN}$ problem is acute due to results from different nuclear [8] and atomic [9] physics experiments which seem to indicate quite different values of $h^{(1)}_{\pi NN}$. Theoretical interpretation of future experiments on more elementary targets should be easier, of course. At the moment neither large nor small $h^{(1)}_{\pi NN}$ option can be excluded and perhaps this constant can be much smaller then so called "best value" [10] ; then the short distance (in comparison with effective π exchange) Parity Violating (p.v.) contributions should be larger than those from the set of "best values" [10] - here a test for importance of such contributions would be photodisintegration of deuteron [4]. Both real and virual photon -initiated effects are of interest and were considered in the literature [1,2,3,4,11,12]. Experimentally these cases correspond to photon and electron -initiated collisions, respectively. A convenient feature of photon (real) initiated collisions is the absence of direct Z exchange between projectile and target so it is a unique situation where p.v. structure of electromagnetic current itself is singled out without further elaborations. On the other hand disentangling virtual photon p.v. contributions from electroproduction seems to be difficult - already at $Q^2 > 0.1 GeV^2$ it's contribution to measured asymmetry is a few percent of the neutral current's contribution [11]. This situation reflects the fact that p.v. interference terms involving only electromagnetic currents must contain additional photon's propagator so extra (α) factor appears when compared with terms which contain electromagnetic and neutral current exchanges (Only for $Q^2 \rightarrow 0$ the additional photon's propagator will dominate - for the π meson electroproduction it takes the place below Q^2 of the order $10^{-3} GeV^2$ [11]). Having this situation in mind we shall limit ourselves to the real photons only in the further discussion so that in the language of ref. [11] we shall confine ourselves to the p.v. electromagnetic and nuclear interaction at $Q^2 = 0$. Model dependent estimates of these interactions lead to asymmetries of the order of 10^{-7} for low energy photoproduction ("best fit " coupling scheme). The Compton amplitude asymmetries have been theoretically estimated in the case of proton target as 10^{-8} in the one-loop approximation of HBχPT [3]. As it was shown [13] that p.v. effects at low energies, ω , do not contribute to the static terms (i.e. they are at least of the order $O(\omega^2)$) in this process, we have to treat these p.v. asymmetries as p.v. spin polarizabilities. Parity conserving (p.c.) spin polarizabilities of nucleon are of theoretical and experimental interest since long time. In particular HBχPT was shown to be rather unpredicdive in calculating these polarizabilities [14]. So in this case apart from uncertainty of couplings this kind of unstability might appear. In general it would be appropriate to make low energy multipole expansion and in such a way to limit a number of adjustable parameters, however, in p.c. analysis it was shown that apart from general theoretical input quite acurate experimental data are needed to find these parameters [15]. We have no comparable data for p.v. analysis at the moment, so it seems sound to try to correlate different p.v. observables via model independent relations such as dispersive sum rules in analogy with spin dependent p.c. domain [16,17]. We shall limit ourselves to the sum rules for forward amplitudes as in this case relations between physically measured cross sections are most transparent. In what follows we shall discuss legitimacy of the relevant dispersion relations, apply them to Compton scattering on proton and deuteron using existing models for p.v. in photoproduction [1,11] and photo- desintegration [4] . Next we shall discuss the problem of subtractions in the context of eventual use of superconvergence assumptions. Tentative predictions made under such assumptions will be formulated. More speculative - but related - problem of a trace of unification in high energy Compton asymptotics might be posed.

2 Dispersive Formulae in Standard Model

2.1 Asymptotic States

As we want to discuss dispersion formulae for collision amplitudes, it is a suitable place to ask in what a degree the usual properties of these amplitudes (existence of asymptotic states and of interpolating local fields) are exhibited in the Standard Model. The asymptotic states have to correspond to Fock space of stable particles so we are left with photon, electron, neutrinos (at least the lightest one), proton and stable atomic ions. Let us mention here that the existence of unstable fields is a source of concern in Quantum Field Theory [18]. Next, each stable particle should correspond to irreducible Poincare (unitary) representation and here trouble appears with charged particles [19]. This is connected with QED infrared radiation and well defined procedure exists in perturbative calculus only. It is the reason that our considerations concerning Compton amplitudes will be limited to the order α in p.c. part and to order α^2 in p.v. part (they are infrared safe and at low energies are αG_F order contributions). Still we are left with the problem of asymptotic states and interpolating fields in QCD part of SM - we shall rely on the results of Oehme: "the analytic properties of physical amplitudes are the same as those obtained on the basis of an effective theory involving only the composite, physical fields" [20] (in other words confinement does not spoil old axiomatic proofs for hadronic interactions [21]).

2.2 Analyticity and Compton Sum Rules for Arbitrary Target

Working in the lowest electroweak order it is reasonable to abandon C, P invariance only and keep in our considerations T-invariance. The strong part of interactions is taken without any approximation. The analyticity of the forward Compton amplitude and crossing properties follow from the typical steps schetched below. We start from LSZ - derived formula for the forward Compton amplitude. For convenience we choose target in lab frame and drop unessential for the further argument contact terms, so that we get [22]

$$S_{fi} = I + i(2\pi)^4 \delta_4(P_f - P_i) M_{fi} \tag{1}$$

$$M_{fi} = e^2 \overline{\epsilon_f^\mu} \epsilon_i^\nu T_{\mu\nu} \tag{2}$$

$$T_{\mu\nu}(q, M, s_f, s_i) = i \int d^4x \, e^{i\omega(x_0 - \vec{n}\vec{x})} \Theta(x_0)\langle M, s_f|[j_\mu(x), j_\nu(0)]|M, s_i\rangle \tag{3}$$

where $\vec{q} = \omega\vec{n}$ is photon's momentum and with z-axis taken in its direction, $\vec{n}$, s_i, s_f is target's initial (final) z-component of spin, ϵ_i, ϵ_f denote initial (final) circular polarization states of photon. Let us choose radiation gauge so only $\mu, \nu = 1, 2$ components of j_μ, j_ν contribute. In what follows we never use parity conservation in the derivations, of course.

The analyticity in the upper complex half-plane of ω follows from the fact that retarded commutator in (3) vanishes for $x_0 < 0$ and , due to causality, vanishes also for $x_0 < |z|$. For $Re\omega > 0$ approaching real axis we get physical Compton amplitude specified by (2). For $Re\omega < 0$ the limiting amplitude can be obtained by applying complex conjugation to (3) and exploiting invariance of matrix elements with respect to rotations; here rotation around y or x-axis by angle π should be used. The result, independent of P,C, T invariances, reads

$$M^{s_f,s_i}_{h_f,h_i}(-Re\omega + i\epsilon) = \overline{M^{s_i,s_f}_{-h_f,-h_i}(Re\omega + i\epsilon)} \tag{4}$$

This is not a customary form of crossing relation,however,adding T invariance demand and using it with suitable translation and rotation we get

$$M^{s_f,s_i}_{h_f,h_i}(-Re\omega + i\epsilon) = \overline{M^{s_f,s_i}_{-h_i,-h_f}(Re\omega + i\epsilon)} \tag{5}$$

In what follows we shall be interested in coherent amplitudes only (i.e. $s_i = s_f$ and $h_i = h_f$), suitable for sum rules as their imaginary parts are proportional to the total cross sections. In this case (4) and (5) are eqiuvalent, so demand of T invariance is not necessary. We shall use abbreviated names f for these amplitudes

$$f_{s,h}(\omega) = M^{s,s}_{h,h}(\omega) \tag{6}$$

so that

$$f_{s,h}(-Re\omega + i\epsilon) = \overline{f_{s,-h}(Re\omega + i\epsilon)} \tag{7}$$

We shall use amplitudes [23] normalized such that, for any target

$$Im f_{s,h}(\omega) = \omega\sigma^T_{s,h}(\omega) \tag{8}$$

The analyticity, crossing (7) and unitarity lead, through Hilbert formulae to the dispersion relations for these amplitudes

$$Re f_{s,h}(\omega) = \frac{1}{\pi} \mathcal{P} \int_{\omega_{th}}^{\infty} \frac{\omega' \sigma_{s,h}^T(\omega')}{\omega' - \omega} d\omega' + \frac{1}{\pi} \int_{\omega_{th}}^{\infty} \frac{\omega' \sigma_{s,-h}^T(\omega')}{\omega' + \omega} d\omega' + (subtr.) \quad (9)$$

On the other hand any amplitude f can be written as

$$f_{s,h} = f_{s,h}^+ + f_{s,h}^- \quad (10)$$

where f^+, f^- are p.c. and p.v., respectively,

$$f_{s,h}^{\pm} = \frac{1}{2}(f_{s,h} \pm f_{-s,-h}) \quad (11)$$

There exist proofs of low energy QED theorems for any target's spin [23,24] up to $O(\omega)$ terms. Explicit proof that p.v. amplitudes of SM are of order $O(\omega^2)$ has been given for spin $\frac{1}{2}$ [13]. I learned from I.B.Khriplovich that this result holds for targets of any spin if one neglects, as we do, T-violation; the reason is that the leading (larger than $O(\omega^2)$) low energy behaviour comes from the pole diagrams and that in this case the only p.v. coupling of real photon involves electric dipole moment [25]. Therefore we shall take for granted that, for any target's spin

$$M_{h_f,h_i}^{s_f,s_i}(\omega \to 0) = \delta_{h_f,h_i} \delta_{s_f,s_i} f_{s_i,h_i}^{(+)LET} + O(\omega^2) = M^{(+)} + O(\omega^2) \quad (12)$$

with $f^{(+)LET}$ known from the Low Energy Theorems [23,24] and

$$M_{h_f,h_i}^{(\pm)s_f,s_i} = \frac{1}{2}(M_{h_f,h_i}^{s_f,s_i} \pm M_{-h_f,-h_i}^{-s_f,-s_i}) \quad (13)$$

so that

$$f_{s,h}^-(\omega)|_{\omega \to 0} = O(\omega^2) \quad (14)$$

Hence for any target in the limit $\omega \to 0$ the ratio

$$A(s_i, h_i) = \frac{\sum_{s_f, h_f}(|M_{h_f,h_i}^{s_f,s_i}|^2 - |M_{h_f,-h_i}^{s_f,-s_i}|^2)}{4 f_{s_i,h_i}^{(+)LET}} = f_{s_i,h_i}^- + O(\omega^4) \quad (15)$$

measures the parity violating part of forward amplitude. This came out to be simple for any target due to diagonal form of M^+ at low energies (comp. (12)).

It will be convenient to consider averaged p.v. amplitudes

$$f_h^{(-)\gamma} = \frac{1}{2S+1} \sum_{s_i} f_{s_i,h}^- \tag{16}$$

and

$$f_s^{(-)tg} = \frac{1}{2}(f_{s,+1}^- + f_{s,-1}^-) \tag{17}$$

These amplitudes are expressed by integrals over relevant differences of the total cross sections (comp. (9)).

$$Ref_h^{(-)\gamma} = \frac{\omega}{\pi}P\int_{\omega_{th}}^{\infty}\frac{\omega'}{\omega'^2 - \omega^2}(\sigma_h^T - \sigma_{-h}^T)d\omega' + (subtr.) \tag{18}$$

where

$$\sigma_h^T = \frac{1}{2S+1}\sum_{s_i}\sigma_{s_i,h}^T \tag{19}$$

and

$$Ref_s^{(-)tg} = \frac{1}{\pi}P\int_{\omega_{th}}^{\infty}\frac{\omega'^2}{\omega'^2 - \omega^2}(\sigma_s^T - \sigma_{-s}^T)d\omega' + (subtr.) \tag{20}$$

where

$$\sigma_s^T = \frac{1}{2}(\sigma_{s,+1}^T + \sigma_{s,-1}^T) \tag{21}$$

2.3 Subtractions vs superconvergence

As the p.v. amplitudes considered in our approximation are infrared safe we are entitled to expect that high energy growth of forward amplitudes will be at most $\omega(ln\omega)^2$ as results from general principles for finite range interactions [21]. Condition (12) means that for any target's spin no arbitrary constants appear in dispersion formulae for $f^{(-)tg}$, $f^{(-)\gamma}$ if subtraction point is taken at $\omega = 0$, therefore

$$Ref_s^{(-)tg} = \frac{\omega^2}{\pi}P\int_{\omega_{th}}^{\infty}\frac{\sigma_s^T - \sigma_{-s}^T}{\omega'^2 - \omega^2}d\omega' = -4\pi\omega^2 a_s^{(-)tg}(\omega) \tag{22}$$

and

$$Ref_h^{(-)\gamma} = \frac{\omega^3}{\pi}P\int_{\omega_{th}}^{\infty}\frac{\sigma_h^T - \sigma_{-h}^T}{\omega'(\omega'^2 - \omega^2)}d\omega' = -4\pi\omega^3 a_h^{(-)\gamma}(\omega) \tag{23}$$

For $\omega \to 0$ eqns.(22, 23) yield sum rules for p.v. forward polarizabilities $a_s^{(-)tg}(\omega = 0)$, $a_h^{(-)\gamma}(\omega = 0)$ defined in analogy with p.c. forward spin polarizabilities [5]

$$a_s^{(-)tg}(0) = \frac{1}{4\pi^2} \int_{\omega_{th}}^{\infty} \frac{\sigma_{-s}^T - \sigma_s^T}{\omega'^2} d\omega' \tag{24}$$

$$a_h^{(-)\gamma}(0) = \frac{1}{4\pi^2} \int_{\omega_{th}}^{\infty} \frac{\sigma_{-h}^T - \sigma_h^T}{\omega'^3} d\omega \tag{25}$$

If we assume superconvergence of the type $\frac{f(z)}{z} \to 0$ at infinity for asymmetric amplitude (23), then the p.v. analogue of DHG [16] is obtained

$$\int_{\omega_{th}}^{\infty} \frac{\sigma_h^T - \sigma_{-h}^T}{\omega'} d\omega' = 0 \tag{26}$$

It is natural to ask for legitimacy and cosequences of such assumption for p.v. amplitude $f^{(-)\gamma}$. Consequences of this formula will be mentioned in the context of model dependent applications in the next section. The check whether at least in perurbative QCD regime relevant contributions to the total cross sections asymptotically cancel in (26) has not yet been done. In the lowest QCD order one should calculate a few different processes which might conspire to give vanishing overall difference of the total cross sections, it might also happen that non perturbative regime plays equal or essential role. We have only checked with dr K.Kurek that a class of hard processes with polarized target (proton), namely those with the production of additional quark antiquark pair via unpolarized photon structure yields non vanishing (in fact slowly rising) contribution to the difference $\sigma_{\frac{1}{2}}^T - \sigma_{-\frac{1}{2}}^T$ unless valence quarks of proton are negligable at asymptotically large photon's energies in hard production. Despite this warning we shall take this liberty and point at consequences of (26) in the next section.

3 Examples

3.1 Proton Target

The p.v. Compton amplitude can be written in c.m.s. as [3]

$$M_{h_f,h_i}^{(-)s_f,s_i}(\vec{k}, \vec{k'}) = \overline{N_{s_f}}[F_1 \vec{\sigma} \cdot (\hat{\vec{k}} + \hat{\vec{k'}}) \vec{\epsilon_i} \cdot \overline{\vec{\epsilon_f}} - F_2(\vec{\sigma} \cdot \overline{\vec{\epsilon_f}} \hat{\vec{k'}} \cdot \vec{\epsilon_i} + \vec{\sigma} \cdot \vec{\epsilon_i} \hat{\vec{k'}} \cdot \overline{\vec{\epsilon_f}})$$

$$- F_3 \hat{\vec{k}} \cdot \overline{\vec{\epsilon}_f} \hat{\vec{k}'} \cdot \vec{\epsilon}_i \vec{\sigma} \cdot (\hat{\vec{k}} + \hat{\vec{k}'}) - i F_4 \vec{\epsilon}_i \times \overline{\vec{\epsilon}_f} \cdot (\hat{\vec{k}} + \hat{\vec{k}'})] N_{s_i} \quad (27)$$

so that

$$f_{\frac{1}{2}}^{(-)p} = 2F_1 = O(\omega^2) \quad (28)$$

$$f_{h=+1}^{(-)\gamma} = -2F_4 = O(\omega^3) \quad (29)$$

The HBχPT analysis [3] provides values of coefficients F_1, F_4

$$F_1|_{\omega \to 0} = -\frac{e^2}{M} (\frac{\omega}{m_\pi})^2 \frac{M}{F_\pi} \frac{g_A h_{\pi NN}^{(1)}}{24\sqrt{2}\pi^2} \quad (30)$$

$$F_4|_{\omega \to 0} = \frac{e^2}{M} (\frac{\omega}{m_\pi})^3 \frac{m_\pi}{F_\pi} \frac{g_A h_{\pi NN}^{(1)} \mu_n}{24\sqrt{2}\pi^2} \quad (31)$$

where $F_\pi = 93$MeV, $g_A = 1.26$, $\mu_n = -1.91$, $h_{\pi NN}^{(1)} \simeq 5 \cdot 10^{-7}$ in the "best fit" parametrization, M is nucleon mass.

On the other hand there are theoretical results for p.v. effects in near threshold photoproduction [1,2,11]. These values could be used in our relations (22, 23). Denoting

$$Re f_{s=\frac{1}{2}}^{(-)p} = -\frac{e^2}{M} \beta^- (\frac{\omega}{m_\pi})^2|_{\omega \to 0} \quad (32)$$

$$Re f_{h=+1}^{(-)\gamma} = -\frac{e^2}{M} \gamma^- (\frac{\omega}{m_\pi})^3|_{\omega \to 0} \quad (33)$$

we have

$$a_{\frac{1}{2}}^{(-)tg}(0) = \frac{\alpha}{M m_\pi^2} \beta^- = 3.2\beta^{(-)}10^{-3}[fm]^3 \quad (34)$$

$$a_1^{(-)\gamma}(0) = \frac{\alpha}{M m_\pi^3} \gamma^- = 4.5\gamma^{(-)}10^{-3}[fm]^4 \quad (35)$$

with

$$\beta^- = -(\frac{e^2}{M m_\pi^2})^{-1} \frac{1}{\pi} \int_{\omega_{th}}^\infty \frac{\sigma_{s=\frac{1}{2}}^T - \sigma_{s=-\frac{1}{2}}^T}{\omega'^2} d\omega' \quad (36)$$

$$\gamma^- = -(\frac{e^2}{M m_\pi^3})^{-1} \frac{1}{\pi} \int_{\omega_{th}}^\infty \frac{\sigma_{h=+1}^T - \sigma_{h=-1}^T}{\omega'^3} d\omega' \quad (37)$$

while β_χ, γ_χ coming from (30,31) are

$$\beta_\chi^- = \frac{M}{F_\pi} \frac{g_A h_{\pi NN}^{(1)}}{12\sqrt{2}\pi^2} \simeq 4 \cdot 10^{-8} \qquad (38)$$

$$\gamma_\chi^- = \mu_n \frac{m_\pi}{F_\pi} \frac{g_A h_{\pi NN}^{(1)}}{12\sqrt{2}\pi^2} \simeq -1 \cdot 10^{-8} \qquad (39)$$

We shall use cross sections from [1,11] in our sum rules, compare consistency and finally, assuming superconvergence (26) discuss posible consequences. Let us start with HBχPT approach to photoproduction; taking dominant at threshold terms in effective lagrangian of reference [1] we integrate them up to $\omega = 200$ MeV i. e. in the region where HBχPT should be reliable. The results are $1 \cdot 10^{-8}$ for β^- and $-4 \cdot 10^{-9}$ for γ^- , to compare with $4 \cdot 10^{-8}$ and $-1 \cdot 10^{-8}$ for β_χ^- and γ_χ^-, respectively. This means that in γp HBχPT calculations [3] contributions from much higher energies have been involved. Indeed, extrapolating threshold behaviour till 1 GeV and inserting into (36), (37) we get values which well compare with results from [3]. However, there is no reason to assume validity of threshold type behaviour in so large region. Therefore we turn to analysis [11] where elaborated Born type exchanges (with resonances and form factors considered) were put together. We shall consider "best fit" predictions contained in figs 11-15 of [11]. Using eqs. (36), (37) we get $-5 \cdot 10^{-9}$ for β^- and $-1 \cdot 10^{-8}$ for γ^-. It is an example of usefulness of measurement not only threshold p.v. photoproduction but low energy Compton asymmetries, too, in future experiments as they can shed some light on the existence of structure at higher energies. In the comparison made above β^- obtained differ by order of magnitude. This is a reflection of quite different behaviour of cross sections. Let us pass now to the superconvergence hypothesis (26). The contributions in the region below 0.55 GeV we calculate from [11] and obtain relation

$$\int_{0.55 GeV}^{\infty} \frac{\sigma_1^T - \sigma_{-1}^T}{\omega'} d\omega' \simeq -30 pb \qquad (40)$$

If we further assume - by analogy with gross features of DHG sum rule saturation - that the necessary contribution comes from the region below 1 GeV we get for average asymmetry in the region $(0.55 - 1$ GeV) a value $(-50$ pb). This might indicate that it is desirable to look for p.v. effects in this region.

3.2   Deuteron Target

We shall consider model by Khriplovich and Korkin [4] for p.v. effects in photodisintegration. Here the difference of cross sections, integrated up to 10 MeV, yields rather small contribution in natural units (i.e. $\frac{\omega}{2.23MeV}$) but compared with that for proton i.e. in units $\frac{\omega}{m_\pi}$ is large:

$$Re f^{(-)\gamma}_{h=+1} = -\frac{e^2}{M_D}\gamma_D^-(\frac{\omega}{m_\pi})^3|_{\omega \to 0} \tag{41}$$

with

$$\gamma_D^- \simeq -2 \cdot 10^{-4} \tag{42}$$

If we use dispersion relation close to disintegration threshold $\omega \simeq 2.23MeV$, we get

$$-Re f^{(-)\gamma}_{+1}(\omega) \cdot (\frac{e^2}{M_D})^{-1}|_{\omega \to 2.23MeV} \simeq -1 \cdot 10^{-8} \tag{43}$$

while extrapolation of (41) would give $-7 \cdot 10^{-10}$. This indicates that p.v. effects are strengthened due to the cusp effect.

4   Concluding remarks

Described above p.v. sum rules are helpful in checking consistency of theoretical approaches (comp. discussion of example in ch.3.1). On the other hand, if superconvergence hypothesis (comp. ch.2.3) were verified, we could claim that p.v. effects in photoproduction should be sizable even far from the threshold (comp. ch.3.1). This subject seems to be a challenge for theoretical studies.

In connection with eventual experimental verifications in future, the cusp enhancement of p.v. effects (comp. ch.3.2) indicates that Compton scattering at inelastic thresholds may be of importance.

Acknowledgements

I would like to thank Professors: Mamoru Fujiwara and Ziemowid Sujkowski for encouragement and discussions on photon induced processes, Youlik Khriplovich for enlightment on LET in SM and Jacques Bros for helpful remarks on analyticity problem in QED.I would also like to thank dr Krzysztof

Kurek for collaboration on high energy p.v. effects . Many thanks to dr Remco Zegers for helpful correspondance on asymmetries.

References

1. J.W.Chen and X.Ji, *Phys. Rev. Lett.* **86**, 4239 (2001).
2. S.-L.Zhu *et al.*, *Phys. Rev.* C **64**, 035502 (2001).
3. P.F.Bedaque and M.J.Savage, *Phys. Rev.* C **62**, 018501 (2001).
   J.W.Chen and T.D.Cohen , *Phys. Rev.* C **64**, 055206 (2001).
4. I.B.Khriplovich and R.V.Korkin, *Nucl. Phys.* A **690**, 610 (2001).
5. J.Ahrens *et al.*, *Phys. Rev. Lett.* **84**, 5950 (2000).
6. M.Fujiwara *et al.*, *Acta Phys.Pol.* B **29**, 141 (1998).
   T.Nakano *et al.*, *Nucl. Phys.* A **629**, 559c (1998).
   M.Fujiwara, *Nucl. Phys. News* **11**, 28 (2001).
7. JLab LOI 00-002, W.van Oers and B.Wojtsekhowski, spokesman.
8. S.A.Page *et al.*, *Phys. Rev.* C **35**, 1119 (1987).
   M.Bini *et al.*, *Phys. Rev.* C **38**, 1195 (1988).
9. C.S.Wood *et al.*, Science **275**, 1759 (1997).
10. B.Desplanques *et al.*, *Ann. Phys.* **124**, 449 (1980).
11. S.P.Li *et al.*, *Ann. Phys.* **143**, 372 (1982).
12. R.M.Woloshyn, Can.J.Phys. **57** 809 (1979).
13. D.J.Almond, *Nucl. Phys.* B **11**, 277 (1969).
    K.J.Kim, N.P.H.Dass, *Nucl. Phys.* B **113**, 336 (1976).
14. K.B.V.Kumar *et al.*, *Phys. Lett.* B **479**, 167 (2000).
15. O.Hanstein *et al.*, *Nucl. Phys.* A **632**, 561 (1998).
16. S.D.Drell, A.C.Hearn *Phys. Rev. Lett.* **16**, 908 (1966).
    S.B.Gerasimow, Yad.Fiz. **2** 598 (1965).
17. S.Ragusa, *Phys. Rev.* D **47**, 3757 (1993).
18. M.J.G.Veltman, Physica **29** 186 (1963).
    W.Beenakker *et al.*, *Nucl. Phys.* B **573**, 503 (2000).
19. D.Buchholz *et al.*, *Phys. Lett.* B **267**, 377 (1991).
    J.Frohlich *et al.*, *Ann. Phys.* **119**, 241 (1979).
20. R.Oehme, Int.J.Mod.Phys.**A10** 1995 (1995).
21. A.Martin, Scattering Theory..., Lecture Notes in Physics Vol.3, Springer-Verlag (1969).
22. C.Itzykson, J.B.Zuber, Quantum Field Theory, ch.5, McGraw-Hill, (1980).
23. A.Pais, *Nuovo Cimento* **53**, 433 (1968).
24. I.B.Khriplovich *et al.*, Sov.Phys.JETP **82** 616 (1996).
25. I.B. Khriplovich, private communication.

GENERALIZED PARTON DISTRIBUTIONS

M. GUIDAL

Institut de Physique Nucléaire, IN2P3, F-91406 Orsay, France
E-mail: guidal@ipno.in2p3.fr

We briefly review in the following the subject of the recently introduced "Generalized Parton Distributions". We first outline the general theoretical formalism, then its relations to experimental observables. We also review the existing and short-term future projects at the various experimental facilities concerned by this physics. Finally, we discuss the recently published data which look extremely promising for the development of this field.

1 Introduction

For the last 20 years, most of what we have learned on the structure of the nucleon has come from the <u>inclusive</u> scattering of high energy leptons on the nucleon in the Bjorken -or "Deep Inelastic Scattering" (DIS)- regime ($Q^2, \nu \gg$ and $x_B = \frac{Q^2}{2M\nu}$ finite). By detecting only the scattered electron, one has already learned a tremendous amount of information : besides having first put in evidence the quark and gluon substructure of the nucleon, one now knows for instance that only 45% of its momentum is carried by the quarks (the remaining being carried by gluons) and that no more than about 25% of the spin of the nucleon originates from the quarks' intrinsic spin.

Now, with the advent of the new generation of high-energy, high-luminosity lepton accelerators combined with large acceptance and high resolution spectrometers, the possibility of fully detecting the final state of the reaction on the hadronic side arises. By looking now at a particular final state, instead of integrating over <u>all</u> final states as is effectively the case when doing *inclusive* scattering, one will access now much more detailed information on the inner structure of the nucleon and has additional lever arm. In particular, the interpretation of *exclusive* photon and meson electroproduction on the nucleon proceeds through the recently introduced "Generalized Parton Distributions" (GPDs) (Refs. [1,2,3]) -also called "Skewed Parton Distributions"- which are a generalization of the standard structure functions accessed in inclusive lepton scattering.

The GPDs contain information on the correlations between quarks (i.e. non-diagonal elements) and on their transverse momentum dependence in the nucleon. As a direct effect of these features, Ji showed [1] for instance that the second moment of these GPDs gives access to the sum of the quark spin and the quark orbital angular momentum to the nucleon spin, which may shed light on the so-called nucleon "spin-puzzle". Most of these informations are not contained in the traditional inclusive parton distributions extracted from inclusive DIS which allows to access only partons densities, i.e. diagonal elements.

In this paper, after briefly outlining the formalism of the GPDs in section 2, we will discuss some general considerations for their experimental study in section 3 : in particular, the relation between GPDs and experimental observables and the need and requirements for a dedicated experimental facility. Finally, in section 4, we will review and comment the first experimental signatures of this physics recently observed by the HERMES and CLAS (at JLab) collaborations.

2 Formalism

A few years ago, Ji [1] and Radyushkin [2] have shown that the leading order pQCD amplitude for Deeply Virtual Compton Scattering (DVCS) in the forward direction can be factorized in a hard scattering part (exactly calculable in pQCD) and a nonperturbative nucleon structure part as is illustrated in Fig.(1-a). In these so-called "handbag" diagrams, the lower blob which represents the structure of the nucleon can be parametrized, at leading order pQCD, in terms of 4 generalized structure functions, the GPDs. These are traditionnally called $H, \tilde{H}, E, \tilde{E}$, and depend upon three variables : x, ξ and t. $x - \xi$ is the longitudinal momentum fraction carried by the initial quark struck by the virtual photon. Similarly, $x + \xi$ relates to the final quark going back in the nucleon after radiating a photon. -2ξ is therefore the longitudinal momentum difference between the initial and final quarks. In comparison to -2ξ which refers to *longitudinal* degrees of freedom, t, the standard squared 4-momentum transfer between the final nucleon and the initial one, contains *transverse* degrees of freedom (so-called "$k_\perp$") as well.

H and E are spin independent, and are often called *unpolarized* GPDs, whereas $\tilde{H}$ and $\tilde{E}$ are spin dependent, and are often called *polarized* GPDs. The GPDs H and $\tilde{H}$ are actually a generalization of the parton distributions measured in deep inelastic scattering. Indeed, in the forward direction, H reduces to the quark distribution and $\tilde{H}$ to the quark helicity distribution measured in deep inelastic scattering. Furthermore, at finite momentum transfer, there are model independent sum rules which relate the first moments of these GPDs to the elastic form factors.

The GPDs reflect the structure of the nucleon independently of the reaction which probes the nucleon. They can also be accessed through the hard

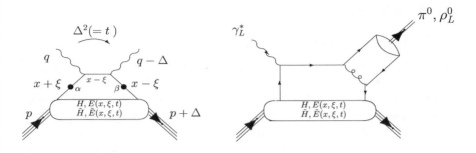

Figure 1. "Handbag" diagrams : a) for DVCS (left) and b) for meson production (right).

exclusive electroproduction of mesons -π^0, ρ^0, ω, ϕ, etc.- (see Fig. (1-b)) for which a QCD factorization proof was given recently [3]. According to Ref. [3], the factorization applies when the virtual photon is longitudinally polarized because in this case, the end-point contributions in the meson wave function are power suppressed. It is shown in Ref. [3] that the cross section for a transversely polarized photon is suppressed by $1/Q^2$ compared to a longitudinally polarized photon. Because the transition at the upper vertices of Fig. (1-b) will be dominantly helicity conserving at high energy and in the forward direction, this means that the vector meson should also be predominantly longitudinally polarized (notation $\rho_L^0, \omega_L, \phi_L$) for a longitudinal photon at QCD leading order and leading twist.

It was also shown in Ref. [3] that leading order pQCD predicts that the vector meson channels (ρ_L^0, ω_L, ϕ_L) are sensitive only to the unpolarized GPDs (H and E) whereas the pseudo-scalar channels ($\pi^0, \eta, ...$) are sensitive only to the polarized GPDs ($\tilde{H}$ and $\tilde{E}$). In comparison to meson electroproduction, DVCS depends at the same time on *both* the polarized and unpolarized GPDs.

Another feature to mention, proper to these handbags diagrams, is the notion of *scaling*. It is predicted that, when asymptotia in Q^2 is reached, the differential cross section $\frac{d\sigma}{dt}$ of these "handbag" mechanisms should show a $\frac{1}{Q^4}$ behavior for DVCS and a $\frac{1}{Q^6}$ behavior for meson production. These Q^2 dependences are strong experimental signatures that the appropriate kinematical regime is reached and are necessary to observe before tempting to interpret data in terms of GPDs. It has been recently an intense effort from the theoretical community to control the corrections (Next to Leading Order, higher twists, ...) to this scaling behavior [4,5].

3 Experimental aspects

3.1 Deconvolution issues

As mentionned in the previous section, the GPDs depend on three variables :
x, ξ and t. However, it has to be realized that only two of these three variables are accessible experimentally, i.e. ξ ($=\frac{x_B}{2-x_B}$, fully defined by detecting
the scattered lepton) and t ($=\Delta^2$, see Fig (1-a), fully defined by detecting
either the recoil proton or the outgoing photon or meson). x however is a
variable which is integrated over, due to the loop in the "handbag" diagrams
(see Fig. (1)). This means that in general a differential cross section will be
proportional to : $\mid \int_{-1}^{+1} dx \frac{H(x,\xi,t)}{x-\xi+i\epsilon} + ... \mid^2$ (where "..." stands for similar terms
for E, $\tilde{H}$, $\tilde{E}$ and $\frac{1}{x-\xi+i\epsilon}$ being the propagator of the quark between the incoming virtual photon and the outgoing photon -or meson-, see Fig. (1)). In
general, one therefore will measure integrals (with a propagator as a weighting
function) of GPDs.

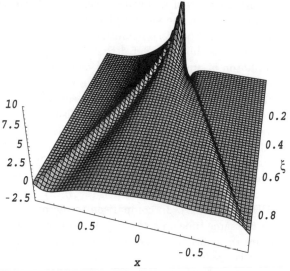

Figure 2. One model for the GPD H as a function of x and ξ for $t=0$. One recognizes for
$\xi=0$ the typical shape of a parton distribution (with the sea quarks rising as x goes to 0,
the negative x part being interpreted as the antiquark contribution). The figure is taken
from Ref. [5].

To illustrate this point, Fig. (2) shows one particular model for the GPD
H as a function of x and ξ (at $t = 0$). One recognizes for $\xi = 0$ a standard

quark density distribution with the rise around $x = 0$ corresponding to the diverging sea contribution. The negative x part corresponds to antiquarks. One sees that the evolution with ξ is not trivial and that measuring the integral over x of a GPD at constant ξ will not uniquely define it.

A particular exception is when one measures an observable proportional to the *imaginary* part of the amplitude (for instance, the beam asymmetry in DVCS which is non-zero in leading order due to the interference with the Bethe-Heitler process, see section 4). Then, because $\int_{-1}^{+1} dx \frac{H(x,\xi,t)}{x-\xi+i\epsilon} = PP(\int_{-1}^{+1} dx \frac{H(x,\xi,t)}{x-\xi}) - i\pi H(\xi,\xi,t)$, one actually measures directly the GPDs at some specific point, $x = \xi$ (i.e., $H(\xi,\xi,t)$).

For mesons, transverse target polarization observables are also sensitive to a different combination of the GPDs, i.e. combinations of the type : $\int_{-1}^{+1} dx \frac{H(x,\xi,t)}{x-\xi} \times E(\xi,\xi,t)$ (the exact formula is more complicated, see for instance Ref. [5,6]). One sees that such transverse spin asymmetries are sensitive to a *product* of the GPDs instead of a sum of their squares as is the case for a typical differential cross section.

It will therefore be a non-trivial (though a priori not impossible) task to actually extract the GPDs from the experimental observables as, to summarize, one actually only accesses in general (weighted) integrals of GPDs or GPDs at some very specific points or product of these two. In absence of any model-independent "deconvolution" procedure at this moment, one will therefore have to rely on some global model fitting procedure.

It should also be added that GPDs are defined for one quark flavor q (i.e. H^q, E^q,...) similar to standard quark distributions. This "flavor" separation will require the measurement of several isospin channels ; for instance, ρ^0 production is proportional to (in a succinct notation) $2/3H^u + 1/3H^d$ while ω production is proportional to $2/3H^u - 1/3H^d$. Similar arguments apply for the polarized GPDs with the $\pi^{0,\pm}$, η,... channels. It can be viewed as an intrinsic richness for mesons channels to allow for such flavor separation.

In summary, it should be clear that a full experimental program aiming at the extraction of the individual GPDs is a broad program which requires the study of several isospin channels and several observables, each having its own characteristics. Only a global study and fit to all this information may allow an actual extraction of the GPDs.

3.2 A dedicated facility

An exploratory study of the GPDs can currently be envisaged at the JLab (E_e=6 GeV), HERMES (E_e=27 GeV) and COMPASS (E_μ=100-200 GeV) facilities in a very complementary fashion, each having its own "advantages" and "disadvantages". The considerations which are relevant for this "exclusive" physics are :

- Kinematical range : it is desirable to span a domain in Q^2 and x_B as large as possible, in particular to test scaling as mentionned in section 2,

- Luminosity : cross sections fall sharply with Q^2 and one has to measure small cross sections,

- Resolution : it is necessary to cleanly identify <u>exclusive</u> reactions. This can be achieved either by a good resolution with the missing mass technique or by detecting <u>all</u> the particles of the final states and thus overdetermining the kinematics of the reaction.

Also, a large acceptance detector is desirable as t and Φ (for asymmetries, studies of decay angular distributions,...) coverages are needed and, more generally, the aim is, as emphasized previously, to measure several channels and kinematic variables simultaneously.

COMPASS, expected to start taking data in 2001, with a 100 to 200 GeV beam has the clear advantage that it is the only facility allowing to reach small x_B (i.e. ξ) at sufficiently large Q^2. However, it suffers from a relatively low luminosity ($\approx 10^{32}cm^{-2}s^{-1}$) and relatively poor resolution to rely on the missing mass technique in order to identify an exclusive reaction (there is a project of overcoming this latter point by adding a recoil detector which would overconstrain the kinematics of the reaction [7]).

HERMES suffers basically from the same issues : relatively low luminosity ($\approx 10^{32-33}cm^{-2}s^{-1}$) and resolution not fine enough to fully select exclusive final states, where, for instance, a typical missing mass resolution of the order of 300 MeV allows the contamination of additional pions into a sample of exclusive events. Here, however, a recoil detector is already under construction which should be operationnal soon and will overcome this issue [8]. HERMES, which has been running since 1996, has the merit of being the first facility to have measured some experimental observables directly relevant to this physics (ρ_L^0 cross sections, DVCS beam asymmetry, exclusive π^+ target asymmetry) as will be discussed in the next section.

JLab (with 6 GeV maximum beam energy in its current running configuration) has the highest luminosity ($\approx 10^{34}cm^{-2}s^{-1}$ for the Hall B large acceptance spectrometer in order to compare fairly with the other two facilities) and very good resolution (a typical missing mass resolution is less than 100 MeV. This good resolution is of course highly correlated with the relatively low energy of the beam). The main drawback of JLab at 6 GeV is obviously the limited kinematical range (at x_B=.3, W > 2 GeV, one cannot exceed Q^2=3.5 GeV^2 for instance). There is also the possibility at JLab to use high resolution arm-spectrometers to study with high precision particular kinematics [9]. Figure 3 summarizes the phase space covered by the current and short-term future projects related to this physics and clearly illustrates the complementarity of each.

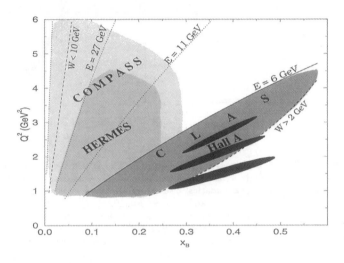

Figure 3. (Q^2, x_B) phase space covered by the current and short-term future experimental projects for DVCS. The figure is taken from Ref. [10].

In spite of all the first "breakthrough" measurements related to the GPD physics that are currently being carried out at these facilities, all these considerations clearly call for a dedicated machine which would combine a high luminosity ($\approx 10^{35-36} cm^{-2} s^{-1}$ desirable) and a high energy ($\approx$ 30 GeV) beam with a good resolution detector (a few tens of MeV for a typical missing mass resolution). The ELFE [11] and JLab upgrade [12] (with a 11 GeV beam energy) projects would be quite well suited for such a physics program devoted to the systematic study of exclusive reactions and the GPDs.

4 First experimental evidences

In this section, we review the first existing experimental data related to GPDs interpretation. Only these past 2 years, have been released by the CLAS

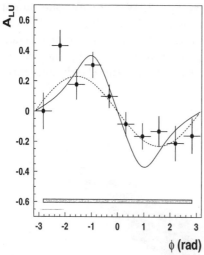 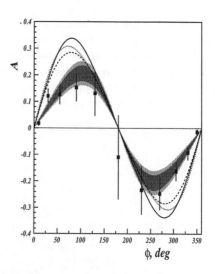

Figure 4. The DVCS beam asymmetry as a function of the azimuthal angle Φ as measured by HERMES [16]. Average kinematics is : $< x >=.11$, $< Q^2 >=2.6$ GeV^2 and $< -t >=.27$ GeV^2. The dashed curve is a $\sin\Phi$ fit whereas the solid curve is the theoretical GPD calculation of Ref. [17].

Figure 5. The DVCS beam asymmetry as a function of the azimuthal angle Φ as measured by CLAS [18]. Average kinematics is : $< x >=.19$, $< Q^2 >=1.25$ GeV^2 and $< -t >=.19$ GeV^2. The shaded regions are error ranges to $\sin\Phi$ and $\sin2\Phi$ fits. Calculations are : leading twist *without* ξ dependence [19,20] (dashed curve), leading twist *with* ξ dependence [19,20] (dotted curve) and leading twist + twist-3 [17] (solid curve).

and HERMES collaborations experimental data precise enough in the relevant kinematical regime. In this paper, we choose to focus on the valence quark region, i.e. W < 10 GeV, where the quark exchange mechanism of Fig. (1) dominates. However, it is to be mentionned that DVCS [13] and vector mesons [14] cross sections at low x_B have also been measured by the H1 and ZEUS collaborations which lend themselves to GPD interpretation through "gluon exchange"-type processes [15].

Clearly, the statistics of all these measurements are still not high enough to allow for a fine binning in the kinematical variables and therefore a precise test of GPD models. Nevertheless, they are very encouraging in the sense that the observed signals, although integrated over quite wide kinematical ranges, are generally compatible (in magnitude and in "shape") with theoretical calculations. It is to be noted by the way that basically all of the calculations accompanying the figures in the following were indeed *predictions* as they were published before the experimental results.

The experimental observables to be discussed are the single spin beam asymmetry (SSA) in DVCS (measured by HERMES and CLAS) and the longitudinal cross section of ρ^0 and ω electroproduction (measured by HERMES). Fig. (4) shows the first measurement of the SSA for DVCS by HERMES with a 27 GeV positron beam. This asymmetry arises from the interference of the "pure" DVCS process (where the outgoing photon is emitted by the nucleon) and the Bethe-Heitler (BH) process (where the outgoing photon is radiated by the incoming or scattered lepton). The two processes are indistinguishable experimentally and interfer. The BH process being purely real and exactly calculable in QED, one has therefore access, through the difference of cross sections for different beam helicities which is sensitive to the imaginary part of the amplitude, to some *linear* combination of the GPDs at the kinematical point $(x = \xi, \xi, t)$ as mentionned in the previous subsection.

The beam asymmetry, which is this latter difference of cross sections divided by their sum, is more straightforward to access experimentally as normalization and systematics issues cancel, at first order, in the ratio. For this asymmetry, a shape close to $\sin\Phi$ (not an exact $\sin\Phi$ shape as higher twists and the Bethe Heitler have some more complex Φ dependence) is expected, where Φ is the standard angle between the leptonic and the hadronic plane. At HERMES, the average kinematics is $< x > = .11$, $< Q^2 > = 2.6$ GeV^2 and $< -t > = .27$ GeV^2 for which an amplitude of .23 for the $\sin\Phi$ moment is extracted from the fit [16]. The discrepancy between the theoretical prediction and the data on Fig. (4) can certainly be attributed on the one hand to the large kinematical range over which the experimental data have been integrated and where the model can vary significantly and, on the other hand, to higher twists corrections not calculated (so far, only twist-3 are under theoretical control for the handbag DVCS process -see Ref. [?]-, the leading twist being twist-2).

Also, the DVCS reaction at HERMES is identified by detecting the scattered lepton (positron) and the outgoing photon from which the missing mass of the non-detected proton is calculated. Due to the limited resolution of the HERMES detector, the selected peak around the proton mass is $-1.5 < M_x < 1.7$ GeV which means that contributions to this asymmetry from nucleon resonant states as well cannot be excluded. Let's recall that a recoil detector aiming at the detection of the recoil proton is projected to be installed at HERMES by 2003 [8]; this will then allow to unambiguously sign the exclusivity of the reaction at HERMES.

This same observable has been measured at JLab with a 4.2 GeV electron beam with the 4π CLAS detector [18]. Due to the lower beam energy compared to HERMES, the kinematical range accessed at JLab is different : $< x > = .19$, $< Q^2 > = 1.25$ GeV^2 and $< -t > = .19$ GeV^2. In this case, the DVCS reaction was identified by detecting, besides the scattered lepton, the recoil proton and then calculating the missing mass of the photon (due to the geometry of the CLAS detector, the outgoing photon which is emitted at forward angles escapes detection). The contamination by $ep \to ep\pi^0$ events can be estimated and subtracted bin per bin, resulting in a rather clean signature of the exclusivity of the reaction.

Figure (5) shows the CLAS measured asymmetry along with theoretical calculations (predictions) which are in fair agreement (the different sign of the CLAS SSA relative to HERMES is due to the use of electron beams in the former case compared to positron beams in the latter). Again, discrepancies can be assigned to the fact that the theory is calculated at a single well-defined kinematical point whereas data has been integrated over several variables and wide ranges. Furthermore, Next to Leading Order as well twist-4 corrections which may be important at these rather low Q^2 values, still need to be quantified.

In the meson sector, the vector meson channel is the most accessible as it allows rather simply to separate the longitudinal from the transverse part of the cross section through its decay angular distribution (we recall, as mentionned in section 2, that only for the longitudinal part of the cross section is the factorization theorem valid at this stage and allows to make interpretations in terms of GPDs). So far, only the ρ^0 channel has yielded sufficient statistics, due to its relatively high cross section, to isolate σ_L. Figure (6) shows the two HERMES points [21] along with the GPD theoretical calculations (predictions). For vector mesons, two mechanisms contribute in two different kinematical regimes : at low W (i.e. large x_B), 2-quark exchange ; at high W (i.e. low x_B), 2-gluon exchange. The quark exchange process can be identified and calculated with the handbag diagrams of figure 1 [19,20]. For meson production, due to presence of the "extra" gluon exchange compared to DVCS, large corrections are expected to the leading order. These corrections can be modelled taking into account $k_\perp$ degrees of freedom [19,20]. At HERMES kinematics, this correction factor is found to be about 3 and allows to predict the magnitude of the cross section.

One way to get rid of such model dependency in the corrections is to look at ratios of cross sections. Indeed, as pointed out by Ref. [3,22], these correction factors are expected to factorize and therefore cancel in ratio. One speaks of "precocious scaling". The HERMES collaboration is about to release the measurement of the ω cross section in the same kinematical range as the ρ^0 cross section, it will be very interesting to compare the $\frac{\omega}{\rho^0}$ ratio to the

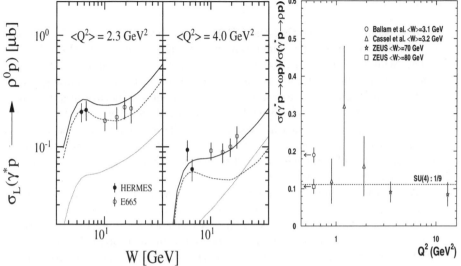

Figure 6. The longitudinal ρ^0 electroproduction cross section in the intermediate W range. The dashed curve represents the quark exchange process calculated in the GPD framework whereas the dotted line represents the 2-gluon exchange process. The solid line is the incoherent sum of the two mechanisms. Calculations are from Refs. [20]. Figure is taken from Ref. [21].

Figure 7. World data for the ρ/ω ratio as a function of Q^2. The figure has been taken and adapted from Ref. [23].

theoretical prediction of the GPD formalism which yields $\approx 1/5$ [19,20], this number, quite model independent, arising basically from the ratio of the u and d quark distributions weighted by known isopin factors. This has to be compared to the well-known SU(3) 1/9 prediction in the low x_B domain. A W (or equivalently x_B) dependence is therefore expected for this ratio. This seems to be already observed with the current world data, see Fig. (7), where one can already distinguish a trend -in spite of quite large error bars- where the low W data are close to $\approx 1/5$ whereas the large W data are closer to 1/9. The preliminary HERMES results tend to confirm this tendency [23].

Similarly, $\frac{\pi^+}{\pi^0}$, $\frac{\pi^0}{\eta}$, $\frac{\rho^+}{\rho^0}$, etc... ratios deserve to be measured as they can be directly compared to leading order and leading twist model independent theoretical predictions in the GPD framework.

5 Conclusion

In conclusion, we believe that the GPDs open a broad new area of physics, by providing a context for understanding <u>exclusive</u> reactions in the valence region (where the quark exchange mechanism dominates) at large Q^2. By "constraining" the final state of the DIS reaction, instead of summing over all final states, one accesses some more fundamental structure functions of the nucleon. These functions provide a unifying link between a whole class of various reactions (elastic and inelastic) and fundamental quantities as diverse as form factors and parton distributions. They allow to access new information on the structure of the nucleon, for instance quark's orbital momentum and, more generally, correlations between quarks.

A full study aiming at the extraction of these GPDs from experimental data requires a new dedicated facility providing high energy and high luminosity lepton beams, equipped with large acceptance and high resolution detectors. First experimental exploratory results from the HERMES and CLAS collaboration provide some evidence that the manifestations of the handbag mechanisms are already observed. This is encouraging and paves the way for a future very rich harvest of hadronic physics and motivates the development of new dedicated projects and facilities.

References

1. X. Ji, Phys.Rev.Lett. **78** (1997) 610; Phys.Rev. **D55** (1997) 7114.
2. A.V. Radyushkin, Phys.Lett. **B380** (1996) 417; Phys.Rev. **D56** (1997) 5524.
3. J.C. Collins, L. Frankfurt and M. Strikman, Phys.Rev. **D56** (1997) 2982.
4. A. Belitsky et al., Phys.Lett. **B437** (1998) 160; Nucl.Phys. **B546** (1999) 114025; Phys.Lett. **B474** (2000) 163.
5. K. Goeke, M. V. Polyakov and M. Vanderhaeghen, Prog.Part.Nucl.Phys. **47** (2001) 401.
6. L. Frankfurt, M. V. Polyakov, M. Strikman and M. Vanderhaeghen, Phys.Rev.Lett. **84** (2000) 2589.
7. N. D'Hose, Letter of intent for COMPASS, private communication.
8. R. Kaiser et al., "A large acceptance recoil detector for HERMES", Addendum to the Proposal DESY PRC 97-07, HERMES 97-032.
9. P.-Y. Bertin and F. Sabatie, spokesperson JLab experiment E00-110.
10. V. Burkert, L. Elhouadrhiri, M. Garcon, S. Stepanyan, spokesperson JLab experiment E01-113.
11. *Elfe : Physics Motivation,*

http://www-dapnia.cea.fr/Sphn/Elfe/Report/ELFE_phys.pdf, to appear as a NUPPEC report and "TESLA: The superconducting electron positron linear collider with an integrated X-Ray laser laboratory", Tehnical Design Report. PT. 6: Appendices. Chapter 4: ELFE: The Electron Laboratory For Europe. By TESLA-N Study Group (M. Anselmino et al.). DESY-01-011FD, DESY-2001-011FD, DESY-TESLA-2001-23FD, DESY-TESLA-FEL-2001-05FD, ECFA-2001-209FD, Mar 2001. 17pp.

12. "White Book", *The Science driving the 12 GeV upgrade of CEBAF*, http://www.jlab.org/div_dept/physics_division/GeV.html

13. L. Favart, DIS 2001 Proceedings, hep-ex/0106067.

14. C. Adloff et al., Eur.Phys.J. **C13** (2000) 371, J. Breitweg et al., Eur.Phys.J. **C6** (1999) 603, J. Breitweg et al., hep-ex/0006013.

15. L. Frankfurt, A. Freund and M. Strikman, Phys.Rev. **D58** (1998) 114001 and Phys.Rev. **D59** (1999) 119901E.

16. A. Airapetian et al., Phys. Rev. Lett. **87** (2001) 182001.

17. N. Kivel, M. V. Polyakov and M. Vanderhaeghen, Phys.Rev. **D63** (2001) 114014.

18. S. Stepanyan et al., Phys.Rev.Lett. **87** (2001) 182002.

19. P.A.M. Guichon and M. Vanderhaeghen, Prog.Part.Nucl.Phys., **41** (1998) 125.

20. M. Vanderhaeghen, P.A.M. Guichon, M. Guidal, Phys.Rev.Lett. **80** (1998) 5064, Phys.Rev. **D60** (1999) 094017.

21. A. Airapetian et al., Eur.Phys.J. **C17** (2000) 389.

22. M. Eides, L. Frankfurt and M. Strikman, Phys.Rev. **D59** (1999) 114025.

23. M. Tytgat, PhD dissertation, Gent University, (2000).

CHIRAL SYMMETRY AND HADRON PROPERTIES IN LATTICE QCD

ANTHONY W. THOMAS

Special Research Centre for the Subatomic Structure of Matter
and Department of Physics and Mathematical Physics,
Adelaide University, Adelaide SA 5005, AUSTRALIA
E-mail: athomas@physics.adelaide.edu.au

We explain some of the challenges and recent discoveries in the struggle to connect hadron properties calculated in lattice QCD at large quark mass to the physical region. We suggest that formal expansions based on effective field theory need to be supplemented by some physical insight before the problem becomes tractable. However, once this is done, some surprisingly accurate results can be extracted.

1 Introduction

With recent developments of novel improved actions, faster computers and better treatments of chiral symmetry, it is reasonable to suggest that lattice QCD may soon deliver on its promise of deep new insights into hadron structure. Indeed, in combination with carefully controlled chiral extrapolation, one can expect to calculate accurate properties of the low mass baryons within a few years.

2 The Role of Chiral Symmetry

A major challenge in performing calculations at realistic quark masses (say 5–8 MeV) is the approximate chiral symmetry of QCD. Goldstone's theorem tells us that chiral symmetry is dynamically broken and that the non-perturbative vacuum is highly non-trivial, with massless Goldstone bosons in the limit $\bar{m} \to 0$ [1]. For finite quark mass these bosons are the three charge states of the pion with a mass $m_\pi^2 \propto \bar{m}$. Although this result strictly holds only for m_π^2 near zero (the Gell-Mann-Oakes-Renner relation), lattice simulations show it is a good approximation for m_π^2 up to 0.8 GeV^2 and we shall use m_π^2 as a measure of the deviation from the chiral limit.

From the point of view of lattice simulations with dynamical quarks (i.e. unquenched) the essential difficulty is that the time taken goes as $\bar{m}^{-3}$, or worse [3]. The state-of-the-art for hadron masses is $\bar{m}$ above 60 MeV, although there is a preliminary result from CP-PACS at about 40 MeV [4]. In general, the quark masses for which simulations currently exist are at masses a factor of 8-20 too high. This means that an increase of computing power to several hundred tera-flops is needed if one is to calculate realistic hadron properties. Even with the current remarkable rate of increase this will take a long time.

Faced with such a serious difficulty physicists (like all other people facing a tough challenge) fall into two classes:

(A) Those who believe that "the cup is half empty":

In this case the emphasis is on the uncertainties associated with having to make big extrapolations to the chiral limit.

(B) Those who believe that "the cup is half full":

In this case we realize that the lattice data obtained so far represents a wealth of information on the properties of hadrons within QCD itself over a range of quark masses. Just as the study of QCD as a function of N_c has taught us a great deal, so the behaviour as a function of $\bar{m}$ can give us great insight into hadronic physics and guide our model building. Furthermore, as a bonus, approach (B) actually leads us to a resolution of the difficulties which led to conslusion (A)!

In light of the brief space available to outline a great deal of evidence, we first summarise the conclusions which emerge from the work of the past three years. We then present just a couple of illustrations of the reasoning which led to these conclusions.

Summary:

- In the region of quark masses $\bar{m} > 60$ MeV or so (m_π greater than typically 400-500 MeV) hadron properties are smooth, slowly varying functions of something like a constituent quark mass, $M \sim M_0 + c\bar{m}$ (with $c \sim 1$).

- Indeed, $M_N \sim 3M, M_{\rho,\omega} \sim 2M$ and magnetic moments behave like $1/M$.

- As $\bar{m}$ decreases below 60 MeV or so, chiral symmetry leads to rapid, non-analytic variation, with:
  $\delta M_N \sim \bar{m}^{3/2}$,
  $\delta \mu_H \sim \bar{m}^{1/2}$,
  $\delta <r^2>_{\text{ch}} \sim \ln \bar{m}$ and
  moments of non-singlet parton distributions $\sim m_\pi^2 \ln m_\pi$.

- Chiral quark models, like the cloudy bag model (CBM) [5], provide a natural explanation of this transition. The scale is basically set by the inverse size of the pion source – the inverse of the bag radius in the CBM.

- When the pion Compton wavelength is smaller than the size of the composite source chiral loops are strongly suppressed. On the other hand, as soon as the pion Compton wavelength is larger than the source one begins to see rapid, non-analytic chiral corrections.

The nett result of this discovery is that one has control over the chiral extrapolation of hadron properties provided one can get at least one data point at pion mass of order 200–300 MeV. This seems feasible with the next generation of supercomputers which should be available within 2–3 years and which will have speeds in excess of 10 tera-flops [6]. This is an extremely exciting possibility in that it will bring forward the scale of realistic calculations of physical hadron properties by a decade or more!

3 Non-Analyticity of Hadron properties

We have already seen that spontaneous chiral symmetry breaking in QCD requires the existence of Goldstone bosons whose masses vanish in the limit of zero quark mass (the chiral limit). As a corollary to this, there must be contributions to hadron properties from Goldstone boson loops. These loops have the unique property that they give rise to terms in an expansion of most hadronic properties as a function of quark mass which are not analytic [7]. As a simple example, consider the nucleon mass. The most important chiral corrections to M_N come from the processes $N \to N\pi \to N$ (σ_{NN}) and $N \to \Delta\pi \to N$ ($\sigma_{N\Delta}$). We write $M_N = M_N^{\text{bare}} + \sigma_{NN} + \sigma_{N\Delta}$. In the heavy baryon limit one has

$$\sigma_{NN} = -\frac{3g_A^2}{16\pi^2 f_\pi^2} \int_0^\infty dk \frac{k^4 u^2(k)}{k^2 + m_\pi^2}. \tag{1}$$

Here $u(k)$ is a natural high momentum cut-off which is the Fourier transform of the source of the pion field (e.g. in the CBM it is $3j_1(kR)/kR$, with R the bag radius [5]). From the point of view of PCAC it is natural to identify $u(k)$ with the axial form-factor of the nucleon, a dipole with mass parameter 1.02 ± 0.08GeV [1].

Totally independent of the form chosen for the ultra-violet cut-off, one finds that σ_{NN} is a non-analytic function of the quark mass. The non-analytic piece of σ_{NN} is independent of the form factor and gives

$$\sigma_{NN}^{LNA} = -\frac{3g_A^2}{32\pi f_\pi^2} m_\pi^3 \sim \bar{m}^{\frac{3}{2}}. \tag{2}$$

This has a branch point, as a function of $\bar{m}$, starting at $\bar{m} = 0$. Such terms can only arise from Goldstone boson loops.

It is natural to ask how significant this non-analytic behaviour is in practice. If the pion mass is given in GeV, $\sigma_{NN}^{LNA} = -5.6m_\pi^3$ and at the physical pion mass it is just -17 MeV. However, at only three times the physical pion mass, $m_\pi = 420\text{MeV}$, it is -460MeV – half the mass of the nucleon. If one's aim is to extract physical nucleon properties from lattice QCD calculations this is extremely important. As we explained earlier, the most sophisticated lattice calculations with dynamical fermions are only just becoming feasible at such low masses and to connect to the physical world one must extrapolate from $m_\pi \sim 500\text{MeV}$ to $m_\pi = 140\text{MeV}$. Clearly one must have control of the chiral behaviour. As shown in Ref. [2], the best fit to recent lattice data for the mass of the nucleon from CP-PACS and UKQCD [4,8], using a form which naively respects the presence of the LNA term:

$$M_N = \alpha + \beta m_\pi^2 + \gamma m_\pi^3, \tag{3}$$

yields a very good fit to the data but the chiral coefficient γ is only -0.761. This should be compared with the value -5.60 required by chiral symmetry. If one insists that γ be consistent with QCD the best fit one can obtain with this form is totally unacceptable.

Such a result is not unexpected. In general, the series expansion generated by an effective field theory is either divergent or asymptotic [10,11] and it would be astonishing if a naive expansion in powers of m_π, in the case of lattice QCD, were to be effective over more than a small range. On the other hand, a recent study [12] of the Euler-Heisenberg problem [13,14,15], which serves as an exactly soluble model with an effective field theory expansion that is is indeed asymptotic, provides enormous encouragement. Dunne et $al.$ [12] found that it was possible to produce an extremely accurate representation of the exact solution using a two parameter, generalised Padé approximant which built in the correct asymptotic behaviour in both the large and small mass limits. This problem is directly analogous (in terms of its asymptotic limits) to the problem of extrapolating charge radii, which we review below and gives great theoretical support to the approach taken there.

In the case of baryon masses, an alternative to the standard (divergent) series expansion of chiral perturbation theory was suggested recently by Leinweber et al. [2]. This approach, which also involves just three parameters, is to evaluate σ_{NN} and $\sigma_{N\Delta}$ with the same ultra-violet form factor, with mass parameter Λ, and to fit M_N as

$$M_N = \alpha + \beta m_\pi^2 + \sigma_{NN}(m_\pi, \Lambda) + \sigma_{N\Delta}(m_\pi, \Lambda), \tag{4}$$

by adjusting α, β and Λ to fit the data. Using a sharp cut-off $(u(k) = \theta(\Lambda - k))$ these authors were able to obtain analytic expressions for σ_{NN} and $\sigma_{N\Delta}$ which reveal the correct LNA behaviour – and next to leading (NLNA) in the $\Delta\pi$ case, $\sigma_{N\Delta}^{NLNA} \sim m_\pi^4 \ln m_\pi$. These expressions also reveal a branch point at

$m_\pi = M_\Delta - M_N$, which is important if one is extrapolating from large values of m_π to the physical value. This approach yields an excellent fit to the lattice data using Eq.(4), with Λ fixed at a value suggested by CBM simulations to be equivalent to a 1 GeV dipole. A small increase in Λ is necessary to fit the lowest mass data point (at $m_\pi^2 \sim 0.1$ GeV^2) well, but for the present the uncertainty in this lowest mass point precludes an accurate determination of Λ. Most importantly, we stress that this fit is obtained with a form which guarantees exactly the right LNA and NLNA behaviour of QCD.

4   Electromagnetic Properties of Hadrons

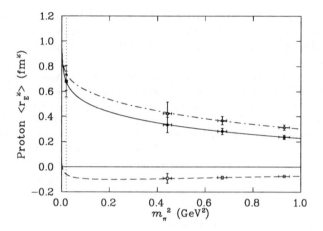

Figure 1. Fits to lattice results for the squared electric charge radius of the proton – from Ref. [17]. Fits to the contributions from individual quark flavors are also shown: the u-quark sector by open triangles and the d-quark sector by open squares. The physical value predicted by the fits are indicated at the physical pion mass. The experimental value is denoted by an asterisk.

It is a completely general consequence of quantum mechanics that the long-range charge structure of the proton comes from its π^+ cloud $(p \to n\pi^+)$, while for the neutron it comes from its π^- cloud $(n \to p\pi^-)$. However it is not often realized that the LNA contribution to the nucleon charge radius goes like $\ln m_\pi$ and diverges as $\bar{m} \to 0$ [16]. This can never be described by a constituent quark model. The latest data from Mainz and Nikhef for the neutron electric form factor shows good agreement with CBM calculations for a confinement radius between 0.9 and 1.0 fm, with the long-range π^- tail of the neutron playing a crucial role [18,19].

While there is only limited (and indeed quite old) lattice data for hadron charge radii, recent experimental progress in the determination of hyperon charge radii has led us to examine the extrapolation procedure for obtaining charge data from the lattice simulations [17]. Figure 1 shows the extrapolation of the lattice data for the charge radius of the proton. The form used, in line with our comments above concerning the Euler-Heisenberg problem, is a two parameter Padé approximant, which builds in the correct LNA behaviour as $m_\pi \to 0$, as well as the correct large mass limit. Clearly the agreement with experiment is much better once the logarithm required by chiral symmetry is correctly included – rather than simply making a linear extrapolation in the quark mass (or m_π^2). Full details of the results for all the octet baryons may be found in Ref. [17].

The situation for baryon magnetic moments is also very interesting. The LNA contribution in this case arises from the diagram where the photon couples to the pion loop. As this involves two pion propagators the expansion of the proton and neutron moments is:

$$\mu^{p(n)} = \mu_0^{p(n)} \mp \alpha m_\pi + \mathcal{O}(m_\pi^2). \tag{5}$$

Here $\mu_0^{p(n)}$ is the value in the chiral limit and the linear term in m_π is proportional to $\bar{m}^{\frac{1}{2}}$, a branch point at $\bar{m} = 0$. The coefficient of the LNA term is $\alpha = 4.4\mu_N \text{GeV}^{-1}$. At the physical pion mass this LNA contribution is $0.6\mu_N$, which is almost a third of the neutron magnetic moment.

Just as for M_N, the chiral behaviour of $\mu^{p(n)}$ is vital to a correct extrapolation of lattice data. One can obtain a very satisfactory fit to some rather old data, which happens to be the best available, using the simple Padé [20]:

$$\mu^{p(n)} = \frac{\mu_0^{p(n)}}{1 \pm \frac{\alpha}{\mu_0^{p(n)}} m_\pi + \beta m_\pi^2} \tag{6}$$

Existing lattice data can only determine two parameters and Eq.(6) has just two free parameters while once again guaranteeing the correct LNA behaviour as $m_\pi \to 0$ **and** the correct behaviour of HQET at large m_π^2. The extrapolated values of μ^p and μ^n at the physical pion mass, $2.85 \pm 0.22\mu_N$ and $-1.90 \pm 0.15\mu_N$ are currently the best estimates from non-perturbative QCD [20]. For the application of similar ideas to other members of the nucleon octet we refer to Ref. [21], and for the strangeness magnetic moment of the nucleon we refer to Ref. [22]. The last example is another case where tremendous improvements in the experimental capabilities, specifically the accurate measurement of parity violation in ep scattering [23], is giving us vital information on hadron structure.

5 Structure Functions

The parton distribution functions (PDFs) of the nucleon are light-cone correlation functions which, in the infinite momentum frame, are interpreted as probability distributions for finding specific partons (quarks, antiquarks, gluons) in the nucleon. At high momentum transfer (Q^2) the dominant component of the PDFs are determined by non-perturbative matrix elements of certain "leading twist" operators. In principle these matrix elements, which correspond to moments of the measured structure functions, contain vital information about the non-perturbative structure of the target. An example is the $\bar{d} - \bar{u}$ asymmetry, predicted [25] on the basis of the nucleon's pion cloud [26], which has been spectacularly confirmed in recent experiments at CERN and Fermilab [27].

Recently the MIT group has performed the first full (unquenched) QCD calculations of non-singlet moments [28]. The moments from the full QCD simulations are very similar to those from the quenched calculations. This is consistent with the suggestions of chiral quark models, like the CBM, and the general conclusions outlined in the Introduction. That is, in the quark mass region currently accessible quark loops are suppressed.

As for the other nucleon properties discussed above, we propose to extraplate the lattice data to the physical pion mass using a formula which is compatible with the LNA structure of the PDFs. This behaviour was derived recently, with the result that the LNA behaviour involved a term in $m_\pi^2 \ln m_\pi$ [29]. For an initial investigation we concentrate on the non-singlet combination of PDFs, $u - d$, in which "disconnected" quark loops cancel. Calculations based on the CBM (which incorporate the LNA chiral structure just discussed) actually produce quite a reasonable description of the behaviour of the moments of the PDFs as a function of quark mass and, most important from the phenomenological point of view, the CBM calculations (for the n'th moment of the PDFs) can be fit with the simple expansion in m_π:

$$\langle x_u^n - x_d^n \rangle = a_n + b_n m_\pi^2 + a_n c_{\mathrm{LNA}} m_\pi^2 \ln \left(\frac{m_\pi^2}{m_\pi^2 + \mu^2} \right) , \qquad (7)$$

where c_{LNA} is model independent.

The scale μ in Eq.(7) is effectively the scale at which the rapid, chiral variation at low m_π turns off. The best fit to the lattice data is obtained with a value $\mu \sim 0.4 - 0.5$ GeV – a very similar scale to that found, for example,

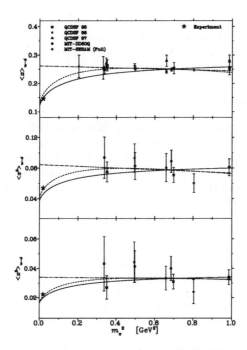

Figure 2. Moments of the $u - d$ quark distribution – from Ref.[30]. The straight (long-dashed) lines are linear fits to the data, while the curves have the correct LNA behavior in the chiral limit. While the solid curves build in the correct heavy quark limit as well as the chiral logarithm, the short-dashed curve uses simply Eq.(7). The star represents the phenomenological values taken from NLO fits.

for the magnetic moments. Clearly, from Fig. 2, Eq.(7) gives a very good description of the lattice data for the first three moments of the non-singlet distribution $d - u$ as well as the experimentally determined moments.

6 Conclusion

We are fortunate to be working in hadron physics at this time. Developments in lattice QCD, especially more powerful computers and improved chiral extrapolations, should finally allow the computation of accurate hadron properties within full QCD, *at the physical quark masses*, within the next five years. In order to carry out these chiral extrapolations we have seen that one needs to go beyond the naive series expansion of traditional chiral perturbation theory, but tremendous progress has been made in this direction. As a result of these advances we can expect new insights into the structure and properties of all hadrons built of light quarks.

Acknowledgements

It is a pleasure to acknowledge the collaborations with many colleagues at CSSM and elsewhere which have led to the results presented here. I would particularly like to express my appreciation to Will Detmold, Gerald Dunne, Emily Hackett-Jones, Derek Leinweber, Wally Melnitchouk, Kazuo Tsushima, Stewart Wright and Ross Young. This work was supported by the Australian Research Council and Adelaide University.

References

1. A. W. Thomas and W. Weise, "The Structure of the Nucleon," *289 pages. Hardcover ISBN 3-527-40297-7 Wiley-VCH, Berlin 2001.*
2. D. B. Leinweber, A. W. Thomas, K. Tsushima and S. V. Wright, Phys. Rev. D **61**, 074502 (2000) [hep-lat/9906027].
3. T. Lippert, S. Gusken and K. Schilling, Nucl. Phys. Proc. Suppl. **83** (2000) 182.
4. S. Aoki *et al.* [CP-PACS-Collaboration], Phys. Rev. D **60**, 114508 (1999) [hep-lat/9902018].
5. S. Theberge, A. W. Thomas and G. A. Miller, Phys. Rev. D **22**, 2838 (1980) [Erratum-ibid. D **23**, 2106 (1980)];
   A. W. Thomas, Adv. Nucl. Phys. **13**, 1 (1984).
6. Lattice Hadron Physics Collaboration proposal, *"Nuclear Theory with Lattice QCD"*, J.W. Negele and N. Isgur principal investigators, March 2000, ftp://www-ctp.mit.edu/pub/negele/LatProp/.
7. S. Weinberg, Physica (Amsterdam) **96 A**, 327 (1979);
   J. Gasser and H. Leutwyler, Ann. Phys. **158**, 142 (1984).
8. C. R. Allton *et al.* [UKQCD Collaboration], Phys. Rev. D **60**, 034507 (1999) [hep-lat/9808016].
9. S. V. Wright, D. B. Leinweber and A. W. Thomas, Nucl. Phys. A **680**, 137 (2000) [nucl-th/0005003].
10. F. J. Dyson, Phys. Rev. **85**, 631 (1952).
11. J. C. Le Guillou and J. Zinn-Justin (Eds.), *Large-Order Behaviour of Perturbation Theory*, (North Holland, Amsterdam, 1990).
12. G. V. Dunne, A. W. Thomas and S. V. Wright, arXiv:hep-th/0110155.

13. W. Heisenberg and H. Euler, Z. Phys. **98**, 714 (1936).

14. V. Weisskopf, Kong. Dans. Vid. Selsk. Math-fys. Medd. XIV No. 6 (1936), reprinted in *Quantum Electrodynamics*, J. Schwinger (Ed.) (Dover, New York, 1958).

15. J.Schwinger, Phys. Rev. **82**, 664 (1951).

16. D. B. Leinweber and T. D. Cohen, Phys. Rev. D **47**, 2147 (1993) [hep-lat/9211058].

17. E. J. Hackett-Jones, D. B. Leinweber and A. W. Thomas, Phys. Lett. B **494**, 89 (2000) [hep-lat/0008018].

18. D. H. Lu, S. N. Yang and A. W. Thomas, Nucl. Phys. A **684**, 296 (2001).

19. D. H. Lu, A. W. Thomas and A. G. Williams, Phys. Rev. C **57**, 2628 (1998) [nucl-th/9706019].

20. D. B. Leinweber, D. H. Lu and A. W. Thomas, Phys. Rev. D **60**, 034014 (1999) [hep-lat/9810005].

21. E. J. Hackett-Jones, D. B. Leinweber and A. W. Thomas, Phys. Lett. B **489**, 143 (2000) [hep-lat/0004006].

22. D. B. Leinweber and A. W. Thomas, Phys. Rev. D **62**, 074505 (2000) [hep-lat/9912052].

23. K. S. Kumar and P. A. Souder, Prog. Part. Nucl. Phys. **45**, S333 (2000).

24. D. B. Leinweber, A. W. Thomas and R. D. Young, Phys. Rev. Lett. **86**, 5011 (2001) [hep-ph/0101211].

25. A. W. Thomas, Phys. Lett. B **126**, 97 (1983).

26. E.M. Henley and G.A. Miller, Phys. Lett. B 251, 453 (1990); A.I. Signal, A.W. Schreiber and A.W. Thomas, Mod. Phys. Lett. A 6, 271 (1991); W. Melnitchouk, A.W. Thomas and A.I. Signal, Z. Phys. A 340, 85 (1991); S. Kumano, Phys. Rev. D43, 3067 (1991); S. Kumano and J.T. Londergan, Phys. Rev. D 44, 717 (1991); W.-Y.P. Hwang, J. Speth and G.E. Brown, Z. Phys. A339, 383 (1991).

27. P. Amaudraz et al., Phys. Rev. Lett. 66, 2712 (1991). A. Baldit et al., Phys. Lett. B 332, 244 (1994).

28. D. Dolgov *et al.*, hep-lat/0011010.

29. A.W. Thomas, W. Melnitchouk and F.M. Steffens, Phys. Rev. Lett. 85, 2892 (2000).

30. W. Detmold, W. Melnitchouk, J. W. Negele, D. B. Renner and A. W. Thomas, Phys. Rev. Lett. **87**, 172001 (2001) [arXiv:hep-lat/0103006]; W. Detmold, W. Melnitchouk and A. W. Thomas, Eur. Phys. J. directC **13**, 1 (2001) [arXiv:hep-lat/0108002].

DETERMINATION OF THE AXIAL COUPLING CONSTANT G_A IN THE LINEAR REPRESENTATIONS OF CHIRAL SYMMETRY

ATSUSHI HOSAKA

Research Center for Nuclear Physics (RCNP), Osaka Univ. Ibaraki 567-0047 Japan
E-mail: hosaka@rcnp.osaka-u.ac.jp

DAISUKE JIDO

Consejo Superior de Investigaciones Científicas, Universitat de Valencia, IFIC, Institutos de Investigación de Peterna, Aptdo. Correos 2085, 46071, Valencia, Spain

MAKOTO OKA

Consejo Superior de Department of Physics, Tokyo Institute of Technology, Meguro, Tokyo 152-8551 Japan

If a baryon field belongs to a certain linear representation of chiral symmetry of $SU(2) \otimes SU(2)$, the axial coupling constant g_A can be determined algebraically from the commutation relations derived from the superconvergence property of pion-nucleon scattering amplitudes. This establishes an algebraic explanation for the values of g_A of such as the non-relativistic quark model, large-N_c limit and the mirror assignment for two chiral partner nucleons. For the mirror assignment, the axial charges of the positive and negative parity nucleons have opposite signs. Experiments of eta and pion productions are proposed in which the sign difference of the axial charges can be observed.

1 Introduction

The axial coupling constant g_A of the nucleon is one of fundamental constants of the nucleon; experimentally, it is $g_A = 1.25$. Sometimes, the fact that g_A is close to unity is considered as an evidence of partially conserved axial vector current (PCAC), and hence it would become one in the limit that the axial current is conserved. This, however, is not correct, and g_A takes any number when chiral symmetry is broken spontaneously [1]. This fact is manifest in the non-linear sigma model. Even in the linear sigma model, g_A can be arbitrary if higher derivative terms are added.

Theoretically, g_A is related to the generators of the chiral group:

$$[Q_V^a, Q_V^b] = i\epsilon_{abc}Q_V^c, \quad [Q_V^a, Q_A^b] = i\epsilon_{abc}Q_A^c, \quad [Q_A^a, Q_A^b] = i\epsilon_{abc}Q_A^c. \quad (1)$$

If chiral symmetry is not spontaneously broken, these commutation relations may be used to determine the value of g_A, the nucleon matrix element of the axial charge operator Q_A^a. When, however, chiral symmetry is spontaneously broken, the operators Q_A^a are not well defined. In his pioneering work, Weinberg used the pion-nucleon matrix elements rather than the axial charges Q_A^a to compute g_A by using Goldberger-Treiman relation and commutation relations among the pion-nucleon coupling matrices X^a [2]. These commutation relations are derived from a super convergence property of pion-nucleon scattering amplitudes; it is a consistency condition between the low momentum expansion and the asymptotic behavior of the amplitudes. In the dispersion theory, Aldler and Weisberger derived a sum rule from the commutation relations [3,4]. Weinberg showed that when continuum intermediate states were saturated by narrow one particle states, the dispersion relations reduce to a set of algebraic equations which are in some cases solved to provide the value of g_A.

From a group theoretical point of view, a closed algebra determines the values of charges. A rather trivial example is the isospin charge of $SU(2)$. Similarly the axial part of the chiral group (a coset $SU(2) \times SU(2)/SU(2)$) when combined with the isospin part determines the axial charge of a linear representation of the full chiral group $SU(2) \times SU(2)$. In this paper we discuss several examples where g_A can be determined by the commutation relations, including the cases corresponding to the non-relativistic quark model, large-N_c limit and the mirror assignment for the two nucleons of chiral partners.

The mirror representation of the chiral group for the nucleon is particularly interesting, since it provides a possibility where positive and negative parity nucleons belong to the same chiral multiplet, showing characteristic behaviors toward chiral symmetry restoration [5]. In the latter part of this paper, we propose experimental method in which we will be able to observe the mirror nucleons in pion and eta productions from the nucleon [6].

2 Algebraic determination of axial charges

The use of the algebraic method was considered by Weinberg long ago [2]. He computed pion-nucleon coupling matrix elements for various linear representations of the chiral group. This method was later used in explaining the reason that the axial charge and magnetic moments of the constituent quarks takes the bare values [9].

It was shown that commutation relations among the pion-nucleon coupling matrices X_a and isospin charges T^a,

$$\left[X^a, X^b\right] = i\epsilon_{abc}T^c, \quad \left[T^a, T^b\right] = i\epsilon_{abc}T^c, \quad \left[T^a, X^b\right] = i\epsilon_{abc}X^c. \quad (2)$$

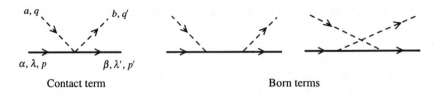

Figure 1. Contact and Born diagrams for the pion-nucleon scatterings.

were derived by considering the asymptotic behavior of the pion-nucleon forward scattering amplitudes when they are computed from a low energy effective lagrangian. Here the matrices X^a and T^a are related to the matrix elements of the axial vector and vector currents,

$$\langle N_\beta \lambda' | \int d^3x \, \tilde{A}_3^a(x) | N_\alpha \lambda \rangle \equiv (X^a)_{\beta\lambda',\alpha\lambda} , \tag{3}$$

$$\langle N_\beta \lambda' | \int d^3x \, V_0^a(x) | N_\alpha \lambda \rangle \equiv (T^a)_{\beta\lambda',\alpha\lambda} . \tag{4}$$

Here α and β are isospin indices of the nucleon, and λ and λ' are the helicities. When the momentum direction is taken along the z-axis, the matrices X^a and T^a are related to the axial charge g_A and isospin charge $g_V = 1$ by $(X^a)_{\beta\lambda',\alpha\lambda} = g_A(\tau^a/2)_{\beta\alpha}(\sigma_3)_{\lambda'\lambda}$, $(T^a)_{\beta\alpha} = g_V(\tau^a/2)_{\beta\alpha}\delta_{\lambda'\lambda}$. The currents appear in the low energy effective lagrangian [1]

$$L_{int} = \frac{2i}{f_\pi^2} V_\mu^a \epsilon_{abc} \pi_b \partial^\mu \pi_c + \frac{1}{f_\pi^2} \tilde{A}_\mu^a \partial^\mu \pi_a , \tag{5}$$

where $\tilde{A}_\mu^a = A_\mu^a - (-f_\pi \partial_\mu \pi^a)$. In (5), the pion-nucleon coupling constant g is replaced by $g_A M / f_\pi$ through the Goldberger-Treiman relation.

The scattering amplitudes computed by using the lagrangian (5) are small momentum expansion around $p = 0$, reproducing the low energy behavior expected from the low energy theorems. However, the large momentum behavior of the amplitudes is not consistent with the lower bound of unitarity. The commutation relations (2) are introduced in order to reproduce the correct asymptotic behavior of the amplitudes [2].

Now using the commutation relations, we can determine the charges, the matrix elements of X^a and T^a. The isospin charge is trivial; it is normalized, $g_V = 1$. Now taking the nucleon matrix elements, one can compute the axial charges (pion-nucleon couplings) X^a. For example, we consider a linear representation $(\frac{1}{2}, 1) \oplus (1, \frac{1}{2})$. The numbers in the parentheses represent isospin values of $SU(2) \otimes SU(2)$. Then the first term of the direct sum, $(\frac{1}{2}, 1)$, is the representation for the chirality plus component $\psi_+ = \frac{1}{2}(1 + \gamma_5)\psi$, and the second term, $(1, \frac{1}{2})$, for the chirality minus component $\psi_- = \frac{1}{2}(1 - \gamma_5)\psi$. The chiral representation $(\frac{1}{2}, 1)$ contains terms of isospin 1/2 (nucleon) and 3/2 (delta) as diagonal combinations of 1/2 and 1.

To be specific, let us consider matrix elements of the commutation relation

$$[X^+, X^-] = -T^3 \tag{6}$$

between $|N\rangle = |IM\rangle = |1/2\ 1/2\rangle$ and $|\Delta\rangle = |IM\rangle = |3/2\ 1/2\rangle$. Here $X^\pm = \mp 1/\sqrt{2}(X_1 \pm iX_2)$. Writing the reduced matrix elements as $\langle N||X||N\rangle \equiv X_N$, $\langle N||X||\Delta\rangle \equiv X_{N\Delta}$ and $\langle \Delta||X||\Delta\rangle \equiv X_\Delta$, we obtain the following three coupled equations:

$$-\frac{1}{3}X_N^2 + \frac{1}{6}X_{N\Delta}^2 = -\frac{1}{2},$$

$$\frac{1}{30}X_\Delta^2 + \frac{1}{12}X_{N\Delta}^2 = \frac{1}{2}, \tag{7}$$

$$2X_N X_{N\Delta} - \sqrt{10}X_\Delta X_{N\Delta} = 0.$$

Solving these coupled equations for $X_{N\Delta} \neq 0$, we find

$$|X_N| = \frac{5}{\sqrt{6}}, \quad |X_\Delta| = \sqrt{\frac{5}{3}}, \quad |X_{N\Delta}| = \frac{4}{\sqrt{3}}. \tag{8}$$

These results lead to the nucleon axial charge $|g_A^N| = 5/3$, which is the value of the non-relativistic quark model. This agreement is not accidental, since in the quark model, the nucleon and delta can be described in the same basis of three quarks, as corresponding to the chiral representation $(1, \frac{1}{2}) \oplus (\frac{1}{2}, 1)$. If $X_{N\Delta} = 0$, there is no coupling between the nucleon and delta, where they belong to separate representations; $N \sim (\frac{1}{2}, 0) \oplus (0, \frac{1}{2})$ and $\Delta \sim (\frac{3}{2}, 0) \oplus (0, \frac{3}{2})$. In this case the nucleon axial charge reduces simply to unity, $g_A = 1$. This explains the result of the linear sigma model.

We can extend this analysis to various cases (representations). Here we show two examples; one is the large-N_c limit and the other is the mirror assignment for parity doublet nucleons. The large-N_c nucleons are represented by the representation $((N_c+1)/4, (N_c-1)/4) \oplus ((N_c-1)/4, (N_c+1)/4)$ [7]. Obviously, this representation contains the isospin states of $1/2 \leq I \leq N_c/2$. The relevant equations corresponding to (7) are recursion equations from $I = 1/2$ to $N_c/2$. The solution gives the nucleon $g_A = (N_c + 2)/3$, which is the result known in the large-N_c quark model [8]. In the mirror assignment, we consider two nucleons of opposite parity (parity doublet). Then, the assignment of the chiral group (I_1, I_2) for the chirality plus and minus components is interchanged for the two nucleons;

$$|1\rangle:\ \psi_+ \sim (\tfrac{1}{2}, 0),\ \psi_- \sim (0, \tfrac{1}{2}),\quad |2\rangle:\ \psi_+ \sim (0, \tfrac{1}{2}),\ \psi_- \sim (\tfrac{1}{2}, 0). \tag{9}$$

Obviously, the absolute values of the axial charges of the two nucleons are unity but with opposite signs. It is also possible to assign the same chiral representation to the two nucleons. This assignment was called the naive assignment. In general, the two representations of (9) can mix in physical nucleons; the chiral eigenstates and mass eigenstates differ. Introducing a mixing angle θ, the axial charges of the two physical nucleons (mass eigenstates) are given in the matrix form (see Eq. (16)

$$g_A = \begin{pmatrix} \cos 2\theta & -\sin 2\theta \\ -\sin 2\theta & -\cos 2\theta \end{pmatrix} . \tag{10}$$

3 Experimental observation of the mirror assignment

Among the above examples, the mirror (as well as the naive) representations were not considered much before. The possibility that the axial charge of the negative parity nucleon can be opposite to the nucleon axial charge was first pointed out by Lee and realized in the form of the linear sigma model by DeTar and Kunihiro. Weinberg also pointed out the two possibilities $g_A = \pm 1$ [9]. At the composite level, the nucleon axial charge should be derived from the underlying theory of QCD. So far, we do not know reliable methods to do so. In this section, therefore, we first present phenomenological properties of the two chiral assignments, the naive and mirror, using the linear sigma model [5]. We then propose an experimental method to observe the two assignments [6].

Let us consider linear sigma models based on the naive and mirror assignments. To do this, meson fields are introduced as components of the representation $(1/2, 1/2)$ of the chiral group, which are subject to the transformation rule: $\sigma + i\vec{\tau} \cdot \vec{\pi} \to g_L(\sigma + i\vec{\tau} \cdot \vec{\pi})g_R^\dagger$.

In the naive assignment, the chiral invariant lagrangian up to order $(\text{mass})^4$ is given by:

$$L_{\text{naive}} = \bar{N}_1 i\partial\!\!\!/ N_1 - g_1 \bar{N}_1 (\sigma + i\gamma_5 \vec{\tau} \cdot \vec{\pi})N_1 + \bar{N}_2 i\partial\!\!\!/ N_2 - g_2 \bar{N}_2 (\sigma + i\gamma_5 \vec{\tau} \cdot \vec{\pi})N_2$$
$$- g_{12}\{\bar{N}_1(\gamma_5 \sigma + i\vec{\tau} \cdot \vec{\pi})N_2 - \bar{N}_2(\gamma_5 \sigma + i\vec{\tau} \cdot \vec{\pi})N_1\} + L_{\text{mes}} , \tag{11}$$

where g_1, g_2 and g_{12} are free parameters. The terms of g_1 and g_2 are ordinary chiral invariant coupling terms of the linear sigma model. The term of g_{12} is the mixing of N_1 and N_2. Since the two nucleons have opposite parities, γ_5 appears in the coupling with σ, while it does not in the coupling with π. The meson lagrangian L_{mes} in (11) is not important in the following discussion.

Chiral symmetry breaks down spontaneously when the sigma meson acquires a finite vacuum expectation value, $\sigma_0 \equiv \langle 0|\sigma|0\rangle$. This generates masses of the nucleons. From (11), the mass can be expressed by a 2×2 matrix in the space of N_1 and N_2. The mass matrix can be diagonalized by the rotated states,

$$
\begin{pmatrix} N_+ \\ N_- \end{pmatrix} = \begin{pmatrix} \cos\theta & \gamma_5 \sin\theta \\ -\gamma_5 \sin\theta & \cos\theta \end{pmatrix} \begin{pmatrix} N_1 \\ N_2 \end{pmatrix} ,
\tag{12}
$$

where the mixing angle and mass eigenvalues are given by

$$
\tan 2\theta = \frac{2g_1}{g_1 + g_2}, \quad m_\pm = \frac{\sigma_0}{2}\left(\sqrt{(g_1 + g_2)^2 + 4g_{12}^2} \pm (g_1 - g_2)\right) .
\tag{13}
$$

In the naive model, since the interaction and mass matrices takes the same form, the physical states, N_+ and N_-, decouple exactly; the lagrangian becomes a sum of the N_+ and N_- parts. [a] Therefore, chiral symmetry imposes no constraint on the relation between N_+ and N_-. The role of chiral symmetry is just the mass generation due to its spontaneous breaking. When chiral symmetry is restored and $\sigma_0 \to 0$, both N_+ and N_- become massless and degenerate. However, the degeneracy is trivial as they are independent; they no longer transform among themselves. The decoupling of N_+ and N_- implies that the off-diagonal Yukawa coupling $g_{\pi N_+ N_-}$ vanishes. This is a rigorous statement up to the order we considered.

Now we turn to the mirror assignment. It is rather straightforward to write down the chiral invariant lagrangian compatible to the mirror transformations:

$$
L_{\text{mirror}} = \bar{N}_1 i \partial\!\!\!/ N_1 - g_1 \bar{N}_1(\sigma + i\gamma_5 \vec{\tau} \cdot \vec{\pi})N_1 + \bar{N}_2 i\partial\!\!\!/ N_2 - g_2 \bar{N}_2(\sigma - i\gamma_5 \vec{\tau} \cdot \vec{\pi})N_2
$$
$$
- m_0(\bar{N}_1 \gamma_5 N_2 - \bar{N}_2 \gamma_5 N_1) + L_{\text{mes}} .
\tag{14}
$$

Here the chiral invariant mass term has been added. Note that in the g_2 term, the sign of the pion field is opposite to that of the g_1 term. This compensates the mirror transformation of N_2. The lagrangian (14) was first formulated by DeTar and Kunihiro.

When chiral symmetry is spontaneously broken, the mass matrix of the lagrangian (14) can be diagonalized by a linear combination similar to (12). The mixing angle and mass eigenvalues are given by

$$
\tan 2\theta = \frac{2m_0}{\sigma_0(g_1 + g_2)}, \quad m_\pm = \frac{\sigma_0}{2}\left(\sqrt{(g_1 + g_2)^2 + 4\mu^2} \pm (g_1 - g_2)\right) ,
\tag{15}
$$

where $\mu = m_0/\sigma_0$. In the mirror model, the interaction term is not diagonalized in the physical basis, unlike the naive model.

---

[a] Small chiral symmetry breaking might induce a small coupling $g_{\pi N_+ N_-}$.

Let us show the axial coupling constants g_A in the mirror model. They can be extracted from the commutation relations between the axial charge operators Q_5^a and the nucleon fields,

$$[Q_5^a, N_+] = \frac{\tau^a}{2}\gamma_5(\cos 2\theta\, N_+ - \sin 2\theta\, \gamma_5 N_-)$$

$$[Q_5^a, N_-] = \frac{\tau^a}{2}\gamma_5(-\sin 2\theta\, \gamma_5 N_+ - \cos 2\theta\, N_-)\,. \tag{16}$$

This implies that g_A's are expressed by a 2×2 matrix whose elements are given by the coefficients of (16), explaining the result given in (10). From this we see that the signs of the diagonal axial charges g_A^{++} and g_A^{--} are opposite. The absolute value is, however, smaller than one in contradiction with experimental value $g_A \sim 1.25$. In the present model, the physical states $N_\pm$ are the superpositions of $N_{1,2}$, whose axial charges are ± 1. This explains why $|g_A^{++}|, |g_A^{--}| < 1$. In the algebraic method, the g_A value can be increased by introducing a mixing with higher representations such as $(1, \frac{1}{2})$.

4 π and η productions at threshold region

In this section, we propose experimental method to study the two chiral assignments. As discussed in the preceding sections, one of the differences between the naive and mirror assignments is the relative sign of the axial coupling constants of the positive and negative parity nucleons. In the following discussions, we identify $N_+ \sim N(939)$ and $N_- \sim N(1535) \equiv N^*$. Strictly, the identification of the negative parity nucleon with the first excited state $N(1535)$ is no more than an assumption. From experimental point of view, however, $N(1535)$ has a distinguished feature that it has a strong coupling with an η meson, which can be used as a filter to observe the resonance. In practice, we observe the pion couplings which are related to the axial couplings through the Goldberger-Treiman relation $g_{\pi N_\pm N_\pm} f_\pi = g_A M_\pm$.

Let us consider π and η productions induced by a pion or photon. Suppose that the two diagrams of (1) and (2) as shown in Fig. 2 are dominant in these process. Modulo energy denominator, the only difference of these processes is due to the coupling constants $g_{\pi NN}$ and $g_{\pi N^* N^*}$. Therefore, depending on their relative sign, cross sections are either enhanced or suppressed. In the pion induced process, due to the p-wave coupling nature, another diagram (3) also contributes substantially.

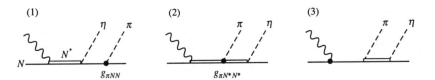

Figure 2. Dominant diagrams for π and η productions. The incident wavy line is either a photon or a pion. For photon induced reactions, the diagrams (1) and (2) are dominant, while for pion induced reactions, the third one (3) also contributes.

Table 1. Parameters used in our calculation.

m_N	m_{N^*}	Γ_{N^*}	$g_{\pi NN}$	$g_{\pi NN^*}$	$g_{\eta NN^*}$	$g_{\pi N^*N^*}$
938	1535	140	13	0.7	2.0	13 (naive)
(MeV)	(MeV)	(MeV)				−13 (mirror)

In actual computation, we take the interaction lagrangians:

$$L_{\pi NN} = g_{\pi NN} \bar{N} i\gamma_5 \vec{\tau} \cdot \vec{\pi} N \,, \quad L_{\eta NN^*} = g_{\eta NN^*} (\bar{N}\eta N^* + \bar{N}^*\eta N) \,,$$
$$L_{\pi NN^*} = g_{\pi NN^*} (\bar{N}\tau \cdot \pi N^* + \bar{N}^*\tau \cdot \pi N)$$
$$L_{\pi N^*N^*} = g_{\pi N^*N^*} (\bar{N}^* i\gamma_5 \tau \cdot \pi N^*) \,. \tag{17}$$

We use these interactions both for the naive and mirror cases with empirical coupling constants for $g_{\pi NN} \sim 13$, $g_{\pi NN^*} \sim 0.7$ and $g_{\eta NN^*} \sim 2$. The coupling constants $g_{\pi NN^*} \sim 0.7$ and $g_{\eta NN^*} \sim 2$ are determined from the partial decay widths, $\Gamma_{N^*(1535)\to\pi N} \approx \Gamma_{N^*(1535)\to\eta N} \sim 70$ MeV, although large uncertainties for the width have been reported. The unknown parameter is the $g_{\pi N^*N^*}$ coupling. One can estimate it by using the theoretical value of the axial charge g_A^* and the Goldberger-Treiman relation for N^*. When $g_A^* = \pm 1$ for the naive and mirror assignments, we find $g_{\pi N^*N^*} = g_A^* m_{N^*}/f_\pi \sim \pm 17$. Here, just for simplicity, we use the same absolute value as $g_{\pi NN}$. The coupling values used in our computations are summarized in Table 1.

Several remarks follow [6]:

- We assume resonance (N^*) pole dominance. This is considered to be good particularly for the η production at the threshold region, since η is dominantly produced by N^*.

- There are altogether twelve resonance dominant diagrams. Due to energy denominator, the three diagrams in Fig. 2 are dominant.

630

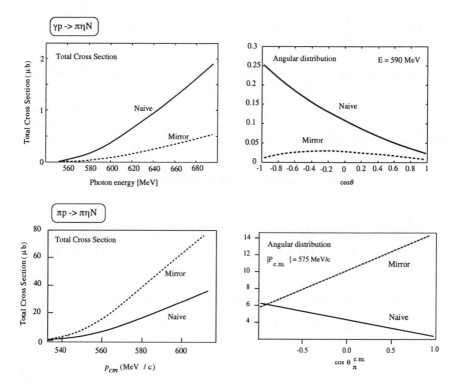

Figure 3. Various cross sections for π and η productions.

- Background contributions, in which two meson (seagull) or three meson vertices appear, are suppressed due to G-parity conservation.

Hence, the processes are indeed dominated by the N^* resonance diagrams.

We show various cross sections for the pion and photon induced processes in Fig. 3. We briefly discuss the results:

1. The total cross sections are of order of micro barn, which are well accessible by the present experiments. In the photon induced process, the diagrams (1) and (2) interfere destructively in the mirror assignment. In the pion induced case, due to the momentum dependence of the initial vertex the third term (3) becomes dominant and the mirror assignment is rather enhanced.

2. In the pion induced reaction, the angular distribution of the final state

pion differs clearly depending on the sign of the $\pi N N$ and $\pi N^* N^*$ couplings.

5 Summary

In this report we have presented an algebraic argument to determine the nucleon axial charge g_A. Assuming that the nucleon belongs to a linear representation of the chiral group, the commutation relations can determine g_A. This explains the nucleon g_A values in the non-relativistic quark model, large-N_c and the mirror nucleons. In order to detect the mirror nucleons, we proposed an experiment of π and η production. Various differential cross sections were computed which can be measured at the facilities such as SPring8 and LNS, Tohoku. Determination of the axial charge is a simple but an interesting question related to chiral symmetry of the nucleon and should be studied further both in theory and experiment.

References

1. S. Weinberg, *The Quantum Theory of Fields*, Cambridge (1996), II p.203.
2. S. Weinberg, Phys. Rev. 177 (1969) 2604.
3. S. Adler, Phys, Rev. 140 (1965) B736.
4. W. Weisberger, Phys, Rev. 143 (1966) 1302.
5. D. Jido, Y. Nemoto, M. Oka and A. Hosaka, Nucl. Phys. A671 (2000) 471.
6. D. Jido, M. Oka and A. Hosaka, Prog. Theor. Phys. 106 (2001) 823; D. Jido, M. Oka and A. Hosaka, Prog. Theor. Phys. 106 (2001) 873.
7. S.R. Beane, Phys. Rev. D59 (1999) 031901.
8. A.V. Manohar, Nucl. Phys. B248 (1984) 19.
9. S. Weinberg, Phys. Rev. Lett. 65 (1990) 1177; 1181.
10. B. W. Lee, *Chiral Dynamics*, Gordon and Breach, New York, (1972)
11. C. DeTar and T. Kunihiro, Phys. Rev. D 39 (1989) 2805.

Study ω and ϕ photoproduction in the nucleon isotopic channels

Q. Zhao

Department of Physics, University of Surrey, Guildford, GU2 7XH, UK
E-mail: qiang.zhao@surrey.ac.uk

We present results for the photoproduction of ω and ϕ meson in the nucleon isotopic channnles. A recently developed quark model with an effective Lagrangian is employed to account for the non-diffractive s- and u-channel processes; the diffractive feature arising from the *natural* parity exchange is accounted for by the t-channel pomeron exchange, while the *unnatural* parity exchange is accounted for by the t-channel pion exchange. In the ω production, the isotopic effects could provide more information concerning the search of "missing resonances", while in the ϕ production, the isotopic effects could highlight non-diffractive resonance excitation mechanisms at large angles.

1 Introduction

The availabilities of high intensity electron and photon beams at JLab (CLAS), ELSA (SAPHIR), MAMI, ESRF (GRAAL), and SPring-8 give accesses to excite nucleons with the clean electromagnetic probes, thus revive the interest in the search of "missing resonances" [1] in meson photo- and electroproduction. Vector meson photoproduction near threshold attracts attentions [2,3,4,5,6] since those "missing resonances" could have stronger couplings to this channel such that their existing signals might be derived.

Taking into account large degrees of freedom in the resonance excitations, experimental efforts for such a purpose is by no means trivial. Historically, experimental data for photoproduction of vector meson in isotopic channels are very sparse. In contrast with the reaction $\gamma + p \rightarrow \omega(\rho^0) + p$, there are only few dottes available for $\gamma + n \rightarrow \rho^- + p$ and $\gamma + p \rightarrow \rho^+ + n$, while experiment on $\gamma + n \rightarrow \omega(\phi) + n$ is not available. The main feature in the latter reactions is that the electric transition vanishes in the nucleon pole terms. Nevertheless, due to the change of the isospin degrees of freedom, interferences between different resonances and processes will change significantly. Nevertheless, in the neutron target reaction, resonances of quark model representation [70, [4]8] are no longer suppressed by the Moorhouse selection rule [7], thus will add more ingredients into the reaction. Briefly, apart from various polarization observables which could provide rich information about the reaction mechanism, coherent study of isotopic reactions could also highlight signals which cannot be seen easily in a single channel.

Our motivation of studying the isotopic production of the ϕ meson is rather different from the ω. Our attention here is paid to the effects from the non-diffractive s- and u-channel ϕ production in polarization observables. The study of isotopic reaction provides us with insights into the non-diffractive processes at large angles (large momentum transfer with $W \approx 2 \sim 4$ GeV). Since the ϕ meson has higher threshold ($E_\gamma \approx 1.57$ GeV), where a large number of resonances can be excited, the non-diffractive resonance effects could exhibit more collectively rather than exclusively from individual resonances. Another relevant interest in ϕ production is to detect strangeness component in nucleons, which could be another important non-diffractive source in the reaction [8,9,10]. Although our study does not take into account such a mechanism, we shall see that our results provide some supplementary information for such an effort.

In this proceeding, a quark model approach to vector meson photoproduction [2,3,11,12] is applied to the photoproduction of ω and ϕ meson off the proton and neutron. Our purpose is to provide a framework on which a systematic study of resonance excitations becomes possible.

2 The model

Our model consists of three processes: (i) vector meson production through an effective Lagrangian; (ii) t-channel pomeron exchange ($\mathcal{P}$) for ω, ρ^0 and ϕ production [13]; (iii) t-channel light meson exchange. Namely, in the ω and ϕ meson photoproduction, the π^0 exchange (π) is taken into account. [a]

The effective Lagrangian for the V-qq coupling is:

$$L_{eff} = \overline{\psi}(a\gamma_\mu + \frac{ib\sigma_{\mu\nu}q^\nu}{2m_q})V^\mu\psi, \tag{1}$$

where ψ ($\overline{\psi}$) denotes the light quark (anti-quark) field in a baryon system; V^μ represents vector meson field (ω, ρ, K^* and ϕ). In this model, the 3-quark baryon system is described by the NRCQM in the $SU(6) \otimes O(3)$ symmetry limit, while the vector meson is treated as an elementary point-like particle which couples to the constituent quark through the effective interaction. Parameter a and b denote the coupling strengths and are determined by experimental data.

The tree level transitions from the effective Lagrangian can be labelled by the Mandelstam variable, s, u, and t:

$$M_{fi} = M_{fi}^s + M_{fi}^u + M_{fi}^t . \tag{2}$$

Given the electromagnetic coupling $H_e = -\overline{\psi}\gamma_\mu e_q A^\mu\psi$, and the effective strong coupling $H_m = -\overline{\psi}(a\gamma_\mu + \frac{ib\sigma_{\mu\nu}q^\nu}{2m_q})V^\mu\psi$, the s- and u-channel with resonance

---

[a] In the ρ^0 production, the σ meson exchange is included.

excitations can be expressed as,

$$M_{fi}^{s+u} = \sum_j \langle N_f|H_m|N_j\rangle\langle N_j|\frac{1}{E_i + \omega_\gamma - E_j}H_e|N_i\rangle$$

$$+ \sum_j \langle N_f|H_e\frac{1}{E_i - \omega_m - E_j}|N_j\rangle\langle N_j|H_m|N_i\rangle, \qquad (3)$$

where ω_γ and ω_m represent the photon and meson energy, respectively. In this expression, all the intermediate states $|N_j\rangle$ are included. A simple transformation for the electromagnetic interaction leads to,

$$M_{fi}^{s+u} = i\langle N_f|[g_e, H_m]|N_i\rangle$$

$$+i\omega_\gamma \sum_j \langle N_f|H_m|N_j\rangle\langle N_j|\frac{1}{E_i + \omega_\gamma - E_j}h_e|N_i\rangle$$

$$+i\omega_\gamma \sum_j \langle N_f|h_e\frac{1}{E_i - \omega_m - E_j}|N_j\rangle\langle N_j|H_m|N_i\rangle, \qquad (4)$$

with $h_e = \sum_l e_l \mathbf{r}_l \cdot \boldsymbol{\epsilon}_\gamma(1 - \boldsymbol{\alpha} \cdot \hat{\mathbf{k}})e^{i\mathbf{k}\cdot\mathbf{r}_l}$, and $g_e = \sum_l e_l \mathbf{r}_l \cdot \boldsymbol{\epsilon}_\gamma e^{i\mathbf{k}\cdot\mathbf{r}_l}$, where $\hat{\mathbf{k}} \equiv \mathbf{k}/\omega_\gamma$ is the unit vector along the photon three-momentum $\mathbf{k}$. We identify the term $i\langle N_f|[g_e, H_m]|N_i\rangle$ as a seagull term M_{fi}^{sg}, and re-define the second and third term as the s- and u-channel, respectively. The tree level diagrams are illustrated in Fig. 1, where a bracket on the seagull term gives a caution about its origin.

In the NRCQM, those low-lying states ($n \leq 2$) have been successfully related to the resonances which can be taken into account explicitly in the formula. For those higher excited states, they can be treated degenerate in the main quantum number n in the harmonic oscillator basis. Detailed description of this approach can be found in Ref. [2,11,3]. The ω and ϕ production are significantly simplified because of isospin conservation. Namely, only isospin 1/2 intermediate resonances will contribute in the reaction. Another advantage for these two reactions is that both particles are charge-neutral. This feature leads to the vanishing of the t-channel vector meson exchange and the seagull term. In the end, only the s- and u-channel (S+U) from Eq. 4 will be the contribution source from the effective Lagrangian. In addition, the Moorhouse selection rule[7] further simplifies the calculations for the proton target reaction. Due to this selection rule, resonances belonging to representation [70,[4] 8] would vanish in the $\gamma p \to N^*$ transitions.

Apart from the S+U from the effective Lagrangian, our model includes the pomeron exchange ($\mathcal{P}$) to account for the diffractive phenomenon in vector

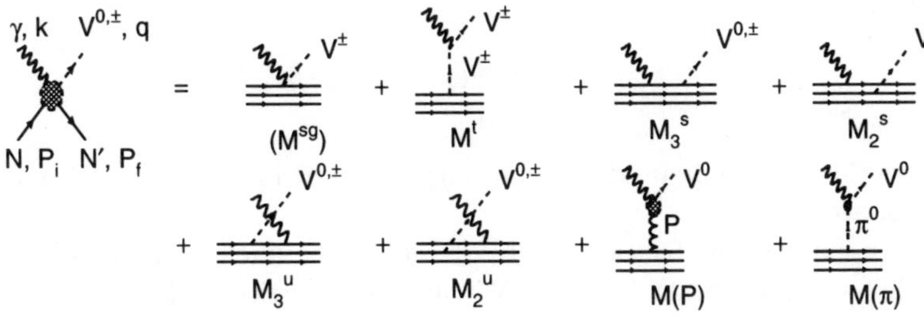

Figure 1: Tree level transition diagrams for the reaction.

meson photoproduction. In this model[13], the pomeron mediates the long range interaction between two confined quarks, and behaves rather like a $C = +1$ isoscalar photon. We summarize the vertices as follows:

(i) pomeron-nucleon coupling:

$$F_\mu(t) = 3\beta_0 \gamma_\mu f(t), \quad f(t) = \frac{(4M_N^2 - 2.8t)}{(4M_N^2 - t)(1 - t/0.7)^2} , \tag{5}$$

where $\beta_0 = 1.27$ GeV^{-1} is the coupling of the pomeron to one light constituent quark; $f(t)$ is the isoscalar nucleon electromagnetic form factor with four-momentum transfer t; the factor 3 comes from the "quark-counting rule".

(ii) Quark-ϕ-meson coupling:

$$V_\nu(p - \frac{1}{2}q, p + \frac{1}{2}q) = f_V M_\phi \gamma_\nu , \tag{6}$$

where f_V is the radiative decay constant of the vector meson, and determined by $\Gamma_{V \to e^+ e^-} = 8\pi \alpha_e^2 e_Q^2 f_V^2 / 3M_V$. A form factor $\mu_0^2/(\mu_0^2 + p^2)$ is adopted for the pomeron-off-shell-quark vertex, where $\mu_0 = 1.2$ GeV is the cut-off energy, and p is the four-momentum of the quark. The pomeron trajectory is $\alpha(t) = 1 + \epsilon + \alpha' t$, with $\alpha' = 0.25$ GeV^{-2}.

The π^0 exchange is introduced with the following Lagrangians for the πNN and $\phi\pi\gamma$ coupling:

$$L_{\pi NN} = -ig_{\pi NN}\overline{\psi}\gamma_5(\boldsymbol{\tau} \cdot \boldsymbol{\pi})\psi . \tag{7}$$

and

$$L_{V\pi^0\gamma} = e_N \frac{g_{V\pi\gamma}}{M_V} \epsilon_{\alpha\beta\gamma\delta} \partial^\alpha A^\beta \partial^\gamma V^\delta \pi^0 \ . \tag{8}$$

where the commonly used couplings, $g^2_{\pi NN}/4\pi = 14$, $g^2_{\omega\pi\gamma} = 3.315$ and $g^2_{\phi\pi\gamma} = 0.143$, are adopted.

An exponential factor $e^{-(\mathbf{q}-\mathbf{k})^2/6\alpha^2_\pi}$ from the nucleon wavefunctions plays a role as a form factor for the πNN and $\phi\pi\gamma$ vertices, where $\alpha_\pi = 300$ MeV is adopted. This factor comes out naturally in the harmonic oscillator basis where the nucleon is treated as a 3-quark system.

3   Results and discussions

Application of this approach to the ω meson photoproduction shows a great promise. A coherent study of the ρ meson photoproduction [2] also highlights the approximately satisfied isospin symmetry between the ω and ρ meson.

In $\gamma p \to \omega p$, the old measurements [14] of the parity asymmetries at $E_\gamma = 2.8$, 4.7 and 9.3 GeV provides very strong constraints on the *natural* ($\mathcal{P}$) and *unnatural*-parity exchange (π). It was shown that at $E_\gamma = 2.8$ GeV, the parity asymmetry has a negative value which suggests the still dominant contribution from the pion exchange. At 4.7 GeV, the *natural*-parity starts to dominate over the cross section, and at 9.3 GeV, the parity asymmetry is completely dominated by the pomeron exchanges. Such an energy evolution of the parity asymmetry above the resonance region constrains the "background" terms, which then can be extrapolated down to the resonance region. In Ref.[3,15], we show that such a scheme is consistent with the experimental data for the parity asymmetries [14], and indeed leads to a reasonable estimation of these "background" terms.

3.1   ω meson photoproduction

Parameters at resonance region for the ω photoproduction are constrained approximately by the SAPHIR data [16]. It shows that $a = -2.72$ and $b' \equiv b - a = -3.42$ give an overall fit of available data.

In Fig. 2, differential cross sections for the isotopic channels are presented at four energy bins, $E_\gamma =$ 1.225, 1.450, 1.675 and 1.915 GeV, and compared with the SAPHIR data [16]. It shows that near threshold the π^0 exchange dominates over the other two processes at small angles, while $\mathcal{P}$ will compete with π with the increasing energies. The exclusive $\mathcal{P}$ and π in the proton and neutron reaction are changed slightly due to the small difference between the proton and neutron mass. The S+U contributions generally dominate at large

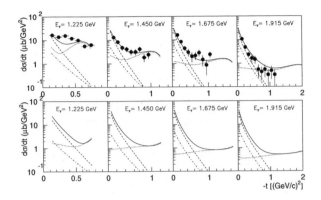

Figure 2: Differential cross section for $\gamma p \to \omega p$ (upper column) and $\gamma n \to \phi n$ (lower column). The dashed, dot-dashed, and dotted curves denote exclusive results for process, S+U, $\mathcal{P}$ and π, respectively; the solid curves represent full model calculation. Data are from Ref.[16].

angles. Compare these two isotopic reactions, we see that the main impact of the change of the isospin space comes from S+U.

Interesting interfering effects among these processes could be seen in polarization observables. In Fig. 3, we present predictions for $\Sigma \equiv (2\rho_{11}^1 + \rho_{00}^0)/(2\rho_{11}^0 + \rho_{00}^0)$, which has been measured by GRAAL[17]. Here, ρ^1 and ρ^0 are density matrix elements in the helicity space[18]. One important feature of Σ is that large asymmetries (deviation from 0) cannot be produced by the exclusive $\mathcal{P}$, π or $\mathcal{P} + \pi$. As shown by the solid curves, large asymmetries are produced by the interferences between the $\mathcal{P} + \pi$ and S+U processes. Interest arises from the comparison of the isotopic reactions. It shows that the S+U produces flattened positive asymmetries in $\gamma n \to \omega n$ near threshold. Specifically, we show the effects of $P_{13}(1720)$ in the Σ, which suggest that isotopic reactions could have outlined more information.

Another interesting observable with polarized photon beam is, $\Sigma_A \equiv (\rho_{11}^1 + \rho_{1-1}^1)/(\rho_{11}^0 + \rho_{1-1}^0)$. As found in Ref.[3], this observable is more sensitive to small contributions from individual resonances. Due to lack of space, we shall only show the calculation of Σ_A for the ϕ production in the next Subsection.

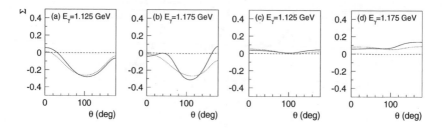

Figure 3: Polarized beam symmetry for the proton [(a)-(b)] and neutron reactions [(c)-(d)] in ω production. The solid, dashed, and dotted curves denote the full model calculation, $\mathcal{P} + \pi$, the full model excluding the $P_{13}(1720)$.

3.2 ϕ meson photoproduction

In the ϕ meson production, the parameter $a = 0.241 \pm 0.105$ and $b' \equiv b - a = 0.458 \pm 0.091$ for the effective ϕ-qq coupling are determined by the fit of old data from Ref. [19] at $E_\gamma = 2.0$ GeV and the recent data from CLAS [20] at 3.6 GeV. These values are much smaller than parameters derived for the ω and ρ production. Qualitatively, the ϕ-qq couplings could be suppressed by the OZI rule. Meanwhile, in the SU(3) symmetry limit, $a_\phi/a_\omega = g_{\phi NN}/g_{\omega NN}$ can be derived, which is consistent with the ratio for an ideal ω-ϕ mixing [21], $g_{\phi NN}/g_{\omega NN} = -\tan 3.7°$.

Using the same parameters derived in $\gamma p \to \phi p$, the cross sections for both $\gamma p \to \phi p$ and $\gamma n \to \phi n$ are calculated at $E_\gamma = 2.0$ GeV (Fig. 4). Significant cross section difference occurs at large angles, where $d\sigma/d\Omega$ for $\gamma n \to \phi n$ is much smaller than for $\gamma p \to \phi p$. A common feature arising from these two reactions is a relatively stronger backward peak from the "background" u-channel nucleon pole term. Similar feature is also found by Laget [22].

In Fig. 5, the isotopic effects of these two reactions are shown for the polarized beam asymmetry $\Sigma_A \equiv (\rho_{11}^1 + \rho_{1-1}^1)/(\rho_{11}^0 + \rho_{1-1}^0) = (\sigma_\parallel - \sigma_\perp)/(\sigma_\parallel + \sigma_\perp)$ at $E_\gamma = 2.0$ GeV, where $\sigma_\parallel$ and $\sigma_\perp$ denote the cross sections for $\phi \to K^+ K^-$ when the decay plane is parallel or perpendicular to the photon polarization vector.

The dashed curves represent results for $\mathcal{P} + \pi$, which deviate from +1 (for pure *natural*-parity exchange) due to the presence of the *unnatural* parity pion exchange. With the S+U, the full model calculation suggests that the large angle asymmetry is strongly influenced by the S+U processes, while the forward angles are not sensitive to them. Interferences between the $\mathcal{P}$ and

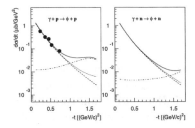

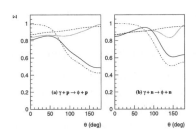

Figure 4: Differential cross section for ϕ production at $E_\gamma = 2.0$ GeV. The dot-dashed, dashed, and solid curves denote the S+U, $\mathcal{P} + \pi$, and full model calculations, respectively, while the dotted curve represents full model calculation excluding the u-channel contribution. Data come from Ref.[19].

Figure 5: Polarized beam symmetry Σ_A for the proton and neutron reactions at $E_\gamma = 2.0$ GeV. Notations are the same as Fig. 4.

S+U can be seen by excluding the π (see the dot-dashed curves). It shows that asymmetries produced by the S+U at forward angles are negligible. Since the pion exchange becomes very small at large angles, we conclude that the large angle asymmetry is determined by the S+U and reflects the isotopic effects. The role played by the s-channel resonances in the two reactions are presented by excluding the u-channel contributions. As shown by the dotted curves, the interferences from the s-channel resonances are much weaker than that from the u-channel, although they are still an important non-diffractive source at large angles.

This feature, which might make it difficult to filter out signals for individual s-channel resonances in the ϕ photoproductions might suggest that the forward angle could be an ideal region for the study of other non-diffractive sources. [b]

It should be noted that no isotopic effects can be seen in Σ and Σ_A if only $\mathcal{P} + \pi$ contribute to the cross section. This is because the transition amplitude of $\mathcal{P}$ is purely imaginary, while that of pion exchange is purely real. In Σ and Σ_A, the sign arising from the $g_{\pi NN}$ will disappear, which is why the dashed curves in Fig. 5 are (almost) the same. It is worth noting that our results for the Σ_A are quite similar to findings of Ref.[23] at small angles, but very different at large angles. In Ref.[23] only the nucleon pole terms for the s- and u-channel

---

[b]For example, it was suggested by Ref.[8,9], the strangeness component in nucleons could produce large asymmetries in beam-target double polarization observable at forward angles, while we find that S+U process only make sense at large angles.[11,12]

processes were included.

4 Conclusions and perspectives

We show that the effective Lagrangian approach in the quark model basis provides an ideal framework for the study of resonance excitations in vector meson photoproduction. The great advantage is that only a limited number of parameters will be introduced. Meanwhile, it permits a systematic and coherent study of different isospin channels, which could be helpful in highlighting model-independent features in the reaction mechanisms. In the $SU(6)\otimes O(3)$ framework, photoproduction of isospin 1 ($\rho^{0,\pm 1}$) and isospin 1/2 ($K^{*\pm}$, K^{*0} and $\overline{K}^{*0}$) can be also studied[2,24].

Conerning the search of signals for $s\bar{s}$ component in the nucleon, the forward angle kinematics might be selective if the findings of Refs.[9,10] are taken into account, since effects from the S+U are negligible. Certainly, since a possible strangeness content has not been explicitly included in this model, the effective ϕ-qq coupling cannot distinguish between an OZI evading ϕNN coupling and a strangeness component in the nucleon. In future study, a more complex approach including the possible strangeness component in the nucleon should be explored.

Concerning the efforts of searching for "missing resonances" (in ω photoproduction)[25] and probing strangeness components (in ϕ)[26,27], it shows that the isotopic channels might be useful for disentangling various processes involved in the reaction mechanism. Experiments using the neutron target could provide more evidence to pin down model-independent aspects for any modelling[28].

Acknowledgments

I thank J.-P. Didelez, M. Guidal, B. Saghai, and J.S. Al-Khalili for collaborations on part of the work. Fruitful discussions with E. Hourany concerning the GRAAL experiment, T. Nakano concerning the SPring-8, and P.L. Cole and D.J. Tedeschi concerning the CLAS experiments, are acknowledged.

1. N. Isgur and G. Karl, *Phys. Lett.* B **72**, 109 (1977); *Phys. Rev.* D **23**, 817 (1981); R. Koniuk and N. Isgur, *Phys. Rev.* D **21**, 1868 (1980); S. capstick and W. Roberts, *Phys. Rev.* D **49**, 4570 (1994).
2. Q. Zhao, Z.-P. Li and C. Bennhold, *Phys. Lett.* B **436**, 42 (1998); *Phys. Rev.* C **58**, 2393 (1998).
3. Q. Zhao, *Phys. Rev.* C **63**, 025203 (2001).
4. Y. Oh, A.I. Titov and T.-S.H. Lee, *Phys. Rev.* C **63**, 025201 (2001).

5. Y. Oh, proceeding of this conference.
6. A.I. Titov, proceeding of this conference.
7. R.G. Moorhouse, *Phys. Rev. Lett.* **16**, 772 (1966).
8. E.M. Henley, G. Krein, and A.G. Williams, *Phys. Lett.* B **281**, 178 (1992).
9. A.I. Titov, Y. Oh, and S.N. Yang, *Phys. Rev. Lett.* **79**, 1643 (1997).
10. A.I. Titov, Y. Oh, S.N. Yang, and T. Morii, *Phys. Rev.* C **58**, 2429 (1998).
11. Q. Zhao, J.-P. Didelez, M. Guidal, and B. Saghai, *Nucl. Phys.* A **660**, 323 (1999).
12. Q. Zhao, B. Saghai and J.S. Al-Khalili, *Phys. Lett.* B **509**, 231 (2001).
13. A. Donnachie and P.V. Landshoff, *Phys. Lett.* B **185**, 403 (1987); *Nucl. Phys.* B **311**, 509 (1989).
14. J. Ballam *et al.*, *Phys. Rev.* D **7**, 3150 (1973).
15. Q. Zhao, proceeding of NSTAR2001 (Mainz), p237, World Scientific, 2001.
16. F.J. Klein, Ph.D. thesis, Univ. of Bonn, Bonn-IR-96-008 (1996); πN Newslett. 14, 141 (1998).
17. J. Ajaka *et al.*, Proceeding of the 14th International Spin Physics Symposium, Osaka, Japan, October 16-21, 2000.
18. K. Schilling, P. Seyboth, and G. Wolf, *Nucl. Phys.* B **15**, 397 (1970).
19. H.J. Besch et al., *Nucl. Phys.* B **70**, 257 (1974).
20. CLAS Collaboration, E. Anciant et al., *Phys. Rev. Lett.* **85**, 4682 (2000).
21. Particle Data Group, D.E. Groom *et al.*, *Euro. Phys. J.* C 15, 1 (2000).
22. J.-M. Laget, *Phys. Lett.* B **489**, 313 (2000).
23. A.I. Titov, T.-S.H. Lee, and H. Toki, *Phys. Rev.* C **59**, R2993 (1999).
24. Q. Zhao, J.S. Al-Khalili and C. Bennhold, *Phys. Rev.* C **64**, 052201(R) (2001).
25. P.L. Cole, proceeding of this conference.
26. T. Nakano, proceeding of this conference.
27. D.J. Tedeschi, proceeding of this conference.
28. J.-P. Didelez, proceeding of this conference, and private communication.

THE ROLE OF LOW MASS NUCLEON RESONANCES IN NEAR THRESHOLD ω MESON PHOTOPRODUCTION

ALEXANDER I. TITOV

Bogoliubov Laboratory of Theoretical Physics, JINR, Dubna 141980, Russia
E-mail: atitov@thsun1.jinr.ru

T.-S. HARRY LEE

Physics Division, Argonne National Laboratory, Argonne, Illinois 60439, U.S.A.
E-mail: lee@theory.phy.anl.gov

We investigate the role of the nucleon resonances with masses $M_{N^*} \leq 1720$ MeV in ω meson photoproduction at low energy with $E_\gamma \leq 1.25$ GeV. The amplitude is calculated within effective Lagrangian approach, which includes, at the tree level, leading t-channel pseudoscalar meson exchange, nucleon Born terms and nucleon resonance excitations. For the later part we use an effective Lagrangians motivated by previous studies of π and η photoproduction. The contributions from the nucleon resonances are found to be significant relative to the all other channels in changing the differential cross sections in a wide interval of t In particular, we suggest that a crucial test of our predictions can be made by measuring single and double spin asymmetries.

The study of the vector meson transition matrix elements $N^* \to NV$, were N^*, N, V are the nucleon resonance, nucleon and vector meson ($V = \omega, \rho, \phi$), respectively, has a deep relation to different aspects of intermediate and high energy physics from resolving so-called "missing" resonance problem[1] up to estimation of in-medium modification of the vector meson properties[2,3]. Electromagnetic production of vector mesons is one of the most promising reactions for study of $N^* \to NV$ couplings experimentally, e.g., at Thomas Jefferson National Accelerator Facility, ELSA-SAPHIR of Bonn, GRAAL of Grenoble, and LEPS of SPring-8. The ω meson photoproduction has some advantage because the strong non-resonant background to the total amplitude is better established as compared with the other vector mesons[4].

The role of the nucleon excitations in ω meson photoproduction at relatively large energy was studied by Zhao *et al.*[5,6], using the SU(6) × O(3) constituent quark model with an effective quark-meson interaction. This model was applied to the relatively large energy region $E_\gamma \geq 1.7$ GeV, where the heavy resonances with $M_{N^*} > 1.9$ GeV are important. In the near threshold region the naive quark model without configuration mixing has some problems because it forbids or suppressed photo excitation $\gamma N \to N^*$ and $N^* \to N\omega$ decay for many of resonances with $M_{N^*} \leq 1.72$

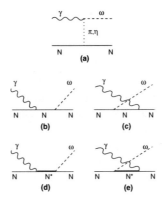

Figure 1. Diagrammatic representation of ω photoproduction mechanisms: (a) (π, η) exchange, (b,c) direct and crossed nucleon terms, (d,e) direct and crossed resonant terms.

GeV [7]. The effect of configuration mixing has been included in recent analysis[8], motivated by the constituent quark model of[9,10]. But since $N^* \to N\omega$ transition is calculated in[10] within $^3P_0$ model for on-shell hadrons, this model can not be applied to calculation of $N^*N\omega$ couplings for the subthreshold resonances. Taking into account these problems, we analyse the low energy photoproduction on the base of the effective Lagrangian approach, where the $N^*N\gamma$ interaction is taken to be similar to that of π, η photoproduction[11,12,13]. The $N^*N\omega$ interaction is derived using the vector dominance model. The found effective Lagrangians and coupling constants might be considered as a starting point for subsequent QCD analysis of the N^*NV transition matrix elements[14]. This study also could get a hint for experimentalists on which spin observables are important for unambiguous study of the resonance excitation. In some sense, our consideration is close to the recent paper[15], but in comparison with it, we use totally covariant and gauge invariant interactions, study different forms of them, and discuss various spin observables

We define the kinematical variables for $\gamma p \to \omega p$ reaction using standard notation. The four-momenta of the incoming photon, outgoing ω, initial nucleon, and final nucleon are denoted as k, q, p, and p' respectively, which defines $t = (p - p')^2 = (q - k)^2$, $s \equiv W^2 = (p + k)^2$, and the ω meson production angle θ by $\cos\theta \equiv \mathbf{k} \cdot \mathbf{q}/|\mathbf{k}||\mathbf{q}|$.

The invariant amplitude can be written as sum

$$I_{fi} = I_{fi}^{PS} + I_{fi}^{N} + I_{fi}^{N^*},$$ (1)

where I_{fi}^{PS}, I_{fi}^{N} and $I_{fi}^{N^*}$ stand for the pseudoscalar (π, η) - meson exchange [Fig.1(a)], direct and crossed nucleon terms [Fig.1(b,c)], and the resonance excitation [Fig.1(d,e)], respectively. The sum of the first two terms represent the non-resonant background. We calculate it, using the model of Ref.[8]. The amplitude $I_{fi}^{N^*}$ will be discussed later.

Together with unpolarized differential cross section, we will discuss the single (i) beam asymmetry Σ_x and (ii) ω meson tensor asymmetry $V_{z'z'}$, and the double beam target spin asymmetry C_{BT}, which are defined as following

$$\Sigma_x = \frac{\text{Tr}\left[I_{fi}\sigma_\gamma^x I_{fi}^\dagger\right]}{\text{Tr}\left[I_{fi} I_{fi}^\dagger\right]} = \frac{d\sigma_\perp - d\sigma_\parallel}{d\sigma_{\text{tot}}},$$ (2)

$$V_{z'z'} = \frac{\text{Tr}\left[I_{fi} I_{fi}^\dagger S_\omega^{zz}\right]}{\text{Tr}\left[I_{fi} I_{fi}^\dagger\right]} = \frac{d\sigma^{s_{z'}^\omega=1} + d\sigma^{s_{z'}^\omega=-1} - 2d\sigma^{s_{z'}^\omega=0}}{\sqrt{2}d\sigma_{\text{tot}}},$$ (3)

$$C_{zz}^{BT} = \frac{d\sigma(\uparrow\downarrow) - d\sigma(\uparrow\uparrow)}{d\sigma(\uparrow\downarrow) + d\sigma(\uparrow\uparrow)},$$ (4)

where subscript $\perp$ ($\parallel$) corresponds to a photon linearly polarized along the $\mathbf{y}$ ($\mathbf{x}$) axis; S_ω^{zz} is the tensor polarization operator for the spin 1 particle and s_z^ω is the ω meson spin projection to the quantization axis $\mathbf{z'}$; the arrows represent the helicities of the incoming photon and the target protons.

The resonant amplitude $I_{fi}^{N_L^*}$ includes contribution of the resonances with mass below of the ω meson production threshold: $M_{N^*} \leq M_N + m_\omega$. In this work we consider contribution of the 8 resonances: $P_{11}(1440)$, $D_{13}(1520)$, $S_{11}(1535)$, $S_{11}(1650)$, $D_{15}(1675)$, $F_{15}(1680)$, $D_{13}(1700)$, $P_{13}(1720)$.

The effective Lagrangian at $\gamma N N^*$ vertex for a spin-$\frac{1}{2}$ particle is chosen by the analogy with the $\gamma N N$ interaction, keeping the only gauge invariant tensor part, since the vector part violates the gauge invariance at $M_{N^*} \neq M_N$. That is the "minimal" covariant interaction, previously used in study of η-photoproduction[13]

$$\mathcal{L}_{\gamma N N^*}^{\frac{1}{2}\pm} = \frac{eg_{\gamma N N^*}}{2(M_{N^*} + M_N)}\bar{\psi}_N \Gamma^{(\pm)}\sigma_{\mu\nu}F^{\mu\nu}\psi_{N^*} + \text{h.c.},$$ (5)

where $F^{\mu\nu} = \partial^\nu A^\mu - \partial^\mu A^\nu$, ψ_N, ψ_{N^*} and A_μ are the nucleon, nucleon resonance and the electromagnetic fields, respectively, $\Gamma^{+(-)} = 1(\gamma_5)$.

For $J = \frac{3}{2}^{\pm}$ we use conventional expression, known from Δ and $N^*(1520)$ excitation in π and η photoproduction [11,13,16,17,18]

$$\mathcal{L}_{\gamma NN^*}^{\frac{3}{2}\pm} = i\frac{eg_{\gamma NN^*}^{(1)}}{M_1^{(\pm)}}\bar{\psi}_{N^*}^{\mu}\gamma^{\lambda}\Gamma^{(\mp)}F_{\lambda\mu}\psi_N$$

$$-\frac{eg_{\gamma NN^*}^{(2)}}{M_2^2}\bar{\psi}_{N^*}^{\mu}\gamma^{\lambda}\Gamma^{(\mp)}F_{\mu\lambda}\partial^{\lambda}\psi_N + \text{h.c.}, \qquad (6)$$

where $M_1^{\pm} \equiv M_{N^*} \pm M_N$, $M_2^2 \equiv (M_{N^*}^2 - M_N^2)/2$, ψ_α is the Rarita-Schwinger spin-$\frac{3}{2}$ baryon field [a].

The effective Lagrangian for the resonances with $J^P = \frac{5}{2}^{\pm}$ is constructed by the analogy with the previous case

$$\mathcal{L}_{\gamma NN^*}^{\frac{5}{2}\pm} = \frac{eg_{\gamma NN^*}^{(1)}}{M_2^2}\bar{\psi}_{N^*}^{\mu\alpha}\gamma^{\lambda}\Gamma^{(\pm)}(\partial_\alpha F_{\lambda\mu})\psi_N$$

$$-i\frac{eg_{\gamma NN^*}^{(2)}}{M_3^{\pm 3}}\bar{\psi}_{N^*}^{\mu\alpha}\gamma^{\lambda}\Gamma^{(\pm)}(\partial_\alpha F_{\mu\lambda})\partial^{\lambda}\psi_N + \text{h.c.}, \qquad (7)$$

where $M_3^{\pm 3} = (M_{N^*}^2 - M_N^2)(M_{N^*} \pm M_N)/2$ and $\psi_{\alpha\beta}$ is the Rarita-Schwinger spin-$\frac{5}{2}$ resonance field. Eq. (7) with $g_{\gamma NN^*}^{(2)} = 0$ for the on-shell baryon is a covariant generalization of non-relativistic expression used in Ref.[3].

The ωNN^* interaction we define using the vector dominance model (VDM) which assumes that the effective $\mathcal{L}_{\omega NN^*}$ Lagrangian has the same form as the corresponding $\mathcal{L}_{\gamma NN^*}$ with substitution

$$A_\mu \to \omega_\mu, \qquad eg \to f_\omega, \qquad f_\omega = 2g_s\gamma_\omega, \qquad f_\omega^{(1,2)} = 2g_s^{(1,2)}\gamma_\omega, \qquad (8)$$

where $\gamma_\omega = 8.53$ and $\gamma_\rho = 2.52$ are corresponding strengths, found from electromagnetic $\omega \to e^+e^-$, $\rho \to e^+e^-$ decays [4].

The scalar strength g_s $(g_s^{(1,2)})$ is related to the proton and neutron strengths as

$$g_s = (g_p + g_n)/2, \qquad g_s^{(1,2)} = (g_p^{(1,2)} + g_n^{(1,2)})/2. \qquad (9)$$

Similarly to the nucleon pole terms, we assume that the γNN^* and ωNN^* vertices in s and u-channels are dressed by form factors $F_B(r^2) = \Lambda_{N^*}^4/(\Lambda_{N^*}^4 + (r^2 - M_{N^*}^2)^2)$, where the choice of cut-off Λ_{N^*} will be discussed later.

---

[a] The dimension constants $M_{1(2)}$ in Eq.(6) are different from that used in Refs [11,13], where $M_1 = M_2 = 2M_N$. Our choice leads to the more simple expression for the helicity amplitudes.

There are many discussions concerning two-component form of the effective Lagrangian (6). Thus, there have been some claims based on the field theoretical arguments, that couplings $g^{(2)}$ should be equal zero[16,19,20]. But on the other hand, the one-component Lagrangian gives strict ratio of $A^{1/2}$ and $A^{3/2}$ helicity amplitudes which often does not correspond to the experimental data. Taking into account this uncertainty, we consider three models or parameter sets for the resonance excitations. The first one (model I) corresponds two-component Lagrangians of (6), (7) with finite $g^{(2)}$. In second case (model II) we use $g^{(2)} = 0$. In both cases we define coupling constants from comparison of calculated and experimentally measured helicity amplitudes, given in[21] with certain error bars, and using their central values. The model III is the model II, but as input we use the values of helicity amplitudes taken within given accuracy to get the best agreement with known experimental data on the ω photoproduction.

Relations between helicity amplitudes and coupling constants for the model I read
$J = \frac{1}{2}^{\pm}$:

$$ eg_a = \frac{2\sqrt{M_{N*} M_N k}}{M_{N*} - M_N} A_a^{\frac{1}{2}}, \tag{10} $$

with

$$ k = \frac{M_{N*}^2 - M_N^2}{2M_{N*}}. \tag{11} $$

$J = \frac{3}{2}^{\pm}$:

$$ eg_a^{(1)} = -\frac{\sqrt{6M_{N*} M_N}}{k} \left(A_a^{\frac{1}{2}} \pm \frac{1}{\sqrt{3}} A_a^{\frac{3}{2}} \right), $$
$$ eg_a^{(2)} = -\frac{\sqrt{6M_{N*} M_N}}{k} \left(A_a^{\frac{1}{2}} - \frac{1}{\sqrt{3}} \frac{M_N}{M_{N*}} A_a^{\frac{3}{2}} \right) \tag{12} $$

$J = \frac{5}{2}^{\pm}$:

$$ eg_a^{(1)} = \frac{\sqrt{40M_{N*} M_N}}{k} \frac{M_{N*}}{M_{N*} \mp M_N} \left(A_a^{\frac{1}{2}} \mp \frac{1}{\sqrt{2}} A_a^{\frac{3}{2}} \right), $$
$$ eg_a^{(2)} = -\frac{\sqrt{40M_{N*} M_N}}{k} \frac{M_{N*}}{M_{N*} \mp M_N} \left(A_a^{\frac{1}{2}} - \frac{1}{\sqrt{2}} \frac{M_{N*}}{M_N} A_a^{\frac{3}{2}} \right), \tag{13} $$

where subscript $a = p, n, s$ denotes proton, neutron, and isoscalar components, respectively.

In the models II, III relations between helicity amplitudes and coupling constants read

$J^P = \frac{3}{2}^{\pm}$

$$eg_a^{(1)} = \mp\sqrt{\frac{24M_{N^*}}{kM_N}}(M_{N^*} \pm M_N)\,A_a^{\frac{1}{2}},\ eg_a^{(1)} = \mp\sqrt{\frac{8M_N}{kM_{N^*}}}(M_{N^*} \pm M_N)\,A_a^{\frac{3}{2}},$$

$J^P = \frac{5}{2}^{\pm}$

$$eg_a^{(1)} = \mp\sqrt{\frac{40M_NM_{N^*}}{kM_N}}\,\frac{M_{N^*}}{M_N}\,A_a^{\frac{1}{2}},\ eg_a^{(1)} = \mp\sqrt{\frac{20M_NM_{N^*}}{kM_N}}\,A_a^{\frac{3}{2}},$$

which results in correlation between helicity-3/2 and 1/2 amplitudes: $A^{3/2}/A^{1/2} = \sqrt{3}M_{N^*}/M_N\ (\sqrt{2}M_{N^*}/M_N)$, for the resonances with $J^P = \frac{3}{2}^{\pm}\ (\frac{5}{2}^{\pm})$. Since this strong correlation is not observed for all resonances, we find $g^{(1)}$ using normalization to the total $N^* \to N\gamma$ decay or to the sum $(A_a^{1/2})^2 + (A_a^{3/2})^2$, taking it's sign to be the sign of the dominant component

$J^P = \frac{3}{2}^{\pm}$

$$eg_a^{(1)} = \mp S_a^{\frac{3}{2}}\sqrt{\frac{24M_{N^*}^3\,M_Nk}{3M_{N^*}^2 + M_N^2}}\left((A_a^{\frac{1}{2}})^2 + (A_a^{\frac{3}{2}})^2\right), \tag{14}$$

$$S_a^{\frac{3}{2}} = \mathrm{sign}(A^{\frac{1}{2}})\theta(h_a) + \mathrm{sign}(A^{\frac{3}{2}})\theta(-h_a),\ h_a = |A_a^{\frac{1}{2}}| - \frac{1}{\sqrt{3}}\frac{M_N}{M_{N^*}}|A_a^{\frac{3}{2}}|;$$

$J^P = \frac{5}{2}^{\pm}$

$$eg_a^{(1)} = \mp S_a^{\frac{5}{2}}\sqrt{\frac{40M_{N^*}^3\,M_N}{k(2M_{N^*}^2 + M_N^2)}}\left((A_a^{\frac{1}{2}})^2 + (A_a^{\frac{3}{2}})^2\right), \tag{15}$$

$$S_a^{\frac{5}{2}} = \mathrm{sign}(A^{\frac{1}{2}})\theta(h_a) + \mathrm{sign}(A^{\frac{3}{2}})\theta(-h_a),\ h_a = |A_a^{\frac{1}{2}}| - \frac{1}{\sqrt{2}}\frac{M_N}{M_{N^*}}|A_a^{\frac{3}{2}}|.$$

The coupling constants for the model I are found by making use of Eqs. (8), (9), (12), (13) and the central values for the helicity amplitudes $A_{p(n)}^{\frac{1}{2}}$, $A_{p(n)}^{\frac{3}{2}}$, listed by the Particle Data Group [21]. For the model II they are calculated with the same input data, but using Eqs. (14), (15). For the model III, the coupling constants are defined using the same expressions as for the model II, but as input we use the values of helicity amplitude within given accuracy to get the best fit to existing data on unpolarized cross section at $E_\gamma = 1.23$ GeV [22] and the beam asymmetry at $E_\gamma = 1.175$ GeV [23]. The corresponding values are listed in Table II of Ref.[24].

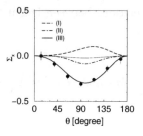

Figure 2. Beam asymmetry as a function of $\omega-$meson production angle at $E_\gamma = 1.175$ GeV for the three models. The thin solid line is the beam asymmetry for the non-resonant background, taken separately. Data are taken from Ref.[23]

The value of cut-off $\Lambda_{N^*} \simeq 0.75$ GeV is fixed by the fitting the unpolarized cross section at $E_\gamma = 1.175$ GeV. The quality of the fit is different for the different models, but the qualitative behaviour of the cross sections are similar.

The striking difference appears in spin variables. In Fig. 2 we show our result for beam asymmetry as a function of ω meson production angle at $E_\gamma = 1.175$ GeV in comparison with the experimental data of Ref.[23]. The non-resonant amplitude, taking alone, leads to zero asymmetry. The full amplitude with resonant part of model I leads to the positive asymmetry. The models II and III result in negative asymmetry but with different absolute value. In all cases the dominant contribution comes from $F_{15}(1680)$ and $D_{13}(1520)$ excitations. We found, that keeping g^2 to be finite (model I) and varying the input helicity amplitudes within the measured accuracy, one can not change the sign of Σ_x (from positive to negative). Fixing incoming parameters, we perform calculation of differential cross sections at other energies and make predictions for spin observables.

Fig. 3 shows our prediction for the differential cross sections for $\gamma p \to p\omega$ reaction as a function of t at $E_\gamma = 1.125$, 1.175, and 1.23 GeV. For convenience, we show contribution from the different channels taken separately. The results are from pseudoscalar-meson exchange (thin solid), direct and crossed nucleon terms (short dashed), N^* excitation (dashed), and the full amplitude (thick solid). One can see, that the total cross section is described by the flat curves, which decrease with $|t|$ much slowly then the pure pseudoscalar exchange, dominant at small $|t|$.

The beam asymmetry Σ_x at three beam energies at c.m.s. is shown in Fig. 4. Interesting, that at $E_\gamma = 1.125$ GeV the absolute value of Σ_x is

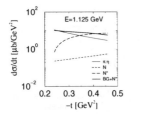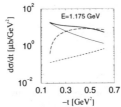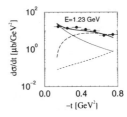

Figure 3. Differential cross sections for $\gamma p \to p\omega$ reaction as a function of t at $E_\gamma = 1.125$, 1.175, and 1.25 GeV. The results are from pseudoscalar-meson exchange (thin solid), direct and crossed nucleon terms (short dashed), N^* excitation (dashed), and the full amplitude (thick solid). Data are taken from Ref.[22].

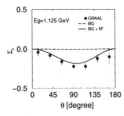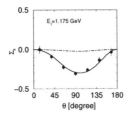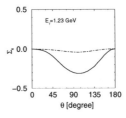

Figure 4. Beam asymmetry at at $E_\gamma = 1.125$, 1.175, and 1.23 GeV as a function of $\omega-$ production angle. The results are from the non-resonant background (BG) (dot-dashed), and the full amplitude (solid). Data are taken from Ref.[23].

smaller than at $E_\gamma = 1.175$, whereas the relative contribution of the resonance amplitude at lower energy is greater.

The Fig. 5 shows our prediction for tensor $V_{z'z'}$ asymmetry calculated in Gottfied-Jackson system. The non-resonant part is close to the value $1/\sqrt{2}$, predicted for pure pseudoscalar exchange[24]. The resonance excitations lead to the strong contribution of longitudinally polarized outgoing ω mesons and decrease of $V_{z'z'}$ up to the negative values. This effect is strengthened with energy.

The double beam-target asymmetry is shown in Fig. 6. For the only non-resonant amplitude one gets positive slightly growing with θ asymmetry. The resonant part changes this behaviour on the qualitative level: the full asymmetry decrease monotonically with θ and reach large negative values at backward production.

650

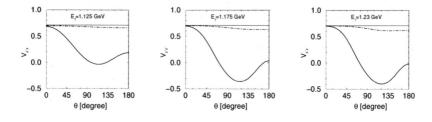

Figure 5. The $\omega-$meson tensor asymmetry at $E_\gamma = 1.125$, 1.175, and 1.23 GeV as a function of $\omega-$ production angle. The thin solid line is contribution from the pseudoscalar exchange amplitude, taken separately. The other notation is the same as in Fig. 4.

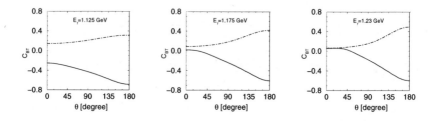

Figure 6. Beam-target asymmetry at $E_\gamma = 1.125$, 1.175, and 1.23 GeV as a function of $\omega-$ production angle. Notation is the same as in Fig. 4.

In summary, we have analysed the manifestation of the low mass resonances in ω photoproduction in the near threshold energy region on the base of the effective Lagrangian approach and the vector dominance model. We found that at $E_\gamma \leq 1.25$ GeV, the dominant contribution comes from $F_{15}(1680)$ and $D_{13}(1520)$ excitations. The beam asymmetry is extremely sensitive to the form of the effective Lagrangian and allows to fix it. The resonance excitation gives strong modification of the single and double spin asymmetries which may be used as a powerful tool for study the resonance dynamics in ω photoproduction. When the energy increases, one has to include the higher mass resonance excitation, similarly to that of Refs.[5,8], which might change considerably our near-threshold predictions because of strong mutual interference of different resonances. The effects due to the initial and final state interactions must be also investigated.

We greatfully acknowledge fruitfull discussions with Yongseok Oh. A.I.T.

thanks for the worm hospitality of the nuclear theory group in Argonne National Laboratory. This work was supported in part by Russian Foundation for Basic Research under Grant No.00-15-96737 and U.S. DOE Nuclear Physics Division Contract No. W-31-109-ENG-38.

References

1. N. Isgur and G. Karl, *Phys. Lett.*B **72**, 109 (1977); *Phys. Rev.*D **23**, 817(E) (1981); R. Koniuk and N. Isgur, *Phys. Rev.*D **21**, 1868 (1980).
2. G.E. Brown and M. Rho, *Phys. Rev. Lett.* **66**, 2710 (1991).
3. B. Friman and H.J. Pirner, *Nucl. Phys.*A **617**, 496 (1997).
4. B. Friman and M. Soyeur, *Nucl. Phys.*A **600**, 477 (1996).
5. Q. Zhao, Z. Li, and C. Bennhold, PLB **436**, 42 (1998);*Phys. Rev.*C **58**, 2393 (1998).
6. Q. Zhao, J.-P. Didelez, M. Guidal, and B. Saghai, *Nucl. Phys.*A **660**, 323 (1999).
7. F.E. Close, *An Introduction to Quark and Partons*, Academic press, London, New York, San Francisco, 1979.
8. Yongseok Oh, A.I. Titov, and T-S.H. Lee, *Phys. Rev.*C **63**, 025201 (2001).
9. S. Capstick, *Phys. Rev.*D **46**, 2864 (1992).
10. S. Capstick and W. Roberts, *Phys. Rev.*D **49**, 4570 (1994).
11. R.M. Davidson, N. C. Mukhopaghyay, and R. S.Wittman, *Phys. Rev.*D **43**, 71 (1991).
12. T. Sato and T.-S. H. Lee, *Phys. Rev.*C **54**, 2660 (1996).
13. M. Benmerrouche, N. C. Mukhopaghyay, and J. F.Zhang, *Phys. Rev.*D **43**, 3237 (1995).
14. D.O. Riska and G.E. Brown, *Nucl. Phys.*A **679**, 577 (2001).
15. M. Post and U. Mosel, *Nucl. Phys.*A **688**, 808 (2001).
16. M.G. Olsson and E.T. Osipovsky, *Nucl. Phys.*B **87**, 399 (1975).
17. H.F. Jones and M.D. Scadron, *Ann. Phys. (N.Y.)* **81**, 1 (1973).
18. S. Nozawa, B. Blankleider, and T.-S. H. Lee, *Nucl. Phys.*A **513**, 459 (1990).
19. L.M. Nath and B.K. Bhattacharyya, ZPC **5**, 9 (1980).
20. I. Blomqvist and G.M. Laget, *Nucl. Phys.*A **280**, 405 (1977)
21. C. Caso, et al., Particle Data Group, *Eur. Phys. J. C* **3**, 1 (1998).
22. F. J. Klein, Ph.D. thesis, Bonn Univ. (1996); SAPHIR Collaboration, F. J. Klein *et al.*, πN Newslett. **14**, 141 (1998).
23. J. Ajaka et al., Proc. 14 Intern. Spin Physics Symposium. AIP Conference Proceedings. V 570, p. 198. Melville, NY, 2001.
24. A. I. Titov and T.-S. H. Lee, to be published.

PHOTOPRODUCTION EXPERIMENTS WITH POLARIZED
HD TARGETS

S. BOUCHIGNY, C. COMMEAUX, J-P. DIDELEZ, G. ROUILLE, M. GUIDAL,
E. HOURANY, R. KUNNE

IN2P3, Institut de Physique Nucléaire, Orsay, France
E-mail: didelez@ipno.in2p3.fr

M. BASSAN, A. D'ANGELO, R. DI SALVO, A. FANTINI, D. MORICCIANI,
C. SCHAERF

INFN sezione di Roma II and Università di Roma "Tor Vergata", Italy

J-P. BOCQUET, A. LLERES, D. REBREYEND

IN2P3, Institut des Sciences Nucléaire, Grenoble, France

P. LEVI SANDRI

INFN, Laboratori Nazionali di Frascati, Italy

F. GHIO, B. GIROLAMI

INFN sezione di Roma I and Istituto Superiore di Sanità, Roma, Italy

V. BELLINI

INFN Laboratori Nazionali del Sud and Università di Catania, Italy

M. CASTOLDI, A. ZUCCHIATTI

INFN sezione di Genova and Università di Genova, Italy

G. GERVINO

INFN sezione di Torino and Università di Torino, Italy

O. BARTALINI

Università di Trento and INFN, Laboratori Nazionali di Frascati, Italy

Within the French-Italian GRAAL collaboration, we have constructed a *Polarized HD Target Factory (HYDILE)*, implanted at IPN Orsay. This project is very similar to the SPHICE one existing at BNL, and developed within the LEGS Spin Collaboration. The expected properties of the HD targets and the results already obtained for the polarization of HD samples at Orsay and BNL are given. New possibilities to measure polarization observables on the proton, the neutron and the deuteron are now opened. Most simple and double polarization measurements for the photoproduction of the low mass pseudoscalar and vector mesons could be performed, by using the high polarization (linear or circular) of the backscattered photon beams and the large effective polarization (longitudinal or tranverse) of the target, resulting essentially, from the good dilution factor. Polarized total cross sections concerning the isovector Drell-Hearn Gerasimov Sum Rule related to the proton neutron difference, could be measured with minimum systematic uncertainties: the target containing both polarized protons and neutrons. By using the available polarized photon beams from LEGS, GRAAL and Spring-8, measurements could be done from the π meson threshold up to 2.5 Gev, where the ϕ meson photoproduction cross section is close to its maximum value. Typical examples are given for GRAAL and Spring-8.

1 Introduction

Polarized HD targets are made of nearly pure HD molecules and have therefore an outstanding "Dilution Factor" (DF), since all nuclear species in the target are polarizable, except for the Al wires necessary for the cooling. Those wires represent at most 20% in weight of the HD material in the target. The H polarization could reach 90% and the D vector polarization exceed 50%. The H and D can be polarized independently and their relative orientation can be either parallel or antiparallel [1].

Figure 1 shows a schematic view of the existing SPHICE (Strongly Polarized Hydrogen ICE) HD target developed by the LEGS Spin Collaboration (LSC) for experiments at BNL [2]. The HYDILE target (HYdrogen Deuterium for Intersecting Laser Electron beams) is very similar to the SPHICE target [3], with however, the foreseen additional possibility of transverse polarization. It is presently under construction within a french-italian (IN2P3-INFN) collaboration, to perform photoproduction experiments using the fully polarized high energy backscattered photon beams of the GRAAL set-up at the European Synchrotron Facility (ESRF) in Grenoble (France).

The present paper gives the status of the HYDILE project and describes some experiments which could be performed at GRAAL and Spring-8, taking advantage of the available energies and of the various configurations for the beam and target polarization states.

SPHIce Target Scheme

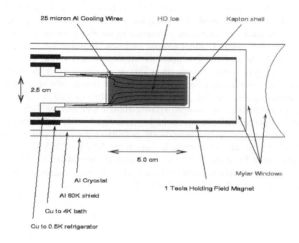

25 micron Al Cooling Wires   HD Ice   Kapton shell

2.5 cm

5.0 cm

Al Cryostat

Al 80K shield

Cu to 4K bath

Cu to 0.5K refrigerator

Mylar Windows

1 Tesla Holding Field Magnet

Figure 1. Schematic view of the SPHICE HD polarized target already produced in US by the LSC collaboration, inside the In-Beam Cryostat (IBC) constructed at IPN-Orsay. Note that the only impurities in the HD material, are thin Al wires, necessary to insure the cooling. They represent at most 20% in weight of the HD content. For 5 cm of HD, the target thickness is 720 mg/cm^2, while the total thickness of mylar and kapton windows is of the order of 50 mg/cm^2. The IBC provides the cooling of the HD sample at a temperature of 1.5 K and the holding field, with a 1 mm thick superconducting coil (NbTi) attached to the 4K ^4He schielding cryostat. The maximum field is 1 Tesla, allowing relaxation times of 5 days for H and D.

2 Polarization-Production Apparatus

Equipment for producing polarized HD targets are now operating at the "Institut de Physique Nuclaire" in Orsay (France). The main piece is a powerful ^3He-^4He Dilution Refrigerator (DR) ordered from Leiden Cryogenics BV, with a heat lift capacity of 12 μW at 10 mK. A circulation rate of 5 mmole/s of ^3He can be sustained, resulting in a cooling power of 3.8 mW at 110 mK. The base temperature measured on the mixing chamber is below 4 mK. The refrigerator is equipped with a 13.5 T superconducting NbSn coil having a bore diameter of 72 mm, a bore length of 53 cm and a field homogeneity of 4.3×10^{-4}.

Solid HD samples are made with a dedicated cryostat, called "Ice Maker" (IM). This is a ^4He cryostat equipped with a 100K-1.5K variable temperature insert and a 2.5 T superconducting coil. This cryostat is also used for the storage and transport of HD polarized targets. Last, to handle the targets between the DR and the IM, a 4.2K telescopic Transfer Cryostat (TC) has been constructed at Orsay. Basically a left-hand right-hand thread mechanism makes it possible to engage, screw or unscrew a target in or from its cold finger, while an 0.5 T superconducting coil provides the required target polarization holding field during the cold transfer.

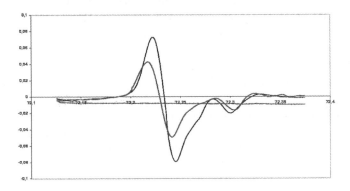

Figure 2. Proton NMR signals, before, during and after the transfer.

3 Polarization Results

The first attempt to polarize a 1 mole HD target has been done at Orsay in October 1999 with the main goal of putting the DR together with its NMR measurement probe into a first full operation cycle. The results of this first test are related in a previous paper [4]. A second polarization run was performed in July 2000, in order to establish the polarization rates and the relaxation times attainable with the present system together with the accuracy of absolute polarization measurements. The corresponding results are given in Ref. [5]. More recently, an other polarization was attempted, in order to demonstrate the ability of the system to transfer polarized samples from the DR to the IM. The HD targets were made from commercially available HD gas with high initial H_2 and D_2 concentrations of respectively 1.1% and 0.5%. Small concentrations of ortho $(o-H_2)$ and para $(p-D_2)$ are necessary for the polarization process, however the above values are at least 10 times too big, compared to ideal concentrations [1]. With such contaminations, relaxation times are short. Nevertheless, we succeeded the transfer of a polarized sample with a loss of polarization of 35%. Figure 2 shows the proton NMR signals, before, during and after the transfer. The polarization rates of protons and deuterons were respectively 60% and 14% before the transfer. With double distilled HD, as available in US from recently working distillation [6], relaxation times as long as 5 days have been achieved and the polarization loss during the transfer, is negligible. The next polarization run to be performed at Orsay will use double distilled HD, in order to reproduce the BNL results in France, before transporting the targets to the GRAAL site at Grenoble, 600 km away from Orsay.

4 Transverse Polarization

In the configuration depicted in Figure 1, only longitudinal polarization of the target is possible, the solenoidal coil providing the holding field, keeps the original polarization axis, either parallel or antiparallel to the incoming photon direction. If saddle coils, as shown schematically in Figure 3, are claming the solenoidal one, it becomes possible, using appropriate variations of the current intensities in the coils, to rotate adiabatically the spins and put them in a transverse direction. It as been shown that the saddle coils alone can produce in the target volume, a transverse field with an homogeneity better than 2.5% and a polar angle with the transverse axis less than 2.5^o, in spite of the constraints imposed by the geometry of the system. This is particularly true if only the limited interaction region between the collimated backscattered beam and the target is considered [7]. Transverse polarization of the target is of major interest, in particular with a linearly polarized beam. Combining all the possible polarization states allows to measure, in the case of pseudoscalar mesons photoproduction, the three single polarization observables: Σ (Beam), $\mathcal{T}$ (Target) and $\mathcal{P}$ (Recoil). Equations (1) to (4), show that the corresponding observables are related to different bilinear combinations of the 4 helicity amplitudes, while the differential cross section is related to their squares.

$$\frac{d\sigma}{d\Omega} = |H_1|^2 + |H_2|^2 + |H_3|^2 + |H_4|^2 \tag{1}$$

$$\Sigma = \frac{1}{P_\gamma} \frac{(N_\perp^\uparrow + N_\perp^\downarrow) - (N_{||}^\uparrow + N_{||}^\downarrow)}{4N_0} = -\Re(H_1 H_4^* - H_2 H_3^*) \tag{2}$$

$$\mathcal{T} = \frac{1}{P_T} \frac{(N_\perp^\uparrow + N_{||}^\uparrow) - (N_\perp^\downarrow + N_{||}^\downarrow)}{4N_0} = -\Im(H_1 H_2^* + H_3 H_4^*) \tag{3}$$

$$\mathcal{P} = \frac{1}{P_\gamma P_T} \frac{(N_\perp^\uparrow + N_{||}^\downarrow) - (N_\perp^\downarrow + N_{||}^\uparrow)}{4N_0} = -\Im(H_1 H_3^* + H_2 H_4^*) \tag{4}$$

H_1 to H_4, are the four helicity amplitudes. The symbols $N_\perp$, $N_{||}$, $N^\uparrow$, $N^\downarrow$, represent the counting rates in the reaction plane, for perpendicular or parallel orientations of the beam polarization and "up" or "down" target polarization respectively. N_0 stands for the unpolarized counting rate. As can be seen, a difficult but important measurement as the recoil polarization, can be accessed by using all "Beam-Target" relative polarization configurations.

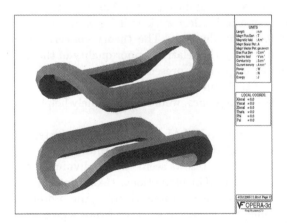

Figure 3. Schematic view of the saddle coils, surrounding the solenoidal coil containing the HD target (see Fig. 1).

5 Physics Prospects

5.1 GRAAL

To complement the beam asymmetries Σ for the η meson photoproduction on proton already measured at GRAAL [8], a typical program could be the continuation of the η meson physics on polarized proton, deuteron and neutron. By looking, near threshold, how well polarization data can be accounted for by the most advanced approach: quark model with an effective Lagrangian [9]. A better determination of the $S_{11}(1535)$ resonance electromagnetic couplings (isoscalar and isovector) and width, would certainly result from such an investigation [10,11]. At higher energies, the sensitivity of polarization observables, in particular target polarization data, to other N^* resonances like the $F_{17}(1990)$ and $G_{17}(2190)$ resonances could be used to determine their properties and unknown ηN branching ratios [12]. On the other hand, the search for missing N^* resonances could be done in all final state channels other than πN, in particular through the ω vector meson photoproduction. In the energy region near the ω meson threshold, this process is dominated by the t-channel *unnatural parity* (π^0) exchange. However, the effect from intermediate excited nucleonic resonances can be seen at large angles. Model study of target polarization observables and density matrix elements [13,14] show a selective sensitivity to the barely established $P_{13}(1900)$ and $F_{15}(2000)$ two stars resonances [15], predicted by Capstick [16] to have a significant ωN branching ratio.

Whereas resonance models work well for η-production, the situation is more difficult for K production. The theory concerning these reactions is essentially hampered by the relatively poor quality of the data. The inclusion in the data bases of precise, presently unexisting, "Σ" beam asymmetries for the $K\Lambda$ and $K\Sigma$ channels, and corresponding target asymmetry data to complete the few existing ones, would be extremely helpful to further improve these models.

The major part of the GRAAL program will be the test of the Drell-Hearn Gerasimov sum rule (DHG) for both proton and neutron [17,18]. If estimations of the proton and neutron results seem to be individually consistent with the respective DHGp and DHGn predictions, taking into account the large uncertainties of the partial wave parametrization, the proton-neutron difference comes out four times too large and with the wrong sign compared to DHG_{p-n} predictions [19]. A first recent experimental investigation of DHGp up to 800 MeV photon energy [20], where saturation of the sum is already significantly achieved, shows that the DHG sum rule is roughly verified for the proton. In which case, the large deviation from DHG_{p-n}, must be attributed to the neutron [21].

It is the aim of the HYDILE polarized HD target, containing both polarized protons and neutrons with the unique GRAAL polarized photon beams, to precisely measure the isovector part of the DHG sum rule, due to the small proton-neutron difference. The identification of individual final state channels by the GRAAL set-up, and the direct comparison of quasi-free reactions on bound, equally polarized proton and neutron from the D nucleus and free proton from the H nucleus are required measurements to enlight the DHG puzzle. Such a technique relating observables on free and bound nucleons has allowed to measure η photoproduction cross sections on the neutron [22].

5.2 Spring-8

a)

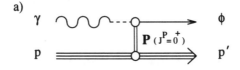

b)

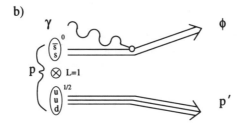

Figure 4. Feynmann graphs of the two principle mechanisms for ϕ photoproduction. a) Diffractive scattering, where the incoming γ mixes into a virtual ϕ that interacts with the nucleon via Pomeron exchange. The Pomeron is a $J^P = 0^+$ object so the ϕ has the γ polarization and the recoil proton has that of the target. b) Knockout of a $^1S_0$ $s\bar{s}$ pair to form the $^3S_1$ $s\bar{s}$ ϕ. The pair is initially coupled to a $^2S_{1/2}$ uud core. The intrinsic parity of the proton requires the coupling to be $L = 1$. The ϕ has the γ polarization and the recoil proton has that of the core.

Investigation of DHG_{p-n}, should of course be pursued at Spring-8, but at the photon energies achievable there, the ϕ meson photoproduction is close

660

to its maximum value. The ϕ is essentially a $s\bar{s}$ pair; therefore, the possibility to "knock-out" a $s\bar{s}$ pair from the target proton to produce a ϕ in the final state and probe the strange content of the nucleon, is extremely attractive. Figure 4 shows the Feynmann graphs of the two principle mechanisms for ϕ photoproduction.

The measurement of the Longitudinal Beam Target Asymmetry (LBTA) observable at forward angles, where the cross section peaks, can be selectively related to the "knock-out" mechanism of a $s\bar{s}$ pair forming the proton wave function [23]. The amplification factor is huge, close to 100, with an 0.25% presence probability of the $s\bar{s}$ pair producing asymmetries as high as 25%. It has been shown, using the quark model approach with an effective Lagrangian for the non-diffractive contribution and a Pomeron exchange for the diffractive contribution, that the contribution of baryonic resonances produces no LBTA at forward angles [24], reinforcing the motivation for this experiment.

6 SUMMARY

The the GRAAL experimental program with the HD polarized target HYDILE will be devoted to the measurement of most beam-target double polarization observables on the neutron for π^0, η and ω mesons. To this aim, reference measurements on the proton must also be done. The DHG sum rules, in particular the DHG(p-n) which requires small systematic errors and the "free neutron" polarized total cross sections remain a priority for GRAAL. Strangeness production on the proton, through the $K^+\Lambda$ and $K^+\Sigma^0$ channels using linearly polarized photons will be pushed as far as possible for both single and double polarization observables. Finally, at Spreing-8, with an energy increase above 2 GeV, the η' physics could be done and the LBTA, measured at forward angles for the ϕ photoproduction, could provide a direct measurement of the strange content of the nucleon.

References

1. A. Honig, Phys. Rev. Lett. **19**, 1009 (1967).
2. A. Honig *et al.*, Nucl. Instr. and Meth. In Phys. Research **356**, 39 (1995).
3. J.P. Didelez, Nucl. Phys. News **Vol 4**, N^o3, 10 (1994).
4. G. Rouille *et al.*, Nucl. Instr. and Meth. In Phys. Research **A 464**, 428 (2001).

5. C. Commeaux *et al.*, Proceedings of the 9th Seminar: Electromagnetic Interactions of Nuclei At Low and Medium Energies, Moscow, September 2000, p 281.
6. S. Whisnant, Contribution to this conference.
7. P. Gara (IPN, Orsay), private communication.
8. J. Ajaka *et al.*, Phys. Rev. Lett. **81**, 1797 (1998).
9. Z. Li and B. Saghai Nucl. Phys. **A644**, 345 (1998).
10. B. Saghai and Z. Li, *Proceedings of the NSTAR2000 Conference "Excited Nucleons and Hadronic Structure", Newport News, USA, February 2000, p 179.* World Scientific, Ed. V. D. Burkert, L. Elouadrhiri, J. J. Kelly and R. C. Minehart.
11. B. Saghai, Contribution to this conference.
12. L. Tiator *et al.*, Phys. Rev. **C 60**, 035210 (1999).
13. Q. Zhao, Phys. Rev. **C 63**, 025203 (2001).
14. Y. Oh, A. I. Titov and T.-S. H. Lee, Phys. Rev. **C 63**, 025201 (2001).
15. *Review of Particle Physics*, The European Physical Journal **C, Vol 3, N^0 1-4**, 613-645 (1998).
16. S. Capstick, Few-Body Systems, **Suppl. 11**, 86 (1999), and references therein.
17. S.D. Drell and A.C. Hearn, Phys. Rev. Lett. **16**, 908 (1966).
18. S.B. Gerasimov, Sov. J. Nucl. Phys. **2**, 430 (1966).
19. A.M. Sandorfi, C.S. Whisnant and M. Khandaker, Phys. Rev. **D 50**, R6681 (1994).
20. J. Ahrens *et al.*, Phys. Rev. Lett. **84**, 5950 (2000).
21. L. Tiator, D. Drechsel and S.S. Kamalov, *Proceedings of the NSTAR2000 Conference "Excited Nucleons and Hadronic Structure", Newport News, USA, February 2000, p 343.* World Scientific, Ed. V. D. Burkert, L. Elouadrhiri, J. J. Kelly and R. C. Minehart.
22. P. Hoffmann-Rothe *et al.*, Phys. Rev. Lett. **78**, 4697 (1997).
23. A.I. Titov, Y. Oh and S.N. Yang, Phys. Rev. Lett. **79**, 1634 (1997).
24. Q. Zhao *et al.*, Nucl. Phys. **A660**, 323 (1999).

COMPTON SCATTERING AND π^+ PHOTO-PRODUCTION AT GRAAL

O. BARTALINI[1,2,3], V. BELLINI[4,5], J.P. BOCQUET[6], M. CAPOGNI[1,7], M. CASTOLDI[8], A. D'ANGELO[1,7], ANNELISA D'ANGELO[1,3], J.P. DIDELEZ[9], R. DI SALVO[1,7], A. FANTINI[1,7], G. GERVINO[10], F. GHIO[11,12], B. GIROLAMI[11,12], M. GUIDAL[9], E. HOURANY[9], I. KILVINGTON[13], R.KUNNE[9], V. KUZNETSOV[1,7,14], A. LAPIK[14], P. LEVI SANDRI[3], A. LLERES[6], D. MORICCIANI[7], V. NEDOREZOV[14], L. NICOLETTI[6,4,5], D. REBREYEND[6], N.V. RUDNEV[15], C. SCHAERF[1,7], M.L. SPERDUTO[4,5], M.C. SUTERA[16,4], A. TURINGE[17], A. ZUCCHIATTI[8]

INVITED PAPER PRESENTED BY D. MORICCIANI[7]

(1) Università di Roma "Tor Vergata", I-00133 Roma, Italy
(2) Università di Trento, I-38100 Trento, Italy
(3) INFN, Laboratori Nazionali di Frascati, I-00044 Frascati, Italy
(4) Università di Catania, I-95123 Catania, Italy
(5) INFN, Laboratori Nazionali del Sud, I-95123 Catania, Italy
(6) Institut des Sciences Nucléaires de Grenoble, 38026 Grenoble, France
(7) INFN, Sezione Roma2, I-00133 Roma, Italy
(8) INFN, Sezione di Genova, I-16146 Genova, Italy
(9) Institut de Physique Nucléaire, 91406 Orsay Cedex, France
(10) INFN, Sezione di Torino and Università di Torino, I-10125 Torino, Italy
(11) Istituto Superiore di Sanità, I-00161 Roma, Italy
(12) INFN, Sezione Roma1, 00185 Roma, Italy
(13) ESRF, Polygone Scientifique, F-38043 Grenoble, France
(14) Institute for Nuclear Research, 117312 Moscow, Russia
(15) Institute of Theoretical and Experimental Physics, 117259 Moscow, Russia
(16) INFN, Sezione di Catania, I-95123 Catania, Italy
(17) Kurchatov Institute of Atomic Energy, 123182 Moscow, Russia

A polarized and tagged gamma ray beam is produced at GRAAL by the Compton scattering of laser light on the high energy electron circulating in the ESRF storage ring. We present results on the beam polarization asymmetries in the Compton Scattering and positive pion photo-production on proton target in the energy region 500-1500 MeV. These very precise results cover the angular range 30-150 degrees, providing stringent constraints to theoretical models.

1   Introduction

Probing of the nucleon with polarized tagged photons provides important information on the spectrum of baryon resonances. Measurements of polarization observables in photon-induced reactions, in particular in meson photo-production, are essential for comparison with recent theoretical models [1,2,3,4]. Asymmetries are much more sensitive than cross sections to the contribution of the less dominating resonances. In facts, while the cross sections are proportional to the sum of the square of the helicity amplitudes, the asymmetries contain the interference between them, thus amplifying the effect of the less contributing resonances. For this reason there is a strong demand for very precise measurements particularly for polarization asymmetry data.

2 GRAAL $\vec{\gamma}-$ beam

The backward Compton scattering of laser light on the high energy electrons circulating in storage rings provides beams of gamma-rays useful for the study of photo-nuclear reactions [5]. After the successful operation of the Ladon beam in Frascati [6], several back-scattered beams have become available for nuclear and particle physics research.

The main characteristics of these beams are the low background of photons of lower energies and the high degree of polarization. Instead of the ($\approx$ 1/k) spectrum of bremsstrahlung beams, Compton-backscattered beams have a quasi-flat spectrum with the maximum intensity at the highest energy.

In the scattering of very relativistic electrons, helicity is a good quantum number and therefore there is little transfer of angular momentum from the electron to the photon: at the highest photon energy, the polarization of the gamma-ray is very close to that of the laser light and can be modified by changing the polarization of the light with optical tools. In this way it is easy to rotate the linear polarization of the beam, to change from linear to circular polarization and to flip the photon spin.

In the case of GRAAL beam two independent sets of data were collected using the green laser line (2.41 eV) and a UV laser line (3.53 eV), allowing to reach a maximum Eγ energy of 1100 MeV and 1500 MeV respectively, allowing the measure a given photo-reaction at same energy but with different polarization degree in order to have a cross check of the measure of this observable.

The energy of the outgoing $\vec{\gamma}$ is measured using a internal tagging system. The tagging detector consists of 128 silicon micro-strips (with a 300 μm pitch) in coincidence with a set of 10 scintillator bars. The spatial resolution of the strips corresponds to an energy resolution of the order of 1% (FWHM) for the scattered photon. The signals from the plastic scintillators are in coincidence with the radio frequency of the ring and are used to start the trigger of the whole experimental acquisition.

3 The GRAAL experimental set-up

The GRAAL apparatus is a large solid angle (almost 4π) detector, as shown in figure 1, for the detection of neutral and charged particles [7].

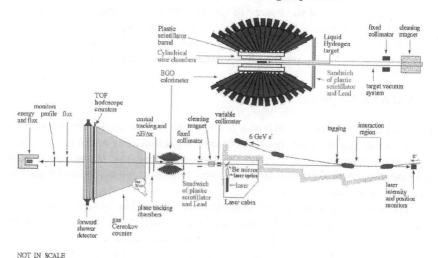

Figure 1. The GRAAL experimental apparatus.

The apparatus has azimuthal φ symmetry and it is divided in three region respect to the ϑ angle:

1) In the forward direction i.e. for polar angles smaller than 25°, the apparatus consists of:

   a) two plane chambers for track reconstruction of charged particles, providing a spatial resolution on track reconstruction of the order of 1 mm;

   b) a double layer scintillator wall of 3×3 m^2 size, providing ΔE and TOF measurements for charged particles with a time resolution of of the order to 600 ps (FWHM);

   c) a shower detector consisting of 16 vertical modules made of layers of lead and plastic scintillators, with a 3 cm thick iron sheet in front of it, providing ΔE and TOF measurement for charged and neutral particles, with a time resolution of of the order to 600 ps (FWHM), with a good efficiency also for photons [8].

2) In the central region i.e. for polar angles between 25° and 155°, the apparatus consists of:

    a) two cylindrical MWPC's for vertex reconstruction of charged particles, providing a spatial resolution on track reconstruction of the order of 1 mm;

    b) a cylindrical barrel of 32 plastic scintillators, for charged-neutral particles discrimination and identification;

    c) an electromagnetic calorimeter, made of 480 BGO crystals, each of 21 r.l. thickness, providing a high energy resolution for γ, $\simeq$ 3% (FWHM) at 1 GeV[9], and a good response to protons with energies up to 200 MeV;

3) In the backward region i.e. for polar angles greater then 155°, two disks of plastic scintillators are separated with a 1 cm of lead between them.

The feasibility of a gas Čerenkov detector was studied [11], the construction was started and it will be installed between the plane chambers and the forward walls for electrons-pion discrimination.

In the forward direction, a beam monitor, composed of three scintillators separated by a gamma converter, constantly measures the incident flux. The efficiency in the detection of photons is measured by a spaghetti calorimeter working at low beam intensities [10]. The measured flux over the entire energy spectrum is about 2×10^6 γ/s with a laser power of 2.8 W.

4   Physics results

We will present in the following some results of beam asymmetries Σ for the reactions: $\vec{\gamma}p \rightarrow \gamma'p$ (Compton scattering) and $\vec{\gamma}p \rightarrow \pi^+ n$ with some theoretical interpretation. They cover the $\vec{\gamma}$ energy range 500 MeV - 1500 MeV and a very wide angular interval, between 30° and 150°. The GRAAL data often extend to unexplored regions, both in energy and in angles, while for the regions where previous results already existed, they considerably reduce the existing errors bars.

4.1   Compton Scattering

We report some preliminary results on measurements of the beam polarization asymmetries in the Compton scattering on the proton in the angular range where both proton and photon are detected inside the BGO (central region). The main background is represented by the π^0 photo-production where one of the two photons from the π^0 decay was lost. The π^0 decay could be symmetric or asymmetric in energy. In the first case the two photons could produce two overlapping clusters in the BGO; in order to optimize the cluster reconstruction different statistical methods have been developed [12]. In the second case, where a low energy photon of π^0 decay can be lost, a comparison between data and Monte Carlo simulation can increase the reconstruction efficiency.

A kinematical fit allows the identification of Compton reaction and the π^0 remaining background. Preliminary results have been extracted for the beam asymmetry Σ in Compton scattering, that are shown in figure 2. GRAAL data (solid square) for $E_\gamma = 710\ MeV$ are in good agreement with Yerevan data [13] (solid triangles) for $E_\gamma = 750\ MeV$. On the other side the comparison of data with the theoretical model from [14] is not yet satisfying. The observed disagreement requires refinements of the model.

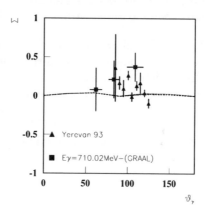

Figure 2. Results for beam asymmetry in $\vec{\gamma}p \to \gamma'p$. See the text for details.

4.2 π^+n reaction

The figure 3 show the GRAAL beam asymmetries in that have already been published [15]. They have been obtained using the green line of the laser. Black triangles and circles are GRAAL results when both the neutron and the positive pion are detected by the BGO ball and when the neutron is detected in the forward shower detector, respectively. They are compared with the results from previous experiments: open circles correspond to the results from Daresbury [16], while open triangles and squares are the results from SLAC [17][18]. GRAAL data cover unexplored regions, especially at higher energies and large angles (backward directions) and the agreement with the previous data is rather good in the common regions. The solid and dotted lines are the WI98 and SM95 solutions of the GW-SAID group (partial wave analysis) [1]. Disagreement with the data are observed in the energy region 850-1000 MeV. Dashed and dashed-dotted lines are the predictions of a unitary isobar model from [19] with and without the insertion of the $S_{11}(1650)$ resonance respectively. The behavior of the two curves clearly illustrates the important role played by this resonance, but the extracted value for its helicity coupling is essentially smaller than it was previously quoted. Further investigations are then requested and a new light can be brought from the extension of data at higher energies.

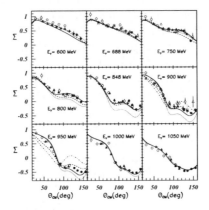

Figure 3. Results for beam asymmetry in $\vec{\gamma} + p \to \pi^+ + n$. See the text for details.

5 Conclusions

The GRAAL apparatus and polarized and tagged $\vec{\gamma}$ ray beam has become operational since several years, allowing the collection of a very high statistics of data. Beam asymmetries and cross sections have been extracted for several channels and other reactions are presently under analysis.

Acknowledgments

We are grateful to our theoretical colleagues who collaborated to the interpretation of our data. We thank the ESRF staff for the stable and reliable operation of the storage ring.

References

1. R.Arndt, I.Strakovsky, and R.Workman,*Bull. Am. Phys. Soc.* **45**, 34(2000); R. Arndt, I. Strakovsky, R. Workman, *Phys. Rev.* **C53** 430 (1996). The SAID solutions and the single pion photoproduction data base are available via telnet gwdac.phys.gwu.edu, user said.
2. B. Saghai and Z. Li, DAPNIA/SPhN-01-04 Submitted to *Eur. Phys. J. A*
3. L. Tiator et al., *Phys. Rev.* **C60**, 35210, (1999)
4. T. Feuster and U. Mosel *Phys. Rev.* **C59**, 460, (1999)
5. D. Babusci et al., *La Rivista del Nuovo Cimento*, **19**, 5, (1996)
6. L. Casano et al., *Laser and Unconventional Optics Journal* **55**, 3, (1975)
7. J. Ajaka et al., *Phys. Rev. Lett.* **81**, 1797, (1998)
8. V. Kouznetsov et al., to be published in *Nucl. Instrum. Methods* **A**
9. P. Levi Sandri et al., *Nucl. Instrum. Methods* **A370**, 396, (1996);
10. Bellini et al., Nucl. Instrum. Methods **A386**, 254, (1997)
11. M. Castoldi et al., Technical Report INFN/TC-99/17
12. A. Zucchiatti et al., *NIM* **A425**, 536, (1999)
13. F.V. Adamian et al., *J. Phys. G* **19**, L193, (1993)
14. A. L'vov et al., *Phys. Rev.* **C55**, 359, (1997)
15. J. Ajaka et al., *Phys. Lett.* **B475**, 372, (2000)
16. P.J. Bussey et al., *Nucl. Phys.* **B154**, 205, (1979)
17. G. Knies et al., *Phys. Rev* **D10**, 2778, (1974)
18. R. Zdarko and E. Dolly, *Nuovo Cimento* **10A**, 10, (1972)
19. D. Drechsel, O. Hanstein, S.S. Kamalov, L. Tiator, *Nucl. Phys* **A645**, 145, (1999). The MAID program is available on the webpage: http://www.kph.uni-mainz.de/MAID/maid.html

NEW (e,e'K+) HYPERNUCLEAR SPECTROSCOPY WITH A HIGH-RESOLUTION KAON SPECTROMETER (HKS)

O. HASHIMOTO

Department of Physics, Tohoku University, Sendai 980-8578, Japan
E-mail: hashimot@lambda.phys.tohoku.ac.jp

Reaction spectroscopy of Λ hypernuclei by the (e,e'K+) reacation with unprecedented quality is under preparation, following the success of Jlab E89-009 experiment, which observied a sub-MeV $^{12}_{\Lambda}$B spectrum in the ^{12}C(e,e'K+)$^{12}_{\Lambda}$B reaction. The experiment, Jlab E01-011, utilizes the "tilt method", in which the scattered electron spectromter is vertically tilted so that 0-degree bremsstrahlung electrons do not enter the spectrometer system. Together with a high-resolution kaon spectrometer(HKS) under construction, reaction spectroscopy of Λ hypernuclei with more than 50 times efficient and twice better resolution(3-400 keV(FWHM)) becomes possible. Hypernuclear spectroscopy by the (e,e'K+) reaction will be extended to the mass region beyond the p-shell region.

1 Introduction

Spectroscopic study of Λ hypernuclei observed significant progress in the past years through meson-induced reactions such as (π^+, K^+) and (K^-, π) at KEK and BNL[1,2]. With advanced spectrometer systems such as the SKS spectrometer at KEK[3], the reaction spectroscopy reached the mass resolution of about 2 MeV(FWHM) and revealed characteristics of hypernuclear single-particle orbits up to Pb and also unique features of Λ hypernuclear states, although the quality of the spectra are still limitted. In addition, energy splittings of spin partner states are intensively studied by high-resolution hypernuclear γ ray spectroscopy, which has become possible only recently thanks to high efficiency Germanium detectors.[4] Reaction and γ-ray spectrsocpy of Λ hypernuclei together plays key roles in the investigation of the strangeness nuclear physics. In the Λ hypernuclear reaction spectroscopy, it is particularly important to explore a way to realized sub-MeV energy resolution, by which the precision information on the behaviour of a Λ hyperon deep inside nuclear medium can be investigated. The (e,e'K+) reaction is one of the most promising reactions which provide an opportunity for high precision spectroscopy.

2 The (e,e'K+) hypernuclear spectroscopy

The (e,e'K+) reaction has a unique characteristic for the hypernuclear spectroscopy among a wide variety of reactions which can be used to produce a strangeness -1 hyperon. Each reaction has its own advantage and potentially plays its role for hypernuclear spectroscopy. However, only the (K^-, π^-) and (π^+, K^+) reactions, which convert a neutron in the target to a Λ hyperon, have been so far intensively used for the spectroscopic investigation. Although the (π^+, K^+) reaction is relatively new compared to the (K^-, π^-) reaction, it is now considered as one of the best reactions for hypernuclear spectroscopic investigation because it favorably populates deeply bound hypernuclear states due to a large momentum transfer. This is in contrast to the (K^-, π^-) reaction, which in general transfers small momentum and thus preferentially excites substitutional states.

Since the momentum transfer of the (e,e'K$^+$) reaction is similar to that of the (π^+, K^+) reaction, it is also expected to preferentially populate high-spin bound hypernuclear states. However, in contrast to reactions with meson beams, the electromagnetic reaction populates spin-flip hypernuclear states as well as non-spin-flip states.[5] The (e,e'K$^+$) reaction excites proton-hole-Λ-particle states which have the configuration $[(l_j)_N^{-1}(l_k)^\Lambda]_J$, by converting a proton to a Λ hyperon. This selectivity is particularly important as it allows us to directly study the spin-dependent structure of Λ hypernuclei.

Experimentally, however, the most important characteristic of the (e,e'K$^+$) reaction could be that significantly better energy resolution can be realized since the reaction is initiated with a primary electron beam of extremely good beam emittance and momentum spread, which is not the case with secondary meson beams. Energy resolution of a few 100 keV is expected, provided an appropriate spectrometer system is available.

Although the (e,e'K$^+$) reaction has many advantages for hypernuclear spectroscopy, hypernuclear production cross sections are much smaller than those by reactions using hadronic beams. The first (e,e'K$^+$) reaction spectroscopy was successfully carried out at Jlab Hall C(E89-009) as reported in this proceedings.[6] However, statistics are quite limitted even though it took almost 2 months to accumulate the data. It is, thus, eagerly waited to significantly upgrade the existing spectrometer system and experimental configuration to overcome the limitation of the (e,e'K$^+$) reaction.[7]

3 New experimental configuration and "Tilt method"

In the previous E89-009 experimental setup, scattered electrons and kaons at angles including zero degrees were detected in order to maximize the number of virtual-photon associated electrons, which have angular distribution of sharply forward peak, and kaons, which have moderate forward angular distribution, as shown in Fig. 1. Particularly, scattered electrons were detected at 0 degrees("0-degree tagging method") where the virtual photon intensity per beam electrons is sharply peaked.

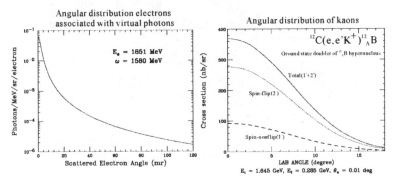

Figure 1. Angular distribution of bremsstrahlung associated electrons and kaons in the $^{12}C(e,e'K^+)^{12}_\Lambda B$ reaction. Note units of angles are msr for electrons and degrees for kaons.

This "0-degree tagging method" allowed one to run the experiment with a low-intensity electron beam and with a thin target. However, in this method, the beam intensity was still limitted by huge background due to bremsstrahlung associated electrons, rate of which amounted up more than a few times 10^8 Hz at the focal plane of the electrons(Enge) spectrometer. Because of that, the yield rate of $^{12}_\Lambda B$ spectrum was low and the signal-to-accidental ratio was also poor.

In the proposed E01-011 experimental configuration[8], which is schematically shown in Fig. 2, scattered electrons and kaons are bent to the opposie directions each other by a splitter magnet as in the case of the E89-009 experiment. However, the electron spectrometer is to be tilted vertically by a small angle in order to avoid dominant backgrounds of electrons due to the bremsstrahlung process as well as Moeller scattering. Since the scattered electrons are momentum dispersed horizontally by the splitter magnet, the Enge spectrometer needs to be tilted vertically. The tilt method works because the angular distribution of the virtual photon associated electrons is less forward peaked compared to that of bremsstrahlung associated electrons. An optimum tilt angle of 4.5 degrees were determined by a GENAT Monte Carlo simulation, taking into account the angular distribution and multiple scattering of virtual photon and bremsstrahlung associated electrons and moeller electrons.

In the present optimized condition, we can run a 100 mg/cm^2 ^{12}C target

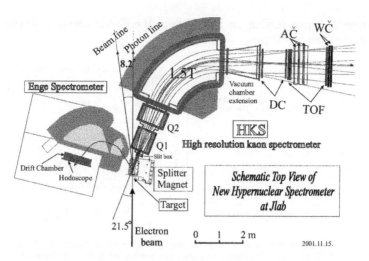

Figure 2. Schematic drawing of the HKS spectrometer system under construction and to be installed at Jlab for the $(e,e'K^+)$ hypernuclear spectroscopy.

with an electron beam intensity of 30 μA. Even then, singles rate at the focal plane of the Enge spectrometer is much smaller than the previous E09-009 expeirment, while hadronic rates of the HKS spectrometer are much higher as shown in Table 1.

Table 1. Expected singles rates for the E01-011 experimetnal configuration with a beam intensity of 30 μA and target thickness of 100 mg/cm^2

Target	Kaon spectrometer(HKS)			e' spectrometer (Enge)	
	π^+ rate (kHz)	K^+ rate (kHz)	p rate (kHz)	e^- rate (MHz)	π^- rate (kHz)
^{12}C	800	0.34	280	2.6	2.8
^{28}Si	800	0.29	240	5.1	2.8
^{51}V	770	0.26	230	6.9	3.0

The momentum acceptances of the two spectrometers are matched with each other to maximize hypernuclear yields, as shown in Fig. 3, where correlation of the electron and kaon momentum corresponding to Λ, Σ and hypernuclear states are indicated.

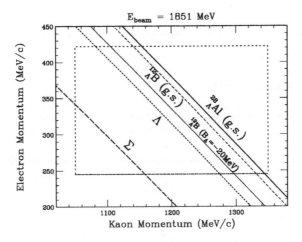

Figure 3. Momentum correlation between kaon arm and electron arm for hyperons and hypernuclei production reaction.

4 The spectrometer system

A plan view of the proposed geometry, the splitter magnet+scttered electron spectrometer(Enge)+high resolution kaon spectrometer (HKS), is already shown in Fig. 2. Both the HKS spectrometer and the Enge spectrometer are positioned at as forward angle as possible. The beam electrons are deflected by the splitter magnet but are to be steered back to the beam dump by a pair of dipole magnets located down stream. The spectrometer system was designed so that it would fit both Hall A and C of Jlab and can be installed in whichever hall available. Schematic layout of the spectrometer system installed in Hall C is shown in Fig. 4 and the spectrometer characteristics are summarized in Table 2.

4.1 High-resolution Kaon spectrometer (HKS)

The HKS spectrometer was designed to achieve 2×10^{-4} momentum resolution at 1.2 GeV/c and 18 msr solid angle acceptance simultaneously, when used with the splitter. A configuration of QQD has been employed in order to have a control over

Table 2. Experimental condition and specification of the proposed hypernuclear spectrometer system

---

Beam Condition

Beam energy	1.8 GeV
Beam momentum stability	1×10^{-4}

General Configuration — Splitter+Kaon spectrometer +Scattered electron spectrometer

High-resolution Kaon spectrometer(HKS spectrometer)

Configuration	QQD and horizontal bend
Central momentum	1.2 GeV/c
Momentum acceptance	± 10 %
Momentum resolution ($\Delta p/p$)	2×10^{-4} (beam spot size 0.1mm assumed)
Solid angle	20 msr with a splitter (30 msr without splitter)
Kaon detection angle	Horizontal : 7 degrees

Scattered electron spectrometer(Enge spectrometer)

Central momentum	0.3 GeV/c
Momentum acceptance	± 20 %
Momentum resolution ($\delta p/p$)	5×10^{-4}
Tilt angle	4.5 degrees vertical
Electron detection angle	Horizontal : 0 degrees Vertical : 2.25 degrees

---

both horizontal and vertical focusing. The spectrometer is placed rotated horizontally by 7 degrees with respect to the beam so that zero degree positive particles, mostly positrons, would not enter the spectrometer. A GEANT simulation was carried out taking into account realistic matter distributions such as a vacuum and chamber windows, detectors and air, and drift chamber position resolutions to study the HKS spectrometer performance. It was shown that expected solid acceptance can be obtaied as shown in 5 and the required energy resolution will be achieved even with a modest chamber resolution of 200 μm.

The detector package of the HKS spectrometer consists of tracking chambers

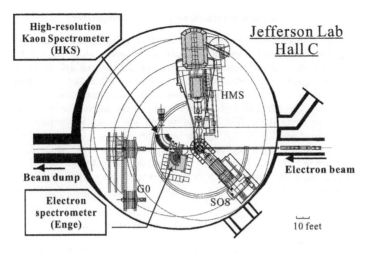

Figure 4. General layout of possible installation of the HKS spectrometer system in Hall C.

and trigger counters. Since the pion and proton rates are expected to be as high as several 100kHZ, it is vital to have Cherenkov counters which have capability to reject those particles in the trigger. To meet the requirements, 3 layers of aerogel Cherenkov counters(n=1.05) and two layers of water Cerenkov counters(n=1.33) are under construction.

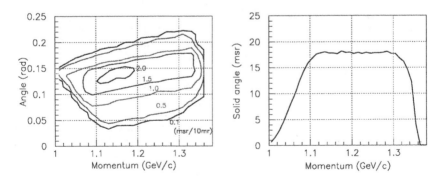

Figure 5. Momentum and angular acceptance of the HKS spectrometer.

4.2 The scattered electron spectrometer(Enge spectrometer)

The Enge spectrometer, which was used in E89-009, will be installed as a spectrometer to analyzes scattered electron momentum. However, since the spectrometer is vertically tilted as already mentioned, the optics of the original spectrometer is modified and higher order terms cannot be neglected. Therefore, measurement of the incident angle as well as the position of electrons at the focal plane becomes vital to achieve The effect of higher order terms in the momentum reconstruction was studied by using RAYTRACE program and was found the required momentum resolution of 5×10^{-4}(FWHM) can be achieved with a thin drift chamber in which multiple scattering is negligible. The expected singles rate at the focal plane is at most a few MHz and is 2 order of magnitude less than the case of E89-009. A hexagonal-cell drift chamber with xx'uu'xx'vv'xx' planes, thus, and hodoscopes meet the requirement.

5 Expected performance of the spectrometer system

Expected hypernuclear mass resolutions for 3 targets together with item-by-item contributions are listed in Table 3. The mass resolution of 3-400 keV (FWHM) is to be achieved with the proposed E01-011 experiment. It is twice better than E89-009.

Table 3. The energy resolution of the HKS system

Item	Contribution to the resolution (keV, FWHM)		
Target	C	Si	V
HKS momentum(2×10^{-4})		216	
Beam momentum($\leq 1 \times 10^{-4}$)		≤ 180	
Enge momentum(5×10^{-4})		150	
K^+ angle resolution	152	64	36
Target (100 mg/cm^2)	≤ 180	≤ 171	≤ 148
Overall	≤ 400	≤ 370	≤ 350

Yield rate of $^{12}_\Lambda$B ground state was evaluated based on the E89-009 result in the ^{12}C(e,e'K$^+$)$^{12}_\Lambda$B reaction and the E89-009 and E01-011 are compared in Table 4. Overall hypernuclear yield rate of the proposed E01-011 is expected to be more than 50 times higher with signal-to-accidental ratio almost 10 times better than E89-009. The considerable improvement is achieved partly because we can use higher intensity beams and thicker targets and partly because the kaon spectrometer has a larger solid angle acceptance. Expected hypernuclear spectrum for a Si target is shown in Fig. 6.

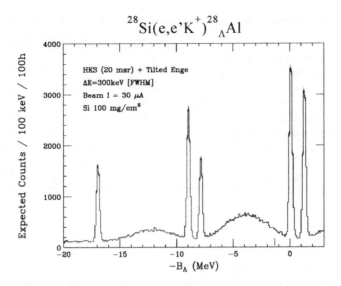

Figure 6. Expected mass spectrum of $^{28}_{\Lambda}$Si for 100 hour beam time with HKS spectrometer system.

6 Summary

The Λ hypernuclear spectroscopy by the (e,e'K$^+$) reaction has great potential to explore strangeness nuclear physics of the S=-1 sector quantitatively. Based on a thorough examination of the previous E89-009 experiment, which made first successful observation of the hypernuclear mass spectrum by the (e,e'K$^+$) reaction, a new experiment(E01-011) which intends to perform precision spectroscopy even beyond the p-shell region and to reveal behaviour of a Λ hyperon in nuclear medium.

With a high-resolution kaon spectrometer(HKS) under construction, which has a large solid angle, and a new "Tilt method", high quality hypernuclear mass spectra will be measured with significantly better resolution, statistics and signal-to-accidental ratio compared witht eh previous E89-009 experiment.

The HKS spectrometer magnet together with the detector packages both for the HKS and Enge spectrometers will be completed in 2003 and the E01-011 experiment will be ready to run.

The present work has been conducted under the support of Monkasho Grant-in-aid for Scientific Research "specially promoted program" and also in collaboration with the Jlab-E01-011 group. Intensive discussions with S.N. Nakamura, Y. Fujii, L. Tang, J. Reinhold and E. Hungerford are particularly acknowledged.

Table 4. Estimated hypernuclear yield gain of E01-011 compared with E89-009

Item	E01-011	E89-009	Gain factor
Virtual photon flux per electron($\times 10^{-4}$)	0.35	4	0.0875
Target thickness(mg/cm^2)	100	22	4.5
e' Momentum acceptance(MeV/c)	150	120	1.2
Kaon survival rate	0.35	0.4	0.88
Solid angle of the K arm	18	5	3.6
Beam current(μA)	30	0.66	45
Estimated yield ($^{12}_{\Lambda}B_{gr}$: counts/h)	61	0.9 (measured)	68

References

1. C. Milner *et. al.*, Phys. Rev. Lett. **54** (1985) 1237; P. H. Pile *et al.*, Phys. Rev. Lett.**66** (1991) 2585.
2. T. Hasegawa *et al.*, Phys. Rev. **C53** (1996) 1210; O. Hashimoto, Proc. Int. Workshop on Strangeness Nuclear Physics, Seoul (World Scientific 1999), p. 116; H. Hotchi *et. al.*, Phys. Rev. **C64** 044302,2001
3. T. Fukuda *et. al.*, Nucl. Instr. Meth. A361 (1995) 485-496.
4. T. Tamura *et. al.*, Phys. Rev. Lett. **84** (2000) 5963. K. Tanida *et. al.*, Phys. Rev. Lett. **86** (2001) 1982-1985.
5. T. Motoba, M. Sotona and K. Itonaga, Prog. Theor. Phys. Suppl. 117 (1994) 123; M. Sotona and S. Frullani, ibid. 151.
6. E. Hungerford, in these proceedings.;
   L. Tang *et. al.*, Proc. of Mesons and Light Nuclei '01
7. Articles in Proceedings of Jlab sponsered workshop"Hypernuclear physics with electromagnetic probes(HYPJLAB99), December 1999, Eds. O. Hashimoto and L. Tang.
8. Jlab proposal E01-011, Spokespersons O. Hashimoto, L. Tang, J. Reinhold and S. Nakamura.

DEVELOPMENT OF A COMPACT PHOTON DETECTOR
FOR ANKE AT COSY JÜLICH

M. BÜSCHER, V. HEJNY, H. R. KOCH, H. MACHNER, H. SEYFARTH AND
H. STRÖHER

Institut für Kernphysik, Forschungszentrum Jülich, Jülich, Germany
E-mail: v.hejny@fz-juelich.de

M. HOEK, R. NOVOTNY AND K. RÖMER

II. Physikalisches Institut, Justus-Liebig Universität Giessen, Giessen, Germany
E-mail: r.novotny@exp2.physik.uni-giessen.de

J. BACELAR AND H. LÖHNER

KVI Groningen, Groningen, The Netherlands
E-mail: h.loehner@kvi.nl

A. MAGIERA AND A. WROŃSKA

Institute of Physics, Jagellonian University Cracow, Cracow, Poland
E-mail: magiera@if.uj.edu.pl

V. CHERNYCHOV

Institute for Theoretical and Experimental Physics, Moscow, Russia
E-mail: chernychov@vitep3.itep.ru

COSY Jülich is a race-track shaped synchrotron which accelerates and cools beams of protons (both polarized and unpolarized) and deuterons with momenta up to 3.6 GeV/c. Those beams are delivered to internal and external target positions for hadron physics experiments. Since magnetic and time-of-flight detectors based on organic scintillators are used in the experimental setups, all measurements are essentially "photon blind". Recent improvements in the performance of high-density inorganic scintillators offer the possibility to design very compact large-acceptance electromagnetic calorimeters with excellent timing and good energy resolution, applicable also for photon energies below 1 GeV. Such a detection system, based on PbWO$_4$, is planned to be built for the internal magnetic spectrometer ANKE at COSY Jülich. The limited space and stray magnetic fields of ANKE place severe boundary conditions, which have to be taken into account for detector layout and the choice of photo sensors.

1 Introduction

Photons are a very sensitive probe to obtain information about hadronic reactions. Nearly all neutral mesons below 1 GeV/c^2 (π^0, η, ω, η', f_0, a_0) have decay branches into multi-photon final states. Since photons are weakly interacting, they escape the production region nearly undisturbed and can be efficiently detected with good energy resolution using modern high-density inorganic scintillators. Such photon detectors with large solid angle have been designed, built and are being successfully operated at many electron and hadron accelerators in the intermediate and high-energy range.

At the Cooler Synchrotron (COSY) of the Forschungszentrum Jülich experiments currently focus on the detection of charged particles with no possibility of a direct identification of single or multiple photon final states. In order to open this new window it is intended to build a dedicated photon detector for use mainly in conjunction with the magnetic spectrometer ANKE [1] installed at an internal target position of COSY. This combination of detectors will allow to perform exclusive experiments by identifying charged and neutral particles simultaneously. Furthermore, a flexible detector design will allow its use at other target stations of COSY or other accelerator facilities as well.

2 Design goals

ANKE (acronym for "Apparatus for Studies of Nucleon and Kaon Ejectiles") is an internal magnetic spectrometer in one of the straight sections of COSY. It consists of three dipole magnets (D1, D2 and D3) to form a chicane, which imposes a closed orbit bump on the circulating beam and allows to momentum analyze the charged reaction products. As shown in figure 1, various detector systems are used for momentum reconstruction (multi-wire proportional chambers, MWPC, and drift chambers, MWDC) and particle identification (energy loss ,ΔE, time-of-flight, TOF, and Cerenkov light, C).

With this existing detector setup, the major constraints for an electromagnetic calorimeter are fixed. Limited space in the target area and the stray magnetic fields are major challenges to be handled. Furthermore, access to the experimental area is limited to a few short maintenance intervals. In order to fulfill individual needs of specific experimental setups (e.g. choice of target, special near-target detectors, etc.), mounting and dismounting of the whole device has to be accomplished within about one week. Thus, the design goals for a photon detector can be summarized as follows:

- large angular acceptance for detection of multi-photon final states,
- fast response for excellent timing ($\gamma\gamma$ coincidences) and high-count rate capability,
- detection of photons with energies substantially below 1 GeV,
- compact design to fit into the available space (total diameter ≤ 75 cm),
- operation of the photo sensors in a magnetic field of $B \leq 0.2$ T,

• modular design for fast installation and de-installation.

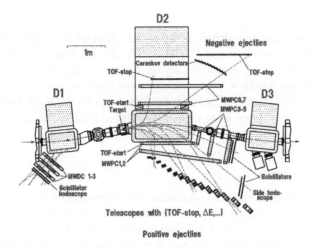

Figure 1. Schematic drawing of the ANKE spectrometer at COSY-Jülich. The beam enters from the left and hits the target in front of D2. Positively and negatively charged reactions products are deflected to the right and left, respectively, where dedicated detection systems are set up to identify and analyze them.

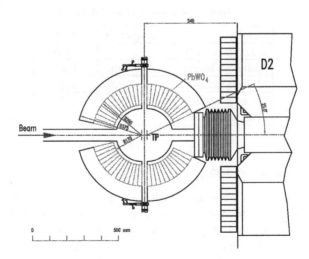

Figure 2. Schematic view of the ball-shaped photon detector in front of the dipole magnet D2 of ANKE. The beam enters from the left. TP indicates the target point.

3 Realization

3.1 General setup

Figure 2 shows a tentative sketch of the photon detector surrounding the target in front of the spectrometer magnet D2. To enable the installation of the new device, the target point has to be shifted slightly upstream compared with figure 1. For the design shown, this results in a minimum opening angle in forward direction of the photon detector of $\Theta=\pm25°$ and a solid angle covered of 93% of 4π. Such a compact layout can only be achieved using modern techniques for detector material and photo sensors.

3.2 Detector material

In recent years $PbWO_4$ has been established as a high-density inorganic scintillator with good energy resolution for photon detection also at intermediate energies [2]. This material is most suitable for the discussed detector, because it allows a very compact design without compromising on energy resolution and timing properties. Considering the energy range of the produced photons, the proposed length of the crystals is 12 cm, i.e. 13.5 radiation lengths X_0. Its main properties are given in table 1.

Table 1. Main properties of $PbWO_4$.

Density	(g/cm^3)	8.28
Index of refraction		2.16
Radiation length	(cm)	0.89
Moliere radius	(cm)	2.20
Decay constant	(ns)	< 20
Peak wavelength	(nm)	420

3.3 Photon response of $PbWO_4$

The photon response of $PbWO_4$ crystals, which during the years have steadily improved in quality, has been investigated in several experiments [3,4]. Nb/La-doped crystals of slightly tapered shape, manufactured und preselected by Bogoroditsk Techno-Chemical Plant (Russia) and by RI&NC (Minsk, Belarus) have been assembled into a 5x5 matrix. The optically polished crystals of 150mm length ($\sim17X_0$) have a quadratic front face ($20.5x20.5mm^2$) and a tapering angle of $\sim0.4^0$. They are individually wrapped in PTFE-foil, coupled with optical grease to photomultiplier tubes (Hamamatsu R4125 or Philips XP1911, covering between 44%

and 35% of the crystal endface) and stacked into a light-tight box, which has been temperature stabilized between 6-10⁰C.

For optimum crystals energy resolutions of $\sigma/E \sim 30\%$ can be achieved for 662 keV photons of ^{137}Cs at room temperature. In the most recent measurements the resolution at 45MeV photon energy has been reduced to $\sigma/E \sim 7.4\%$. The overall energy dependence is shown in figure 3 and can be parametrized by $\sigma/E \sim 1.41\%/E_\gamma^{1/2} + 0.9\%$.

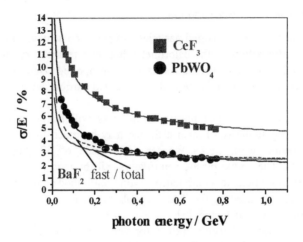

Figure 3. Energy response of the PbWO₄ array in comparison to CeF₃ and BaF₂ (TAPS).

Several tests of the time resolution give an upper limit of $\sigma_t < 130$ps, a value which allows photon/particle separation even for compact setups with flight paths well below 1m.

3.4 Charged particle detection with PbWO₄

In general, clean photon detection in this energy regime relies in addition on an efficient particle/photon discrimination or even on a particle spectroscopy to allow a complete kinematical characterization of the studied reaction channels. Therefore, the previous investigations of the photon response of PbWO₄ has been extended to the detection of charged hadrons with kinetic energies up to 1.2 GeV. The tested PbWO₄-matrix was identical to the one used in measurements with monoenergetic photons up to 800MeV (see above).

As a results it was found that the energy resolution for protons up to kinetic energies, which are completely stopped in the detector material (E_p =365 MeV), can be parametrized by $\sigma/E = 0.97\%/E_p^{1/2} + 3.33\%$ (E_p given in GeV), i.e. $PbWO_4$ shows similar or even better energy resolution for hadrons for electromagnetic probes. However, the simultaneously obtained response for charged pions indicates a strong quenching of the scintillation light initiated by protons relative to pions. This effect could be related to a sensitivity of the scintillation mechanism to the ionization density of different probes, but requires more quantitative studies and a detailed characterization of the used crystal quality.

3.5 Charged particle discrimination

The discrimination against charged particles can partly be achieved by exploiting the time-of-flight information of the calorimeter but, due to the very short flight path from the target, additional techniques have to be applied. Depending on specific experimental needs, two distinct methods for charged particle detection are foreseen:

1. A ball-shaped array of fast plastic scintillators will be put inside the photon detector. The granularity of such a device may be much lower than that of the photon detector itself, i.e. a "soccer ball" made of 32 pieces will fit the requirements. The single plastic scintillators will be read out using wavelength-shifting optical fibers in combination with multi-anode photomultipliers. This method is already used by the TAPS collaboration for its individual, hexagonal charged-particle veto detectors [5].
2. A near-target silicon-detector system for vertex/spectator detection is currently under development for ANKE [6]. It will fit into the inner part of the photon detector and will provide the possibility to measure spectator protons (i.e. using the deuteron as an effective neutron target) and to reconstruct reaction vertices when extended targets (like the polarized gas target with a storage-cell [7]) are used. Recently, first experiments with a spectator-detector prototype (ω production in pn-collisions) have been carried out at ANKE.

3.6 Photo sensors

In order to exploit the fast timing properties of $PbWO_4$ it is necessary to use photo multipliers rather than slow conventional photodiodes or avalanche photodiodes (APDs). However, due to the stray magnetic field in front of the dipole ($B \leq 0.2$ T), standard phototubes cannot be installed. Instead, special fine-mesh tubes have to be used. They are basically insensitive to magnetic fields parallel to the multiplier axis (up to $B \approx 1$ T). Thus, only the perpendicular component B_{perp} has to be shielded. In addition, effects by the ramping of the magnets during the acceleration phase of COSY have to be managed.

Tests with an unshielded Hamamatsu R5505 phototube showed that the gain already drops at $B_{perp} = 15$ mT. Shielding the tubes with soft iron of 1mm and 2mm wall thickness, the operating range can be extended up to 90 mT and beyond 120 mT, respectively. To withstand the stray field of 200 mT, a shielding of 3 mm thickness has been chosen and tested in the experimental environment of ANKE. It has been shown that the magnet ramping has no persistent effect on the response of the phototubes; a stable operation during the data-taking cycle could be achieved.

To some extent, the dimensions of the R5505 tube ($d = 25.6 \pm 0.7$ mm, $l = 75$ mm including base) will influence the layout of the modules. They will have a total length of $l = 20$ cm fixing the inner radius of the photon detector to $R_i = 17$ cm. The total diameter corresponds to a solid angle of 13 msr and, consequently, the total number of channels will be about 800.

3.7 High voltage supply, readout and trigger

The limited space in the target area in combination with the large number of channels prevents the installation of individual high-voltage supply cables. The preferred method, which is currently under investigation, would use bases with on-board miniaturized Cockroft-Walton generators employing Greinacher voltage multipliers. Such devices have been developed, for example, for WA98 (CERN) [8], the HADES time-of-flight system[1] and the WASA detector at CELSIUS[2]. Such a solution would also be preferable in view of the temperature dependence of the light output of PbWO$_4$: the power consumption of these devices can be one order of magnitude lower than that of standard dividers.

For readout and trigger electronics, modern integrated concepts will be chosen, i.e. a solution with several complete channels on one board or even performing digitization already at the multiplier. Electronic modules of the first type are currently developed by the TAPS collaboration. Presumably, one version of these VME boards will have 8 complete channels including CFD, LED, TDC and QDC. This fast and compact VME solution in combination with Multiplicity Coincidence Units, developed in parallel for versatile triggering, seems to be well suited for application in this project.

4 Summary

The main properties of the proposed photon detector at ANKE are summarized in table 2.

---

[1] See http://hp.uif.cas.cz/hades
[2] See http://www.tsl.uu.se/wasa

Table 2. Overview of the main detector properties.

Solid angle	(% of 4π)	93
Granularity	(msr)	13
Material		PbWO$_4$
Crystal length	(cm / X$_0$)	12 / 13.5
Time resolution σ_t	(ps)	130
Energy resolution σ_E at 500 MeV	(%)	3
Invariant mass resolution FWHM	(MeV)	23 (π^0)
		40 (η)
		50 (ω)
Number of channels		~ 800

The COSY Program Advisory Committee (PAC) has accepted the proposal for a photon detector at COSY in November 2000. R&D concerning photo-tubes, high-voltage supply and read-out is currently going on. It is planned to have a final decision on the detector layout in spring/summer 2002. Taking into account delivery and assembling times, the detector should be ready for installation at ANKE in fall 2004.

Consequently, the first experiments are planned for end of 2004. As accepted by the PAC, these will cover η and ω production off the proton and neutron from threshold up to the maximum COSY energy ($T_p = 2.65$ GeV) and the study of charge symmetry breaking in the reaction d+d $\rightarrow$ ^{4}He+π^0 [9]. Investigations of neutral scalar meson production (a$_0$/f$_0$) are also foreseen [10].

Acknowledgements

The research has been performed within the ANKE-, Crystal Clear- and TAPS-Collaboration and has been supported by BMBF (Bundesministerium für Bildung, Wissenschaft, Forschung und Technologie), DFG (Deutsche Forschungs-gemeinschaft) and FZJ (Forschungszentrum Jülich).

References

1. Barsov, S. et al., *Nucl. Instr. And Meth. A* **462** (2001) p. 364.
2. Novotny, R. et al., *IEEE Trans. on Nucl. Sc.* **47** (2000) p. 1499.
3. Novotny, R. et al., *IEEE Trans. on Nucl. Sc.* **44** (1997) p. 447.
4. Mengel, K. et al., *IEEE Trans. on Nucl. Sc.* **45** (1998) p. 681.
5. Janssen, S. et al., *IEEE Trans. on Nucl. Sc.* **47** (2000) p. 798.

6. Barsov, S. et al., In *Annual Report Institut für Kernphysik, Forschungszentrum Jülich* (2000) p. 26.
7. Emmerich, R. et al., In *Annual Report Institut für Kernphysik, Forschungszentrum Jülich* (2000) p. 32.
8. Neumaier, S. et al., *Nucl. Instr. and Meth. A* **360** (1995) p. 593.
9. Bacelar, J. et al., A photon detector for COSY, *COSY proposal* **83.2** (2000).
10. Büscher, M. et al., Investigation of neutral scalar mesons with ANKE, *COSY proposal* **97** (2001).

THE COMMISSIONING OF THE HALL-B BEAMLINE OF JEFFERSON LAB FOR COHERENTLY PRODUCING A BEAM OF LINEARLY-POLARIZED PHOTONS

P.L. COLE,* J. KELLIE, F.J. KLEIN, K. LIVINGSTON, J.A. MUELLER, J.C. SANABRIA, AND D.J. TEDESCHI, COSPOKESPERSONS OF G8

*University of Texas at El Paso, El Paso, TX 79968;
Jefferson Lab, Newport News, VA 23606
E-mail: cole@jlab.org

FOR THE CLAS COLLABORATION

The set of experiments forming the g8 run took place this past summer (6/04/01 – 8/13/01) in Hall B of Jefferson Lab. These experiments made use of a beam of linearly-polarized photons produced through coherent bremsstrahlung and represent the first time such a probe has been employed at Jefferson Lab. Among the several new and upgraded Hall-B beamline devices commissioned prior to the production running of g8a were the photon tagger, the coherent bremsstrahlung facility (goniometer + an instrumented collimator), a photon profiler, and the PrimEx dipole + pair spectrometer telescopes. We essentially commissioned a new beamline for photon running in Hall B. The scientific purpose of g8 is to improve the understanding of the underlying symmetry of the quark degrees of freedom in the nucleon, the nature of the parity exchange between the incident photon and the target nucleon, and the mechanism of associated strangeness production in electromagnetic reactions. With the high-quality beam of the tagged and collimated linearly-polarized photons and the nearly complete angular coverage of the Hall-B spectrometer, we seek to extract the differential cross sections and polarization observables for the photoproduction of vector mesons and kaons at photon energies ranging between 1.1 and 2.25 GeV. For the first phase of g8, i.e. g8a, we collected approximately 1.8 billion triggers for $1.75 \leq E_{\bar{\gamma}} \leq 2.25$ GeV. In this paper, we report on the results of the commissioning of the beamline devices for the g8a run.

1 Motivation

From our set of experiments on vector-meson photoproduction for the g8 run,[1,2,3] we seek to study the evolution of the spin density matrix elements as functions of the Mandelstam variables, s and t, in the effort to extract spin-parity information on the underlying baryon resonances which decay through the ρ or ω channel and the nature of the parity exchange for ϕ production near threshold. With linearly-polarized photons, one has access to nine independent spin density matrix elements. Combinations of which correspond to the vector-meson polarization,[a] the beam asymmetry Σ, the parity asymmetry P, and beam-vector meson double polarization observables. In the case of s-channel helicity conservation or natural parity exchange ($J^\pi = 0^+, 2^+$), the decay angular distribution is given by $P_\gamma \sin^2\theta \cos 2\Psi$,[b] which corresponds to $\Sigma = P = 1$. Whereas for unnatural-parity exchange ($J^\pi = 0^-$), the parity asymmetry flips to -1. Linearly-polarized photons therefore serve as a parity filter. It is by extracting these density matrix elements as functions of the Mandelstam variables, s and t, that we shall obtain a *model-independent* pool of data. These density matrix elements form the meeting ground between theory and experiment. Through their models, theorists predict the helicity amplitudes, and from the decay angular distribution data, experimentalists extract the *bilinear combinations of these helicity amplitudes* or the *density matrix elements*. Accurate and precise measurements of the evolution of these density matrix elements over a wide range of s and t will constrain the theory, and thereby give insight into the underlying production mechanisms. For recent information on vector meson production, we refer the reader to the papers[4,5] in these conference proceedings

In the case of the hyperon production, employing the probe of linearly-polarized photons, together with measuring the polarization of the recoil hyperon, gives access to the beam asymmetry Σ and the beam-recoil double polarization observables $O_{x'}$, $O_{y'}$ and $O_{z'}$. These data from the g8 run,[6] coupled with the CLAS-g1[7] data from unpolarized and circularly-polarized photon beams, will provide an almost complete description of the $\gamma p \to K^+ \Lambda$ channel and will facilitate a *model-independent* analysis towards identifying the intermediate resonances.

At present there exist conflicting interpretations of the $\gamma p \to K^+\Lambda$ cross section data from SAPHIR.[8] A calculation from Mart and Bennhold[9] interprets the structure at $W = 1900$ to be the result of one of the 'missing' resonances, $[D_{13}]_3(1960)$. Saghai and collaborators,[10,11,12] have taken another tack; in their approach they include off-shell effects and Λ^*-exchange and take only the known N^* resonances into account. The measurement of polarization observables is necessary for settling this issue of the presence or absence of the $[D_{13}]_3(1960)$.

---

[a]By means of the decay of the spin-1 vector mesons into the spin-0 pseudoscalars, one gains access to the the tensor polarization components T^2_0, $T^2_{\pm 1}$, and $T^2_{\pm 2}$. Additional information on the polarization observables can be found in Refs.[13,14]

[b]Here, Ψ is the angle between the direction of the photon polarization vector and the azimuth of the decay plane.

2 Brief Theoretical Background on Coherent Bremsstrahlung

The technique of obtaining linearly-polarized photons has been successfully employed at SLAC[16] and Mainz.[17] Detailed discussions of the underlying theory of coherent bremsstrahlung can be found in the papers listed in Ref.[18]

For an electron of energy E_0 to radiate a photon of energy k requires that the momentum $\vec{q}_e$ be transferred to a nucleus in the crystal lattice. The minimum momentum transfer, q_{min}, – for each individual reciprocal lattice vector – scales with increasing photon energy. Only the lattice vectors for which q_{min} is less than q_e may contribute to the coherent process.

$$q_{min} = \frac{m_e^2 c^3}{2E_0} \frac{x}{1-x} \le q_e, \tag{1}$$

where x is the fractional photon energy, k/E_0. A discontinuity will arise for $q_e > q_{min}$; thus demarcating the coherent edge for a given lattice vector. In the limit of small goniometer angles, for a diamond crystal we can write[19]

$$\theta_v(k-l) - \theta_v(k+l) = \frac{\sqrt{2}a m_e^2 c^3}{4\pi\hbar} \frac{1}{E_0}\left(\frac{x}{1-x}\right)$$
$$= [57.0 \text{ GeV} - \text{mrad}] \cdot \frac{1}{E_0}\left(\frac{x}{1-x}\right). \tag{2}$$

Here, (h, k, l) define the Miller indices of the reciprocal lattice vector and a is the length of the fundamental cell. In Fig. 1 we plot the theoretical lattice map for the [100] plane of diamond once it is perpendicular to the direction of the electron beam.

The angular distribution of the coherent and incoherent bremsstrahlung photons are very different. Whereas the polar angular distribution of the incoherent bremsstrahlung photons is *independent* of the photon energy

$$\frac{dN(\theta)}{d\theta} = \frac{\theta}{(1+\theta^2)^2}, \tag{3}$$

the emission angle of the coherent bremsstrahlung photons *decreases* with increasing photon energy ($x \le x_d$)

$$\theta^2 = \frac{1-x}{x} \cdot \frac{x_d}{1-x_d} - 1. \tag{4}$$

Here, x_d is the maximum fractional photon energy for a given setting of crystal. We can make use of the physical fact that the emission angle of the coherent bremsstrahlung photon is correlated with its energy to enhance the

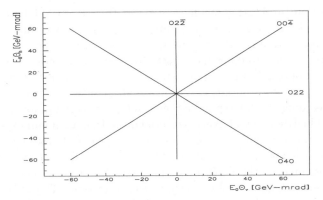

Figure 1. Theoretical lattice map for the [100] plane of diamond, which is approximately perpendicular to the electron beam.

polarization of the beam of photons. We can then extract the spectral peak by tightly collimating the beam and thereby reduce the incoherent background, which further serves to increase the degree of polarization.

3   Hall-B Beamline for Producing Linearly-Polarized Photons

The g8 group of experiments employed the coherent bremstrahlung facility, which was commissoned at the beginning of the g8a run (see Fig. 2). To

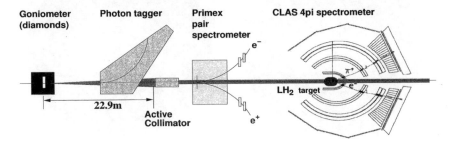

Figure 2. Layout of the Hall-B beamline for the coherent bremsstrahlung facility.

produce polarized photons of energy 2 GeV from an incident 5.7-GeV electron beam requires that the angle of incidence between the reciprocal lattice vector and the electron beam be aligned to approximately 1 μrad. We employed the GWU goniometer[20] to align the 50-μm diamond radiator. The technique of Glasgow University for aligning the crystal by means of a series of scans is an extension of Lohman[21] and will be detailed in an upcoming NIM article.[22] The alignment procedure entails executing small angular movements of the crystal and recording the corresponding photon tagger spectrum for each shift in θ_h and θ_v. The [100] crystal axis is set at 60 mrad from its nominal position and is swept through a 360° cone in azimuth. The procedure for aligning the crystal is shown in Fig. 3, where on the left we show a simulation of a scan that is not yet aligned; the θ_h-θ_v offset indicates the degree that the [100] plane is not perpendicular to the electron beam direction. The scan on the right is real data. The dark radial ridges of these plots trace the energy of the coherent peak as the angle between the beam and the face of the crystal varies. As can be seen in the left hand figure, the orientation of the crystal with respect to the beam is found by fitting a template (shown with dashed lines) composed of 8 lines, spaced 45° apart. The final scan is close to a perfect 4-fold symmetry; this shows that the crystal is very well aligned to the beam.

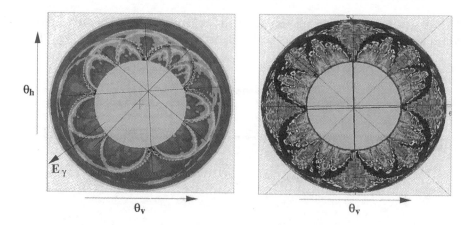

Figure 3. (Left) Simulated scan illustrating the alignment procedure by fitting a template. Here the face of the crystal, i.e. the [100] plane, is offset in θ_v and θ_h with respect to the direction of the incident electron beam. (Right) Final scan taken during a data scan run. We see that the crystal is aligned to the desired degree of precision.

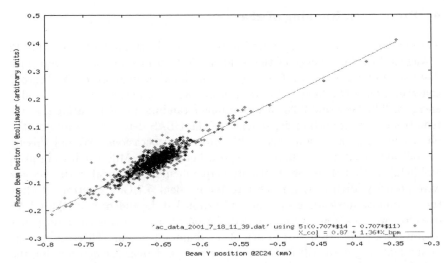

Figure 4. Correlation of the vertical beam position as determined by the embedded collimator scintillators and the beam position monitor. The electron beam position monitor is located 11 m upstream of the collimator, in front of the tagger magnet (2C24). The data were collected on 18 July 2001.

The photon spectrum, as measured by the tagger,[23] is the key diagnostic tool for aligning the reciprocal lattice vectors of the diamond radiator with respect to the incident electron beam to allow for coherent bremsstrahlung production. The focal plane of the photon tagger consists of 384 $\frac{1}{3}$-overlapping plastic scintillators spanning the energy range of 20 to 95% of the incident electron energy. Each counter thus behaves as an independent detector. From the scan plot on the right in Fig. 3 (see also Fig. 5), we can see that the photon tagger performed very well indeed.

An instrumented collimator[24,25] of aperture 2 mm was installed in the Hall-B beamline downstream of the tagger magnet at a distance of 22.9 m from the diamond radiator. By comparing the asymmetries obtained online from the active elements of the collimator with the electron beam position monitors, we see that the collimator was sensitive to beam shifts to better than 50 μm (see Fig. 4).

4 Production Running of g8a

The g8a production running period took place in the summer of 2001 (July 12 – August 13). The energy of the incident electron beam on the 50-μm thick diamond radiator was 5.7 GeV, with a nominal current of 7 nA. For our coherent bremsstrahlung data, we ran at two separate coherent peak edge energies: 2.07 GeV and 2.25 GeV. For understanding and delineating the systematics of the azimuthal dependence of the CLAS detector, we rotated the photon-beam polarization axis by 90° on several occasions. We did this by periodically rotating the diamond crystal between the Miller indices of [02$\bar{2}$] and [022], so that these two mutually perpendicular reciprocal lattice vectors were properly aligned with respect to the incident 5.7 GeV electron. To further eliminate misleading or 'built-in' azimuthal dependences of the CLAS detector, we took several unpolarized-photon runs employing the amorphous 50-μm thick carbon radiator in lieu of the diamond crystal. These incoherent bremsstrahlung data will furthermore be our yardsticks for determining the polarization of the beam. These complementary data sets will aid us in understanding our azimuthal dependences and thereby will serve to reduce the systematic uncertainties in the differential cross sections of the hyperons and vector mesons.

In Fig. 5 we plot the normalized photon spectrum before and after the collimator. Here, *normalized* means that the spectrum obtained with the diamond radiator is divided by the reference spectrum from an amorphous carbon radiator of thickness 50 μm, and where we have set the baseline to the value of 100. The uncollimated data were collected online with the free-running scalers. The collimated spectrum is derived from TDC hit patterns with the random background subtracted out. This preliminary spectrum results from a first pass through the data with a rough timing calibration. Both data sets, i.e. uncollimated and collimated, are fit with the anb code.[26] We see excellent agreement between the data and the fits, and based upon this calculational evidence, we claim we have succeeded in producing a beam of linearly-polarized photons with a high degree of linear polarization of up to a maximum of 84%.

5 Summary

With the g8a run, we were able to show 'proof of principle' that we can coherently produce a tagged and collimated beam of linearly-polarized photons from a 50-μm diamond radiator. The maximum degree of polarization exceeds 80%. For the first phase of g8, i.e. g8a, we collected approximately

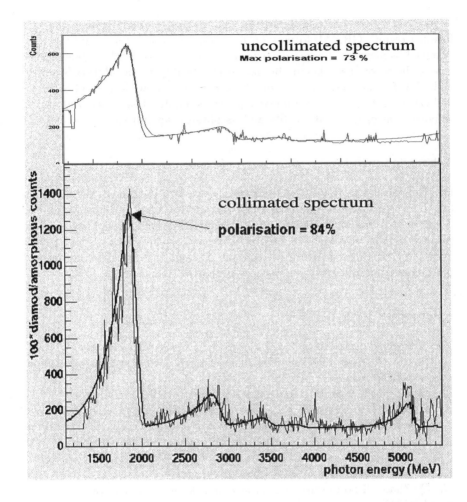

Figure 5. Normalized photon spectra before and after collimation. We remark that both plots have the same vertical scale; this highlights the enhancement of the spectral peak due to the collimation of the photon beam. The fits to both the uncollimated and collimated normalized photon spectra are from the anb code.[26]

1.8 billion triggers, which, after our data cuts and analysis, should give us well over 100 times the world's data set for ρs and ωs in the energy range of $1.75 \leq E_{\bar{\gamma}} \leq 2.25$ GeV. We remark that the g8a data set is completely complementary to those of SPring-8 and GRAAL. SPring-8 strengths lie in the very forward region, GRAAL focuses on neutral-particle detection, and CLAS is designed to measure charged particles in the more central polar-angle regime. The addition of the coherent bremsstralung facility to the Hall-B beamline will allow for the cross comparison of the JLab results with those of SPring-8 and GRAAL. In summary, a linearly-polarized photon beam represents a real enhancement of the JLab facility and its physics capabilities.

Acknowledgments

P.L. Cole wishes to thank Dr. T. Hotta and Prof. T. Nakano of the Research Center for Nuclear Physics of the University of Osaka for their hospitality and is grateful to Prof. M. Fujiwara and the organizers of EMI2001 for inviting him to this thought-provoking conference. This work is supported through grants from the U.S. Department of Energy, the National Science Foundation, and the Engineering and Physical Sciences Research Council of the U.K.

References

1. "Photoproduction of ρ Mesons from the Proton with Linearly Polarized Photons," JLab E94-109, P.L. Cole, J.P. Connelly, and K. Livingston, cospokespersons.
2. "Photoproduction of ϕ Mesons with Linearly Polarized Photons," JLab E98-109, D.J. Tedeschi, P.L. Cole, and J.A. Mueller, cospokespersons.
3. "Photoproduction of ω Mesons off Protons with Linearly Polarized Photons," JLab E99-013, F.J. Klein and P.L. Cole, cospokespersons.
4. A. Titov and T.-S. H. Lee, "The Role of Low Mass Nucleon Resonances in Near Threshold ω Meson Photoproduction," *EMI2001*.
5. Q. Zhao, "Photoproduction of ω and ϕ Mesons in the Quark Model," *EMI2001*.
6. "Photoproduction of Associated Strangeness using a Linearly Polarized Beam of Photons," CLAS Approved Analysis 2001-02, J.C. Sanabria, J. Kellie, and F.J. Klein, cospokespersons.
7. "Electromagnetic Production of Hyperons," JLab E89-004, R. Schumacher, spokesperson.
8. SAPHIR Collaboration, M.Q. Tran *et al.*, Phys. Lett. B 445 (1998) 20.
9. T. Mart and C. Bennhold, Phys. Rev. C 61 (2000) 012201.
10. W.-T. Chiang, F. Tabakin, T.-S. H. Lee, and B. Saghai, Phys. Lett. B 517 (2001) 101.

11. Z. Li and B. Saghai, Nucl. Phys. A 644 (1998) 345;
    Z. Li, B. Saghai, T. Ye, and Q. Zhao, *work in progress.*
12. B. Saghai, "Missing Resonance Search via Meson Electromagnetic Production," *EMI2001.*
13. K. Schilling, P. Seyboth, and G. Wolf, Nucl. Phys. B 15 (1970) 397;
    D. Schildknecht and B. Schrempp-Otto, Nuovo Cim. 5A, (1971) 183.
14. C.G. Fasani, F. Tabakin, and B. Saghai, Phys. Rev. C 46 (1992) 2430;
    B. Saghai and F. Tabakin, Phys. Rev. C 53 (1996) 66;
    M. Pichowsky, C. Savkli, and F. Tabakin, Phys. Rev. C 53 (1996) 593;
    C. Savkli, F. Tabakin, and S.N. Yang, Phys. Rev. C 53 (1996) 1132;
    B. Saghai and F. Tabakin, Phys. Rev. C 55 (1997) 917;
    W.M. Kloet, W.-T. Chiang, and F. Tabakin, Phys. Rev. C 58 (1998) 1086.
15. F.J. Klein, Ph.D. thesis, Bonn-IR-96-008 (1996);
    F.J. Klein, πN Newsletters, No. 14 (1998) 141.
16. W. Kaune *et al.*, Phys. Rev. D 11 (1975) 478.
17. D. Lohman *et al.*, Nucl. Inst. Meth. A (1994) 343; J. Peise, M.S. Thesis, Universität Mainz (1989) ; D. Lohman, M.S. Thesis, Universität Mainz (1992); F. Rambo, M.S. Thesis, Universität Göttingen (1995); A. Schmidt, M.S. Thesis, Universität Mainz (1995).
18. F.H. Dyson and H. Überall, Phys. Rev. 99 (1955) 604;
    H. Überall, Phys. Rev. 103 (1956) 1055;
    H. Überall, Phys. Rev. 107 (1957) 223;
    H. Überall, Z. Naturforsch. 17a (1962) 332;
    G. Diambrini-Palazzi, Rev. Modern Phys. 40 (1968) 611;
    U. Timm, Fortschr. Phys. 17 (1969) 765;
    R. Rambo *et al.*, Phys. Rev. C 58 (1998) 489.
19. D. Luckey and R.F. Schwitters, Nucl. Instr. and Meth. 81 (1970) 64.
20. NSF Award 9724489, W.J. Briscoe, K.S. Dhuga, and L.C. Maximon, co-PIs, The George Washington University, Washington, D.C., 20052.
21. D. Lohman, M. Schumacher *et al.*, Nucl. Instr. and Meth. A 343 (1994) 494.
22. K. Livingston, to be submitted to Nucl. Instr. and Meth. A.
23. D. Sober *et al.*, Nucl. Instr. and Meth. A 343 (2000) 263.
24. P.L. Cole, "The IPN-Orsay/UTEP Instrumented Collimator," August (1999) 17pp. *unpublished.*
25. A. Puga, M.S. thesis, University of Texas at El Paso, "Calibration of the UTEP/Orsay Instrumented Collimator via the LabVIEW-Benchtop Data Acquisition System," December, 2001.
26. A. Natter, *ANB code*, URL:
    http://www.pit.physik.uni-tuebingen.de/ ~natter/software/brems-anb.html

LIST OF PARTICIPANTS

Laith J.   ABU-RADDAD
RCNP,
Osaka University
10-1 Mihogaoka, Ibaraki
Osaka 567-0047   JAPAN
laith@rcnp.osaka-u.ac.jp

Tatsuya   ADACHI
Dept. of Physics,
Osaka University
1-1 Machikaneyama, Toyonaka
Osaka 560-0043   JAPAN
adachi@lns.sci.osaka-u.ac.jp

Hidetoshi   AKIMUNE
Dept. of Physics,
Konan University
8-9-1 Okamoto, Higashinada-ku, Kobe
Hyogo 658-8501   JAPAN
akimune@konan-u.ac.jp

Sam M.   AUSTIN
NSCL,
Michigan Satae University
East Lansing
MI 48824-1321   USA
austin@nscl.msu.edu

Jose   BACELAR
Kernfysisch Versneller Instituut
Zernikelaan 25
9747 AA Groningen
NETHERLANDS
bacelar@kvi.nl

Nicola   BIANCHI
INFN - Frascati
Via E. Fermi 40,
C.P. Box 13, Frascati
I-00044-Roma   ITALY
bianchi@hermes.desy.de

William J.   BRISCOE
Dept. of Physics,
The George Washington University
825 21st Street, NW, Washington DC
20052   USA
briscoe@gwu.edu

Daniel S.   CARMAN
Dept. of Physics,
Ohio University
Athens
OH 45701 USA
carman@ohio.edu

Gordon D.   CATES
Dept. of Physics,
University of Virginia
Physics Building, PO Box 400714,
Charlottesville VA 22904-4714   USA
cates@virginia.edu

Philip L .   COLE
Dept. of Physics,
University of Texas at El Paso
El Paso
Texas 79968   USA
cole@jlab.org

Schin   DATÉ
JASRI/SPring-8
1-1-1 Kouto, Mikazuki-cho,
Sayo-gun, Hyogo 679-5198
JAPAN
schin@spring8.or.jp

Rachele   DI SALVO
Dipartimento di Fisica,
Universita' di Roma Tor Vergata
I-00133 – Roma
ITALY
rachele.disalvo@roma2.infn.it

Jean-P. DIDELEZ
Institut de Physique Nucleaire
91406 ORSAY Cedex
FRANCE

didelez@ipno.in2p3.fr

Hiroyasu EJIRI
RCNP,
Osaka University
10-1 Mihogaoka, Ibaraki
Osaka 567-0047 JAPAN
ejiri@rcnp.osaka-u.ac.jp

Hisako FUJIMURA
RCNP,
Osaka University
10-1 Mihogaoka, Ibaraki
Osaka 567-0047 JAPAN
fujimura@rcnp.osaka-u.ac.jp

Yoshitaka FUJITA
Dept. of Physics,
Osaka university
1-1 Machikaneyama, Toyonaka
Osaka 560-0043 JAPAN
fujita@lns.sci.osaka-u.ac.jp

Moshe GAI
Dept. of Physics,
University of Connecticut
U3046 2152 Hillside Rd., Storrs
CT 06269-3046 USA
gai@uconn.edu

Sergo B. GERASIMOV
Bogoliubov Lab. of
Theoretical Physics, JINR,
Dubna, Moscow region
RU-141980 RUSSIA
gerasb@thsun1.jinr.ru

Yu-Bing DONG
Institute of High Energy Phsyics,
Academia Sinica
P. O. B. 918(4-1)
Beijing 100039 P. R. CHINA
dongyb@mail.ihep.ac.cn

Manuel J. FIOLHAIS
Dept. de Fisica,
Universidade de Coimbra
P-30004-516 Coimbra
PORTUGAL
tmanuel@teor.fis.uc.pt

Hirohiko FUJITA
Dept. of Physics,
Osaka University
1-1 Machikaneyama, Toyonaka
Osaka 560-0043 JAPAN
hfujita@lns.sci.osaka-u.ac.jp

Mamoru FUJIWARA
RCNP,
Osaka University
10-1 Mihogaoka, Ibaraki
Osaka 567-0047 JAPAN
fujiwara@rcnp.osaka-u.ac.jp

Haiyan GAO
Massachusetts Institute of Technology
26-413 MIT, 77 Massachusetts Ave.,
Cambridge
MA 02139 USA
haiyan@mit.edu

Ronald GILMAN
Rutgers University /Jefferson Lab
MS12H/A123, Jefferson Lab, 12000
Jefferson Ave, Newport News
VA, 23693 USA
gilman@jlab.org

Michel GUIDAL
Institut de Physique Nucleaire
Bat 100 - M052
91406 ORSAY Cedex
FRANCE
guidal@ipno.in2p3.fr

Keigo HARA
RCNP,
Osaka University
10-1 Mihogaoka, Ibaraki
Osaka 567-0047 JAPAN
hara@rcnp.osaka-u.ac.jp

Takeo HASEGAWA
Faculty of Engineering,
Miyazaki University
1-1, Gakuen Kibanadai Nishi
Miyazaki 889-2192 JAPAN
hasegawa@phys.miyazaki-u.ac.jp

Osamu HASHIMOTO
Dept. of Physics,
Tohoku University
Aoba-ku, Sendai
Miyagi 981-8578 JAPAN
hashimot@lambda.phys.tohoku.ac.jp

Takehito HAYAKAWA
Japan Atomic Energy Research Institute
Shirakata-shirane 2-4, Tokai-mura
Ibaraki 319-1195
JAPAN
hayakawa@jball4.tokai.jaeri.go.jp

Tomokatsu HAYAKAWA
Dept. of Physics,
Osaka University
1-1 Machikaneyama, Toyonaka
Osaka 560-0043 JAPAN
hayakawa@km.phys.sci.osaka-u.ac.jp

Kenneth H. HICKS
Dept. of Physics,
Ohio University
Athens
OH 45701 USA
hicks@oak.cats.ohiou.edu

Douglas W. HIGINBOTHAM
Massachusetts Inst. of Tech.
(Jlab branch)
12000 Jefferson Ave., MS 12H,
Newport News VA 23606 USA
doug@jlab.org

Seung-Woo HONG
Dept. of Physics,
Sungkyunkwan University
Suwon, 440-746
Rep. of KOREA
swhong@skku.ac.kr

Hisashi HORIUCHI
Dept. of Physics,
Kyoto University
Kitashirakawa-Oiwake-cho, Sakyo-ku
Kyoto 606-8502 JAPAN
horiuchi@ruby.scphys.kyoto-u.ac.jp

Charles J. HOROWITZ
Dept. of Physics,
Indiana University
Bloomington
IN 47401 USA
horowitz@iucf.indiana.edu

Atsushi HOSAKA
RCNP,
Osaka University
10-1 Mihogaoka, Ibaraki
Osaka 567-0047 JAPAN
hosaka@rcnp.osaka-u.ac.jp

Tomoaki   HOTTA
RCNP,
Osaka University
10-1 Mihogaoka, Ibaraki
Osaka 567-0047   JAPAN
hotta@rcnp.osaka-u.ac.jp

Ed V.   HUNGERFORD
Dept. of Physcs,
University of Houston
4800 Calhoun, Houston
TX 77204   USA
hunger@uh.edu

Kiyomi   IKEDA
RIKEN,
RI beam factory
2-1 Hirosawa, Wako
Saitama 351-0198   JAPAN
k-ikeda@postman.riken.go.jp

Ken'ichi   IMAI
Dept. of Physics,
Kyoto University
Kitashirakawa-Oiwake-cho, Sakyo-ku
Kyoto 606-8502   JAPAN
imai@nh.scphys.kyoto-u.ac.jp

Takahisa   ITAHASHI
RCNP,
Osaka University
10-1 Mihogaoka, Ibaraki
Osaka 567-0047   JAPAN
itahasi@rcnp.osaka-u.ac.jp

Takahiro   IWATA
Dept. of Physics,
Nagoya University
Furo-chou, Chikusa-ku, Nagoya
Aichi 464-8602   JAPAN
iwata@kiso.phys.nagoya-u.ac.jp

Kees   De JAGER
Jeffereson Laboratory
12000 Jefferson Avenue,
Newport News
VA 23606   USA
kees@jlab.org

Richard T.   JONES
University of Conneticuit
U3046 2152 Hillside Road, Storrs
CT 06269-3046
USA
richard.t.jones@uconn.edu

Toshitaka   KAJINO
National Astronomical Observatory
2-21-1 Osawa, Mitaka
Tokyo 181-8588
JAPAN
kajino@nao.ac.jp

Jun'ichiro   KAMIYA
RCNP,
Osaka University
10-1 Mihogaoka, Ibaraki
Osaka 567-0047   JAPAN
kamiya@rcnp.osaka-u.ac.jp

Jirota   KASAGI
Lab. of Nuclelar Science,
Tohoku University
1-2-1 Mikamine, Taihaku-ku, Sendai
Miyagi 982-0826   JAPAN
kasagi@lns.tohoku.ac.jp

Takahiro   KAWABATA
RCNP,
Osaka University
10-1 Mihogaoka, Ibaraki
Osaka 567-0047   JAPAN
takahiro@rcnp.osaka-u.ac.jp

Keigo KAWASE
RCNP,
Osaka University
10-1 Mihogaoka, Ibaraki
Osaka 567-0047 JAPAN
kawase@rcnp.osaka-u.ac.jp

H. KIM
Inst. of Phys. and Applied Phys.,
Yonsei University
134 Sinchon-dong, Seodaemun-gu
Seoul 120-749 Rep. of KOREA
hung@phya.yonsei.ac.kr

Miho KOMA-TAKAYAMA
RCNP,
Osaka University
10-1 Mihogaoka, Ibaraki
Osaka 567-0047 JAPAN
takayama@rcnp.osaka-u.ac.jp

Karlheinz LANGANKE
Institute for Physics and Astronomy,
University of Aarhus
Ny Munkegade
DK-8000 Aarhus C DENMARK
langanke@ifa.au.dk

Noriaki MAEHARA
RCNP,
Osaka University
10-1 Mihogaoka, Ibaraki
Osaka 567-0047 JAPAN
maehara@rcnp.osaka-u.ac.jp

Yuichiro MANABE
RCNP,
Osaka University
10-1 Mihogaoka, Ibaraki
Osaka 567-0047 JAPAN
manabe@rcnp.osaka-u.ac.jp

K. KIM
Dept. of Physics,
Sungkyunkwan University
Suwon, 440-746
Rep. of KOREA
kyungsik@color.skku.ac.kr

Yasuyuki KITAMURA
RCNP,
Osaka University
10-1 Mihogaoka, Ibaraki
Osaka 567-0047 JAPAN
yasuyuki@rcnp.osaka-u.ac.jp

Shunzo KUMANO
Dept. of Physics,
Saga University
Honjo-1
Saga 840-8502 JAPAN
kumanos@cc.saga-u.ac.jp

Leszek W. LUKASZUK
Soltan Institute for Nuclear Studies
Z-d P8, Hoza 69
00-681 Warsaw
POLAND
lukaszuk@fuw.edu.pl

Hiroyuki MAKII
RCNP,
Osaka University
10-1 Mihogaoka, Ibaraki
Osaka 567-0047 JAPAN
makii@rcnp.osaka-u.ac.jp

Nobuyuki MATSUOKA
RCNP,
Osaka University
10-1 Mihogaoka, Ibaraki
Osaka 567-0047 JAPAN
matsuoka@rcnp.osaka-u.ac.jp

704

Tsutomu MIBE
RCNP,
Osaka University
10-1 Mihogaoka, Ibaraki
Osaka 567-0047 JAPAN
mibe@rcnp.osaka-u.ac.jp

Shizu MINAMI
Dept. of Physics,
Osaka University
1-1 Machikaneyama, Toyonaka
Osaka 560-0043 JAPAN
minami@km.phys.sci.osaka-u.ac.jp

Kenji MISHIMA
RCNP,
Osaka University
10-1 Mihogaoka, Ibaraki
Osaka 567-0047 JAPAN
mishima@rcnp.osaka-u.ac.jp

Yoshiyuki MIYACHI
Dept. of Physics,
Tokyo Institute of Technology
2-12-1, Meguro-ku
Tokyo 152-8551 JAPAN
miyachi@nucl.phys.titech.ac.jp

Kazuya MIYAGAWA
Dept. of Applied Physics,
Okayama Univ. of Science
1-1 Ridai-cho
Okayama 700-0005 JAPAN
miyagawa@dap.ous.ac.jp

Peter MOHR
Institut fuer Kernphysik,
TU Darmstadt
Schlossgartenstr. 9
D-64289 Darmstadt GERMANY
mohr@ikp.tu-darmstadt.de

Dario MORICCIANI
INFN - Sezione Roma2
via della Ricerca Scientifica, n.1
I-00133-Roma
ITALY
dario.moricciani@roma2.infn.it

Toshiyuki MORII
Faculty of Human Development,
Kobe University
3-11 Tsurukabuto, Nada-ku, Kobe
Hyogo 657-8501 JAPAN
morii@kobe-u.ac.jp

Hiko MORITA
Saporo Gakuin University
Bunkyo-dai 11, Ebestu
Hokkaido 069-8555
JAPAN
hiko@earth.sgu.ac.jp

Tohru MOTOBAYASHI
Dept. of Physics,
Rikkyo University
3 Nishi-Ikebukuro, Toshima
Tokyo 171-8501 JAPAN
motobaya@rikkyo.ne.jp

Kentaro MUKAIDA
Dept. of Physics,
Osaka University
1-1 Machikaneyama, Toyonaka
Osaka 560-0043 JAPAN
mukaida@km.phys.sci.osaka-u.ac.jp

Yasuki NAGAI
RCNP,
Osaka University
10-1 Mihogaoka, Ibaraki
Osaka 567-0047 JAPAN
nagai@rcnp.osaka-u.ac.jp

Takashi NAGAO
Dept. of Physics,
Osaka University
1-1 Machikaneyama, Toyonaka
Osaka 560-0043   JAPAN
nagao@km.phys.sci.osaka-u.ac.jp

Takashi NAKANO
RCNP,
Osaka University
10-1 Mihogaoka, Ibaraki
Osaka 567-0047   JAPAN
nakano@rcnp.osaka-u.ac.jp

Masaharu NOMACHI
Dept. of Physics,
Osaka University
1-1 Machikaneyama, Toyonaka
Osaka 560-0043   JAPAN
nomachi@lns.sci.osaka-u.ac.jp

Karapet OGANESSYAN
INFN - LNF
Via E. Fermi 40,
C.P. Box 13, Frascati (RM)
I-00044   ITALY
kogan@lnf.infn.it

Yongseok OH
Dept. of Physics,
Yonsei University
Seoul 120-749
Rep. of KOREA
yoh@phya.yonsei.ac.kr

Takaharu OTSUKA
Dept. of Physics,
University of Tokyo
Hongo, Bunkyo-ku
Tokyo 113-0033   JAPAN
otsuka@phys.s.u-tokyo.ac.jp

Kohsuke NAKANISHI
RCNP,
Osaka University
10-1 Mihogaoka, Ibaraki
Osaka 567-0047   JAPAN
nakanisi@rcnp.osaka-u.ac.jp

Vladimir G.   NEDOREZOV
Institute for Nuclear Research
Prospect of 60-th October Anniversary,
7A
117312 Moscow   RUSSIA
Vladimir@cpc.inr.ac.ru

Takashi NUMATA
Dept. of Physics,
Osaka University
1-1 Machikaneyama, Toyonaka
Osaka 560-0043   JAPAN
numata@km.phys.sci.osaka-u.ac.jp

Kazuyuki OGATA
RCNP,
Osaka University
10-1 Mihogaoka, Ibaraki
Osaka 567-0047   JAPAN
kazuyuki@rcnp.osaka-u.ac.jp

Grant V.   O'RIELLY
Physics Dept.,
The George Washington University
825 21st Street, NW,
Washington DC 20052   USA
orielly@gwu.edu

Gerassimos G.   PETRATOS
Physics Dept.,
Kent State University
Kent
OH 44242   USA
gpetrato@kent.edu

Chary RANGACHARYULU
Dept. of Physics,
University of Saskatchewan
116 Science Place, Saskatoon,
Saskatchewan S7N5E2 CANADA
chary@sask.usask.ca

Achim RICHTER
Institut fuer Kernphysik,
TU Darmstadt
Schlossgartenstr. 9
D-64289 Darmstadt GERMANY
richter@ikp.tu-darmstadt.de

Hiroyuki SAGAWA
Center for Mathematical Science,
University of Aizu
Aizu-Wakamatsu
Fukushima 965-8580 JAPAN
sagawa@u-aizu.ac.jp

Bijian SAGHAI
Commissariat a l'Energie Atomique -
Saclay
Bat. 703, DAPNIA/SPhN, CEA/Saclay
91191 Gif-sur-Yvette Cedex FRANCE
bsaghai@cea.fr

Atsushi SAKAGUCHI
Graduate School of Science,
Osaka University
1-1 Machikaneyama, Toyonaka
Osaka 560-0043 JAPAN
sakaguch@phys.sci.osaka-u.ac.jp

Emiko SANO
RCNP,
Osaka University
10-1 Mihogaoka, Ibaraki
Osaka 567-0047 JAPAN
esano@rcnp.osaka-u.ac.jp

Tohru SATO
Graduate School of Science,
Osaka University
1-1 Machikaneyama, Toyonaka
Osaka 560-0043 JAPAN
tsato@phys.sci.osaka-u.ac.jp

Olaf SCHOLTEN
Kernfysisch Versneller Instituut
Zernikelaan 25
9747 AA Groningen
NETHERLANDS
scholten@kvi.nl

Tatsushi SHIMA
RCNP,
Osaka University
10-1 Mihogaoka, Ibaraki
Osaka 567-0047 JAPAN
shima@rcnp.osaka-u.ac.jp

Yoshihiro SHIMBARA
Dept. of Physics,
Osaka University
1-1 Machikaneyama, Toyonaka
Osaka 560-0043 JAPAN
shimbara@lns.sci.osaka-u.ac.jp

Youhei SHIMIZU
RCNP,
Osaka University
10-1 Mihogaoka, Ibaraki
Osaka 567-0047 JAPAN
yshimizu@rcnp.osaka-u.ac.jp

Toshiyuki SHIZUMA
Japan Atomic Energy Research Institute
Shirakata-shirane 2-4, Tokai-mura
Ibaraki 319-1195
JAPAN
hayakawa@jball4.tokai.jaeri.go.jp

Paul A. SOUDER
Dept. of Physics,
Syracuse University
201 Physics Building, Syracuse
NY 13244-1130 USA
souder@suhep.phy.syr.edu

Hans STROEHER
Forschungszentrum Jülich,
Institut für Kernphysik
Postfach 1913
D-52425 Jülich GERMANY
h.stroeher@fz-juelich.de

Kazuko SUGAWARA-TANABE
Otsuma Women's University
2-7-1 Karakida, Tama
Tokyo 206-8540
JAPAN
tanabe@otsuma.ac.jp

Yorihito SUGAYA
Graduate School of Science,
Osaka University
1-1 Machikaneyama, Toyonaka
Osaka 560-0043 JAPAN
sugaya@rcnp.osaka-u.ac.jp

Ayumu SUGITA
RCNP,
Osaka University
10-1 Mihogaoka, Ibaraki
Osaka 567-0047 JAPAN
sugita@rcnp.osaka-u.ac.jp

Mizuki SUMIHAMA
RCNP,
Osaka University
10-1 Mihogaoka, Ibaraki
Osaka 567-0047 JAPAN
sumihama@km.phys.sci.osaka-u.ac.jp

Toshio SUZUKI
Dept of Physics,
Nihon University
Sakurajosui 3-25-40, Setagaya-ku
Tokyo 156-8550 JAPAN
suzuki@chs.nihon-u.ac.jp

Keiji TAKAHISA
RCNP,
Osaka University
10-1 Mihogaoka, Ibaraki
Osaka 567-0047 JAPAN
takahisa@rcnp.osaka-u.ac.jp

Masayoshi TANAKA
Kobe Tokiwa Jr. College
2-6-2 Ohtani-cho, Nagata-ku, Kobe
Hyogo 653-0838
JAPAN
tanaka@rcnp.osaka-u.ac.jp

Isao TANIHATA
RIKEN,
RI beam factory
2-1 Hirosawa, Wako
Saitama 351-0198 JAPAN
tanihata@rikaxp.riken.go.jp

David J. TEDESCHI
Dept. of Phys. and Astronomy,
University of South Carolina
Columbia
SC 29208 USA
tedeschi@sc.edu

Anthony W. THOMAS
CSSM,
University of Adelaide
Adelaide
SA 5005 AUSTRALIA
athomas@physics.adelaide.edu.au

A. TITOV
Bogolyubov Laboratory of
Theoretical Physics, JINR
Dubna, Moscow Region
141980 Moscow   RUSSIA
atitov@thsun1.jinr.ru

Hiroshi   TOKI
RCNP,
Osaka University
10-1 Mihogaoka, Ibaraki
Osaka 567-0047   JAPAN
toki@rcnp.osaka-u.ac.jp

S. TOMII
Dept. of Physics,
Osaka University
1-1 Machikaneyama, Toyonaka
Osaka 560-0043   JAPAN
tomii@km.phys.sci.osaka-u.ac.jp

Kazuo   TSUSHIMA
CSSM,
University of Adelaide
Adelaide
SA 5005   AUSTRALIA
ktsushim@physics.adelaide.edu.au

Makoto   UCHIDA
Dept. of Physics,
Kyoto University
Kitashirakawa-Oiwake-cho, Sakyo-ku
Kyoto 606-8502   JAPAN
uchida@ne.scphys.kyoto-u.ac.jp

Hitoshi   UEDA
RCNP,
Osaka University
10-1 Mihogaoka, Ibaraki
Osaka 567-0047   JAPAN
ueda@rcnp.osaka-u.ac.jp

Mikhail G.   URIN
Moscow Engineering Physics Institute
Kashirskoe shosse 31
115409 Moscow
RUSSIA
urin@theor.mephi.ru

Hiroaki   UTSUNOMIYA
Dept. of Physics,
Konan University
8-9-1 Okamoto, Higashinada-ku, Kobe
Hyogo 658-8501   JAPAN
hiro@konan-u.ac.jp

J.   van de WIELE
Institut de Physique Nucleaire
CNRS-IN2P3 Universite Paris-Sud
F 91406 ORSAY Cedex
FRANCE
VANDEWI@IPNO.IN2P3.FR

W.T.H.   van OERS
Dept. of Phys. and Astronomy,
University of Manitoba/TRIUMF
TRIUMF 4004 Wesbrook Mall,
Vancouver B.C. V6T 2A3   CANADA
vanoers@triumf.ca

P.   von NEUMANN-COSEL
TU Darmstadt,
Institut fur Kernphysik
Schlossgartenstr. 9
D-64289 Darmstadt   GERMANY
vnc@ikp.tu-darmstadt.de

Henry R.   WELLER
TUNL and Duke University
TUNL Box 90308, Duke University
Durham
NC 27708-0308   USA
weller@tunl.duke.edu

C. S.   WHISNANT
Physics Department,
James Madison University
MSC 7702, Miller Hall, Room 113,
James Madison University,
Harrisonburg VA 22807   USA
whisnacs@jmu.edu

Hirohito   YAMAZAKI
Lab. of Nuclear Science,
Tohoku University
Mikamine 1-2-1, Taihaku-ku, Sendai
Miyagi 982-0826   JAPAN
yamazaki@lns.tohoku.ac.jp

Masaru   YOSOI
Dept. of Physics,
Kyoto University
Kitashirakawa-Oiwake-cho, Sakyo-ku
Kyoto 606-8502   JAPAN
yosoi@ne.scphys.kyoto-u.ac.jp

Qiang   ZHAO
Dept. of Physics,
University of Surrey
Guildford, Surrey
GU2 7XH   UK
qiang.zhao@surrey.ac.uk

R.   WOLOSHYN
TRIUMF
4004 Wesbrook Mall, Vancouver
V6T 2A3
CANADA
rwww@triumf.ca

Masanobu   YAHIRO
Dept. of Physics,
University of the Ryukyus
Senbaru, Nishihara-cho, Okinawa,
903-0213   JAPAN
yahiro@sci.u-ryukyu.ac.jp

Remco G. T.   ZEGERS
Advanced Photon Resaerch Center,
JAERI
8-1 Umemidai, Kizu-cho, Souraku-gun
Kyoto 619-021   JAPAN
zegers@spring8.or.jp